# Progress in Optical Science and Photonics

Volume 20

The purpose of the series Progress in Optical Science and Photonics is to provide a forum to disseminate the latest research findings in various areas of Optics and its applications. The intended audience are physicists, electrical and electronic engineers, applied mathematicians, biomedical engineers, and advanced graduate students.

T. Stewart McKechnie

# General Theory of Light Propagation and Imaging Through the Atmosphere

Second Edition

T. Stewart McKechnie
St. Andrews, Scotland, UK

ISSN 2363-5096 ISSN 2363-510X (electronic)
Progress in Optical Science and Photonics
ISBN 978-3-030-98830-2 ISBN 978-3-030-98828-9 (eBook)
https://doi.org/10.1007/978-3-030-98828-9



This Springer imprint is published by the registered company Springer Nature Switzerland AG
The registered company address is: Gewerbestrasse 11, 6330 Cham, Switzerland

*... [Telescopes] cannot be so formed as to take away that confusion of Rays which arises from the Tremors of the Atmosphere. The only Remedy is a most serene and quiet Air, such as may perhaps be found on the tops of the highest Mountains above the grosser Clouds.*

*Isaac Newton*

**Note**

The epigraph is quoted from Book 1, Part 1 of Sir Isaac Newton's great work "Opticks: or a Treatise of the Reflections, Refractions, Inflections and Colours of light," (1730). Newton refers to diffraction effects as the "inflection of light." The first edition of "Opticks" was published in 1704, the second of Newton's great works, "Principia" being the first.

**Reference**

Newton, I. (1730). Opticks: Or a treatise of the reflections, refractions, inflections and colours of light (4th ed.). London: William Innys at the Weft-End of St. Paul's.

*The emergence of truth:*

*"First, it is ridiculed.*
*Second, it is violently opposed.*
*Third, it is accepted as being self-evident."*

*Arthur Schopenhauer*
*(1788–1860)*

*To John, Heather, and Iain*

# Preface

As an undergraduate in the mid-1960s, I studied mathematical physics at the Tait Institute of Mathematical Physics, the University of Edinburgh.[1] Having developed an interest in astronomy and built telescopes as a teenager, after graduating I was able to pursue these interests as a postgraduate student in the Optics Section of the Department of Physics at Imperial College, London. With Prof. Walter T. Welford as mentor, Dr. Michael E. Barnett as internal Ph.D. Adviser, and surrounded by noteworthy postgraduate research colleagues such as J. Christopher Dainty and Gareth Parry, I found myself in a particularly stimulating optics research environment. It was also my privilege to have Prof. Joseph W. Goodman as my external Ph.D. examiner in 1974; the Ph.D. thesis topic, the statistical properties of laser speckle, led naturally to research in the subject area of this treatise: Light Propagation and Imaging through the Atmosphere.

Between 1974 and 1976, while holding a postdoctoral research position at Imperial College, I was able to show that in certain crucial ways the atmosphere behaves as though it were flattened into a single scattering layer. This made it possible to apply in the area of atmospheric scattering several key theoretical results and measurement techniques[2] known and used in the area of light scattering from rough surfaces.

These findings opened up novel approaches. One in particular enabled calculation of the average size distribution of the turbulence structure in atmospheric paths from readily measurable properties of the intensities in point object images formed by telescopes observing over these paths; unresolved stars of course make ideal point objects. Initial measurement data obtained from images of the bright star, Vega (using the Yapp 36-in. reflector at the Royal Greenwich Observatory, Herstmonceux Castle), indicated that turbulence structure in the atmosphere was significantly

[1] The Tait Institute at the University of Edinburgh, named after Peter Guthrie Tait, Professor of Natural Philosophy at the University, a school and lifelong friend of James Clerk Maxwell. Courses attended by the author included some given by future Nobel laureate, Dr. Peter W. Higgs. Neither the author nor his class colleagues had any inkling that Higg's particle physics research contribution at the time would carry such enormous moment.

[2] Many of these results and techniques had just been published in the book, *Laser speckle and related phenomena*, edited by J. C. Dainty, the first edition of which appeared in 1975.

smaller, on average, than assumed in the prevailing Kolmogorov understanding—an understanding developed by the Soviet mathematician, Andrey N. Kolmogorov (1903–1987), and others.

Taken at face value, the smaller measured turbulence structure led directly to the controversial postulation from a Kolmogorov perspective (McKechnie 1976) that central cores should appear frequently in star images formed by large ground-based astronomical telescopes. It was further postulated that the largest telescopes at that time (4–5 m class), if equipped with diffraction-limited optics, should be capable of delivering—through these cores—angular resolution at least an order of magnitude better than the ~ 1-arcsec resolution customarily obtained from large ground-based instruments. However, due to the lack of mathematical rigor in that early work, I and no doubt others familiar with it, such as J. C. Dainty and A. Gologly, considered it both incomplete and inconclusive.

Some years later, between 1989 and 1992, while providing optics support to the Falcon Nuclear Laser program at Sandia National Laboratories, Albuquerque, New Mexico, I was at last able to return to the subject of light propagation and imaging through the atmosphere. During that period, I was finally able to achieve the mathematical rigor that had been lacking previously. This resulted in a number of published papers that more resolutely advanced the same basic ideas developed more than a decade earlier at Imperial College. My interest in the subject has continued ever since, constantly drawing me back to work on various other aspects of this multifaceted subject. In 2010, I began the task of consolidating the various papers published over the years into a single document: the 2015 first edition of this book. In late 2019, I began authoring this second edition. With little else to distract me during the 2020/21 COVID-19 pandemic, progress was rapid.

The book discusses fundamental problems inherent in Kolmogorov formulations, problems that include dimensional inconsistencies. It also provides an extensive list of observational evidence documented by various researchers and observational astronomers over the years that contradicts basic assumptions underpinning Kolmogorov theory. By rigorous scientific standards, the existence of even one piece of sustainable observational evidence contradicting these assumptions ought to be enough to undermine the general validity of the theory; but the fact that instances are found everywhere we look should surely signal the entire demise of the theory. Yet, in the absence of any better alternative, most researchers simply ignored the problems and continued to use Kolmogorov theory as though it were still fully viable. From an objective standpoint, it is clear that a more precise and more broadly encompassing theory is needed to describe light propagation and imaging through Earth's turbulent atmosphere.

This book lays out just such a theory. It also discusses in detail the many ways in which the predictions that emerge from this theory differ from those of Kolmogorov theory. Applications for the new theoretical formulation are found in the fields of laser communications, high-energy laser (HEL) beam propagation, imaging through the atmosphere with large ground-based astronomical telescopes—including next-generations Extremely Large Telescopes—and Earth surveillance imaging from satellites in space.

The theory is based on a generalized atmosphere rather than one whose properties are assumed, a priori, to be Kolmogorov. For this and other reasons outlined in the text, we label the theory a "general theory." Although the turbulence properties in the atmospheric observing paths are initially regarded as unknown, they are nonetheless encrypted in the properties of point object or star images formed by large telescopes. The encryption is complicated because phase information associated with the field disturbances in the images is lost when only intensity is recorded. Nevertheless, to decipher the essential turbulence information merely requires measuring certain key properties of the image intensities and inserting the measured intensity data into the appropriate mapping equations. Procedures for making the necessary star image measurements are described in detail.

The data is used initially to calculate certain crucial—and in fact decisive—properties of the amplitude and phase of the image-forming light waves in the telescope pupil. Once these wave properties have been established, we can then calculate from them, not only the essential properties of the turbulence in the atmospheric path, but all other meaningful properties of the complex amplitudes and intensities in point object images. These image properties include both monochromatic and polychromatic properties, as well as the various speckle and core and halo properties.

The generalized formulation upon which the measurement procedures are based ensures non-partisan measurement outcomes. Thus, turbulence properties extracted from the measured data are free of bias either toward or against any one type of turbulence, Kolmogorov or other. To counter any impression that the book is too critical of Kolmogorov theory, it is emphasized here that, in circumstances where the atmospheric turbulence were in fact "Kolmogorov," the measurement data should faithfully bear witness to that fact.

The overall mathematical approach relies only on Maxwell's electromagnetic wave equations. There is no requirement for any knowledge, perceived or otherwise, of the physics of turbulence. (Maxwell's equations have been universally accepted as valid for more than one hundred years and have become an essential part of our understanding of the universe and the laws that govern it.) The formulation laid out is dimensionally robust and naturally provides consistency between the properties of turbulence in atmospheric observing paths and telescope image properties—essential attributes that are not shared by Kolmogorov formulations.

Before concluding here, it is my pleasure to thank Prof. J. Christopher Dainty for his interest, support, and objective inputs, dating all the way back to when I first became involved in the subject area at Imperial College. It is also my pleasure to thank the late Prof. Roger F. Griffin, for generously sharing with me so much of the practical observational knowledge acquired by him over hundreds of nights (spread over many decades) while observing with many of the world's largest telescopes. I also owe a huge debt of gratitude to Dr. Scott A. Sallberg whose extensive knowledge of the subject area, combined with a more than generous allocation of his precious time, contributed countless helpful ideas and suggestions, while at the same time identifying myriads of errors, small and large, that with the passage of time had become invisible to me. For any errors that still remain—and I am sure there are a few—I take sole responsibility.

Gratitude is also due to Sandia National Laboratories for supporting this work in the early 1990s and, in particular, to my colleague and friend there, Dr. Daniel R. Neal, for his support and long-term interest in this work. I also owe a debt of gratitude to my youngest son, Iain, who is now at Heriot Watt University, Edinburgh, also studying mathematical physics. He frequently came to the rescue when all seemed lost. Younger minds have no equal when it comes to computer keyboard skills and manipulating text editing programs.

Perhaps a nod of appreciation is also due to the redoubtable Sir Patrick Moore (1923–2012) whose "Sky and Night" television program and book, *The Amateur Astronomer* (Moore 1958), helped kindle my interest in telescopes and astronomy more than 60 years ago.

And by no means least, I am forever enormously indebted to my dear wife, Alison, for her infinite patience and devotion during the many hours it took to write the two editions of this book. Without her indulgence and support, neither would have been possible.

St. Andrews, Scotland, UK T. Stewart McKechnie

## References

McKechnie, T. S. (1976). Cores in star images. *JOSA*, *66*(6), 635.
Moore, P. (1958). The amateur astronomer (2nd ed.). London: Lutterworth Press.

# Contents

# In Memoriam

John Campbell Brown
OBE, FRSE (1947–2019)
10th Astronomer Royal for Scotland

Photograph by courtesy of Mike Peel. http://www.mikepeel.net/

The author had the privilege of knowing Prof. John Brown through our mutual association with the Astronomical Society of Glasgow. Straightaway, I was captivated by his passion for astronomy and the unique mix of energy and excitement he brought to the subject wherever he ventured. After the Society's monthly meetings, he and I would often indulge our mutual fascination by talking "astronomy" until late into the evening. John was an inspiring and supportive friend, an honest man who recognized the need for transparency and truth in science.

I felt particularly honored when John invited me to present a seminar at his prestigious Physics and Astronomy School at the University of Glasgow. The title of the Talk (May 24, 2018): "Attaining the Ultimate Resolution Limits of Large Ground-Based Astronomical Telescopes, Including Extremely Large Telescopes (ELTS)."*

Professor Brown held posts continuously in the Glasgow School after his first appointment as Research Assistant in 1968. His career advanced rapidly. In 1984, he was promoted to chair of astrophysics. In 1995, he was appointed Astronomer Royal for Scotland, and in 1996, he was appointed Regius Professor of Astronomy, holding the latter two positions until his passing in 2019.

It was my earnest hope that John would write a foreword to this second edition of the book, knowing that anything he wrote would surely add weight, sparkle, and flair. Alas, that was not to be. His sudden passing while staying on his beloved Isle of Skye came as a great shock to those who knew him, leaving a void not easily filled. Sadly, we may never see his likes again.

*Talk Slides: TalkSlides24May2018.pdf.

# About the Author

**T. Stewart McKechnie** though born in Glasgow, most of my upbringing took place in Edinburgh where we moved as a family—to 10 Observatory Road—when I was four years of age. I attended South Morningside Primary School, transferring at the age of twelve to George Watson's College for secondary schooling. From there, I went to Edinburgh University, obtaining an honors degree in Mathematical Physics. After briefly working for an optics company in Glasgow, I attended Imperial College, London, obtaining a master's degree in Applied Optics followed by a Ph.D. in the subject area of light scattering. The next six years were divided between postdoctoral research positions at Imperial College and Loughborough University.

In 1981, an 18-month optics consultancy opportunity took me across the Atlantic to the USA. I continued to live and work in the USA as an Optics Consultant for the next 32 years, obtaining US citizenship in 1994 and thus becoming a dual UK/US citizen. While in the USA, I met and married Alison, from Kent, England, who was temporarily working in New York. We have three children, John, Heather, and Iain. In 2013, three years after I retired to focus on writing the first edition of this book, we relocated as a family back to the UK. As an undergraduate, I had the honor of representing Edinburgh and the East of Scotland at cricket as an opening bowler, and while at Imperial College earned tennis colors playing for the college team.

Portrait of the 12-year-old author painted in the summer of 1956. Having just gained entry to George Watson's College, Edinburgh, the author wears the maroon and white striped school tie, just as worn previously by the artist himself, Henry Raeburn Dobson (1901–1985)

Author as a teenager and his home-built six-inch Newtonian reflector set up in the backyard at 10 Observatory Road, Edinburgh. Photograph taken shortly after "first light" in March 1962, six months before enrolling at Edinburgh University. Additional telescope details are given in Footnote 10 (Chap. 10). Photograph by courtesy of high school colleague and friend, Iain E. Stark

# Chapter 1
# History of the Telescope and Its Remarkable Contribution to Scientific Discovery (and the 400-Year Journey from Galileo to a Rigorous General Theory of Imaging Through Earth's Turbulent Atmosphere)

**Abstract** The General Theory presented in this book primarily relates to imaging with telescopes. It is therefore fitting that the first chapter should provide a brief history of the telescope, its physical evolution over the centuries and, more recently, the development of a rigorous understanding of image formation through Earth's turbulent atmosphere. It is also fitting to recount the astonishing advancements in scientific understanding that have accompanied the evolution of this singular instrument. Following Galileo's epochal discoveries in 1610, Ptolemy's thousand-year-old geocentric model of the universe began to crumble. Today, some four centuries later, we find ourselves living in an infinitely more bewildering universe—a universe—governed by relativistic laws at large scales and quantum mechanical laws at small scales. With construction having already begun on a new generation of Extremely Large Telescopes (ELTs) and with continuing advances anticipated in Adaptive Optics (AO), when first light is achieved with these prodigious instruments in the next few years, we will surely enter a new golden age of the telescope. The images harvested by these instruments may turn out no less epochal than those first images observed by Galileo. Not for the first time, the telescope may lead us into a new golden age of scientific discovery.

Two of the telescopes used by Galileo Galilei (1564–1642) to make his epochal discoveries of 1609 and 1610 survive to this day and can be seen at the Museo Galileo in Florence, the city where Galileo lies buried. One of them consists of two lenses which Galileo ground himself, spaced apart by about one meter. The diameter of the front objective lens is about 37 mm, the diameter of the negative rear lens about 22 mm. Nowadays, this type of telescope configuration is referred to as "Galilean." Despite the modest quality of its optics and its meager 20 × magnification, Galileo used this "perspicillum"[1] to view the heavens with a level of clarity unimagined by all preceding generations of naked-eye observers. As word quickly spread, and others saw the same images, the foundations of scientific and religious thought were shaken

[1] Galileo used the name perspicillum to describe his instrument in the Starry Messenger.



T. S. McKechnie, *General Theory of Light Propagation and Imaging Through the Atmosphere*, Progress in Optical Science and Photonics 20,
https://doi.org/10.1007/978-3-030-98828-9_1

as never before. Ptolemy's geocentric model of the universe[2] in which the Earth is supposed fixed, while everything else rotates about it, now faced challenges more threatening than any encountered previously. There would be a prolonged struggle, but ultimately, the heliocentric model developed by Nicolaus Copernicus (1473–1543)—by which the planets (now including the downgraded Earth) rotate around a stationary Sun—would be the only one to survive.

In time, Earth's bewildering status downgrade would be followed by no less of a downgrade in the Sun's status. The Sun would come to be seen as an unremarkable star—just one of many billions—voyaging silently in the outer suburbs of the Milky Way galaxy, tens of trillions of miles from its nearest stellar kin. Even the Milky Way would eventually come to be seen as just one of many billions of galaxies—albeit a particularly large one—existing in a universe vastly greater in size.

Today, some four hundred years after Galileo, magnificent telescopes, such as the 2.4-m Hubble Space Telescope, the 6.5-m James Webb Space Telescope (JWST), and the twin 10-m Keck telescopes, allow astronomers to see objects at incredible distances. Light from the farthest of these objects—some $10^{23}$ miles away—is thought to have been traveling through the cosmos since shortly after the birth of the universe, some 13.8 billion years ago. By the time Sun and Earth came into being, the light we see today from the more distant objects had already been traveling for some 8 or 9 billion years.

Images produced over the centuries by successive generations of ever larger and more powerful telescopes have awed and perplexed no less than those so famously seen by Galileo. Ideas inspired by images delivered by these instruments have enabled scientists and thinkers to construct a staggeringly sophisticated understanding of the universe: a universe now seen as governed by quantum and relativistic laws, one in which energy and mass occur in discrete packages and have an equivalence, where nothing is perfectly certain, and where distances distort, time mutates, and light is bent by gravity; a universe in which a big bang once occurred, followed by an expansion that not only persists to this day but even seems to be accelerating. Such is the universe we live in. It could hardly be any more extraordinary. In this historical introduction, we trace the remarkable contribution of the telescope to the development of our present understanding of the universe and the scientific laws that govern it.

The invention of eyeglasses in the late thirteenth century preceded telescopes. The Italian, Salvino D'Armate, produced the first wearable pair in 1284 and is credited with the invention. By 1300, they were widely available in European cities such as Venice and Florence. At about the same time, the promise of lens use beyond mere eyeglasses was foreseen by Roger Bacon (1214–1294): "… we can dispose bodies in such order with respect to the eye and the objects, that the rays shall be refracted and bent towards any place we please, so that … from an incredible distance we may

[2] The Hellenistic astronomer and mathematician Claudius Ptolemy (90–168) who lived most of his life in Alexandria, Egypt, laid out the geocentric model of the cosmos in his main astronomical work, the Almagest, a greatly influential treatise that represented the culmination of centuries of work by Hellenic and Babylonian astronomers, in particular the Greek philosopher, Aristotle (384 BC–322 BC). Ptolemy's geocentric ideas were widely accepted throughout Europe and elsewhere for over a thousand years.

read the smallest letters and number the smallest particles of dust and sand by reason of the greatness of the angle under which we may see them."[3]

Despite this uncanny foresight, the invention of the first telescope still had to wait another 300 years. In 1608, a Flemish spectacle-maker, Hans Lippershey (1570–1619), applied for a telescope patent. It was refused and he is not credited with the invention, but he is credited with making telescopes widely known throughout Europe. Hearing of them in May 1609, Galileo began building his own telescopes, with various sizes and magnifications. In October 1609, he turned his instruments to the heavens and thus began the prolific association of the telescope and scientific advancement, an association that has forever changed man's understanding of the universe and his place in it.

By late 1610, Galileo had observed craters and mountains on the Moon, the phases of Venus, the "lobes" of Saturn, the Milky Way as a collection of stars, sunspots, and the four "Galilean" moons of Jupiter. His March 1610 publication of Sidereus Nuncius—usually translated into English as Starry Messenger—attracted worldwide attention. His 1612 notebook recording of a "star" close to Jupiter that appeared to move relative to the other stars even turned out to be the planet Neptune, glimpsed more than two hundred years before its official "discovery" in 1846. The name "telescope" was coined at a banquet for Galileo in 1611 by Prince Cesi at the suggestion of Greek mathematician, Giovanni Demisiana.

The "horns of Venus" had been documented in ancient times, as had sunspots and a least one of Jupiter's moons. Nonetheless, Galileo was able to observe these sights with an unprecedented level of sharpness. He saw the slow formation of sunspots, and by their movement, he could see that the Sun was rotating. This observation and his observation of the full planetary phases of Venus had momentous impact. Both were grave setbacks to Aristotelians, essentially ruling out the Ptolemaic geocentric system and instead providing compelling evidence for the Copernican heliocentric model. Father Benedetto Castelli, a monk at Montecassino and Galileo's favorite pupil, surmised that these discoveries (Tauber, 1982, p. 137) "must convince the most obstinate." Galileo, however, saw things more realistically: "It would not be enough even if the stars came down to Earth to bring witness themselves."

Convicted of grave suspicion of heresy by the church in 1633, Galileo was ordered imprisoned, a sentence later commuted to house arrest, which he endured for the rest of his life. Four years before he died, the young English poet John Milton (1608–1674) visited him at his villa just outside of Florence, finding him "in darkness, and with dangers compassed round." But Galileo may have gotten off lightly: For similar transgressions in 1660, the Roman Inquisition declared Giordano Bruno (Mason, 2008) "an impenitent, pertinacious, and obstinate heretic," sufficient reason to have him burned at the stake.

Galileo stands apart from the standards of his time: "I do not feel obliged to believe that the same god who has endowed us with sense, reason and intellect has intended us to forgo their use." He understood the supremacy of mathematics: "Philosophy is

[3] This and other quotes in this section are taken from "History of the Telescope," King (1955) and encyclopedia Wikipedia.

written in this grand book, the universe ... It is written in the language of mathematics, and its characters are triangles, circles, and other geometric figures ..." Galileo's masterpiece, "Discourse on the Two New Sciences," which was smuggled out of Italy and published in the Netherlands in 1636, laid the foundations of modern physics. Albert Einstein (1879–1955) held Galileo in highest regard: "Propositions arrived at by purely logical means are completely empty as regards reality. Because Galileo recognized this, and particularly because he drummed it into the scientific world, he is the father of modern physics— indeed, of modern science altogether."

The year 1609 turned out to be doubly auspicious when Johannes Kepler (1571–1630), the German-born astronomer and mathematician and assistant to the Danish nobleman Tycho Brahe (1546–1601), published two radical laws of planetary motion based on his analysis of Brahe's precise astronomical observations.[4] His third law followed in 1619. Kepler's empirically deduced laws were solidified later by Isaac Newton (1642–1727) who showed them to be direct consequences of the inverse square law of gravity.

In 1676, the Danish astronomer Ole Romer (1644–1710) established that light has a finite, though extremely fast speed, his measured value—about 220,000 km per second—based on observations of the motions and eclipse times of Jupiter's moons, particularly those of Io, which can vary by tens of minutes depending on the annual, 300–500 million mile change in Earth–Jupiter separation. Despite the 26% error in his measurement, his technique was sound; the error was later attributed to inaccurate planetary orbital data. All of these developments contributed to a momentous seventeenth century scientific revolution.

Galileo's first telescope was a refractor. By combining positive and negative lenses, his instrument was able to provide upright images. Later, positive eye lenses became more common as they naturally provide better aberration correction. (Inverted images are only mildly inconvenient for astronomical objects.) The reflector telescope had to wait until 1663, when one was described in Optica Promota by the Scottish mathematician and astronomer, James Gregory (1638–1675). Gregory never actually built his telescope,[5] though subsequently Robert Hooke (1635–1703) and others did. Isaac Newton constructed the first working reflector in 1669. It had a 2- inch- diameter primary mirror which he stopped down to about 1.3 inch; it had a focal length of about 6.25 inch and magnified objects about 35 times.

Newtonian telescopes became the standard observing instrument for a century and a half despite reports in 1672 from M. de Berce of a reflector design by an obscure Frenchman, Laurent Cassegrain (1629–1693), in which a convex secondary mirror is suspended above the center of the primary concave mirror. The Cassegrain design

[4] Tycho Brahe's comprehensive measurements of the locations of the stars and planets were made to an unprecedented accuracy of about 1.5 arc min. In this pre-telescope era, his measurement instruments comprised long sighting poles.

[5] The James Gregory Telescope, named after Gregory, is the largest working optical telescope in the UK. The instrument—a 37-inch Schmidt–Cassegrain design that saw first light in 1962—is located just above sea level near the School of Physics and Astronomy at St. Andrews University and is still used frequently. Gregory also discovered the diffraction grating by passing light through bird feathers and observing the patterns produced. He observed the splitting of sunlight into its component colors about one year after Newton had done so with a prism.

as it became known caused great controversy but, rejected by Newton and criticized by Christian Huygens (1629–1695), interest in it soon faded.

About a hundred years later, Jesse Ramsden (1735–1800) praised the design for its cancelation of spherical aberration. As word gradually spread of the advantages, Cassegrain telescopes became more fashionable at the expense of Newtonians. Most large professional telescopes built since the beginning of the twentieth century are based either on Cassegrain designs or on later Ritchey-Chrétien variations thereof.

Because reflector telescopes are simpler to construct than refractors, they facilitated rapid growth in aperture size. By the end of the eighteenth century, in 1789, William Herschel (1738–1822)—best known for his 1781 discovery of Uranus—had constructed a 48-inch reflector, an instrument that would remain the world's largest until 1845 when Lord Rosse (William Parsons, 1800–1867) completed an enormous 72-inch instrument, the "Leviathan of Parsonstown."

New opportunity arose for refractors in 1729 with the invention of the achromatic doublet by Englishman Chester Moor Hall (1703–1771). Yet, over the next hundred years or so refractor apertures barely increased in size. It took improvements in optical fabrication methods and glass quality to make the crucial difference. Beginning around 1840—when the largest refractor aperture was only about 10 inch—refractor aperture sizes grew steadily, culminating in the Yerkes 40-inch refractor which saw first light in 1897.

This instrument, the creation of American astronomer George Ellery Hale (1868–1938) with funds provided by the Chicago railway magnate Charles Tyson Yerkes, still remains the world's largest refractor. But despite its considerable size, this instrument still did not surpass in size Lord Rosse's 72-inch instrument. Soon, with the coming of the twentieth century, even larger reflector telescopes would come to dominate, leaving refractor telescopes—at least for now—as a colorful part of the rich history of the telescope.

Before the Yerkes instrument was completed, Hale already had plans for even larger reflector telescopes. In 1908, his 60-inch reflector saw first operation on Mt. Wilson and within a few years, with funds supplied by the Los Angeles businessman John D. Hooker and the Carnegie Institution, he began planning the famous 100-inch instrument. The latter instrument, which saw first light on Mt. Wilson in 1917, would hold the honor of being the world's largest until 1948 when it was surpassed by the 200-inch Mt. Palomar telescope—yet another telescope orchestrated by Hale, though it was one he would not live to see commissioned.

In 1975, the Palomar telescope was surpassed in size, though not in optical performance, by the 6-m (236-in.) BTA-6 Soviet Union telescope.[6] Beginning around 1990, the monolithic primary mirrors that were used in these and other large telescopes lost favor, interest having shifted to much thinner and lighter primary mirror designs, such as the segmented design used by the 10-m Keck telescopes and the thin meniscus design used by the twin 8-m Gemini telescopes.

[6] Translated into English, BTA stands for Bolshoi Telescope Altazimuth. This instrument pioneered the use of an altazimuth mount with a computer-controlled de-rotator, now a standard feature in large astronomical telescopes.

The success of the Keck segmented mirror principle has strongly influenced design thinking for the even larger, next-generation Extremely Large Telescopes (ETTs), including the 25-m Giant Magellan Telescope, the Thirty Meter Telescope (TMT), and the 39-m European Extremely Large Telescope (E-ELT). Wave theory, and the phenomenon of diffraction, put an end to the alluring prospect of limitless resolution that Herschel had once thought possible by simply increasing magnification, but not until after protracted delay.

Newton and his contemporaries were only familiar with longitudinal waves, as in sound waves. To Newton, the "transversality" of light, as implied by the discovery of polarization by Huygens, constituted an insuperable objection to wave theory. Faced with Newton's formidable opposition, wave theory gained little ground for nearly one hundred years. But attitudes began to change at the beginning of the nineteenth century.

In 1801, Thomas Young (1773–1829) enunciated the principle of interference and explained the colors in thin films. In 1822, he commented "on some lines of light and shade" that appeared around small objects viewed in a microscope, interpreting these in terms of wave theory. Joseph von Fraunhofer (1787–1826) and John Herschel (1792–1871), son of William, commented on the spurious disk and the surrounding system of faint rings. Both recognized, as did Young, that these were wavelike phenomena.

In 1818, Augustin-Jean Fresnel (1788–1827) combined Huygens' intuitive secondary wavelet principle with the principle of interference and was able to accurately describe a number of diffraction effects. In 1834, the Astronomer Royal, Sir George Airy (1801–1892), used wave theory to give a power series expression for the light distribution in a star image formed by a diffraction-limited telescope with circular aperture. His expression described the image—now called an Airy pattern—as having a bright central disk surrounded by a system of bright and dark rings. In 1882, Gustav R. Kirchhoff (1824–1887) placed the Huygens Fresnel principle on a more solid mathematical footing by deriving it from the Fresnel–Kirchhoff formula. With wave theory now well established, astronomers now faced the disappointing reality: Telescope resolution was after all fundamentally limited; magnification beyond a certain degree was pointless.

In 1880, Lord Rayleigh (John William Strutt, 1842–1919) proposed an angular resolution criterion for telescopes based on point sources and telescopes with circular apertures.[7] According to the Rayleigh criterion, the just-resolved condition is attained when the twin components of a two-point-object—such as a binary star— are separated so that the geometrical image of one lies on the first dark ring of the other. This criterion gives rise to the widely used formula for the angular resolution of a telescope, expressed in radians by $1.22\ \lambda/D$. Though defined somewhat arbitrarily, Rayleigh's criterion has become firmly established as the standard metric for assessing telescope resolution.

[7] Rayleigh's criterion for telescope resolution is similar to one he proposed in 1879 for spectrograph resolution.

In 1855–1856, James Clerk Maxwell (1831–1879) demonstrated that electric and magnetic fields travel through space in the form of waves. In 1862, he calculated the propagation speed of these waves based on electric and magnetic data measured in the laboratory. Finding it to be approximately the same as the speed of light and recognizing the enormous significance, he commented, "We can scarcely avoid the conclusion that light consists of the transverse undulations of the same medium which is the cause of electric and magnetic phenomena." In 1864, he published his earth-shaking paper, "Dynamical Theory of the Electromagnetic Field." That same year, in a spoken presentation of this paper to the Royal Society of London, he commented: "We have strong reason to conclude that light itself—including radiant heat and other radiation, if any—is an electromagnetic disturbance in the form of waves propagated through the electro-magnetic field according to electro-magnetic laws."

Maxwell's theory—the first field theory—was developed at a time when scientists were more familiar with notions such as action at a distance. His new theory left most of his contemporaries puzzled and suspicious, including influential ones such as George Airy who still felt that a proper understanding of electromagnetic phenomena was more likely to come from a dynamical approach of the kind Maxwell had devised earlier. To Airy and other contemporaries, Maxwell's field theory was a step backward. For the next 20 years, Maxwell's theory lay dormant.

There were no significant developments until 1887, when the German physicist, Heinrich Hertz (1857–1894), with a cleverly designed laboratory experiment, detected signals produced by a spark gap. This was the first definitive evidence of electromagnetic waves. Things moved quickly thereafter, but alas this was now eight years after Maxwell's death. Remarking on the long struggle of Maxwell's electromagnetic theory to gain acceptance, Born and Wolf[8] observed, "It seems to be a characteristic of the human mind that familiar concepts are abandoned only with the greatest reluctance, especially when a concrete picture of phenomena has to be sacrificed."

The year 1887 was also the year Michelson and Morley carried out their famous ether-drift measurement experiment in a laboratory at Case Western University, Ohio. Using a device invented by Michelson—later known as an interferometer—the measured drift was found to be much slower than the 30 km/s needed to account for Earth's motion round the Sun.[9] With experimental error taken into account, perplexingly it seemed that ether-drift velocity might be zero. With the speed of light, denoted by c, now seeming to be unaffected by the speed of the observer, Newton's classical mechanics fell into disarray. Order was restored in 1905 by Einstein who made sense of both Maxwell's equations and the Michelson–Morley experiment by his breathtaking special theory of relativity (Einstein, 1905).

The central axiom of Einstein's theory is that the speed of light c is the same for all observers, making it a fundamental characteristic of the universe. Einstein had succeeded in combining space and time into what is now known as space–time. The

[8] "Principles of Optics," Max Born and Emil Wolf, fourth edition (1975).

[9] The Sun's orbital speed around the galaxy—about 275 km/sec—is an order of magnitude greater than Earth's orbital speed around the Sun.

consequences were non-intuitive and hard to comprehend. Distances and time could no longer be considered absolute; they now depended on the relative speed of the observer.

Of course, c was already central to Maxwell's electromagnetic equations. It was simply the gearing ratio between space and time, defined in his equations as the ratio of the electromagnetic and electrostatic units of charge. Though relativity theory devalued Newton's stock, Maxwell's stock instantly appreciated. Remarkably, his equations remained unchanged when subjected to the transform equations of Hendrik Lorentz (1853–1928, equations that describe the compression of space and time at relativistic speeds and which form a central part of Einstein's theory. Thus, Maxwell's equations found themselves at the very core of special relativity (Mahon, 2004) thereby elevating them to an even higher level of primacy.[10]

On the centennial of Maxwell's birthday, Einstein stated: "Since Maxwell's time, physical reality has been thought of as represented by continuous fields, and not capable of any mechanical interpretation. This change in the conception of reality is the most profound and the most fruitful that physics has experienced since the time of Newton." Maxwell's equations set the stage, not only for the arrival of relativity theory, but also for quantum mechanics. Remarkably, his equations survived the rigors of both these momentous theories unscathed.

The construction of ever larger telescopes during the twentieth century promoted astounding advancements in cosmological and scientific understanding. In 1925, only eight years after the 100-inch telescope saw first light, measurement data obtained from this instrument by Edwin Hubble (1889–1953) culminated in his landmark paper indicating that the Milky Way was not, after all, the whole universe. Distance measurements of several nebulae including the Andromeda nebula [using the period–luminosity relation for Cepheid variable stars discovered in 1912 by Henrietta Leavitt (1868–1921)] indicated to Hubble that these objects lay one hundred thousand times further away than the nearest stars (Voller, 2012). Staggeringly, Hubble explained nebula as being other galaxies, lying far beyond the Milky Way.

Einstein anticipated an expanding universe in his 1916 general theory of relativity.[11] However, in 1917, thinking that the universe must be constant and unchanging, he refined his theory, fudging the calculations by the ad hoc addition of a "cosmological constant." In 1929, in a second landmark paper, again based on measurements made with the 100-inch telescope, Hubble showed a linear correspondence between the distance and speed of recession of nearby galaxies. This remarkable relation not only provided an observational basis for an expanding universe, it also gave the expansion rate—now known as the Hubble constant.

The "expanding universe" idea had been proposed earlier in 1927 by Georges Lemaitre (1894–1966) who had also concluded that a "creation-like" event must

[10] Maxwell identified the speed of light c as a fundamental constant in his unified electromagnetic theory. Einstein identified c as a fundamental constant in his special theory of relativity, a theory that unified space and time into so-called space–time.

[11] Einstein, A. (1915) Berl. Sitz. 778, 779, 831, 844. (1916) Ann. D. Physik, (4) 49, 769. Einstein's general theory of relativity unified his 1905 special theory of relativity with Newton's laws of gravity.

have occurred. Einstein, while initially disagreeing with Lemaitre, reversed his position after seeing Hubble's 1929 data; thereafter, he gave unreserved support to the expanding universe idea. Subsequently, Einstein referred to his 1917 cosmological constant as his "biggest blunder." However, in yet one more twist, nowadays cosmologists suspect the existence of other physical entities that have influences akin to Einstein's cosmological constant. Referred to as dark energy and dark matter, it now seems that these two entities account for 95% of the cosmic density, 73% and 22%, respectively. Despite their immense presence, because neither emits light and they cannot therefore be seen directly by telescopes, little is known about them.

The name given to the creation event—the big bang—was coined in 1949 by Sir Fred Hoyle (1915–2001). Ironically, Hoyle was passionately opposed to big bang Theory, preferring instead the alternative steady state theory. First conjectured in 1928 by Sir James Jeans (1877–1946), the steady state theory was revived in 1948 by Hoyle, Thomas Gold, Hermann Bondi, and others. The two competing theories drove research, most of which ultimately supported the big bang theory. The discovery in 1964 of the cosmic microwave background, with its collected frequencies sketching out a blackbody curve, convinced most cosmologists that some sort of big bang event must have occurred.

Einstein's general theory of relativity predicted the deflection of light by gravity—a phenomenon now called gravitational lensing. The first concrete evidence of the phenomenon was obtained by Arthur Eddington (1882–1944) who, in 1919, armed with several small telescopes, led a total solar eclipse expedition to the Principe Island off the west coast of equatorial Africa. His confirmation that stars lying in the direction of the Sun appeared shifted by roughly the amount predicted (about 1 arcsecond) made Einstein instantly famous. But many remained skeptical. Could Eddington (Berman, 2009) really have made such precise observations of 5th magnitude stars in the Hyades cluster looking through blurry air and through the solar corona with a 4-inch telescope?

Mainstream acceptance of general relativity theory did not come until later—between 1960 and 1975—when astronomers and physicists advanced understanding of relativistic cosmology via concepts such as black holes and quasars. Since 1979, hundreds of gravitationally distorted images have been observed. The most famous of these, discovered by John Huchra (1948–2010) and now referred to as Einstein's Cross, is a spectacular example, the image constituted by four separate images of a distant quasar imaged by the gravity of a foreground galaxy—a galaxy now referred to as Huchra's lens.[12]

Without realizing the enormous significance, in 1802, William Wollaston (1766–1828) noticed dark lines in the solar spectrum. In 1817, these were rediscovered by Joseph von Fraunhofer (1787–1826) who accurately cataloged more than 600 of them. Neither Fraunhofer nor his scientific contemporaries could explain these 'Fraunhofer lines.' Nor were they explained later by Maxwell's electromagnetic theory. One hundred years would elapse before a valid explanation emerged.

[12] The quasar lies some 8 billion light-years distant; the foreground galaxy—Huchra's lens—lies a mere 440 million light-years away.

The basis of quantum theory is the 1900 supposition by Max Planck (1858–1947) that electromagnetic energy is emitted in quantized packets. Energy can only arise in integer multiples of the elementary unit, h v, where h is Planck's constant (introduced in 1899) and v is the radiation frequency. Considered by Planck as "only a formal assumption ... actually I did not think much about it ..." the assumption turned out to be entirely incompatible with classical physics and is regarded as the birth of quantum physics.

In 1905, Einstein explained the photoelectric effect – the emission of electrons from metals and other substances when light strikes their surfaces. Einstein's explanation was closely related to Planck's 1900 work on energy quantization, where light is composed of discrete particles of energy—named photons in 1926. The present-day understanding of light is one of wave–particle duality, in which light particles and waves are seen as possessing certain properties of both but cannot be regarded as solely one or the other.

Despite the strong and mounting evidence for quantization, other physicists such as Rayleigh, Jeans, and Lorentz set Planck's constant to zero to align with classical physics. But Planck knew that it had a precise, nonzero value: "I am unable to understand Jeans' stubbornness—he is an example of a theoretician as should never be existing ... So much the worse for the facts if they are wrong." Later, Max Born would write about Planck: "... his belief in the imperative power of logical thinking based on facts was so strong that he did not hesitate to express a claim contradicting all traditions, because he had convinced himself that no other resort was possible.

Planck expected that wave mechanics would soon render his creation—quantum theory—unnecessary. However, further work only cemented his theory, even against "his and Einstein's revulsions." Planck's struggle against older views would go on for many years, causing him to observe (Tolstoy, 1981):

> "A new scientific truth does not triumph by convincing its opponents and making them see the light, but rather its opponents eventually die, and a new generation grows up that is familiar with it."

In a pithier version of the same sentiment he stated

> "Science advances one funeral at a time."

Einstein, as did others, disliked the probabilistic nature of quantum mechanics. He considered it incomplete, thinking that there must be some underlying determining mechanism, or some local hidden variable. In a letter to his lifelong friend, Max Born, he famously wrote, "God does not play dice with the universe." Nonetheless, the present understanding would seem to indicate the contrary.

## 1.1 Telescope Imaging Through Earth's Turbulent Atmosphere

With the enormous growth of telescope apertures in the nineteenth and twentieth centuries, it became increasingly clear that telescope resolution was more limited by atmospheric turbulence than by diffraction from the telescope aperture. According to the Rayleigh criterion (1.22 λ/D), at visible wavelengths a 100-inch telescope should resolve to about 1/20th of an arcsecond and a 200-inch telescope to about 1/40th of an arcsecond. Disappointingly, the resolution levels obtained by the Mount Wilson 100-inch and Mt. Palomar 200-inch instruments, 0.5–1.0 arcsecond, were lower than theoretical limits by almost two orders of magnitude.

Imaging through the turbulent atmosphere with large ground-based telescopes remained poorly understood until the latter part of the twentieth century. Earlier in the century, understanding had been notional, lacking credible mathematical foundation. A significant breakthrough came in 1964 with Hufnagel and Stanley's (1964) solution to the Helmholtz equation—of which we will learn more in Chap. 3.

They solved the equation for the case of an initially plane light wave propagating through a randomly fluctuating refractive index field. Plane waves, of course perfectly describe light waves from distant stars as they arrive at the top of the atmosphere, while a randomly varying refractive index field perfectly describes—for our purposes—Earth's turbulent atmosphere.

## 1.2 Kolmogorov Theory (Mid-1960s)

Recognizing that there was no need to solve the Helmholtz equation for the amplitude and phase of the light waves explicitly, Hufnagel and Stanley instead solved directly for the atmospheric modulation transform function, or atmospheric MTF. This key function provides useful—though ambiguous[13]—information for calculating the blurring effect of the atmosphere on star images and, by extension, the blurring effect on more complicated objects, such as star clusters, the Moon, and planets.

Hufnagel and Stanley applied their new result to light waves propagating through an atmosphere constituted by Kolmogorov turbulence. From the resulting atmospheric MTF they were able to predict how star images might appear at visible wavelengths in large ground-based telescopes for that type of turbulence. The predicted images were indeed similar to the seeing-disc star images familiar to astronomers at that time. Imaging through the atmosphere had finally yielded—or so it appeared—to mathematical description. Hufnagel and Stanley's important contribution formed the basis of what became known as Kolmogorov theory, named after the Russian turbulence physicist, Andrey N. Kolmogorov (1903–1987).

Kolmogorov theory is based on the postulation that atmospheric turbulence[14] always comprises the same relative proportions of structure sizes, irrespective of

---

[13] The same atmospheric MTF can arise from many different types of atmospheric turbulence spectra as shown in Chap. 2 (Sect. 2.1). To unambiguously identify the correct type requires use of a more general function the two-point two-wavelength correlation function (Chap. 6, Sect. 6.3)—an expression for which was developed by the Author in 1975 though not published until 1991 (McKechnie 1991a, 1991b, Eq. 13).

[14] Atmospheric turbulence can be considered a catch-all term that refers essentially to the random temperature- and pressure-related refractive index variations in the atmosphere.

geographical location, altitude, or the prevailing seeing conditions; only the turbulence strength is permitted to vary within these parameter spaces. The smallest turbulence structures are postulated to measure less than one centimeter across and the largest many tens of meters across, with the intermediate sizes disposed in certain fixed, self-similar proportions—the entirety known as Kolmogorov turbulence.

In the mid-1960s when Kolmogorov theory was developed, it was widely believed that resolution through the atmosphere with large ground-based telescopes would rarely be better than 0.5- to 1-arcsecond. In accordance with that belief, Kolmogorov theory saw no need to prescribe telescope optical tolerances in terms of a precise diffraction-based standard, such as wave front error. Instead, it chose to adopt the same crude, geometrical optics-based standard that had been used earlier in the century to build the 100-inch and 200-inch instruments. According to that standard, large ground-based telescopes merely had to be 'intrinsically' capable—that is, in the absence of atmosphere—of concentrating most of the light in a star image into a ~ 0.5-arcsecond diameter image patch. But, what if this ~ 0.5-arcsecond telescope performance prescription was mistaken? Or what if any of the one-size-fits-all Kolmogorov turbulence assumptions were mistaken? With the lack of substantive answers to questions like these, could it be justly claimed that Kolmogorov theory was scientifically sound and trustworthy?

Despite these and other unresolved issues, the incorporation of Hufnagel and Stanley's solution of the Helmholtz equation into Kolmogorov theory gave the theory considerable weight. That weight, together with aggressive promotion by the theory's originators and bristling intolerance of any criticism of the mysteriously conjured Kolmogorov turbulence assumptions, helped secure the theory's widespread acceptance and use.[15]

But alas, the adoption by Kolmogorov theory of the ~0.5-arcsecond telescope optical performance standard ensured that images delivered by large ground-based astronomical telescopes constructed in the 1970s and 80s would be no sharper than those delivered by telescopes built half a century earlier. With the development of adaptive optics still several decades off in the future, there was no near-term prospect of ground-based telescopes delivering resolution significantly better than the supposed 0.5- to 1-arcsecond ceiling.

[15] Kolmogorov theory was developed by a loosely connected group of individuals, principally, A. N. Kolmogorov, V. I. Tatarski, R. E. Hufnagel, N. R. Stanley and D. L. Fried. Each made separate contributions, leading to something of a patchwork theory sewn together by – as fate would have it – dimensionally inconsistent mathematics (Appendix I). As others embraced the theory, its development effectively became 'open source.' A notable contribution was made by R. J. Noll (Noll, 1976) who established expressions for the Zernike polynomial coefficients for Kolmogorov turbulence.

## 1.3 Origins of the New General Theory

Such was the stalled—or perhaps stifled—situation in 1974 when the Author's research activities broadened from light scattering off single surfaces to continuous light scattering and imaging over extended atmospheric paths. In the latter area, two initiatives were soon identified that might help improve understanding:

(1). Take rigorous account of diffraction by the telescope optics, and
(2). Avoid making arbitrary, a priori, assumptions about the properties of atmospheric turbulence, Kolmogorov or other.

A mathematical formulation supporting these initiatives was developed by mid-1975 and mapping equations were established that enabled turbulence properties over atmospheric observing paths to be calculated directly from measurable properties of star images. The inverse equations were also established.

The new general theory was first field-tested in September 1975 during a seven-night observing trip to the Royal Greenwich Observatory (RGO), Herstmonceux, UK. To the Author's astonishment, the star image measurement data consistently indicated turbulence properties significantly different from those postulated in Kolmogorov theory. Most of the turbulence structure was found lying in the much smaller size range—from about 1 to 50 cms.

If the mapping equations were correct and the atmospheric turbulence properties calculated from them were representative of turbulence at other observatory sites, the consequences for ground-based observational astronomy would be both staggering and profound.

Calculations based on the foremost telescope in the world at that time—the Mt. Palomar 200-inch telescope—indicated that if that instrument were to observe through an atmosphere similar to the (non-Kolmogorov) atmosphere lying above the RGO site, star images obtained by it would no longer necessarily have the familiar seeing-disc appearance. While a portion of the light energy in the image would indeed scatter into a feature similar to a seeing-disc, referred to now as a halo, the remaining light portion would image into a more highly resolved feature, known as an image core.

Star images comprising both features are known as core and halo images. The new theory categorizes this type of image as the most general type of star image, with halo-only images (i.e., seeing-disc images) and core-only images now merely special cases in the short- and long-wavelength extremes.

As indicated in Chaps. 9 and 10, star image cores formed by large ground-based telescopes have exactly the same angular size and shape as star images formed by these instruments in the absence of atmosphere. Intriguingly, for telescopes with circular apertures constructed to diffraction-limited precision, image cores arise as ideal Airy patterns. At longer wavelengths, the Airy pattern cores naturally broaden in proportion to the wavelength. They also brighten exponentially, ultimately becoming the dominant features in star images. For aberrated telescopes, cores become distorted, often showing comatic or astigmatic characteristics. For severely aberrated telescopes, cores may even become unrecognizable, their light energy now spread so widely and thinly that it is no longer distinguishable from the halo.

Contrary to all previous understanding and practice, to fully exploit the astonishingly high resolution levels offered by image cores, large ground-based telescopes would now have to be constructed to diffraction-limited precision. A 200-inch instrument built to such precision would now be expected to deliver 0.05-arcsecond resolution at the near-IR wavelength, 1 $\mu$m, and even 0.025-arcsecond resolution at visible wavelengths in pristine, sub-arcsecond seeing.

To advance such revolutionary new propositions in the mid-1970s carried significant risk. What if there was an error in the mapping equations, or perhaps an error in the experimental procedure? To avoid potentially career ending humiliation and embarrassment, it was only prudent to proceed cautiously. A draft manuscript (McKechnie, 1976a, 1976b) showing the development of the mapping equations was circulated in 1975 to a select number of individuals.[16] No obvious errors having been identified, the material was presented at the 1976 annual meeting of the Optical Society of America, held in February of that year in Monterey, CA. The following excerpt from the talk abstract (McKechnie, 1976a, 1976b) captures the radical, albeit tentative conclusions:

> ... in the Author's opinion, the observations of R. F. Griffin,[17] made using the Hale 200-inch and the Mt. Wilson 100-inch telescopes are in such close agreement with the predictions of the [new] theory that it is almost certain that this observer has seen central cores. By optimizing some of the factors that influence the brightness of the core, it may be possible to use this phenomenon as a tool for improving the resolution of large telescopes.

The Talk was received with a mixture of incredulity and bemusement. It was nonetheless reassuring that nothing new or unexpected was offered by those audience members who rose to the defense of Kolmogorov theory. One prominent Kolmogorov advocate flatly dismissed the possibility of image cores, arguing that the complete absence of photographic evidence was surely proof enough that the claimed phenomenon was illusory. In the words of Alber Guinon,

> When everyone is absolutely against you, it means you are absolutely wrong – or absolutely right.

## 1.4 Publication of the New General Theory (1989/90)

With progress on the new theory interrupted by life's vagaries, job changes and relocation to the United States, it would take the Author a further twelve years—until 1989—to develop rigorous proofs of the cornerstone elements in the mapping

[16] Copies of the manuscript, "Light propagation through the atmosphere and imaging in astronomy," were circulated to J. C. Dainty and others at Imperial College along with several specialist mathematicians.

[17] Laterally, Roger F. Griffin (1935 – 2021) was Emeritus Professor of observational astronomy, Cambridge University, Cambridge, UK.

equations. Now confident that the equations were correct, the need for caution had finally lifted. It was time to publish the new theory, (McKechnie, 1990, 1991a, 1991b, 1992) its startling content evident in the following excerpt from the 1990 paper Abstract:

> In night seeing conditions at typical telescope sites, the optimum wavelength lies in the range, 1.0 to 2.5 micrometers. For a 5-meter telescope [the largest in existence, circa 1990] it should be possible to obtain resolution of the order 0.05 to 0.15 arc seconds routinely in this wavelength range. However, to facilitate such high levels of resolution, the telescope must be diffraction limited.

New theories rarely please everyone. By challenging previous understanding and long-cherished beliefs, they can cause discomfort, suspicion, and bruised egos. But if the new theory directly contradicts an established understanding with which many have become familiar and grown to trust, and some may have used—perhaps injudiciously—as the basis of high-profile development initiatives, the reaction can be considerably more fraught.

Those whose careers and prestige owe directly to their association with the development and/or exploitation of the old theory, now face the cheerless prospect of embarrassment and tarnished reputations. In such circumstances, rather than embrace the new theory, denial might offer a more comfortable option. If the more influential and invested of those affected choose the latter path, and subordinates fall into line, scientific advancement can find itself obstructed by an injured and dangerous foe who may resort to any strategy, fair or unfair, within the generous bounds afforded by unprincipled self-interest. Thomas Gold (1989) has written about the behavioral dynamics in such situations.

> "If a large proportion of the scientific community in one field is guided by the herd instinct, then they cannot adopt another viewpoint since they cannot imagine that the whole herd will swing around at the same time. It is merely the planning of the situation. Even if everybody were willing to change course, nobody individually will be sure that he will not be outside the herd when he does so. Perhaps if they could do it as neatly as a flock of starlings, they would. So this inertia-producing effect is a very serious one. It is not just the herd instinct in the individuals that you have to worry about, you have to worry about how it is augmented by the way in which science is handled. If support from peers, if moral and financial consequences are at stake, then overall staying with the herd is the successful policy for the individual who is depending on these, but it is not the successful policy for the pursuit of science."

Resistance to new scientific theories is nothing new; the more disruptive the theory the fiercer the resistance. History abounds with tales of initial rejection of new theories. Earlier in the chapter, we saw how Copernicus, Galileo, Maxwell, Planck, and even Einstein had to contest and endure decades of resistance to their new theories; in some instances mainstream acceptance only came posthumously.

Publication of the new general theory (circa 1989/90) met dogged resistance from various journal referees and even journal officers, quagmiring the process with a seemingly endless succession of objections and ploys, some bizarre, some plainly mischievous. One referee submitted mathematics showing how Kolmogorov theory 'could readily explain' the 0.1-arcsec star images shown in one of the papers—a submission hastily withdrawn when a square root error was identified in his mathematics. Undismayed, he and a colleague repeated the same misinformative claim in a subsequent rebuttal article in which mathematics was conspicuously absent.

The resistance caused an excruciating nineteen-month publication delay for one paper, while another suffered an even more excruciating thirty-month delay. Though trying, these delays wrought no concessions. The published papers stood firm on each of the radical and disruptive predictions.

But as well as opposition, there was also encouragement and support, most notably from the distinguished US scientist, Prof. Charles H. Townes (1915 – 2015). Townes shared the 1964 Nobel Prize in physics for the invention of the laser[18] and, subsequently, from 1966 to 1970, directed the US Government Science and Technology Advisory Committee for the momentous Apollo lunar landing program—whose success in July 1969 many consider mankind's most magnificent achievement.

During a 1991 presentation by the Author (McKechnie, 1991a, 1991b) on the new theory, Townes happened to be present in the audience. After the presentation, he approached, introduced himself and, in the conversation that followed, spoke of his own misgivings about Kolmogorov turbulence assumptions.

At the time, Townes was leading a team from the University of California, Berkley, making independent atmospheric measurements (Bester et al., 1992). When one of the team members and paper co-author, Dr. William C. Danchi, was later asked for his views on Kolmogorov turbulence, he responded categorically (Danchi, 2001):

> We found that you see Kolmogorov turbulence only when the seeing conditions are poor, which are not conditions for observational astronomy. Under good seeing conditions the atmosphere is non-Kolmogorov.

In the summer of 1989, optics personnel at the US Air Force Research Laboratory (AFRL), Albuquerque, New Mexico, expressed interest in obtaining tangible proof of star image cores. To that end, they invited astronomers, K. Hege and D. W. McCarthy (Steward Observatory, Arizona), to attend a seminar[19] given by the Author: "Obtaining diffraction limited images at near infrared wavelengths using large ground based telescopes." Details were spelled out at the seminar of a definitive imaging experiment that would either prove, or disprove, the existence of image cores. (Particulars of the experiment are given in Chap. 10.)

---

[18] Townes also acted as science adviser to every US president from Harry S. Truman to William J. Clinton.

[19] Seminar presentation (September 26, 1989) by the Author, "Obtaining diffraction limited images at near infrared wavelengths using large ground based telescopes." AFRL (Bldg. 400), Albuquerque, New Mexico, USA. (Attendees: D. W. McCarthy, Jr., K. Hege, Steward Observatory, Tucson, G. Loos, B. Venet, AFRL, R. Haddock, Lentec Corp.

## 1.5 Definitive Confirmation of Cores in Star Images

The experiment was carried out in late-1989 using the 5.6-m Multi Mirror Telescope (MMT), Mt. Hopkins, Arizona. That particular instrument was chosen because one of its mirrors (Mirror E) was known to have extremely high optical quality. Almost immediately, 0.1-arcsecond star image cores were confirmed at the near-IR imaging wavelength, 2.2 μm, even though visible seeing was an unremarkable 1.25-arcsecond.

So finally, after decades of argument and controversy, here were actual recorded images of star image cores. No longer could cores be regarded as "illusory." Soon afterwards, similar 0.1-arcsecond star image cores were obtained (McKechnie, 1992) using the 3.8-m Mayall instrument, Kitt Peak, AZ. The latter images can be viewed in Chap. 10 (Figs. 10.10 and 10.11).

### *1.5.1 The UKIRT 3.8-m Instrument and Star Image Cores*

In July 1990, star images obtained by the 3.8-m Mayall telescope were shown at a seminar presentation given by the Author and hosted by Mr. Eli Etad at the Royal Observatory Edinburgh (ROE), the administrative headquarters of the 3.8-m United Kingdom InfraRed Telescope (UKIRT), Mauna Kea, Hawaii. Astonished and intrigued by the 0.1-arcsec image cores they had just seen, ROE astronomers were soon attempting to harvest similar images using UKIRT (but now using 13 × extra magnification so that the pixel spacing in their high resolution camera was effectively reduced from 0.65 arcsec to 0.05 arcsec, small enough to resolve the tiny image cores.)

Sadly, their efforts were entirely fruitless. Even star images obtained at the longer and therefore less challenging wavelength, 3.4 μm, failed to show image cores. In September 1990, when Dr. Colin Humphries (ROE) telephoned with the disappointing news, there was little one could say in consolation. The problem pointed directly to UKIRT's optics.

UKIRT was constructed in the 1970s (1979 first light) as a low-cost instrument, built to relaxed, Kolmogorov-based optical tolerance prescriptions. But, since the instrument was still capable of delivering the 'full' 1-arcsecond resolution 'permitted' by the atmosphere, it was widely celebrated as an engineering cost/performance triumph. However, following its troubling failure in 1990 to deliver near-IR image cores similar to those obtained by the 3.8-m Mayall instrument—despite average Mauna Kea seeing being 40% better than Kitt Peak seeing—the celebrations abruptly ended.

It was now painfully clear to all concerned that the UKIRT instrument was grossly underperforming, even when compared to rival instruments that themselves were far from diffraction-limited. By the fall of 1990, substantial funding (£650,000) had been obtained by ROE to upgrade UKIRT's optics.

Once the upgrades were finally completed in 1998, UKIRT had been transformed from a crude "light bucket" instrument into a magnificent, near diffraction-limited instrument. According to the UKIRT website (Hawarden, 1998) the instrument was now delivering the sharpest images of any large ground-based telescope in the world — images claimed as

> "close to the limit set by the wave nature of light (the diffraction limit) at wavelengths of 2 μm or longer ..." [The best UKIRT images now rivaled] "those of the Hubble Space Telescope's NICMOS infrared imager for sharpness."

Except for the most intransigent, it was now incontrovertibly clear that imaging with large ground-based telescopes was a far different—and much more favorable—proposition than previously recognized. Following its mid-1960s development, Kolmogorov theory had 'justified' and encouraged the construction of cheap, under-performing ground-based telescopes. But now, more than thirty years later, it had become shockingly clear that, rather than helping ground-based astronomers deliver sharper images, this flawed theory had instead brought progress to an abrupt halt.

## 1.6 Horace Babcock and Adaptive Optics

In 1953, Horace W. Babcock (1912–2003)[20] proposed the prescient and startling new idea of using adaptive optics to correct atmospherically induced image distortions [Babcock, 1953]. Alas, another forty years would elapse before technological advances were able to transform his ideas into practical AO systems.

As described in later chapters, the new general theory applies equally to telescopes with and without AO capability. But that did not impress Dr. Babcock, at least not initially. In a December 1990 newspaper article about the new general theory,[21] "Looking Sharp on Earth, Ground-Based Telescopes Can Be as Good as Hubble, Physicist Says," Babcock's solicited assessment was nothing if not blunt:

> "Bad seeing is caused by the atmosphere – there is no doubt about it… McKechnie simply doesn't know the score."

Some six months later, at the May 1991 Annual Meeting of the American Astronomical Society, Seattle, WA., Dr. Babcock happened to be in the audience during a presentation by the Author (McKechnie, 1991a, 1991b) and saw for the first time the 0.1-arcsecond star image cores obtained from the Mayall instrument. Afterwards,

---

[20] From 1964 to 1978, Horace Babcock was director of Palomar Observatories and thus had authority over two of the largest telescopes in the world at that time, the 200-inch Mt. Palomar and the 100-inch Mt. Wilson instruments.

[21] Albuquerque Journal, Mon Dec 17, 1990, Article Headline: "Looking Sharp on Earth, Ground-Based Telescopes Can Be as Good as Hubble, Physicist Says." To balance the Article, the Science Editor Author, Rex Graham, telephone-interviewed several astronomers, including Dr. Babcock.

to his credit, Dr. Babcock was first to his feet to acknowledge the new images, volunteering that the resolution achieved in them was.

> "entirely new to my experience."

Carl E. Sagan (1934–1996) the celebrated American astronomer and cosmologist commented on the age-old problem of inertial resistance to new scientific theories:

> "That the Earth is at the center of the universe was once obvious. The truth may be puzzling or counterintuitive; it may contradict deeply held prejudices. But, as the history of both science and politics has amply demonstrated, preferring comfortable error to the hard truth is, sooner or later, disastrous."

## 1.7 Pivotal Equation Yielded by the New General Theory

While authoring his widely-acclaimed book, "A brief history of time" (Hawking, 1988) Stephen W. Hawking (1942–2018) was advised that every mathematical equation he included would halve the book's sales. Though mostly heeding the advice, Hawking could not resist including Einstein's momentous mass-energy equivalence equation,

$$E = mc^2$$

Had the Author been similarly restricted to just one equation, it would undoubtedly have been the following (McKechnie, 1991a, 1991b, Eq. 77):

$$\lambda_{opt} = 2\pi\sigma,$$

where $\lambda_{opt}$ denotes the so called 'optimum wavelength' and the crucial parameter, $\sigma$, denotes the root mean square (rms) Optical Path Difference (OPD) variation imprinted by the turbulent atmosphere on the image-forming light wave portions collected by the telescope.

The equation's modest appearance belies the fact that it provides pivotal insight into light propagation and imaging through the atmosphere in several crucial ways. First, it identifies the optimum wavelength, $\lambda_{opt}$, at which any large ground-based telescope will deliver maximum possible resolution in given seeing conditions. The equation therefore identifies the center of what is now popularly referred to as the 'sweet spot' wavelength region. By use of AO to shrink $\sigma$, the optimum wavelength also shrinks, providing a proportional resolution improvement.

Second, whenever natural guide stars are used as reference objects, the equation identifies the wavelength region best suited for maximizing tracking and image stabilization precision. Third, it identifies the optimum wavelength for laser beam propagation through the atmosphere, including High Energy Laser (HEL) beams.

In the latter application, a telescope projects an image of a point-object—albeit an extraordinarily potent point-object—on to a distant target. Once a representative $\sigma$

value has been established for the entire end-to-end optical beam path, the equation identifies the unique wavelength, $\lambda_{opt}$, that comes to the sharpest focus and thus delivers maximum possible irradiance flux (Watts/m$^2$) at the target.

By applying AO pre-corrections to the outwardly directed HEL beam, $\sigma$ can be minimized but as discussed more fully in Chap. 16, it cannot be reduced to zero. The lowest realistically achievable value of $\sigma$ identifies the HEL wavelength region for delivering maximum beam irradiance, or beam potency, at the target. Use of HEL wavelengths substantially different from $\lambda_{opt}$—whether longer or shorter—can decimate target irradiance levels, catastrophically reducing the effectiveness of the system.

Following the equation's development in 1989, its troubling implications to the future AirBorne Laser (ABL) program were instantly clear. Though the ABL program (1996–2012) would not formally begin for another seven years, the HEL wavelength had already been set at 1.315 μm by the selection of the oxygen iodide laser (COIL).

For the envisaged long-range ABL engagement scenarios, the equation indicated that the 1.315-μm wavelength choice could hardly have been any less suitable. It not only indicated failure of the program from the outset, it spelled failure by a disturbingly large margin (Sect. 16.9.1.1).

## 1.8 Future Direction of Ground-Based Observational Astronomy

Nowadays, most large ground-based astronomical telescopes (e.g., Keck and Keck II) are equipped with AO systems, enabling them to achieve near diffraction-limited resolution at near-IR and longer wavelengths. But since there is no theoretical reason whatsoever (Chap. 14) why diffraction-limited images cannot be obtained at visible wavelengths (Sect. 14.6) or any other wavelength transmitted by the atmosphere,[22] plainly the full potential of AO has yet to be fully unlocked. At visible wavelengths, the Keck instruments should ultimately be able to deliver 0.01-arcsecond, or 10-milli-arcsecond (mas) resolution.

With improvements in AO technology combined with the immense size of the various ELT instruments now under construction, the time fast approaches when it should be routinely possible to resolve down to 3 mas (Chap. 14, Sect. 14.11). William Herschel may not, after all, have been entirely wrong: By constructing large enough telescopes—equipped with AO systems—there may not be any theoretical limit to the powers of our instruments.

[22] The UV atmospheric transmission cut-off occurs at about 0.3 μm.

## 1.9 Final Destination!

For thousands of years, philosophers and thinkers have suspected that the universe comprises more than just twinkling points of light and blurry specks. Today, after the increasingly astonishing discoveries made by successive generations of larger and more powerful telescopes, physicists and cosmologists have constructed an understanding of the universe that is both staggering and bewildering. The fodder for future generations of physicists and cosmologists will be the discoveries made by future telescopes. These discoveries may in time lead to an understanding of our universe that is even more extraordinary than the one we have come to know.

Present-day physicists and scientists are pursuing two major objectives: the formulation of a single cohesive quantum gravity model that resolves present-day inconsistencies and describes the entire universe from subatomic to astronomical scales, and the development of a grand unification theory that combines all of the universe's four forces. Some encouragement can be taken from the fact that less than fifty years ago there were five.[23] From these two partial theories, it may ultimately be possible to construct a complete unified theory.

Though it is far from guaranteed that any such theory will ever be accomplished, advances in understanding will surely occur. Some may lead in entirely unexpected directions. Some versions of quantum gravity based on string theory predict that our universe is just a tiny corner of creation, just one of a much larger number of universes forming a larger entity called a multiverse.

Projecting ahead, say, one hundred or so years into the future, it is difficult to imagine what man's understanding of the universe will be at that time. One view has already been voiced by J. B. S. Haldane: "My own suspicion is that the universe is not only queerer than we suppose, but queerer than we can suppose." The only certainty is that a fascinating journey of discovery lies ahead as ever more magnificent and spectacular images are provided by ever larger and more powerful telescopes. These images will naturally help scientists and cosmologists continue to peel back, layer by layer, the physical reality that is our universe.

Will humanity finally attain the ultimate goal, "a complete unified theory?" That is by no means certain but, as mused by Stephen Hawking, "If we do, it should in time be understandable in broad principle by everyone, not just a few scientists. Then we shall all, philosophers, scientists, and just ordinary people, be able to take part in the discussion of why we and the universe exist. If we find an answer to that, it would be the ultimate triumph of human reason, for then we would know the mind of God."

---

[23] M.A. Salam (1926–1996) took up the position of professor of physics at Imperial College in 1957. In 1979, he along with S.L. Glashow and S. Weinberg was awarded the Nobel Prize in physics for contributions to the theory of the unified weak and electromagnetic interaction between elementary particles."

## References


Babcock, H. (1953). Possibility of compensating astronomical seeing. *Astronomical Society of the Pacific, 65*, 229.

Berman, R. (2009). Strange Universe. *Astronomy, 37*(5), 16.

Bester, M., Danchi, W. C., Degiacomi, C. G., Greenhill, L. J., & Townes, C. H. (1992). Atmospheric fluctuations: Empirical structure functions and projected performance of future instruments. *The Astrophysical Journal, 392*, 357–374.

Danchi, W. C. (2001, December 27). Private communication. Albuquerque, NM.

Einstein, A. (1905). On the electrodynamics of moving bodies. *Annalen Der Physik, 17*, 891–916.

Gold, T. J. (1989). *Journal of Science Exploration, 3*(2), 103–112.

Hawarden, T. (1998, August 20). *The United Kingdom infrared telescope*. Retrieved from http://www.jach.hawaii.edu/UKIRT/public/tele-descrip.html

Hawking, S. W. (1988). *A brief history of time*. Bantam Press. A division of Transwworld Publishers Ltd.

Hufnagel, R. E., & Stanley, N. R. (1964). Modulation transfer function associated with image transmission through turbulent media. *Journal of the Optical Society of America, 54*, 52–61.

King, H. C. (1955). *History of the telescope*. Sky Publishing Corp.

Mahon, B. (2004). *The man who changed everything (The life of James Clerk Maxwell)*. Wiley.

Mason, M. (2008, September 8). *Discover Magazine*. Retrieved from the September 2008 issue, published online September 8, 2008, from http://discovermagazine.com/2008/sep/06-burned-at-the-stake-for-believing-in-science

McKechnie, T. S. (1976a). *Light propagation through the atmosphere and imaging in astronomy* (personal communication to J. C. Dainty et al.).

McKechnie, T. S. (1976b). Cores in star images. *Journal of the Optical Society of America, 66*, 635.

McKechnie, T. S. (1990). Diffraction limited imaging using large ground-based telescopes. In *Proceedings of SPIE, V. 1236, Symposium on astronomical telescopes and instrumentation for the 21st century* (pp. 164–178), 11–17 February 1990

McKechnie, T. S. (1991a). Light propagation through the atmosphere and the properties of images formed by large ground-based telescopes. *Journal of the Optical Society of America A, 8*, 346–365.

McKechnie, T. S. (1991b). Focusing infrared laser beams on targets in space without using adaptive optics. In *Proceedings of SPIE, Propagation of high energy laser beams through the Earth's atmosphere* (Vol. 1408, pp. 119–135).

McKechnie, T. S. (1992). Atmospheric turbulence and the resolution limits of large ground-based telescopes. *Journal of the Optical Society of America A, 9*, 1937–1954.

Noll, R. J. (1976). Zernike polynomials and atmospheric turbulence. *Journal of the Optical Society of America, 66*, 207–211.

Tauber, G. E. (1982). *Man and the cosmos* (p. 137). Greenwich House.

Tolstoy, I. (1981). *James Clerk Maxwell: A biography*. Canongate.

Voller, R. L. (2012). The man who measured the cosmos. *Astronomy, 20*(1), 52–57.

# Chapter 2
# Introduction

**Abstract** Soon after the new 'general theory' described in this book was submitted for publication in July 1989, imaging experiments facilitated by the US Air Force began tentatively confirming that much higher resolution levels were indeed possible using large ground-based telescopes. Star image cores measuring 0.1-arcsecond across were recorded at the near-IR wavelength, 2 μm. (Though 0.05-arcsecond image cores were also anticipated at wavelength, 1 μm, to display them in their full glory would have required diffraction-limited telescopes.) These developments added to the growing dissatisfaction with the widely used, but increasingly doubted Kolmogorov theory. The lax optical tolerances prescribed by that theory for large ground-based telescopes had led to an entire generation of underperforming telescopes built in the 1970s and 1980s. But increasingly these prescriptions were being quietly abandoned within the astronomical community in favor of much tighter prescriptions. The result was hugely improved resolution. Plainly, an uncomfortably large gap had developed between Kolmogorov understanding and actual reality. The new theory smoothly closes that gap. In addition to providing a more comprehensive and precise understanding of imaging through the atmosphere with large telescopes (with or without AO), the new theory also finds applications in the areas of laser communications and high energy laser beam propagation.

In common with other books dealing with the subject of light propagation and imaging through the atmosphere, this book treats atmospheric turbulence as a stochastic, or random, process. Consequently, extensive use is made of the methods of statistical optics, constructed around the well-tested mathematical formulations of diffraction theory, coherence theory, Fourier optics, and even geometrical optics when appropriate. The fundamental point of departure between this book and other books on the subject is that the treatment is based on a generalized atmosphere rather than one which is assumed, a priori, to be characterized by Kolmogorov turbulence.

The book lays out a general theory governing light propagation and imaging through this generalized atmosphere, avoiding the inflexible—and now widely doubted—Kolmogorov turbulence assumptions. In this way, it provides a more complete and precise account of the subject, one that naturally accommodates all



T. S. McKechnie, *General Theory of Light Propagation and Imaging Through the Atmosphere*, Progress in Optical Science and Photonics 20,
https://doi.org/10.1007/978-3-030-98828-9_2

the different atmospheric turbulence conditions that occur from instant to instant at all observatory sites, as well as from one site to another. The theory applies to imaging through the atmosphere with any large astronomical telescope, including next generation Extremely Large Telescopes (ELTs). The theory also applies to laser communications, high-energy laser (HEL) beam propagation, and Earth surveillance imaging from satellites in space.

The mathematical approach relies solely on Maxwell's electromagnetic wave equations. There is no requirement for any knowledge, perceived or otherwise, of the physics of turbulence. Maxwell's equations, which are set out in Chap. 3 and further discussed in Appendix A, have been universally accepted as valid for more than one hundred years and are an essential part of our understanding of the universe and the laws governing it. Though elegant and concise, these landmark equations contain the entire blueprint for describing all optical phenomena, including diffraction, polarization, interference, coherence, color and—of prime importance to this book—"light propagation and imaging through the atmosphere." The theory laid out in the book applies generically to all forms of electromagnetic radiation, including X-ray, microwave, radar, and radio waves. However, the primary applications to which the book limits consideration are at visible and IR wavelengths. The theory set out in the book was originally published around 1989/90. At that time, Kolmogorov theory had been in existence for about twenty-five years and was enjoying widespread use. But it was also facing increasing criticism from observational astronomers and atmospheric experimentalists. Kolmogorov theory's inadequacies were becoming increasingly evident; skepticism towards it was mounting.

We saw in the last chapter how Kolmogorov theory had embraced the geometrical optics-based notion that star images formed by large ground-based telescopes would generally appear in the form of seeing discs, rarely measuring less than 0.5- to 1-arcsecond across. Consistent with this pessimistic expectation—or so it was reasoned at the time—the optical performance of large ground-based telescopes was considered acceptable as long as they were intrinsically capable—that is, in the absence of atmosphere—of concentrating most of the light in a star image into a ~0.5 arcsecond diameter patch. The new general theory predicted resolution levels almost an order of magnitude higher than those just indicated. But, to fully realize these new levels would now require telescopes built to diffraction-limited precision. By themselves, these two crucial differences represent a paradigm change in understanding. But the matter does not simply end there. Other equally significant and readily exploitable differences emerge from the new theory. But before listing and describing them, it would be helpful if the reader had at least a top-level understanding of the principal underlying cause.

## 2.1 Principal Cause of Differences Between Kolmogorov Theory and the New General Theory

The schematic in Fig. 2.1 shows two initially plane light waves from a distant star after passing through the turbulent atmosphere. Both waves show significant phase disruption. One wave (solid line) has supposedly encountered small-scale sinusoidal turbulence structure with periodicity 0.3 m. The other (dashed line) has encountered five-times larger sinusoidal turbulence structure with periodicity 1.5 m. Since both waves identically explore the same range of wave front gradients, −0.5-arcsecond to +0.5-arcsecond, geometrical optics principles suggest that large telescopes would focus both waves into nominally similar 1-arcsecond star images. But as we shall discover, the whole story is much less simplistic and infinitely more intriguing.

The principles of Fourier analysis tell us that any continuous wave, regardless of its shape, may be represented as the sum of its constituent sinusoidal spatial frequency components. Thus, the waves shown in Fig. 2.1 may be regarded simply as two of the basic 'building blocks' of actual randomly corrugated light waves. Because of their widely different periodicities, the two waves allow us to examine in isolation the different effects of small- and large-scale turbulence structure on star images formed by large telescopes.

Because the imaging wavelength is just as important as turbulence structures size, we shall also explore image appearance for two well-separated wavelengths, $\lambda = 0.55\ \mu m$ and $\lambda = 2.2\mu m$, as well as for the two different turbulence structure sizes. The image intensity sections in Fig. 2.2 therefore correspond to the four possible combinations of the two turbulence structure sizes and the two wavelengths.

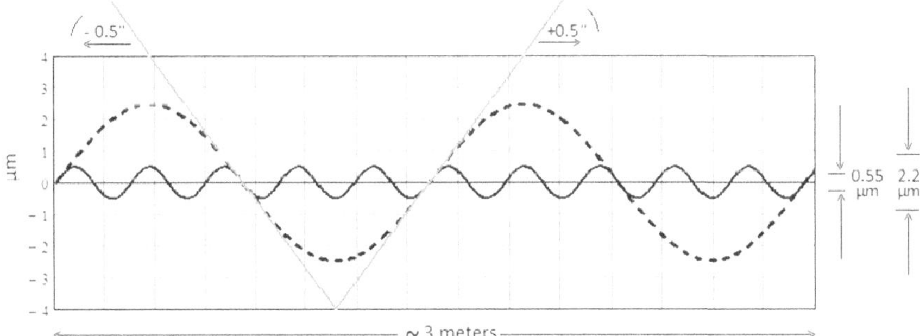


**Fig. 2.1** Schematic showing the effect on initially plane light waves of small- and large-scale turbulence structure in the atmosphere. The wave (in solid line) has encountered sinusoidal turbulence structure with periodicity 0.3 m. The other wave (in dashed line) has encountered five-times larger structure with sinusoidal periodicity 1.5 m. Notice that both waves explore the identical gradient angle range, −0.5-arcsecond to + 0.5-arcsecond, as depicted by the (orange) construction lines which tangentially touch both waves as they cross the horizontal axis where they attain maximum gradient. (Note the ~160,000-times scale expansion on the vertical axis to distinguish the wave corrugations more easily.)

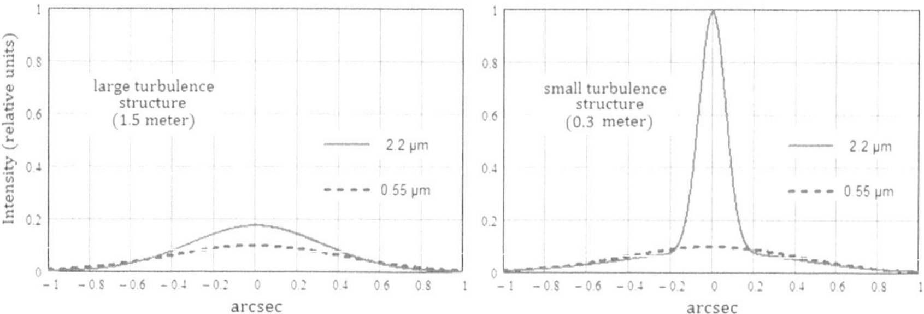


**Fig. 2.2** Star image intensity sections formed by a 5 m (200-inch) telescope observing in 1-arcsecond visible seeing conditions. The images on the left correspond to large, 1.5 m, sinusoidal turbulence structure. The images on the right correspond to smaller, 0.3 m, structure. Upon rotation, all four image sections enclose the same volume and thus the same amount of light energy

### *2.1.1 Visible and IR Star Images for Large Turbulence Structure*

Plainly, the peak to valley (P-V) depth range, 5.0-μm, explored by the wave corresponding to the larger turbulence is larger than either of the two wavelengths considered, 0.55 μm and 2.2 μm. Consequently, for either wavelength, the phase[1] variations will be uniformly distributed (approximately) over the entire primary phase angle range, $-\pi$ to $+\pi$. Thus, the star images at both wavelengths will be nominally identical, ~1-arcsecond seeing disc images, as indicated by the two image sections in Fig. 2.2 (Left).[2]

### *2.1.2 Visible and IR Star Images for Small Turbulence Structure*

The P-V depth range, 1.0-μm, explored by the wave corresponding to the smaller turbulence is again significantly larger than the 0.55-μm imaging wavelength. Consequently, the wave front phases will again be uniformly distributed over the primary phase angle range, $-\pi$ to $+\pi$. Again, the telescope will form a ~1-arcsecond seeing disc star image, as shown in Fig. 2.2 (dashed line, right).

However, for the longer, 2.2-μm, wavelength, the 1.0-μm P-V corrugation depth range is now considerably smaller than the wavelength. Consequently, the phase angle range explored by this wave occupies only a fraction of the primary phase angle range, a condition indicating only partial phase disruption.

In effect, this means that a certain fraction of the light energy is fully disrupted and is therefore imaged by the telescope into a seeing disc-like feature, known as a

[1] The phase angle, or phase, should not be confused with the wave front gradient angle.

[2] The small difference between the image sections need not concern us at this stage.

halo, while the remaining light fraction remains entirely undisrupted and images into a highly resolved feature, known as an image core. The final star image is simply the sum of these two image features, known as a core-and-halo image (Fig. 2.2, solid line, right).

In Chap. 10, core appearance is shown to be identical to that of a star image formed by the telescope in the absence of atmosphere. Ignoring the effect of the central obstruction, for a 200-inch telescope constructed to diffraction-limited accuracy, the image cores at 2.2-$\mu$m would be ideal Airy patterns.

The star image sections corresponding to the larger turbulence structure (Fig. 2.2, left) show only marginal resolution improvement at the longer imaging wavelength, 2.2 $\mu$m. However, the sections corresponding to the smaller turbulence structure (Fig. 2.2, Right) show significantly larger resolution improvement at the longer wavelength.

The existence, or absence, of star image cores at any given wavelength depends on the 'average' size of the turbulence structure in the atmospheric path. Large average size indicates less likelihood of finding image cores at any given wavelength. Unsurprisingly, Kolmogorov turbulence structure, with its surmised tens-of-meters outer scale limit, naturally implies large average structure size. Consequently, if a 200-inch telescope were observing through a Kolmogorov atmosphere in typical ~1-arcsecond visible seeing, there would be little prospect of image cores appearing at any wavelength shorter than the far-IR wavelength, 10 $\mu$m.

However, if the same 200-inch instrument were observing through an atmosphere consisting of the smaller turbulence structure measured at RGO, again assuming ~1-arcsecond visible seeing, image cores would be expected routinely at wavelengths as short as 1 $\mu$m. In sub-0.5 arcsecond visible seeing, cores would likely appear at visible wavelengths. In such superb seeing conditions, the Airy pattern image cores formed by a diffraction-limited 200-inch instrument would provide 0.025-arcsecond resolution.

## 2.2 Significant Features of the New General Theory

(1) **Generality of the new theory with respect to atmospheric turbulence**

The new theory makes no a-priori assumptions about the properties of atmospheric turbulence, Kolmogorov or other. Instead it deals in terms of a generalized atmosphere whose initially unknown turbulence properties can be established, on site and as needed, from readily-measurable properties of star images formed by large telescopes.

(2) **Other areas of generality of the new theory**

Diffraction effects, phase variation, amplitude scintillation and refractive index dispersion are all rigorously dealt with, as are isotropic and non-isotropic atmospheric turbulence. Turbulence strength and structure size distributions are allowed to vary independently, both temporally and in all three (x,y,z) spatial dimensions. Telescopes of all sizes are considered, aberrated or diffraction-limited, with or without central obstructions, and with or without AO.

(3) **Significantly higher resolution levels expected from large telescopes**

Core-and-halo star images—previously regarded as improbable and even illusory–are now regarded as the most general type of star image. Seeing disc and core-only star images are now simply special cases in the short- and long-wavelength limits. With five-meter class telescopes (large by 1989/90 standards) star image cores were expected at all near-IR and longer wavelengths, even in unexceptional ~1-arcsecond visible seeing conditions. Cores at wavelengths, 1.1-$\mu$m and 2.2-$\mu$m, were expected to deliver five- to ten-times higher resolution than before. With the even larger ELT instruments, even larger resolution improvement are anticipated.

(4) **Optimum wavelengths for delivering highest possible telescope resolution**

Optimum wavelengths are now anticipated ($\lambda opt = 2 \pi \sigma$) where large telescopes achieve highest theoretically possible resolution. For a diffraction-limited 5 m instrument observing in ~ 1-arcsecond visible seeing conditions, the anticipated optimum wavelength range, 1.1 $\mu$m to 2.2 $\mu$m, offers the possibility of resolution between 0.05 arcsecond and 0.1-arcsecond. Nowadays, the wavelength region straddling the optimum wavelength is referred to as the'sweet-spot' imaging region.

(5) **Optimum wavelengths for focusing laser beams through the atmosphere:**

Similarly, when laser beams—High Energy Laser (HEL) or other—are focused through the atmosphere, optimum wavelengths can again be identified that enable maximum possible irradiance flux (Watts/$m^2$) concentration at the beam focus (assuming the same amount of beam power at all wavelengths). Non-optimum wavelengths needlessly decimate flux levels and/or the lethality range of the system.

(6) **Use of star image cores as reference objects rather than light energy centroids**

Cores in natural guide star images are shown to be more stable reference objects than light energy centroids. The continuously 'boiling' speckle in the halo portion of a star image introduces angular jitter into the image centroid. Because image cores are much less affected, their use as reference objects enables more precise, diffraction-limited image stabilization and AO image correction.

(7) **Exploiting isoplanatic properties of star image cores for increased sky coverage**

The twin cores in binary star images are expected to remain angularly locked together for binary separations of at least several arcminutes. (For comparison, individual speckle pairs, and hence individual image centroids only remain locked for angular separations of a few arcseconds.) By using cores in natural guide star images as reference objects, rather than light energy centroids, diffraction-limited image quality can be obtained over much wider sky areas. Solid-angle sky coverage may increase by three or four orders of magnitude.

(8) **Characterization of atmospheric observing paths and/or laser beam paths**

By measuring key properties of star images (or images of other types of point-objects) obtained by any large telescope at a suitably chosen wavelength, fully definitive object-to-image beam path characterizations can be obtained. All meaningful properties of star images formed by the same telescope over the same path can then be calculated for all other wavelengths in the extended optical wavelength range, 0.3–1000 $\mu$m. These properties include speckle properties, core-and-halo properties, and image intensity envelops.

(9) **Characterization of AO-corrected beam paths**

The above path characterization procedure equally applies to AO-equipped telescopes. The only procedural difference is that the AO system would now operate during image characterization. Again, by measuring key properties of star images obtained at a suitably chosen wavelength, all meaningful properties of star images formed by the same telescope/AO combination, including ELT instruments, can be calculated for all other wavelengths in the extended optical range, 0.3–1000 $\mu$m.

(10) **Scaled laboratory image simulations**

Scaling equations consistent with the new theory enable accurate laboratory simulations of star images, or images of extended objects, formed by large telescopes in the field. By using scaled optics to represent the telescope and (prescription) scaled diffuser screens to represent the atmosphere, meter-size telescope diameters can readily scale down to millimeter diameters. Wavelength scaling is also possible. For example, images obtained in the field at IR wavelengths can be simulated in the laboratory at—more convenient—visible wavelengths, while faithfully preserving all speckle, core and halo, and other image properties.

## 2.3 Book Content Preview

Such is the embrace of Maxwell's equations, the task undertaken in this book of describing light propagation and imaging through atmospheric turbulence amounts to applying these equations under applicable boundary conditions and then using established mathematical techniques to find solutions. The burden was greatly reduced by many solutions already existing in the literature, e.g., the Kirchhoff diffraction integral. In those instances where the desired solutions could not be found in the literature, solutions were obtained using the well-tested mathematical formulations listed earlier in the chapter. If Maxwell's electromagnetic equations are indeed incontrovertible, the primacy of mathematics[3] ensures that appropriate solutions will provide true and faithful descriptions of actual physical behavior.[4]

Maxwell's equations underpin the analysis given in Chaps. 5 and 6 of the propagation behavior of infinitely extensive, initially plane (i.e., collimated) light waves, considered to have originated from a distant unresolved star. (Because such objects closely approximate ideal point-objects, the terms "unresolved stars" and "point-objects" are sometimes used interchangeably.) Chapter 5 deals with the propagation behavior of light waves after scattering by a single thin atmospheric layer. Chapter 6 deals with the propagation behavior when the waves are continuously scattered over an extended atmospheric path, precisely modeled by a conterminous stack comprising an infinite number of limitingly thin layers.

The refractive index distribution function of the atmosphere, $n(x,y,z,\lambda)$, is the essential function that controls how light waves propagate over extended atmospheric paths. This function describes, in effect, a four-dimensional (three spatial and one temporal) scalar field, where the refractive indices are considered to vary randomly and continuously throughout. Thus, the refractive index field may be considered as a randomly varying field whose properties are best described statistically.

Once the light waves emerge from the extended atmospheric path, having now been scattered and disrupted by the turbulence structures encountered in the path, Maxwell's equations are used, in effect, a second time to describe (Chap. 7) the focusing and imaging of these light waves by telescopes. In this way, mapping relationships are established between the properties of point-object images formed by telescopes observing through the atmosphere and the properties of the turbulence structure in the atmospheric observing paths. The formulation is dimensionally robust and naturally provides consistency between the atmospheric turbulence properties and the properties of telescope images—both crucially important attributes that are not equally enjoyed by Kolmogorov formulations.

Though the turbulence properties in the atmospheric observing paths are initially regarded as unknown, they are nonetheless encrypted in the properties of point-object images formed by large telescopes observing over these paths. The encryption is complicated because phase information is lost when only image intensities are recorded.

Nonetheless, the essential turbulence properties can be deciphered by first measuring certain key properties of the intensities in point-object images and then inserting the measured data into the appropriate mapping equations.

As described in Chap. 8, the measured intensity data are initially used to calculate the two-point two-wavelength correlation function of the complex amplitudes of the light waves as they emerge from the atmospheric path, capturing the function in effect at the instant the waves arrive in the entrance pupil of the telescope. We denote

[3] The philosopher, David Hume (1711–1776), stated "nothing can be proved except in mathematics. Much of what we accept as fact is mere conjecture. Arguably, mathematics is the only exact science.".

[4] Exceptions arise when there are multiple solutions, with some having no obvious real-world interpretation. It is usually clear when this occurs, and such solutions are simply ignored.

this correlation function by $S(x', y', x, y, \lambda', \lambda)$ while noting that it contains all essential statistical information needed to accomplish all of our theoretical and practical objectives. This crucially important—and in fact decisive—function is defined as the unit-normalized form of the function $\langle U(x', y', \lambda') \cdot U^*(x, y, \lambda)\rangle$, where $U(x, y, \lambda)$ denotes the complex amplitude of the light waves arriving in the pupil plane of the telescope, $\lambda$ is the wavelength, and the $\langle\cdot\rangle$ brackets denote ensemble average. Brackets of this sort represent the expectation operator where the expectation value is the ensemble average. Once function $S(x', y', x, y, \lambda', \lambda)$ has been established, we can then use it to calculate, not only the essential properties of the turbulence in the atmospheric path, but all other meaningful properties of the complex amplitudes and intensities in point-object images, including monochromatic and polychromatic properties as well as the various speckle and core and halo properties.

In the special case where the two wavelengths coalesce (i.e., $\lambda' \rightarrow \lambda$), function $S(x', y', x, y, \lambda', \lambda)$ degenerates into the complex coherence factor,[5] which is the unit-normalized form of function $\langle U(x', y', \lambda) \cdot U^*(x, y, \lambda)\rangle$. The complex coherence factor is simply the autocorrelation function, or self-correlation function, of the complex amplitude distribution. To distinguish this important complex coherence factor from others that arise in the book—such as the complex coherence factor that arises from the complex amplitude distribution in the telescope image plane—our practice will be to refer to this function by its more familiar title, the atmospheric modulation transfer function, or atmospheric MTF. We denote this function by $M(x', y', x, y, \lambda)$. Though this function is less general than $S(x', y', x, y, \lambda', \lambda)$, it is nonetheless important; it essentially sets the resolution limits for telescopes observing over atmospheric paths.

The generalized formulation used in Chap. 8 to convert measured star image properties into atmospheric turbulence properties ensures non-partisan outcomes. Thus, the various other star image properties and turbulence structure properties deduced from the measurements are free of bias toward any one particular type of turbulence, Kolmogorov or other. If the turbulence properties happen to be those of Kolmogorov turbulence, the measurements should indeed show this; the general formulation given in this book naturally includes the Kolmogorov formulation as a special case.

The image of an unresolved star is simply the intensity point-spread function (PSF) of the telescope/atmosphere imaging combination. As discussed in Chaps. 9 and 10, the most general type of star image exhibits core and halo structure. Images comprising only cores or only halos are regarded in the treatment as degenerate cases. The surface plot images in Fig. 2.3 show the average intensity distribution for all three types of star image, each calculated for a 3 m diffraction-limited telescope imaging at the near-IR wavelength, 2.2 μm, in average, 1-arcsec, visible seeing conditions.

Star images consisting only of cores are routinely formed by small backyard telescopes at visible wavelengths; often these images appear as Airy patterns. Core and

[5] The complex coherence factor is closely related to the mutual coherence function. As discussed in Chap. 7 (Sect. 3.8), for quasi-monochromatic light—with which we will deal frequently—the mutual coherence function degenerates into the complex coherence factor.

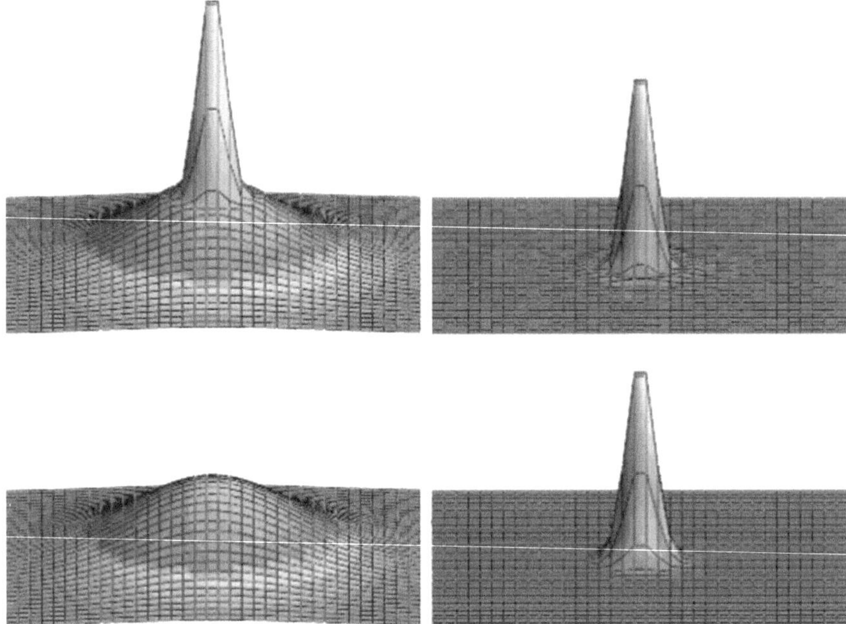

**Fig. 2.3** Star image intensity envelopes. as obtained from a 3 m diffraction-limited telescope at the near-IR wavelength, 2.2 μm in average 1-arcsec visible seeing. At this wavelength a core and halo image would likely arise (Top left). Bottom left: The halo shown in isolation. Top right: The anticipated Airy pattern core. Bottom right: The Gaussian approximation to the Airy pattern core

halo structure can be seen in images formed by larger telescopes at visible wavelengths in excellent (sub-0.5 arcsecond) seeing conditions. At near-IR and longer wavelengths, they can be seen in almost any reasonable seeing conditions.

At visible wavelengths in mediocre or poor seeing conditions, large telescopes generally deliver star images comprising only a halo (Bottom left in Fig. 2.3) commonly known as a seeing disc star image. Star image properties are controlled by the imaging wavelength, the telescope optics (by the aperture size and shape and the instrument's aberration characteristics), and by the average size and strength of the turbulence structure in the atmospheric path. Image properties also depend on whether short- or long-exposure versions of the image are recorded. By choosing to either correct or not correct image motion during star image acquisition, the resulting image naturally falls into one or other category.

Once the amplitude and intensity PSFs of the atmosphere/telescope combination have been established, it is a simple matter to determine the image properties of more complicated types of object; such objects are known as extended objects. Subject to restrictions imposed by the isoplanatic angle properties of the telescope and the isoplanatic angle limits imposed by the atmosphere, the intensity distribution describing the final image of an extended object can be obtained by convolution of the appropriate object illumination function with the appropriate amplitude or intensity

PSF of the telescope/atmosphere combination. Whether the convolution operation is carried out in terms of one or other of these two fundamental PSFs, or indeed in terms of some combination of the two, depends on the state of coherence of the illumination. For distant astronomical objects, the illumination is usually incoherent. For actively illuminated objects, such as satellites in low Earth orbit, the illumination can be either partially coherent or fully coherent.

To properly account for diffraction, the formulation developed in the book makes extensive use of the methods of Fourier optics. Many of the key equations are therefore integral equations, and many of these are two-dimensional Fourier transform equations (Titchmarsh, 1948). In an attempt to simplify things, Chaps. 13 and 18 reduce these equations down to simple, more easily manageable approximate equations.

The Chap. 13 simplification is made by approximating the average intensity distribution in the most general type of star image—the core and halo image—as the sum of two best-fit Gaussian functions, a narrow function to represent the core and a wider one to represent the halo. For the degenerate core-only and halo-only image types, it is only necessary to use one or other of the two Gaussian functions to represent the star image. While this modeling scheme might at first appear crude, the Airy pattern core and its Gaussian approximation shown in Fig. 2.3 in the top- right and bottom-right images are almost indistinguishable.

In new Chap. 18, the most general type of star image—the core and halo image—is modelled even more precisely as the sum of three best-fit Gaussian functions. This representation provides a uniquely detailed picture of the underlying physics of star image formation. Because the accuracy of the formulation increases as the diameter of the telescope, the Chap. 18 formulation may be found particularly useful for dealing with star images formed by ELT instruments.

Applications of the formulae given in Chaps 13 and 18 include generating detailed seeing logs in parallel with, and thus not interfering with, the primary imaging task of the observing telescope. These seeing logs go beyond the usual practice of recording visible seeing via quantities such as the full-width half-maximum (FWHM) angular size of star images (Chromey, 2010). The envisaged seeing logs maintain a more complete seeing record that includes not only the integrated strength of the turbulence in the atmospheric path but also the turbulence structure function. Other equally simple back-of-the-envelope formulae can be used to calculate from the seeing log data the properties of star images formed at any wavelength in the entire (UV to far-IR) wavelength range. Seeing logs of the type envisaged, built up hour-by-hour and site-by-site should gradually lead to a broader and more complete picture of the characteristics of atmospheric turbulence.

A second application of the simplified formulae is the rapid identification of optimum wavelengths. For imaging applications (Chap. 10) maximum resolution can be obtained at these wavelengths. Optimum wavelengths are particularly important when dealing with laser beam propagation through the atmosphere. Applications include laser communication and HEL weapon systems. With regard to the latter application, Chap. 16 describes how optimum wavelengths enable HEL weapon systems to achieve maximum lethality range and maximum irradiance potency at the target.

Often, the propagating laser beam comes to focus within the atmosphere. In such cases, the propagation analysis given in Chaps. 5 and 6 for collimated light requires modification. Chapter 16 therefore examines the propagation behavior of convergent and divergent light beams as they travel through the atmosphere. Measurement techniques are described for characterizing either beam path type. As well as establishing the turbulence properties in these paths, optimum wavelengths can also be identified.

Chapter 14 examines the resolution limits of telescopes observing through the atmosphere. Telescopes with and without AO are considered. Suitable optical tolerance specifications are developed for both cases, consistent with allowing the highest theoretically possible resolution levels. In general, telescope resolution depends on the atmospheric seeing quality and the aberrations of the telescope (including defocus). For telescopes equipped with AO, resolution also depends on the efficacy of the AO system.

Throughout the book, telescope resolution is characterized by the instrument's ability to resolve two-point objects, such as binary stars. Various two-point resolution criteria will be discussed. Rayleigh's original criterion (defined by the angle subtended by the radius of the first dark ring of an Airy pattern) gives rise to the familiar angular resolution limit, expressed in radians by $1.22\ \lambda/D$. Strictly, this criterion only applies to diffraction-limited telescopes with circular apertures, where the images of the individual point-objects are indeed Airy patterns.

For other types of telescope, such as those with non-circular apertures, aberrated telescopes, or telescopes with central obstructions, where the star images may radically differ from Airy patterns, a broader more flexible resolution criterion is required. We therefore choose to deal in terms of an extended version of Rayleigh's criterion (Born & Wolf, 2003), where the two-point object is considered just-resolved when the intensity in the center of the image falls to a value 26.5% below the level attained in the two bright peaks. It is no coincidence that, when the latter version of Rayleigh's criterion is applied to diffraction-limited telescopes with circular apertures, it provides a just-resolved angular limit identical to that given by Rayleigh's original angular resolution limit, $1.22\ \lambda/D$ .

Chapter 14 also deals with optical tolerance specifications for large ground-based telescopes. For any given telescope, the appropriate tolerance specification depends on the envisaged imaging application as well as on the intended imaging wavelengths. Lax tolerances can be adopted if halo-only images are considered adequate. Tighter tolerances are required if the intention is to resolve image cores and thus obtain diffraction-limited resolution. For telescopes equipped with AO, the tolerance specifications for the telescope optics are closely tied to the performance of the AO system. Significant fixed telescope aberrations can be tolerated if the AO system is capable of correcting them.

An analysis given in Chap. 14 shows that even allowing for the fact that AO systems do not correct intensity scintillation in the image-forming waves, an AO system that near-perfectly corrects the phase errors alone should still be able to deliver star images with Strehl intensities greater than 0.8 at all wavelengths transmitted by the atmosphere. (It is noted that Strehl intensity values greater than 0.8 imply substantially diffraction-limited image quality.) Thus, at the shortest wavelength transmitted by the atmosphere, about 0.3 μm, and for todays' largest telescopes—the Keck 10 m instruments—theoretically possible resolution from these gigantic instruments is about 0.008-arcsecond, or 8 milli-arcsecond. This level of resolution is about 7x higher than presently obtained by the Keck instruments at 2.2 μm, and about 4x higher than delivered at 0.3 μm by the Hubble Space Telescope.

For the even larger ELT instruments, theoretically possible resolution increases with the diameters and approaches 0.002 arcsecond (2 mas) for the largest, 39 m, E-ELT at the shortest transmitted wavelength, 0.3 μm. Considerable technological progress and many decades may elapse before this extremely high resolution level is even approached. Chap 14 includes a discussion of more realistic near-term achievable resolution levels. The discussion is illustrated by computer generated images showing how single and binary stars will likely appear through ELT instruments at visible and IR wavelengths. The images are shown for currently achievable levels of AO correction as well as for modestly higher levels that may be achievable within a decade of first light.

Chapter 15 deals with scaled optical simulators whose purpose is to produce simulated images in the laboratory with properties identical to those delivered by large telescopes in the field. An image simulator, referred to as the Space Shuttle image simulator, is described in detail. This particular simulator was constructed and used in 1989 (McKechnie, 1990) at a time when NASA faced the potentially disastrous problem of thermal insulation tiles accidentally detaching from the space shuttle body; it was designed to provide image simulations at both visible and IR wavelengths with properties closely replicating those produced by an actual 4 m ground-based telescope observing the space craft as it passed directly overhead. The actual images obtained from this simulator are shown in Fig. 15.6. These images clearly demonstrate that, whereas missing tiles would probably not be identifiable in images obtained at visible wavelengths, they would be readily identifiable in images obtained at the near-IR wavelength, 2.5-μm, a wavelength deemed to be close to the "optimum wavelength" for a 4 m ground-based telescope in the typical seeing conditions considered.

Newly added Chapter 18 provides a mathematical tool kit specifically designed for day-to-day use with the forthcoming generation of ELT instruments. The tool kit equations are entirely analytic and provide a convenient and precise platform for characterizing (AO-corrected) ELT imaging paths, star image appearance, and other properties of star images formed by these huge instruments. The equations are developed from a uniquely powerful and precise star image model based on the sum of three Gaussian functions (rather than the two used in Chap. 13).

## 2.4 Kolmogorov Theory

Turbulence occurs in all fluids, but particularly in low-viscosity fluids such as air. Achieving a satisfactory understanding of turbulence, with its chaotic and stochastic characteristics, has proven to be one of the most stubborn and difficult problems of physics. A commonly used model for describing atmospheric turbulence traces back to ideas developed in the 1940s by the Russian mathematician, A.N. Kolmogorov. According to the "Kolmogorov model" Kolmogorov (1941a, 1941b), Tatarski (1961), the outer scale size limit of the turbulence structure is set by the physical size scales of the driving mechanisms which, in the case of the atmosphere, include wind and thermal convection currents. The intermediate structure scales are considered self-similar over a broad range of scales in the so-called inertial subrange.

The atmospheric MTF, M(x', y', x , y , λ), for any given atmospheric path is the crucial function that, when taken in combination with the Optical Transfer Function (OTF) of the observing telescope, describes the image quality and resolution that can be obtained by that telescope observing over that path. In a landmark mid-1960s paper, Hufnagel and Stanley (1964) significantly advanced understanding by showing how the atmospheric MTF can be calculated in terms of certain optical path difference (OPD) integrals obtained over the atmospheric path. Their analysis indicated the counterintuitive, yet correct, result that this important function can be calculated without having explicit knowledge of either the amplitude or the phase of the light waves arriving at the telescope.

Seeking to build further on their important result, Hufnagel and Stanley adopted the Kolmogorov atmospheric turbulence model and produced an expression for the atmospheric MTF for turbulence of that type. Their deduction of the average intensity envelope in the image of an unresolved star effectively set the resolution limits for telescopes observing through Kolmogorov turbulence. But by electing to "bypass" certain difficulties faced by Kolmogorov theory and by bypassing without comment a number of contradictions and inconsistencies in the turbulence assumptions, Hufnagel and Stanley now found themselves heading along a poorly charted and much less certain path.

Examination of Kolmogorov theory reveals that it contains certain physical inconsistencies and dimensional contradictions. The various problems are discussed in detail in Chap. 6 and Appendix I. The dimensional problem is readily evident. If the Kolmogorov 2/3-power law structure function were to hold true over an infinite range of turbulence structure scale sizes, from zero to infinity (meters), the dimensions of the structure function would be that of $\text{Length}^{-2/3}$, while the dimensions of the structure constants, $C_n$, would be that of $\text{Length}^{-1/3}$, and there would be no dimensional contradictions. However, as soon as departures occur from the 2/3-power law structure function—as required by Kolmogorov formulations at the inner and outer scale limits (c.f., Sect. 6.1.2.1)—to accommodate these departures and restore overall dimensional consistency the dimensions of the structure constants would (somehow?) have to change.

We now face an absurd problem. Because the dimensions of $C_n$ are fixed at the outset, there is no available mechanism for restoring dimensional consistency outside of the inertial subrange. A practical consequence of this problem is that measured values of the structure constants, $C_n$, would now depend on the length of the measurement baseline relative to the supposed outer scale limit, $L_0$. Measured $C_n$ values would tend to be smaller when longer measurement baselines are used and larger for shorter baselines. Such behavior inevitably leads to anomalous and inconsistent atmospheric path characterizations—a problem that has persistently plagued Kolmogorov theory.

Yet another problem arises when we examine the quantity known as the coherence parameter, or Fried parameter, denoted by $r_0$ (Fried, 1966). This parameter relates specifically to Kolmogorov turbulence. It has useful conceptual value for quantifying seeing at visible wavelengths, but problems arise when it is applied to IR wavelengths. At these longer wavelengths, $r_0$ supposedly grows as the 6/5-power of wavelength and the resolution of large telescopes improves slowly, as the -1/5th power of wavelength. The improved resolution manifests itself by the gradual narrowing of seeing disc star images with increasing wavelength. But nowhere in this scheme is there any indication of image cores developing. Since $r_0$ is a geometrical optics-based quantity, it cannot account for diffraction and therefore cannot properly account for star image cores.

In the 1960s, Hufnagel and Stanley's lead in incorporating Kolmogorov assumptions into their expression for the atmospheric MTF was soon followed by others. Though Kolmogorov theory had yet to demonstrate that it could survive rigorous scientific scrutiny, in the absence of any better theory, it was passively accepted as a valid theory describing imaging through the atmosphere with large telescopes. Yet not everyone was aboard with this theory. Many were far from convinced, particularly observational astronomers.

Reports persistently surfaced of image properties quite different from those predicted by Kolmogorov theory: for example, cores in star images (Griffin, 1973) and too little image motion (Woolf et al., 1982). We might also recall from Chap. 1 the scathing response from William Danchi[6] when asked to paraphrase his views on Kolmogorov turbulence. Nobel laureate, Richard P. Feynman (1918–1988), has described turbulence as "the most important unsolved problem of classical physics." Disagreement and controversy have surrounded the application of Kolmogorov theory to imaging through the atmosphere ever since its development in the 1960s. The continued use of this theory may owe more to the forceful insistence of its promoters than to actual experimental confirmation.

An extensive list of observational evidence contradicting Kolmogorov assumptions is given in Appendix I. As noted in the Preface, to merely show that one piece of sustainable observational evidence contradicts these assumptions should, by rigorous scientific standards, be enough to undermine the general validity of the theory. But, in this instance, there is such an abundance of evidence that it should surely signal the final demise of the theory. Yet many scientists and researchers simply ignore the problems and continue using Kolmogorov theory as though it were still viable.

[6] Danchi was a member of the turbulence measurement team led by Charles H. Townes. The team used an 11 m baseline interferometer on Mount Wilson to deduce the turbulence properties from the optical path differences measured at different separations. Their findings can be found in several papers, including "Atmospheric fluctuations: Empirical structure functions and projected performance for future instruments," as listed in the references under Bester et al. (1992).

From an objective viewpoint, a more exact and more widely embracing theory is needed to describe light propagation and imaging through Earth's turbulent atmosphere. Such a theory is set out in the chapters that follow. While the approach allows for the possibility of Kolmogorov turbulence, it also allows for the possibility of any other kind of turbulence, in particular, the non-Kolmogorov kinds measured by various researchers, including Coulman et al.[7] and the Author, further details of which can be found in Chap. 8 and Appendices C and I.

### 2.4.1 Kolmogorov Theory and Its Damage Legacy

The trust placed in Kolmogorov theory for so many years had some extremely damaging, even if unintended consequences. The theory gave rise to unduly coarse optical tolerance prescriptions for large ground-based telescopes, which were readily adopted by astronomers and telescope builders. They were even reaffirmed as recently as 1991 (Martin et al., 1991). The argument was both simple and enticing: Why waste scarce resources building ultra-precise telescopes when the atmosphere is only going to ruin image quality anyway? The 1970s and 80s saw the construction of a generation of large ground-based astronomical telescopes built to coarse Kolmogorov optical tolerance prescriptions. For twenty years, these instruments—princes of their era—represented the forefront of ground-based observational astronomy. Only later did it become clear that, by their radically poor imaging performance, they were short-changing astronomy and the astronomers who used them.

By the late 1980s, skepticism towards Kolmogorov optical prescriptions for large ground-based telescopes had grown to the point where astronomers and telescope builders simply went ahead anyway and began constructing ground-based telescopes to higher optical standards. The 3.6 m New Technology Telescope (NTT) saw first light in March 1989. By virtue of its active optics (Wilson, 2003), this instrument immediately delivered much sharper images. Roger Griffin, Emeritus Professor of Observational Astronomy, Institute of Astronomy, University of Cambridge, who for many years was a voice-in-the-wilderness advocate for higher quality telescope optics observed wryly (Griffin, 1990): "Isn't it amazing, then, … that advantage of their capabilities can be taken rather often, in the case of the NTT on the very first night it was used?"

To account for image cores at visible to mid-IR wavelengths with large telescopes, a large fraction of the total atmospheric turbulence energy must reside in sizes smaller than 1 m—a conclusion directly contrary to Kolmogorov turbulence assumptions. Thus, between 1965 and 1990, by eschewing the possibility of image cores at these wavelengths, Kolmogorov adherents were also eschewing the 5–10 times resolution

[7] Coulman et al. made extensive turbulence structure size measurements over a multi-year period at observatory sites in France, Chile, and the USA. They consistently measured significantly smaller turbulence structure than assumed in Kolmogorov theory. Their measured turbulence spectrum is remarkably similar to the one measured by the Author at RGO in 1975/76.

improvement that these features would, in due course, make routinely possible, even in unremarkable 1-arcsecond visible seeing conditions. Further, by denying the existence of image cores, there was a needless delay in exploiting the unique advantages of these features over star image centroids for tracking, image stabilization, and AO image correction.

## 2.5 The New General Theory

The new approach to light propagation and imaging through the atmosphere set out in this book avoids all of the various problems faced by Kolmogorov theory. It does this in various ways. The most crucial is that it makes no a priori assumptions whatsoever about the properties of atmospheric turbulence. In particular, the treatment makes no assumptions about the form of the atmospheric turbulence structure function, Kolmogorov or other. The formulism treats the structure function as unknown initially, its functional form, or forms, only to be attached later, based on on-site measurements of key intensity properties of point-object, or unresolved star, images.

Analysis provided in Chaps. 10 and 17 shows that, by using the core in a star image as the reference feature—rather than the time-honored image centroid—tracking, image stabilization and AO image correction can be carried out to significantly higher levels of precision. This advantage is further leveraged by the larger isoplanatic angles associated with star image cores which, as discussed in Chap. 17, offer the possibility of much larger sky coverage areas from the limited number of natural stars in the sky bright enough to be used as reference objects.

As stated earlier, the new theory is dimensionally consistent, it derives entirely from Maxwell's electromagnetic wave equations, and does not depend on any detailed understanding of the mechanisms of atmospheric turbulence. The theory applies to the imaging of light waves arising from any type of object—astronomical, spaceborne, or Earth-bound. It also applies to telescopes of all types and sizes, as well as telescopes with and without AO.

If we are to believe Danchi (Danchi, 2001) a single turbulence structure function simply does not apply to all observing conditions. Nor would any such function likely apply to all observing sites. The new approach set out in the book allows for all possible types of structure function. The measurement procedures enable these functions to be obtained from on-site star image measurements, as needed, in whatever form/s they might take. The measurement procedures also allow for the possibility of non-isotropic turbulence structure functions, such as might arise from the effects of wind shear.

The star image measurement procedures enable any residual aberrations of AO-equipped telescopes to be separated into fixed and dynamical components, thus allowing these two kinds of residual aberration to be analyzed and minimized separately. Application of these procedures (Chaps. 13 and 18) should assist large telescopes, such as the Keck instruments and the forthcoming generation of ELT instruments, inch forward towards their ultimate resolution potentials. Despite the astonishingly large, 39 m diameter of the largest ELT, in quiescent atmospheric conditions there is no theoretical reason why, given sufficient time and resources, this colossal instrument should not deliver near diffraction-limited images (i.e., Strehl intensity $\geq 0.8$) at all visible and IR wavelengths transmitted by the atmosphere.

For telescopes equipped with AO systems, the properties of the OPD fluctuation in the waves arriving in the telescope pupil are not, per se, the properties that determine the final image properties. In this case, the final image properties are determined by the properties of the residual OPD fluctuation in the image-forming wave fronts after the AO system has corrected these OPD fluctuations to the best of its ability. Thus, the final imaging performance of any AO-equipped telescope depends at least as much on the efficacy of the AO system as it does on the prevailing seeing conditions. With AO-system performance now so critical and, with different AO systems performing with different levels of efficacy, there is even less likelihood of finding blanket descriptions of image properties delivered by different instruments.

The new theory laid out in this book also offers fresh opportunity for progress in the area of laser beam propagation through the atmosphere. As described in Chap. 16 (Sect. 16.9.1.1), the choice of the HEL wavelength, 1.315 μm, for the Airborne Laser (ABL) program was a particularly poor choice. The equation, $\lambda opt = 2\ \pi\ \sigma$, is used in that section to show that more optimum wavelength choices were available, use of which would have resulted in an order of magnitude increase in irradiance potency at the target. But, perhaps more troublingly, at the envisioned hundreds of kilometer ABL standoff distances, even the most optimum HEL wavelength would still only have created irradiance potency levels at the target two orders of magnitude less than required levels.

The projected performance numbers used to justify and launch the ABL program were evidently very different from the despairingly poor performance numbers just indicated. A speaking campaign was mounted by the Author in 1990 (and sustained over the entire life of the ABL program) to alert the ABL community of the dire outlook facing the program should it continue along the lines proposed. The warnings went unheeded.

## References

Bester, M., Danchi, W. C., Degiacomi, C. G., Greenhill, L. J., & Townes, C. H. (1992). Atmospheric fluctuations: Empirical structure functions and projected performance of future instruments. *The Astrophysical Journal, 392*, 357–374.

Born, M., & Wolf, E. (2003). *Principles of optics* (7th ed., revised). Cambridge University Press.

Chanan, G., Troy, M., Dekens, F., Michaels, S., Nelson, J., & Mast, T. (Jan 1998). Phasing the mirror segments of the Keck telescopes: The broadband phasing algorithm. *Applied Optics, 37*(1), 140–155.

Chromey, F. R. (2010). *To measure the sky: An introduction to observational astronomy*. Cambridge University Press.

Coulman, C. E., & Vernin, J. (1991). Significance of anisotropy and the outer scale of turbulence for optical and radio seeing. *Applied Optics, 30*, 118–126.

Coulman, C. E., Vernin, J., Coqueugniot, Y., & Caccia, J. L. (1988). Outer scale of turbulence appropriate to modeling refractive index structure profiles. *Applied Optics, 27*, 155–160.

Danchi, W. C. (27 Dec 2001). Private communication.

Fried, D. L. (1966). Optical resolution through a randomly inhomogeneous medium for very long and very short exposures. *JOSA, 56*, 1372–1379.

Griffin, R. F. (1973). On image structure, and the value and challenge of very large telescopes. *Observatory, 93*, 3–8.

Griffin, R. F. (May 1990). Giant telescopes, tiny images. *Sky and Telescope*, 469.

Hufnagel, R. E., & Stanley, N. R. (1964). Modulation transfer function associated with image transmission through turbulent media. *JOSA, 54*, 52–61.

Kolmogorov, A. N. (1941a). The local structure of turbulence in incompressible viscous fluid for very large Reynolds numbers. *Doklady Akademii Nauk SSSR, 30*, 301–305.

Kolmogorov, A. N. (1941b). Dissipation of energy in locally isotropic turbulence. *Doklady Akademii Nauk SSSR, 32*, 16–18.

Martin, B., Hill, J. M., & Angel, R. (1991). The new ground-based optical telescopes. *Physics Today, 44*, 22–30.

McKechnie, T. S. (1990). Diffraction limited imaging using large ground-based telescopes. In *Proceedings of SPIE, V. 1236, Symposium on Astronomical Telescopes and Instrumentation for the 21st Century*, February 11–17 (pp. 164–178).

Ratcliffe, J. A. (1956). Some aspects of diffraction theory and their applications to the ionosphere. *Reports on Progress in Physics, 19*, 188–287.

Tatarski, V. I. (1961). *Wave propagation in a turbulent medium*. McGraw-Hill Book Co., Inc.

Titchmarsh, E. C. (1948). *Introduction to the theory of Fourier integrals* (2nd ed.). Clarendon Press.

Wilson, R. N. (2003). The history and development of the ESO active optics system. *The Messenger, 113*. ESO.

Woolf, N. J., McCarthy, D. W., & Angel, J. R. (1982). Performance of the MMT VII: Image shrinkage in sub-arc second seeing at the MMT and 2.3 meter telescopes. In: L. D. Barr & G. Burbridge (Eds.), Advanced technology optical telescopes, Proc. SPIE. Eng. (Vol. 332, pp. 50–56).

# Chapter 3
# Terms, Definitions, and Theoretical Foundations

**Abstract** This chapter provides basic information about the essential property of air that controls the behavior of light waves as they propagate through the atmosphere: refractive index. Quantities used to describe light waves are introduced, such as the complex amplitude and intensity; and Maxwell's electromagnetic equations are set out along with their solution for light propagating in an inhomogeneous medium—the atmosphere. Because atmospheric refractive index is a stochastic process, a statistical approach is needed to describe light propagation and imaging through the atmosphere. Statistical functions are introduced that characterize disrupted light waves and the disrupted state of images formed from these waves by large telescopes (e.g., the autocorrelation function of the complex amplitude). The terminologies used to describe the coherence properties of light are also introduced, and practical definitions are given for frequently used hypothetical entities such as "quasimonochromatic" light and "point-object." Scalar diffraction theory is used; vector diffraction theory is only used when polarization effects make a difference.

As with all other forms of 'electromagnetic radiation, light exhibits the properties of both waves and particles; it is emitted and absorbed in tiny packets called photons. The word "waves" is a loosely defined term that will frequently be used to describe the state of electromagnetic field disturbances; "wavefronts" is another such term that may be considered as describing the shape of electromagnetic waves when frozen at an instant of time. Scalar wave diffraction theory will generally be used in the book to describe the field disturbance as light waves propagate through the atmosphere; it will also be used to describe how optical systems, such as telescopes, bring the collected wave portions to a focus in the image plane. In fact, there is only one brief exception to the use of scalar diffraction theory; in Chap. 11, Sect. 11.8, vector waves and vector diffraction theory are used to treat the statistical properties of speckle when the telescope optics cause either complete, or partial, depolarization of the image-forming light waves.

Complex amplitude is the primary quantity that will be used to describe the light fields. Complex amplitude is a scalar function that accounts for both the phase and amplitude fluctuations associated with electromagnetic waves. Complex amplitude



T. S. McKechnie, *General Theory of Light Propagation and Imaging Through the Atmosphere*, Progress in Optical Science and Photonics 20,
https://doi.org/10.1007/978-3-030-98828-9_3

is the time-independent part of the electromagnetic field. The intensity, which is proportional to the squared modulus of the complex amplitude, is the quantity that allows electromagnetic waves to be detected. The Helmholtz equation—an equation that arises directly from Maxwell's electromagnetic equations—will be used for wave propagation in homogeneous free space. A closely related equation will be used for wave propagation through the inhomogeneous free space that primarily concerns us: Earth's atmosphere.

## 3.1 Air Refractive Index

The refractive index of air depends mainly on temperature and pressure, both of which vary strongly with altitude. Humidity is a lesser influence. Increased humidity lowers air density[1] and hence lowers the refractive index. For standard air—that is, dry air at sea level at temperature 15 °C and pressure 1013.25 mbar (1 mbar = 100 Pa) containing 0.045 % by volume of carbon dioxide—the refractive index $n_S$ is given by the formula (Birch & Downs, 1994),

$$n_S = 1 + \frac{8342.54 + 2406147 \cdot \left(130 - \frac{1}{\lambda_o^2}\right)^{-1} + 15998 \cdot \left(38.9 - \frac{1}{\lambda_o^2}\right)^{-1}}{10^8} \quad (3.1)$$

where the vacuum wavelength, $\lambda_o$, is expressed in microns (μm). This formula is based on observations made over the wavelength range, 0.2–2.0 μm.

In the visible wavelength range, 0.405–0.705 μm, Birch and Downs give the following convenient approximation for $n_S$ where discrepancies are claimed to be less than $1.4 \times 10^{-8}$,

$$n_S = 1 + 0.0472326 \cdot \left(173.3 - \frac{1}{\lambda_o^2}\right)^{-1}. \quad (3.2)$$

For $\lambda_o = 0.55\mu m$, 3.1 gives the refractive index as 1.000277838 and 3.2 gives the nearly identical value 1.000277848.

For air at temperature $T$ in °C and pressure $P$ in mbar, the refractive index $n_{TP}$ is given by

$$n_{TP} = 1 + (n_S - 1) \cdot \frac{P \cdot \left[1 + P \cdot (60.1 - 0.972 \cdot T) \cdot 10^{-8}\right]}{960.9543 \cdot (1 + 0.003661 \cdot T)}. \quad (3.3)$$

---

[1] The molecular weights of oxygen ($O_2$) and nitrogen ($N_2$) are 32 and 28, respectively. Assuming air to be a 20%:80% mix ratio, the weight of an "average" air molecule is 28.8. Water molecules ($H_2O$) are lighter, having a molecular weight of 18. Since (by Avogadro's hypothesis) the density of a gas is proportional to the molecular weight, moist air is less dense than dry air.

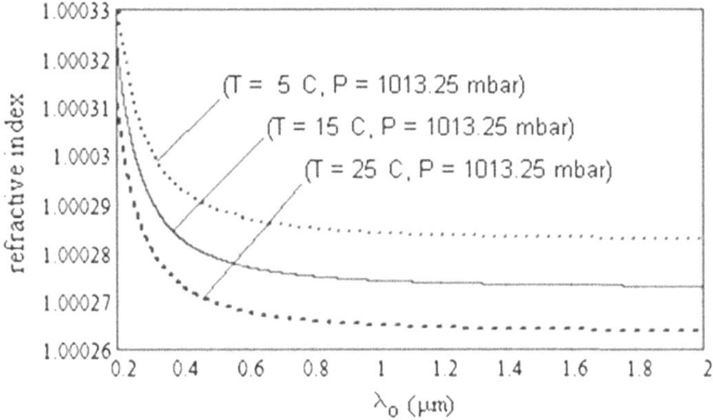


**Fig. 3.1** Refractive index versus wavelength for standard dry air ($T = 15$ °C, $P = 1013.25$ mbar). The index is also shown for temperatures 5 and 25 °C

Figure 3.1 shows how refractive index varies with wavelength for standard dry air over the entire range of validity of 3.1. The figure also shows the refractive index variation with wavelength at temperatures ±10 °C (above and below) standard temperature.

### 3.1.1 *Air Temperature and Altitude*

From sea level to about 11 km, the temperature of standard air reduces approximately linearly with altitude as shown in Fig. 3.2; the data used to produce this figure was obtained from the 1962 standard atmosphere graph of geometric altitude against temperature. Between 11 km and about 20 km, temperature remains approximately constant at 217 K. Above 20 km to about 100 km, temperatures oscillate over the range, 190–320 K.

### 3.1.2 *Air Pressure and Altitude*

Air pressure decreases with increasing altitude. Air pressure at a given observatory site provides a measure of the atmospheric mass lying above the site. At altitude $h$, atmospheric pressure $P$ is given in mbar[2] by the formula

$$P = P_o \cdot \left(1 - \frac{L \cdot h}{T_o}\right)^{\frac{g \cdot M}{R \cdot L}}. \tag{3.4}$$

[2] For SI units, the following conversions can be used: one bar = 100,000 Pa (Pa); 1 mbar = 100 Pa.

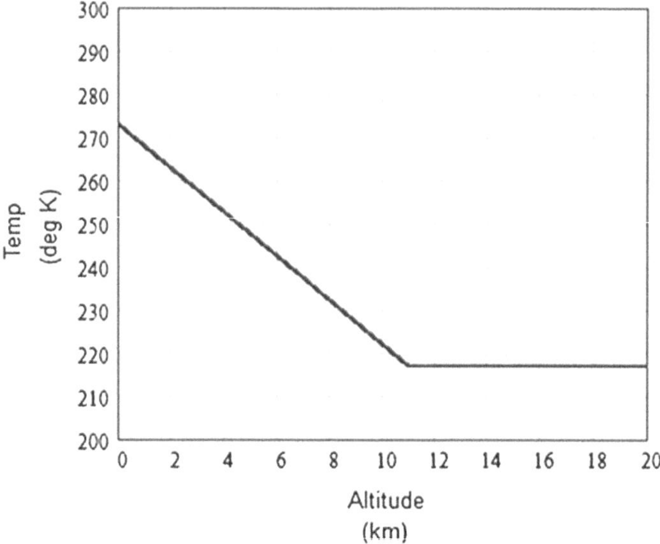


**Fig. 3.2** Atmospheric temperature variation with altitude (derived from US standard atmosphere graph of geometric altitude against temperature)

where

$P_o = 1013.25$ mbar (sea-level standard pressure)
$L = 0.0065 \cdot \frac{K}{m}$ (temperature lapse rate with altitude)
$T_o = 288.15$ K (sea-level standard temperature)
$g = 9.80665 \cdot \frac{m}{s^2}$ (acceleration due to gravity)
$M = 0.0289644 \ \frac{\text{kg}}{\text{mol}}$ (molar mass of dry air).
$R = 8.31447 \ \frac{J}{(\text{mol K})}$ (universal gas constant).

Relative pressure defined by the ratio $P/P_o$ is plotted against altitude in Fig. 3.3.

### 3.1.3 *Integrated Optical Path Difference Over the Entire Atmospheric Depth*

The integrated optical path difference (OPD) over a vertical atmospheric path through the zenith depends on the air pressure and temperature profiles over the path. For an observatory site at altitude *h*, the integrated OPD is given by[3]

[3] OPD often refers to optical path differences for two different paths through an optical system. Here, we use it to indicate the optical path difference between the path through the atmosphere and the same path in the absence of atmosphere.

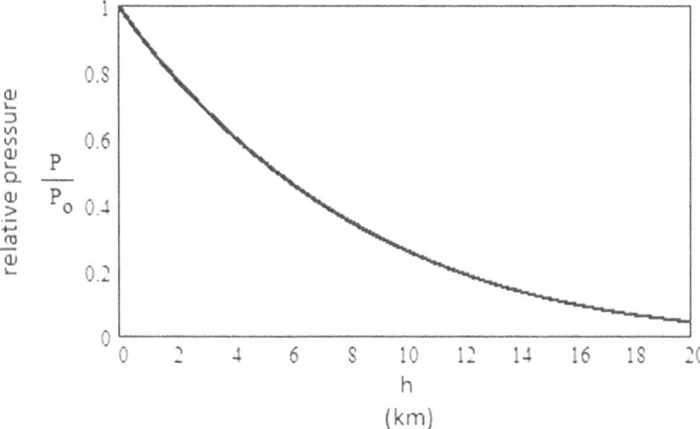


**Fig. 3.3** Atmospheric pressure (relative to that at sea level) versus altitude for standard temperature and pressure conditions ($T = 15°$ C, $P = 1013.25$ mbar)

$$OPD(h) = \int_h^\infty \left[n\left(h'\right) - 1\right] \cdot dh'. \tag{3.5}$$

Assuming standard temperature and pressure at sea level, with pressure versus altitude as given by 3.4, and a temperature lapse rate of 0.0065 K/m, the variation of OPD($h$) with altitude is as shown in Fig. 3.4.

From sea level, the integrated OPD for a vertical path through the atmosphere amounts to about 2.35 m. For comparison, a 4.57 m thick crown glass window

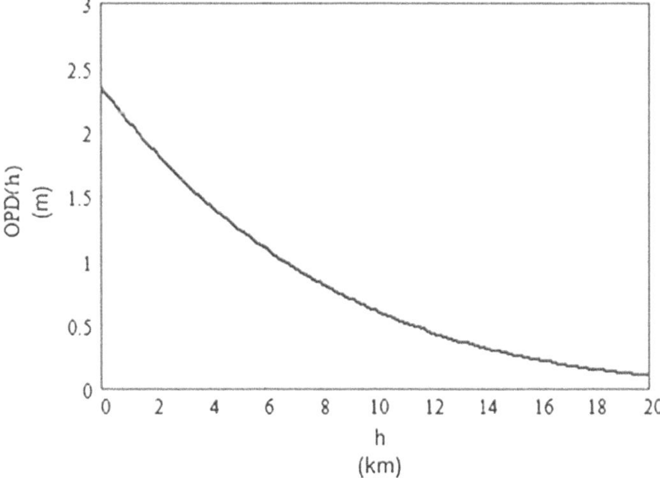


**Fig. 3.4** Integrated OPD versus altitude for vertical atmospheric paths (with standard temperature and pressure conditions assumed at sea level)

($n$~1.515) introduces a comparable amount of integrated OPD. On Mauna Kea, HI (4200 m altitude), where many of the world's largest telescopes are located, the integrated OPD for a vertical path is about 1.39 m, about 40.8 % less than the integrated OPD value measured from sea level.

For slanted paths through the atmosphere, it may readily be seen that the values given for OPD(h) by 3.5 would have to be multiplied by the quantity 1/cos (ZA) where ZA is the zenith angle of the path.

### 3.1.4 *Effect of Humidity*

The refractive index of water vapor is less than that of air. The refractive index of moist air is therefore less than that of dry air—this in spite of the widely held, but mistaken, notion that that moist air is "heavier" than dry air (cf., Footnote 1, p. 34). Over the visible spectral range, 0.405–0.644 μm, the refractive index of moist air, $n_{TPM}$, is given by the formula

$$n_{TPM} = n_{TP} + P_M \cdot \left(3.7345 - 0.0401 \cdot \frac{1}{\lambda_o^2}\right) \cdot 10^{-8}, \qquad (3.6)$$

where $P_M$ is the partial pressure due to the water vapor expressed in mbar, the total pressure still being $P$mbar. Birch and Downes claim that this expression is valid for conditions "not deviating very much" from standard laboratory air conditions ($T$ = 15 $^0$C, $P$ 1013.25 mbar).

For moisture-saturated air at standard temperature and pressure, the partial pressure due to water vapor, $P_M$, is about 23 mbar. Figure 3.5 shows how refractive index reduces for moisture-saturated air.

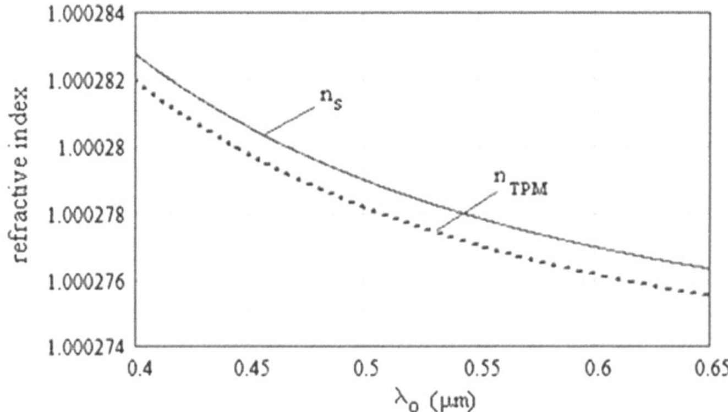


**Fig. 3.5** Refractive index of standard dry air (calculated from 3.1) and for air at 100% humidity (calculated from 3.6) at standard temperature and pressure ($T$ = 15 °C, $P$ = 1013.25 mbar)

### 3.1.5 *Effect of Dispersion*

The refractive index dispersion of air can be expressed by the $V$ -value, a measure defined in terms of the refractive indices at three traditional Fraunhofer spectral lines: the sodium D-line (0.5893 μm),[4] the hydrogen C-line (0.6563 μm), and hydrogen F-line (0.4861 μm). Denoting the refractive indices at these wavelengths by $n_D$, $n_C$, and $n_F$, the $V$-value is expressed by

$$V - \text{value} = \frac{n_D - 1}{n_F - n_C}. \tag{3.7}$$

Table 3.1 gives the refractive indices of standard dry air at the three standard wavelengths (calculated from 3.1). Inserting these values in the above formula gives the $V$ -value of standard dry air as 89.4, a value indicating relatively low dispersion. For comparison, achromatic doublets used in standard binoculars typically consist of a low-dispersion crown element with $V$-value ~60 and a high-dispersion flint element with V -value ~36.

### 3.1.6 *Random Variables Associated with Atmospheric Turbulence*

When temperature, pressure, and moisture content are constant throughout a given air volume, for most practical purposes the refractive index may be considered constant throughout that volume. An air volume of this kind may be described as optically homogeneous. When light waves travel through such a volume, the phases remain undisrupted. Vacuum space is perfectly optically homogeneous.

Unlike vacuum space, Earth's atmosphere behaves as though it were optically inhomogeneous. The inhomogeneity derives from a number of sources, including those just mentioned—temperature-, pressure-, and moisture-level fluctuations—all of which are capable of producing random fluctuations of the refractive index. When

**Table 3.1** Refractive index of standard dry air ($T = 15\,^0$C, $P = 1013.25$ mbar) for the hydrogen C—and F-lines and the sodium D-line

| Wavelength (μm) | Refractive index |
|---|---|
| 0.4861 | 1.0002794 |
| 0.5893 | 1.0002771 |
| 0.6563 | 1.0002763 |

[4] The so-called sodium D-line is actually a doublet centered on wavelength 0.58929 μm.

light waves propagate through an inhomogeneous volume, the phases of the waves—and ultimately the amplitudes too—can suffer significant disruptions. Whereas a great many parameters are required to describe all of the characteristics of the atmosphere, the propagation behavior of light waves through the atmosphere is ultimately governed by just one crucial parameter: refractive index. Thus, for the purposes of this book where we are concerned with light propagation through the atmosphere, any randomly varying mechanism in the atmosphere that affects refractive index may be regarded as a contributor to "atmospheric turbulence."

Weather systems can span hundreds of kilometers; the associated temperature and pressure fluctuations can produce huge refractive index variations over the extent of these systems. From one locality to another within a typical weather system, it is not unusual to find pressure variations of the order $\Delta p/p \approx 0.05$; such variations produce corresponding refractive index variations of the order $\Delta n \approx (n - 1)\, 0.05$.[5] The gradients associated with refractive index variations of this magnitude cause measurable angular shifts in telescope images; the time constants of the image shifts are typically of the order of hours or days. For future extremely large telescopes (ELTs), with diameters of several tens of meters, the effects of nonzero second- and higher-order derivatives of these large-scale refractive index variations will become increasingly evident since both can cause significant phase distortions over the correspondingly larger image-forming wave portions collected by these massive instruments. The effect on the images would be similar to that of mild, though slowly varying, telescope aberrations.

In Kolmogorov theory (Hufnagel & Stanley, 1964), the term "atmospheric turbulence" refers to atmospheric structures postulated to lie in a size range bounded by the so-called inner and outer scale limits. However, in this book, we consider the term as having a broader connotation. The term no longer merely refers to turbulence structures limited to the Kolmogorov size range; it is now considered to include all physically possible scale sizes—that is, sizes ranging from less than one millimeter at one extreme to tens, or even hundreds of kilometers at the other. In the general formulation laid out, there is no need to suppose either an inner or an outer scale limit to turbulence structure size. Moreover, we no longer consider that the term speaks only for the effect of temperature on the refractive index field as is assumed in Kolmogorov theory; now it speaks for the effects of all mechanisms that affect the refractive index. These naturally include the effects of pressure- and moisture-level fluctuations, in addition to those of temperature.

The absence of an in-depth understanding of how the various controlling parameters affect the atmospheric refractive index field—a fact noted by Hufnagel and

[5] A 5% change in weather pressure, say from 1000 to 1050 mbar, produces an OPD change of about 12 cm for vertical propagation paths (an OPD amount equivalent to about 220,000 visible light wavelengths). While OPD variations of this magnitude commonly occur over ground baselines of the order of hundreds of kilometers, they do not cause measurable amounts of scintillation on the ground below.

Stanley (1964)[6]—does not present a problem to the general theory laid out in this book. There is no need to know details of how these parameters interact together to determine this field. The essential statistical properties of the refractive index field are all ultimately encoded in the statistical properties of point-object images formed by telescopes. By simply measuring certain key properties of the intensities in these images, we can calculate the essential properties of the refractive index field from the measured data.

The first significant steps toward understanding the mechanisms of turbulence came from the English mathematician and physicist, Lewis Fry Richardson (1881–1953), who carried out many terrestrial atmospheric turbulence experiments and pioneered modern mathematical techniques for forecasting weather (Richardson, 1922). The dimensionless Richardson number is named after him; in aviation, this number gives a rough indication of expected air turbulence. According to Richardson's understanding, air motions driven by wind and convection cause the formation of eddies and vortices which interfere with one another, spinning off smaller secondary structures, the process continuing in a downward cascade until a broad spectrum of turbulence structure sizes develops.[7]

In steady-state mode, the turbulence field comprises a hierarchy of structures whose sizes can range over several orders of magnitude. In the grip of the fluid's molecular viscosity, the smallest structures dissipate their kinetic energy in the form of heat. As long as a constant rate of energy is supplied by the driving mechanisms, which naturally include wind and convection, one would naturally expect the statistical distribution of structure sizes in the turbulent field to remain constant over time.

At night, however, after the Sun has set and the energy supply to the driving mechanisms abruptly stops, and yet with the same natural tendency persisting of larger turbulence structures cascading down into smaller ones, an entirely different type of turbulence spectrum likely arises—one characterized by the turbulence energy residing in structures whose average size is now smaller than before.

The issue of turbulence structure size is of fundamental importance to light propagation and imaging through the atmosphere. As we shall see in Chap. 12 and other chapters, the image characteristics that result when the average turbulence structure size in the atmospheric path is small are dramatically different from those resulting from turbulence with larger average structure size. For the relatively large average turbulence structure sizes assumed in Kolmogorov formulations, where sizes range from less than a centimeter to many tens of meters, the resolution outlook for large ground-based astronomical telescopes not equipped with adaptive optics (AO) is poor. However, if the average turbulence structure size is significantly smaller

[6] These controlling factors include the rate of energy per unit mass dissipated by viscous friction, the average shear rate of the wind, and the average vertical gradient of the potential temperature. According to Hufnagel and Stanley, "each of these quantities is a function of the other two" and "there appears to be no satisfactory theory for predicting their functional interdependence, except for the lowest few hundred meters of the atmosphere.".

[7] Richardson (1922) famously summarized turbulence in the rhyming verse: "*Big whirls have little whirls that feed on their velocity, and little whirls have lesser whirls and so on to viscosity.*".

than assumed in Kolmogorov formulations, an entirely different and altogether more optimistic resolution outlook presents itself.

The refractive index fluctuation associated with atmospheric turbulence solely determines how light waves are affected as they propagate through the atmosphere. As we have just seen, the refractive index—which at any single instant in time may be considered as a three-dimensional scalar field—is determined by the combined effect of a number of underlying variables, of which pressure and temperature are the most important. Because these variables are random variables, the refractive index is itself a random variable (Papoulis, 1965). As time unfolds, the refractive index field evolves, but in a random chaotic manner rather than deterministically. Random refractive index fluctuations disrupt the phases of propagating light waves. The severity of the disruption depends on both the strength of the atmospheric turbulence and the length of the atmospheric path.

Distances over which the temperature and pressure—and hence the refractive index—remain substantially correlated provide a measure of the size of the atmospheric turbulence structures. The distribution of turbulence structure size—and hence the distribution of refractive index structure size—is crucial to determining how turbulence affects propagating light waves. Typically, for turbulence structures in the size range, 0–10 m, the random temperature changes from one turbulence structure to neighboring structures is small, of the order 0.01 °C. (Such temperature fluctuations are of course superposed on much larger variations associated with weather systems and altitude where, as indicated by Fig. 3.2, from sea level to an altitude of 11 km average temperature varies by more than 50 °C.) Though temperature changes of the order 0.01 °C may seem small, because they occur over relatively short distances (perhaps of the order of 10 cm), they produce disproportionately large changes in the wavefront gradients. For telescopes with apertures larger than about 20 cm, these turbulence-related phase disruptions can severely degrade image quality. Ignoring dispersion effects, which are relatively small anyway, the phase fluctuation associated with a given amount of OPD fluctuation increases in inverse proportion to the wavelength. Thus, shorter wavelengths are generally more seriously disrupted by turbulence than longer wavelengths.

### *3.1.7 Astronomical Refraction*

Astronomical refraction refers to the angular displacement of astronomical objects from their true position because of refraction by the Earth's atmosphere. A similar effect, known as terrestrial refraction, occurs when atmospheric refraction causes objects seen through terrestrial telescopes, such as mountains, aircraft, and ships, to appear displaced from their true positions.

Consider Fig. 3.6 where a light ray from a star is seen passing through the atmosphere before finally arriving at an observer located at the point, O, on Earth's surface. The analytic approach used here closely follows that previously given by Smart (1958). Due to the small (<80 km) height of the atmosphere compared to Earth's

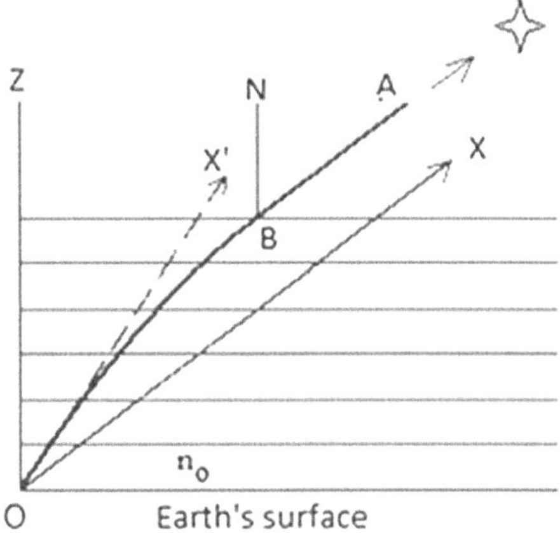


**Fig. 3.6** Refraction through Earth's atmosphere where the atmosphere is approximated by a large number of horizontal layers

radius (6370 km), for moderate zenith distances the atmosphere may be considered to act like a system of horizontal layers where the density—and hence the refractive index—of the layers progressively increases between the uppermost layer and the lowest surface layer. With this simplified model, the distribution of refractive index with altitude does not affect the overall refraction angle of rays by the time they finally reach the ground; this angle is entirely determined by the refractive index of the lowest layer in contact with the ground, as is now shown.

AB indicates the direction of the ray incident on the highest atmospheric layer. If we draw OX parallel to BA, angle ZOX is the true zenith distance, or zenith angle, of the star; we denote this angle by ZA. In the figure, OX'' gives the direction of the ray reaching the observer in the final element of the ray path. Thus, OX′ is the direction in which the observer sees the star. We refer to ZOX′ as the observed zenith angle, denoting it by ZA′.

By applying Snell's law of refraction, it can be seen that for any particular layer, say the *j*th layer, the following relation holds:

$$\sin(ZA) = n_j \cdot \sin(ZA'_j), \tag{3.8}$$

where $n_j$ is the refractive index of the *j*th layer and $ZA'_j$ is the zenith angle of the ray in that layer. For the layer nearest the ground, where $j = 0$, we can write

$$\sin(ZA) = n_0 \cdot \sin(ZA'). \tag{3.9}$$

Atmospheric refraction always makes stars appear closer to the zenith than if there were no atmosphere. The observed position always lies in the vertical great circle through the true position.

The angle of refraction, *R*, is defined as the angle through which the star's observed direction is displaced from its true direction. This angle is given by

$$R = ZA - ZA'. \quad (3.10)$$

Equations 3.9 and 3.10 allow us to write

$$\sin(R + ZA') = n_0 \cdot \sin(ZA'). \quad (3.11)$$

Expanding $\sin(R + ZA')$ in the form $\sin(R) \cdot \cos(ZA') + \cos(R) \cdot \sin(ZA')$ and noting that $R$ is a small angle, 3.11 may be seen to reduce to

$$R = (n_0 - 1) \cdot \tan(ZA'). \quad (3.12)$$

This formula may be regarded as valid for zenith angles up to about 50°, though it still gives fairly good approximations up to about 70°. At larger angles, the formula breaks down; it is no longer acceptable to ignore the curvature of the atmospheric layers.

It may be noted that angle $R$ is always positive. Thus, the true zenith distance is always greater than the observed zenith distance. $ZA'$ is the quantity actually measured in an observation. By using 3.12, we may calculate the true zenith distance, ZA, and thus eliminate the effect of astronomical refraction from the observed zenith distance.

The refractive index of the ground layer, $n_0$, depends on the altitude of the observing site and on the temperature, pressure, and moisture level at the site. It also depends on the observation wavelength. Equations 3.1–3.4 and 3.6 may be used to obtain appropriate values for $n_0$.

#### 3.1.7.1 Typical Angles of Refraction

By inserting the refractive index values indicated in Table 3.1 for standard dry air at sea level ($T$ = 15 °C, $P$ = 1013.25 mbar) for a star at observed zenith angle, $ZA' = 45^0$, the angles of refraction (calculated from 3.12) are as indicated in Table 3.2.

Unless a compensating prism is used, the angular spread indicated in the table means that star images formed by any telescope observing at the zenith angle, 45°, will be stretched in the direction toward the zenith into about a 1-arcsec-wide rainbow.

**Table 3.2** Angles of refraction at sea level for P = 1013.25 mbar) for a star at observed zenith angle, 45°

| Wavelength (μm) | Angle of refraction (arcsec) |
|---|---|
| 0.4861 | 57.63 |
| 0.5893 | 57.16 |
| 0.6563 | 56.99 |

### 3.1.8 Atmospheric Extinction

Atmospheric extinction refers to the dimming of stars and other astronomical objects caused by light scattering by Earth's atmosphere. While most of the light loss at visible wavelengths is caused by Rayleigh scattering (Strutt J., "On the light from the sky, its polarization and colour," 1871), Mie scattering (Stratton, 1941) also contributes significantly. Both Rayleigh and Mie scattering are elastic scattering processes; thus, no wavelength changes are involved. Raman scattering—an inelastic form of scattering that does involve wavelength changes—also occurs, but in amounts several orders of magnitude smaller than Rayleigh and Mie scattering. For most practical purposes, the contribution of Raman scattering to extinction may be ignored (Gardiner, 1989).

Rayleigh scattering is caused by air molecules and small dust particles in the atmosphere, with scattering strength showing a $1/\lambda^4$ dependence on wavelength, $\lambda$. Larger dust particles in the atmosphere, with sizes $> \lambda$, more likely cause Mie scattering. The fact that blue wavelengths are more strongly scattered than red—for both Rayleigh and Mie scattering—accounts for blue skies and red sunsets (Lilienfeld, 2004).

Atmospheric extinction is usually quantified (Ridpath, 1989) by the change of magnitude caused by the scattering losses. At visible wavelengths in clear sky viewing conditions, a star near the zenith observed from sea level is typically dimmed by about 0.2 visual magnitudes, which corresponds to the brightness pull-down factor, 0.832 ($= 2.5119^{-0.2}$) where 2.5119 is Pogson's ratio (cf., Sect. 14.5). Extinction losses are larger for stars lying further from the zenith; these losses may be approximately calculated using the same simple atmospheric model shown previously in Fig. 3.6.

In terms of that model, it may readily be shown that the magnitude loss, $\Delta m$, of a star at arbitrary zenith angle, ZA, consistent with the 0.2-magnitude loss for a through-zenith observing path, is given by

$$\Delta m = -\frac{\ln\left[2.5119^{-\left[\frac{0.2}{\cos(ZA)}\right]}\right]}{\ln(2.5119)}. \tag{3.13}$$

Figure 3.7 shows atmospheric extinction plotted against zenith angle according to the above formula. If haze is present, extinction levels can be considerably higher than shown in the figure. On the other hand, if the observing site is at high altitude rather than at sea level, the losses may be significantly less. Because the atmospheric model used to produce 3.13 ignores Earth's curvature (cf., atmospheric refraction in Sect. 3.1.7), this equation loses accuracy for zenith angles greater than about 70°. For a horizontal path through the entire atmosphere, the equation indicates that extinction is total, with no light at all reaching the observer. An accurate calculation—one that takes proper account of Earth's curvature—shows that a horizontal path through the atmosphere passes through an air mass about 40 times that of a vertical path. Based on the 0.2-magnitude loss for the path through zenith, losses for horizontal paths

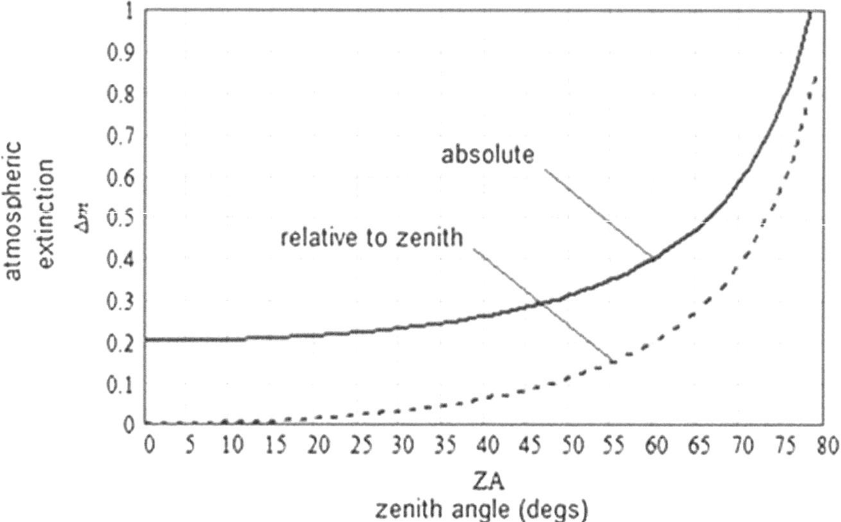


**Fig. 3.7** Atmospheric extinction versus zenith angle in typical clear sky astronomical viewing conditions. The plotted quantity, $\Delta m$, represents the brightness loss caused by the atmospheric path measured in visual magnitudes. The plot in *firm line* shows the actual dimming loss. The plot in *dotted line* indicates the dimming loss relative to the loss over the zenith path. The latter plot is useful for estimating the magnitudes of variable stars using comparison stars that lie at other zenith angles

would in fact be limited to about 8 magnitudes (i.e., ~ $40 \times 0.2$). This loss is enough to render even the brightest star, Sirius (mag $= -1.46$), invisible to the naked eye, though not quite enough to do the same for Venus (mag $= -4.4$ at mean elongation) or Jupiter (mag $= -2.7$ at mean opposition).

## 3.2 Point-Objects

Throughout the book, the object most frequently dealt with is the single point-object. Ideally, a point-object is a light source subtending an infinitesimally small angle with respect to the imaging system. Since actual light sources must have some physical extent, in practice it is not possible to realize an ideal point-object. In this book, therefore, a point-object is simply considered as an object whose angular subtense is much smaller than the minimum size that can be resolved by the optical system. Unresolved stars can make nearly ideal point-objects. However, whereas a large star such as Betelgeuse (1/20th arcsec angular subtense) might make an ideal point-object for a small telescope, it might not be so ideal for very large telescopes that are capable of resolving it.

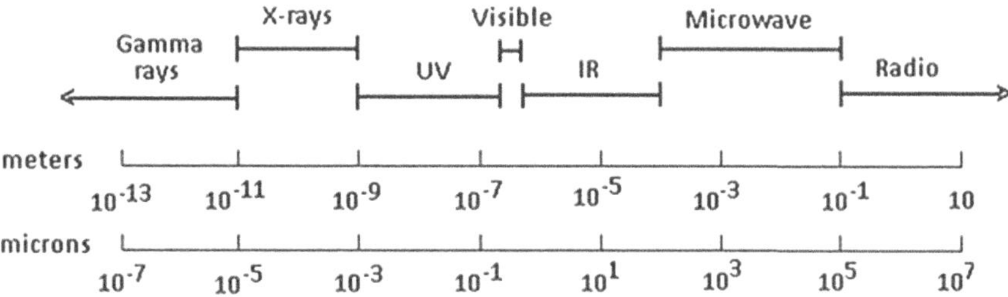


**Fig. 3.8** The electromagnetic spectrum. The numbers indicate the wavelength, expressed in both meters and microns

## 3.3 The Electromagnetic Spectrum

The electromagnetic spectrum is represented schematically in Fig. 3.8. The spectral portion relevant to this book comprises the upper end of the UV spectrum, the entire visible spectrum, and the IR spectral bands transmitted by the atmosphere. Wavelengths shorter than 0.3 μm are largely blocked by the atmosphere as well as by standard glasses. The effective wavelength range considered in this book therefore lies between about 0.3 and 1000 μm. Reflector telescopes can provide images over this entire range as long as suitable optical coatings are used on the mirror surfaces.

The infrared spectrum is usually divided into regions referred to as near-infrared, mid-infrared, long-wavelength infrared, and far-infrared. However, there is no universal agreement about the boundaries.

Table 3.3 gives the boundary conventions used in this book. The type of detector used principally determines the classification of individual IR wavelengths. Appropriate detector types are indicated in the table. Cooled detectors are usually necessary at IR wavelengths. Since detector technology is considered outside the scope of this book, for more information on this extremely important technological area we merely refer the reader to the literature and books such as those by Rogalski (2011) and Henini and Razeghi (2002). For infrared system engineering, the reader is referred to the time-honored reference book by Hudson (1969).

Due to absorption by the atmosphere, primarily by water vapor and $CO_2$, only certain IR wavelength bands reach the ground. A typical atmospheric transmission plot is shown in Fig. 3.9. The major transmission bands are commonly referred to by the alphabetical characters indicated in Table 3.4.

## 3.4 Quasi-Monochromatic Light

In practice, even the narrowest spectral line has a nonzero width. The term quasi-monochromatic refers to light comprised of such a narrow spread of wavelengths that our experiments cannot distinguish differences between them. Thus, for all practical purposes, quasi-monochromatic light is light that behaves as though it were constituted by a single monochromatic emission.

**Table 3.3** Wavelength band nomenclature for UV through IR wavelengths and appropriate detector types

| Wavelength name | Abbreviation | Wavelength range (μm) | Detector types |
|---|---|---|---|
| Vacuum ultraviolet | VUV | 0.01–0.2 | Blocked by atmosphere |
| Middle ultraviolet | MUV | 0.2–0.3 | Blocked by glass |
| Ultraviolet B | UVB | 0.28–0.315 | $SiO_2$ |
| Ultraviolet A | UVA | 0.315–0.4 | $SiO_2$ |
| Visible | Vis | 0.40–0.75 | $SiO_2$, human eye |
| Near-infrared | NIR | 0.75–3.0 | $SiO_2$, InGaAs, PbS |
| Mid-infrared | MIR | 3–8 | InSb, HgCdTe, PbSe, PtSi |
| Long-wavelength infrared | LWIR | 8–15 | HgCdTe, microbolometers |
| Far-infrared | FIR | 15–1000 | Doped $SiO_2$ |

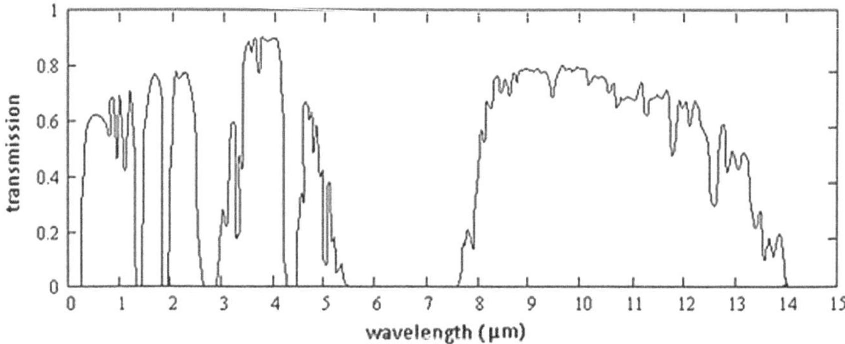


**Fig. 3.9** Atmospheric transmission windows typical of a vertical propagation path

**Table 3.4** Infrared atmospheric transmission windows

| Alphabetical character | Wavelength range (μm) |
|---|---|
| I | 0.7–1.0 |
| J | 1.0–1.4 |
| H | 1.5–1.8 |
| K | 2.0–2.4 |
| L | 3.0–4.0 |
| M | 4.5–5.5 |
| N | 7.5–14.0 |
| Q | 17–25 |
| Z | 28–40 |

## 3.5 Amplitude and Phase of Light Waves Disrupted by Turbulence

Atmospheric turbulence in the inertial subrange is caused by the mixing of warm and cold air in the atmosphere, the driving forces being wind, convection, and the thermodynamic changes caused by water vapor as it alternately evaporates and condenses. The random temperature and pressure fluctuations associated with these mechanisms produce random refractive index fluctuations. After passage through turbulent air (with its associated inhomogeneous refractive index), initially plane waves (such as might have originated from distant stars) or spherical waves (such as might have originated from objects in low Earth orbit) soon acquire random wave front corrugations (Fig. 3.10). In either case, these wave corrugations can be described by the phase fluctuations over the wavefronts. As these phase-disrupted waves continue forward propagation, some portions of the wavefronts converge, while others diverge, causing amplitude fluctuations to develop. The light waves that finally reach the telescope are generally characterized by random fluctuations of both phase and amplitude, the latter usually referred to as scintillation.

Intensity detectors, such as the eye or charge-coupled device (CCD) arrays, respond to the intensity scintillation but not to the phase fluctuation. Scintillation patterns formed by light from the brightest star, Sirius, are shown in Fig. 3.11. The strong scintillation seen in this figure is because Sirius was low in the sky when the images were recorded (zenith angle $\approx 70^\circ$). As indicated in the figure, patterns

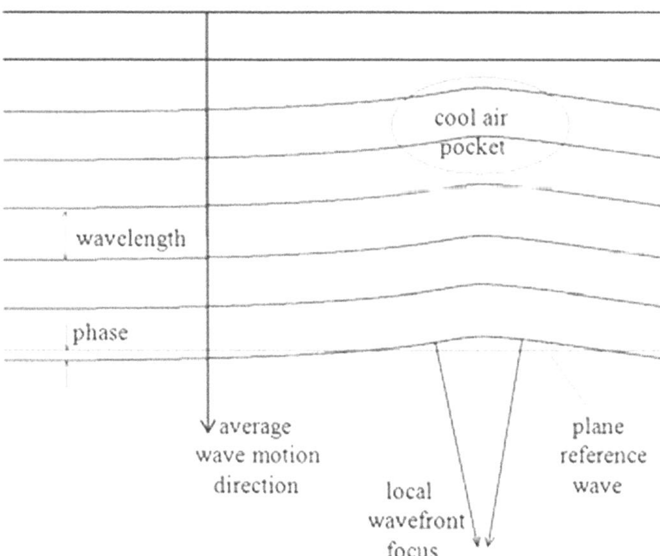


**Fig. 3.10** Pockets of cool air (higher refractive index) cause convergence, or focusing, of local wavefront portions. Warm air pockets (lower refractive index) cause divergence of the wavefront

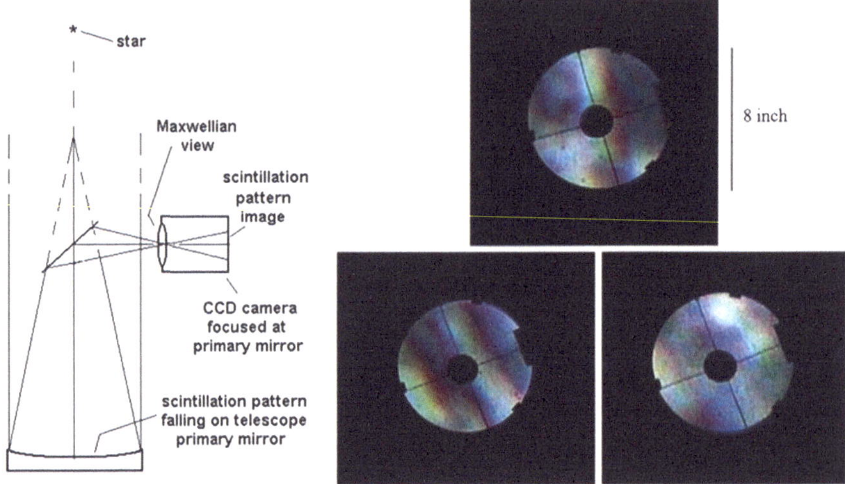


**Fig. 3.11** Scintillation patterns formed by light from the brightest star, Sirius, falling on the primary mirror of the author's 8-inch Newtonian reflector (pictured from the Maxwellian view). Light patterns like these falling on a naked-eye observer looking directly up at the star would result in him/her seeing wild fluctuations in the star's apparent brightness—familiar to everyone as twinkling. Twinkling is generally accompanied by color variations, mirroring the color variations evident in the patterns

like these are obtained by pointing the telescope directly at the star, while pointing and focusing the camera at the primary mirror. The view so obtained is sometimes referred to as the Maxwellian view.

Anyone who has ever seen a star twinkle is already familiar with the phenomenon of scintillation. Scintillation is caused primarily by upper atmospheric layers. As light waves pass through these layers, the waves acquire phase fluctuations that gradually transform into a mixture of both amplitude and phase fluctuation by the time the waves reach the ground below. To a lesser extent, layers at intermediate altitudes also contribute to scintillation, while layers near the ground make almost no contribution.

High-altitude air motion causes scintillation patterns, such as shown in Fig. 3.11, to move across the telescope aperture. Sometimes the movement rate is slow—perhaps just a few km/hour—but, when the jet stream lies overhead, patterns like these can flit across the aperture at speeds in excess of 160 km/h (100 mph). Turbulence churning can also cause evolution of these patterns as they move across the aperture. The time-constant of scintillation is largely determined by turbulence structure size and the wind speed in the upper layers of the atmosphere.

Decorrelation rates typically lie in the range, 1 Hz–1 kHz. In contrast, decorrelation rates for the phase fluctuation of the light waves reaching the ground are determined by wind speed and turbulence structure size at all altitudes, not just high altitudes. Phase fluctuation rates are also strongly wavelength dependent; they reduce (roughly) in inverse proportion to the wavelength. Phase decorrelation rates are therefore likely to be quite different from those of scintillation. Further discussion of scintillation is given in Chaps. 5 and 6. A quantitative analysis of scintillation caused by typical turbulence structure is given in Appendix H.

## 3.6 The Atmosphere Considered as a Stochastic Process

The refractive index field associated with the atmosphere is entirely responsible for modifying the complex amplitude in propagating light waves. Atmospheric turbulence and its associated refractive index field may both be considered as stochastic processes. These processes exist in four-dimensional space, three spatial dimensions and one time dimension. Even if the initial starting conditions in the atmosphere are known precisely, future evolution of n(x,y,z,t) cannot be predicted exactly. A certain level of indeterminacy allows for an uncountable number of possible future outcomes. The values taken by the refractive index field, n(x,y,z,t), continually change in a random manner, both spatially and temporally. It is appropriate, therefore, to describe fields of this type in statistical terms, with the field characteristics described by probability distributions.

For widely separated locations in either space or time, the values taken by n(x,y,z,t) are likely to be poorly correlated; for closer separations, the values likely show higher levels of correlation. In addition to these correlation properties, for any one particular location in space and time, both the spatial and temporal fluctuations of the refractive index field, n(x,y,z,t), likely share common attributes; for example, they may both have approximately the same mean value or the same variance.

The entire family of possible atmospheric refractive index distributions comprises an ensemble. This leads to the concept of ensemble averages. In this book, when the term "average" is used, usually the ensemble average is implied. Many different properties of the refractive index field can be described in terms of ensemble averages: for example, the mean and the variance. The average size and shape of the refractive index structures can also be described by ensemble averages; both relate directly to the average size and shape of the turbulence structures.

First-order statistical properties relate to properties at a single location in either space or time. These properties are entirely described by probability density distributions, or probability density functions (PDFs). Meaningful first-order statistical properties may be derived from PDFs, such as the mean and the variance.[8] Second-order statistical properties refer to properties at two separated locations in either space or time. These properties are described by joint-probability density functions. In the spatial domain, second-order statistical properties contain meaningful information about the size and shape of the refractive index structures. In the time domain, they contain information about temporal change rates of the refractive index structures.

[8] Some authors consider the variance as a second-order statistical property.

### 3.6.1 *Spatial and Temporal Stationarity and the Ensemble Average*

Atmospheric conditions can change significantly from day to day and even over the course of a day. However, over meaningfully long-time intervals, or over meaningfully large expanses of atmosphere, average values can remain substantially invariant. This leads to the notion of extended regions of space and time (perhaps both at once) over which the statistical properties remain substantially stationary (Davenport & Root, 1958). Within these regions, meaningful estimates of ensemble average quantities, such as the mean and the variance, can be obtained from either space or time-averages. A stochastic process is said to be ergodic in a temporal sense if the ensemble average statistical properties can be deduced from a sufficiently long sample realization of the process. Similarly, the process is ergodic in a spatial sense if the statistical properties can be obtained from a sufficiently extensive spatial area.

Space-average estimates are obtained from measurements made at a single instant of time over many spatial locations. Time-average estimates are obtained from measurements made at a single location in space over extended periods of time. While we might casually refer to the "ensemble average," in practice such an average is usually obtained as either a space-average or a time-average estimate.

### 3.6.2 *Standard Error and Standard Deviation*

The measurement accuracy of an estimate can be obtained from the standard error of the measured sample. Assuming an unbiased estimator—that is, a quantity that does not exhibit estimator bias—and assuming that the estimate is based on a large enough measured sample, standard error increasingly tends toward the standard deviation of the ensemble. Standard error is usually estimated by dividing the sample mean of the deviations by the square root of the sample size.

### 3.6.3 *Autocovariance and Autocorrelation Functions, the Variance, and Rms*

Different fields of study define autocorrelation differently. In some fields, the term is used interchangeably with the term autocovariance. In this section, therefore, to eliminate any possibility of ambiguity, we give mathematical definitions describing exactly what these two terms represent; we will rigorously adhere to these definitions and usages throughout the text. The autocorrelation function and auto covariance function are closely connected and convey similar information. Such differences as there might be arise solely from how these functions are normalized. We also define the terms variance and rms, quantities that relate directly to the autocovariance function at zero lag.

Given a real random process, $F(x)$, the autocovariance function is the covariance of the function $F(x)$ relative to an x-shifted version of itself. The $x$ argument may represent either a spatial or a temporal argument, as appropriate. The autocovariance function $F(x)$ at two shifted locations, $x$ and $x'$, can be written in the form, $\langle (F(x') - \langle F(x') \rangle) \cdot (F(x) - \langle F(x) \rangle) \rangle$, where the $\langle \cdot \rangle$ brackets denote the ensemble average and $\langle F(x') \rangle$ and $\langle F(x) \rangle$ represent the mean values of the process at $x'$ and $x$, respectively. The variance, or second central moment, is given by $\sigma^2(x) = \langle F^2(x) \rangle - \langle F(x) \rangle^2$.

In this book, to normalize the autocovariance function to unity, we can accomplish this task by simply dividing by the product term, $\sigma\left(x'\right) \cdot \sigma(x)$.

We refer to the resulting function as the autocorrelation function, formally defining it by $\frac{\langle (F(x') - \langle F(x') \rangle) \cdot (F(x) - \langle F(x) \rangle) \rangle}{\sigma(x') \cdot \sigma(x)}$.

If $F(x)$ constitutes a stationary process, the mean value and the variance (and all other functions of the process) are then independent of $x$. The variance, $\sigma^2(x)$, may then be written simply as $\sigma^2$. The autocorrelation function in this case can then be written in the form, $\frac{\langle F(x') \cdot F(x) \rangle - \langle F(x) \rangle^2}{\sigma^2}$, where $\langle F(x) \rangle$ takes the same value for all $x$.

In general, the autocorrelation function of such a process takes values in the range, [0, 1], with 0 indicating zero correlation and 1 indicating perfect correlation.

The root-mean-square deviation, or root-mean-square error, $\sigma$, is simply the square root of the variance, $\sigma^2$. The root-mean-square deviation is usually abbreviated to "rms." For an unbiased estimator, the rms is known as the standard error. An example of a real stationary process can be found in the intensity distribution in the scintillated image-forming waves arriving in the pupil of a telescope after propagation over an extended atmospheric path.

When the function $F(x)$ constitutes a complex random process, the autocovariance function, the autocorrelation function, the variance, and rms are all defined in a similar way to their real function counterparts. For a stationary complex random process with zero mean, the autocorrelation function can be written in the form, $\frac{\langle F(x') \cdot F^*(x) \rangle}{\sigma^2}$, where $F^*$ indicates the complex conjugate. An example of such a process might be found in the complex amplitude distribution over image forming waves arriving in the pupil of a telescope after propagation over an extended atmospheric path.

In general, the autocorrelation function takes values in the range, [−1, 1], with 0 indicating zero correlation, 1 indicating perfect correlation, and − 1 indicating perfect anti-correlation.

It might be an appropriate moment to observe that the autocorrelation function of a random process is a second-order statistical property of the process. Similarly, the structure function of a random process is also a second-order statistical property of the process.

The autocorrelation function of a random process and the structure function of that process are closely related; the one can be derived from the other. As indicated in the next section, the refractive index distribution of the atmosphere is a prime example of a random process. The autocorrelation function of the atmospheric refractive index distribution and the structure function of that distribution are closely related; the actual relationship is given later, in Sect. 6.5.1.

### 3.6.4 The Atmospheric Refractive Index Field

The atmospheric refractive index field is a random process in both space and time which, in its most general form, we denote by n(x,y,z,t). However, because of the rapid speed that light waves travel through the atmosphere, for typical atmospheric paths the refractive index field can generally be regarded as frozen in time for the travel duration. In this book, therefore, we generally deal in terms of the time-invariant refractive index field, denoted simply by n(x,y,z).

#### 3.6.4.1 The Normalized Fluctuating Part of the Refractive Index Field

For the purposes of this book, atmospheric inhomogeneity primarily concerns with the fluctuations of the atmospheric refractive index field, n(x,y,z). It is therefore convenient to define the normalized fluctuating part of this field, N(x,y,z), as follows:

$$N(x, y, z) = \frac{n(x, y, z) - \langle n(x, y, z)\rangle}{\langle n(x, y, z)\rangle}, \tag{3.14}$$

where $\langle n(x, y, z)\rangle$ is the ensemble average value of n(x,y,z).

It is also clear from the above expression that

$$\langle N(x, y, z)\rangle = 0. \tag{3.15}$$

It might be observed here that, since the refractive index of air at sea level is only about 1.00028, the highest value that can be attained by any specific value of $N$ is generally less than 0.00028. Thus, the approximations used in later chapters, and which were also made by Hufnagel and Stanley (1964), $\langle N\rangle << 1$ and $\langle N^2\rangle \approx 0$, are fully justified.

## 3.7 Scalar Diffraction Theory

Before discussing scalar diffraction theory, it is worth briefly discussing what is meant by the term diffraction. Diffraction refers to various phenomena that occur when waves encounter obstacles. For light waves, it refers to the apparent bending of light around apertures. Similar effects occur when light waves travel through inhomogeneous media where the refractive index varies randomly.

Feynman (1963) has observed: "No one has ever been able to define the difference between interference and diffraction satisfactorily. It is just a question of usage, and there is no specific, important physical difference between them." Feynman suggests that when there are only a few sources, say two, we call it interference, as in a Young's slit experiment. But when there are a large number of sources, such as when there are numerous sources of Huygens's secondary wavelets, the process is called diffraction.

### 3.7.1 Scalar Diffraction Theory Applied to Atmospheric Propagation

Scalar wave diffraction theory is used almost exclusively throughout this book to describe light propagation through the atmosphere. This classical optics approach ignores polarization effects and, strictly, is only valid in the small-angle scattering limit, $\sin(\theta) \approx \theta$. The small-angle scattering approximation is easily justified for our applications. Over path lengths of the order of the entire atmospheric depth, light deflection angles are generally no greater than a few arcseconds.

In contrast, backscattering from both clouds and aerosol distributions can involve multiple scattering events that produce much larger scatter angles. Since depolarization can occur in such cases and vector diffraction theory would then be required to describe the behavior, the book does not deal in any depth with such large-angle scattering processes.

### 3.7.2 Scalar Diffraction Theory Applied to Telescope Imaging

The use of scalar diffraction theory greatly simplifies the analysis of telescope imaging. However, the application is only strictly valid in two cases: (1) for telescopes that are truly diffraction-limited for light of any initial state of polarization and (2) when we restrict to the paraxial region where all polarization states are treated equally regardless of telescope aberrations. Thus, to justify the use of scalar diffraction theory, we must restrict its application to axially symmetric telescopes with $F$/numbers $\geq 4$. We must also restrict to field angles of less than a few degrees. Most refractor telescopes and most large reflector telescopes of Cassegrain or Ritchey-Chretien designs operate at $F$/numbers $\geq 4$ and their fields of view are generally less than a few degrees.

However, if we were to depart significantly from the paraxial region, as might happen with faster telescopes with $F$/numbers $<4$ or with telescopes with wide fields of view ($>$ a few degrees), we would then be obliged to use generalized refractive indices (Born & Wolf, 2003). Generalized refractive indices depend not just on the refractive indices of the various elements (and coating materials) that comprise the telescope optical train, they also depend on the sines and cosines of the angles of incidence (AOI) of the light beams as they reflect off and refract through the elements. The overall effect is that the telescope pupil function becomes polarization-dependent, which can cause significant phase and amplitude differences between the complex amplitudes in each of the two orthogonal polarization directions. The large phase angle difference that can develop at high angles of incidence for light waves at the S- and P-polarizations is illustrated in Fig. 3.12; the plot in this figure was calculated for a standard silver mirror coating with a silicon dioxide ($SiO_2$) protective layer.

For certain commonly used types of telescope, such as Newtonian reflectors or telescopes with Coude paths where the imaging rays reflect off mirror surfaces at

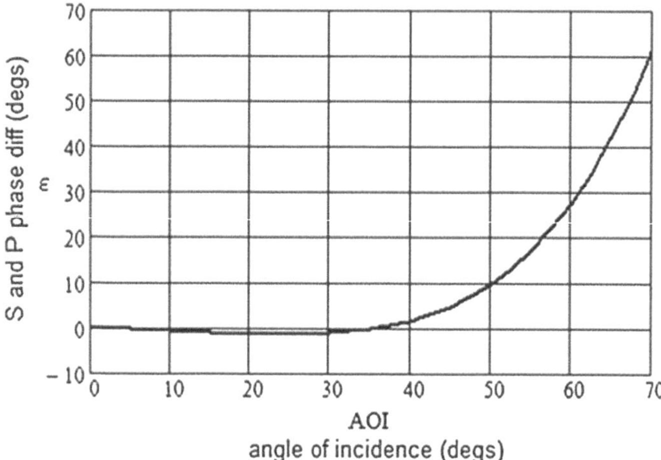


**Fig. 3.12** Phase difference at wavelength 0.55 μm between S- and P-polarized light beams for a silver (Ag) reflective coating with silicon dioxide protective layer. For AOI angles less than 10° (as would generally be the case for axially symmetric telescopes with $F/\#s \geq 4$) the phase differences for this typical telescope mirror coating are negligible

large (typically 45°) angles of incidence, it may simply not be possible to ignore polarization effects. In such instances, half-wave and quarter-wave plates can sometimes be used to compensate, or eliminate, path length differences at the two orthogonal polarizations. Compensation can also be effected by use of custom optical coating designs. However, these compensation strategies generally only work over limited wavelength bands and over limited AOI ranges.

If polarization effects cannot safely be neglected, vector field representations must be used, such as provided by Jones matrices and other equivalent formulations. For more information on vector field representations, the reader is referred to Born and Wolf (2003). Further discussion of polarization effects in this book is limited to Sect. 11.8 where we consider the effect of depolarizing telescopes on the statistical properties of speckle in star images.

### 3.7.3 Monochromatic Light Fields

Monochromatic light fields can be described by the real scalar field function, $V^r(x, y, z, t)$, which may be considered to represent either the electric field or the magnetic field. Function $V^r(x, y, z, t)$ can be expressed in the form,

$$V^r(x, y, z, t) = A(x, y, z) \cdot \cos[\omega \cdot t - \varphi(x, y, z)], \tag{3.16}$$

where both the field amplitude $A(x, y, z)$ and the phase angle $\varphi(x, y, z)$ are real scalar functions of position. Angular frequency $\omega$ is related to the optical frequency $\nu$ by the relation

$$\nu = \frac{\omega}{2 \cdot \pi}. \tag{3.17}$$

### 3.7.4 Analytic Signal

Notation is greatly simplified by using exponential rather than trigonometric functions. Therefore, rather than dealing with the real scalar function for monochromatic fields, $V^r(x, y, z, t)$, we choose to deal in terms of the complex function, $V(x, y, z, t)$, which we refer to as the analytic signal (Gabor, 1946). In general, the complex valued analytic signal can be written in terms of the real and imaginary parts of the field as follows,

$$V(x, y, z, t) = V^r(x, y, z, t) + i \cdot V^i(x, y, z, t), \tag{3.18}$$

where $V^i(x, y, z, t)$ is the imaginary part of the field.

For the analytic signals encountered in scalar diffraction theory (the diffraction theory primarily used in this book), the quantities $V^{\mathrm{r}}$ and $V^{\mathrm{i}}$ are identical apart from a $\pi/2$ phase angle difference. The real part of the analytic signal may be considered as representing either the electric vector $E$ or the magnetic vector $H$. [In the more general case of vector fields, $V^{\mathrm{r}}$ and $V^{\mathrm{i}}$ are related via Hilbert transforms (Born & Wolf, 2003; Bracewell, 1978).

### 3.7.5 Complex Amplitude

For monochromatic light, the analytic signal, $V(x, y, z, t)$, can be separated into spatial and temporal components,

$$V(x, y, z, t) = U(x, y, z) \cdot \exp(-i \cdot \omega \cdot t). \tag{3.19}$$

The crucial function $U(x, y, z)$ describes the time-independent part of the field and is referred to as the complex amplitude. This function—which in some texts is referred to as the phasor—will be used frequently throughout the book; it may be considered as representing the spatially varying part of either the electric field vector $E$ or the magnetic field vector $H$. These two field vectors are discussed further in Sect. 3.9.1. Complex amplitude may be subdivided into the amplitude and phase terms, $A(x,y,z)$ and $\varphi(x,y,z)$, both of which are entirely real,

$$U(x, y, z) = A(x, y, z) \cdot \exp[i \cdot \varphi(x, y, z)]. \tag{3.20}$$

For light fields consisting of the sum of many monochromatic contributions, the term $\exp(-i \cdot \omega \cdot t)$ in 3.19 occurs as a common factor. Since no useful purpose is served in this book by carrying forward such time-dependence, henceforth the term $\exp(-i \cdot \omega \cdot t)$ will be omitted. For our applications, the essential physics of light wave propagation through the atmosphere and the imaging of the wave portions collected by telescopes can be wholly represented by the time-independent part of the field—the complex amplitude, $U(x, y, z)$, as defined above by 3.20.

### 3.7.6 *Intensity*

Intensity is the quantity that allows electromagnetic energy to be detected. At the visible and IR wavelengths that are of primary interest in this book, optical frequencies are extremely high ($\upsilon \approx 10^{14}$Hz). Since intensity detectors (such as the eye, photographic plates, CCD arrays, and IR focal plane arrays) have response rates many orders of magnitude slower than this, in practice, measured intensity is invariably obtained as the average of the disturbance over a very large number of temporal cycles. Consequently, measured intensities are time-independent so that, apart from a nonessential constant, the intensity $I(x, y, z)$ may be expressed in terms of the complex amplitude in the form,

$$I(x, y, z) = U(x, y, z) \cdot U^*(x, y, z) = |U(x, y, z)|^2, \tag{3.21}$$

where $*$ denotes complex conjugate. Using 3.20 to substitute for, $(x, y, z)$, intensity may also be expressed by

$$I(x, y, z) = A^2(x, y, z). \tag{3.22}$$

Though the complex amplitude is usually calculated first, the intensity distribution in the image is usually the desired final description of the image.

### 3.7.7 *Irradiance*

Irradiance may be loosely considered as a measure of intensity. However, the term is assigned a very specific meaning in this book; it is considered as having the dimensions of Watts per square meter (Electro-Optics Handbook, 1974).

### 3.7.8 Polychromatic Light Fields

Generalization from monochromatic light fields to polychromatic fields is readily accomplished by invoking the linear superposition principle for electromagnetic fields (Goodman, 2004). According to this principle, the polychromatic field is simply the sum of the individual monochromatic components that make up the disturbance. As we shall see in Chap. 11, to calculate the statistical properties of polychromatic images, it is necessary to know the degree of correlation between the complex amplitudes at all pairs of wavelengths included in the polychromatic disturbance.

## 3.8 Coherence Terminology

In this section, we provide a brief summary list of the terminologies that are used to describe coherence-related functions. A fuller list is provided in Appendix B. Since we generally refer to the complex amplitude in defined planes, such as the telescope pupil plane or the telescope image plane, where the same $z$-value is implied, we choose here to omit $z$ from the argument lists.

The mutual coherence function (Born & Wolf, 2003) describes the correlation between the complex amplitudes in waves at two separate points in space at time separation $\tau$. We denote this function[9] by $\Gamma\left(u', v', u, v, \lambda, \tau\right)$ and define it in terms of the analytic signal by.

$$\Gamma\left(u', v', u, v, \lambda, \tau\right) = V\left(u', v', \lambda, t + \tau\right) \cdot V^*(u, v, \lambda, t). \tag{3.23}$$

For quasi-monochromatic light waves (Sect. 3.7), $\Gamma\left(u', v', u, v, \lambda, \tau\right)$ differs inappreciably from $\Gamma\left(u', v', u, v, \lambda, 0\right)$. Thus, we may write

$$\begin{aligned}\Gamma\left(u', v', u, v, \lambda, \tau\right) &= \Gamma\left(u', v', u, v, \lambda, 0\right) \\ &= V\left(u', v', \lambda, t\right) \cdot V^*(u, v, \lambda, t).\end{aligned} \tag{3.24}$$

For notational brevity, the function $\Gamma\left(u', v', u, v, \lambda, 0\right)$ is usually denoted by $J\left(u', v', u, v, \lambda\right)$, so that we can write

$$J\left(u', v', u, v, \lambda\right) = \Gamma\left(u', v', u, v, \lambda, 0\right). \tag{3.25}$$

Recalling 3.19, we see that $J\left(u', v', u, v, \lambda\right)$ is simply the autocovariance function of the complex amplitude,

[9] $\Gamma\left(u', v', u, v, \lambda, \tau\right)$ is essentially the same function as the mutual coherence function which Born *and Wolf denote by $\Gamma_{1,2}(\tau)$. They denote the mutual intensity function by $J_{1,2}$ and the complex degree of coherence by $\mu_{1,2}$.

$$J(u', v', u, v, \lambda) = U(u', v', \lambda) \cdot U^*(u, v, \lambda). \tag{3.26}$$

Function $J(u', v', u, v, \lambda)$ is commonly referred to as the mutual intensity function.

Frequent use is made of the unit-normalized form of $J(u', v', u, v, \lambda)$. We denote this function by $\mu(u', v', u, v, \lambda)$, formally defining it by

$$\mu(u', v', u, v, \lambda) = \frac{J(u', v', u, v, \lambda)}{\sqrt{J(u', v', u', v', \lambda)} \cdot \sqrt{J(u, v, u, v, \lambda)}} = \frac{J(u', v', u, v, \lambda)}{\sqrt{I(u', v', \lambda)} \cdot \sqrt{I(u, v, \lambda)}}. \tag{3.27}$$

Function $\mu(u', v', u, v, \lambda)$ can also be referred to as the complex coherence factor.

## 3.9 Free-Space Propagation

The term free space often refers to a perfect vacuum—a lossless medium, devoid of all particles and free of all currents and charges. In this book, a slightly wider definition is used, one that includes air, glasses, and other dielectric materials. In the assumed absence of charges and currents, the electromagnetic properties of these other types of media are not significantly different from vacuum properties, except that the refractive index now takes values greater than unity. Denoting the speed of light in vacuum by $c$, light travels through such a free-space medium at a speed, v, given by

$$\mathrm{v} = \frac{c}{n}, \tag{3.28}$$

where $n$ is the refractive index of the medium.

Two wavelengths are now distinguished, $\lambda_0$ and $\lambda$. Each corresponds to light of exactly the same optical frequency, $\nu$. For the former, the wavelength is that of the light while propagating through vacuum; for the latter, the wavelength is that of the light while propagating through a medium of refractive index $n$. These two wavelengths are defined as follows:

$$\lambda_0 = \frac{c}{\nu}, \tag{3.29}$$

$$\lambda = \frac{\mathrm{v}}{\nu} = \frac{c}{n \cdot \nu}, \tag{3.30}$$

from which it is clear that they are related by

$$\lambda = \frac{\lambda_0}{n}. \tag{3.31}$$

Two quantities, known as wave numbers, $k_0$ and $k$, are now defined:

$$k_0 = \frac{2 \cdot \pi}{\lambda_0} = \frac{\omega}{c}, \tag{3.32}$$

$$k = \frac{2 \cdot \pi}{\lambda} = \frac{n \cdot \omega}{c}, \tag{3.33}$$

where $\omega$ is the angular frequency of the light, defined previously by 3.17. It follows from 3.32 and 3.33 that $k$ and $k_0$ are related by

$$k = n \cdot k_0. \tag{3.34}$$

### 3.9.1 Maxwell's Electromagnetic Wave Equations

Electromagnetic wave propagation is governed by Maxwell's equations[10] which hold at every point where the physical properties of the medium are continuous. These equations may be expressed as follows (Born & Wolf, 2003):

$$curl\ H - \frac{1}{c} \cdot \frac{\partial D}{\partial t} = \frac{4 \cdot \pi}{c} \cdot j, \tag{3.35}$$

$$curl\ E + \frac{1}{c} \cdot \frac{\partial B}{\partial t} = 0, \tag{3.36}$$

$$divD = 4 \cdot \pi \cdot \rho, \tag{3.37}$$

$$divB = 0, \tag{3.38}$$

where $E$ is the electric field vector, $H$ is the magnetic field vector, $B$ is the magnetic induction vector, $D$ is the electric displacement vector, $j$ is the electric current density, $\rho$ is the electric charge density, and $c$ is the speed of light in vacuum. The field vectors $E$, $H$, $D$, and $B$ are functions of both location (x,y,z) and time $t$.

The four equations just given must be used in conjunction with material equations that describe the properties of the medium. In their most general form, the material equations are complicated, but if the medium is either at rest or only moving slowly

[10] Further discussion of these crucial equations is given in Appendix A which also contains background information about James Clerk Maxwell, the man, and his other contributions toscience.

(compared to the speed of light) and if the medium is isotropic, they may be written in the form

$$j = \sigma \cdot E, \tag{3.39}$$

$$D = \varepsilon \cdot E \tag{3.40}$$

$$B = \mu \cdot H, \tag{3.41}$$

where $\sigma$ is specific conductivity, $\varepsilon$ is the relative dielectric constant (or permittivity), and $\mu$ is the relative magnetic permeability, each of them normalized by the free-space values.

In this book, we primarily concern with non-conducting materials, which may also be referred to as dielectrics, such as air or glass. These materials cannot support currents, and we assume the absence of charges. Thus, we may set $\sigma = 0$ and $j = 0$. The properties of such materials are then entirely described by $\varepsilon$ and $\mu$.

If the materials, or media, are assumed homogeneous on the scale of the optical wavelengths, it can be shown (Born & Wolf, 2003) that possible solutions[11] for the electric and magnetic field vectors, $E$ and $H$, have the form

$$\nabla^2 E - \frac{\varepsilon \cdot \mu}{c^2} \cdot \frac{\partial^2 E}{\partial t^2} = 0, \tag{3.42}$$

$$\nabla^2 H - \frac{\varepsilon \cdot \mu}{c^2} \cdot \frac{\partial^2 H}{\partial t^2} = 0, \tag{3.43}$$

where $\nabla^2$ is the Laplace operator, or Laplacian, defined by

$$\nabla^2 = \frac{\partial^2}{\partial x^2} + \frac{\partial^2}{\partial y^2} + \frac{\partial^2}{\partial z^2}. \tag{3.44}$$

Equations 3.42 and 3.43 are recognized as standard equations describing wave motion, which suggest that electromagnetic waves travel at a speed given by

$$\mathrm{v} = \frac{c}{\sqrt{\varepsilon \cdot \mu}}. \tag{3.45}$$

Recalling 3.28, we see that the refractive index of the non-conductive media and materials considered here is given by

[11] Other solutions such as standing waves are also possible but these are not discussed in this book.

$$n = \sqrt{\varepsilon \cdot \mu}, \tag{3.46}$$

a relation known as Maxwell's formula.

For vacuum, where $\mu = 1$ and $\mu = 1$, the refractive index takes the value unity for all frequencies. Thus, the propagation velocity of electromagnetic waves (3.45) in vacuum space takes the same value, $c$, for all frequencies.

For air and glass, $\mu$ takes values close to unity. Thus, for these two media refractive index is given approximately by

$$n \approx \sqrt{\varepsilon}. \tag{3.47}$$

More generally, $\varepsilon$ is a function of frequency. For materials with low dispersion and simple chemical structure such as air, 3.47 is often found to be reasonably accurate. For example, for air at yellow wavelengths (Born & Wolf, 2003) measured values show $n = 1.000294$ and $\sqrt{\varepsilon} = 1.000295$. However, for glasses and liquids the formula can become widely inaccurate. For example, for water $n \approx 1.33$ and $\sqrt{\varepsilon} = 9$. Because of the unreliability of 3.47, we shall avoid using it. Instead, we use directly measured values refractive index values.

### 3.9.2 *Helmholtz Equation*

If we consider a monochromatic scalar wave and separate the analytic signal into its spatial and temporal components (cf., 3.19), the complex amplitude $U$ may be obtained by solving the time-independent forms of 3.42 and 3.43. The result is commonly known as the Helmholtz equation (Born & Wolf, 2003) which may be written in the form,

$$\left(\nabla^2 + \left(\frac{n\ \omega}{c}\right)^2\right) \cdot U = 0, \tag{3.48}$$

where $\omega$ is the angular frequency of the light as defined earlier by 3.17.

### 3.9.3 *Solutions for Infinitely Extensive Plane Waves*

#### 3.9.3.1 Solution for a Homogeneous Medium

For a homogeneous medium, the refractive index, $n$, is constant everywhere throughout the medium. For an infinitely extensive, monochromatic plane wave of angular frequency, $\omega$, a solution to the Helmholtz equation for the complex amplitude can be given in the form,

$$U = A_o \cdot \exp\left[-i \cdot \left(\frac{n \cdot \omega}{c} \cdot z\right)\right], \tag{3.49}$$

where $A_o$ may be considered as the wave amplitude which in this case is constant. It is worth again noting that in scalar wave theory the complex amplitude $U$ may be considered as either the electric field vector $E$ or the magnetic field vector $H$.

#### 3.9.3.2 Solution for Earth's Inhomogeneous Atmosphere

For Earth's inhomogeneous atmosphere, the refractive index field is a random process. As per Sect. 3.6.4, we denote this field at any arbitrary instant of time by n(x,y,z). Assuming that the atmospheric refractive index fluctuations only occur over distances much larger than the visible and IR wavelength scales that concern us, wave propagation can again be described by the Helmholtz equation, but where we now replace the constant refractive index, $n$, by the randomly varying refractive index, n(x,y,z).

As indicated by Hufnagel and Stanley, the solution of 3.48 in this case (for an initially plane, infinitely extensive wave) is the same as that given by 3.49, except that now $A_o$ is no longer constant; instead, it arises as a complex randomly varying function that obeys the equation,

$$\nabla^2 A_o(x, y, z) + 2 \cdot i \cdot k \cdot \frac{\partial A_o(x, y, z)}{\partial z} + 2 \cdot k^2 \cdot N(x, y, z) \cdot A_o(x, y, z) = 0, \tag{3.50}$$

where terms containing $N^2$ have been neglected. (It is important to note here that function, $A_o\ (x, y, z)$, is not the same as, and should not be confused with, the entirely real function, $A(x, y, z)$, that was used previously (cf., 3.20) to denote the amplitude portion of the complex amplitude.)

Though there are no known analytical solutions to the 3.50, Hufnagel and Stanley have shown that, if the sole purpose is simply to obtain an expression for the atmospheric modulation transfer function (MTF), there is no need to solve this equation explicitly for $A_o\ (x, y, z)$. Likewise, for the purposes of this book, where we are primarily concerned with the atmospheric MTF, $M(x', y', x, y, \lambda)$ and the more general two-point two-wavelength correlation function, $S(x', y', x, y, \lambda', \lambda)$, again there is no need to solve 3.50 explicitly for $A_o(x, y, z)$.

## 3.10 Mathematical Notations and Quantity Dimensions

The mathematical notation used in the different chapters is listed in tables at the end of each chapter, Table 3.5 being the appropriate table for this chapter. The dimensional

**Table 3.5** Mathematical notation used in this chapter along with the SI dimensional units of the individual quantities

| Symbol | Quantity | Dimensions |
|---|---|---|
| $\lambda$ | Wavelength | m |
| $n$ | Refractive index | "1" |
| $N$ | Normalized fluctuating part of the refractive index | "1" |
| $P$ | Air pressure | $kg \cdot s^{-2} \cdot m^{-1}$ |
| $T$ | Air temperature | "1" |
| $h$ | Altitude | m |
| OPD | Optical path difference | m |
| ZA | Zenith angle | "1" |
| $R$ | Astronomical refraction angle | "1" |
| $\Delta m$ | Atmospheric extinction (in magnitudes) | "1" |
| $V$ | Analytic signal | kg m $s^{-3}$ $amp^{-1}$ |
| $U$ | Complex amplitude | kg m $s^{-3}$ $amp^{-1}$ |
| $A$ | Amplitude | kg m $s^{-3}$ $amp^{-1}$ |
| $\phi$ | Phase | "1" |
| $I$ | Intensity | $kg^2$ $m^2$ $s^{-6}$ $amp^{-2}$ |
| $\nu$ | Optical frequency | s − 1 |
| $\omega$ | Angular optical frequency | s − 1 |
| $c$ and $v$ | Speed of light in vacuum and air | m s − 1 |
| $k$ and $k_o$ | Wavenumbers in air and vacuum | m − 1 |
| $J$ | Mutual intensity function | $kg^2$ $m^2$ $s^{-6}$ $amp^{-2}$ |
| $\mu$ | Complex coherence factor | "1" |
| $E$ | Electric field vector | kg m $s^{-3}$ $amp^{-1}$ |
| $H$ | Magnetic field vector | $kg^{-1}$ $amp^{-1}$ |

Dimensionless quantities are indicated by "1"

units of individual quantities are also indicated in the tables using the international system (SI) of units, a modern form of the metric system. In this system, the fundamental units of mass, length, and time, are kilograms (kg), meters (m), and seconds (s). Electrical quantities, such as the electric field strength of an electromagnetic wave also make use of the electrical unit, Ampere, which we will abbreviate to "amp," and thus avoid confusion with "A" which is used throughout the book to denote the real amplitude of a light wave.

Whereas the common practice is generally adopted of expressing the complex amplitude and the intensity in dimensionless forms, the actual dimensions of these functions are as indicated in Table 3.5. Throughout the book, when dimensionless forms are used to express these two important quantities, a note is generally made to the effect that a multiplier term with appropriate dimensions has been omitted.

## References

Birch, K. P., & Downs, M. J. (1994). Correction to the updated Edlén equation for the refractive index of air. *Metrologia, 31*, 315.
Born, M., & Wolf, E. (2003). *Principles of optics* (7th ed., revised). Cambridge University Press.
Bracewell, R. N. (1978). *The Fourier transform and its applications*. McGraw-Hill.
Davenport, W. B., & Root, W. L. (1958). *An introduction to the theory of random signals and noise*. McGraw-Hill.
Electro-Optics Handbook. (1974). *Electro-optics handbook*, (pp. 17601–5688). Burle Industries, Inc.
Feynman, R. (1963). *Lectures in physics* (Vol. 1, pp. 30–31). Addison Wesley Publishing Company.
Gabor, D. (1946). Theory of communication. Part 1: The analysis of information. *Journal of the Institution of Electrical Engineers-Part III: Radio and Communication Engineering, 93*(26), 429.
Gardiner, D. (1989). *Practical Raman spectroscopy*. Springer.
Goodman, J. W. (2004). *Introduction to Fourier optics* (3rd ed.). Roberts and Company Publishers.
Henini, M., & Razeghi, M. (2002). *Handbook of infrared detection technologies*. Elsevier Science Inc.
Hudson, R. D. (1969). *Infrared systems engineering*. Wiley.
Hufnagel, R. E., & Stanley, N. R. (1964). Modulation transfer function associated with image transmission through turbulent media. *JOSA, 54*(1), 52–61.
Lilienfeld, P. (2004). A blue sky history. *Optics and Photonics News, 15*(6), 32–39.
Papoulis, A. (1965). *Probability, random variables and stochastic processes*. McGraw-Hill.
Richardson, L. (1922). *Weather prediction by numerical process*. Cambridge University Press (reprinted by Dover Publications, 1965).
Ridpath, I. (1989). *Norton's 2000.0, star atlas and reference handbook* (18th ed.). Longman Group UK Ltd.
Rogalski, A. (2011). *Infrared detectors* (2nd ed.). CRC Press.
Smart, W. M. (1958). *Foundations of astronomy, sixth impression*. Longmans, Green and Co.
Stratton, J. A. (1941). *Electromagnetic theory*. McGraw-Hill.

# Chapter 4
# Diffraction

**Abstract** This chapter introduces the subject of diffraction—the key mechanism that determines how light propagates through the atmosphere and comes to final focus in the telescope image plane. The origin and basis of the Fresnel-Kirchhoff diffraction formula is described; this formula derives directly from Maxwell's equations. Solutions to the formula are given in three regions, all of which are used in the subsequent light propagation analysis: the geometrical optics region, the near-field Fresnel region and the far-field Fraunhofer region. The distribution of light energy in the Fraunhofer region describes the final image formed by the telescope. Optical system terminologies used to describe optical systems—telescopes in particular—are introduced (e.g., optical axis, telescope objective, central obstruction, and telescope pupil function). Ray terminologies are also introduced (e.g., principal ray, marginal ray). The amplitude and intensity point-spread functions of telescopes are defined. Linear superposition, convolution, isoplanaticity, and coherence are described for dealing with extended objects.

To properly describe how light propagates through the atmosphere and how images are formed by telescopes, it is essential to account for the effects of diffraction. While Maxwell's equations are the ultimate recourse for the treatment of diffraction given in this book, the Fresnel–Kirchhoff diffraction formula derives directly from these equations (by application of Kirchhoff's boundary conditions and Green's theorem[1] to the Helmholtz equation), and we consider this crucial diffraction formula as ultimately valid for dealing with diffraction wherever it arises.

In this chapter, therefore, the Fresnel–Kirchhoff diffraction formula will be used to develop expressions describing the complex amplitude and intensity distributions in point-object images formed by telescopes in the absence of atmosphere. These two important distributions are of course referred to as the amplitude and intensity

[1] George Green (1793–1841) was a self-taught English mathematical physicist. In "An essay on the application of mathematical analysis to the theories of electricity and magnetism," published in 1828, he introduced what are now referred to as Green's functions and Green's theorem (Born and Wolf 2003). His theory formed a crucial part of the foundation for later work on electromagnetic theory by James Clerk Maxwell.



T. S. McKechnie, *General Theory of Light Propagation and Imaging Through the Atmosphere*, Progress in Optical Science and Photonics 20,
https://doi.org/10.1007/978-3-030-98828-9_4

point spread functions (PSFs) of the telescope. Whereas the intensity PSF is given uniquely by the squared modulus of the amplitude PSF, the amplitude PSF is not uniquely determined by the intensity PSF. Thus, for general imaging applications, it is usually necessary to first obtain the amplitude PSF. The amplitude PSF and the intensity PSF associated with an imaging system, such as a telescope, may be considered as the two fundamental building blocks for creating images of extended objects.

The image intensity distribution of an extended object can be calculated from the convolution of these PSFs with the object illumination function. The same object can have different appearances depending on the state of coherence of the illumination. (Witness the smooth appearance of a white sheet of paper when it is illuminated by an incandescent light source and the speckled appearance when illuminated by a laser.) Thus, we shall consider illumination in all three basic coherence categories: incoherent illumination, coherent illumination, and partially coherent illumination.

For each category, expressions are developed for the image intensity for an arbitrarily shaped extended object. The effects of diffraction are fully taken into account by the expressions. Diffraction patterns formed by telescopes can take an infinite variety of forms depending on the imaging wavelength, telescope aperture shape, telescope aberrations, central obstructions, object shape, and the coherence properties of the illumination. Figures are included at the end of the chapter to illustrate some of the diffraction patterns that can arise for a point-object; they show the familiar Airy pattern as well as the closely related patterns that arise for telescopes with central obstructions. The same figures also show diffraction patterns arising for a two-point object. It will be apparent that even this—the simplest of all extended objects—can have a bewildering number of image apparitions depending on the state of coherence of the illumination. Because images can equally be regarded as diffraction patterns, the terms "image" and "diffraction pattern" are often used interchangeably.

The discussion given in this chapter is restricted to telescopes imaging in the absence of atmosphere. This may be seen as a starting platform from which to explore (in Chaps. 5 and 6) more complicated diffraction effects that arise when the turbulent atmosphere is taken into account. For large telescopes viewing through the atmosphere, the images of point-objects such as stars (and also extended objects such as binary stars) generally break up into speckle patterns. As the atmosphere moves and churns, these patterns continuously evolve.

## 4.1 Diffraction by an Aperture

The crucial Fresnel–Kirchhoff diffraction formula has its origins in the Huygens–Fresnel theory—a theory developed by Fresnel by combining Huygens' intuitively developed principle of secondary wavelets with Fresnel's application of Young's interference principles. Whereas the Huygens–Fresnel theory successfully explains diffraction in many imaging applications, the Fresnel–Kirchhoff diffraction formula provides a more robust mathematical platform for treating diffraction problems. Born and Wolf (2003) and Goodman (2004) have given accounts of the origins of this formula and how it relates to the earlier Huygens– Fresnel principle and ultimately to Maxwell's equations via the rigorous analysis given by the German theoretical physicist, Arnold J.H. Sommerfeld (1868–1951) (Sommerfeld, 1896).

In applying the Fresnel–Kirchhoff diffraction formula in this book to diffraction from an aperture, it is assumed that the size of the aperture is always significantly larger than the size of the light wavelength and, to avoid unnecessary complication in this chapter, it is also assumed that the refractive index of the medium in which the light travels is nominally unity, irrespective of whether the medium considered might be air or vacuum.

At visible and IR wavelengths, the aperture-size requirement is satisfied by telescopes with apertures even as small as 1 mm and thus satisfied by even the smallest backyard telescope. For larger telescopes, such as used by professional astronomers where the aperture sizes are measured in meters, the requirement is always comfortably satisfied. The Fresnel-Kirchhoff diffraction formula facilitates mathematical descriptions of a wide variety of diffraction effects. In particular, it enables calculation of the field distribution in the image of a point-object.

### 4.1.1 Fresnel Number

The Fresnel number, which we denote by $F$, is a dimensionless number frequently used in diffraction theory. It is loosely defined by

$$F \approx \frac{L^2}{Z \cdot \lambda}, \tag{4.1}$$

where the parameter $L$ is the characteristic size of the aperture (e.g., diameter), $Z$ is the propagation distance beyond the aperture, and $\lambda$ is the wavelength.

There are three main solutions to the Fresnel–Kirchhoff diffraction formula; all three are used in the propagation analyses given in Chaps. 5 and 6. The solution adopted for any particular application is determined by the magnitude of the Fresnel number, $F$, in the space in which the diffraction occurs. There are three principal Fresnel number regions:

- $F >> 1$ Geometrical optics,
- $F \geq 1$ Fresnel near-field diffraction, and
- $F << 1$ Fraunhofer far-field diffraction.

The geometrical optics region ($F >> 1$) holds for short distances beyond the diffracting aperture. Rectilinear wave propagation occurs in this regime, with the light waves traveling in straight lines and showing little evidence of diffraction.

Geometrical optics principles can be applied in this regime. At larger distances from the aperture, where $F$ takes smaller values, two other types of solution are generally used: the Fresnel near-field solution and the Fraunhofer far-field solution.

### 4.1.2 Fresnel–Kirchoff Diffraction Formula

Figure 4.1 shows a monochromatic light source located at $P$ producing spherical light waves that spread uniformly in all directions. An infinitely extensive opaque screen lies to the right of $P$. The screen contains an aperture that allows a portion of the light from $P$ to pass through. It is assumed that light portions lying beyond the extent of the aperture are entirely blocked and hence do not contribute to the diffraction field to the right of the aperture.

The Fresnel–Kirchhoff diffraction formula, or integral, allows us to calculate the complex amplitude, which we denote by $U$, at the arbitrarily chosen point, $P'$, in the region to the right of the aperture. With the notation indicated in Fig. 4.1, the formula may be expressed in the form,

$$U(P') = -\frac{i \cdot C}{2 \cdot \lambda} \cdot \iint\limits_{aperture} \frac{e^{i \cdot k \cdot (r+s)}}{r \cdot s} \cdot [\cos(\bar{n}, \bar{r}) - \cos(\bar{n}, \bar{s})] \cdot dS, \tag{4.2}$$

where $C$ is a constant that sets the field units, $r$ is the distance between points $P$ and $Q$ (i.e., the modulus of $\bar{r}$), $s$ is the distance between points $Q$ and $P'$ (i.e., the modulus of $\bar{s}$), and d$S$ is an elemental area in the aperture plane. The cosines of the

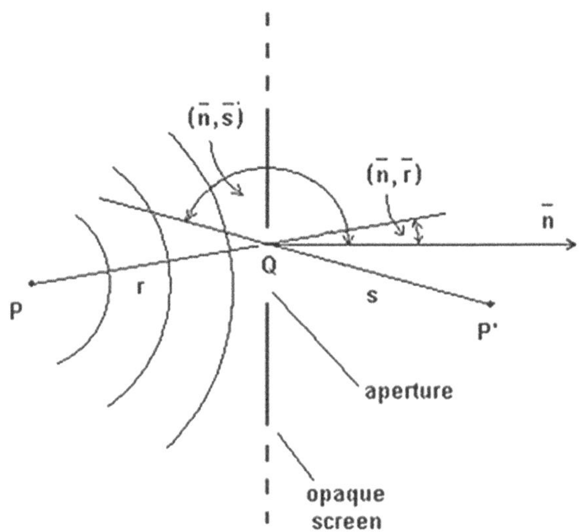


**Fig. 4.1** Geometrical basis of the Kirchhoff–Fresnel diffraction formula (4.2)

angles between the outward normal, $\overline{n}$, and the vectors, $\overline{r}$ and $\overline{s}$ are represented by $\cos(\overline{n}, \overline{r})$ and $\cos(\overline{n}, \overline{s})$, respectively.

As indicated at the top of the chapter, because it was rigorously shown by Sommerfeld that the Fresnel–Kirchhoff diffraction formula derives directly from Maxwell's equations, in this book we consider this formula (4.2) as being a sufficiently rigorous basis for describing diffraction whenever it arises.

### *4.1.3 Fresnel Near-Field Diffraction*

Generally, the complex amplitude is evaluated in specific planes of interest, such as the telescope image plane or the telescope pupil plane. Since the $z$ value corresponding to these planes is always known, there is no need to carry the $z$-dependence forward. Henceforth, therefore, the complex amplitude will be represented only in terms of its $x$ and $y$ spatial arguments by $U(x,y)$. The notation used in this section is the same as that used by Goodman (2004).

The geometry used to describe the diffraction of light by an aperture is shown in Fig. 4.2. Denoting the complex amplitude distribution in the $(x_1,y_1)$ aperture plane on the left of the figure by $U(x_1, y_1)$, we now wish to evaluate the Fresnel–Kirchhoff diffraction formula in a near-field observation plane, lying a distance Z to the right of the aperture in the $(x_0,y_0)$ plane. Compared to the wavelength scale and the aperture size, the Z value is assumed large enough to be consistent with Fresnel numbers falling in the range, $F \geq 1$.

Suppose now that a point source of monochromatic light of wavelength, $\lambda$, located as shown at an arbitrary point, $P_1$, in the $(x_1,y_1)$ aperture plane emits spherical light

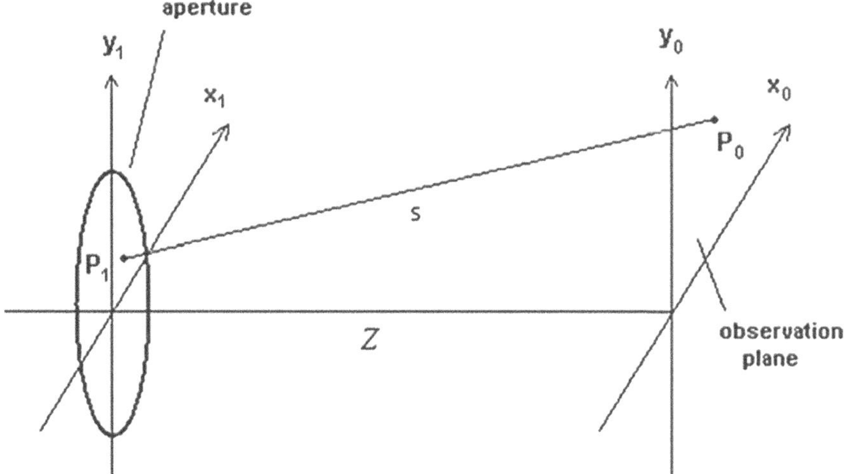


**Fig. 4.2** Diffraction geometry

waves; we are particularly interested in those spherical wave portions that travel toward the right. The exact distance, $s$, between point, $P_1$, in the aperture plane and a second arbitrary point $P_0$ in the $(x_0,y_0)$ plane of observation to the right of the aperture is given by

$$s = \left[Z^2 + (x_0 - x_1)^2 + (y_0 - y_1)^2\right]^{1/2}. \tag{4.3}$$

Using a binomial expansion for the square root, we may adequately approximate $s$ by the first two terms of the expansion:

$$s \cong Z \cdot \left[1 + \frac{1}{2} \cdot \left(\frac{x_0 - x_1}{Z}\right)^2 + \frac{1}{2} \cdot \left(\frac{y_0 - y_1}{Z}\right)^2\right]. \tag{4.4}$$

This approximation greatly simplifies the Fresnel–Kirchhoff diffraction formula (4.2); it allows the complex amplitude in the plane of observation to be written in the form

$$U(x_0, y_0) = \frac{-i \cdot \exp(i \cdot k \cdot Z)}{\lambda \cdot Z} \cdot \int_{-\infty}^{\infty} \int_{-\infty}^{\infty} U(x_1, y_1) \cdot \exp\left\{\frac{i \cdot k}{2 \cdot Z} \cdot \left[(x_0 - x_1)^2 + (y_0 - y_1)^2\right]\right\} \cdot dx_1 \cdot dy_1 \tag{4.5}$$

By expanding the quadratic terms in the exponent, the above equation may be expressed in the form,

$$U(x_0, y_0) = \frac{-i \cdot \exp(i \cdot k \cdot Z)}{\lambda \cdot Z} \cdot \exp\left[i \cdot \frac{k}{2 \cdot Z} \cdot \left[x_0^2 + y_0^2\right]\right] \int_{-\infty}^{\infty} \int_{-\infty}^{\infty} U(x_1, y_1) \times \exp\left[i \cdot \frac{k}{2 \cdot Z} \cdot \left(x_1^2 + y_1^2\right)\right] \cdot \exp\left[-i \cdot \frac{k}{Z} \cdot (x_0 \cdot x_1 + y_0 \cdot y_1)\right] \cdot dx_1 \cdot dy_1 \tag{4.6}$$

It may be noted that the term inside the double integral is the Fourier transform of the quantity $U(x_1, y_1) \cdot \exp\left[i \cdot k/(2 \cdot Z) \cdot \left(x_1^2 + y_1^2\right)\right]$, where the term $\exp\left[i \cdot k/(2 \cdot Z) \cdot \left(x_1^2 + y_1^2\right)\right]$ is recognized as having the same form as the integrand in a Fresnel integral (Magnus et al., 1966).

By neglecting the higher order terms in the binomial exponent, a limit is imposed on the range of usefulness of 4.5 and 4.6. On first appearances, the range would appear to be set by

$$Z^3 >> \frac{\pi}{4 \cdot \lambda} \cdot \left[(x_0 - x_1)^2 + (y_0 - y_1)^2\right]^2_{\max}, \tag{4.7}$$

which is consistent with the condition assumed earlier, $F \geq 1$. However, for many applications, the principle of stationary phase facilitates significant enlargement of the range of usefulness.

### 4.1.4 Stationary Phase

Though 4.7 is a sufficient condition, it is not a necessary one. For distances small enough to violate the condition, the primary contributions to the integral arise only from points in the vicinity $x_1 \approx x_0$ and $y_1 \approx y_0$. At optical wavelengths, beyond the so-called stationary phase region, the quantity $k/(2 \cdot Z)$ takes extremely large values. This causes the phases produced by the quadratic phase factor in 4.6 to oscillate rapidly. As a result, the net contributions to the diffraction integral from regions beyond the stationary phase region are often entirely negligible (as are contributions from higher order phase terms from the same regions). Thus, the Fresnel near-field diffraction integrals given by 4.5 and 4.6 can generally be used over a greatly expanded and more useful $Z$ range than indicated by 4.7; this result will be used later in Chap. 5. A fuller discussion of stationary phase has been given by Born and Wolf (2003).

### 4.1.5 Fraunhofer Far-Field Diffraction

By imposing the even more stringent restriction,

$$Z >> \frac{1}{2} \cdot k \cdot (x_1^2 + y_1^2)_{\max}, \tag{4.8}$$

a more concise version of the Fresnel–Kirchhoff diffraction formula arises, known as the Fraunhofer approximation,

$$\begin{aligned} U(x_0, y_0) = {} & \frac{-i \cdot \exp(i \cdot k \cdot Z)}{\lambda \cdot Z} \cdot \exp\left[i \cdot \frac{k}{2 \cdot Z} \cdot \left(x_0^2 + y_0^2\right)\right] \\ & \times \int_{-\infty}^{\infty} \int_{-\infty}^{\infty} U(x_1, y_1) \cdot \exp\left[\frac{-2 \cdot \pi \cdot i}{\lambda \cdot Z} \cdot (x_0 \cdot x_1 + y_0 \cdot y_1)\right] \cdot dx_1 \cdot dy_1. \end{aligned} \tag{4.9}$$

Apart from the multiplicative terms outside the integral, this expression is simply the Fourier transform of the field distribution in the plane of the aperture, a distribution referred to hereafter as the aperture distribution, $K(x_1, y_1)$. When the intensity is calculated from 4.9 by obtaining the quantity $U \cdot U^*$ (cf., 3.21), the phase terms preceding the integral exactly cancel and thus do not affect the calculated intensity. Therefore, as we continue forward, it will be convenient to omit these phase terms, thus enabling the Fraunhofer approximation to be written in the concise form,

$$\begin{aligned} U(x_0, y_0) = {} & \frac{-i}{\lambda \cdot Z} \cdot \int_{-\infty}^{\infty} \int_{-\infty}^{\infty} U(x_1, y_1) \cdot \\ & \exp\left[-i \cdot \frac{k}{Z} \cdot (x_0 \cdot x_1 + y_0 \cdot y_1)\right] \cdot dx_1 \cdot dy_1. \end{aligned} \tag{4.10}$$

Using 3.21, the intensity, $I(x_0, y_0)$, can be obtained from the above expression for the complex amplitude as follows:

$$I(x_0, y_0) = \frac{1}{\lambda^2 \cdot Z^2} \cdot \left| \int_{-\infty}^{\infty} \int_{-\infty}^{\infty} U(x_1, y_1) \cdot \exp\left[-i \cdot \frac{k}{Z} \cdot (x_0 \cdot x_1 + y_0 \cdot y_1)\right] \cdot dx_1 \cdot dy_1 \right|^2 \tag{4.11}$$

#### 4.1.5.1 Range of Applicability of Fraunhofer Approximations

The restriction indicated by 4.8 appears severe. However, provided that the wave in the aperture (Fig. 4.2) is spherical and converges toward the plane of observation—as might be arranged by use of a lens or other type of optical element—Fraunhofer diffraction patterns may be observed at distances much closer to the diffracting aperture. This permits use of 4.10 to calculate the complex amplitude distribution in the image of a point-object, such as formed in the image plane of a telescope. The squared modulus of this complex amplitude distribution provides the image intensity distribution given by 4.11.

## 4.2 Optical System Terminology

For general background reading about optical principles and optics terminology, the reader is referred to the book, "Optics," by Hecht (1998). But for convenience, in this section, we offer a brief synopsis of some of the key terms that are used in this book when discussing telescopes and other optical systems.

### *4.2.1 Telescopes, Telescope Objectives, and Eyepieces*

There are three main kinds of telescope. Refractor telescopes use lenses to form magnified images as shown in Fig. 4.3; reflector telescopes use mirrors to form the images as shown in Fig. 4.4; and catadioptric telescopes use a combination of mirrors and lenses to form the images.

The telescope objective element is the optical element that gathers light from the object and brings it to a focus in a plane referred to as the primary image plane. The focused image in this plane is a real image.[2] For reflector telescopes, the telescope objective is a concave mirror, referred to as the primary mirror. Generally, this

[2] If a screen is placed at the location of a real image, the image will be seen by looking directly at the screen.

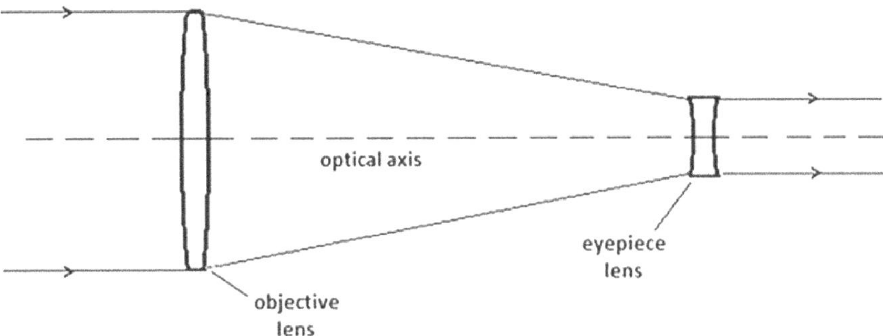


**Fig. 4.3** Galilean refractor telescope

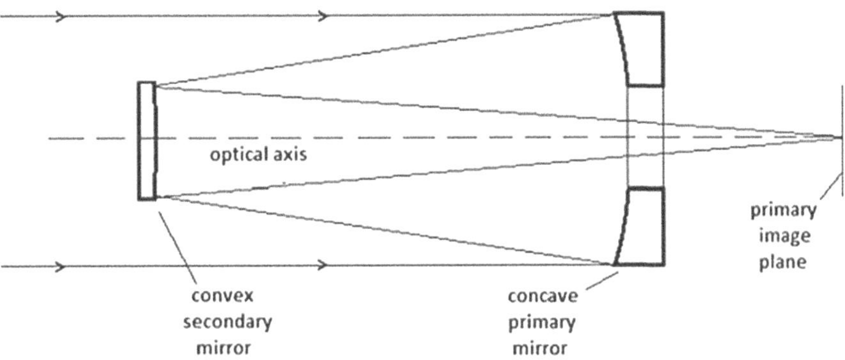


**Fig. 4.4** Cassegrain reflector telescope

mirror has an aspheric shape to correct for spherical aberration. Cassegrain reflectors (Fig. 4.4) also have a smaller convex mirror lying in front of the primary mirror, referred to as the secondary mirror. Light arriving at this mirror from the primary mirror is reflected back through a hole in the center of the primary mirror to form an image in a plane referred to as the primary image plane. The secondary mirror is usually aspheric. In Ritchey–Chretien designs, which may be considered as variants of the Cassegrain design, the reflective surfaces of the primary and secondary mirrors are both hyperboloids of revolution, with the surface shape parameters chosen in a specific combination that corrects coma as well as spherical aberration. Because all wavelengths reflect in exactly the same way—that is, the angle of reflection equals the angle of incidence—reflector telescopes are not affected by chromatic aberrations. For refractor telescopes, the telescope objective takes the form of a lens, often a doublet or even a triplet, designed to correct not only chromatic aberration but spherical aberration and coma as well.

The eyepiece is a lens—often compound—placed in the vicinity of the real image. Eyepieces can have either positive or negative focal lengths. If negative, they are placed just in front of the real image and, if positive, they are placed just behind it. The function of the eyepiece is to magnify the image perceived by the eye. The angular magnification of a telescope is determined by the ratio of the focal length of the objective to that of the eyepiece. Denoting these two focal lengths by $f_{\text{Objective}}$ and $f_{\text{Eyepiece}}$, respectively, angular magnification is given by

$$Magnification = \frac{f_{Objective}}{f_{Eyepiece}} \tag{4.12}$$

The eyepiece used in a Galilean telescope has negative power which results in upright images. Telescopes of this type are mostly used for terrestrial viewing applications. In astronomy, where image orientation is less critical, the use of positive eyepieces is more common. Positive eyepieces, first described by Kepler in 1611, produce inverted images, but in exchange provide better aberration correction. Positive eyepieces place the "exit pupil" (described in the next section) in a more convenient location with respect to the eye, thus affording much wider fields of view. The Galileo telescope mentioned in the Historical Introduction, despite giving only 20 times magnification, only allows one quarter of the Moon's diameter to be seen without repointing the instrument.

Nowadays, images formed by large astronomical telescopes are rarely viewed directly by eye, so that no eyepiece is necessary. Instead, the images are recorded by an imaging device such as a charge-coupled device (CCD) array, which can be located either in the primary image plane or in a secondary image plane, with relay lenses and/or mirrors used to provide the desired image magnification.

### *4.2.2 Aperture Stops, Pupils, Conjugate Distances, Focal Lengths, and F/Numbers*

The physical stop that limits the cross section of the image-forming wavefronts is known as the aperture stop. The entrance pupil is the image of the aperture stop formed by the optical components preceding it. For many types of reflector telescopes,[3] the primary mirror constitutes the aperture stop. The entrance pupil and the aperture stop are then one of the same. For off-axis image locations, sometimes the image-forming light bundles are clipped by the rims of optical components in planes other than that of the aperture stop. Such clipping is referred to as vignetting.

The image of the aperture stop formed by the optical components that follow the aperture stop (such as the eyepiece) is known as the exit pupil. To directly view the image, the eye is generally placed at the exit pupil. With a positive eyepiece, the exit pupil lies just after the eyepiece in a space accessible to the eye. For Galilean telescopes, the exit pupil lies in the space in front of the eyepiece and is not therefore accessible to the eye. For such telescopes, the eye is simply placed as close to the

[3] Such telescopes do not include Schmidt–Cassegrain telescopes where the front aspheric corrector plate generally acts as the aperture stop.

exit pupil as permitted by the eyepiece hardware. A more complete discussion of aperture stops, entrance pupils, etc., has been given by Born and Wolf (2003).

Because we are primarily interested in diffraction, details of the locations and sizes of the aperture stop, the entrance pupil, and the exit pupil need not unduly concern us. The quantity that does primarily concern us is the $F$-number of the optical system in image space, which we denote by $F\,\#_I$. Our purposes can mostly be served by considering the optical system as though it were a "thin lens." The aperture stop, the entrance pupil, and the exit pupil are then coincident in a plane which we loosely refer to as the pupil plane. We refer to the diameter of the image-forming beam in this plane as the pupil diameter $D$. For most reflecting telescopes, $D$ is, in effect, the clear aperture diameter of the primary mirror. The effective focal length of the telescope is denoted by $f$.

### 4.2.3 Light Rays and Ray Terminology

The optical axis is an imaginary line that defines the path along which light propagates through the optical system as shown in Fig. 4.5. In many commonly used optical systems, such as telescopes, microscopes, and cameras, there is rotational symmetry about the optical axis.

A meridional ray, or tangential ray, is a ray confined to the plane containing the system's optical axis and the object point from which these rays originate.

A skew ray is a ray that does not propagate in a plane that contains both the object point and the optical axis. Skew rays do not cross the optical axis and are not parallel to it.

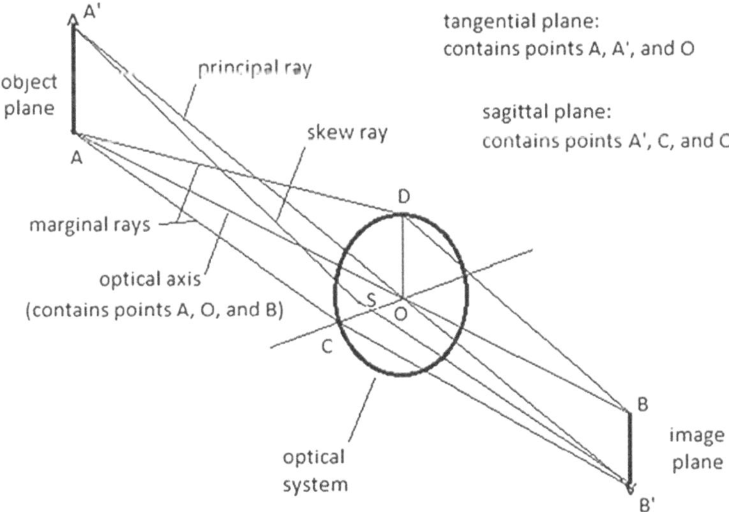


**Fig. 4.5** Optical system ray and other terminologies

A marginal ray is a ray that starts at the point where the object crosses the optical axis and touches the edge of the aperture stop of the system. This ray crosses the optical axis again at the location where the image forms. The distance of the marginal ray from the optical axis at the locations of the entrance and exit pupils determines the sizes of these pupils since both are images of the aperture stop.

The principal ray, or chief ray, is the meridional ray that starts at the edge of the object and passes through the center of the aperture stop. This ray crosses the optical axis at the locations of the entrance and exit pupils. The distance between the principal ray and the optical axis in the plane of the image determines the size of the image.

A sagittal ray, or transverse ray, from an off-axis object point is a ray that propagates in the plane perpendicular to the meridional plane and contains the principal ray. Sagittal rays intersect the pupil along a line that is perpendicular to the meridional plane for the ray's object point; these rays pass through the optical axis. A paraxial ray is a ray that makes a small angle with the optical axis of the system. Paraxial rays lie close to the axis throughout the system. Such rays can be modeled using paraxial approximations. For well-corrected optical systems, non-paraxial rays come to the same point of focus as paraxial rays. Therefore, for such systems, paraxial calculations carry a broader significance.

A finite ray, or real ray, is a ray that is traced without making paraxial approximations.

The marginal and chief rays together define the Lagrange invariant. This quantity provides a measure of the light energy throughput of the optical system.

### 4.2.4 Objects at Finite Distances

With reference to the thin-lens optical system shown in Fig. 4.6, for an object lying at finite distance, $L$, the conjugate distance in image space, $L'$, can be obtained by the use of paraxial approximations (Born & Wolf, 2003) in the form

$$L' = \frac{f \cdot L}{L - f}. \tag{4.13}$$

Defining the lens $F$/number in the usual way,

$$F/\# = f/D, \tag{4.14}$$

the $F$/number in image space, which we denote by $F/\#_I$ is given by

$$F/\#_I = \frac{L'}{D} = F/\# \cdot \frac{L}{L - f} \tag{4.15}$$

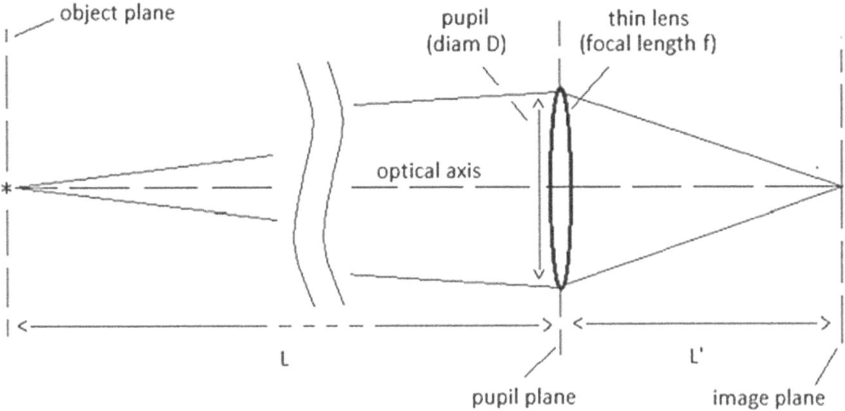


**Fig. 4.6** Optical system notation and terminology

### 4.2.5 *Objects at Infinite Distances*

For objects at infinity, the conjugate distance in image space, L', can be obtained from 4.13 by allowing $L \to \infty$. Thus,

$$L' = f \tag{4.16}$$

For this case, 4.15 reduces to

$$F/\#_I = F/\#. \tag{4.17}$$

For all practical purposes, stars and galaxies can be considered located at infinity. But even for considerably closer objects, such as planets and satellites, usually $L \gg f$, in which case 4.16 and 4.17 are still adequate approximations. Therefore, expressions given in this book describing diffraction phenomena are, by default, set out in terms of the quantities, $D, f$, and $F/\#$. However, in cases where the object is close enough to the optical system to make a significant difference—as is the case in Chap. 16 which deals with lasers beams focused on objects within the atmosphere—it is important to use the correct quantities, $D$, $L'$, and $F/\#_I$. The two latter quantities can be calculated using 4.13 and 4.15.

### 4.2.6 *Pupil Functions*

The pupil function of an optical system is a complex function that describes both the amplitude transmission and the aberrations of the optical system. The pupil function can be considered to act in the pupil plane where it modifies both the transmitted amplitude and the phase of image-forming waves. Because the pupil function is generally wavelength dependent, we denote it by $K(x_1, y_1, \lambda)$.

For an aberrated telescope with a uniformly transmitting aperture, the pupil function may be expressed in terms of the wavefront aberration of the telescope, $W(x_1, y_1, \lambda)$,

$$\begin{aligned} K(x_1, y_1, \lambda) &= \exp\left[-\frac{2\cdot\pi\cdot i\cdot W(x_1, y_1, \lambda)}{\lambda}\right] \text{ within the aperture,} \\ &= 0 \qquad \text{otherwise.} \end{aligned} \tag{4.18}$$

### 4.2.6.1 For Circular Apertures

For circularly symmetric telescopes, it is convenient to deal in terms of radial coordinates. The radial coordinate in the telescope pupil plane, $r_1$, can be defined in terms of the Cartesian coordinates in that plane:

$$r_1 = \left(x_1^2 + y_1^2\right)^{\frac{1}{2}}. \tag{4.19}$$

For a uniformly transmitting, aberration-free telescope with circular aperture and circular central obstruction, the pupil function is entirely real valued and may be expressed by

$$\begin{aligned} K(r_1, \lambda) &= 1 \text{ for} \frac{d}{2} \leq r_1 \leq \frac{D}{2}, \\ &= 0 \qquad \text{otherwise,} \end{aligned} \tag{4.20}$$

where $D$ is the diameter of the telescope objective and $d$ is the diameter of the central obstruction. If there is no central obstruction, we simply set $d$ to zero.

For an aberrated version of the same telescope where the aberrations are again circularly symmetric, the pupil function $K$ takes complex values that may be expressed in terms of the wavefront error of the telescope, $W(r_1, \lambda)$, by

$$\begin{aligned} K(r_1, \lambda) &= \exp\left[-\frac{2\cdot\pi\cdot i\cdot W(r_1, \lambda)}{\lambda}\right] \text{ for} \frac{d}{2} \leq r_1 \leq \frac{D}{2}, \\ &= 0 \qquad \text{otherwise.} \end{aligned} \tag{4.21}$$

For reflector telescopes and chromatically corrected refractor telescopes, the wavefront error can remain constant over broad wavelength ranges. For such telescopes, wavefront error may then be considered independent of wavelength and may be denoted simply by $W(r_1)$.

## 4.3 The Amplitude Point Spread Function

The amplitude in the image of a point-object formed by telescopes and other optical systems is referred to as the amplitude point spread function, which we denote by $PSF_A$. For a telescope of focal length, $f$, we may replace Z by $f$ in 4.10 to obtain

$$PSF_A(x_0, y_0) = \frac{-i}{\lambda \cdot f} \times \int_{-\infty}^{\infty} \int_{-\infty}^{\infty} K(x_1, y_1, \lambda) \cdot \exp\left[\frac{-2 \cdot \pi \cdot i}{\lambda \cdot f} \cdot (x_0 \cdot x_1 + y_0 \cdot y_1)\right] \cdot dx_1 \cdot dy_1 \quad (4.22)$$

### 4.3.1 *For Diffraction-Limited Telescopes with Circular Apertures*

For telescopes with uniformly transmitting circular apertures and circularly symmetric aberrations, the amplitude PSF may be written in terms of the Hankel transform,

$$PSF_A(r_0) = \frac{-2 \cdot \pi \cdot i}{\lambda \cdot f} \cdot \int_0^{\infty} K(r_1, \lambda) \cdot J_0\left[\frac{2 \cdot \pi \cdot r_0 \cdot r_1}{\lambda \cdot f}\right] \cdot r_1 \cdot dr_1, \quad (4.23)$$

where $r_0 \left(= \sqrt{x_0^2 + y_0^2}\right)$ denotes the radial distance parameter here which should not be confused with the coherence parameter, also denoted by $r_0$ that was previously introduced in Sect. 2.1.

If the telescope is also free of aberrations, the appropriate pupil function is as given previously by 4.20. Inserting that pupil function into the above equation for an unobstructed telescope (i.e., d = 0) leads to the unit-normalized amplitude PSF (Born & Wolf, 2003),

$$PSF_{A,Airy}(r_0) = \frac{2 \cdot J_1\left(\frac{\pi \cdot D \cdot r_0}{\lambda \cdot f}\right)}{\left(\frac{\pi \cdot D \cdot r_0}{\lambda \cdot f}\right)}, \quad (4.24)$$

where $J_1(\cdot)$ is the first-order Bessel function of the first kind (Abramowitz and Stegun, 1970). This particular amplitude point spread function was first derived in the form of an expansion by Airy (1835).

## 4.4 The Intensity Point Spread Function

The intensity point spread function of a telescope, which we denote by $\mathrm{PSF_I}$, may be obtained as the squared modulus of the amplitude PSF. Equation 4.22 allows $\mathrm{PSF_I}$ to be written in the un-normalized form,

$$PSF_I(x_0, y_0) = \frac{1}{\lambda^2 \cdot f^2} \times \left| \int_{-\infty}^{\infty} \int_{-\infty}^{\infty} K(x_1, y_1, \lambda) \cdot \exp\left[ \frac{-2 \cdot \pi \cdot i}{\lambda \cdot f} \cdot (x_0 \cdot x_1 + y_0 \cdot y_1) \right] \cdot dx_1 \cdot dy_1 \right|^2 . \quad (4.25)$$

The unit-normalized form may be written as follows:

$$PSF_I(x_0, y_0) = \frac{\left| \int_{-\infty}^{\infty} \int_{-\infty}^{\infty} K(x_1, y_1, \lambda) \cdot \exp\left[ \frac{-2 \cdot \pi \cdot i}{\lambda \cdot f} \cdot (x_0 \cdot x_1 + y_0 \cdot y_1) \right] \cdot dx_1 \cdot dy_1 \right|^2}{\left[ \int_{-\infty}^{\infty} \int_{-\infty}^{\infty} |K(x_1, y_1, \lambda)| \cdot dx_1 \cdot dy_1 \right]^2}, \quad (4.26)$$

where the most general type of telescope pupil function, $K(x_1, y_1, \lambda)$, was defined previously by 4.18.

### 4.4.1 The Airy Pattern

For diffraction-limited telescopes with circular apertures, the intensity PSF is the familiar Airy pattern, the unit-normalized form of which may be obtained from 4.24 in the form,

$$PSF_{I,Airy}(r_0) = \left[ \frac{2 \cdot J_1\left( \frac{\pi \cdot D \cdot r_0}{\lambda \cdot f} \right)}{\left( \frac{\pi \cdot D \cdot r_0}{\lambda \cdot f} \right)} \right]^2 . \quad (4.27)$$

Telescope diameter $D$ and focal length $f$ occur together in the above expression as the ratio, $D/f$. This accounts for the significance of the $F$/number of the optical system referred to previously in Sect. 4.2. Airy patterns can readily be seen by viewing star images at high magnification through small telescopes. An Airy pattern is shown in Fig. 4.7 (left). For telescopes with diameters larger than about 15 cm (6 in.), the effects of atmospheric turbulence degrade the Airy patterns, ultimately fragmenting them into random speckle patterns. The statistical properties of these speckle patterns are examined in Chap. 11.

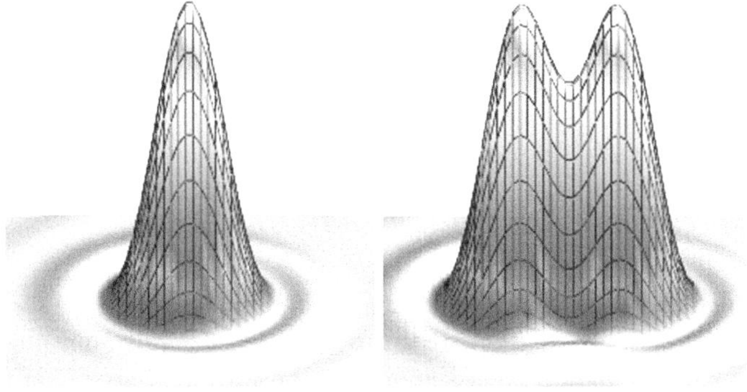

**Fig. 4.7** Diffraction pattern images. *Left* Airy pattern. *Right* Incoherently illuminated two-point object at the just-resolved separation according to the Rayleigh criterion

An Airy pattern is a circularly symmetric pattern, so that it may be fully described by a central section. Defining the image-space angular coordinate, $\theta$, by

$$\theta = \frac{r_0}{f}, \tag{4.28}$$

the unit-normalized Airy pattern may be expressed in terms of $\theta$ by

$$PSF_{I,Airy}(\theta) = \left[\frac{2 \cdot J_1\left(\frac{\pi \cdot D \cdot \theta}{\lambda}\right)}{\left(\frac{\pi \cdot D \cdot \theta}{\lambda}\right)}\right]^2. \tag{4.29}$$

## 4.5 Strehl Intensity

The Strehl intensity, or Strehl ratio, is named after Karl Wilhelm Andreas Strehl (1864–1940), a German physicist, mathematician, and astronomer. Strehl intensity is defined (Strehl, 1895, 1902) as the ratio of intensity in the center of a point-object image formed by an actual—in general aberrated—optical system to the intensity in the center of the image formed by an aberration-free version of the same optical system. Strehl intensity is generally wavelength dependent. We therefore denote it here by $SI(\lambda)$. For telescopes with uniformly transmitting pupils and wavefront aberrations denoted by $W(x_1, y_1, \lambda)$, the following expression for Strehl intensity may be deduced from 4.26,

$$SI(\lambda) = \frac{\left|\int\int_{Aperture} \exp\left[-\frac{2 \cdot \pi \cdot i \cdot W(x_1, y_1, \lambda)}{\lambda}\right] \cdot dx_1 \cdot dy_1\right|^2}{\left|\int\int_{Aperture} dx_1 \cdot dy_1\right|^2}. \tag{4.30}$$

For a diffraction-limited telescope, Strehl intensity, $SI(\lambda)$ , takes its maximum value, unity. For aberrated telescopes, Strehl intensity takes values in the range, $0 \leq SI(\lambda) \leq 1$.

### 4.5.1 Expressed in Terms of Rms Wavefront Error

Although telescope wavefront error characteristics might not be known at every location in the telescope pupil, often the rms wavefront error over the pupil, which we denote by $\sigma_T$, is known. By assuming that the wavefront errors are randomly distributed over the telescope aperture and that their probability density function (PDF) distribution is approximately Gaussian,[4] the Strehl intensity of the telescope at wavelength $\lambda$ is given by

$$SI(\lambda) = \exp\left[-\left(\frac{2 \cdot \pi \cdot \sigma_T}{\lambda}\right)^2\right]. \tag{4.31}$$

If the wavefront error over the telescope pupil, $W(x_1, y_1, \lambda)$, is precisely known, the rms quantity, $\sigma_T$, can be obtained from the relation

$$\sigma_T = \sqrt{\langle W^2(x_1, y_1, \lambda)\rangle - \langle W(x_1, y_1, \lambda)\rangle^2}, \tag{4.32}$$

where the $\langle\cdot\rangle$ brackets indicate the average obtained over the telescope pupil.

Often, wavefront error is expressed as a zero-mean function (i.e., $\langle W(x_1, y_1, \lambda)\rangle = 0$. In such instances, 4.32 reduces to the simpler form

$$\sigma_T = \sqrt{\langle W^2(x_1, y_1, \lambda)\rangle}. \tag{4.33}$$

### 4.5.2 For Circularly Symmetric Images

For telescopes with circular apertures, circular central obstructions, and circularly symmetric aberrations, the intensity PSF has circular symmetry. For such telescopes, 4.30 reduces to the form

$$SI(\lambda) = \frac{64}{\left(D^2 - d^2\right)^2} \cdot \left|\int_{d/2}^{D/2} r_0 \cdot \exp\left[-\frac{2 \cdot \pi \cdot i \cdot W(r_0, \lambda)}{\lambda}\right] \cdot dr_0\right|^2 \tag{4.34}$$

[4] The Gaussian PDF is frequently assumed when calculating Strehl intensity from rms wavefront error (Malacara 1992). For general applications, it would be difficult to justify any other PDF choice.

where D is telescope diameter, d is the central obstruction diameter, and $W(r_0, \lambda)$ is wavefront error.

#### 4.5.2.1 Cases Where Wavefront Error Does not Depend on Wavelength

For reflector telescopes, wavefront error is independent of wavelength. In spite of this, Strehl intensity can still have strong wavelength dependence because of the $1/\lambda$ dependence of the phase errors that control it, as indicated by the exponent term in 4.34. Generally, Strehl intensity is highest at longer wavelengths, with a gradual reduction (not necessarily monotonic) occurring at shorter wavelengths.

#### 4.5.2.2 For Diffraction-Limited Telescopes

For a diffraction-limited telescope in perfect focus (i.e., $W(r_0) = 0$), as may readily be confirmed by evaluating 4.34 for this case, Strehl intensity takes its highest possible value, unity. Thus,

$$SI(\lambda)_{Diffrn-Limited} = 1. \tag{4.35}$$

## 4.6 Rayleigh Resolution Criterion

According to Rayleigh's original two-point resolution criterion (Born & Wolf, 2003), a criterion that applies to diffraction-limited telescopes with circular apertures and two incoherently illuminated point-objects of equal brightness, the points are just resolved when the two individual Airy pattern images are separated by a distance equal to the radius of the first dark ring of either Airy pattern. Referring to 4.29, the first dark zero occurs when the Bessel function argument takes the value 3.832. Thus, the just-resolved angle according to this criterion, $\vartheta_R$, is given in radians by

$$\vartheta_R = \frac{3.832 \cdot \lambda}{\pi \cdot D} = \frac{1.22 \cdot \lambda}{D}. \tag{4.36}$$

## 4.7 Images of Extended Objects

It is assumed throughout this book that when light fields overlap, they add together linearly. This superposition property (Goodman, 2004) leads to significant analytical simplifications. It also allows accurate and meaningful representations of the imaging behavior of optical systems.

### 4.7.1 Superposition Property

The superposition property allows images of extended objects to be expressed by a superposition integral that has the general form

$$g_2(x_2, y_2) = \int_{-\infty}^{\infty} \int_{-\infty}^{\infty} g_1(\xi, \eta) \cdot PSF(x_2, y_2, \xi, \eta) \cdot d\xi \cdot d\eta, \quad (4.37)$$

where $PSF(x_2, y_2, \xi, \eta)$ is the PSF of the imaging system, $g_1(\xi, \eta)$ is the illumination function over the object, and $g_2(x_2, y_2)$ is the illumination distribution over the image. The superposition enacts in different ways depending on the state of coherence of the illumination. For coherent illumination, the optical system behaves linearly with respect to complex amplitude. For incoherent illumination, the system behaves linearly with respect to intensity. For partially coherent illumination, a hybrid calculation is required where final image appearance depends on the complex coherence factor of the illumination,[5] a quantity discussed further in Chap. 7, Sect. 3.8, and by Born and Wolf (2003).

Thus, in the general form expressed above by 4.37, we have purposely not specified whether the quantity that has to be linearly superposed is the complex amplitude or the intensity. Clarification is given in Sects. 4.7.5–4.7.7 where specific superposition integrals are given for illuminations with different states of coherence.

### 4.7.2 Nonlinear Optical Phenomena

Nonlinear optical phenomena only occur in light fields where the intensity is extremely high. Such intensities may be created by pulsed lasers. Nonlinear behaviors occur when the light field amplitudes are large compared to interatomic field amplitudes. In this regime, photon–particle interactions take place that cause the superposition property to break down. Because such extreme field conditions are not considered in any of the applications discussed in this book, in proceeding forward we simply assume that the superposition property holds rigorously and without exception.

### 4.7.3 Isoplanaticity

Image regions over which the functional shape of the intensity PSF remains invariant are known as isoplanatic regions. Sometimes, the entire image plane can be considered an isoplanatic region. An optical system that is diffraction limited over the

[5] The complex coherence factor is also referred to sometimes as the complex degree of coherence when discussing illumination coherence properties.

entire field of view would be one such example. More generally, isoplanaticity is restricted to subregions of the image plane known as isoplanatic patches. An isoplanatic patch is usually found in the image plane region close to the optical axis, but isoplanatic patches can also arise in off-axis regions as long as the intensity PSF remains approximately invariant over these regions. If the size of the isoplanatic patch significantly exceeds the size of the intensity PSF, the imaging system is said to be spatially invariant within the isoplanatic patch. As a point source moves in the object plane within an isoplanatic patch, its image moves correspondingly, with the image shape—the intensity PSF of the optical system—remaining unchanged.

When space invariance applies over usefully large areas of the image, the superposition integral (4.37) within these areas takes on a particularly simple form, referred to as a convolution integral.

### 4.7.4 Convolution Integrals

When an optical system forms the image, each source point in the object plane gives rise to its own individual complex amplitude distribution in the image plane. Within an isoplanatic patch, where the PSF remains spatially invariant, the image may be described by the convolution of the PSF with the illumination distribution over the object,

$$g_2(x, y) = \int_{-\infty}^{\infty} \int_{-\infty}^{\infty} g_1(\xi, \eta) \cdot PSF(x - \xi, y - \eta) \cdot d\xi \cdot d\eta, \tag{4.38}$$

where $PSF(x - \xi, y - \eta)$ represents the PSF of the optical system. In the general form represented by 4.38, again we have not yet specified whether complex amplitude or intensity is the quantity to be added linearly.

For the image to constitute a reasonably faithful replication of the object, the PSF of the imaging system must be compact. Image appearance most closely replicates the object appearance in the limit where the PSF approximates a two- dimensional Dirac delta function, where we can write,

$$PSF(x - \xi, y - \eta) = \delta(x - \xi, y - \eta). \tag{4.39}$$

Replacing the PSF in 4.38 by this delta function, we find that,

$$g_2(x, y) = g_1(x, y), \tag{4.40}$$

indicating that the image exactly replicates the object.

For all nonzero wavelengths, diffraction dictates a certain minimum point spread function size. Thus, though the image may reasonably resemble the object, the resemblance can never be perfect. For astronomical imaging applications, by simply building ever-larger diffraction-limited telescopes (equipped with adaptive optics) the angular size of the point spread function can be reduced without limit, as controlled by the $1/D$ dependence seen in 4.36. Though the intensity PSFs of such telescopes may never quite reach the Dirac delta function limit, there is no theoretical limit to how close we can approach that limit by simply building larger telescopes, equipped with efficiently working AO systems.

### 4.7.5 Images of Coherently Illuminated Extended Objects

When the illumination is fully coherent, optical systems behave linearly with respect to the complex amplitude. Therefore, images obtained under this type of illumination must be calculated by summing together the complex amplitude distributions corresponding to the various individual source points that constitute the object (with care taken to account for the initial phases of the illumination at each individual source point).

For a coherently illuminated extended object lying entirely within an isoplanatic patch, the final complex amplitude in the image may be calculated by convolution of the complex amplitude distribution of the illumination in the object plane with the amplitude PSF of the optical system. The final complex amplitude distribution in the image, $U(x, y)$, may then be written as the convolution integral,

$$U(x, y) = \iint PSF_A(x - x', y - y') \cdot U\left(x', y'\right) \cdot dx' \cdot dy' \qquad (4.41)$$

where $PSF_A$ is the amplitude point spread function of the optical system and $U(x, y)$ is the complex amplitude distribution associated with the object illumination. The intensity distribution in the final image, $I(x, y)$, may be obtained by obtaining the squared modulus of the complex amplitude distribution in the form,

$$I(x, y) = \left| \iint PSF_A(x - x', y - y') \cdot U\left(x', y'\right) \cdot dx' \cdot dy' \right|^2 . \qquad (4.42)$$

### 4.7.6 Images of Incoherently Illuminated Extended Objects

For incoherent illumination, optical systems behave linearly with respect to intensity. Thus, the final image intensity distribution is given directly by the following convolution integral,

$$I(x,y)=\iint PSF_I(x-x',y-y')\cdot I\left(x',y'\right)\cdot dx'\cdot dy' \tag{4.43}$$

where $PSF_I$ is the intensity point spread function of the optical system and $I\left(x',y'\right)$ is the intensity distribution associated with the object illumination. For extended self-luminous astronomical objects, including resolved stars, binary stars, planets, and other object types, the object illumination can generally be considered incoherent.[6]

### 4.7.7 *Images of Partially Coherently Illuminated Extended Objects*

Partially coherent illumination as its name suggests is constituted by illumination that is part-coherent and part-incoherent. For this kind of illumination, the image intensity distributions are calculated separately for the coherent and incoherent light portions in the way described in the two preceding subsections. The intensity distribution in the final image is then obtained by summing the two resulting intensity distributions.

## 4.8 Images of Two-Point Objects

A two-point object is the simplest type of extended object. The complex amplitude and intensity distributions in the (diffraction pattern) image that arises from a single-point object were given previously by 4.24 and 4.29. An expression for the intensity distribution in the image of a two-point object illuminated by light of arbitrary state of coherence can be obtained from 4.24 and 4.29 by adhering to the convolution rules for incoherent, coherent, and partially coherent illumination set out in Sects. 4.7.5–4.7.7 above. In this case, the object illumination function can be written as the sum of two (separated) Dirac delta functions (cf., 4.39). Denoting the resulting intensity distribution by $I(x,y)$, we may express this distribution in the form,

$$I(x,y)=\left[\frac{2\cdot J_1\left(\frac{\pi\cdot D\cdot\sqrt{(x+\Delta x/2)^2+y^2}}{\lambda\cdot f}\right)}{\frac{\pi\cdot D\cdot\sqrt{(x+\Delta x/2)^2+y^2}}{\lambda\cdot f}}\right]^2+\left[\frac{2\cdot J_1\left(\frac{\pi\cdot D\cdot\sqrt{(x-\Delta x/2)^2+y^2}}{\lambda\cdot f}\right)}{\frac{\pi\cdot D\cdot\sqrt{(x-\Delta x/2)^2+y^2}}{\lambda\cdot f}}\right]^2$$

[6] Occasional exceptions are conceivable but of doubtful practical significance. Large plasma-filled regions in space, if energized by a nearby star or other object, could possibly provide just the right conditions for amplifying stray light photons. Such regions act like giant, single-pass laser cavities, which could in principal produce powerful coherent laser beam pulses traveling in the same direction as the initial triggering photon.

$$+2\cdot|\mu|\cdot\cos(\varphi)\cdot\frac{2\cdot J_1\left(\frac{\pi\cdot D\cdot\sqrt{(x+\Delta x/2)^2+y^2}}{\lambda\cdot f}\right)}{\frac{\pi\cdot D\cdot\sqrt{(x+\Delta x/2)^2+y^2}}{\lambda\cdot f}}\cdot\frac{2\cdot J_1\left(\frac{\pi\cdot D\cdot\sqrt{(x-\Delta x/2)^2+y^2}}{\lambda\cdot f}\right)}{\frac{\pi\cdot D\cdot\sqrt{(x-\Delta x/2)^2+y^2}}{\lambda\cdot f}} \tag{4.44}$$

where the $\Delta$x is the separation of the two object points which, without loss of generality, lie along the x-axis, $\mu$ is the complex coherence factor of the illumination at the two object points, with $|\mu|\cdot\cos(\varphi)$ being the real part and $\varphi$ the appropriate phase angle between the coherent portion of the illuminations at the two object points, $J_1(\cdot)$ is the first-order Bessel function, D is the telescope diameter, and f is the focal length.

For incoherent illumination (i.e., $|\mu| = 0$), only the first two terms in 4.44 contribute. These terms can be considered to be the individual intensity distributions arising separately from each of the two-point objects. As noted previously, for incoherent illumination the intensities sum linearly.

For fully coherent illumination (i.e., $|\mu| = 1$), the complex amplitudes sum linearly as long as proper account is taken of any initial phase angle difference $\varphi$ between the illuminations at the two object points. As we shall see in Sect. 4.8.2, for fully coherent illumination the appearance of a coherently illuminated two- point object radically depends on the value taken by $\varphi$.

For partially coherent illumination (i.e., $0 < |\mu| < 1$), the illuminations are treated as part-coherent and part-incoherent. For the coherent part, the complex amplitudes are summed linearly prior to obtaining the intensity. For the incoherent part, the individual intensities are summed linearly. The final image is then the sum of the intensity distributions produced by the coherent and incoherent illumination portions.

### *4.8.1 Incoherently Illuminated Two-Point Objects*

The image seen in Fig. 4.7 (left) shows the Airy pattern image of a single-point-object formed by a diffraction-limited optical system with an unobstructed circular aperture. The image to the right shows the appearance of an incoherently illuminated two-point object (i.e., $\mu = 0$) where the two points are separated by the just-resolved separation according to the Rayleigh criterion. Images like these are seen routinely when single and binary stars are viewed through small telescopes with good quality optics and circular apertures.

### 4.8.2 Coherently Illuminated Two-Point Objects

Figure 4.8 shows the appearance of the same two-point object, separated by exactly the same amount, when the points are illuminated coherently. This kind of illumination corresponds to $|\mu| = 1$. Image appearance in this case crucially depends on the relative phase of the illuminations at the two points, $\varphi$ (Considine, 1966).

For equal phase coherent illumination, $\varphi = 0$, the left image in Fig. 4.8 shows the points as no longer resolved. However, for anti-phase coherent illumination. $\varphi = \pi$, the right image shows that the points are now perfectly resolved. For the latter condition, the intensity always falls to zero along a line bisecting the image points, no matter how close the point separation. Thus, a super-resolution condition exists where Rayleigh's resolution criterion no longer applies.[7] By setting up this coherent anti-phase illumination condition, even the smallest imaginable telescope is capable of resolving even the most closely separated two-point object (Cook & Mountain, 1978; McKechnie, 1973).

Self-luminous astronomical objects such as binary stars give rise to mutually incoherent illumination. Unfortunately, with this kind of illumination there is no possibility of observing super resolution behavior. Thus, subject to the comments made in Footnote 6, in astronomical applications the coherent anti-phase illumination condition would seem to have little practical value. For these applications, Rayleigh's criterion remains the most realistic measure of two-point resolution.

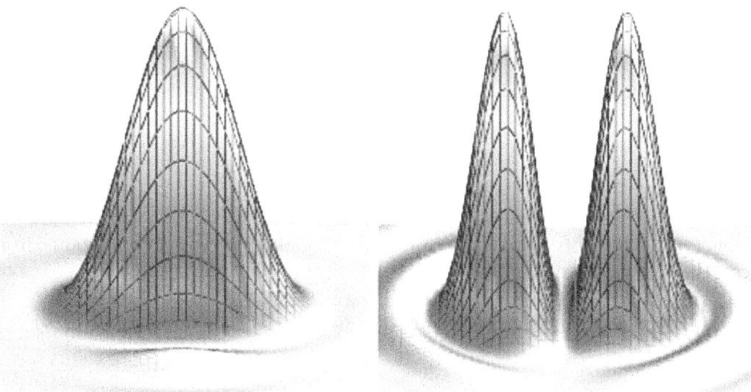

**Fig. 4.8** Diffraction pattern images of a two-point object for *Left* fully coherent in-phase illumination and *Right* for fully coherent anti-phase illumination. The two points are separated by exactly the same amount for these images. The actual separation corresponds to the Rayleigh limit for incoherent illumination $(1.22 \cdot \lambda/D)$

[7] Rayleigh was only familiar with incoherent light sources.

## 4.9 Stellar Speckle Patterns

For telescopes with apertures larger than about 15 cm (6 in.) observing at visible wavelengths, wavefront degradation caused by atmospheric turbulence breaks up the Airy pattern into speckle patterns, as shown in Fig. 4.9. Such patterns are referred to as stellar speckle patterns, and they are found routinely in images formed at visible wavelengths by large astronomical telescopes.[8] As the atmosphere moves and churns, the individual speckle structures in these patterns continually evolve. As discussed in Sect. 11.4.10.1, the average size of a speckle in a star image is determined by the intensity PSF formed by a diffraction-limited version of the telescope; thus, average speckle size does not depend on telescope aberrations (Dainty, 1984).[9] For a 10-m telescope imaging at visible wavelengths, the angular size of the individual speckles is about 0.012 arcsec (12 mas). In stellar speckle patterns, the speckle is distributed over the entire extent of the "seeing disk." The angular size of this disk—typically 0.5–1.5 arcsec—is determined by the angular blur patch caused by the atmosphere. For large telescopes, star images may consist of hundreds or even thousands of individual speckles. For the planned 39-m European Extremely Large Telescope, at visible wavelengths the speckle pattern images of stars will typically comprise one

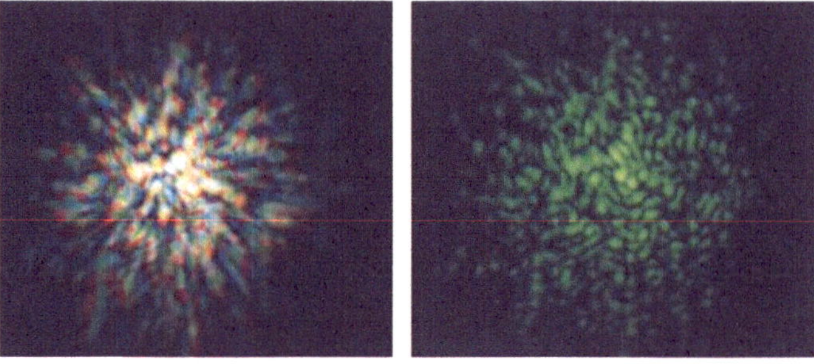

**Fig. 4.9** Instantaneous star images showing stellar speckle patterns formed by a 2-m telescope in ~ 1-arcsec visible seeing conditions. *Left* Polychromatic. *Right* The same image seen through a narrowband green filter

[8] In 1856, more than one hundred years after Newton first proposed the remedy of "serene and quiet air … on the tops of the highest mountains" for seeing the stars more clearly through Earth's atmosphere, Charles Piazzi Smyth (1819–1900), petitioned the Admiralty for a grant of £500 to take a telescope to the mountain tops of Tenerife and test whether Newton had been correct. The funding was granted and Smyth duly completed the mission (Piazzi Smyth, 1858). After moving to the Royal Observatory, Edinburgh in 1846 to take up his appointment as the second Astronomer Royal for Scotland, he was appalled by the poor observing conditions he found there. While nothing immediately came of his efforts, starting around the beginning of the twentieth century, the siting of large astronomical telescopes "on the tops of the highest mountains" has become standard practice.

[9] This certainly holds true for telescope aberrations of several waves or less and therefore holds for most large astronomical telescopes. It does not hold, however, for gross aberrations (cf., Sect. 11.4.10.3).

hundred thousand individual speckles. The properties of stellar speckle patterns are analyzed and discussed in Chap. 11.

## 4.10 Effect of Central Obstruction on Telescope Point Spread Functions

Most large reflecting telescopes have circular apertures and circular central obstructions. When a diffraction-limited telescope of this type (with diameter $D$ and central obstruction diameter $d$) observes in the absence of atmosphere, the intensity distribution in the image of a point-object such as an unresolved star (i.e., the intensity PSF of the telescope is given by

$$I(\vartheta) = 4 \cdot \left\{ \left[ \frac{J_1\left(\frac{\pi \cdot D \cdot \vartheta}{\lambda}\right)}{\frac{\pi \cdot D \cdot \vartheta}{\lambda}} - \left(\frac{d}{D}\right)^2 \cdot \frac{J_1\left(\frac{\pi \cdot d \cdot \vartheta}{\lambda}\right)}{\frac{\pi \cdot d \cdot \vartheta}{\lambda}} \right] \cdot \frac{1}{1 - \left(\frac{d}{D}\right)^2} \right\}^2. \tag{4.45}$$

As can be seen in Fig. 4.10, the central disks of the intensity PSFs produced by a centrally obstructed telescope are narrower than for the unobstructed version of the same telescope. Thus, it may be concluded that there is some resolution benefit in having a central obstruction. However, even for the relatively large obstruction ratios used to generate the plots in this figure, $d/D = 0.4$ and $d/D = 0.9$, only modest resolution improvements are evident. For a limitingly large central obstruction where

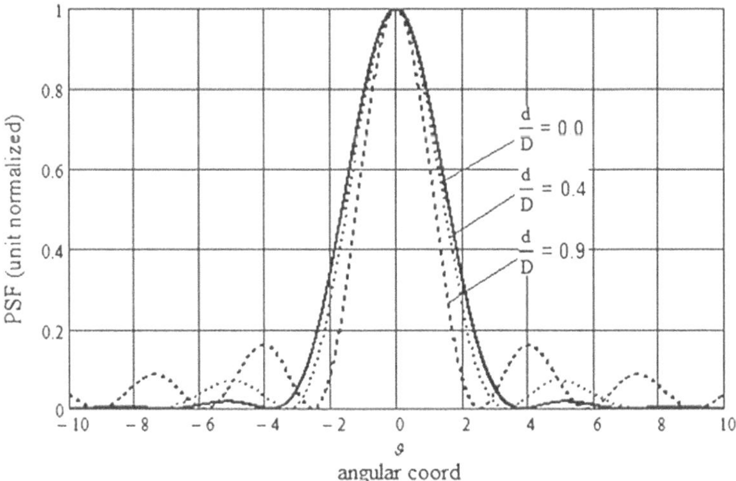


**Fig. 4.10** Effect of central obstructions on the point spread function formed by diffraction-limited telescopes with circular apertures. The intensity PSF for the unobstructed case, $d/D = 0$ is of course an Airy pattern

$d \rightarrow D$, the first dark ring occurs at $0.766 \cdot \lambda/D$, indicating that the maximum resolution improvement factor that can be obtained by centrally obstructing the telescope aperture is about 1.6 (i.e., 1.22/0.766).

A more complete account of the effect of central obstructions on image quality results in a less favorable commentary than the one just given. As discussed later in Chap. 13 and illustrated in Fig. 13.6, central obstructions cause increasingly large fractions of the light energy in the central disk to transfer outwards into the surrounding rings. For extended objects, such as the planets or the Moon, the overall wider intensity PSFs resulting from large central obstructions cause additional image blurring and reduced image contrast. For general applications, large central obstructions should be avoided.

## 4.11 Mathematical Notation Used in This Chapter

The mathematical notation used in this chapter is indicated in Table 4.1.

**Table 4.1** Mathematical notation used in this chapter along with the SI dimensional units of the individual quantities

| Symbol | Quantity | Dimensions |
|---|---|---|
| $\lambda$ | Wavelength | m |
| $F$ | Fresnel number | "1" |
| $Z$ | Propagation distance beyond aperture | m |
| $L$ | Nominal aperture size (e.g., diameter) | m |
| $k$ | Wavenumber in air | $m^{-1}$ |
| $C$ | Constant that sets the complex amplitude units | kg m $s^{-3}$ $amp^{-1}$ |
| $U$ | Complex amplitude | "1" or as set by $C$ units |
| $I$ | Intensity | "1" or as set by $C^2$ units |
| $\mu$ | Complex coherence factor | "1" |
| $D$ | Telescope diameter | m |
| $d$ | Central obstruction diameter | m |
| $f$ | Telescope focal length | m |
| $F/\#$ | Telescope $F$/number | "1" |
| $L$ and $L'$ | Conjugate distances in object and image spaces | m |
| $K$ | Telescope pupil function | "1" |
| $W$ | Telescope wavefront error | m |

(continued)

Table 4.1 (continued)

| Symbol | Quantity | Dimensions |
|---|---|---|
| $\sigma_T$ | Rms telescope wavefront error | m |
| $PSF_A$ | Telescope amplitude point spread function | "1" |
| $PSF_I$ | Telescope intensity point spread function | "1" |
| SI | Telescope Strehl intensity | "1" |
| $g_1$ and $g_2$ | Illumination function in object and image spaces | "1" |
| $\theta$ | Radial angular coordinate in image space | "1" |
| $\theta_R$ | Rayleigh angular resolution limit | "1" |

Dimensionless quantities are indicated by "1". Since only the relative variation of the complex amplitude and the intensity generally matters in this book, the constant, $C$, is often set, in effect, to the dimensionless value, "1"

## References


Abramowitz, M., & Stegun, I. A. (1970). *Handbook of mathematical functions*. U.S. Government Printing Office.

Airy, G. B. (1835). *Transactions of the Cambridge Philosophical Society, 5*, 283.

Born, M., & Wolf, E. (2003). *Principles of Optics* (7th ed., revised). Cambridge: Cambridge University Press.

Considine, P. S. (1966). Effects of coherence on imaging systems. *Journal of Optical Society of America, 56*, 1001.

Cook, D. K., & Mountain, G. D. (1978). The effect of phase angle on the resolution of two coherently illuminated points. *Optical and Quantum Electronics*, 179–180.

Dainty, J. C. (1984). Laser speckle and related phenomena (Vol. 9). In J. C. Dainty (Ed.), *Topics in applied physics*. Springer.

Goodman, J. W. (2004). *Introduction to fourier optics* (3rd ed.). Roberts & Company Publishers.

Hecht, E. (1998). *Optics* (3rd ed.). Addison Wesley Longman.

Magnus, W., Oberheffinger, F., & Soni, R. P. (1966). *Formulas and theorems for the special functions of mathematical physics*. Springer.

Malacara, D. (1992). *Optical shop testing*. Wiley.

McKechnie, T. S. (1973). The effect of condenser obstruction on the two-point resolution of a microscope. *Optica Acta, 19*, 729–737.

Piazzi Smyth, C. (1858). *Teneriffe, an astronomer's experiment: Or specialities of a residence above the clouds*. Retrieved December 7, 2014, from http://books.google.co.uk/books?id=TmsPAAAAYAAJ&printsec=frontcover&dq=charles+smyth+teneriffe&redir_esc=y#v=onepage&q&f=false

Sommerfeld, A. (1896). Mathematische theorie der diffraction.

Strehl, K. (1895). Aplanatische und fehlerhafte Abbildung im Fernrohr. *Zeitschrift Für Instrumentenkunde, 15*, 362–370.

Strehl, K. (1902). Über Luftschlieren und Zonenfehler. *Zeitschrift Für Instrumentenkunde, 22*, 213–217.

# Chapter 5
# Wave Propagation After Scattering by a Thin Atmospheric Layer

**Abstract** This chapter deals with the propagation behavior of infinitely extensive plane light waves after passing through, and being scattered by, a (hypothetical) thin inhomogeneous atmospheric layer. The propagation behavior is examined in the near-field Fresnel space beyond the layer which, at this preliminary stage, we consider homogeneous free space. The behavior of the crucially important two-point two-wavelength correlation function of the complex amplitudes is examined as the waves propagate beyond the layer. The insights and results obtained in this chapter provide a foundation for the analysis given in the next chapter, which deals with wave propagation over extended atmospheric paths. When the two wavelengths coalesce, the function degenerates into the atmospheric MTF. For any atmospheric path, the latter function determines the resolution that can be obtained when observing over that path. The development of scintillation is examined in the space below the layer, the extent to which it occurs depends crucially on the size of the turbulence structures in the layer.

In this chapter, we propagate infinitely extensive initially plane light waves through a thin inhomogeneous atmospheric layer and study the behavior of the scattered waves in two specific regions: First, in the geometrical optics region, a region lying immediately below the layer characterized by Fresnel numbers in the range, $F \gg 1$, where rectilinear wave propagation occurs and where phase fluctuations acquired by the waves have not yet developed into amplitude fluctuations; and second, in the Fresnel near-field region, a region that ranges further from the layer characterized by Fresnel numbers, $F \geq 1$, where the waves generally display both phase and amplitude fluctuations, the latter responsible for intensity scintillation.

Two important statistical functions are derived for the scattered light waves in the space below the layer: (1) the complex coherence factor, $M(x', y', x, y, \lambda)$,[1] a function that describes the degree of correlation between the complex amplitudes for monochromatic waves at two separate locations along the wavefronts; and the more

[1] This function may be considered to be the unit-normalized form of the mutual intensity function, and in classical coherence theory, it is referred to as the mutual intensity (Born & Wolf, 2003).



T. S. McKechnie, *General Theory of Light Propagation and Imaging Through the Atmosphere*, Progress in Optical Science and Photonics 20,
https://doi.org/10.1007/978-3-030-98828-9_5

general function, the two-point two-wavelength correlation function $S(x_1, y_1, x_2, y_2, \lambda_1, \lambda_2)$,[2] a function that describes the degree of correlation between the complex amplitudes in waves at two different wave lengths, $\lambda_1$, $\lambda_2$ and two separate locations along the wavefronts, $(x_1, y_1)$ and $(x_2, y_2)$. The latter function has been discussed by Fante (1978, 1981, 1985, 1986) and McKechnie (1991).

The important previously known result will also be derived that the complex coherence factor, $M$, for disrupted waves of infinite lateral extent conserves perfectly as the waves propagate beyond the layer in the near-field region in homogeneous free space. In contrast, we shall find that the two-point two-wavelength correlation function, $S$, for disrupted waves generally undergoes some degree of modification as the waves propagate beyond the layer in the same homogeneous free-space region. However, exceptions occur in certain special cases where the function is found to approximately conserve over usefully long propagation distances.

Because of the stationarity assumption (Sect. 3.6.1) for the refractive index field, n(x, y, z), it follows that many functions derived from n(x, y, z)—including the aforementioned complex coherence factor, $M$, and the two-point two-wavelength correlation function, $S$—are also spatially stationary. Consequently, the values taken by the functions, $M$ and $S$, depend only on the separation of the two points considered rather than on the absolute locations of the two points. Thus, these two important functions may be represented in terms of spatial argument differences in the forms $M(x_1 - x_2, y_1 - y_2, \lambda)$ and $S(x_1 - x_2, y_1 - y_2, \lambda_1, \lambda_2)$.

The degenerate form of $S(x_1 - x_2, y_1 - y_2, \lambda_1, \lambda_2)$ in the limit $\lambda_1 \to \lambda_2 = \lambda$ is of course $M(x_1 - x_2, y_1 - y_2, \lambda)$. Thus, the propagation behavior of the latter function is fully described by the propagation behavior of the former. However, in spite of the risk of duplication, because $M$ is such an important function we choose to give separate expressions for both functions as the light waves propagate beyond the scattering layer.

Discussion is limited in this chapter to the propagation of initially plane waves after scattering by the single thin atmospheric layer. Light waves arriving from distant unresolved stars obviously closely approximate plane waves. For closer objects, however, such as satellites in low Earth orbits, the arriving waves display small, though significant amounts of spherical curvature. Such curvature can be represented by quadratic phase terms. But, since these may easily be removed later by simply refocusing the telescope, the propagation analysis given in this chapter for initially plane waves also applies more broadly to waves displaying small amounts of spherical curvature.

At the end of the chapter, in Sect. 5.7, we examine quantitatively the development of scintillation over atmospheric paths as a function of turbulence structure size. For typical atmospheric paths, it is found that structure sizes in the approximate range, 0–10 m, are overwhelmingly responsible for its development; for all practical purposes, larger structures make zero contribution to scintillation. The study provides quantitative insight into a fundamental issue related to scintillation: the issue of when

[2] In speckle statistics theory, this function has been referred to by Parry as the generalized spectral correlation function (Dainty, 1984).

we are allowed to apply rectilinear wave propagation principles and when we are not. A basic understanding of this issue is essential before beginning the analysis in Chap. 6 of wave propagation over extended atmospheric paths.

## 5.1 Characterizing Atmospheric Paths and Telescopes by MTFs and OTFs

Hufnagel and Stanley (1964) showed that the (average) atmospheric optical transfer function (OTF) for an atmospheric path is given by the complex coherence factor for that path.[3] A consequence of the stationarity assumption for the refractive index field, n(x, y, z), which was made by Hufnagel and Stanley and which is also made in this book, is that temporal instant-to-instant image shifts caused by atmospheric inhomogeneity always average to zero.

Random tilts associated with wavefronts that have propagated over atmospheric paths give rise to corresponding random phase terms in the instantaneous, or short-exposure, atmospheric OTF. Thus, because the short-exposure atmospheric OTF can take complex values, it is not generally the same as the entirely real-valued short-exposure atmospheric modulation transfer function (MTF). Because we are primarily concerned in this book with either the average short-exposure atmospheric OTF or the long-exposure atmospheric OTF (which is an averaged quantity anyway), the phase terms in both of these atmospheric OTFs always average to zero. Thus, because these functions are exactly identical to their corresponding moduli, in this book, we can refer to them interchangeably as either atmospheric OTFs or atmospheric MTFs; generally, we will use the term atmospheric MTF.

As discussed later in Chap. 7, the resolution achieved by a telescope observing through the atmosphere is determined by the product of the atmospheric MTF and the telescope OTF. Because the telescope OTF and the telescope MTF are not generally the same, these two functions cannot be used interchangeably. Most large astronomical telescopes introduce aberrations into the imaging wavefronts, especially for wavefronts arising from off-axis field locations. An aberration such as coma distorts the image away from its true location by an amount proportional to the local field angle. Shifts of this sort manifest themselves as phase terms in the telescope OTF, causing this function to be complex-valued and therefore different (in general) from the entirely real-valued telescope MTF.

[3] Hufnagel and Stanley use the term mutual coherence factor. However, they define it in a way that is identical to the function referred to in this book as the complex coherence factor.

## 5.2 The Atmospheric Refractive Index

Since the refractive index of air depends on wavelength—a phenomenon referred to as dispersion—a complete propagation analysis must necessarily take account of this phenomenon. However, to simplify the analysis in this chapter, we temporarily disregard dispersion. It will be re-introduced later, in Chap. 6, when we deal with waves that have propagated over extended atmospheric paths, which we represent by a continuous stack comprising a very large, perhaps infinite, number of thin layers of the kind dealt with in this chapter. The final expressions developed in Chap. 6 for the functions, $M$ and $S$, will take proper account of dispersion.

As indicated in Chap. 2, the refractive index field, n(x, y, z), is considered to behave as a random process. The properties of such a process are best described by probability density functions and averages. The normalized fluctuating part of the refractive index field, N(x, y, z), which was defined previously by 3.14, is also a random process.

As discussed in Chap. 3, the ensemble average of any quantity related to a random process refers to the average of that quantity obtained over the ensemble of all possible apparitions. For the various light propagation and imaging applications considered in this book, the statistics of the atmospheric refractive index field, n(x, y, z) will be assumed to be stationary and ergodic, both temporally and spatially. Consequently, it will be assumed that meaningful ensemble average estimates of the refractive index properties can be obtained from either time- or space-average estimates. The same holds true for a number of other functions deriving from n(x, y, z), including the normalized fluctuating part of the refractive index field, N(x, y, z); the integrated optical path difference (OPD) over the atmospheric path; the complex coherence factor for the path, $M$; and the two-point two-wavelength correlation function for the path, $S$. In this book, wide-sense stationarity of the refractive index field is not considered necessary, nor would the assumption of such be physically realistic. Our only real requirement in this regard is that the refractive index field should remain substantially statistically stationary, either over long-enough time periods or over wide-enough regions of space, to enable us to obtain accurate estimates of all meaningful statistical quantities from either the time-average estimate or the space-average estimate.

## 5.3 Wave Propagation in the Geometrical Optics Region

In this section, we examine the propagation behavior of infinitely extensive plane waves after they have passed through a thin (inhomogeneous) atmospheric scattering layer of the type depicted in Fig. 5.1. The propagation region examined in this case corresponds to a region characterized by Fresnel numbers (cf., Sect. 4.1.1) in the range $F \gg 1$. The immediate effect of the refractive index fluctuations within the layer is to disrupt the phase of the transmitted light waves. Such disruption is represented schematically in the figure by the corrugations seen in the wave exiting the layer.

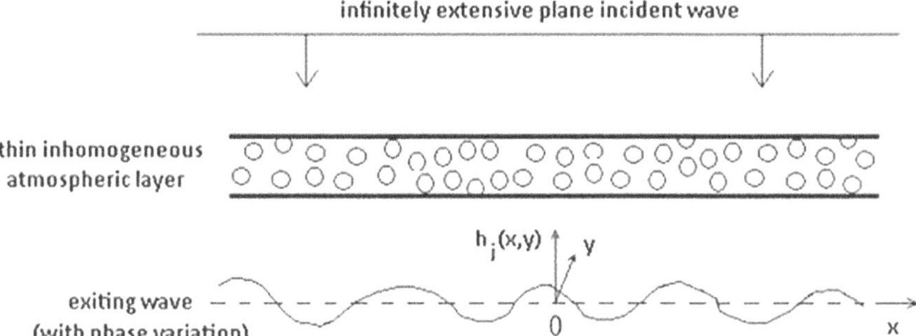


**Fig. 5.1** Schematic showing a plane wave passing through a thin atmospheric scattering layer. The refractive index fluctuations of the inhomogeneous air contained within the layer imprint random phase fluctuations on the exiting wave

The space below the layer in which the disrupted waves are about to propagate is considered a homogeneous free space.

### *5.3.1 Optical Path Difference*

While we only consider a single scattering layer in this chapter, by referring to this layer as the "*j*th layer," we gain the notational flexibility needed in the next chapter when many such thin layers are stacked together to form, in effect, an extended atmospheric scattering path. The optical path difference fluctuation, $OPD_j(x, y)$, introduced by the *j*th layer may be given in terms of the refractive index field, n(x, y, z), by

$$OPD_j(x, y) = \int_{z_j}^{z_{j+1}} n(x, y, z) \cdot dz. \tag{5.1}$$

where $z_j$ and $z_{j+1}$ denote the *z*-coordinates of the upper and lower planes bounding the layer. The normalized fluctuating part of the OPD fluctuations (cf., Sect. 3.6.4.1), which we denote by $h_j(x, y)$, may be expressed as follows:

$$h_j(x, y) = \frac{OPD_j(x, y) - \langle OPD_j(x, y)\rangle}{\langle OPD_j(x, y)\rangle}, \tag{5.2}$$

Equation 5.1 allows $h_j(x, y)$ to be written in the form

$$h_j(x, y) = \int_{z_j}^{z_{j+1}} \left[\frac{n(x, y, z) - \langle n(x, y, z)\rangle}{\langle n(x, y, z)\rangle}\right] \cdot dz, \tag{5.3}$$

were we note that, since n(x, y, z) is a real random variable, $h_j(x, y)$ is also a real random variable. By using 3.14, $h_j(x, y)$ may be expressed in the more concise form

$$h_j(x, y) = \int_{z_j}^{z_{j+1}} N(x, y, z) \cdot dz. \tag{5.4}$$

From 3.15 and 5.4, it is evident that $h_j(x, y)$ is also a zero-mean function. Thus, we can write

$$\langle h_j(x, y) \rangle = 0. \tag{5.5}$$

#### 5.3.1.1 Variance of the OPD Fluctuation

The variance of the OPD fluctuation introduced by the thin atmospheric layer, which we denote by $\sigma_j^2$, is given by

$$\sigma_j^2 = \langle h_j^2(x, y) \rangle, \tag{5.6}$$

where $\sigma_j^2$ may be considered as a measure of the strength of the refractive index fluctuation—and hence the turbulence strength—within the layer.

#### 5.3.1.2 Autocorrelation Function of the OPD Fluctuation

Because $h_j(x, y)$ is assumed to be spatially stationary, the unit-normalized autocorrelation of this function, which we denote by $\rho_j(\xi, \eta)$, may be expressed in the form,

$$\rho_j(\xi, \eta) = \frac{\langle h_j(x + \xi, y + \eta) \cdot h_j(x, y) \rangle}{\langle h_j^2(x, y) \rangle}. \tag{5.7}$$

The autocorrelation function, $\rho_j(\xi, \eta)$, may be considered as a measure of the average size and shape of the refractive index structures in the layer, and hence, this function relates directly to the size and shape of the turbulence structures contained in the layer.

### *5.3.2 Phase Angle of the Exiting Wave*

The phase angle, or phase, of the wave exiting the layer, $\varphi_j(x, y)$, may be expressed in terms of the OPD fluctuation introduced by the layer, $h_j(x, y)$, by

**Table 5.1** Refractive index of standard dry air at sea level for the hydrogen C- and F-lines and the sodium D-line

| Wavelength (μm) | Refractive index |
|---|---|
| 0.4861 | 1.0002794 |
| 0.5893 | 1.0002771 |
| 0.6563 | 1.0002763 |

Refractive indices calculated from 3.1

$$\varphi_j(x, y) = \frac{2 \cdot \pi \cdot h_j(x, y)}{\lambda}, \tag{5.8}$$

where $h_j(x, y)$ was previously given by 5.4 and where $\lambda$ is the wavelength in the homogeneous medium below the layer. If this space were vacuum space, $\lambda$ in the above expression could be replaced by $\lambda_0$. Since $\varphi_j(x, y)$ is a function of the random variable, $h_j(x, y)$, it must itself be a random variable.

Air refractive index n(x, y, z) takes its highest values at sea level where air density is highest. But, as can be seen from Table 5.1, for visible wavelengths, the refractive index of standard dry air even at sea level never exceeds the approximate value 1.0003. For IR wavelengths, or at any altitude above sea level, the refractive index generally takes even smaller values. Consequently, to an accuracy of 0.03% or better, air and vacuum wavelengths are approximately equal. Thus, in this book, it is often acceptable to use the approximation,

$$\lambda \approx \lambda_0. \tag{5.9}$$

### 5.3.3 *Complex Amplitude of the Exiting Wave*

The complex amplitude $U_j(x, y, \lambda)$ of the wave exiting the layer is also a random variable, one that may be expressed in terms of the wave amplitude $A_j(x, y)$ and phase angle $\varphi_j(x, y)$ by

$$U_j(x, y, \lambda) = A_j(x, y) \cdot \exp\left(i \cdot \varphi_j(x, y)\right), \tag{5.10}$$

where $A_j(x, y)$ and $\varphi_j(x, y)$ are both real-valued.

By using 5.8, $U_j(x, y, \lambda)$ may be expressed in terms of the OPD fluctuation by

$$U_j(x, y, \lambda) = A_j(x, y) \cdot \exp\left(\frac{2 \cdot \pi \cdot i \cdot h_j(x, y)}{\lambda}\right) \tag{5.11}$$

The forms of 5.10 and 5.11 tacitly assume rectilinear wave propagation (Sect. 4.1.1). Over the (assumed small) thickness of the atmospheric layer, phase fluctuations that develop in the waves simply do not have sufficient distance in which to convert into significant amounts of scintillation. Thus, in the rectilinear wave propagation region of the homogeneous free space immediately below the layer, we consider that both the wave amplitude, $A_j(x, y)$, and the phase, $\varphi_j(x, y)$, substantially conserve.

The rectilinear wave propagation assumption is obviously valid for small distances just below the layer. But exactly how large are these distances and what are the governing factors? The answer turns out to be slightly more complicated than one might first suspect. As we shall see later in Sect. 5.7, the answer depends partly on the magnitude of the OPD fluctuation introduced by the layer and partly on the lateral structure size of the OPD fluctuations. In that section, we explore in detail the conditions under which we can and cannot apply rectilinear wave propagation principles.

### 5.3.4 *The Two-Point Two-Wavelength Correlation Function for Exiting Waves*

The two-point two-wavelength correlation function, $S_j(x_1 - x_2, y_1 - y_2, \lambda_1, \lambda_2)$ describes the degree of correlation between the complex amplitudes of the light waves exiting the layer at two arbitrary locations, $(x_1, y_1)$ and $(x_2, y_2)$, and at two arbitrary wavelengths, $\lambda_1$ and $\lambda_2$. This function may be expressed in terms of the complex amplitudes of the waves exiting the layer by

$$S_j(x_1 - x_2, y_1 - y_2, \lambda_1, \lambda_2) = \frac{\langle U_j(x_1, y_1, \lambda_1) \cdot U_j^*(x_2, y_2, \lambda_2)\rangle}{\left[\left\langle \left|U_j(x_1, y_1, \lambda_1)\right|^2\right\rangle \cdot \left\langle \left|U_j(x_2, y_2, \lambda_2)\right|^2\right\rangle\right]^{\frac{1}{2}}}, \tag{5.12}$$

Equation 5.11 allows $S_j(x_1 - x_2, y_1 - y_2, \lambda_1, \lambda_2)$ to be expressed in terms of $h_j(x_1, y_1)$ and $h_j(x_2, y_2)$ by

$$S_j(x_1 - x_2, y_1 - y_2, \lambda_1, \lambda_2) = \left\langle \exp\left\{2 \cdot \pi \cdot i \cdot \left[\frac{h_j(x_1, y_1)}{\lambda_1} - \frac{h_j(x_2, y_2)}{\lambda_2}\right]\right\}\right\rangle. \tag{5.13}$$

By using the variable transformation equations,

$$x_1 - x_2 = \xi, \tag{5.14}$$

$$y_1 - y_2 = \eta, \tag{5.15}$$

$$x = x_2, \tag{5.16}$$

$$y = y_2, \tag{5.17}$$

Equation 5.13 may be written in the form

$$S_j(\xi, \eta, \lambda_1, \lambda_2) = \left\langle \exp\left\{2 \cdot \pi \cdot i \cdot \left[\frac{h_j(x+\xi, y+\eta)}{\lambda_1} - \frac{h_j(x, y)}{\lambda_2}\right]\right\}\right\rangle. \tag{5.18}$$

This expression is the desired expression for $S_j(\xi, \eta, \lambda_1, \lambda_2)$. In the next chapter, when we propagate light waves over extended atmospheric paths, expressions of this sort arise for each of the many layers in the path model.

#### 5.3.4.1 Explicit Form for $S_j(\xi, \eta, \lambda_1, \lambda_2)$ for Gaussian Distributed OPD Fluctuation

While a more explicit form of 5.18 is not required at this stage of our analysis, we choose to give one anyway so that we can illustrate the behavior of the function, $S_j(\xi, \eta, \lambda_1, \lambda_2)$ as the waves propagate forward. But before we can give such an expression, we must first choose a suitable form for the probability density function (PDF) of the OPD fluctuation. For convenience, we chose the Gaussian form,

$$PDF(h_j) = \frac{1}{\sigma_j \cdot \sqrt{\pi}} \cdot \exp\left[-\frac{h_j^2}{\sigma_j^2}\right], \tag{5.19}$$

where $\sigma_j^2$ was previously given by 5.6. With this PDF choice, the averaging operation indicated by the $\langle\cdot\rangle$ brackets in 5.18 may be carried out analytically, yielding the following explicit expression for $S_j(\xi, \eta, \lambda_1, \lambda_2)$:

$$S_j(\xi, \eta, \lambda_1, \lambda_2) = \exp\left\{-2 \cdot \pi^2 \cdot \sigma_j^2 \cdot \left[\frac{1}{\lambda_1{}^2} + \frac{1}{\lambda_2{}^2} - \frac{2 \cdot \rho_j(\xi, \eta)}{\lambda_1 \cdot \lambda_2}\right]\right\}, \tag{5.20}$$

where $\rho_j(\xi, \eta)$ was defined previously by 5.7. For the thin atmospheric layers considered thus far, the Gaussian PDF assumption may not necessarily be valid. However, it is important to note here that the extended atmospheric path analysis that ultimately concerns us (Chap. 6) depends only on the validity—of which we can be sure—of 5.18. Again, it should be emphasized that the only reason for including 5.20 here is to give a specific functional form for $S_j(\xi, \eta, \lambda_1, \lambda_2)$ that we can use later (cf., Sect. 5.3.7) to gain graphical appreciation of the behavior of this function.

### 5.3.5 Complex Coherence Factor for Exiting Waves

As previously discussed, the complex coherence factor, $M_j(\xi, \eta, \lambda)$, is simply the degenerate case of the two-point two-wavelength correlation function, $S_j(\xi, \eta, \lambda_1, \lambda_2)$ in the limit where $\lambda_1 \rightarrow \lambda_2 \rightarrow \lambda$. For light waves exiting the thin layer shown in Fig. 5.1, $M_j(\xi, \eta, \lambda)$ may be obtained from 5.18 in the form,

$$M_j(\xi, \eta, \lambda) = \left\langle \exp\left\{2 \cdot \pi \cdot i \cdot \left[\frac{h_j(x + \xi, y + \eta) - h_j(x, y)}{\lambda}\right]\right\}\right\rangle. \tag{5.21}$$

By again assuming that the OPD fluctuation is Gaussian distributed (cf., 5.19), the complex coherence factor can be shown to take the explicit form (Beckmann & Spizzichino, 1963),

$$M_j(\xi, \eta, \lambda) = \exp\left\{\frac{-4 \cdot \pi^2 \cdot \sigma_j^2}{\lambda^2} \cdot \left[1 - \rho_j(\xi, \eta)\right]\right\}. \tag{5.22}$$

Again, the Gaussian assumption used here may or may not be valid; our long term goals can be achieved without this assumption. We simply use it here, without prejudice, so that we can produce an explicit expression for $M_j(\xi, \eta, \lambda)$ that can be evaluated to provide plots (Sect. 5.3.7) that nominally illustrate the shape characteristics of this function. However, in the next chapter, where we deal with extended atmospheric paths comprised of a large number of thin layers of the type dealt with here, at that point we shall be able to invoke the central limit theorem, and that alone will justify making the Gaussian PDF assumption. We shall then find that the complex coherence factor that arises for these extended paths, $M(\xi, \eta, \lambda)$, has a form (cf., 6.52) closely similar to that given above by 5.22.

### 5.3.6 Illustrative Plots of the Complex Coherence Factor

To produce illustrative plots of the complex coherence factor $M_j$ from 5.22, specific values must be attached to $\sigma_j$ and explicit functionalities must be attached to the autocorrelation function of the OPD fluctuation, $\rho_j(\xi, \eta)$. For convenience here and without prejudice to the functional form, or forms, which might arise from actual field measurements of the kind described later in Chaps. 8 and 13, we choose $\rho_j(\xi, \eta)$ to have the circular Gaussian form,

$$\rho_j(\xi, \eta) = \exp\left[-\left(\frac{\xi^2 + \eta^2}{w_o^2}\right)\right], \tag{5.23}$$

where $w_o$ is the 1/e half-width of the function.

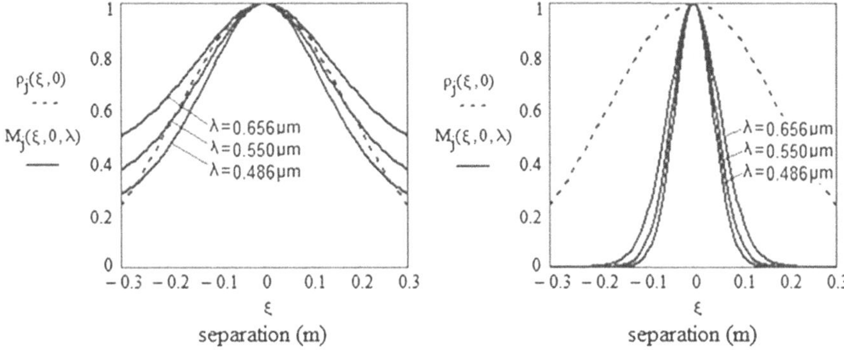


**Fig. 5.2** Sections through the complex coherence factor, $M_j(\xi, 0, \lambda)$, for light waves scattered by a thin atmospheric layer at the three visible wavelengths indicated. Plots generated from 5.22 with $\rho_j(\xi, 0)$ given by 5.23 where $w_o = 0.25$ m. *Left* Weaker turbulence where $\sigma_j = 0.1$ μm. *Right* Stronger turbulence where $\sigma_J = 0.35$ μm

Figure 5.2 shows central sections $\eta = 0$ through the complex coherence factor, $M_j(\xi, 0, \lambda)$ with 5.22 and 5.23 used to calculate the plotted data. The autcorrelation function of the OPD fluctuation, $\rho_j(\xi, 0)$, is also shown in the figure. The $\sigma_j$, $w_o$ and $\lambda$ values used to calculate the plots are indicated in the figure itself and in the figure caption. The two $\sigma_j$ values used may be considered to represent in a relative sense weak and strong turbulence.

For the infinitely extensive light waves considered, it is shown in Sect. 5.6 that no matter how far the waves propagate beyond the scattering layer in the (assumed) homogeneous free space below the layer, the complex coherence factor conserves perfectly. Thus, the complex coherence factors shown in Fig. 5.2 for the waves exiting the layer apply equally to the waves after they propagate beyond the layer, no matter how large the propagation distance might be.

### 5.3.7 *Illustrative Plots of the Two-Point Two-Wavelength Correlation Function*

Figure 5.3 shows central sections ($\eta = 0$) through the two-point two-wavelength correlation function, $S_j(\xi, 0, \lambda_1, \lambda_2)$, for initially plane waves as the waves emerge from the thin scattering layer. The autocorrelation function of the OPD fluctuations used to calculate the plots is again given by 5.23. For this case, $S_j(\xi, 0, \lambda_1, \lambda_2)$ is entirely real-valued. Plot sets are shown for (relatively) weak and strong turbulence, as quantified by the $\sigma_j$ values indicated in the figure caption, for three wavelength pair combinations, blue/green, green/red, and blue/red, where 0.486, 0.550, and 0.656 μm are the actual wavelengths used. Figure 5.4 shows the two-wavelength single-point correlation function, $S_j(0, 0, \lambda_1, \lambda_2)$ plotted against $\lambda_2$ for three fixed $\lambda_1$ values. This function may be referred to as the spectral correlation function.

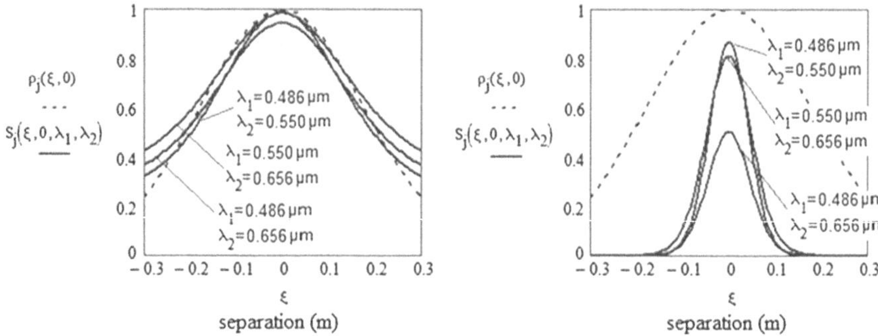


**Fig. 5.3** The two-point two-wavelength correlation function sections, $S_j(\xi, 0, \lambda_1, \lambda_2)$, for light waves scattered by a thin atmospheric layer at the three visible wavelength pairs indicated. Plots generated from 5.20 with $\rho_j(\xi, 0)$ given by 5.23 where $w_o = 0.25$ m. *Left* Weaker turbulence where $\sigma_J = 0.1$ μm. *Right* Stronger turbulence where $\sigma_J = 0.35$ μm. Note that the *blue* and *red* wavelengths are the least correlated

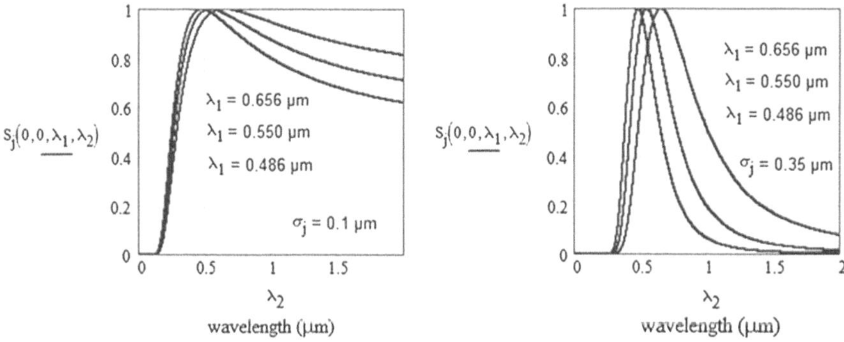


**Fig. 5.4** The single point correlation function $S_j(0, 0, \lambda_1, \lambda_2)$ plotted against $\lambda_2$ for light waves scattered by a thin atmospheric layer for the three $\lambda_1$ values indicated. Plots generated from 5.20 with $\rho_j(\xi, 0)$ given by 5.23 where $w_o = 0.25$ m. *Left* Weaker turbulence ($\sigma_j = 0.1$ μm). *Right* Stronger turbulence ($\sigma_j = 0.35$ μm). Note that the correlation goes to unity when the two wavelengths coalesce (i.e., $\lambda_1 = \lambda_2$)

## 5.4 Near-Field Propagation of the Complex Amplitude

In this section, we make use of the Fresnel near-field propagation equation to develop expressions for the complex amplitudes associated with infinitely extensive light waves at two different wavelengths as the waves propagate in the Fresnel near-field region of the homogeneous free space below the layer. The expressions developed will be used as starting points in the next section when we examine how the two-point two-wavelength correlation function modifies as the waves propagate beyond the layer.

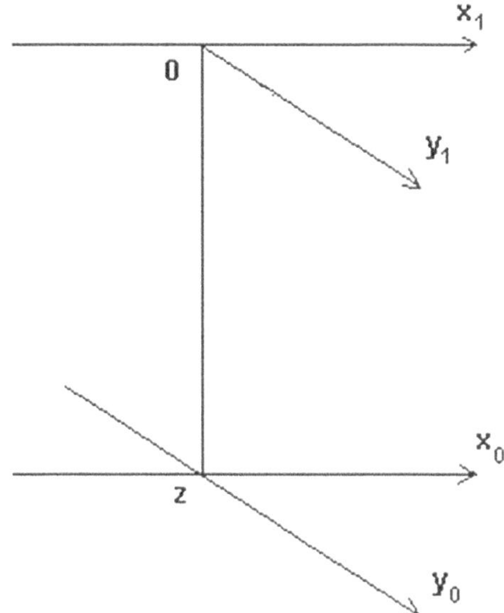


**Fig. 5.5** Coordinate systems used for wave propagation in the Fresnel near-field region

The complex amplitudes of the waves in the upper $(x_1, y_1)$ coordinate plane shown in Fig. 5.5 are denoted, $U(x_1', y_1'\lambda_1)$ and $U(x_1, y_1, \lambda_2)$. It is assumed here that, prior to arriving at this plane, these waves may have previously encountered atmospheric turbulence so that their phases and amplitudes are already randomly disrupted; we denote the complex amplitudes of the waves when they arrive at the lower $(x_0, y_0)$ coordinate plane by $U(x_0', y_0', \lambda_1)$ and $U(x_0, y_0, \lambda_2)$, respectively.

The Fresnel near-field propagation Eq. (4.6) allows the complex amplitudes in the lower plane to be expressed in terms of the corresponding complex amplitudes in the upper plane as follows:

$$\begin{aligned} U(x_0, y_0, \lambda_1) = & \frac{-i \cdot \exp(2 \cdot \pi \cdot i \cdot z/\lambda_1)}{\lambda_1 \cdot z} \\ & \times \exp\left[\frac{\pi \cdot i}{\lambda_1 \cdot z} \cdot \left[x_0^2 + y_0^2\right]\right] \int_{-\infty}^{\infty} \int_{-\infty}^{\infty} U(x_1, y_1, \lambda_1) \\ & \times \exp\left\{\frac{\pi \cdot i}{\lambda_1 \cdot z} \cdot \left[x_1^2 + y_1^2\right]\right\} \\ & \times \exp\left[\frac{-2 \cdot \pi \cdot i}{\lambda_1 \cdot z} \cdot \left[x_0 \cdot x_1 + y_0 \cdot y_1\right]\right] \cdot dx_1 \cdot dy_1, \end{aligned} \tag{5.24}$$

$$U\left(x_0', y_0', \lambda_2\right) = \frac{-i \cdot \exp(2 \cdot \pi \cdot i \cdot z/\lambda_2)}{\lambda_2 \cdot z}$$
$$\times \exp\left[\frac{\pi \cdot i}{\lambda_2 \cdot z} \cdot \left[x_0'^2 + y_0'^2\right]\right] \int_{-\infty}^{\infty} \int_{-\infty}^{\infty} U\left(x_1', y_1', \lambda_2\right)$$
$$\times \exp\left\{\frac{\pi \cdot i}{\lambda_2 \cdot z} \cdot \left[x_1'^2 + y_1'^2\right]\right\}$$
$$\times \exp\left[\frac{-2 \cdot \pi \cdot i}{\lambda_2 \cdot z} \cdot \left[x_0' \cdot x_1' + y_0' \cdot y_1'\right]\right] \cdot dx_1' \cdot dy_1'. \tag{5.25}$$

## 5.5 Near-Field Propagation of the Two-Point Two-Wavelength Correlation Function

Since the random refractive index field in the region above the upper $(x_1, y_1)$ coordinate plane is assumed spatially stationary, we anticipate that the two-point two-wavelength correlation function of the disrupted light waves arriving in this coordinate plane will also be spatially stationary. Thus, we may denote the spatial arguments of this function as difference pairs, in the form $S_1(x_1' - x_1, y_1' - y_1, \lambda_1, \lambda_2)$. Similarly, when the waves arrive in the lower $(x_0, y_0)$ plane, we denote the function in this plane by the spatially stationary form, $S_0\left(x_0' - x_0, y_0' - y_0, \lambda_1, \lambda_2\right)$.

Functions $S_1(x_1' - x_1, y_1' - y_1, \lambda_1, \lambda_2)$ and $S_0\left(x_0' - x_0, y_0' - y_0, \lambda_1, \lambda_2\right)$ may be expressed in terms of the complex amplitudes in the two planes in the standard unit-normalized forms,

$$S_1\left(x_1' - x_1, y_1' - y_1, \lambda_1, \lambda_2\right) = \frac{\left\langle U\left(x_1', y_1', \lambda_1\right) \cdot U^*(x_1, y_1, \lambda_2)\right\rangle}{\left[\left\langle I\left(x_1', y_1', \lambda_1\right)\right\rangle \cdot \left\langle I(x_1, y_1, \lambda_2)\right\rangle\right]^{\frac{1}{2}}}, \tag{5.26}$$

and

$$S_0\left(x_0' - x_0, y_0' - y_0, \lambda_1, \lambda_2\right) = \frac{\left\langle U\left(x_0', y_0', \lambda_1\right) \cdot U^*(x_0, y_0, \lambda_2)\right\rangle}{\left[\left\langle I\left(x_0', y_0', \lambda_1\right)\right\rangle \cdot \left\langle I(x_0, y_0, \lambda_2)\right\rangle\right]^{\frac{1}{2}}}. \tag{5.27}$$

By using 5.24–5.27, function $S_0$ may be expressed in terms of function $S_1$ as follows:

$$S_0\left(x_0' - x_0, y_0' - y_0, \lambda_1, \lambda_2\right) = \frac{\exp[2 \cdot \pi \cdot i \cdot z \cdot (1/\lambda_1 - 1/\lambda_2)]}{\lambda_1 \cdot \lambda_2 \cdot z^2}$$
$$\times \exp\left[\frac{\pi \cdot i}{z} \cdot \left(\frac{x_0'^2 + y_0'^2}{\lambda_1} - \frac{x_0^2 + y_0^2}{\lambda_2}\right)\right]$$

$$\times \int_{-\infty}^{\infty}\int_{-\infty}^{\infty}\int_{-\infty}^{\infty}\int_{-\infty}^{\infty} S_1(x_1' - x_1, y_1' - y_1, \lambda_1, \lambda_2)$$
$$\times \exp\left[\frac{\pi \cdot i}{z} \cdot \left(\frac{x_1'^2 + y_1'^2}{\lambda_1} - \frac{x_1^2 + y_1^2}{\lambda_2}\right)\right]$$
$$\times \exp\left[\frac{-2 \cdot \pi \cdot i}{z} \cdot \left(\frac{x_0' \cdot x_1' + y_0' y_1'}{\lambda_1} - \frac{x_0 \cdot x_1 + y_0 \cdot y_1}{\lambda_2}\right)\right]$$
$$\cdot dx_1' \cdot dx_1 \cdot dy_1' \cdot dy_1. \tag{5.28}$$

By using the coordinate transformation equations

$$x_1' - x_1 = \xi_1., \tag{5.29}$$

$$y_1' - y_1 = \eta_1, \tag{5.30}$$

$$x_0' - x_0 = \xi_0, \tag{5.31}$$

$$y_0' - y_0 = \eta_0. \tag{5.32}$$

and noting that the Jacobian determinant, or simply the Jacobian, of these transformations is unity, 5.28 may be expressed in the form,

$$S_0(x_0 + \xi_0, x_0, y_0 + \eta_0, y_0, \lambda_1, \lambda_2) = \frac{\exp[2 \cdot \pi \cdot i \cdot z \cdot (1/\lambda_1 - 1/\lambda_2)]}{\lambda_1 \cdot \lambda_2 \cdot z^2}$$
$$\times \exp\left[\frac{\pi \cdot i}{z} \cdot \left(\frac{x_0^2 + 2 \cdot \xi_0 \cdot x_0 + \xi_0^2}{\lambda_1} - \frac{x_0^2}{\lambda_2} + \frac{y_0^2 + 2 \cdot \eta_0 \cdot y_0 + \eta_0^2}{\lambda_1} - \frac{y_0^2}{\lambda_2}\right)\right]$$
$$\times \int_{-\infty}^{\infty}\int_{-\infty}^{\infty} S_1(\xi_1, \eta_1, \lambda_1, \lambda_2) \cdot \exp\left[\frac{\pi \cdot i}{z} \cdot \left(\frac{\xi_1^2 + \eta_1^2}{\lambda_1}\right)\right] \cdot \exp\left[\frac{-2 \cdot \pi \cdot i}{z} \cdot \left(\frac{\xi_1 \cdot x_0 + \xi_1 \cdot \xi_0 + \eta_1 \cdot y_0 + \eta_1 \cdot \eta_0}{\lambda_1}\right)\right]$$
$$\times \left\{\int_{-\infty}^{\infty}\int_{-\infty}^{\infty} \exp\left[\frac{\pi \cdot i}{z} \cdot \left(x_1^2 + y_1^2\right) \cdot (1/\lambda_1 - 1/\lambda_2)\right] \cdot \exp\left[\frac{-2 \cdot \pi \cdot i \cdot x_1}{z} \cdot \left(\frac{x_0 + \xi_0 - \xi_1}{\lambda_1} - \frac{x_0}{\lambda_2}\right)\right] \cdot dx_1 \cdot dy_1\right\}$$
$$\times \exp\left[\frac{-2 \cdot \pi \cdot i \cdot y_1}{z} \cdot \left(\frac{y_0 + \eta_0 - \eta_1}{\lambda_1} - \frac{y_0}{\lambda_2}\right)\right] \cdot d\xi_1 \cdot d\eta_1 \tag{5.33}$$

The double integral term in curly brackets in the above equation may be recognized as the Fourier transform of a Gaussian function. After carrying out the Fourier transform analytically, this term is found to reduce to the form,

$$\left\{\frac{i \cdot \lambda_1 \cdot \lambda_2 \cdot z}{(\lambda_2 - \lambda_1)} \cdot \exp\left[\frac{-\pi \cdot i}{z} \cdot \left(\frac{\lambda_1 \cdot \lambda_2}{\lambda_2 - \lambda_1}\right) \cdot \left(\frac{x_0 + \xi_0 - \xi_1}{\lambda_1} - \frac{x_0}{\lambda_2} + \frac{y_0 + \eta_0 - \eta_1}{\lambda_1} - \frac{y_0}{\lambda_2}\right)\right]\right\}.$$

By substituting the above term into 5.33 and canceling terms in the exponent, 5.33 finally simplifies to the form,

$$
\begin{aligned}
S_0(\xi_0, \eta_0, \lambda_1, \lambda_2) = & \frac{i \cdot \exp[2 \cdot \pi \cdot i \cdot z \cdot (1/\lambda_1 - 1/\lambda_2)]}{z \cdot (\lambda_2 - \lambda_1)} \\
& \times \int_{-\infty}^{\infty} \int_{-\infty}^{\infty} S_1(\xi_1, \eta_1, \lambda_1, \lambda_2) \\
& \cdot \exp\left[\frac{-\pi \cdot i}{z \cdot (\lambda_2 - \lambda_1)} \cdot \left((\xi_1 - \xi_0)^2 + (\eta_1 - \eta_0)^2\right)\right] \\
& \cdot d\xi_1 \cdot d\eta_1 \qquad (5.34)
\end{aligned}
$$

The right-hand side of this equation may be recognized as a Fresnel transform. The occurrence of the spatial arguments in the transform kernel in the forms, $(\xi_1 - \xi_0)$ and $(\eta_1 - \eta_0)$ confirms that $S_0$ is indeed spatially stationary as we had anticipated earlier.

Numerical evaluations of 5.34 show that, in general, both the modulus and phase of $S_0(\xi_0, \eta_0, \lambda_1, \lambda_2)$ modify during propagation. The phase term outside the integral varies rapidly as the propagation distance, $z$, increases. The integral term also gives rise to phase fluctuations, although these occur at a much slower rate.

### *5.5.1 Cases Where the Function Conserves*

In the limit $z \cdot (\lambda_2 - \lambda_1) \to 0$, it may be shown (Hoskins, 1979) that

$$
\begin{aligned}
& \exp\left[\frac{-\pi \cdot i}{z \cdot (\lambda_2 - \lambda_1)} \cdot \left((\xi_1 - \xi_0)^2 + (\eta_1 - \eta_0)^2\right)\right] \\
& \quad \to -i \cdot z \cdot (\lambda_2 - \lambda_1) \cdot \delta(\xi_1 - \xi_0, \eta_1 - \eta_0). \qquad (5.35)
\end{aligned}
$$

where $\delta$ denotes the two-dimensional Dirac delta function. The above limit may be approached in different ways as we now discuss.

#### 5.5.1.1 Conservation in the Limit Z → 0

In this limit, 5.34 reduces to the form

$$
S_0(\xi_0, \eta_0, \lambda_1, \lambda_2) = i \cdot \exp[2 \cdot \pi \cdot i \cdot z \cdot (1/\lambda_1 - 1/\lambda_2)] \cdot S_1(\xi_0, \eta_0, \lambda_1, \lambda_2). \qquad (5.36)
$$

By taking the modulus of the right-hand side of this expression, the rapidly varying phase term disappears and we find that

$$
|S_0(\xi_0, \eta_0, \lambda_1, \lambda_2)| = |S_1(\xi_0, \eta_0, \lambda_1, \lambda_2)|, \qquad (5.37)
$$

which indicates that the modulus of the two-point two-wavelength correlation function conserves perfectly in the $z \rightarrow 0$ regime.

#### 5.5.1.2 Conservation in the Limit $(\lambda_2 - \lambda_1) \rightarrow 0$

As the limit $(\lambda_2 - \lambda_1) \rightarrow 0$ is approached, function $|S_0(\xi_0, \eta_0, \lambda_1, \lambda_2)|$ is found to conserve with good accuracy even for extremely large propagation distances. Quantitative insight into the conservation behavior of this function based on actual measurements is given in Chap. 8. We will also discover in that chapter that this conservation behavior provides us with a powerful set of analytical tools that allow atmospheric paths to be characterized in terms of certain readily measurable polychromatic properties of point-object, or unresolved star, images.

At the limit itself $\lambda_2 = \lambda_1$ the two-point two-wavelength correlation function degenerates into the complex coherence factor. This function is found to perfectly conserve no matter how far the waves propagate. This case is of such importance that it will be discussed in more detail in Sect. 5.6.

#### 5.5.1.3 Conservation of the Modulus of the Enclosed Volume

Although function $S_0(\xi_0, \eta_0, \lambda_1, \lambda_2)$ does not generally conserve perfectly during propagation, it can be shown (McKechnie, 1991) that the modulus of the volume contained under $S_0(\xi_0, \eta_0, \lambda_1, \lambda_2)$ does conserve perfectly. Thus, we may write

$$\left| \int_{-\infty}^{\infty} \int_{-\infty}^{\infty} S_0(\xi_0, \eta_0, \lambda_1, \lambda_2) \cdot d\xi_0 \cdot d\eta_0 \right| = \left| \int_{-\infty}^{\infty} \int_{-\infty}^{\infty} S_1(\xi_1, \eta_1, \lambda_1, \lambda_2) \cdot d\xi_1 \cdot d\eta_1 \right|. \tag{5.38}$$

### 5.5.2 *General Case of Non-Conservation of the Function*

In this section, we examine via 5.34 the propagation behavior of the two-point two-wavelength correlation function, $S_1(\xi_1, \eta_1, \lambda_1, \lambda_2)$ for waves scattered by the thin atmospheric layer (Fig. 5.1) in the supposed homogeneous free space below the layer. In fact, we choose to study the behavior of the modulus of the function, $|S_1(\xi_1, \eta_1, \lambda_1, \lambda_2)|$, rather than the function itself to eliminate the rapid phase fluctuations that occur as the waves propagate (cf., 5.34). Though these phase fluctuations are important, eliminating them here avoids unnecessary confusion.

To study the propagation behavior quantitatively, we must first choose an explicit form for $S_1(\xi_1, \eta_1, \lambda_1, \lambda_2)$. Again, without prejudice, we choose the function that arises from the thin atmospheric layer described previously by 5.20 and 5.23:

$$S_1(\xi_1, \eta_1, \lambda_1, \lambda_2) = \exp\left\{-2 \cdot \pi^2 \cdot \sigma_j^2 \cdot \left[\frac{1}{\lambda_1^2} + \frac{1}{\lambda_2^2} - \frac{2 \cdot \exp\left[-\left(\frac{\xi_1^2+\eta_1^2}{w_o^2}\right)\right]}{\lambda_1 \cdot \lambda_2}\right]\right\}. \tag{5.39}$$

The propagation behavior of $|S_1(\xi_1, \eta_1, \lambda_1, \lambda_2)|$ depends critically on the initial state of the function. In the next three subsections, we explore the behavior for three important cases: (1) as a function of wavelength difference, $(\lambda_1 - \lambda_2)$, (2) as a function of the variance of the OPD fluctuation introduced by the thin layer, $\sigma_j^2$, and (3) as a function of the sizes of the turbulence structures contained in the layer as described by the width, $w_o$, of the OPD autocorrelation function, $\rho_j(x, y)$.

#### 5.5.2.1 Propagation Dependence on Wavelength Difference

Wavelength difference $(\lambda_2 - \lambda_1)$ clearly influences the propagation behavior of $S_0(\xi_0, \eta_0, \lambda_1, \lambda_2)$ as can be seen from the plots in Fig. 5.6. To isolate the effect of wavelength difference, all plots in this figure are based on the same turbulence strength value, $\sigma_j = 0.35\ \mu$m, and the same average turbulence structure size value, $w_o = 0.25$ m.

While the $\sigma_j$ and $w_o$ values just indicated have been chosen somewhat arbitrarily for illustration purposes, it is noted here that they are nonetheless representative of the OPD fluctuations for typical nighttime vertical paths through the entire atmosphere

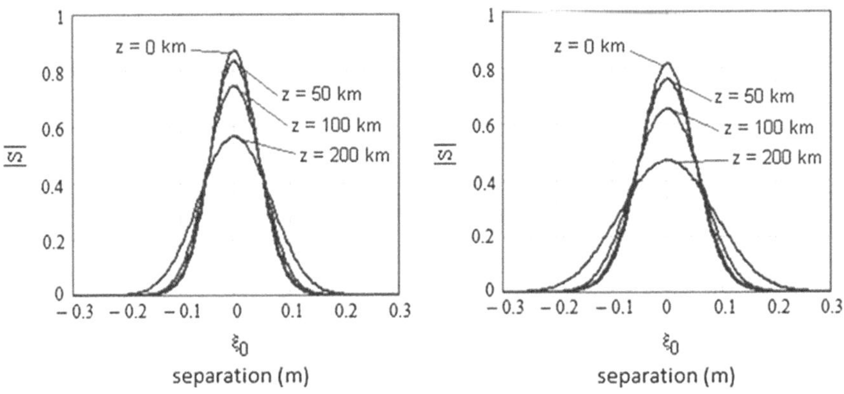


**Fig. 5.6** Central sections showing $|S_1(\xi_1, \eta_1, \lambda_1, \lambda_2)|$ for various propagation distances, $z$, for the wavelengths pairs: *Left* $\lambda_1 = 0.486\ \mu$m and $\lambda_2 = 0.550\ \mu$m (i.e., $\lambda_2 - \lambda_1 = 0.064\ \mu$m); *Right* $\lambda_1 = 0.550\ \mu$m and $\lambda_2 = 0.656\ \mu$m (i.e., $\lambda_2 - \lambda_1 = 0.106\ \mu$m). Other parameters values used in the calculations: $\sigma_j = 0.35\ \mu$m, $w_o = 0.25$ m (plots calculated using 5.34 and 5.39)

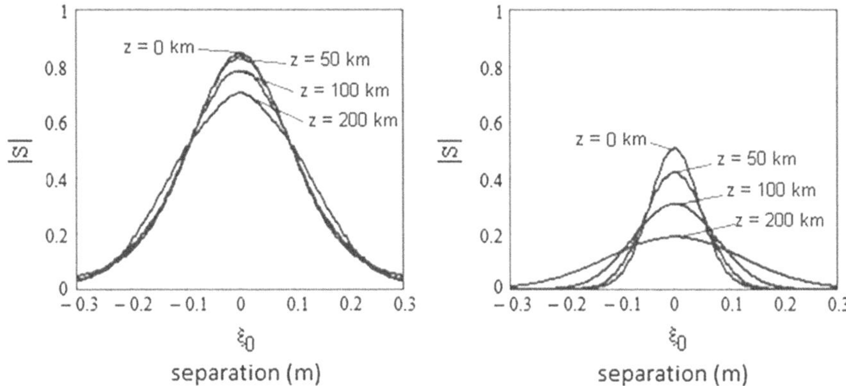


**Fig. 5.7** Central sections showing $|S_0(\xi, 0, \lambda_1, \lambda_2)|$ behavior during propagation for various propagation distances, $z$, for: *Left* relatively weak turbulence ($\sigma_j = 0.175$ μm); and *Right* stronger turbulence ($\sigma_j = 0.35$ μm). Other parameter values used in the calculations: $w_o = 0.25$ m, $\lambda_1 = 0.486$ μm, $\lambda_2 = 0.656$ μm (plots calculated using 5.34 and 5.39)

over wavefront areas of the order of size collected by the Mayall 3.8-m telescope (cf., Chap. 10, Sect. 10.3.5).

#### 5.5.2.2 Propagation Dependence on rms OPD Fluctuation

The rms OPD fluctuation, $\sigma_j$, of the waves as they exit the layer (Fig. 5.1) strongly influences the behavior of $|S_0(\xi_0, \eta_0, \lambda_1, \lambda_2)|$ during propagation. The plot sets in Fig. 5.7 show the propagation behavior for two $\sigma_j$ values, 0.175 and 0.35 μm, for the same average turbulence structure size $w_0 = 0.25m$, and the same two wavelengths, $\lambda_1$ and $\lambda_2$. The plots plainly show that larger $\sigma_j$ values cause more severe modifications of the function $|S_0(\xi_0, \eta_0, \lambda_1, \lambda_2)|$ during propagation.

#### 5.5.2.3 Propagation Dependence on Turbulence Structure Size

Turbulence structure size also exerts strong influence on the propagation behavior of $|S_0(\xi_0, \eta_0, \lambda_1, \lambda_2)|$. The plots in Fig. 5.8 show the behavior for two $w_o$ values, 0.175 and 0.50 μm, corresponding to (relatively) small and large average turbulence structure sizes. To isolate the effect of structure size, all plots in this figure were calculated using the same turbulence strength value, $\sigma_j$, and the same wavelength pair, $\lambda_1$ and $\lambda_2$.

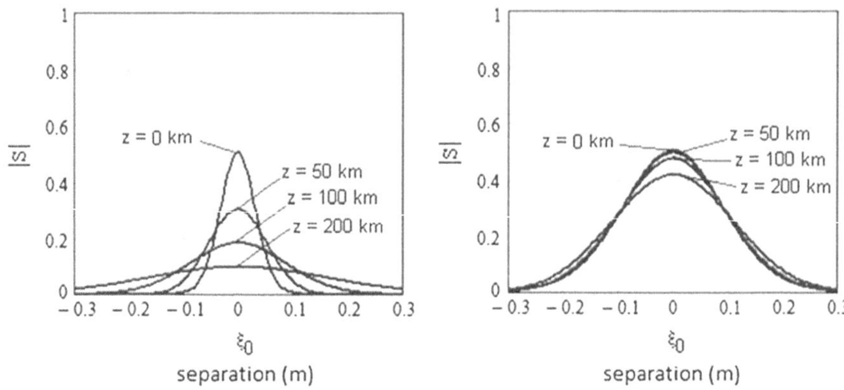


**Fig. 5.8** Central sections showing $|S_0(\xi, 0, \lambda_1, \lambda_2)|$ during propagation for various propagation distances, $z$, after scattering by: *Left* relatively small average turbulence structures ($w_o = 0.175$ m); and *Right* larger average turbulence structures ($w_o = 0.5$ m). Other parameter values: $\sigma_j = 0.35$ μm, $\lambda_1 = 0.486$ μm, and $\lambda_2 = 0.656$ μm (plots calculated using 5.34 and 5.39)

#### 5.5.2.4 The Single-Point Two-Wavelength Correlation Function

Figure 5.9 shows that the function $|S_0(0, 0, \lambda_1, \lambda_2)|$ decays gradually with increasing propagation distance; at limitingly large distances, the function ultimately falls to zero. The case shown ($\sigma_j = 0.35$ μm and $w_o = 0.25$ m) is again representative of the OPD fluctuations for typical nighttime vertical paths through the entire atmosphere over wavefront areas of the order of size collected by the 3.8-m Mayall telescope. The function can be seen to substantially conserve for about the first 20 km, comparable

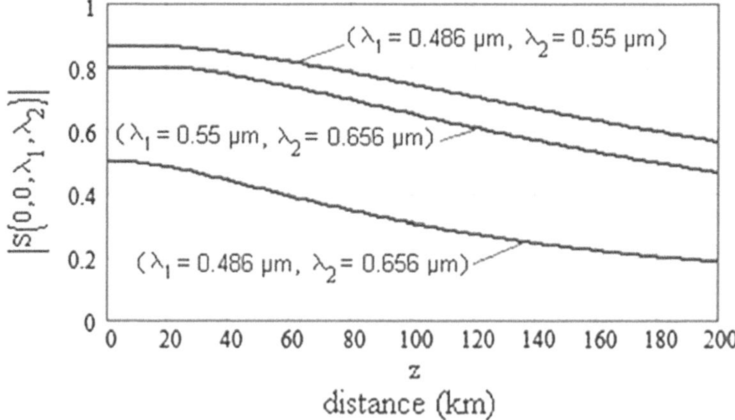


**Fig. 5.9** Plots illustrating propagation behavior of the modulus of the single-point, two-wavelength correlation function $|S_0(0, 0, \lambda_1, \lambda_2)|$ for the three visible wavelength pairs indicated. Other parameter values: $\sigma_j = 0.35$ μm, $w_0 = 0.25$ m (plots calculated using 5.34 and 5.39).

to the distance from the sea level to the top of the atmosphere. The plots shown correspond to the three different visible wavelength pair combinations indicated in the figure.

## 5.6 Near-Field Propagation of the Complex Coherence Factor

In the limit, $\lambda_1 \rightarrow \lambda_2 \rightarrow \lambda$, 5.35 allows 5.34 to reduce to the form

$$S_0(\xi_0, \eta_0, \lambda, \lambda) = S_1(\xi_0, \eta_0, \lambda, \lambda). \tag{5.40}$$

In this limit, $S_0(\xi_0, \eta_0, \lambda, \lambda)$ degenerates into the complex coherence factor, so that the above equation may be written in the alternative form,

$$M_0(\xi_0, \eta_0, \lambda) = M_1(\xi_0, \eta_0, \lambda), \tag{5.41}$$

Equation 5.41 indicates that, for scattered light waves of infinite lateral extent propagating in the Fresnel near-field region of a homogeneous medium, the complex coherence factor conserves perfectly during propagation. Since the Fresnel near-field region for infinitely extensive light waves is itself infinitely extensive, 5.41 indicates that the complex coherence factor conserves perfectly no matter how large the propagation distance. This important result was first established for radio waves by Ratcliffe (1956).[4]

## 5.7 Development of Scintillation After Light Scattering by a Thin Layer

In this section, we explore how scintillation develops in initially plane, infinitely extensive light waves after scattering by a thin atmospheric layer of the kind depicted in Fig. 5.1. Our objective in this exercise is to gain an overall understanding of the dependence of scintillation on the various controlling parameters. We continue to assume that the space below the layer is homogeneous free space. The light waves arriving at the layer are again considered as having originated from a distant unresolved star and, as before, we assume that the refractive index distribution in the layer varies randomly and continuously. To simplify the analysis in this section, we choose to ignore refractive index dispersion.

Although we describe the waves as "infinitely extensive," for most practical atmospheric propagation applications, the waves merely have to extend over distances no

[4] J. A. Ratcliffe was the research adviser at Cambridge University to R. N. Bracewell.

larger than about one hundred meters to justify this description. We also consider the scattering layer to be located at the top of the atmosphere, thus affording the scattered waves maximum opportunity to develop scintillation by the time they reach the ground below. For most practical purposes, the "top of the atmosphere" may be considered to lie at an altitude of about 20 km; from previous Fig. 3.3, we saw that less than 5% of the atmospheric mass lies above this altitude. Because the plane waves arriving at the layer are assumed to have infinite lateral extent, the entire atmospheric volume below the layer may be considered as a near-field propagation region, thus justifying our use of the Fresnel near-field diffraction formula (4.6) to propagate the light waves in the homogenous space below the layer.

Whereas we can be confident that we are dealing with a near-field diffraction problem here (i.e., $F \geq 1$), in Chap. 6 when we propagate light waves over extended atmospheric paths, we shall discover that it is also important to know the extent of the geometrical optics region—or regions as it turns out—within which we can make rectilinear wave propagation approximations. A geometrical optics region may be considered as a subset of the near-field region, in which the more stringent requirement, $F \gg 1$, is imposed. As we shall see shortly, the depth of the geometrical optics region depends on the size of the turbulence structures contained in the layer rather than on the (much larger) lateral extent of the layer itself.

### *5.7.1 Dependence of Scintillation on Turbulence Scale Sizes in the Layer*

In principle, the OPD fluctuation introduced by an atmospheric layer (Fig. 5.1) may be Fourier decomposed into its constituent sinusoidal spatial frequency components, where each component relates to the turbulence structure of a specific size. For the present investigation, however, it makes better sense to deal in terms of the spatial period of the OPD fluctuation rather than the spatial frequency, since the size of any given turbulence structure contained in the layer directly relates to this period. (The spatial period is of course simply the inverse of the spatial frequency.) Thus, for the remainder of Sect. 5.7, we choose to represent turbulence structure of a particular size by the spatial period, $L$, corresponding to that size. While we are interested in turbulence structures of all sizes, from a few millimeters to tens of kilometers, we shall find that the most significant range in regard to the development of scintillation is the range set by $0m \leq L \leq 100m$.

To quantify scintillation strength as a function of turbulence structure size, we first consider the scattering layer to be composed of a sinusoidal turbulence structure with spatial period $L$. By calculating scintillation strength on the ground below the layer for various sampled $L$ values, we can generate plots of scintillation strength versus turbulence structure size. To baseline these plots so that they provide meaningful comparisons between the scintillation strengths corresponding to the different sampled $L$ values, we scale the OPD fluctuation amplitude for each sampled $L$ value so that the diffracted waves in the homogeneous space below the layer are all scattered through the same nominal fixed angle, which we denote by *B*. To avoid unnecessary complication, only a one-dimensional analysis is given.

The scatter angle of the waves emerging from the layer, *B*, can readily be made independent of the spatial period, *L*, by expressing the sinusoidal OPD fluctuation introduced by the layer, $h_j(x)$, by

$$h_j(x) = \frac{B \cdot L}{4 \cdot \pi} \cdot \sin\left(\frac{2 \cdot \pi \cdot x}{L}\right), \tag{5.42}$$

where the amplitude of the sinusoid is given by $B \cdot L/(4 \cdot \pi)$ and $L$ is the spatial period of the sinusoid.

The OPD fluctuation, $h_j(x)$, expressed by 5.42, may also be regarded as representing the sinusoidal wavefront undulations of an initially plane wave immediately after the wave emerges from the layer (cf., Fig. 5.1). The property of 5.42 that allows it to express $h_j(x)$ in a way that results in the scatter angle being independent of *L* may be appreciated by differentiating both sides of 5.42 with respect to $x$, thus obtaining

$$\frac{dh_j(x)}{dx} = \frac{B}{2} \cdot \cos\left(\frac{2 \cdot \pi \cdot x}{L}\right). \tag{5.43}$$

Irrespective of the value chosen for *L*, it is clear from this expression that the wavefront gradients are all constrained to lie in the same interval, $-B/2$ to $B/2$, where the full range of wavefront gradients is simply angle *B*. In the geometrical optics limit, wavefronts generated from 5.42 are all scattered through the same fixed angle *B*. This property is illustrated by the wavefront plots shown in Fig. 5.10 for two different *L* values.

The complex amplitude of the scattered waves as they emerge from the layer can be written in the dimensionless form,

$$U_0(x, \lambda) = \exp\left[\frac{2 \cdot \pi \cdot i}{\lambda} \cdot \frac{B \cdot L}{4 \cdot \pi} \cdot \sin\left(\frac{2 \cdot \pi \cdot x}{L}\right)\right] = \exp\left[\frac{B \cdot L \cdot i}{2 \cdot \lambda} \cdot \sin\left(\frac{2 \cdot \pi \cdot x}{L}\right)\right]. \tag{5.44}$$

As the waves propagate downward in the homogeneous near-field region below the layer, the complex amplitudes in this region may be calculated using the Fresnel diffraction formula (c.f., 4.6). Thus

$$U(x, \lambda, Z) = \int_{-\infty}^{\infty} U_0(x', \lambda) \cdot \exp\left(\frac{\pi \cdot i}{\lambda \cdot Z} \cdot x'^2\right) \cdot \exp\left(\frac{-2 \cdot \pi \cdot i}{\lambda \cdot Z} \cdot x \cdot x'\right) \cdot dx', \tag{5.45}$$

where *Z* may be considered as the propagation distance measured from the layer to the ground below.

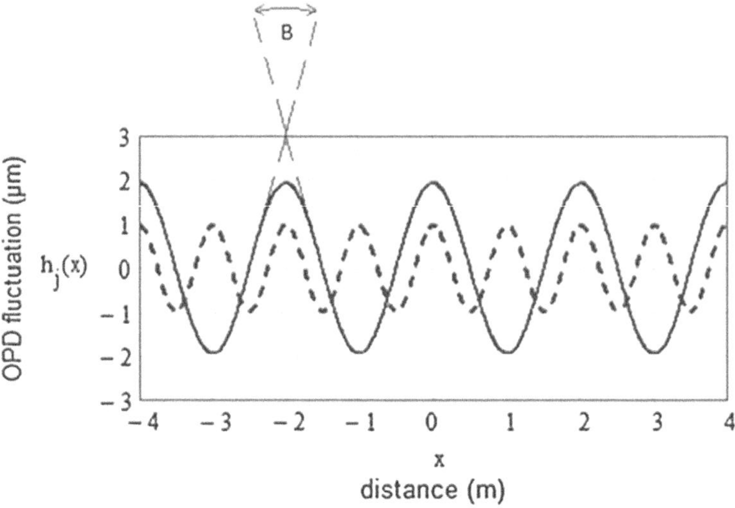


**Fig. 5.10** Plots of the sinusoidal OPD fluctuation introduced by a scattering layer. Because both plots were generated using 5.42, the scatter angle, $B$, is independent of the sinusoidal period, $L$. Parameter values used for the plots: $B = 2.5$ arcsec, and $L = 1$ m and 2 m

The intensity fluctuation—i.e., the scintillation—that develops in the propagating waves is given (in dimensionless form) by the squared modulus of the right-hand side of 5.45. By using 5.44 to express the initial complex amplitude, $U_0(x)$, the intensity fluctuation in the space below the layer can be calculated from the equation,

$$I(x, \lambda, Z) = \left| \int_{-\infty}^{\infty} \exp\left[\frac{B \cdot L \cdot i}{2 \cdot \lambda} \cdot \sin\left(\frac{2 \cdot \pi \cdot x'}{L}\right)\right] \cdot \exp\left(\frac{\pi \cdot i}{\lambda \cdot Z} \cdot x'^2\right) \cdot \exp\left(\frac{-2 \cdot \pi \cdot i}{\lambda \cdot Z} \cdot x \cdot x'\right) \cdot dx' \right|^2. \tag{5.46}$$

Plainly, the complex amplitude and intensity that arise in near-field planes below the scattering layer must display a fundamental cyclical fluctuation with spatial period, $L$. Consequently, a full description of the scintillation characteristics of the scattered wave may be obtained by limiting inspection of these characteristics to within the extent of a single spatial period. By searching for the maximum and minimum intensities that arise over such a period, and by denoting these by $I_{max}$ and $I_{min}$, scintillation strength may be quantified in terms of the intensity modulation factor, MF, or visibility, which we define here in the standard way (Born & Wolf, 2003) by

$$MF = \frac{I_{max} - I_{min}}{I_{max} + I_{min}}. \tag{5.47}$$

With the above definition for MF, it is clear that $0 \leq MF \leq 1$. It is also clear that the extreme values attained by MF, 0 and 1, correspond, respectively, to zero scintillation, where $I_{Min} = I_{Max}$ and fully modulated scintillation where $I_{Min} \ll I_{Max}$.

### 5.7.2 Dependence of Scintillation on the Various Controlling Parameters

In this section, we examine how scintillation strength on the ground below the layer (as expressed by MF) depends on the various controlling parameters, these being: turbulence structure size, *L*; propagation distance, *Z*; scatter angle, *B*; and wave-length, λ. We are particularly interested in the dependence of scintillation on turbulence structure size as described by the quantity, *L*.

Figures 5.11 and 5.12 show scintillation strength, MF, plotted against turbulence structure size, *L*, where 5.46 and 5.47 were used to calculate the plot data.[5] The plots in Fig. 5.11 show scintillation strength on the ground below the layer for two layer altitudes, Z = 3 km and Z = 10 km, and for the visible and IR wavelengths, 0.5 μm, 5 μm, and 25 μm, in seeing conditions corresponding to the scatter angles, B = 1 arcsec, B = 5 arcsec, and B = 25 arcsec. These seeing conditions correspond, respectively, to typical nighttime astronomical seeing, typical daytime astronomical seeing, and atrociously poor seeing that would rarely occur in practice.

Figure 5.12 shows in more detail how scintillation strength depends on layer altitude, with three layer altitudes considered Z = 3 km, Z = 10 km, and Z = 25 km. The plots in this figure are shown for visible light at wavelength 0.5 μm in seeing conditions corresponding to B = 1 arcsec and B = 5 arcsec. The plots corresponding to *Z* = 25 km give some indication of behavior over longer propagation distances, such as associated with long slanted atmospheric paths.

As demonstrated in Figs. 5.11 and 5.12, scintillation strength, MF, cuts off once the turbulence structure period, *L*, exceeds a certain size. Whereas the cutoff point is largely (but not entirely) independent of wavelength, it shows strong dependence on both the seeing angle, *B*, and the layer altitude, *Z*. The unmistakable message conveyed by the various plots in these figures is that scintillation is overwhelmingly attributable to turbulence structure sizes falling in the relatively small size range, $1\,\text{cm} \leq L \leq 50\,\text{m}$.

The above message may be considered ultra-conservative for a number of reasons: First, layer altitude, *Z*, for some of the plots has been set to the extreme value, 25 km. Second, the scattering strength of the layer, as indicated by the parameter,

[5] To accurately evaluate 5.46, the integral limits must be chosen wide enough to capture all sensible contributions to the complex amplitude on the ground below the scattering layer. Appropriate limits depend on *L*, *Z*, *B*, and λ. They also depend on the issue of stationary phase, discussed previously in Sect. 4.1.4. Appropriate limits can be established on a case-by-case basis by varying the limits and examining convergence.

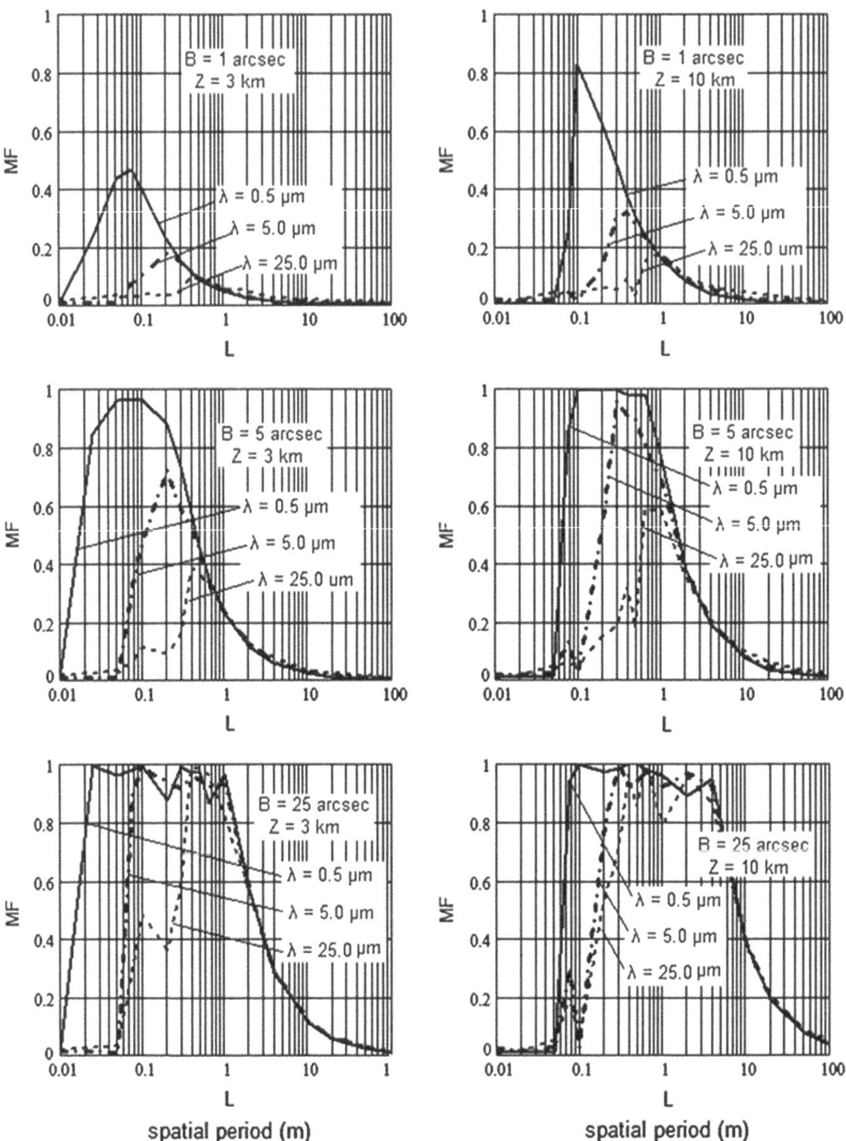


**Fig. 5.11** Scintillation strength on the ground, MF, plotted against the spatial period, *L*, of sinusoidal turbulence structure contained in a thin high-altitude atmospheric layer. The plots are shown for the visible and IR wavelengths indicated, while the layer altitudes correspond to the *Z* values indicated; angle *B* provides a rough measure of visible seeing quality

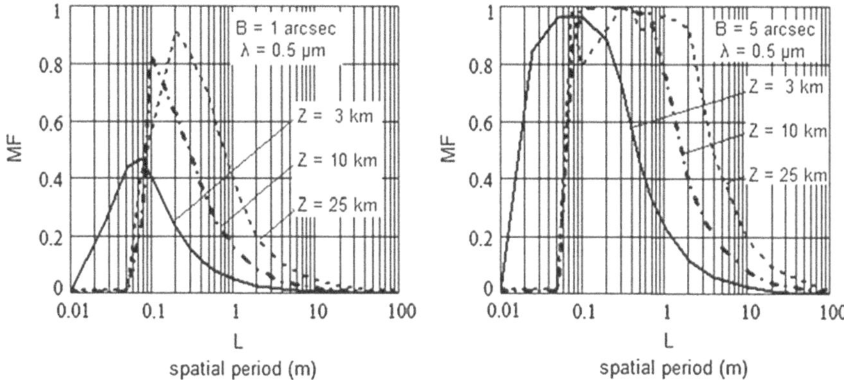


**Fig. 5.12** Scintillation strength on the ground for visible light at 0.5 μm plotted against the spatial period, *L*, of sinusoidal turbulence structure contained in a thin high-altitude atmospheric layer. Layer altitude corresponds to the *Z* values indicated; angle *B* provides a rough measure of visible seeing quality

*B*, has been set in some cases to the extreme value, 25 arcsec. Third, scatter angles normally associated with an entire atmospheric path have been entirely invested in a single (high altitude) scattering layer. Fourth, it is supposed that the entire scatter angle arises from turbulence structure at just a single spatial period component, *L*, whereas in reality, the entire scatter angle arises from the integrated effect of structures corresponding to a broad range of spatial periodicities. Adding to the conservative perspective here, it is noted that actual $C_n^2(z)$ atmospheric turbulence strength distributions (cf., Appendix C, Figure C2) indicate that most of the turbulence strength lies in the first two kilometers above the ground, with only a small fraction lying above the altitude, 5 km.

#### 5.7.2.1 Turbulence Structure Size Range that Effectively Controls Scintillation

When all of these conservative factors are taken into account, it may safely be concluded that actual scintillation strengths that arise from turbulence structure of a given size are always significantly less than the exaggerated amounts indicated in Figs. 5.11 and 5.12. The "exaggeration factor"—particularly in regard to the plots corresponding to Z = 25 km and B = 25 arcsec—no doubt accounts for these plots showing fully saturated scintillation (i.e., MF ≈ 1) in certain spatial period ranges. The same exaggeration factor also indicates that the relatively small amount of scintillation indicated for atmospheric structures larger than about 10 m may, for all practical purposes, be considered negligible.

For the vast majority of atmospheric paths, it can be concluded that the turbulence structure size range that significantly contributes to the development of scintillation is confined to the subrange, $1\text{cm} \leq L \leq 10$ m. It may therefore be concluded that any turbulence structures corresponding to $L > 10$ m, or there-abouts, essentially do not contribute to scintillation.

### 5.7.3 Effective Fresnel Numbers for Atmospheric Paths

Fresnel numbers associated with the diffraction of light waves are generally defined in terms of the size of the aperture that limits the lateral extent of the waves (Sect. 4.1.1). For present purposes, however, it will be helpful to broaden the "Fresnel number" concept to one where we consider the waves limited, not by actual physical apertures, but instead by the size of coherence patches associated with the waves.

After an initially plane wave has been scattered by atmospheric turbulence, the size scale over which both the phase and amplitude retain some measure of coherence may be expressed in terms of the coherence parameter, or Fried parameter, $r_0$. At the arbitrary wavelength, $\lambda$, where the seeing quality at that wavelength is set by angle $B$, $r_0$ is given approximately by

$$r_0 \approx \frac{\lambda}{B}. \tag{5.48}$$

By again making the simplifying assumption that all of the scattering occurs in a single layer at altitude, $Z$, we can define an "effective Fresnel number" for the scattered light waves in the homogeneous space below the layer. Denoting this Fresnel number by $F_{\mathrm{E}}$, we may express it in terms of $r_0$ as follows (cf., 4.1):

$$F_E \approx \frac{r_0^2}{Z \cdot \lambda}. \tag{5.49}$$

Using 5.48 to substitute for $r_0$ in 4.49, we obtain the desired expression for $F_{\mathrm{E}}$,

$$F_E \approx \frac{\lambda}{B^2 \cdot Z}. \tag{5.50}$$

Equation 5.50 captures some—but not all—of the major dependencies of the effective Fresnel numbers associated with atmospheric paths. These numbers increase with increasing wavelength, $\lambda$. They also increase as seeing improves (i.e., as angle $B$ reduces) and reduce in inverse proportion to the layer altitude, $Z$. For any given atmospheric path, the effective Fresnel number given by 5.50 provides us with rough quantitative insight into whether or not scintillation is likely to develop over that path.

For instance, 5.50 could help decide whether or not we might expect to see scintillation if we looked directly by eye (i.e., at wavelength $\lambda \approx 0.55\,\mu$m) at a star lying near the zenith in average nighttime visible seeing conditions (i.e., $B \approx 1.5$arcsec. By making the simplifying assumption that all of the atmospheric turbulence is retained in a single scattering layer, which we (somewhat arbitrarily) assume here to lie at altitude $Z \approx 5$ km, 5.50 generates the effective Fresnel number, $F_E \approx 2$. This number is consistent with near-field propagation (which merely requires $F_E \geq 1$) and therefore indicates that we might expect to see at least some measure of scintillation; the number certainly falls well short of the requirement for zero scintillation, where the entire atmospheric path would have to lie within the geometrical optics region (which requires $F_E \gg 1$).

But now, suppose that an IR viewing scope were used to observe the same star at the mid-IR wavelength, 5 μm, where we assume that the collection diameter of the instrument is about the same as that of the human eye. Inserting the values; $\lambda \approx 5\,\mu\text{m}$, $\text{B} \approx 1.5$ arcsec, and $Z \approx 5$ km into 5.50 gives the effective Fresnel number, $F_\text{E} \approx 20$. This relatively high Fresnel number indicates that the observation path now lies substantially within the geometrical optics region; in this case, we would no longer expect to see any significant amount of scintillation.

The effective Fresnel numbers given by 5.50 clearly offer useful insight into whether or not scintillation is likely to develop over certain types of atmospheric paths. However, because the effective Fresnel numbers given by that equation do not take into account all of the controlling factors, they are not fully definitive; in particular, they tell us nothing about how scintillation depends on turbulence structure size. To account for this crucially important factor, we must approach the problem from a slightly different direction.

#### 5.7.3.1 Effective Fresnel Number that Accounts for Turbulence Structure Size

We saw in Sect. 5.7.2 that, even after propagation over extremely long atmospheric paths, initially plane waves scattered by large (>10 m) turbulence structures do not develop significant amounts of scintillation within the extent of the paths.

Thus, as far as such large-scale structure is concerned, any physically possible atmospheric path may be considered entirely contained within the geometrical optics propagation region, and the effect of such structure on propagating light waves may then be adequately modeled using the simplest of wave propagation principles: rectilinear wave propagation. From these considerations, it would be natural to expect that the effective Fresnel numbers associated with atmospheric paths comprised solely of such large-scale turbulence structure would be consistent with the geometrical optics region condition, $F_\text{E} \gg 1$. We now consider how to express the quantity, $F_\text{E}$, in a way that does indeed reflect this condition.

Figure 5.13 shows a sinusoidal wave with a large spatial period, $L$. It follows from rectilinear wave propagation principles that, to produce the scatter angle, $B$, the

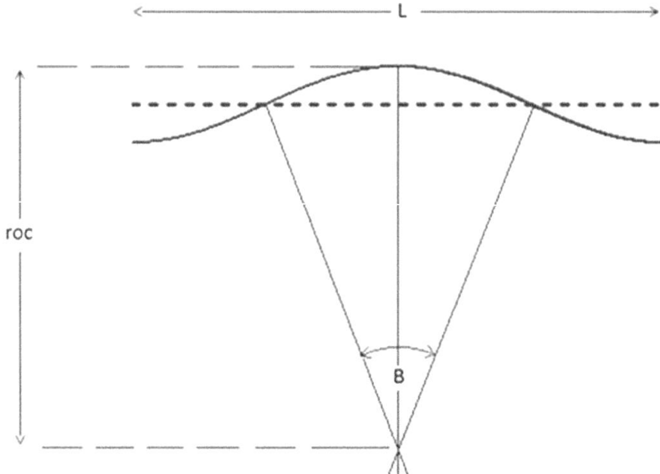


**Fig. 5.13** The radius of curvature in the center of a sinusoidal wave portion, roc, approximately determines where the central wave portion comes to a focus and where we could expect to see strong scintillation. The distance, roc, scales linearly with the turbulence structure period, *L*, and inversely as the wave scatter angle, *B*

radius of curvature of the central portion of the wave, which we denote by roc, must be approximately given by[6]

$$roc \approx \frac{L}{2 \cdot B}. \tag{5.51}$$

If the wave shown in Fig. 5.13 is now allowed to propagate downward, we would expect strong scintillation to develop over any propagation distance, Z, comparable in size to roc (i.e., Z $\approx$ roc). For smaller propagation distances (i.e., Z < roc), milder scintillation would be expected, and, for extremely small propagation distances (i.e., $Z \ll roc$), we would not expect to see any scintillation at all.

Thus, the ratio, roc/Z, may be seen as providing a useful metric for gauging whether or not we deal with the geometrical optics propagation region. By using 5.51, this ratio may be expressed in the form,

$$\frac{roc}{Z} \approx \frac{L}{2 \cdot B \cdot Z}. \tag{5.52}$$

It may be observed here that the dimensionless quantity, roc/Z, provides a propagation path classification scheme similar to that provided by actual Fresnel numbers (cf., Sect. 4.1.1), namely:

[6] By the theorem of intersecting chords, we obtain a slightly different radius of curvature, $roc \approx \pi \cdot L/(8 \cdot B)$, but for the rough analysis given here, there is no need to concern with such a small difference.

- $roc/Z \gg 1$ geometrical optics region (no significant scintillation expected),
- $roc/Z \geq 1$ Fresnel near-field diffraction region (scintillation possible)
- $roc/Z \ll 1$ Fraunhofer far-field diffraction region

Figure 5.14 shows the quantity roc/Z (calculated from 5.52) plotted against turbulence structure size *L*, for representative seeing conditions and over the representative atmospheric path lengths as set by the *B* and *Z* values indicated in the figure. As is clearly evident from these plots, the turbulence structures corresponding to $L > 10$ m produce the outcome, roc/Z $\geq 10$. Because large roc/Z values like this indicate that we are dealing with the geometrical optics region, it may be inferred that turbulence structures larger than about 10 m do not give rise to significant amounts of scintillation. Reassuringly, this outcome is entirely consistent with the scintillation strength data shown in Figs. 5.11 and 5.12.

The lack of wavelength dependence in 5.52 stems from our use of geometrical optics where, in effect, the wavelength is assumed to be smaller than the amplitude of the OPD fluctuation. While the application of geometrical optics principles can sometimes lead to over-simplistic and incorrect outcomes, the use here is apparently justified because the outcome is the same as that of the more rigorous diffraction analysis given in Sects. 5.7.1 and 5.7.2.

Summarizing here, the effective Fresnel number expressed by roc/Z (cf., 5.52) evidently shows a dependence on turbulence structure size (as quantified by *L*) but does not show wavelength dependence. In contrast, the effective Fresnel number defined by 5.50 does not show any dependence on turbulence structure size but does show wavelength dependence. Whereas the effective Fresnel numbers given by 5.50 and 5.52 can both be used to indicate whether or not we deal with the geometrical optics region, individually these quantities can only be considered as partial metrics.

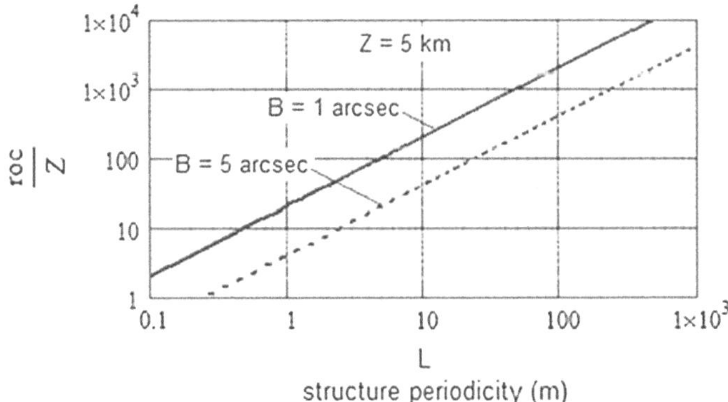


**Fig. 5.14** The quantity roc/Z (cf., 5.52) may be considered as being an effective Fresnel number that helps determine the likelihood of scintillation developing over a propagation path. The plots are shown for representative atmospheric propagation paths. We can see that the condition $roc/Z \gg 1$ is obeyed for all large turbulence structure in the size range, $L > 10$ m, thus indicating that we may treat light wave disruptions caused by such large turbulence structures according to the principles of rectilinear wave propagation

Just as the effective Fresnel number given by 5.50 does not provide specific information about turbulence structure size, the form of 5.49 indicates that neither does the coherence parameter, $r_0$. Thus, we see that the coherence parameter, $r_0$, cannot be regarded as anything more than a partial metric. Further discussion of the limitations of the coherence parameter as an atmospheric path characterization metric can be found in Chap. 12 (Sect. 12.3) and Appendix A.

#### 5.7.3.2 The Geometrical Optics Region for Large-Scale Turbulence Structures

The ultra-conservative assumptions underlying many of the plots in Figs. 5.11 and 5.12 allow us to draw the following important conclusion:

In general, when light waves propagate through the atmosphere—even in the worst possible seeing conditions and even over the longest physically possible atmospheric paths—any turbulence structures that might exist in the path corresponding to a spatial period, $L$, larger than about 10 m simply do not have the ability to create any significant amount of scintillation. Thus, as far as turbulence structures larger than about 10 m are concerned, even the longest atmospheric propagation path may be considered as entirely contained within the geometrical optics region.

This conclusion is further reinforced in Appendix H, where we examine light scattering from individual spherical turbulence structures. The conclusion is considered sufficient justification in this book for treating light propagation through all atmospheric turbulence structures larger than about 10 m according to the simplest of all propagation principles: rectilinear wave propagation. (It might be noted that such treatment is entirely consistent with Hufnagel and Stanley's general expression for the atmospheric MTF, an expression given in terms of OPD path integrals.)

Based on the analysis given in this section (Sect. 5.7) and also on the separate analysis given in Appendix H, it is concluded that the overwhelming majority of scintillation arising from physically realistic atmospheric paths arises from turbulence structure in the size range, 0.01 m–10 m, and in fact by far the largest portion arises from the much smaller subrange, 0.05m–1 m. In Chap. 6, where we propagate light waves over extended atmospheric paths, rectilinear wave propagation principles will indeed be used to deal with any large (>10 m) turbulence structure that might lie in the atmospheric path.[7] For all physically realistic atmospheric paths, the effect of any large (>10 m) turbulence structure can therefore be accurately modeled by simply adding yet another phase screen to the path—one that introduces the appropriate amount of OPD fluctuation into the propagating waves.

---

[7] As mentioned previously in Sects. 2.1 and 3.1.6, in this book, the term "turbulence structure" is used in a slightly broader sense than normal; it is understood to include atmospheric structures of all possible sizes that can affect the atmospheric refractive index field. Thus, the term includes extremely large atmospheric structures, such as constituted by weather systems.

## 5.8 Mathematical Notation Used in This Chapter

The mathematical notation used in this chapter is indicated in Table 5.2.

**Table 5.2** Mathematical notation used in this chapter along with the SI dimensional units of the individual quantities

| Symbol | Quantity | Dimensions |
|---|---|---|
| $\lambda$ | Wavelength | m |
| (x, y) | Cartesian coordinate system in plane perpendicular to the direction of light travel | m |
| $F$ | Fresnel number | "1" |
| $U$ | Complex amplitude | "1" |
| $A$ | Wave amplitude | "1" |
| $\varphi$ | Wave phase | "1" |
| $I$ | Intensity | "1" |
| $M$ | Atmospheric MTF | "1" |
| $S$ | The two-point two-wavelength correlation function of complex amplitudes | "1" |
| $n$ | Refractive index | "1" |
| $N$ | Normalized fluctuating part of refractive index | "1" |
| OPD | Optical path difference | m |
| $h$ | Normalized fluctuating part layer OPD fluctuation | m |
| PDF(h) | Probability density function of layer OPD fluctuation | $m^{-1}$ |
| $\sigma$ | rms OPD fluctuation | m |
| $\rho$ | Autocorrelation function of OPD fluctuation | "1" |
| $L$ | Spatial period of sinusoidal wave | m |
| $B$ | Scatter angle of sinusoidal wave | "1" |
| MF | Intensity modulation factor | "1" |
| $Z$ | Propagation distance from scattering layer | m |
| $r_0$ | Fried coherence parameter | m |
| $F_E$ | Effective Fresnel number | "1" |
| roc | Wavefront radius of curvature | m |

Dimensionless quantities are indicated by "1"

## References

Beckmann, P., & Spizzichino, A. (1963). *The scattering of electromagnetic waves from rough surfaces*. New York: Pergamon.

Born, M., & Wolf, E. (2003). *Principles of optics* (7th ed., revised). Cambridge University Press.

Dainty, J. C. (1984). Laser speckle and related phenomena. In J. C. Dainty (Ed.), *Topics in applied physics* (Vol. 9). Springer.

Fante, R. L. (1978). Multiple-frequency mutual coherence functions for a beam in a random medium. *IEEE Transactions on Antennas and Propagation, AP-26*, 621–623.

Fante, R. L. (1981). Two-position, two-frequency mutual coherence function in turbulence. *Journal of the Optical Society of America, 71*, 1446–1451

Fante, R. L. (1985). In E. Wolf (Ed.), *Progress in optics* (Vol. 22). Amsterdam: North-Holland. Fante, R. L. (1986). Generalized coherence functions for propagation in a random medium. *Journal of the Optical Society of America, A3*, 1326–1327

Fante, R. L. (1986). Generalized coherence functions for propagation in a random medium. *Journal of the Optical Society of America, A3*, 1326–1327

Hoskins, R. F. (1979). *Generalized functions*. Horwood.

Hufnagel, R. E., & Stanley, N. R. (1964). Modulation transfer function associated with image transmission through turbulent media (pp. 52–61).

McKechnie, T. S. (1991). Propagation of the spectral correlation function in a homogeneous medium. *Journal of the Optical Society of America A, 8*, 339–345.

Ratcliffe, J. A. (1956). Some aspects of diffraction theory and their applications to the ionosphere. *Reports on Progress in Physics, 19*, 188–287.

# Chapter 6
# Wave Propagation Over Extended Atmospheric Paths

**Abstract** This chapter examines the behavior of infinitely extensive plane waves as they propagate over extended atmospheric paths. The analysis draws on results from the previous chapter for scattering by thin atmospheric layers. Using a coherency matrix approach for the refractive index field, the crucial result is established that extended atmospheric paths can be represented by finite stacks of mutually uncorrelated random phase screens. By analytically propagating light waves through these stacks, expressions are developed for the atmospheric MTF and the two-point two-wavelength correlation function of the complex amplitudes of the waves emerging from the stacks. The expressions are given in terms of the rms OPD fluctuation over the path, $\sigma$, and the autocorrelation function of that fluctuation, $\rho(\xi, \eta)$. These two measures provide the essential statistical information about turbulence structure in the path. They also provide the essential statistical properties of the complex amplitudes and intensities in star images formed by telescopes observing over the path.

In this chapter, we propagate two initially plane light waves at wavelengths, $\lambda_1$ and $\lambda_2$ over an extended inhomogeneous atmospheric path while keeping track of the key statistical properties of the waves as they become increasingly disrupted. The treatment draws on some of the basic results developed in the previous chapter for scattering by a single thin atmospheric layer. As the waves emerge at the end of the extended atmospheric path, their disrupted state can be sufficiently described for our purposes by two statistical functions of the complex amplitudes: the complex coherence factor, M, and the two-point two-wavelength correlation function, S. Following Hufnagel and Stanley (1964), we shall refer to the complex coherence factor for an atmospheric path, M, as the atmospheric modulation transfer function (MTF) for that path.[1]

[1] In later chapters, when other complex coherence factors arise, for example, the coherence factor that describes the correlation of the complex amplitudes at two different locations in telescope images, we will continue to use the term, complex coherence factor.



T. S. McKechnie, *General Theory of Light Propagation and Imaging Through the Atmosphere*, Progress in Optical Science and Photonics 20,
https://doi.org/10.1007/978-3-030-98828-9_6

General expressions are developed for both M and S. They are considered "general" for several reasons. First, they allow for turbulence structure of all possible types including, but not limited to, Kolmogorov turbulence. Second, they take into account all mechanisms capable of influencing the atmospheric refractive index field, including temperature, pressure, and wavelength dispersion. Third, they allow for turbulence structure of all sizes, from the smallest to the largest conceivable sizes. Fourth, they allow for the possibility of turbulence structure properties varying over the span of the atmospheric path. Fifth, they allow for the possibility of non-isotropic as well as isotropic turbulence structure.

Function M restricts the scattering analysis to monochromatic light, while function S allows us to analyze both monochromatic and polychromatic light scattering behavior. Function M may be regarded as the degenerate case of function S in the limit where the two wavelengths coalesce, i.e., $\lambda_1 \rightarrow \lambda_2 \rightarrow \lambda$.

The expressions developed for M and S take account of both the phase and amplitude fluctuation of the scattered light waves. Amplitude fluctuation directly relates to intensity scintillation, or simply scintillation. The expressions for M and S ultimately derive from just one crucial property of the atmosphere: that its effect on propagating plane light waves is solely determined by the atmospheric refractive index field, n(x, y, z).

A general expression for the atmospheric MTF for extended atmospheric paths was first developed by Hufnagel and Stanley in their (1964) landmark paper; they did not address the more general function, S. Their atmospheric MTF expression (4.6 in their paper) is considered "general" here because it takes account of turbulence structure of all types, including Kolmogorov turbulence. Hufnagel and Stanley indicate that their expression only applies in the limit of small-angle scattering, but give no quantitative indication of what constitutes a "small-angle" in this context.[2] In Sect. 6.2.2.1 we explore this issue in some detail and establish whether typical atmospheric paths do indeed fall into the small-angle scattering category. The analysis deals with both the scattering angles associated with individual turbulence structures in the atmospheric path and the overall scattering angle that develops over the entire path.

As one would hope, the Author's general expression for the atmospheric MTF (developed in Sect. 6.4) turns out to be identical to Hufnagel and Stanley's general expression. But from that point onwards, Hufnagel and Stanley proceed in one direction and the Author in another.

Hufnagel and Stanley assume the validity of Kolmogorov turbulence assumptions and give an expression for the atmospheric MTF for that particular kind of turbulence. In the altogether different approach taken by the Author, no a priori assumptions are made about the atmospheric turbulence properties. While these properties are

[2] Hufnagel and Stanley assume weak scattering by the individual turbulence structures in the atmospheric path (an assumption that is soundly justified) because the largest value of the fluctuating part of the atmospheric refractive index field, N, defined previously by 3.14, is always $\ll 1$. But, as discussed in Appendix H, Hufnagel and Stanley do not address the other equally important matter of the accumulated scatter angle that develops when many such structures are encountered over very long atmospheric paths.

initially treated as unknown, they can be established whenever needed from readily measurable properties of star images.

## 6.1 Atmospheric MTF Expressions Developed by Hufnagel and Stanley

In their 1964 paper, Hufnagel and Stanley essentially develop two expressions for the atmospheric MTF. The first is a general expression that holds true for any type of atmospheric turbulence structure. The second is a version tailored specifically to Kolmogorov turbulence. They obtain their general expression by using as their starting point an approximate version of the Helmholtz equation for light propagating in an inhomogeneous medium (c.f., 3.48) in which the Laplacian operator term, $\partial^2 U/\partial z^2$, is omitted. Thus, the starting-point equation used by Hufnagel and Stanley can be written in the form,

$$\frac{\partial^2 U}{\partial x^2} + \frac{\partial^2 U}{\partial y^2} + \left(\frac{n(x, y, z) \cdot \omega}{c}\right)^2 \cdot U \approx 0, \tag{6.1}$$

where U denotes the complex amplitude of the propagating wave, n(x,y,z) is the atmospheric refractive index field, ω is angular frequency, and c is the speed of light in vacuum space.

An approximate solution of 6.1 can be obtained after first obtaining the function $A_o(x, y, z)$, which was defined previously by 3.50, by solving the following approximate form of that equation,

$$\left(\frac{\partial^2}{\partial x^2} + \frac{\partial^2}{\partial y^2}\right) A_o(x, y, z) + 2 \cdot i \cdot k \cdot \frac{\partial A_o(x, y, z)}{\partial z} + 2 \cdot k^2 \cdot N(x, y, z) \cdot A_o(x, y, z) = 0, \tag{6.2}$$

where we see that the term, $\partial^2 A_o/\partial z^2$ , has been omitted and we recall that N(x,y,z) and k were previously defined by 3.14 and 3.34, respectively, in terms of optical path difference (OPD) line integrals of the type $\int$N(x, y, z) · dz.

Confirmation of the validity of these line integrals as an intermediate step to calculating the atmospheric MTF is provided in Sect. 6.2 where it is rigorously shown that identical OPD line integrals arise from two different, yet closely related atmospheric path models. The first is an ultimately precise model based on an infinitely large number of infinitesimally thin atmospheric layers stacked together. Because the individual layers are infinitesimally thin, each may be considered as behaving like a phase screen. However, phase correlations between at least some of these phase screens (especially near neighbors that share common turbulence structures) make a propagation analysis through such a model extremely difficult.

The second path model derives directly from the first but is based on a finite number of uncorrelated random phase screens. This model can be seen as only an approximation to the first model, albeit a very close approximation, but the mathematics of light propagation through such a model turns out to be much more tractable. As we shall see, the approximations made in establishing the equivalence of the two random phase screen models (particularly in regard to the development of scintillation) can be quantified and, since these approximations likely correspond to Hufnagel and Stanley's approximation, $\partial^2 A_o/\partial z^2 \approx 0$, they appear to affirm its validity.

### 6.1.1 Hufnagel and Stanley's General Expression for the Atmospheric MTF

We begin by reproducing the general expression developed by Hufnagel and Stanley (4.6 in their paper) for the atmospheric MTF, which they denote by M $(p_1, p_2)$:

$$M(p_1, p_2) = \langle A(p_1, z) \cdot A^*(p_2, z)\rangle$$
$$= \left\langle \exp\left\{ i \cdot k \cdot \int_0^z \left[N(p_1, z') - N(p_2, z')\right] \cdot dz' \right\}\right\rangle, \qquad (6.3)$$

where $k = 2 \cdot \pi/\lambda$, and the vectors, $p_1$ and $p_2$, represent two arbitrary points in a Cartesian coordinate plane set up at the end of the path at right angles to the propagation direction. This plane may be considered as the (x, y) pupil plane of the observing telescope, where the z-direction is the propagation direction.

Function N(p, z) was defined previously in Chap. 3 (3.14) as the normalized fluctuating part of the refractive index field, n(p, z); in Hufnagel and Stanley's notation, this function can be written as $N(p, z) = (n(p, z) - \langle n(p, z)\rangle)/\langle n(p, z)\rangle$. The crucially important result given by 6.3 is that the atmospheric MTF can be determined by OPD line integrals over the atmospheric path of the form, $\int_0^z \left[N(p_1, z') - N(p_2, z')\right] \cdot dz'$. By making the reasonable assumption that the values taken by these integrals are Gaussian distributed,[3] Hufnagel and Stanley then obtain the result,

$$M(p_1, p_2) = \exp\left\{ -\frac{1}{2} \cdot k^2 \cdot \int_0^z dz_1 \int_0^z \langle[N(p_1, z_1) - N(p_2, z_1)]\right.$$
$$\left. \cdot[N(p_1, z_2) - N(p_2, z_2)] \cdot dz_2\rangle\right\} \qquad (6.4)$$

They then showed that the integrand in the above equation can be expressed in terms of four quantities of the form

[3] The Gaussian assumption here follows from the central limit theorem. For atmospheric paths of any significant length, it would be difficult to justify any distribution other than Gaussian.

$$D_N = \left\langle \left[ N\left(p_i, z_j\right) - N(p_k, z_l) \right]^2 \right\rangle, \tag{6.5}$$

where i, j, k, and l can each take either of the values, 1 or 2. Hufnagel and Stanley refer to the quantity $D_N$ as "the structure function" of the random refractive index field.

Most researchers in the field of atmospheric propagation, the author included, acknowledge the validity of Hufnagel and Stanley's expression for the atmospheric MTF given above by 6.4. The small-angle scattering assumptions,[4] where for all practical purposes, both $\langle N^2 \rangle$ and $\partial^2 U/\partial z^2$ may be considered to be zero (cf., Appendix H). Otherwise, their expression amounts to a general expression that holds true for any type of turbulence spectrum. It will be shown in Sect. 6.4.3 that the general expression that we develop separately for the atmospheric MTF fully accords with their general expression.

### 6.1.2 *Hufnagel and Stanley's Kolmogorov-Based Expression for the Atmospheric MTF*

To develop their Kolmogorov-based expression for the atmospheric MTF from their general expression, Hufnagel and Stanley make a number of simplifying physical assumptions and mathematical approximations. By examining these in detail in the next three subsections, we discover that some are troublingly fragile, while others are simply unrealistic. These issues, combined with several others identified in Appendix I, lead us to conclude that Hufnagel and Stanley's Kolmogorov-based expression for the atmospheric MTF is best viewed as simply an initial attempt to quantify this important function and that a more general formulation is needed to express it more exactly—of the type developed in this book.

#### 6.1.2.1 Parametric Simplification of the Refractive Index Structure Function

Hufnagel and Stanley contend that the "condition of local isotropy and lateral stationariness for the random process for N implies that $D_N$ depends functionally on the separation distance Δr between the pair of $r = (p, z)$ points with a parametric (slowly varying) dependence on the mean z value" justifies writing $D_N$ in the parametric form given by their 5.6 which, for convenience, we now reproduce:

$$D_N = D_N(\Delta r, \overline{z}), \tag{6.6}$$

[4] On p. 55 of their paper, Hufnagel and Stanley employ the small-angle assumption to justify use of the lateral Laplacian approximation, $\nabla^2 A \approx \partial^2 A/\partial x^2 + \partial^2 A/\partial y^2$, in place of the more exact three-dimensional form, $\nabla^2 A \approx \partial^2 A/\partial x^2 + \partial^2 A/\partial y^2 + \partial^2 A/\partial z^2$.

where Δr and $\overline{z}$ are defined by Hufnagel and Stanley (5.7 and 5.8 in their paper) as follows:

$$\Delta r = \left[ |p_i - p_k|^2 + (z_j - z_l)^2 \right]^{\frac{1}{2}}, \tag{6.7}$$

and

$$\overline{z} = \frac{1}{2}\left(z_j + z_l\right). \tag{6.8}$$

But it is readily apparent that parametric variables, Δr and $\overline{z}$ cannot be treated as independent variables as inferred by 6.6. The parameters, Δr and $\overline{z}$, are related to one other through their separate dependencies on $z_j$ and $z_l$ as indicated by 6.7 and 6.8. To consider Δr and $\overline{z}$ as independent variables is neither mathematically correct nor does it properly reflect the underlying physics. Adding to the problems here, the quantity $\overline{z}$ defined by 6.8 can only be considered as having useful physical meaning for closely separated $z_j$ and $z_l$ pairs; for larger separations, the quantity loses useful meaning.

While the 6.6 representation of $D_N$ results in huge simplifications in the subsequent mathematics, they come at considerable cost. By adopting the 6.6 parametric form, Hufnagel and Stanley effectively restrict consideration to refractive index correlations over limited local regions of the atmospheric path where Δr is small compared to the overall path length, z.[5]

To justify 6.6, the sizes of the largest atmospheric turbulence structures capable of affecting the refractive index are obliged to be significantly smaller than the path length itself.[6] Thus, Kolmogorov formulations are obliged to assume that atmospheric turbulence structure has an "outer scale limit," $L_0$. The size of $L_0$ is not firmly specified and little is known about how the refractive index structure function rolls off in the vicinity of this limit. Some might regard $L_0$ as simply a fudge factor.

One can readily argue that, in general, there is no conveniently small outer scale limit. Large thermal plumes rising from hot spots or hot areas[7] on the ground can extend over distances of many hundreds of meters, while weather systems (cyclones and anticyclones) can extend over distances of tens, or even hundreds, of kilometers. When large "atmospheric structures" like these are present—as they routinely are—one would expect to find refractive index correlations over distances of at least the order of the atmospheric depth (~10 km). However, such large structure sizes are

[5] The exact form for $D_N$ provided by Hufnagel and Stanley by their 5.5 plainly indicates a dependence on the degree of correlation between the refractive indices at all location pairs, $(p_i, z_j)$ and $(p_k, z_l)$, As discussed in upcoming Sect. 6.2.1, another way of considering function, $D_N$, is that it takes values determined by the joint probability density function of the refractive index at the two field locations, $(p_i, z_j)$ and $(p_k, z_l)$.

[6] Plainly, the distances over which the refractive indices in the atmospheric path show significant. correlations are related to the sizes of the turbulence structures in the path.

[7] Hot or cold areas on the ground can be small or large. They are caused by non-uniformity of the absorption/reflection characteristics of the ground and also the local ground inclination to the Sun.

inconsistent with the Kolmogorov requirement (Sect. 6.1.2.1) for an outer scale limit small enough to justify treating $\Delta r$ and $\overline{z}$ as independent parameters.

Thus, we see that the basic foundations of Kolmogorov theory preclude the possibility of extremely large "turbulence structure" features in the atmosphere for which generating mechanisms are known to exist. The temperature and pressure gradients associated with weather systems can cause refractive index fluctuations over the extent of these systems of about $\Delta n/(n-1) \approx 0.05$ (cf., Sect. 3.1.6). For weather systems measuring hundreds of kilometers across, refractive index changes of this magnitude can cause measurable angular shifts in telescope images. Though such shifts rarely exceed 0.1 arcsec, they are nonetheless significant; their time constants can be of the order of hours or days.

Second, and higher, derivatives of the refractive index variations associated with weather systems and thermal plumes can also give rise to significant phase corrugations, with lateral sizes that can extend over hundreds of meters or even many tens of kilometers. With the new generation of extremely large telescopes (ELTs) now under construction whose diameters range in size from 25 m to 39 m, and with the baselines of future telescope interferometer arrays likely to grow well beyond the 90-m baseline of the Keck I and II telescope-pair, it will become increasingly important to account for the phase corrugations that arise from these very large atmospheric systems.

#### 6.1.2.2 Consequences of the Parametric Simplification

The final expression for the atmospheric MTF given by Hufnagel and Stanley for Kolmogorov turbulence (via their 7.2 and 7.6) can be written the form

$$M(p) = \exp\left\{ -\frac{1}{2} \cdot k^2 \cdot 2.91 \cdot p^{\frac{5}{3}} \cdot \int_0^z C_n^2(z') \cdot dz' \right\}, \tag{6.9}$$

where $p = p_1 - p_2$ and $C_n(z')$ is the refractive index structure constant at altitude $z'$. The term, $p^{5/3}$, represents the wavefront structure function for Kolmogorov turbulence. (The wavefront structure function arises as the integral of the refractive index structure function, which is postulated to be $p^{2/3}$ for Kolmogorov turbulence.)

By separating the parameters, $p$ and $z'$ as in 6.9, Hufnagel and Stanley in effect consider that the OPD contributions to the atmospheric MTF from widely separated z-locations in the atmospheric path are uncorrelated and thus may be added in quadrature. This accounts for the term, $p^{5/3} \cdot \int_0^z C_n^2(z') \cdot dz'$, in the exponent; this term has the dimensions of Length$^2$, the same dimensions as the variance of the integrated OPD fluctuation. But now let us consider a very short horizontal path. By assuming that $C_n$ remains constant over the entire length of the path, Hufnagel and Stanley give the atmospheric MTF for such a path by their 7.3 which, for convenience, we reproduce here in the form,

$$M(p) = \exp\left\{-\frac{1}{2} \cdot k^2 \cdot 2.91 \cdot p^{\frac{5}{3}} \cdot L \cdot C_n^2\right\}, \tag{6.10}$$

where L is the path length.

Examination of the exponent in this equation indicates that, as the length of the path increases, the width[8] of the atmospheric MTF reduces in proportion to $L^{-3/5}$ Because the average image envelope formed by large telescopes is determined by the Fourier transform of the atmospheric MTF and because the respective widths of any given function and its Fourier transform are inversely related (see Chap. 7, Sect. 7.7), we might expect the angular width of this envelop[9] to increase roughly in proportion to $L^{3/5}$.

However, let us now consider a path so limitingly short that the refractive indices at the two ends of the path are almost perfectly correlated. Naturally, the length of such a path would have to be smaller than the size of the smallest turbulence structure in the path, a restriction that limits us to barely useful path lengths. However, this extreme case allows us to identify another problem faced by Kolmogorov formulations. Later, by relaxing the path length restriction a little, we show that these formulations continue to suffer from the same problem even when longer, more practically meaningful path lengths are considered.

When the refractive indices are highly correlated over the entire length of the path, we might expect the width of the atmospheric MTF for such paths to reduce in proportion to $L^{-1}$. At the same time, we might expect the average image size to grow in proportion to L (i.e., inversely as the width of the atmospheric MTF). Dimensional considerations show that the atmospheric MTF in this case arises in the form,

$$M(p) \approx \exp\left\{-\frac{c_o}{2} \cdot k^2 \cdot 2.91 \cdot (p \cdot L \cdot C_n)^{\frac{6}{5}}\right\}, \tag{6.11}$$

where $c_o$ is a dimensionless constant whose precise value need not concern us here. Equation 6.11 provides us with an intuitively correct expression for the atmospheric MTF for very short atmospheric paths where the structure constant, Cn, remains constant over the entire length of the path. However, evidently this expression is at odds with the expression actually given by Hufnagel and Stanley (cf., 6.10). Thus, we conclude that Hufnagel and Stanley's expression does not correctly describe behavior for short atmospheric paths.

In the more practically useful case where the paths are still short, but not limitingly short—say with a length of between 10 m and 100 m—it would be unrealistic to expect the refractive indices at the two ends of the path to be fully correlated. However, one might reasonably expect that the refractive indices at the two ends of the path would be at least partially correlated, the correlation attributable to any

---

[8] The width of the atmospheric MTF may be expressed by giving the p value at which M(p) takes some suitable prescribed value, such as 0.5.

[9] To a first approximation, the width of the Fourier transform of a function is inversely related to the width of the function itself.

large turbulence structures in the path whose size is comparable to the length of the path. For such paths, a hybrid formulation is needed to properly account for the atmospheric MTF, a formulation that would comprise some appropriate blend of the terms seen on the right-hand sides of 6.10 and 6.11.[10]

But now let us again revisit the case of very long atmospheric paths. In particular, let us consider an entire vertical path through the atmosphere, where it is supposed that extremely large atmospheric structures are present in the path, perhaps of the type associated with weather systems that can extend laterally over distances comparable to the atmospheric depth. The central assumption in Kolmogorov formulations—that the outer scale turbulence limit, $L_0$, is significantly smaller than the path length—prevents these formulations from taking any account of the effects of such large-scale atmospheric structures. Thus, we conclude that the expression for the atmospheric MTF given by Hufnagel and Stanley (6.9) only provides partial accounting. The expression does not properly account for short atmospheric paths and nor does it account for the effects of very large atmospheric structures that might be present in longer atmospheric paths.[11]

#### 6.1.2.3 Disregard of Pressure- and Moisture-Level Fluctuations in the Atmosphere

Hufnagel and Stanley maintain in Sect. 6.6 of their paper that only the temperature fluctuation associated with atmospheric turbulence has any significant capacity to alter the refractive index, and thus, they consider the effects of all other mechanisms, such as random pressure- and moisture-level fluctuations, as negligible. According to Hufnagel and Stanley, the refractive index structure function, $D_N$, is entirely determined by the temperature structure function, $D_\vartheta$. For convenience, we now reproduce the relationship they give between $D_N$ and $D_\vartheta$ (6.2 in their paper):

$$D_N(\Delta r, \bar{z}) \approx 10^{-12} \cdot \left[\frac{\rho(\bar{z})}{\rho_o}\right]^2 \cdot D_\vartheta(\Delta r, \bar{z}), \tag{6.12}$$

[10] The problem described here is analogous to the problem that arises in imaging theory when we only consider imaging for the case of incoherent illumination while entirely ignoring the no less important cases where the illumination is either partially or fully coherent.

[11] One reviewer expressed justifiable concern that the analysis used to develop general expressions for functions, M and S, did not seem to take account of correlations between the refractive indices at widely separated locations in the atmospheric path. Consequently, the analysis was expanded to its present form to clearly demonstrate (Sect. 6.2) that correlations over all possible distance scales are indeed taken into account. Ironically, though Kolmogorov-based formulations suffer from the same deficiency identified by the reviewer, Kolmogorov theory has not been held to account on this issue.

where $\Delta r$ and $\overline{z}$ were previously defined by 6.7 and 6.8, and $\rho(\overline{z})$ denotes the (deterministic) air pressure at altitude $\overline{z}$ and $\rho_o$ denotes the air pressure at sea level (i.e., where $\overline{z} = 0$).[12]

According to the theory postulated by Kolmogorov and others for isotropic turbulence, the temperature structure function can be written in the form (corresponding to 6.3 in Hufnagel and Stanley's paper),

$$D_N(\Delta r, \overline{z}) = C_{\vartheta}^2(\overline{z}) \cdot \Delta r^{\frac{2}{3}}, \tag{6.13}$$

where $C_{\vartheta}$ is the temperature structure constant. Hufnagel and Stanley express the postulated relationship between the refractive index structure constants, $C_n$, and the temperature structure constants, $C_{\vartheta}$, in the form,

$$C_n(\overline{z}) \approx 10^{-6} \cdot \left(\frac{\rho(\overline{z})}{\rho_o}\right) \cdot C_{\vartheta}(\overline{z}). \tag{6.14}$$

By combining 6.12, 6.13, and 6.14, the refractive index structure function can be written in the final form given by Hufnagel and Stanley,

$$D_N(\Delta r, \overline{z}) = C_n^2(\overline{z}) \cdot \Delta r^{\frac{2}{3}} \tag{6.15}$$

Because random pressure fluctuations certainly exist in a dynamically turbulent atmosphere, it might have been prudent if Kolmogorov formulations had taken such fluctuations into account. The same also applies to moisture-level fluctuations, which can arise when mixing occurs at boundaries between moist and dry air masses.

Thus, the standard expression for the refractive index structure function used in Kolmogorov formulations (6.15 above) does not take account of a number of mechanisms with a capacity to affect the atmospheric refractive index field. It might be observed here that the atmospheric turbulence measurement procedures described in Chaps. 8 and 13—that go hand-in-hand with the general formulations developed later (Sects. 6.3 and 6.4) for both the atmospheric MTF, M, and the two-point two- wavelength correlation function, S—naturally take into account the effects of temperature-, pressure-, moisture-level fluctuations, and in fact all other mechanisms that might possibly influence the atmospheric refractive index field.

[12] It might be observed here that the dependence expressed by the term, $[\rho(\overline{z})/\rho_o]^2$, in 6.12 takes no account of the temperature variations that occur at different altitudes. As may be seen by referring to Fig. 3.2, temperature typically varies with altitude over the range 217 K – 273 K. Such temperature variations cause ~ 20% variations in the refractive index quantity, (n – 1).

**Table 6.1** Tabulated values of the atmospheric MTF, $M(p)$, for Kolmogorov turbulence over vertical atmospheric paths in 1-arscec visible seeing conditions (where 6.17 was used to calculate the values)

| Atmospheric MTF $M(p)$ | Wavelength ($\lambda$) | | |
|---|---|---|---|
| | 0.5 μm | 2.2 μm | 4.8 μm |
| M(0 cm) | 1.00 | 1.00 | 1.00 |
| M(5 cm) | 0.361 | 0.949 | 0.989 |
| M(10 cm) | $4.0 \times 10^{-2}$ | 0.846 | 0.966 |
| M(20 cm) | $3.5 \times 10^{-5}$ | 0.589 | 0.895 |
| M(50 cm) | $3.0 \times 10^{-21}$ | 0.087 | 0.599 |
| M(100 cm) | $7.5 \times 10^{-66}$ | $4.3 \times 10^{-4}$ | 0.0197 |

#### 6.1.2.4 Exploring the Refractive Index Structure Function at Large Separations

For vertical paths through the atmosphere, the atmospheric MTF can be obtained from 6.9 by setting the upper integral limit, z, to infinity. Thus,

$$M(p) = \exp\left\{ -\frac{2 \cdot \pi^2}{\lambda^2} \cdot 2.91 \cdot p^{\frac{5}{3}} \cdot \int_0^{\infty} C_n^2(z') \cdot dz' \right\}, \tag{6.16}$$

where the term, $p^{5/3}$, corresponds to the wavefront structure function used in Kolmogorov formulations.

In Hufnagel and Stanley's illustrative numerical example (Sect. 6.7 in their paper), where 6-arcsec full-width-half-maximum (FWHM) daytime visible seeing conditions are considered, they indicate that the $C_n^2$ integral in the exponent of 6.16 above takes the value $6 \times 10^{-11}$ cm$^{1/3}$. It can be inferred directly from this that in typical (1-arcsec) nighttime seeing conditions, the $C_n^2$ integral must take the smaller value $3.03 \times 10^{-12}$ cm$^{1/3}$.[13]

Inserting the latter value into 6.16 leads to the following expression for the atmospheric MTF for a vertical path through the atmosphere in 1-arcsec nighttime seeing conditions:

$$M(p) = \exp\left\{ -\frac{p^{\frac{5}{3}}}{\lambda^2} \cdot 1.74 \cdot 10^{-10} \cdot \mathrm{cm}^{\frac{1}{3}} \right\}. \tag{6.17}$$

Tabulated values of $M(p)$ obtained from this relation are given in Table 6.1 for a selection of visible and IR wavelengths. According to Table 6.1, in 1-arcsec seeing conditions at the visible wavelength, 0.5 μm, $M(10\,\mathrm{cm}) = 0.04$ while, for $p > 20$ cm, $M(p) < 3.5 \times 10^{-5}$. Thus, for visible wavelengths, apart from the requirement

[13] According to Kolmogorov theory, if the seeing improves (i. e., FWHM seeing reduces) by a certain factor, the $C_n^2$ integral reduces in proportion to the 5/3-power of that factor. For the six-times improvement considered here, the $C_n^2$ integral reduces by the factor, 65/3 ≈ 19.8.

that function $M(p)$ should fall substantially to zero for $p \geq 20\,\text{cm}$, the precise functionality of $M(p)$ for $p \geq 20\,\text{cm}$ is of little consequence; one can readily conceive of many functionalities other than $p^{5/3}$ that also cause $M(p)$ to fall substantially to zero for argument values in this size range. (We might also note that since the Kolmogorov functionality, $p^{5/3}$, supposedly rolls off in the vicinity of the turbulence outer scale limit, $L_0$, this by itself tells us that functionalities other than $p^{5/3}$ must ultimately develop for large enough p values.)

For IR wavelengths, where the quantity $k = 2 \cdot \pi/\lambda$ takes smaller values than it does for visible wavelengths, function $M(p)$ must take correspondingly larger values for any given $p$ value. According to the table, at the mid-IR wavelength, 4.8 μm, M(20 cm) = 0.197. Thus, if for any reason the wavefront structure function happens to depart from the $p^{5/3}$ functionality used in Kolmogorov formulations, the problem would more likely reveal itself at IR wavelengths where the function can be explored [via M($p$)] out to much larger $p$ values. As discussed in Chap. 8 (Sect. 8.3.2), the most plausible explanation for the existence of image cores at visible and near-IR wavelengths – and Kolmogorov theory's failure to predict these features – is that departures from the supposed 5/3-power law wavefront structure function are often evident at $p$ values of less than one meter.[14]

The difficulties and deficiencies outlined above represent just a sampling of the problems faced by Kolmogorov formulations. In Appendix I, other problems are also identified, including dimensional inconsistencies in the formulations. While some researchers continue to reaffirm their faith in Kolmogorov theory (Tatarski & Zavorotny, 1993), no satisfactory mathematics has yet been presented to explain away any of the above problems.

## 6.2 Layered Model Representations of Extended Atmospheric Paths

In this section, a number of mathematical results and physical concepts are developed that act as foundations in subsequent Sects. 6.3, 6.4 and 6.6, where we propagate light waves over extended atmospheric paths and develop general expressions for the two-point two-wavelength correlation function, S, and the atmospheric MTF, M. Because this foundation material is so crucial, the analysis and discussion given in this section are rigorous and detailed.

[14] It might also be noted that the quantity, $k = 2 \cdot \pi/\lambda$, that arises in the atmospheric MTF, M, is replaced in the two-point two-wavelength correlation function, S (cf., Sect. 6.3), by the more general quantity, $2\pi(1/\lambda_1 - 1/\lambda_2)$. The latter quantity can take relatively small values even at visible wavelengths when the two wavelengths, $\lambda_1$ and $\lambda_2$, are closely separated. Thus, to accurately establish the functionality of the wavefront structure function out to large values of the argument, p, the behavior is best explored by studying either the behavior of the atmospheric MTF at IR wavelengths or the behavior of the two-point two-wavelength correlation function at closely separated wavelength pairs at either visible or IR wavelengths, as convenient.

For readers who would prefer to avoid all of this rigor, they can indeed do this by skipping directly to Sect. 6.3, but only if they are comfortable in accepting the essential results and conclusions developed in this section which may be summarized as follows:

> Any atmospheric path can be precisely modeled by slicing the path into a limitingly large number of infinitesimally thin slices, or layers, each of which may be considered to act as though it were a random phase screen. However, because the OPD fluctuation introduced by at least some of the various phase screens in this model will inevitably show correlations, particularly for closely separated phase screen pairs, a second path model will be developed that closely approximates the first path model, but which comprises only a certain finite number of uncorrelated random phase screens.
>
> In the analysis given, both of these phase screen models are rigorously shown to share two crucially important properties: (1) The sum of the squares of the OPD fluctuations arising from the various individual phase screens in each of the two path models are identical (cf., later 6.24) and (2) the integrated OPD fluctuations arising from each of the two path models are also identical (cf., later 6.25). In addition, it is shown that, in the "small-angle" scattering limit (a limit that is shown to apply even for the longest conceivable atmospheric path), for all practical purposes the two path models produce identical scattering characteristics, in terms of both phase and amplitude (scintillation). Additionally, such negligibly small differences as there might be in the scattering behaviors of the two path models will be shown to relate directly to the "small-angle" approximation made by Hufnagel and Stanley when they omit the term $\partial^2 U)/(\partial z^2)$ from the Helmholtz equation (cf., 6.1).
>
> While it is clear that extended atmospheric paths can be exactly modeled by the first of the two phase screen models, unfortunately the mathematics of light propagation through such a model is not easily tractable. On the other hand, while extended atmospheric paths are seen as only being approximately modeled by the second of the two phase screen models, the approximation will be seen to be extremely accurate. Of no less importance, the mathematics of light propagation through this type of model turns out to be readily tractable. Consequently, the propagation analysis given in Sect. 6.3 and subsequent sections will be given in terms of the second of the two phase screen path models, constituted by a certain finite number of uncorrelated random phase screens.

For those who would prefer to see the detailed justification of the above results and conclusions, they should simply continue reading. Otherwise, they may skip directly to Sect. 6.3.

Plainly, any extended atmospheric path can be modeled by slicing the path—as though it were a loaf of bread—into a stack of thin atmospheric layers, as indicated schematically in Fig. 6.1. By choosing a small enough layer thickness, the effect of individual layers on propagating light waves increasingly approximates that of a random phase screen—that is, a device that introduces OPD fluctuation into traversing light waves without causing any immediate change in the scintillation characteristics of the waves.

By choosing a small enough layer thickness, clearly the atmospheric path can be modeled to any required level of precision as a discrete stack of random phase screens, where each stack member may be considered located in the center of the layer that it represents. It might be observed here that phase screen models of this type can be used to model vertical, horizontal, and slanted atmospheric paths, as

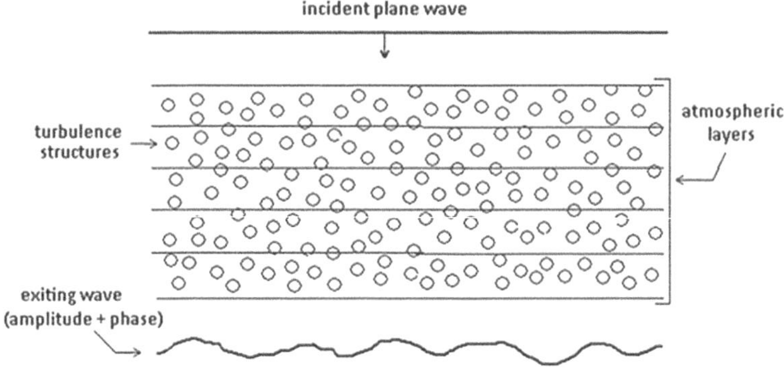


**Fig. 6.1** Layered model of the atmosphere for a vertical atmospheric path. The same type of model also applies to horizontal and slanted paths

required. Modeling of atmospheric paths using uncorrelated random phase screen stacks has been described previously (McKechnie, 1976a, 1976b, 1991b).

But as the layers become limitingly thin, as required to achieve ultimate modeling accuracy, neighboring layers—and perhaps even more widely separated layers as well—increasingly find themselves sharing common turbulence structures. Consequently, the OPD fluctuations introduced by the various random phase screens in the model become increasingly correlated. Plainly, the distances over which correlations arise relate directly to the sizes of the turbulence structures in the path. Correlations between the OPD fluctuations introduced by even one pair of random phase screens in the stack greatly complicate the analysis. However, as we shall discover in the next section (Sect. 6.2.1), such complication can be avoided. In that section, matrix algebra methods (Horn & Johnson, 1985) and (Murdoch, 1957) are used to show that the path model just described can be replaced by a second random phase screen stack model, equivalent to the first in its scattering effect but now comprised of a certain finite number of uncorrelated random phase screens; the latter model turns out to be much more mathematically tractable.

As just indicated, two crucial results will be established in Sect. 6.2.1: (1) It will be rigorously shown that the OPD path integrals arising from either of the two path models are identical, and (2) it will be rigorously shown that the sum of the squares of the OPD fluctuations introduced by the various random phase screens in the two path models are also identical. However, it will also become apparent that there are subtle differences between the two models. Whereas the first model (which in its limiting form comprises an infinite number of infinitesimally thin random phase screens) enables atmospheric paths to be modeled with ultimate precision, the second model (which comprises only a finite number of uncorrelated random phase screens) will be seen as only an approximation—albeit a very good approximation.

The actual approximations used to show that the light scattering effects of the two models are essentially the same are described in detail in Sect. 6.2.2. As just indicated, these approximations are shown to relate directly to the approximation made by Hufnagel and Stanley when they omit the term $\partial^2 U)/\partial z^2$ from the Helmholtz equation (cf., 6.1).

The actual "finite" number of uncorrelated random phase screens in the second of the two path models turns out to be unimportant. We shall find that the task of developing general expressions for the atmospheric MTF, M, and the two-point two-wavelength correlation function, S, is satisfactorily accomplished just as long as the number is large enough (say > 6) to allow invocation of the central limit theorem, a theorem that we use later to allow us to make the hugely simplifying assumption that the integrated OPD fluctuation over the atmospheric path is Gaussian distributed.

Hufnagel and Stanley's general expression for the atmospheric MTF essentially depends on the OPD line integrals obtained over the entire length of the atmospheric path. Because the OPD path integrals arising from either one of our two phase screen path models turn out to be identical, and because the first of these models corresponds to an ultimately precise representation of the path—one that is exactly equivalent to the continuous path considered by Hufnagel and Stanley—it might reasonably be expected that the general expressions for the atmospheric MTF that arise from either of these path models will be identical to the general expression developed by Hufnagel and Stanley (cf., 6.4). As we shall see in due course (Sect. 6.4) this does indeed turn out to be the case.

### 6.2.1 *Two Equivalent Random Phase Screen Atmospheric Path Models*

We begin by considering the first of the two path models, one that comprises an extremely large number, $n_L$, of thin atmospheric layers as depicted schematically in Fig. 6.1. By choosing a small enough layer thickness, clearly the effect of each individual layer on propagating light waves can be approximated to any required level of accuracy by considering the layer as though it were a random phase screen. However, because some layers inevitably share common turbulence structures, we anticipate that the OPD fluctuations introduced by at least a subset of the $n_L$ phase screens will show some level of correlation. As previously mentioned, any such correlations greatly complicate the analysis.

In seeking to avoid such complication, the crucial result will be established in this section that the integrated OPD fluctuations that arise from this extremely precise path model arise equally from a second, mathematically related path model comprised of only a certain finite number of random phase screens, where the OPD fluctuations introduced by the individual phase screens in this model are now all mutually uncorrelated.

For the precise path model consisting of a very large number, $n_L$, of random phase screens, we denote (following the practice adopted in Chap. 5) the OPD fluctuation introduced by the jth phase screen member of the stack by $h'_j(x, y)$, where $j = 1....n_L$ and where without loss of generality the various $h'_j(x, y)$ are defined as zero-mean functions. Also, for notational simplicity in this section, we abbreviate $h_j^{'}(x, y)$ to $h_j^{'}$.

The OPD fluctuations introduced by any arbitrarily chosen pair of these phase screens will generally show at least some degree of correlation. For the jth and kth members of the phase screen stack, the degree of correlation is proportional to the quantity $\langle h_j^{'} \cdot h_k^{'} \rangle$. Correlations are likely to be highest for neighboring phase screens which inevitably share at least some common turbulence structures. However, when larger structures are also present in the atmospheric path, one can also envisage OPD correlations occurring between more widely separated random phase screen pairs.

The OPD fluctuation introduced by each phase screen in the model can be described by an $n_L$—element column matrix $\left[h'\right]$ with elements $h_1^{'}, h_2^{'} ... h_{n_L}^{'}$ as follows:

$$\left[h'\right] = \begin{bmatrix} h'_1 \\ h'_2 \\ \vdots \\ h'_{n_L} \end{bmatrix}. \tag{6.18}$$

Correlations between the OPD fluctuations associated with all the different phase screen combinations can be concisely expressed by the real symmetric matrix $\left[J'\right]$,[15]

$$\left[J'\right] = \left\langle \left[h'\right] \cdot \left[h'\right]^T \right\rangle = \begin{bmatrix} \langle h'_1 \cdot h'_1 \rangle & \langle h'_1 \cdot h'_2 \rangle & \cdots & \langle h'_1 \cdot h'_{n_L} \rangle \\ \langle h'_2 \cdot h'_1 \rangle & \langle h'_2 \cdot h'_2 \rangle & \cdots & \langle h'_2 \cdot h'_{n_L} \rangle \\ \cdots & \cdots & \cdots & \cdots \\ \langle h'_{n_L} \cdot h'_1 \rangle & \langle h'_{n_L} \cdot h'_2 \rangle & \cdots & \langle h'_{n_L} \cdot h'_{n_L} \rangle \end{bmatrix}, \tag{6.19}$$

where the symbol T denotes the transpose operation.

Suppose now that a linear transformation is carried out on the column matrix, $\left[h'\right]$, using the $n_L \times n_L$ transformation matrix $[TM]$, thus yielding a new column matrix $[h]$. Thus, we may write as follows:

$$[h] = [TM] \cdot \left[h'\right], \tag{6.20}$$

In matrix algebra theory (Strang 2009), for a real symmetric matrix $\left[J'\right]$ there exists a unitary transformation matrix $[TM_o]$ which diagonalizes $\left[J'\right]$, a result that allows us to write as follows:

[15] The entries of a symmetric matrix are symmetric with respect to the main diagonal (top left to bottom right). If the various entries are denoted by $a_{j,k}$, then $a_{j,k} = a_{k,j}$.

$$[J] = [TM_o] \cdot [J'] \cdot [TM_o^T] = \begin{bmatrix} \lambda_1 & & & 0 \\ & \lambda_2 & & \\ & & \ddots & \\ 0 & & & \lambda_{n_L} \end{bmatrix}, \tag{6.21}$$

where $[J]$ is the diagonalized form of $[J']$ and $\lambda_1$. $\lambda_2$ ....., $\lambda_{n_L}$ are real non negative eigenvalues of [J′].

Since matrix $[TM_o]$ is orthogonal, its inverse is equal to its transpose, and thus

$$[TM_o^T] \cdot [TM_o] = [TM_o] \cdot [TM_o^T] = \begin{bmatrix} 1 & & & 0 \\ & 1 & & \\ & & \ddots & \\ 0 & & & 1 \end{bmatrix}. \tag{6.22}$$

By applying the transformation matrix $[TM_o]$ to [h′], we can create a new set of field components, $h$, all of which are mutually uncorrelated and again all are defined as zero-mean functions. Because $[TM_o^T] \cdot [TM_0]$ is simply the identity matrix, the sum of the squares of the new field components remains unchanged after the transformation. For the real symmetric matrix with which we deal, $J'$, the sum of the diagonal entries—the trace—is equal to the sum of the eigenvalues. Thus, we can write as follows:

$$\sum_{k=1}^{n_L} \langle h'^2_k \rangle = \sum_{k=1}^{n_L} \langle h^2_k \rangle = \sum_{k=1}^{n_L} \lambda_k. \tag{6.23}$$

If the number of layers, or phase screens, $n_L$, in the first path model is (as supposed) extremely large, many of the eigenvalues simply turn out to be zero.

Therefore, let us now suppose that only n of the eigenvalues are nonzero and that all are distinct. The eigenvectors [h] corresponding to each of these eigenvalues must then be orthogonal. [For a real symmetric matrix, the eigenvectors are always orthogonal (Strang 2009).] The number of nonzero eigenvalues, n, in effect determines the maximum number of uncorrelated random phase screens that can be used to model the atmospheric path.[16]

#### 6.2.1.1 Properties of the Integrated OPD Fluctuation for the Two Path Models

Because all eigenvalues, $\lambda_k$, are zero for k > n, we may write as follows:

[16] To model an atmospheric path in the laboratory, it would make practical sense to use only a few uncorrelated random phase screens, i.e., a number considerably less than n. As we shall see later in Sect. 6.5, if we are only interested in modeling the functions, M and S, a single random phase screen can be sufficient.

$$\sum_{k=1}^{n_L} \langle h'^2_k \rangle = \sum_{k=1}^{n_L} \lambda_k = \sum_{k=1}^{n} \lambda_k = \sum_{k=1}^{n} \langle h^2_k \rangle. \tag{6.24}$$

Equation 6.24 indicates the crucially important result that—regardless of whether the atmospheric path is precisely modeled using a very large number, $n_L$, of correlated phase screens or by the much smaller number, n, of uncorrelated phase screens—the variance of the integrated OPD fluctuation over the entire propagation path arises equally as the sum of the variances associated with the individual random phase screens in either of the two path models.

The effective number of uncorrelated random phase screens used to represent a given atmospheric path, n, has a value roughly given by dividing the length of the path by the average size of the turbulence structures in the path. For vertical atmospheric paths from ground to space, that number could lie somewhere in the approximate range, $10^3$–$10^5$, while for steeply slanted paths even larger values could arise. However, it should be emphasized here that, for the limited purpose in this chapter of establishing general expressions for the atmospheric MTF and the two-point two-wavelength correlation function, the exact number turns out to be inconsequential; whatever the number might be for any given atmospheric path, it merely has to be large enough to justify our later invocation, in Sect. 6.3, of the central limit theorem.

For extremely short paths of perhaps just a few centimeters, n might take a value close to unity. This tells us, in effect, that such a path may be represented by a single random phase screen, an outcome that would of course disqualify us from using the powerful central limit theorem. However, if the distribution of the OPD fluctuation introduced by such a phase screen happened to be approximately Gaussian anyway, there would be no need to invoke this theorem. It might be noted here that, at some point of another, most atmospheric propagation analyses, including that of Hufnagel and Stanley (1964), rely on the assumption that the integrated OPD fluctuation is Gaussian distributed.

Denoting the integrated OPD fluctuation introduced by the entire atmospheric path by H, and noting that the matrix transformation is a linear operation, the following result can also be seen to hold,

$$H = \sum_{k=1}^{n_L} h'_k = \sum_{k=1}^{n} h_k, \tag{6.25}$$

where $h'_k$ refers to the model with the limitingly large number of random phase screens and $h_k$ refers to the model with the finite number of mutually uncorrelated random phase screens.

It is now a suitable moment to abandon the abbreviated notations adopted at the beginning of Sect. 6.2.1 for the OPD fluctuations introduced by the various random phase screens in the two models, $h'_k$ and $h_k$, and re-adopt the full forms, $h_k(x, y)$ and $h'_k(x, y)$. In the limit, $n_L \to \infty$, it is clear that

$$\sum_{k=1}^{n_L} h'_k(x, y) \to \int_0^z N(x, y, z') \cdot dz' \tag{6.26}$$

and hence, by using 6.25, we may write as follows:

$$H(x, y) = \sum_{k=1}^{n_L} h'_k(x, y) = \sum_{k=1}^{n} h_k(x, y) = \int_0^z N(x, y, z') \cdot dz'. \tag{6.27}$$

The OPD path integrals underlying the general expression for the MTF developed by Hufnagel and Stanley first appear in their 4.6 in the form, $\int_v^z N(x, y, z') \cdot dz'$. Since the vector, p, used by these authors corresponds to our Cartesian coordinate point (x, y), Hufnagel and Stanley's OPD line integrals equate directly to the OPD function used in this paper, H(x, y). Thus, 6.27 indicates, in effect, that the integrated OPD fluctuations arising from either of our two random phase screen path models are identical to the integrated OPD fluctuations arising from Hufnagel and Stanley's continuous atmospheric path model.

Noting that the h(x, y) and $h'(x, y)$ are both zero-mean functions, it follows that the mean value of H(x, y) is also zero. Thus, 6.24 allows us to express the variance of the integrated OPD fluctuation, $\langle H(x, y)^2 \rangle$, as follows:

$$\langle H(x, y)^2 \rangle = \sum_{k=1}^{n_L} \langle h'_k(x, y)^2 \rangle = \sum_{k=1}^{n} \langle h_k(x, y)^2 \rangle. \tag{6.28}$$

Denoting the variance of H(x, y) by $\sigma^2$, we can now formally write

$$\sigma^2 = \langle H(x, y)^2 \rangle. \tag{6.29}$$

The unit-normalized autocorrelation function of H(x, y), which we denote by $\rho(\xi, \eta)$, may be written

$$\rho(\xi, \eta) = \frac{\langle H(x + \xi, y + \eta) \cdot H(x, y) \rangle}{\langle H(x, y)^2 \rangle}. \tag{6.30}$$

Using 6.29 and 6.30, the autocovariance function of H(x, y) may be written in the form[17]

$$\langle H(x + \xi, y + \eta) \cdot H(x, y) \rangle = \sigma^2 \cdot \rho(\xi, \eta). \tag{6.31}$$

[17] The variance of the integrated OPD fluctuation, $\sigma^2$, may be regarded as the value taken by the autocovariance function at zero lag.

We have now seen that the function, H(x, y), its variance, $\sigma^2$, and its autocorrelation function, $\rho(\xi, \eta)$, all arise identically from either of our two random phase screen path models. Also, because it turns out that the two crucial functions that primarily concern us—the atmospheric MTF, $M(\xi, \eta, \lambda)$, and the two-point two-wavelength correlation function, $S(\xi, \eta, \lambda_1, \lambda_2)$—are solely determined by $\sigma^2$ and $\rho(\xi, \eta)$ for any given wavelength choices, we could use either one of the two path models to develop general expressions for $M(\xi, \eta, \lambda)$ and $S(\xi, \eta, \lambda_1, \lambda_2)$. However, because the model comprised of the finite number of uncorrelated random phase screens is so much more mathematically tractable, the development of these expressions in Sects. 6.3 and 6.4 will in fact be carried out using that model.

#### 6.2.1.2 Locations of the Random Phase Screens in the Two Path Models

For the path model comprising a limitingly large number of correlated random phase screens, plainly the locations of the individual phase screens in that model correspond exactly to the locations of the thin atmospheric layers represented by these phase screens. However, for the model comprised of a finite number of uncorrelated random phase screens, the OPD contribution made by each individual phase screen is now drawn from an extended path portion whose length naturally relates to the size of the turbulence structures represented by that phase screen. Thus, the locations of the individual phase screens in the second of our two models are now less-precisely defined. While this might at first appear problematic, we will conveniently discover in Sects. 6.3 and 6.4 that the matter is rendered entirely moot by the fact that, as far as functions $M(\xi, \eta, \lambda)$ and $S(\xi, \eta, \lambda_1, \lambda_2)$ are concerned, the precise location of the individual phase screens in either of the two atmospheric path models turns out to be inconsequential.

Since the evaluation of integrals of the type indicated by 6.3 involves only multiplication and addition operations, the associative and commutative properties of these operations [when applied to real-valued operands such as $N(x, y, z')$ allow us to calculate the integral summations using any convenient summation order. This generous property grants us, in effect, the freedom to rearrange the order of the individual random phase screens in our atmospheric path model without compromising the two functions that concern us here, $M(\xi, \eta, \lambda)$ and $S(\xi, \eta, \lambda_1, \lambda_2)$. Thus, as far as these two crucial functions are concerned, the order of arranging the individual phase screens in either of the two equivalent random phase screen path models is inconsequential.

#### 6.2.1.3 Representing an Entire Atmospheric Path by a Single Equivalent Phase Screen

For the limited purpose of calculating functions, $M(\xi, \eta, \lambda)$ and $S(\xi, \eta, \lambda_1, \lambda_2)$, the various random phase screens comprising either of our two atmospheric path models may even be regarded as squashed together into what then becomes, in effect, a single random phase screen. For any given atmospheric path, such a phase screen can be referred to as the equivalent phase screen (EPS) for that path. EPS representations of atmospheric paths are discussed in more detail in Sect. 6.5.

### *6.2.2 Properties of the Phase Screens in the Uncorrelated Random Phase Screen Path Model*

The principle assumption made by Hufnagel and Stanley (1964) in developing their general expression for the atmospheric MTF is that wave propagation behavior can be adequately described by the lateral Laplacian operator; on p. 55 of their paper they write, "Under the assumption of small scattering angles ... the lateral Laplacian is used ...." Thus, they omit the term, $\partial^2 U/\partial z^2$, from their analysis. They also assume, with good justification that $\langle N(x, y, z)^2 \rangle \approx 0$.

The "small scattering angle" assumption made by Hufnagel and Stanley is made in one form or another in all atmospheric propagation analyses known to the author. For the assumption to be valid, the overall scatter angle that builds up over the entire atmospheric path must always remain suitably small; the actual angle is controlled by the combined effects of the following: (1) the scattering angles arising from the individual turbulence structures in the atmospheric path and (2) the overall length of the path.

Infinitely long atmospheric propagation paths are inconsistent with small-angle scattering. Even if only mild levels of turbulence are present in an atmospheric path, it is clear that, if that path is long enough, the overall scatter angle that develops must eventually exceed the small-angle scattering limit. (Scatter angle increases approximately as the square root of the number of randomly distributed scattering structures encountered in the path.) Therefore, for the small-angle scattering assumption to remain valid, restrictions must be imposed, not only on the scattering strengths of the individual scattering structures contained in the path, but also on the overall length of the path itself.

Appendix H examines the scattering angles caused by atmospheric turbulence structures for various representative structure sizes and scattering strengths. As demonstrated quantitatively in that appendix, for typical atmospheric paths the scattering strengths of individual turbulence structures, whether small or large, invariably lie comfortably within the small-angle requirement. However, as far as path lengths are concerned, it turns out that we are not obliged to impose any arbitrary path length restrictions; suitable restrictions are naturally imposed anyway, a consequence of the atmospheric thickness being much smaller than Earth's radius.[18]

---

[18] Air pressure at 15-km altitude is about 10 times less than sea-level air pressure (cf., Fig. 3.3). Therefore, in rough terms, if the effective atmospheric thickness is considered to be about 15 km, the longest physically possible atmospheric path is the path travelled by a beam of light projected from an altitude of 15 km, directed so that it grazes the ground at the horizon and then rises again to 15 km. Noting that Earth's radius is about 6370 km, the length of such a path calculates out at

As noted earlier, the development of expressions for functions, $M(\xi, \eta, \lambda)$ and $S(\xi, \eta, \lambda_1, \lambda_2)$, in Sects. 6.3 and 6.4 is based on the atmospheric path model comprised of a finite stack of uncorrelated random phase screens. It was shown in Sect. 6.2.1 that the integrated OPD fluctuations that arise from this model are identical to those that arise both from the precise path model comprising an infinitely large number of correlated random phase screens and from Hufnagel and Stanley's OPD path integrals. But to be suitably comprehensive for our purposes, the chosen path model must also possess one other crucial property: It must allow for the free development of scintillation, a phenomenon that assuredly occurs over typical atmospheric paths as can be attested by anyone who has ever seen a star twinkle.

Referring back to 6.25, evidently the limitingly large number of OPD contributions, $h_j^{'}(x, y)$, drawn from the first (extremely precise) path model must sum together in groups to generate the finite number of OPD contributions, $h_j(x, y)$, that define the second (uncorrelated random phase screen) path model. To permit such grouping obliges us to make certain approximations. Not surprisingly, these approximations relate closely to the "small-angle scattering" approximations made by Hufnagel and Stanley.

Equation 6.25 is a linear equation expressing the sums of OPD contributions. Because OPD contributions can equally be interpreted as phase contributions (the phase and OPD fluctuations are linked via the light wavelength), this equation might appear to indicate that the entire atmospheric path can be considered contained within the geometrical optics region and thus cannot be reconciled with the development of scintillation. However, in this instance, appearances are deceptive; as we shall see shortly, by making readily justifiable approximations, the linear addition of OPD contributions expressed by 6.25 can be made fully consistent with the development of scintillation.[19]

We now formally lay out our understanding of the properties of atmospheric turbulence structure in relation to propagating light waves and identify how these properties relate to the properties of the individual phase screens in our uncorrelated random phase screen atmospheric path model. At the same time, we identify the crucial approximations that reconcile this model with the issue of scintillation.

1. It is self-evident that the distance spans over which the refractive indices in atmospheric paths exhibit correlations relate directly to the sizes of the random turbulence structures contained in these paths. Thus, for small turbulence structures, the distance spans over which correlations occur are correspondingly small; for larger structures, the spans are correspondingly larger.

about 883 km. For ground-based astronomical telescopes sited at sea level, considerably shorter atmospheric paths arise: the shortest path (through zenith) is about 15 km; for high altitude observing sites, zenith paths can be as short as 10 km.

[19] The same holds true in Hufnagel and Stanley's analysis for the atmospheric MTF where OPD integrals, contrary to first appearances, are also entirely consistent with the development of scintillation.

2. It is assumed that the OPD contributions arising from turbulence structures of significantly different sizes are uncorrelated.[20] Thus, we infer that the distance span over which any individual uncorrelated random phase screen in the atmospheric path model draws its unique OPD contributions is comparable in size to the size of the turbulence structures represented by that phase screen.

   (Note that we must allow for the possibility of an extremely large range of turbulence structure sizes. Some may be smaller than 1 cm, others as large as many kilometers.[21]
3. Twinkling increases as zenith angle increases because of the increased atmospheric path length. To observe strong twinkling, we must generally observe stars lying at lower elevations (cf., Fig. 3.11). Thus, the development of any significant amount of scintillation typically requires large propagation distances—distances of the order of several kilometers. It can be surmised from this that, to effect the smallest perceptible change in the scintillation characteristics of a propagating light wave, requires a distance of at least a few tens of meters. Thus, we surmise that, for typical atmospheric paths, the scintillation characteristics of propagating light waves remain largely invariant over distances of the order of a few tens of meters.
4. For small turbulence structure in the approximate size range, 0 m – 10 m, Item (3) above indicates that the immediate effect of such structure on initially plane waves (even if these waves already exhibit some measure of scintillation as a result of having traversed the earlier portion of the atmospheric path) is simply that of adding an additional phase term to the complex amplitudes associated with these waves. Thus, we surmise that the immediate effect of any given turbulence structure in this relatively small size range may be approximated by an appropriately chosen random phase screen located in the vicinity of the plane closest to that turbulence structure. In effect here, we invoke the principle of rectilinear wave propagation, the simplest of propagation principles, one that generally only applies in the geometrical optics limit.

   Of course, the additional phase introduced by any phase screen representing turbulence structure in the 0–10 m size range must ultimately lead to the development of additional scintillation in downstream path portions than would otherwise have developed in the absence of that phase screen. We might note here that the wave propagation analysis given presently in Sect. 6.3 does indeed allow for the free development of scintillation in all path portions downstream of the various uncorrelated random phase screens in the path model.

---

[20] While we assume that these OPD contributions are uncorrelated, they are not necessarily statistically independent; strong turbulence at one structure size is likely to be matched by correspondingly strong turbulence at other sizes. However, for our purposes, it is sufficient that the OPD contributions are merely uncorrelated.

[21] Clearly, the OPD contribution of an individual random phase screen—one that represents turbulence structures of a certain size range lying in a certain path neighborhood—will be poorly correlated with the OPD contribution made by another random phase screen that draws its contribution from similar-sized turbulence structures in another path neighborhood, separated from the first by a distance significantly greater than the turbulence structure sizes considered.

5. For large turbulence structures lying in the approximate size range, greater than or equal to 10 m (which includes the largest structure sizes that could possibly exist in the atmosphere, some perhaps measuring many kilometers across), as has been demonstrated quantitatively in Chap. 5, Sect. 5.7, and Appendix H, even the longest physically possible atmospheric path does not generally provide sufficient distance for the additional phase contributions arising from these large turbulence structures to develop into significant amounts of scintillation.

   Thus, we see that the effect of large ($\geq$ 10 m) turbulence structures on propagating light waves may again be modeled by simply adding an appropriate random phase screen to the path model. We might observe here that we have again invoked the principle of rectilinear wave propagation.[22] But in this case, there is a significant difference: We can now justifiably conclude that the effect of adding such a phase screen makes no significant difference to the scintillation characteristics that develop over any physically possible atmospheric path. For all practical purposes, we may consider that any scintillation that does develop over the path remains the same whether or not such a phase screen is included in the path.

   Thus, we conclude that the scintillation characteristics that finally develop over any atmospheric path are overwhelmingly determined by the smaller ($\leq$ 10 m) turbulence structures in the path, with by far the largest scintillation contributions arising from structure sizes in the smaller subrange, 0.01–1 m.

The concept that large-scale turbulence structures do not contribute significant amounts of scintillation over extended path lengths is neither original nor unusual. When interferometric testing is carried out over extended test paths, it is generally assumed that the interference fringes are entirely caused by phase differences that accrue over the test path. Usually, no consideration is given to the fact that small amounts of scintillation must ultimately develop from these phase differences in downstream portions of the test path; it is simply assumed that the amount of scintillation that develops from these phase differences is too small to significantly affect fringe appearance.

It might also be noted here that a random phase screen representing the OPD contributions of large turbulence structures ($\geq$ 10 m) could, in principle, be subdivided into any number (small or large, as convenient) of fully correlated secondary phase screens. The only firm requirement for such subdivision is that the OPD contributions from the various secondary screens should exactly add up to the OPD contribution of the parent phase screen. For the limited purpose of calculating functions $M(\xi, \eta, \lambda)$ and $S(\xi, \eta, \lambda_1, \lambda_2)$, there are no strict requirements in regard to the placement of such secondary phase screens in the atmospheric path; they may be distributed in any convenient order.

[22] The rectilinear wave propagation principles that, in effect, have been separately justified here for propagation through both small-scale (<10 m) and large-scale ($\geq$10 m) turbulence structures are entirely consistent with Hufnagel and Stanley's apparent use of the same principles when their general expression for the atmospheric MTF is found to depend only on the OPD line integrals over the propagation path.

The approximation, $\partial^2 U/\partial z^2$, made by Hufnagel and Stanley in developing their general expression for the atmospheric MTF is hardly mathematically transparent. It is far from clear how this approximation affects the quantitative accuracy of their atmospheric MTF expression for typical atmospheric paths. In this regard, since the approximations used to develop our general expression for the atmospheric MTF (Sects. 6.3 and 6.4) are readily quantifiable, they are readily transparent. We might also observe that because the general expression that we develop for the atmospheric MTF turns out to be identical to Hufnagel and Stanley's general expression, the quantifiable approximations that are used in its development (cf., Appendix H) can be seen as equivalent to the approximation made by Hufnagel and Stanley, $\partial^2 U/\partial z^2 \approx 0$.

#### 6.2.2.1 "Small-Angle" Scattering

The understanding of atmospheric scattering laid out in Items (1)–(5) is a consequence of scattering over any physically realistic atmospheric path truly being small-angle scattering. A common attribute of Items (4) and (5) is that the amount of convergence, or divergence, effected by the lens-like behavior of any given turbulence structure is entirely negligible over propagation distances comparable to the size of the structure itself. This attribute enables the following criterion to be established for assessing the accuracy of the "small-angle scattering" approximation as it applies to actual atmospheric propagation paths:

By assuming "small-angle scattering" in atmospheric propagation applications, it is assumed in effect that, as light waves[23] propagate through any given turbulence structure encountered in any physically realistic propagation path, the additional phase introduced into the waves by that structure, irrespective of its size and scattering strength, causes no significant additional amount of scintillation to develop as the waves propagate through the structure itself. Stated another way, it is assumed that the scintillation characteristics of light waves as they emerge from any given turbulence structure are approximately the same as they would have been had that turbulence structure not been present in the path in the first place.

The above criterion may be seen to take account of the dual requirements (1) that all individual turbulence structures in the atmospheric path cause only insignificantly small scatter angles and (2) that the overall scatter angle that develops over the path remains within suitably small limits (say $< 1$ degree) for even the longest physically possible path lengths.

The approximations implicit in the above "small-angle scattering" criterion may be considered as physical analogues of the "small-angle scattering" approximation made by Hufnagel when they omit the term, $\partial^2 U/\partial z^2$, from the wave propagation equation (cf., 6.1).[24] Though Hufnagel and Stanley do not expressly discuss path

---

[23] The light waves considered here are assumed to have originated from initially plane waves that gradually develop increasingly strong phase and amplitude fluctuations as they propagate over the atmospheric path.

[24] Hufnagel and Stanley also omit terms in $N(x, y, z)^2$, but since this approximation is exceptionally accurate, it presents no cause for concern. The quantity, $N(x, y, z)$, was defined previously by 3.14

length restrictions, such restrictions are nonetheless implied by their approximation. As propagation distance, z, increases without limit, the integrated error accruing from the omitted term, $\partial^2 U/\partial z^2$, must also increase without limit. Ultimately, if the path length is long enough, it would no longer be safe to ignore this omission.

### 6.2.3 *Effect of Individual Random Phase Screens on Transmitted Light Waves*

For our purposes, to fully describe the immediate effect of any individual uncorrelated random phase screen in the atmospheric path model on a traversing light wave, it is only necessary to specify the additional OPD contribution introduced by that phase screen.[25] For the arbitrarily chosen jth phase screen, where the OPD contribution is given by $h_j(x, y)$ (cf., 5.2), the complex amplitude associated with an incident plane wave as it emerges from this phase screen may be expressed by

$$U_j(x, y, \lambda) = \exp\left\{2 \cdot \pi \cdot i \cdot \left[\frac{h_j(x, y)}{\lambda}\right]\right\}. \tag{6.32}$$

In the more general case where an initially plane wave has been disrupted (in terms of both phase and amplitude) by earlier phase screens encountered prior to its arrival at the jth phase screen, we use a slightly modified version of the above equation. Denoting the complex amplitude of the wave as it arrives at the jth phase screen by the general function, $B_{j-1}(x, y)$, the complex amplitude of the wave emerging from the phase screen in this case may be expressed in the form,

$$U_j(x, y, \lambda) = B_{j-1}(x, y) \cdot \exp\left\{2 \cdot \pi \cdot i \cdot \left[\frac{h_j(x, y)}{\lambda}\right]\right\}. \tag{6.33}$$

## 6.3 General Expression for the Two-Point Two-Wavelength Correlation Function

In this section, we propagate two initially plane waves at wavelengths, $\lambda_1$ and $\lambda_2$, through the uncorrelated random phase screen atmospheric path model depicted in Fig. 6.2 and, as we do so, develop an expression for the two-point two-wavelength

in terms of the atmospheric refractive index, n(x, y, z). Because $n(x, y, z) - 1 < 0.0003$ at sea level and even smaller at altitude, $N(x, y, z)^2 - 1 < 10^{-7}$ for all values of x, y, and z.

[25] By "immediate effect," we refer only to the additional phase acquired by the wave immediately after it emerges from the random phase screen. Subsequently, of course, in downstream path portions, this additional phase could result in additional amounts of scintillation.

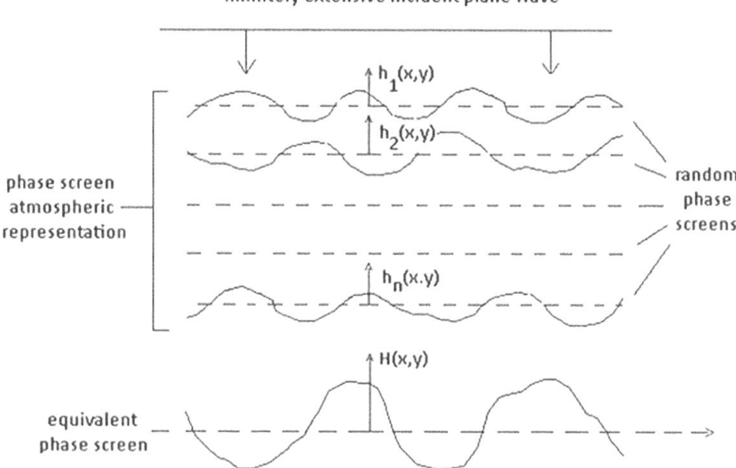


**Fig. 6.2** Sectional view through the multiple phase screen representation of the atmosphere and the equivalent phase screen

correlation function, $S(\xi, \eta, \lambda_1, \lambda_2)$, for the scattered waves that emerge at the end of the path. At each stage in the propagation, we remain mindful of the understanding and approximations set out above in Sect. 6.2.2 (under Items 1–5) in regard to the behavior of the individual uncorrelated random phase screens in the path model.

Once the expression for $S(\xi, \eta, \lambda_1, \lambda_2)$ has been developed in this section, it is a trivial matter in Sect. 6.4 to provide an expression for the atmospheric MTF; such an expression may be obtained by simply coalescing the two wavelengths, $\lambda_1$ and $\lambda_2$, to a single wavelength, $\lambda$, so that function $S(\xi, \eta, \lambda_1, \lambda_2)$ degenerates into function $M(\xi, \eta, \lambda)$.

It might be noted that the analysis given in this chapter, as well as applying to initially plane waves, also applies to light waves that have gentle spherical curvatures. Wavefront curvatures like this arise when we observe less-distant objects, such as satellites in low Earth orbits. It is a simple matter to deal with such curvature; at any time, its effect can be nulled to zero by simply refocusing the telescope. However, for objects that are much closer, such as objects lying within the atmosphere itself, more extreme wavefront curvatures can arise that produce substantial amounts of beam divergence over the atmospheric path itself, amounts that cannot be properly compensated by simply refocusing the telescope; a modified version of the propagation analysis becomes necessary, one that is given later, in Chap. 16.

To avoid overcomplicating the analysis in this section, we ignore wavelength dispersion by considering that the OPD fluctuations introduced by the various random phase screens in the path model are independent of wavelength. Later, in Sect. 6.6, dispersion will be reintroduced and properly taken into account.

Recalling 5.18, once the waves have propagated through the first phase screen depicted in Fig. 6.2, the two-point two-wavelength correlation function for the waves exiting this phase screen may be expressed by the form (cf., 6.32),

$$S_1(\xi, \eta, \lambda_1, \lambda_2) = \left\langle \exp\left\{2 \cdot \pi \cdot i \cdot \left[\frac{h_1(x+\xi, y+\eta)}{\lambda_1} - \frac{h_1(x, y)}{\lambda_2}\right]\right\}\right\rangle, \quad (6.34)$$

where $h_1(x, y)$ is the OPD fluctuation introduced by the phase screen. The spatial arguments of function $S_1$ reflect the fact that this function is spatially stationary, a consequence of the refractive index field, n(x, y, z), having previously been assumed spatially stationary (Sect. 5.2).

We now suppose that the waves emerging from the first phase screen propagate to the second phase screen. Since all of the turbulence strength in the atmospheric path is, in effect, compressed into the various individual phase screens in our atmospheric path model, the spaces between the phase screens may be considered homogeneous free spaces. Plainly, by the time the waves arrive at the second screen, the phase fluctuations imprinted by the first phase screen will have undergone at least some small measure of depletion into scintillation. However, since our only immediate objective here is to discover the form of the two-point two-wavelength correlation function at the instant the waves arrive at the second phase screen, we do not in fact need specific details of the complex amplitudes of the waves as they arrive at this phase screen, and thus, we do not need to know specific details of either the phase or the amplitude fluctuation of the waves as they arrive at this phase screen.

To allow for both phase and amplitude (scintillation) to develop freely as the waves propagate in the space between the first and second phase screens, we proceed by denoting the complex amplitudes of the waves arriving at the second phase screen by what are essentially two unknown functions, $B_1(x, y, \lambda_1)$ and $B_1(x, y, \lambda_2)$. As the waves at each of the two wavelengths emerge from the second phase screen, their complex amplitudes may then be expressed (cf., 6.33) in terms of $B_1(x, y, \lambda_1)$ and $B_1(x, y, \lambda_2)$ by the respective forms,

$$B_1(x, y, \lambda_1) \cdot \exp\left[\frac{2 \cdot \pi \cdot i \cdot h_2(x, y)}{\lambda_1}\right] \quad \text{and} \quad B_1(x, y, \lambda_2) \cdot \exp\left[\frac{2 \cdot \pi \cdot i \cdot h_2(x, y)}{\lambda_2}\right].$$

The two-point two-wavelength correlation function associated with the waves as they emerge from the second phase screen, which we denote by $S_2(\xi, \eta, \lambda_1, \lambda_2)$, may then be written as follows:

$$S_2(\xi, \eta, \lambda_1, \lambda_2) = \left\langle \begin{array}{c} B_1(x+\xi, y+\eta, \lambda_1) \cdot \exp\left[\frac{2 \cdot \pi \cdot i \cdot h_2(x+\xi, y+\eta)}{\lambda_1}\right] \\ \cdot B_1^*(x, y, \lambda_2) \cdot \exp\left[\frac{-2 \cdot \pi \cdot i \cdot h_2(x, y)}{\lambda_2}\right] \end{array} \right\rangle. \quad (6.35)$$

Now, it is known (Papoulis, 1965) that for two uncorrelated random variables, say X and Y, the expected value of the product, $X \cdot Y$, is identical to the product of the expected values of the two individual parts. Thus, we may write as follows:

$$\langle X \cdot Y \rangle = \langle X \rangle \cdot \langle Y \rangle. \tag{6.36}$$

Because any two functions of random uncorrelated variables are themselves random uncorrelated variables, it follows that the following two terms that appear in 6.35,

$$B_1(x+\xi, y+\eta, \lambda_1) \cdot B_1^*(x, y, \lambda_2)$$

and

$$\exp\left\{2 \cdot \pi \cdot i \cdot \left[\frac{h_2(x+\xi, y+\eta)}{\lambda_1} - \frac{h_2(x, y)}{\lambda_2}\right]\right\},$$

must be uncorrelated. (These two terms derive as functions of two separate uncorrelated random phase screens.) This crucial property, taken together with the result given by 6.36, allows the right-hand side of 6.35 to be expressed in the form,

$$\begin{aligned} S_2(\xi, \eta, \lambda_1, \lambda_2) = &\left\langle B_1(x+\xi, y+\eta, \lambda_1) \cdot B_1^*(x, y, \lambda_2)\right\rangle \\ &\cdot \left\langle \exp\left\{2 \cdot \pi \cdot i \cdot \left[\frac{h_2(x+\xi, y+\eta)}{\lambda_1} - \frac{h_2(x, y)}{\lambda_2}\right]\right\}\right\rangle. \end{aligned} \tag{6.37}$$

In Chap. 5, Sect. 5.5.1, it was shown that, for any suitably close pair of wavelengths, $\lambda_1$ and $\lambda_2$, the two-point two-wavelength complex coherence factor associated with infinitely extensive waves, regardless of the magnitude of their phase and amplitude disruptions, approximately conserves during propagation in a homogeneous medium. Therefore, because the interstitial spaces between the random phase screens in our atmospheric path model are considered homogeneous free spaces—for which the two-point two-wavelength correlation function was shown (cf., Sect. 5.5.1.2) to conserve in the limit of small $(\lambda_1 - \lambda_2)$— we may write as follows:

$$\begin{aligned} &\left\langle B_1(x+\xi, y+\eta, \lambda_1) \cdot B_1^*(x, y, \lambda_2)\right\rangle = \\ &\qquad \left\langle \exp\left\{2 \cdot \pi \cdot i \cdot \left[\frac{h_1(x+\xi, y+\eta)}{\lambda_1} - \frac{h_1(x, y)}{\lambda_2}\right]\right\}\right\rangle. \end{aligned} \tag{6.38}$$

It is important to observe here that, while we have no specific knowledge of the complex amplitudes of either of the two waves arriving at the second phase screen, $B_1(x, y, \lambda_1)$ and $B_1(x, y, \lambda_2)$, the above result indicates that we can nonetheless quantify the function $\left\langle B_1(x+\xi, y+\eta, \lambda_1) \cdot B_1^*(x, y, \lambda_2)\right\rangle$ in terms of the OPD fluctuation associated with the waves that emerged from the first random phase screen. By combining 6.37 and 6.38, function $S_2(\xi, \eta, \lambda_1, \lambda_2)$ may now be written in the form,

$$S_2(\xi, \eta, \lambda_1, \lambda_2) = \left\langle \exp\left\{2 \cdot \pi \cdot i \cdot \left[\frac{h_1(x+\xi, y+\eta)}{\lambda_1} - \frac{h_1(x, y)}{\lambda_2}\right]\right\}\right\rangle \cdot \left\langle \exp\left\{2 \cdot \pi \cdot i \cdot \left[\frac{h_2(x+\xi, y+\eta)}{\lambda_1} - \frac{h_2(x, y)}{\lambda_2}\right]\right\}.\right\rangle \quad (6.39)$$

If we continue propagating the light waves in the same manner through the remaining uncorrelated random phase screens in the path model, the final two-point two-wavelength correlation function arises in the form,

$$S(\xi, \eta, \lambda_1, \lambda_2) = \left\langle \exp\left\{2 \cdot \pi \cdot i \cdot \left[\frac{h_1(x+\xi, y+\eta)}{\lambda_1} - \frac{h_1(x, y)}{\lambda_2}\right]\right\}\right\rangle \times \left\langle \exp\left\{2 \cdot \pi \cdot i \cdot \left[\frac{h_2(x+\xi, y+\eta)}{\lambda_1} - \frac{h_2(x, y)}{\lambda_2}\right]\right\}\right\rangle \times \cdots \times \cdots \times \exp\left\{2 \cdot \pi \cdot i \cdot \left[\frac{h_n(x+\xi, y+\eta)}{\lambda_1} - \frac{h_n(x, y)}{\lambda_2}\right]\right\}. \quad (6.40)$$

where we recall that n is the (finite) number of uncorrelated random phase screens in the path model (Fig. 6.2).

We might observe here that, by not imposing any constraints on the properties of the complex amplitude functions of the sort, $B_j(x, y, \lambda_1)$ and $B_j(x, y, \lambda_2)$, our analysis allows scintillation to develop freely as the waves propagate through the model. Despite this, the final form developed for $S(\xi, \eta, \lambda_1, \lambda_2)$ shows no specific dependence on scintillation. Evidently, for purposes of calculating function $S(\xi, \eta, \lambda_1, \lambda_2)$, scintillation does not factor into the analysis.

A similar conclusion was noted by Hufnagel and Stanley in regard to their general expression for the atmospheric MTF—the atmospheric MTF simply being a degenerate case of $S(\xi, \eta, \lambda_1, \lambda_2)$. In the words of these authors, "in determining the average coherence function, the effects of diffraction exactly cancel the effects of scintillation, when the medium is random and laterally, statistically stationary."

According to 6.40, the two-point two-wavelength correlation function, $S(\xi, \eta, \lambda_1, \lambda_2)$, that finally arises from our atmospheric path model can be expressed directly in terms of the individual OPD fluctuations, $h_j(x, y)$, introduced by each of the uncorrelated random phase screens in the model. However, though knowledge of the individual $h_j(x, y)$ would obviously be useful, this amount of information turns out to be vastly greater than the amount actually needed to fulfill our present limited objective of developing a general expression for $S(\xi, \eta, \lambda_1, \lambda_2)$.

Because the individual terms within the various sets of averaging brackets in 6.40 are all mutually uncorrelated, we may use the result stated previously by 6.36 to justify removal of all but the outer set of averaging brackets. Thus, we may express $S(\xi, \eta, \lambda_1, \lambda_2)$ in the much simpler form,

$$S(\xi, \eta, \lambda_1, \lambda_2) = \left\langle \exp\left\{2 \cdot \pi \cdot i \cdot \left[\frac{\sum_{j=1}^{n} h_j(x+\xi, y+\eta)}{\lambda_1} - \frac{\sum_{j=1}^{n} h_j(x, y)}{\lambda_2}\right]\right\}\right\rangle. \tag{6.41}$$

We might again observe that, as discussed earlier in Sect. 6.2.1.2, because of the associative and commutative properties of the addition operator when applied to real operands [as constituted here by the various OPD contributions $h_j(x, y)$], the order of adding terms in the two summations in the above equation is inconsequential. As far as function $S(\xi, \eta, \lambda_1, \lambda_2)$ is concerned, this effectively means that the order of laying out the individual random phase screens in the path model is also inconsequential. The implications of this important result are discussed further in Sect. 6.5.

The integrated OPD fluctuation over the atmospheric path, which we denote by H(x, y), may be expressed in terms of the individual phase screen contribution, $h_j(x, y)$, as follows (cf., 6.27):

$$H(x, y) = \sum_{j=1}^{n} h_j(x, y). \tag{6.42}$$

Because the individual $h_j(x, y)$ are all zero-mean functions (cf., 5.5), it follows that function H(x, y) must itself be a zero-mean function. Thus, we may formally write.

$$\langle H(x, y)\rangle = 0. \tag{6.43}$$

Equation 6.42 allows 6.41 to be expressed more concisely in terms of H(x, y) as follows:

$$S(\xi, \eta, \lambda_1, \lambda_2) = \left\langle \exp\left\{2 \cdot \pi \cdot i \cdot \left[\frac{H(x+\xi, y+\eta)}{\lambda_1} - \frac{H(x, y)}{\lambda_2}\right]\right\}\right\rangle. \tag{6.44}$$

Equation 6.44 infers that, for the purpose of calculating $S(\xi, \eta, \lambda_1, \lambda_2)$, only the final integrated amount of OPD fluctuation carries any significance; thus, there is no need to know details of the OPD contributions introduced by any of the individual phase screens. This outcome is of course entirely consistent with Hufnagel and Stanley's expression for calculating the atmospheric MTF based on OPD path integrals.

As noted previously in Sect. 6.2.1.1, though the individual OPD contributions, $h_j(x, y)$, may or may not be Gaussian distributed, for any given extended atmospheric path, the first two fundamental theorems of probability theory—the law of large numbers and the central limit theorem (Middleton, 1960)—obliges the distribution of their sum, H(x, y), to closely approximate a Gaussian distribution.

The variance, $\sigma^2$, and the autocorrelation function, $\rho(\xi, \eta)$, of function H(x, y) were formally defined earlier by 6.29 and 6.30. With the assumption that H(x, y) is indeed Gaussian distributed, it follows that functions, H(x, y)/$\lambda_1$ and H (x, y)/$\lambda_2$, are also Gaussian distributed, which allows (Beckmann & Spizzichino, 1963) 6.44 to be re-expressed in terms of the variance, $\sigma^2$, and the autocorrelation function, $\rho(\xi, \eta)$, of the integrated OPD fluctuation, H(x, y), by the form,

$$S(\xi, \eta, \lambda_1, \lambda_2) = \exp\left\{-2 \cdot \pi^2 \cdot \sigma^2 \cdot \left[\frac{1}{\lambda_1^2} + \frac{1}{\lambda_2^2} - \frac{2 \cdot \rho(\xi, \eta)}{\lambda_1 \cdot \lambda_2}\right]\right\}. \tag{6.45}$$

Equation 6.45 is the desired general expression at this stage for the two-point two-wavelength correlation function for extended atmospheric paths. Whereas the expression does not include the effects of wavelength dispersion, as we shall discover in Sect. 6.6, it is a relatively simple matter to generalize the above expression to include these effects.

In subsequent chapters, we shall find that function $S(\xi, \eta, \lambda_1, \lambda_2)$ determines all of the important statistical properties of point-object images formed by telescopes for both monochromatic and polychromatic light. We shall also find in Chaps. 8, 13 and 18 that 6.45 provides us with a crucial mathematical tool that allows the key statistical properties of the integrated OPD fluctuation over the propagation path—that is, $\sigma^2$ and $\rho(\xi, \eta)$—to be obtained from readily measurable properties of the intensities in point-object, or unresolved star, images.

### *6.3.1 Case of Isotropic Turbulence*

For isotropic turbulence, it is convenient to define the radial coordinate in the telescope pupil plane, ε, thus

$$\varepsilon = (\xi^2 + \eta^2)^{\frac{1}{2}}. \tag{6.46}$$

For this type of sturbulence, 6.45 may be written in the circularly symmetric form,

$$S(\varepsilon, \lambda_1, \lambda_2) = \exp\left\{-2 \cdot \pi^2 \cdot \sigma^2 \cdot \left[\frac{1}{\lambda_1^2} + \frac{1}{\lambda_2^2} - \frac{2 \cdot \rho(\varepsilon)}{\lambda_1 \cdot \lambda_2}\right]\right\}. \tag{6.47}$$

### *6.3.2 The Functional Form When $\rho(\xi, \eta)$ is Gaussian*

Actual field measurements of the autocorrelation function of the integrated OPD fluctuation, $\rho(\xi, \eta)$, will speak for themselves and these will of course supersede any functional form that we might assume here. For the time being, however, irrespective of whether we deal with an actual measured function or whether the function has been established in some other way, it is convenient to proceed by simply approximating $\rho(\xi, \eta)$ by a best-fit Gaussian function.

In Appendix C (Fig. C.3), the functional form of $\rho(\xi, \eta)$ deduced from the turbulence structure outer scale limits measured by Coulman et al. (1988) closely.

resembles a Gaussian function. We also find in Chap. 14 (Sect. 14.10) that the Gaussian approximation for $\rho(\xi, \eta)$ yields computer-generated image simulations of the 0.11-arcsec binary brown dwarf system, CFBDSIR 1458 + 10, that are remarkably similar to actual images of this binary system obtained by the Keck II telescope. These two outcomes may be seen as providing at least some measure of justification here for representing the $\rho(\xi, \eta)$ in terms of a best-fit Gaussian function.

#### 6.3.2.1 Non-isotropic Turbulence

In this case, we use an elliptical Gaussian function to approximate function $\rho(\xi, \eta)$. Thus, we write as follows:

$$\rho(\xi, \eta) = \exp\left[-\left(\frac{\xi^2}{w_{ox}^2} + \frac{\eta^2}{w_{oy}^2}\right)\right]. \tag{6.48}$$

where $w_{ox}$ and $w_{oy}$ are the 1/e half-widths of the function in the x- and y-directions which are of course aligned with the $\xi$- and $\eta$-directions. Inserting the above expression for into 6.45 gives $S(\xi, \eta, \lambda_1, \lambda_2)$ in the form,

$$S(\xi, \eta, \lambda_1, \lambda_2) = \exp\left\{-2 \cdot \pi^2 \cdot \sigma^2 \cdot \left[\frac{1}{\lambda_1^2} + \frac{1}{\lambda_2^2} - \frac{2}{\lambda_1 \cdot \lambda_2} \cdot \exp\left[-\left(\frac{\xi^2}{w_{ox}^2} + \frac{\eta^2}{w_{oy}^2}\right)\right]\right]\right\} \tag{6.49}$$

where we see that three quantities, $\sigma^2$, $w_{ox}$, and $w_{oy}$, are needed to describe the integrated effect of the turbulence structure in the propagation path.

#### 6.3.2.2 Isotropic Turbulence

In this case, we can set $w_{ox} = w_{oy} = w_o$, which allows 6.48 to simplify to the form,

$$\rho(\xi, \eta) = \exp\left[-\left(\frac{\varepsilon^2}{w_o^2}\right)\right]. \tag{6.50}$$

Using the above equation to substitute for $\rho(\varepsilon)$ in 6.47 gives $S(\varepsilon, \lambda_1, \lambda_2)$ in the form

$$S(\varepsilon, \lambda_1, \lambda_2) = \exp\left\{-2 \cdot \pi^2 \cdot \sigma^2 \cdot \left[\frac{1}{\lambda_1^2} + \frac{1}{\lambda_2^2} - \frac{2}{\lambda_1 \cdot \lambda_2} \cdot \exp\left[-\left(\frac{\varepsilon^2}{w_o^2}\right)\right]\right]\right\}, \tag{6.51}$$

where we see that now only two parameters, $\sigma^2$ and $w_o$, are needed to describe the integrated effect of turbulence structure in the propagation path.

## 6.4 General Expression for the Atmospheric MTF

The two-point two-wavelength correlation function, $S(\varepsilon, \lambda_1, \lambda_2)$, for an atmospheric path describes in statistical terms the overall scattering effect of that path on both monochromatic and polychromatic light waves. However, in instances where only monochromatic properties are of interest, these properties are more concisely described by the atmospheric MTF, $M(\xi, \eta, \lambda)$, a function that may be obtained from 6.45 as the degenerate case where $\lambda_1 \to \lambda_2 \to \lambda$:

$$M(\xi, \eta, \lambda) = \exp\left\{\frac{-4 \cdot \pi^2 \cdot \sigma^2}{\lambda^2} \cdot [1 - \rho(\xi, \eta)]\right\}. \tag{6.52}$$

The quantity, $[1 - \rho(\xi, \eta)]$, may be recognized as the unit-normalized wavefront structure function. This function is discussed further in Sect. 8.4.

### *6.4.1 Case of Isotropic Turbulence*

For this type of turbulence, 6.52 may be written in the circularly symmetric form,

$$M(\varepsilon, \lambda) = \exp\left\{\frac{-4 \cdot \pi^2 \cdot \sigma^2}{\lambda^2} \cdot [1 - \rho(\varepsilon)]\right\} \tag{6.53}$$

### *6.4.2 Functional Forms When $\rho(\xi, \eta)$ is Gaussian*

#### 6.4.2.1 Non-Isotropic Turbulence

The atmospheric MTF in this case may be obtained by combining 6.48 and 6.52:

$$M(\xi, \eta, \lambda) = \exp\left\{\frac{-4 \cdot \pi^2 \cdot \sigma^2}{\lambda^2} \cdot \left[1 - \exp\left[-\left(\frac{\xi^2}{w_{ox}^2} + \frac{\eta^2}{w_{oy}^2}\right)\right]\right]\right\}, \tag{6.54}$$

where again three parameters, $\sigma^2$, $w_{ox}$, and $w_{oy}$, are needed to describe the integrated effects of turbulence structure in the propagation path.

#### 6.4.2.2 Isotropic Turbulence

In this case, by combining 6.50 and 6.53, we obtain the circularly symmetric form

$$M(\varepsilon,\lambda) = \exp\left\{\frac{-4\cdot\pi^2\cdot\sigma^2}{\lambda^2}\cdot\left[1-\exp\left[-\left(\frac{\varepsilon^2}{w_o^2}\right)\right]\right]\right\}, \tag{6.55}$$

where just two quantities, $\sigma^2$ and $w_o$, are needed to describe the integrated effects of turbulence structure in the propagation path.

### 6.4.3 *Comparison of the General Expression to that of Hufnagel and Stanley*

We now show that the general expression given above by 6.52 for the atmospheric MTF is identical to Hufnagel and Stanley's widely accepted general expression. We first begin with 6.44, which is a direct antecedent of 6.52. In the limit, $\lambda_1 \to \lambda_2 \to \lambda$, we can obtain the atmospheric MTF from that equation in the form,

$$M(\xi,\eta,\lambda) = \left\langle\exp\left\{\frac{2\cdot\pi\cdot i}{\lambda}\cdot[H(x+\xi,y+\eta)-H(x,y)]\right\}\right\rangle. \tag{6.56}$$

Recalling 6.27, the above expression for the atmospheric MTF may be written in the form,

$$M(\xi,\eta,\lambda) = \left\langle\exp\left\{\frac{2\cdot\pi\cdot i}{\lambda}\cdot\int_0^z\left[N\left(x+\xi,y+\eta,z'\right)-N(x,y,z'\right]\cdot dz'\right\}\right\rangle. \tag{6.57}$$

This expression for the atmospheric MTF is given in a form that allows direct comparison with the general expression given by Hufnagel and Stanley (1964). For convenience, we now reproduce Hufnagel and Stanley's 4.6 using the same notation as these two authors,[26]

$$M(p_1,p_2,\lambda) = \left\langle A^*(p_2,z)\cdot A(p_1,z)\right\rangle$$

[26] It might be noted that Hufnagel and Stanley place the complex conjugate symbol on the first-listed complex amplitude term, $A^*(p_z, z)$, whereas our practice throughout the book has been to place it on the second-listed complex amplitude.

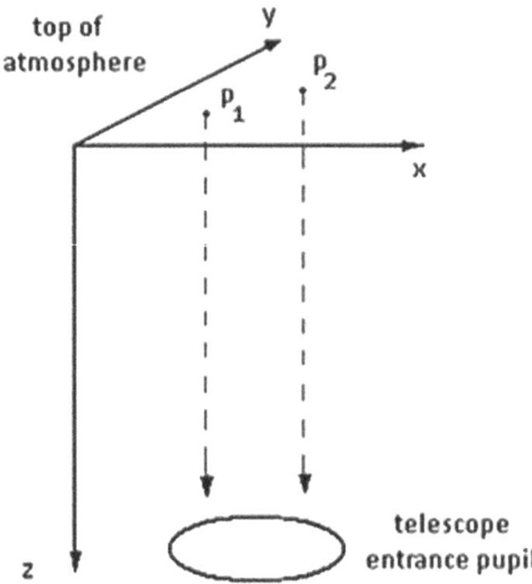


**Fig. 6.3** To calculate the atmospheric MTF, $M(p_1, p_2, \lambda)$, OPD line integrals are calculated over the atmospheric paths indicated by the dashed/arrowed lines

$$= \left\langle \exp\left\{ i \cdot k \cdot \int_0^z \left[ N\left(p_2, z'\right) - N\left(p_1, z'\right) \right] dz' \right\} \right\rangle. \tag{6.58}$$

The paths of integration for the two line integrals in the above equation are depicted in Fig. 6.3. Noting that $k = 2\pi/\lambda$, 6.57 and 6.58 are now seen as being exactly equivalent, thus establishing the equivalence of the general expression developed for the atmospheric MTF in this chapter (6.52) and that developed previously by Hufnagel and Stanley.

## 6.5 Equivalent Phase Screen Representation of an Atmospheric Path

The expressions given for the two-point two-wavelength correlation function, $S(\xi, \eta, \lambda_{1,}, \lambda_2)$, and the atmospheric MTF, $M(\xi, \eta, \lambda)$, by 6.45 and 6.52 are immediately seen as identical to the expressions that would have arisen had the light waves been scattered by a single random phase screen, one that imprints OPD fluctuation, $H(x, y)$, on incident plane waves. Thus, for the two crucial functions that we concern with in this book, $M(\xi, \eta, \lambda)$ and $S(\xi, \eta, \lambda_{1,}, \lambda_2)$, any extended atmospheric path may be considered[27] as behaving as though it were a single random phase screen

[27] When the EPS idea was first proposed in a private communication (McKechnie 1976a) and conference presentation (McKechnie 1976b, c), the reaction was mostly one of disbelief. To the majority of those exposed to the idea, the representation seemed over-simplistic. Nonetheless, the sound mathematical basis of this representation has resulted in its becoming an established part of

(Fig. 6.2). We refer to such an Equivalent Phase Screen as the EPS for that atmospheric path. The essential property of any EPS is that it introduces OPD fluctuations whose variance, $\sigma^2$, and autocorrelation function, $\rho(\xi, \eta)$, are the same as those of the integrated OPD fluctuations over the atmospheric path.

The quantity, $\sigma^2$, provides a measure of the integrated strength of the atmospheric turbulence structures in the path, while function $\rho(\xi, \eta)$ provides a measure of the average size of the turbulence structures in the path. The EPS concept allows us to make use of a wide range of theoretical results and practical measurement techniques contained in the extensive literature dealing with light scattering from rough surfaces (Dainty, 1984). Some of these are exploited in Chap. 8 where techniques are described for measuring $\sigma^2$ and $\rho(\xi, \eta)$. Others are exploited in Chap. 11 where expressions are given for a wide range of properties of star image speckle patterns—stellar speckle patterns—formed by large ground-based telescopes.

Whereas EPS representations of extended atmospheric paths facilitate accurate descriptions of the statistical properties of the complex amplitudes and intensities in point-object images, they do not necessarily allow accurate descriptions of the properties of images of extended objects. For such objects, image properties also depend on the isoplanatic properties of the atmosphere, a subject dealt with later, in Chap. 17.

### 6.5.1 Relationship Between $\rho(\xi, \eta)$ and the Refractive Index Structure Function, $D_N$

It might be observed here that, since both the autocorrelation function of the integrated OPD fluctuation, $\rho(\xi, \eta)$, and the refractive index structure function, $D_N$, provide information about the average turbulence structure sizes present in atmospheric paths, a mathematical relationship must exist between these two functions. Details of this relationship are provided later, in Chap. 8 (Sect. 8.4).

### 6.5.2 Location of the Equivalent Phase Screen in the Atmospheric Path

The conservation behavior of $M(\xi, \eta, \lambda)$ and $S(\xi, \eta, \lambda_1, \lambda_2)$ for light waves of infinite lateral extent as they propagate in a homogeneous medium, discussed previously in Chap. 5 (Sect. 5.5.1), tells us that the EPS can be located anywhere in the near-field with respect to the telescope. For ground-based astronomical telescopes, it is sometimes convenient to consider the EPS as located just in front of the telescope. However, it may equally be considered to lie at any other location in the atmospheric

conventional understanding of light propagation and imaging through the atmosphere (Roggemann and Welsh 1996).

path. Irrespective of the supposed location of the EPS in the atmospheric path, the resulting atmospheric MTF, $M(\xi, \eta, \lambda)$, is always that given by 6.52; similarly, the two-point two-wavelength correlation function, $S(\xi, \eta, \lambda_1, \lambda_2)$, is always that given by 6.45 (albeit with the wavelength restrictions discussed in Chap. 5, Sect. 5.5.1).

### 6.5.3 *Complex Amplitude Properties Arising from an Equivalent Phase Screen*

If the EPS is considered located just in front of the telescope, clearly the waves entering the telescope will only show phase fluctuations, scintillation not yet having had opportunity to develop. On the other hand, if the EPS is considered located at a significant distance from the telescope (e.g., at an altitude of 10 km), the waves entering the telescope will exhibit amplitude fluctuations (scintillation) in addition to the phase fluctuations. These conflicting behaviors indicate that, in general, the EPS representation does not allow accurate modeling of the complex amplitude. Consequently, use of the EPS representation in this book is restricted to applications relating only to certain statistical properties of the complex amplitude and the intensity, in particular, those described by functions, $M(\xi, \eta, \lambda)$ and $S(\xi, \eta, \lambda_1, \lambda_2)$.

Hufnagel and Stanley (1964) previously noted that it is not always necessary to know explicit details of the complex amplitude, commenting that while the complex amplitude is "of obvious interest, it is important to recall that for the present purpose (of calculating the atmospheric MTF), it is necessary only to compute a statistically averaged quantity. The central result of this paper is that the average coherence function M given by (2.3) can be calculated easily and exactly without ever solving explicitly for V and A."[28]

Although EPS representations generally do not permit accurate modeling of the complex amplitude, in certain instances accurate modeling may still actually be possible with this representation. Included among these are the following:

(1) imaging with ground-based astronomical telescopes at visible wavelengths when scintillation is negligible, as is often the case for stars lying within about 45° of the zenith, and (2) imaging with ground-based astronomical telescopes at infrared wavelengths where scintillation levels are usually significantly less that at visible wavelengths.

[28] In the notation of Hufnagel and Stanley, V is the complex scalar representation of the electromagnetic wave. The modulus of V is denoted by A. The equation referred to in the quoted passage, (2.3), refers to Hufnagel and Stanley's equation.

### 6.5.4 *Properties of the OPD Fluctuation Created by an Equivalent Phase Screen*

For the uncorrelated random phase screen model of an atmospheric path used in our propagation analysis (Sect. 6.2.1), it may be shown using 5.6, 5.7, 6.29 and 6.30 that

$$\sigma^2 = \sum_{j=1}^{n} \sigma_j^2, \tag{6.59}$$

and

$$\rho(\xi,\eta) = \frac{1}{\sigma^2} \cdot \sum_{j=1}^{n} \sigma_j^2 \cdot \rho_j(\xi,\eta). \tag{6.60}$$

Equation 6.59 states that the variance of the OPD fluctuations introduced by the EPS, $\sigma^2$, arises as the sum of the variances of the OPD contributions from each of the uncorrelated random phase screens in the atmospheric path model. The quantity $\sigma^2$ is a unique and unambiguous quantity that governs a number of key properties of telescope images. Techniques for obtaining $\sigma^2$ from star image intensity measurements are described in Chaps. 8, 13 and 18.

Generally, phase fluctuations associated with light waves that have propagated over extended atmospheric paths do not have any unique significance. As the waves propagate forward and scintillation gradually develops, a corresponding amount of phase depletion likely occurs. Consequently, the rms phase fluctuation associated with actual light waves would generally be smaller than $\sigma$. Since the task of keeping track of such phase depletion is comparable to the (almost impossible) task of keeping track of the complex amplitude, in this book we avoid such complication by limiting consideration to the phase fluctuations associated with the scintillation-free waves that emerge from the EPS (which directly relate to the integrated OPD fluctuation over the atmospheric path). For our purposes, it is sufficient to know that the two crucial functions that arise from these waves, $M(\xi, \eta, \lambda)$ and $S(\xi, \eta, \lambda_1, \lambda_2)$, are identical to the functions that arise from the phase-disrupted and scintillated waves that emerge at the end of the actual atmospheric path itself.

Plainly, the widths of the individual $\rho_j(\xi, \eta)$ functions associated with the various uncorrelated random phase screens in the atmospheric path model relate directly to the average size of the turbulence structures represented by the individual random phase screens in the model. Equation 6.60 indicates that the autocorrelation function of the integrated OPD fluctuation $\rho(\xi, \eta)$ arises as the weighted average of the individual $\rho_j(\xi, \eta)$ functions, the weighting factors being the individual variances, $\sigma_j^2$. Consequently, once $\rho(\xi, \eta)$ has been established, in effect, we have established the average turbulence structure size distribution over the entire propagation path. Techniques for obtaining $\rho(\xi, \eta)$ from readily measurable properties of the intensities in point-object images are described in Chaps. 8, 13 and 18.

## 6.6 General Expressions for M and S that Include Dispersion

In this section, more general expressions are given for $M(\xi, \eta, \lambda)$ and $S(\xi, \eta, \lambda_1, \lambda_2)$ that take account of the refractive index dispersion of air. The convention adopted thus far is that $\sigma$, $\rho(\xi, \eta)$, $H(x, y)$ and the refractive index, n, all relate to wavelength $\lambda$. By now adopting the supplementary conventions that $\sigma_{\lambda_k}$, $\rho_{\lambda_k}(\xi, \eta)$, and $H_{\lambda_k}(x, y)$ relate to wavelengths, $\lambda_k$, where k = 1 and 2, the following relationships can be shown to hold between the integrated OPD fluctuations at the three wavelengths, $\lambda$, $\lambda_1$ and $\lambda_2$[29]:

$$H(x, y) = \left(\frac{n-1}{n_{\lambda_1}-1}\right) \cdot H_{\lambda_1}(x, y) = \left(\frac{n-1}{n_{\lambda_2}-1}\right) \cdot H_{\lambda_2}(x, y). \tag{6.61}$$

By combining the above relations with previous 6.29 and 6.30, the following relationships are also seen to hold:

$$\sigma = \left(\frac{n-1}{n_{\lambda_1}-1}\right) \cdot \sigma_{\lambda_1} = \left(\frac{n-1}{n_{\lambda_2}-1}\right) \cdot \sigma_{\lambda_2} \tag{6.62}$$

and

$$\rho(\xi, \eta) = \rho_{\lambda_1}(\xi, \eta) = \rho_{\lambda_2}(\xi, \eta). \tag{6.63}$$

Thus, to take dispersion into account, clearly function, $S(\xi, \eta, \lambda_1, \lambda_2)$, which was previously given by 6.45, must take the more general form,

$$S(\xi, \eta, \lambda_1, \lambda_2) = \exp\left\{-2 \cdot \pi^2 \cdot \left[\frac{\sigma_{\lambda_1}^2}{\lambda_1^2} + \frac{\sigma_{\lambda_2}^2}{\lambda_2^2} - \frac{2 \cdot \sigma_{\lambda_1} \cdot \sigma_{\lambda_2} \cdot \rho(\xi, \eta)}{\lambda_1 \cdot \lambda_2}\right]\right\}. \tag{6.64}$$

By using 6.62 and 6.63, the above relation may be written in the alternative form,

$$S(\xi, \eta, \lambda_1, \lambda_2) = \exp\left\{-2 \cdot \pi^2 \cdot \frac{\sigma^2}{(n-1)^2} \cdot \left[\frac{(n_{\lambda_1}-1)^2}{\lambda_1^2} + \frac{(n_{\lambda_2}-1)^2}{\lambda_2^2}\right] \cdot \right.$$
$$\left. - \frac{2 \cdot \left(n_{\lambda_1}-1\right) \cdot \left(n_{\lambda_2}-1\right) \cdot \rho(\xi, \eta)}{\lambda_1 \cdot \lambda_2}\right\}. \tag{6.65}$$

[29] Equation 6.61 is based on the assumption that the refractive index dispersion ratios, $(n-1)/\left(n_{\lambda_1}-1\right)$ and $(n-1)/\left(n_{\lambda_2}-1\right)$ are both invariant with respect to temperature and pressure and therefore invariant over the entire atmospheric path. Appropriate values for these ratios may be obtained from the refractive index values for standard air (cf., 3.1).

As per Footnote 30, appropriate values of the refractive indices, $n$, $n_{\lambda_1}$ and $n_{\lambda_2}$ may be obtained by using the values for standard air at sea level, given previously in Chap. 3 by 3.1.

The atmospheric MTF, $M(\xi, \eta, \lambda)$, may be obtained directly from 6.65 as the limiting case where $\lambda_1 \to \lambda_2 \to \lambda$. Thus, it may be written,

$$M(\xi, \eta, \lambda) = \exp\left\{\frac{-4 \cdot \pi^2 \cdot \sigma^2}{\lambda^2} \cdot [1 - \rho(\xi, \eta)]\right\}. \quad (6.66)$$

It should come as no surprise to find that the above expression for $M(\xi, \eta, \lambda)$ is identical to the expression given previously by 6.52. It was implicitly assumed in the development of that earlier expression that the value attached to $\sigma$ referred specifically to the value at wavelength, $\lambda$, thus rendering the dispersion issue moot.

## 6.7 Mathematical Notation Used in This Chapter

The mathematical notation used in this chapter is indicated in Table 6.2.

**Table 6.2** Mathematical notation used in this chapter along with the SI dimensional units of the individual quantities

| Symbol | Quantity | Dimensions |
|---|---|---|
| $\lambda$ | Wavelength | m |
| (x, y) | Cartesian coordinate system in plane perpendicular to direction of light travel | m |
| $(\xi, \eta)$ | Cartesian coordinate system in plane perpendicular to direction of light travel | m |
| $\varepsilon$ | Radial coordinate in plane perpendicular to direction of light travel, $\varepsilon = \sqrt{\xi^2 + \eta^2}$ | m |
| M | Atmospheric MTF | "1" |
| S | Two-point two-wavelength correlation function of the complex amplitude | "1" |
| U | Complex amplitude | "1" |
| A | Amplitude | "1" |
| $n$ | Refractive index | "1" |
| N | Normalized fluctuating part of refractive index | "1" |
| OPD | Optical path difference | m |
| $k$ | Wave number in air | $m^{-1}$ |
| $p$ | Distance between points $p_1$ and $p_2$ as per H&S | m |
| $D_N$ | Refractive index structure function | "1" |

(continued)

**Table 6.2** (continued)

| Symbol | Quantity | Dimensions |
|---|---|---|
| $C_\vartheta$ | Temperature structure constant | $m^{-1/3}$ |
| $C_n$ | Refractive index structure constant | $m^{-1/3}$ |
| $\frac{\rho(\bar{z})}{\rho_o}$ | Average air pressure at altitude relative to sea level (as per H&S paper) | "1" |
| $L_0$ | Turbulence structure outer scale limit | m |
| $L$ | Horizontal path length (as per H&S paper) | m |
| $[h]$ | OPD fluctuation column matrix | m |
| $[J']$ | Real symmetric OPD correlation matrix | $m^2$ |
| $[J]$ | Diagonalized OPD correlation matrix | $m^2$ |
| $[TM]$ | Transformation matrix | "1" |
| $[TM_0]$ | Unitary transformation matrix | "1" |
| $\lambda_k$ | kth eigenvalue | $m^2$ |
| $h_j$ | OPD fluctuation introduced by jth phase screens | $m$ |
| $H$ | Summed OPD over all phase screens in path | $m$ |
| $\sigma$ | rms of summed OPD, H | $m$ |
| $\rho$ | Autocorrelations function of summed OPD, H | "1" |
| $B_{j-1}$ | Complex amplitude of wave arriving at jth phase screen | "1" |

Dimensionless quantities are indicated by "1"

## References


Beckmann, P., & Spizzichino, A. (1963). *The scattering of electromagnetic waves from rough surfaces*. Pergamon.

Coulman, C. E., Vernin, J., Coqueugniot, Y., & Caccia, J. L. (1988). Outer scale of turbulence appropriate to modeling refractive index structure profiles. *Applied Optics, 27*, 155–160.

Dainty, J. C. (1984). Laser speckle and related phenomena. In J. C. Dainty (Ed.), Topics in applied physics (Vol. 9). Berlin: Springer.

Horn, R. A., & Johnson, C. R. (1985). *Matrix analysis*. Cambridge University Press. ISBN 978–0–521–38632–6.

Hufnagel, R. E., & Stanley, N. R. (1964). Modulation transfer function associated with image transmission through turbulent media. *JOSA, 54*, 52–61.

McKechnie, T. S. (1976a) *Light propagation through the atmosphere and imaging in astronomy*. Private Communication to J. C. Dainty et al. London.

McKechnie, T. S. (1976b). *Cores in Star Images*. *JOSA, 66*(6), 635.

McKechnie, T. S. (1976c). Image plane speckle in partially coherent illumination. *Optical and Quantum Electronics, 8*, 61–67.

McKechnie, T. S. (1991). Light propagation through the atmosphere and the properties of images formed by large ground-based telescopes. *JOSA A, 8*, 346–365.

Middleton, D. (1960). *Introduction to statistical communication theory*. McGraw Hill Book Co.

Murdoch, D. C. (1957). *Linear algebra for undergraduates*. Wiley.

Papoulis, A. (1965). *Probability, random variables and stochastic processes*. McGraw-Hill.

Roggemann, M. C., & Welsh, B. M. (1996). *Imaging through turbulence*. Boca

Strang, G. (2009). *Introduction to linear algebra* (4th ed.). Wellesley-Cambridge Press. (Raton: CRC Press).

Tatarski, V. I., & Zavorotny, V. U. (1993). Atmospheric turbulence and the resolution limits of large ground-based telescopes: Comment. *JOSA, A10*, 2410–2417.

# Chapter 7
# Properties of Point-Object Images Formed by Telescopes

**Abstract** This chapter provides descriptions of the key statistical properties of point-object images formed by telescopes observing over extended atmospheric paths. These properties include the average intensity envelopes for long- and short-exposure images and the statistical properties of speckle in instantaneous images. Descriptions are given for both monochromatic and polychromatic images. All of the properties can be expressed in terms of the telescope pupil function, the rms OPD fluctuation over the extended atmospheric path, $\sigma$, and the autocorrelation function of that fluctuation, $\rho(\xi, \eta)$. Because image intensity envelopes are generally wavelength dependent, telescope resolution is also wavelength dependent. Expressions are also provided for the average intensity envelops of images of extended objects. The image of such an object can take an infinite variety of different forms depending on the coherence properties of the illumination.

Expressions are developed in this chapter for the correlation functions that control and describe the key statistical properties of both monochromatic and polychromatic point-object images formed by telescopes of arbitrary size observing through the atmosphere. Unresolved stars of course closely approximate ideal point-objects. Expressions are given for the average intensity envelopes in the images, the shapes and widths of which control telescope resolution through the atmosphere. These expressions form the starting point of more detailed discussion of the characteristics of star image envelopes given in subsequent Chaps. 9 and 10. Expressions are also given for the functions that control the statistical properties of the stellar speckle patterns that arise in star images formed by large telescopes. These expressions form the starting point of a detailed examination of the properties of stellar speckle patterns given in Chap. 11.

The most general of the correlation functions considered in this chapter is the two-point two-wavelength correlation function of the complex amplitudes in point-object images. The expression developed for this function takes account of the atmospheric path via the two-point two-wavelength correlation function of the complex amplitudes, S, of the light waves arriving in the telescope pupil, for which expressions were previously developed in Chap. 6 (6.45 and 6.65). The expression also takes account



T. S. McKechnie, *General Theory of Light Propagation and Imaging Through the Atmosphere*, Progress in Optical Science and Photonics 20,
https://doi.org/10.1007/978-3-030-98828-9_7

of the two-point two-wavelength correlation function associated with the telescope optics, for which expressions will be developed in this chapter. Two versions of point-object images are considered in this and following chapters: the so-called long- and short-exposure versions. In Chap. 8, methods are described for characterizing atmospheric paths in terms of key statistical properties of the integrated optical path difference (OPD) fluctuation, $H(x, y)$, associated with the image-forming light waves in the telescope pupil, namely the variance, $\sigma^2$, and autocorrelation function, $\rho(\xi, \eta)$. Depending on whether or not image wander is corrected during the measurements, the measured data naturally provide either the long- or short-exposure versions of $\sigma^2$ and $\rho(\xi, \eta)$. By taking the appropriate $\sigma^2$ and $\rho(\xi, \eta)$ combination together with the pupil function of the observing telescope, expressions can be obtained for the average intensity envelopes in either long- or short-exposure point-object images.

## 7.1 Long- and Short-Exposure Images of Point-Objects

Figure 7.1 shows a light wave, assumed to be of infinite lateral extent, arriving in the pupil plane of a telescope after having propagated over an extended atmospheric path. The wave may be considered as having originated from a point-object at infinity.

Two dashed straight lines are also shown in the figure. One represents an imaginary plane wave that best fits the instantaneous, infinitely extensive incoming wave. The location in the image plane where the telescope brings this imaginary plane wave to a focus corresponds to the location of the long-exposure image. The other dashed line corresponds to a plane wave that best fits the finite, instantaneous wave portion collected by the telescope at the instant captured in the figure. The focus location of this imaginary wave corresponds to the location of the instantaneous, or short-exposure, image. This location generally wanders randomly in the image plane; not unexpectedly, the mean location of the short-exposure image corresponds to the location of the long-exposure image. The average short-exposure image can be obtained by summing individual short-exposure images, with image wander subtracted prior to each addition. The long-exposure image may be obtained by summing the same set of images without subtracting image wander. In general, instantaneous short-exposure images, and also the average short-exposure image, provide higher levels of resolution than long-exposure images.

Several factors combine together to determine the overall wander characteristics of star images. As indicated in Fig. 7.1, atmospheric turbulence is an important contributor. Telescope shake and tracking errors also contribute. The blurring effect of image wander arising from all sources can result in long-exposure images providing significantly lower resolution levels than the average short-exposure image. In fact, in windy conditions, telescope shake can be the dominant cause of image wander; it can cause image blur patches measuring several arcsec across.

Because image wander is easily corrected, and because so many of its causes are entirely unrelated to atmospheric turbulence, in this book—where we are generally

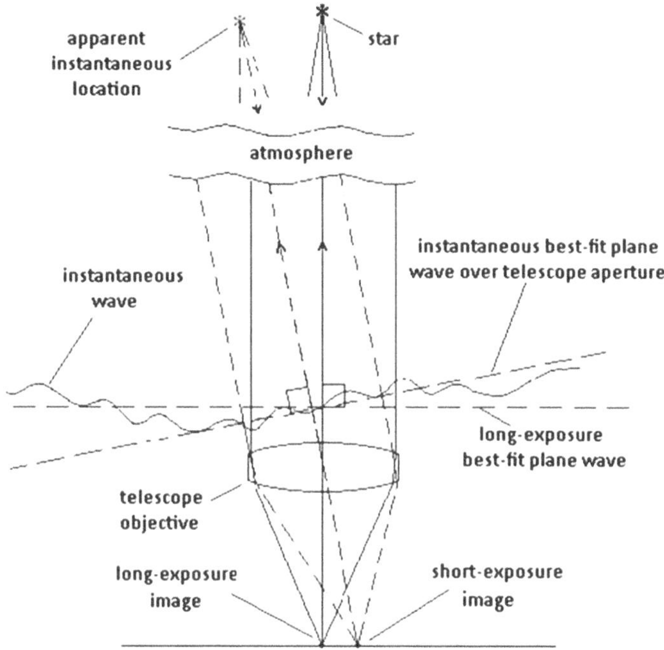


**Fig. 7.1** Instantaneous wave arriving in the pupil of a telescope. The average departure behavior of the instantaneous wave (solid line) from the two best-fit plane reference wavefronts shown (dashed lines) determine the respective properties of the long- and short-exposure images of a point-object such as a distant unresolved star. Generally, both the phase and amplitude of such a wave exhibit random fluctuations

concerned with achieving the highest levels of image quality and telescope resolution—we shall generally deal in terms of the average short-exposure image. While long-exposure images without image wander correction might be the simplest type of image to record, as a general rule this type of image should be avoided.

## 7.2 Telescope Coordinate Systems

Figure 7.2 shows an $(x, y)$ Cartesian coordinate system located in the telescope pupil plane. Two other Cartesian coordinate systems are also shown in the image plane of the telescope: the Cartesian $(u, v)$ system is shown on the left, and the angular coordinate system $(\alpha, \beta)$ is shown on the right. (Because telescope resolution is usually expressed in terms of angle, it is often convenient to use angular coordinates in the telescope image plane.) Angles $\alpha$ and $\beta$ may be considered as the direction angles in the telescope image plane (not to be confused with the direction cosines of these angles). In the small-angle limit, $\alpha$ and $\beta$ are related to $u$ and $v$ as follows:

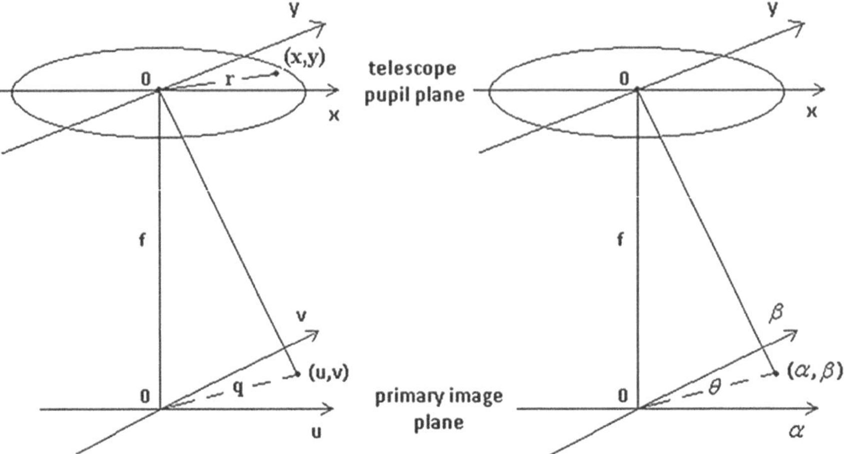


**Fig. 7.2** Telescope coordinate systems. Left Cartesian coordinate systems referred to the telescope pupil and image planes. Right Alternative angular coordinate system ($\alpha$, $\beta$) in the image plane

$$\alpha = \frac{u}{f}, \tag{7.1}$$

$$\beta = \frac{v}{f}, \tag{7.2}$$

where f is the focal length of the telescope, which is tacitly assumed here to be focused at infinity. (For targets at finite distances, where telescope refocusing would be necessary, f would be replaced in the above equations by the conjugate distance in image space as discussed previously in Chap. 4, Sect. 4.2.4.)

For circularly symmetric images (such as Airy patterns), it is often convenient to deal in terms of the radial coordinate, q, defined by

$$q = \left(u^2 + v^2\right)^{\frac{1}{2}}. \tag{7.3}$$

The corresponding angular coordinate, $\vartheta$, is similarly defined by

$$\vartheta = \left(\alpha^2 + \beta^2\right)^{\frac{1}{2}} = \frac{q}{f}. \tag{7.4}$$

It might be noted that, in Chap. 4, $\vartheta$ was defined (cf., 4.28) with respect to image plane coordinates ($x_0$, $y_0$). From now on, we supersede that earlier definition by considering angle $\vartheta$ as re-defined by 7.4 as the radial angular coordinate in the ($\alpha$, $\beta$) telescope image plane.

## 7.3 The Complex Amplitude in an Instantaneous Point-Object Image

Using Fig. 7.2 coordinate systems, the instantaneous complex amplitude in the image of a point-object, $U(u, v, \lambda)$, may be given in terms of the complex amplitude of the wave arriving in the telescope pupil, $B(x, y, \lambda)$, by the Fraunhofer diffraction integral (4.10),

$$U(u, v, \lambda) = \frac{1}{\lambda} \cdot \int_{-\infty}^{\infty} \int_{-\infty}^{\infty} B(x, y, \lambda) \cdot K(x, y, \lambda) \cdot \exp\left[\frac{-2 \cdot \pi \cdot i}{\lambda \cdot f} \cdot (u \cdot x + v \cdot y)\right] \cdot dx \cdot dy, \tag{7.5}$$

where $K(x, y, \lambda)$ is the telescope pupil function (introduced previously in Sect. 4.2.6). The intensity transmission function over the pupil is given by $K(x, y, \lambda) \cdot K^*(x, y, \lambda)$. For an aberration-free telescope, $K(x, y, \lambda)$ is entirely real-valued and may be regarded simply as the amplitude transmission factor. For an aberrated telescope, $K(x, y, \lambda)$ generally takes complex values.

Complex amplitude in the image can equally be expressed in terms of the angular coordinates

$$U(\alpha, \beta, \lambda) = \frac{1}{\lambda} \cdot \int_{-\infty}^{\infty} \int_{-\infty}^{\infty} B(x, y, \lambda) \cdot K(x, y, \lambda) \cdot \exp\left[\frac{-2 \cdot \pi \cdot i}{\lambda} \cdot (\alpha \cdot x + \beta \cdot y)\right] \cdot dx \cdot dy. \tag{7.6}$$

For an aberration-free telescope with a circular pupil of diameter D, the pupil function can be expressed in the form

$$K(x, y, \lambda) = \begin{cases} 1 & \text{for } \left(x^2 + y^2\right) \le \left(\frac{D}{2}\right)^2, \\ 0 & \text{otherwise.} \end{cases} \tag{7.7}$$

## 7.4 Telescope OTFs and MTFs

The image of a sinusoidal intensity pattern formed by an optical system is itself a sinusoidal intensity pattern. In general, the pattern arising in the image is a demodulated, phase-shifted version of the original object pattern. Optical systems may be considered as acting like spatial frequency filters (Goodman, 2004). The optical transfer function (OTF) describes how sinusoidal patterns in object space transform to sinusoidal patterns in image space as a function of the pattern's spatial frequency.

In general, the OTF is a complex-valued function. The modulus of the OTF, referred to as the modulation transfer function (MTF), describes the modulation of the image pattern relative to the modulation of the object pattern. The phase angle associated with the OTF describes the lateral shifting of the sinusoidal patterns in the image relative to their true Gaussian image plane locations.

Figure 7.3 shows a fully modulated sinusoidal object intensity pattern along with the corresponding sinusoidal intensity pattern in the image. The modulation factor associated with the optical system, MF, may be calculated from the maximum and minimum intensities attained by the sinusoidal pattern formed in the image, $I_{max}$ and $I_{min}$, using the standard formula (cf., 5.47),

$$MF = \frac{I_{\max} - I_{\min}}{I_{\max} + I_{\min}}, \tag{7.8}$$

where MF takes values in the range, $0 \leq MF \leq 1$. The MTF of an optical system describes the variation of the real quantity, MF, as a function of the spatial frequency of the sinusoidal pattern.

The phase angle associated with the OTF, which describes lateral shifts of sinusoidal image patterns, also depends on the spatial frequency. The OTF and MTF are related as follows:

$$OTF = MTF \cdot \exp(i \cdot phaseshift). \tag{7.9}$$

For incoherently illuminated extended objects, the intensity distribution over the object can be Fourier-decomposed into a number (perhaps infinitely large) of sinusoidal components. The OTF enables transformation of each Fourier component represented in the object into a corresponding intensity component in the image. The superposition property (Sect. 4.7.1) may then be applied to calculate the final image.

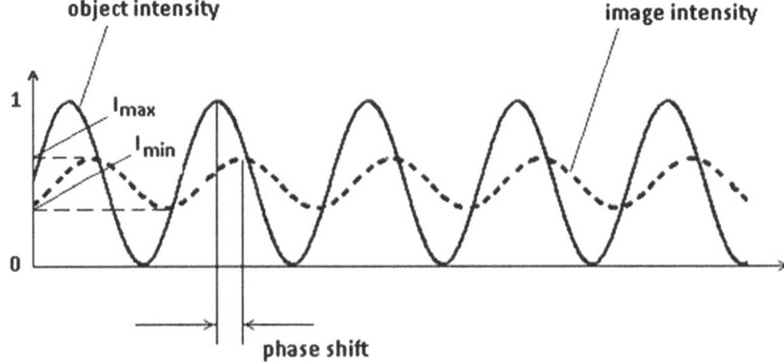


**Fig. 7.3** The OTF of an imaging system describes both the intensity modulation and phase response in the image of a sinusoidal object intensity pattern as a function of the spatial frequency

### 7.4.1 *Telescope OTF and MTF for Incoherent Illumination*

Incoherent illumination can generally be assumed for astronomical imaging and is therefore the most common type of illumination considered in this book. For notational brevity, we denote the OTF of the telescope by $M_T$. For the less frequently occurring case of coherent illumination discussed in Sect. 7.4.2, we attach an additional subscript and denote the telescope OTF by $M_{T.Coh}$. For incoherent illumination, it can be shown (Born & Wolf, 2003) that the OTF of a telescope is given by

$$M_T\left(\frac{\xi}{\lambda \cdot f}, \frac{\eta}{\lambda \cdot f}, \lambda\right) = \frac{\int_{-\infty}^{\infty}\int_{-\infty}^{\infty} K(x+\xi, y+\eta, \lambda) \cdot K^*(x, y, \lambda) \cdot dx \cdot dy}{\int_{-\infty}^{\infty}\int_{-\infty}^{\infty} K(x, y, \lambda) \cdot K^*(x, y, \lambda) \cdot dx \cdot dy}, \tag{7.10}$$

where the OTF arguments, which have the dimensions of spatial frequency (i.e.,$Length^{-1}$), are seen to be scaled versions of the coordinates in the telescope pupil plane.

The MTF may be obtained from the above equation by obtaining the modulus of both sides. Thus,

$$\left|M_T\left(\frac{\xi}{\lambda \cdot f}, \frac{\eta}{\lambda \cdot f}, \lambda\right)\right| = \left|\frac{\int_{-\infty}^{\infty}\int_{-\infty}^{\infty} K(x+\xi, y+\eta, \lambda) \cdot K^*(x, y, \lambda) \cdot dx \cdot dy}{\int_{-\infty}^{\infty}\int_{-\infty}^{\infty} K(x, y, \lambda) \cdot K^*(x, y, \lambda) \cdot dx \cdot dy}\right|, \tag{7.11}$$

#### 7.4.1.1 OTFs and MTFs for Diffraction-Limited Telescopes with Circular Apertures

The pupil function of an aberration-free telescope with circular aperture was given previously by 7.7. Because the function is both circularly symmetric and entirely real, the telescope OTF must then also be circularly symmetric and entirely real. Thus, for this case, telescopes OTF and MTF are identical, with 7.10 and 7.11 both reducing to the form,

$$\left|M_T\left(\frac{\varepsilon}{\lambda \cdot f}, \lambda\right)\right| = M_T\left(\frac{\varepsilon}{\lambda \cdot f}, \lambda\right) = \frac{2}{\pi} \cdot \left[\cos^{-1}\left(\frac{\varepsilon}{D}\right) - \frac{\varepsilon}{D} \cdot \sqrt{1-\left(\frac{\varepsilon}{D}\right)^2}\right], \tag{7.12}$$

where the radial distance in the telescope pupil plane, $\varepsilon$, was defined previously by 6.46. The radial distance in the pupil may be transformed into the corresponding spatial frequency in the image plane, $\nu$, using the relation,

$$\nu = \frac{\varepsilon}{\lambda \cdot f}. \tag{7.13}$$

By using the above transformation, the MTF given by 7.12 may be expressed in the alternative form,

$$|M_T(\nu, \lambda)| = M_T(\nu, \lambda)$$
$$= \frac{2}{\pi} \cdot \left[ \cos^{-1}\left(\frac{\lambda \cdot f \cdot \nu}{D}\right) - \frac{\lambda \cdot f \cdot \nu}{D} \cdot \sqrt{1 - \left(\frac{\lambda \cdot f \cdot \nu}{D}\right)^2} \right]. \tag{7.14}$$

Figure 7.4 shows the telescope MTF calculated from the above equation. Also shown in the same figure is an MTF plot for an aberrated version of the same telescope, this plot reflecting the general characteristics of an MTF for an aberrated optical system where the values attained are always less than or equal to those of an unaberrated version of the optical system.

**Effect of Central Obstructions**

Central obstructions have the effect of pulling down modulation in the intermediate band of spatial frequencies. For a diffraction-limited telescope with circular aperture of diameter, D, and central obstruction diameter, d, the pupil function may be written in the form,

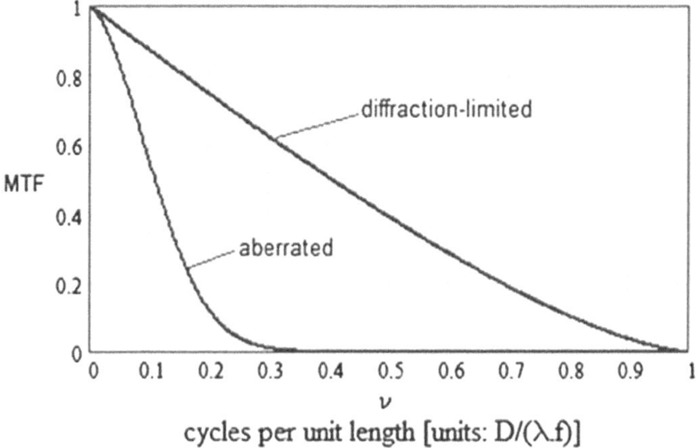


**Fig. 7.4** MTF response of telescopes with circular apertures for incoherent illumination for diffraction-limited and aberrated telescopes

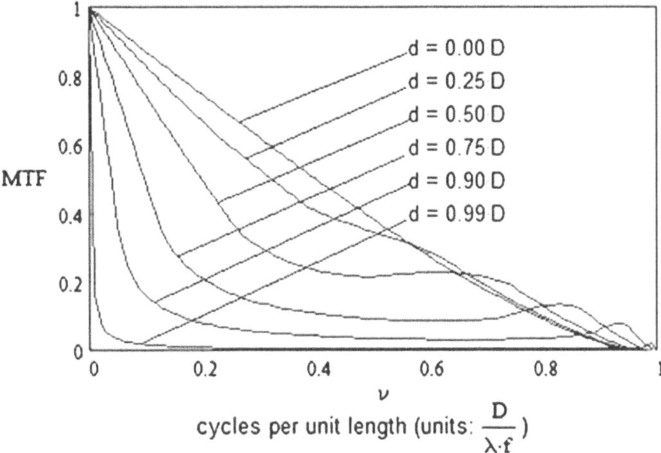


**Fig. 7.5** MTF and OTF are identically equal for incoherent illumination for a diffraction-limited telescope with circular aperture and various central obstruction diameters, d (where d is expressed as indicated in terms of the telescope diameter D)

$$K(x, y, \lambda) = 1 \quad \text{for} \quad \left(\frac{d}{2}\right)^2 \leq \left(x^2 + y^2\right) \leq \left(\frac{D}{2}\right)^2, \\ = 0 \qquad \text{otherwise} \tag{7.15}$$

Figure 7.5 shows the MTF for various central obstruction sizes, where 7.11 and 7.15 were used to calculate the plots. (The MTF and OTF are identical for diffraction-limited optical systems).

Significant MTF degradation can be seen for $> 0.25 \cdot D$. In the limit, as $d \to D$, some modulation remains near $v = 0$ and there is even some modulation enhancement in the neighborhood of the cutoff, but otherwise the MTF is seen to collapse toward zero. If there is any choice in the matter, large central obstructions should be avoided.

### MTF Cutoff Frequency for Incoherent Illumination

The frequency at which the MTF ultimately goes to zero is referred to as the cutoff frequency. Inspection of the above equation indicates that this frequency, which we denote by $\nu_{C,Incoh}$, is given by

$$\nu_{C,Incoh} = \frac{D}{\lambda \cdot f}. \tag{7.16}$$

### Angular Resolution at the MTF Cutoff Frequency for Incoherent Illumination

For incoherent illumination, the angular resolution limit for an aberration-free telescope with circular pupil, $\vartheta_{C,Incoh}$, is given in radians in terms of the cutoff frequency by

$$\vartheta_{C,Incoh} = \frac{1}{\nu_{C,Incoh} \cdot f}. \tag{7.17}$$

Using 7.16, this angular resolution limit may also be expressed in the form,

$$\vartheta_{C,Incoh} = \frac{\lambda}{D}. \tag{7.18}$$

Though this limit is seen to be 1.22 times smaller than the Rayleigh resolution limit given previously by 4.36, this disparity does not amount to a fundamental contradiction. At the somewhat arbitrarily chosen Rayleigh limit separation for a two-point object (previously illustrated in Fig. 4.7), some residual modulation can be seen in the image of the "just-resolved" two-point object. At the MTF cutoff frequency, modulation falls entirely to zero.

### *7.4.2 Amplitude Transfer Function of a Telescope for Coherent Illumination*

For a coherently illuminated object, the complex amplitude rather than the intensity is the quantity that sums linearly in the image. The frequency response of the telescope would then be significantly different from that just indicated for incoherent illumination. By using suitably scaled arguments, Born and Wolf (2003) show how the amplitude transfer function for coherent illumination, $M_{T,Coh}$, can be expressed in terms of the telescope pupil function, $K(\xi, \eta, \lambda)$, by

$$M_{T,Coh}\left(\frac{\xi}{\lambda \cdot f}, \frac{\mu}{\lambda \cdot f}, \lambda\right) = K(\xi, \eta, \lambda). \tag{7.19}$$

The subscript, Coh, is used here to discriminate the amplitude transfer function of the telescope for coherent light from the telescope OTF for incoherent light. As mentioned earlier, for astronomical imaging, the illumination is invariably incoherent. The incoherent OTF, $M_T$, therefore arises more frequently than the coherent amplitude transfer function, $M_{T,Coh}$.

An expression for the modulus of the amplitude transfer function of a telescope for coherent illumination can be obtained directly from 7.19 in the form,

$$\left|M_{T,Coh}\left(\frac{\xi}{\lambda \cdot f}, \frac{\mu}{\lambda \cdot f}, \lambda\right)\right| = |K(\xi, \eta, \lambda)|. \tag{7.20}$$

Equation 7.19 indicates that the telescope amplitude transfer function for coherent illumination is sensitive to aberrations. However, 7.20 indicates the remarkable result that the modulus of the function is insensitive to aberrations, no matter how severe these might be.

#### 7.4.2.1 The Function for Diffraction-Limited Telescopes with Circular Apertures

The pupil function for this case was previously given by 7.7. By inserting this function into 7.19 and 7.20, the amplitude transfer function and the modulus of this function are seen to be identical:

$$\left|M_{T,Coh}\left(\frac{\varepsilon}{\lambda \cdot f}, \lambda\right)\right| = M_{T,Coh}\left(\frac{\varepsilon}{\lambda \cdot f}, \lambda\right) = 1 \quad \text{for} \quad \varepsilon \leq \frac{D}{2},$$
$$= 0 \qquad \text{otherwise} \tag{7.21}$$

The above results may also be expressed in terms of the spatial frequency coordinate $\nu$ (cf., 7.13):

$$\left|M_{T,Coh}(\nu, \lambda)\right| = M_{T,Coh}(\nu, \lambda) = 1 \quad \text{for} \quad \nu \leq \frac{D}{2 \cdot \lambda \cdot f},$$
$$= 0 \quad \text{otherwise,} \tag{7.22}$$

from which we see that the cutoff frequency, denoted by $\nu_{C,Coh}$, is given by

$$\nu_{C,Coh} = \frac{D}{2 \cdot \lambda \cdot f}. \tag{7.23}$$

We might note here that the cutoff frequency is exactly half that found for incoherent illumination (cf., 7.16). Figure 7.6 shows the amplitude transfer function for coherent illumination, calculated using 7.22. Modulation is seen to be perfectly preserved out to the cutoff frequency, at which it falls abruptly to zero.

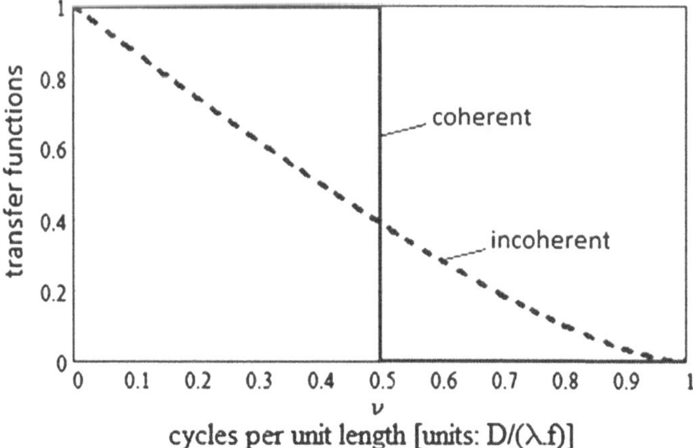


**Fig. 7.6** Amplitude transfer function (solid line) for a telescope with circular aperture for coherent illumination. Unlike the MTF for incoherent illumination (dotted line) for a diffraction-limited telescope, the amplitude response for coherent illumination is not affected by telescope aberrations

#### 7.4.2.2 "Anomaly" Between the Cutoff Frequencies for Coherent and Incoherent Light

Though it might at first seem anomalous that the frequency response cutoffs seen in Fig. 7.6 for coherent and incoherent illumination differ by a factor of 2, this difference has a ready explanation. For coherent illumination, complex amplitude is the quantity that adds linearly in the image. If we suppose that the image comprises a fully modulated sinusoidal complex amplitude pattern described by $0.5 \cdot [1 + \sin(x)]$, a function that takes values in the range, 0–1, the final image intensity is given by $0.25 \cdot [1 + \sin(x)]^2$, which also takes values in the range, 0–1. However, the trigonometric identities,

$$[1 + \sin(x)]^2 = 1 + 2 \cdot \sin(x) + \sin(x)^2 = 1 + 2 \cdot \sin(x) + \cos(x)^2 - \cos(2 \cdot x), \tag{7.24}$$

indicate that, among other contributions, the final image intensity contains the term, $-\cos(2 \cdot x)$, which indicates a sinusoidal intensity component at a spatial frequency twice that of the complex amplitude, $0.5 \cdot [1 + \sin(x)]$. Thus, whether the object is illuminated coherently or incoherently, as intuition might have already suspected, in either case the final image intensity displays spatial frequencies up to the same absolute cutoff limit, $D/(\lambda \cdot f)$.

**Effect of central obstructions**

For coherent illumination, central obstructions, such as found in most large reflector telescopes, cause complete loss of response over a certain range of low spatial frequencies. For a uniformly transmitting aperture of diameter, D, and central obstruction of diameter, d, the pupil function may be expressed in the form

$$\begin{aligned} K(x, y, \lambda) &= \exp[i \cdot \varphi(x, y, \lambda)] \quad \text{for} \quad \left(\frac{d}{2}\right)^2 \leq \left(x^2 + y^2\right) \leq \left(\frac{D}{2}\right)^2 \\ &= 0 \qquad \text{otherwise} \end{aligned} \tag{7.25}$$

where the phase aberrations of the telescope are expressed by $\varphi(x, y, \lambda)$.

For an aberration-free telescope (i.e., $\varphi(x, y, \lambda) = 0$) the amplitude transfer function and its modulus are identical (as well as being circularly symmetric here) so that, in terms of the spatial frequency parameter,, we can write

$$\begin{aligned} \left|M_{T,Coh}(\nu, \lambda)\right| = M_{T,Coh}(\nu, \lambda) &= 1 \quad \text{for} \quad \frac{d}{2 \cdot \lambda \cdot f} \leq \nu \leq \frac{D}{2 \cdot \lambda \cdot f}, \\ &= 0 \qquad \text{otherwise.} \end{aligned} \tag{7.26}$$

Figure 7.7 shows the response calculated from this equation for the case, $= 0.2 \cdot D$, the plot shown applying equally to aberrated and unaberrated telescopes.

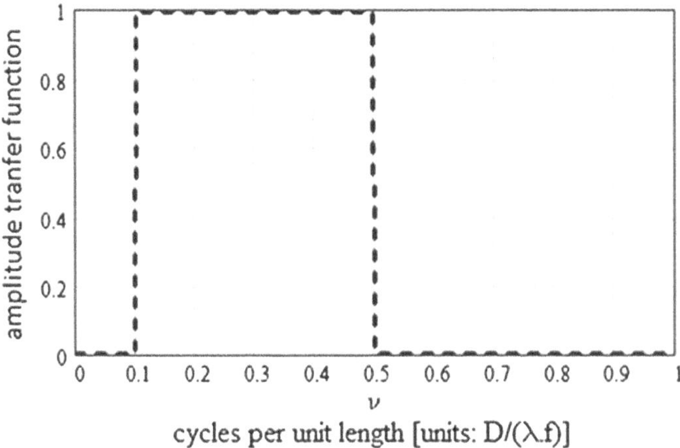


**Fig. 7.7** The modulus of the amplitude transfer function for coherent illumination for a telescope with circular aperture and circular central obstruction for the case $d = 0.2 \cdot D$. The modulus does not depend on telescope aberrations

## 7.5 Two-Point Two-Wavelength Correlation Function of the Complex Amplitudes in the Image

The two-point two-wavelength autocovariance function of the complex amplitudes in the image of a point object, $\langle U(u', v', \lambda_1) \cdot U^*(u, v, \lambda_2)\rangle$, determines the key statistical properties of the complex amplitudes and intensities in the image of this, the simplest type of object. (Without loss of generality, we can consider the respective mean values of the complex amplitudes, $\langle U(u', v', \lambda_1)\rangle$ and $\langle U(u, v, \lambda_2)\rangle$ to be zero.) Function $\langle U(u', v', \lambda_1) \cdot U^*(u, v, \lambda_2)\rangle$ may be considered as a generalized form of the mutual intensity function of the complex amplitudes in the image. Using 7.5, this function may be expressed in terms of the complex amplitudes in the telescope pupil plane as follows:

$$\begin{aligned}\langle U(u', v', \lambda_1) \cdot U^*(u, v, \lambda_2)\rangle &= \frac{1}{\lambda_1 \cdot \lambda_2} \\ &\times \int_{-\infty}^{\infty}\int_{-\infty}^{\infty}\int_{-\infty}^{\infty}\int_{-\infty}^{\infty} \langle B(x', y', \lambda_1) \cdot B^*(x, y, \lambda_2)\rangle \cdot K(x', y', \lambda_1) \cdot K^*(x, y, \lambda_2) \\ &\times \exp\left[\frac{-2 \cdot \pi \cdot i}{f} \cdot \left(\frac{u' \cdot x' + v' \cdot y'}{\lambda_1} - \frac{u \cdot x + v \cdot y}{\lambda_2}\right)\right] \cdot dx \cdot dy \cdot dx' \cdot dy',\end{aligned} \tag{7.27}$$

where $K(x', y', \lambda_1)$ and $K(x, y, \lambda_2)$ denote the telescope pupil functions at the two wavelengths. In general, these functions are both complex; both the amplitude and phase may vary independently.

Because the atmospheric refractive index field was previously assumed spatially stationary (Sect. 5.2), the unit-normalized form of $\left\langle B\left(x', y', \lambda_1\right) \cdot B^*(x, y, \lambda_2)\right\rangle$ must then be identical to the two-point two-wavelength correlation function, $S\left(x' - x, y' - y, \lambda_1, \lambda_2\right)$, which arose in Chap. 6. Thus, we may write

$$S(x' - x, y' - y, \lambda_1, \lambda_2) = \frac{\langle B(x', y', \lambda_1) \cdot B^*(x, y, \lambda_2)\rangle}{[\langle B(x', y', \lambda_1) \cdot B^*(x', y', \lambda_1)\rangle \cdot \langle B(x, y, \lambda_2) \cdot B^*(x, y, \lambda_2)\rangle]^{\frac{1}{2}}}. \tag{7.28}$$

Function $S(x' - x, y' - y, \lambda_1, \lambda_2)$ may be calculated from the integrated OPD fluctuation over the propagation path, $H(x, y)$, using either 6.44 or 6.45.

Combining 7.27 and 7.28, we may write $\langle U\left(u', v', \lambda_1\right) \cdot U^*(u, v, \lambda_2)\rangle$ in the form

$$\begin{aligned}\left\langle U\left(u', v', \lambda_1\right) \cdot U^*(u, v, \lambda_2)\right\rangle &= \frac{1}{\lambda_1 \cdot \lambda_2} \\ &\times \int_{-\infty}^{\infty}\int_{-\infty}^{\infty}\int_{-\infty}^{\infty}\int_{-\infty}^{\infty} S\left(x' - x, y' - y, \lambda_1, \lambda_2\right) \cdot K\left(x', y', \lambda_1\right) \cdot K^*(x, y, \lambda_2) \\ &\times \exp\left[\frac{-2 \cdot \pi \cdot i}{f} \cdot \left(\frac{u' \cdot x' + v \cdot' y'}{\lambda_1} - \frac{u \cdot x + v \cdot y}{\lambda_2}\right)\right] \cdot dx \cdot dy \cdot dx' \cdot dy'.\end{aligned} \tag{7.29}$$

By using the coordinate transformations,

$$x' = x + \xi, \tag{7.30}$$

$$y' = y + \eta, \tag{7.31}$$

Equation 7.29 may be re-expressed in the form,

$$\begin{aligned}\left\langle U\left(u', v', \lambda_1\right) \cdot U^*(u, v, \lambda_2)\right\rangle &= \frac{1}{\lambda_1 \cdot \lambda_2} \cdot \int_{-\infty}^{\infty}\int_{-\infty}^{\infty}\int_{-\infty}^{\infty}\int_{-\infty}^{\infty} S(\xi, \eta, \lambda_1, \lambda_2) \cdot K(x + \xi, y + \eta, \lambda_1) \cdot K^*(x, y, \lambda_2) \\ &\times \exp\left\{\frac{-2 \cdot \pi \cdot i}{f} \cdot \left[x \cdot \left(\frac{u'}{\lambda_1} - \frac{u}{\lambda_2}\right) + y \cdot \left(\frac{v'}{\lambda_1} - \frac{v}{\lambda_2}\right) + \xi \cdot \frac{u'}{\lambda_1} + \eta \frac{v'}{\lambda_1}\right]\right\} \cdot dx \cdot dy \cdot d\xi \cdot d\eta.\end{aligned} \tag{7.32}$$

Regrouping terms in the above equation gives the desired expression for $\langle U(u', v', \lambda_1) \cdot U^*(u, v, \lambda_2)\rangle$ :

$$\left\langle U\left(u', v', \lambda_1\right) \cdot U^*(u, v, \lambda_2)\right\rangle = \frac{1}{\lambda_1 \cdot \lambda_2} \cdot \int_{-\infty}^{\infty}\int_{-\infty}^{\infty} S(\xi, \eta, \lambda_1, \lambda_2)$$

$$\times \left\{ \int_{-\infty}^{\infty} \int_{-\infty}^{\infty} K(x+\xi, y+\eta, \lambda_1) \cdot K^*(x, y, \lambda_2) \right.$$
$$\left. \times \exp\left[\frac{-2 \cdot \pi \cdot i}{f} \cdot \left[x \cdot \left(\frac{u'}{\lambda_1} - \frac{u}{\lambda_2}\right) + y \cdot \left(\frac{v'}{\lambda_1} - \frac{v}{\lambda_2}\right)\right]\right] \cdot dx \cdot dy \right\}$$
$$\times \exp\left[\frac{-2 \cdot \pi \cdot i}{\lambda_1 \cdot f} \cdot \left(u' \cdot \xi + v' \cdot \eta\right)\right] \cdot d\xi \cdot d\eta. \tag{7.33}$$

### 7.5.1 *Characterizing the Influence of the Telescope Optics*

The influence of the telescope optics on the two-point two-wavelength correlation function of the complex amplitudes in the image is expressed by the double-integral term in the curly brackets in 7.33. The unit-normalized form of this term, which we denote by $S_T\left(u', v', u, v, \xi, \eta, \lambda_1, \lambda_2\right)$, may be written

$$S_T\left(u', v', u, v, \xi, \eta, \lambda_1, \lambda_2\right) =$$
$$\frac{\int_{-\infty}^{\infty} \int_{-\infty}^{\infty} K(x+\xi, y+\eta, \lambda_1) \cdot K^*(x, y, \lambda_2) \cdot \exp\left\{\frac{-2 \cdot \pi \cdot i}{f} \cdot \left[x \cdot \left(\frac{u'}{\lambda_1} - \frac{u}{\lambda_2}\right) + y \cdot \left(\frac{v'}{\lambda_1} - \frac{v}{\lambda_2}\right)\right]\right\} \cdot dx \cdot dy}{\left[\int_{-\infty}^{\infty} \int_{-\infty}^{\infty} K(x, y, \lambda_1) \cdot K^*(x, y, \lambda_1) \cdot dx \cdot dy \cdot \int_{-\infty}^{\infty} \int_{-\infty}^{\infty} K(x, y, \lambda_2) \cdot K^*(x, y, \lambda_2) \cdot dx \cdot dy\right]^{\frac{1}{2}}}. \tag{7.34}$$

Function, $\langle U\left(u', v', \lambda_1\right) \cdot U^*(u, v, \lambda_2)\rangle$, given above by 7.33, may now be expressed more concisely in terms of $S_T\left(u', v', u, v, \xi, \eta, \lambda_1, \lambda_2\right)$:

$$\left\langle U\left(u', v', \lambda_1\right) \cdot U^*(u, v, \lambda_2)\right\rangle =$$
$$\frac{1}{\lambda_1 \cdot \lambda_2} \cdot \int_{-\infty}^{\infty} \int_{-\infty}^{\infty} S(\xi, \eta, \lambda_1, \lambda_2) \cdot S_T\left(u', v', u, v, \xi, \eta, \lambda_1, \lambda_2\right)$$
$$\times \exp\left[\frac{-2 \cdot \pi \cdot i}{\lambda_1 \cdot f} \cdot \left(u' \cdot \xi + v' \cdot \eta\right)\right] \cdot d\xi \cdot d\eta \tag{7.35}$$

It might be noted that the normalizations of 7.33 and 7.35 can be seen to be slightly different, so that although the shapes of the functions given by these equations are the same, the magnitudes are different. The issue of normalization is addressed in the next section when we give the unit-normalized form of the function—i.e., the autocorrelation function. Although function $S(\xi, \eta, \lambda_1, \lambda_2)$ is spatially stationary as discussed previously,[1] the consequence of $S_T\left(u', v', u, v, \xi, \eta, \lambda_1, \lambda_2\right)$ not being spatially stationary is that the telescope image properties are generally non-stationary.

[1] The spatial stationarity of S follows as a consequence of the spatially stationarity assumption made in Sect. 3.6.1 for the refractive index field, n(x, y, z).

The variation of the average intensity over the image is the most obvious manifestation of such non-stationary behavior. Radial structure in polychromatic stellar speckle images (cf., Fig. 4.9) is yet another manifestation, further discussion of which is given in Chap. 11 (Sect. 11.5.1.1.

### 7.5.2 Unit-Normalized Form of the Function

It is useful at this point to define the unit-normalized form of $\langle U(u', v', \lambda_1) \cdot U^*(u, v, \lambda_2)\rangle$, which we denote by $\mu(u', v', u, v, \lambda_1, \lambda_2)$:

$$\mu(u', v', u, v, \lambda_1, \lambda_2) = \frac{\langle U(u', v', \lambda_1) \cdot U^*(u, v, \lambda_2)\rangle}{[\langle I(u', v', \lambda_1)\rangle \cdot \langle I(u, v, \lambda_2)\rangle]^{\frac{1}{2}}}. \tag{7.36}$$

An explicit expression for the numerator in the above equation was given previously by 7.33. Explicit expressions for the average intensity quantities in the denominator, $\langle I(u', v', \lambda_1)\rangle$ and $\langle I(u, v, \lambda_2)\rangle$ can also be obtained from 7.33 as follows:

$$\begin{aligned}\langle I(u', v', \lambda_1)\rangle &= \langle U(u', v', \lambda_1) \cdot U^*(u', v', \lambda_1)\rangle \\ &= \frac{1}{\lambda_1^2} \cdot \int_{-\infty}^{\infty}\int_{-\infty}^{\infty} M(\xi, \eta, \lambda_1) \cdot \left[\int_{-\infty}^{\infty}\int_{-\infty}^{\infty} K(x, y, \lambda_1) \cdot K^*(x, y, \lambda_1) \cdot dx \cdot dy\right] \\ &\quad \times \exp\left[\frac{-2 \cdot \pi \cdot i}{\lambda_1 \cdot f} \cdot (u' \cdot \xi + v' \cdot \eta)\right] \cdot d\xi \cdot d\eta,\end{aligned} \tag{7.37}$$

$$\begin{aligned}\langle I(u, v, \lambda_2)\rangle &= \langle U(u, v, \lambda_2) \cdot U^*(u, v, \lambda_2)\rangle \\ &= \frac{1}{\lambda_2^2} \cdot \int_{-\infty}^{\infty}\int_{-\infty}^{\infty} M(\xi, \eta, \lambda_2) \cdot \left[\int_{-\infty}^{\infty}\int_{-\infty}^{\infty} K(x, y, \lambda_2) \cdot K^*(x, y, \lambda_2) \cdot dx \cdot dy\right] \\ &\quad \times \exp\left[\frac{-2 \cdot \pi \cdot i}{\lambda_2 \cdot f} \cdot (u \cdot \xi + v \cdot \eta)\right] \cdot d\xi \cdot d\eta\end{aligned} \tag{7.38}$$

where we note that $S(\xi, \eta, \lambda_1, \lambda_1) = M(\xi, \eta, \lambda_1)$ and $S(\xi, \eta, \lambda_2, \lambda_2) = M(\xi, \eta, \lambda_2)$.

### 7.5.3 The Function at a Single Point in the Image

In the limit, where $u' \to u$ and $v' \to v$, function $\langle U(u', v', \lambda_1) \cdot U^*(u, v, \lambda_2)\rangle$ degenerates into the two-wavelength autocovariance function of the complex amplitudes at a single point in the image. Using 7.33 as our starting point, we can see how this function can be written in the form,

$$\begin{aligned}\langle U(u,v,\lambda_1)\cdot U^*(u,v,\lambda_2)\rangle &= \frac{1}{\lambda_1\cdot\lambda_2}\cdot\int_{-\infty}^{\infty}\int_{-\infty}^{\infty} S(\xi,\eta,\lambda_1,\lambda_2)\\ &\times\Bigg\{\int_{-\infty}^{\infty}\int_{-\infty}^{\infty} K(x+\xi,y+\eta,\lambda_1)\cdot K^*(x,y,\lambda_2)\\ &\times\exp\left[\frac{-2\cdot\pi\cdot i}{f}\cdot(x\cdot u+y\cdot v)\cdot\left(\frac{1}{\lambda_1}-\frac{1}{\lambda_2}\right)\right]\cdot dx\cdot dy\Bigg\}\\ &\times\exp\left[\frac{-2\cdot\pi\cdot i}{\lambda_1\cdot f}\cdot(u\cdot\xi+v\cdot\eta)\right]\cdot d\xi\cdot d\eta. \end{aligned} \tag{7.39}$$

In this limit, the unit-normalized version of the function, $\mu(u,v,u,v,\lambda_1,\lambda_2)$, can be written as (cf., 7.36),

$$\mu(u,v,u,v,\lambda_1,\lambda_2) = \frac{\langle U(u,v,\lambda_1)\cdot U^*(u,v,\lambda_2)\rangle}{[\langle I(u,v,\lambda_1)\rangle\cdot\langle I(u,v,\lambda_2)\rangle]^{\frac{1}{2}}}. \tag{7.40}$$

Function $\mu(u,v,u,v,\lambda_1,\lambda_2)$ is commonly referred to as the spectral correlation function.

## *7.5.4 The Spectral Correlation Function at the Center of a Point-Object Image*

An expression for $\langle U(0,0,\lambda_1)\cdot U^*(0,0,\lambda_2)\rangle$ may be obtained from 7.39 by setting u = 0, v = 0:

$$\begin{aligned}\langle U(0,0,\lambda_1)\cdot U^*(0,0,\lambda_2)\rangle &= \frac{1}{\lambda_1\cdot\lambda_2}\cdot\int_{-\infty}^{\infty}\int_{-\infty}^{\infty} S(\xi,\eta,\lambda_1,\lambda_2)\\ &\times\left[\int_{-\infty}^{\infty}\int_{-\infty}^{\infty} K(x+\xi,y+\eta,\lambda_1)\cdot K^*(x,y,\lambda_2)\cdot dx\cdot dy\right]\cdot d\xi\cdot d\eta. \end{aligned} \tag{7.41}$$

#### 7.5.4.1 Two-Wavelength Correlation Coefficient

The spectral correlation function at the center of the image, $\mu(0,0,0,0,\lambda_1,\lambda_2)$, follows directly from the above expression in the form,

$$\mu(0,0,0,0,\lambda_1,\lambda_2)$$
$$= \int_{-\infty}^{\infty}\int_{-\infty}^{\infty} S(\xi,\eta,\lambda_1,\lambda_2)$$
$$\times \left[\int_{-\infty}^{\infty}\int_{-\infty}^{\infty} K(x+\xi,y+\eta,\lambda_1)\cdot K^*(x,y,\lambda_2)\cdot dx\cdot dy\right]\cdot d\xi\cdot d\eta$$
$$\times \frac{1}{\left\{\int_{-\infty}^{\infty}\int_{-\infty}^{\infty} M(\xi,\eta,\lambda_1)\cdot\left[\int_{-\infty}^{\infty}\int_{-\infty}^{\infty} K(x+\xi,y+\eta,\lambda_1)\cdot K^*(x,y,\lambda_1)\cdot dx\cdot dy\right]\cdot d\xi\cdot d\eta\right\}^{\frac{1}{2}}}$$
$$\times \frac{1}{\left\{\int_{-\infty}^{\infty}\int_{-\infty}^{\infty} M(\xi,\eta,\lambda_2)\cdot\left[\int_{-\infty}^{\infty}\int_{-\infty}^{\infty} K(x+\xi,y+\eta,\lambda_2)\cdot K^*(x,y,\lambda_2)\cdot dx\cdot dy\right]\cdot d\xi\cdot d\eta\right\}^{\frac{1}{2}}}. \quad (7.42)$$

Function $\mu(0, 0, 0, 0, \lambda_1, \lambda_2)$ may be regarded as the correlation coefficient for the complex amplitudes in the center of the image at wavelengths $\lambda_1$ and $\lambda_2$. In general, this coefficient is complex; the values taken by its modulus are constrained as follows:

$$0 \le |\mu(0, 0, 0, 0, \lambda_1, \lambda_2)| \le 1. \quad (7.43)$$

When $\mu$ takes the value zero, no correlation exists between the complex amplitudes at the two wavelengths. A positive real value for $\mu$ infers that as one complex amplitude increases, the other also increases. A negative real value for $\mu$ infers that as one complex amplitude increases, the other also increases, but with opposite phase. Complex values infer phase angle differences other than 0 or $\pi$.

## 7.6 Complex Coherence Factor of the Complex Amplitude in the Image

When the two wavelengths coalesce, $\lambda_1 \to \lambda_2 \to 0$, the two-point two-wavelength correlation function degenerates, becoming the autocovariance function of the complex amplitude in the image. This function is simply the function, $J\left(u', v', u, v, \lambda\right)$, which was previously defined by 3.26. Function $J\left(u', v', u, v, \lambda\right)$ may be referred to as the mutual intensity function associated with the complex amplitude in the image.

An explicit form for $J\left(u', v', u, v, \lambda\right)$ may be obtained from 7.33 by setting $\lambda_1 = \lambda_2 = \lambda$. For this case, we note that $S(\xi, \eta, \lambda_1, \lambda_2)$ degenerates into $M(\xi, \eta, \lambda)$, so that we may write

$$J\left(u', v', u, v, \lambda\right) = \frac{1}{\lambda^2} \cdot \int_{-\infty}^{\infty}\int_{-\infty}^{\infty} M(\xi, \eta, \lambda) \cdot \exp\left[\frac{-2\cdot\pi\cdot i}{\lambda\cdot f} \cdot \left(u'\cdot\xi + v'\cdot\eta\right)\right]$$

$$\times \left\{ \int_{-\infty}^{\infty} \int_{-\infty}^{\infty} K(x+\xi, y+\eta, \lambda) \cdot K^*(x, y, \lambda) \right.$$

$$\left. \times \exp\left[ \frac{-2 \cdot \pi \cdot i}{\lambda \cdot f} \cdot \left( x \cdot \left( u' - u \right) + y \cdot \left( v' - v \right) \right) \right] \cdot dx \cdot dy \right\} \cdot d\xi \cdot d\eta. \quad (7.44)$$

The unit-normalized mutual intensity function, i.e., the complex coherence factor, which we denote by $\mu\left(u', v', u, v, \lambda\right)$, is given by the unit-normalized form,

$$\mu\left(u', v', u, v, \lambda\right) = \frac{J\left(u', v', u, v, \lambda\right)}{\left[\langle I(u', v', \lambda)\rangle \cdot \langle I(u, v, \lambda)\rangle\right]^{\frac{1}{2}}}. \quad (7.45)$$

Equation 7.44 can be used to give explicit expressions for the function, $J\left(u', v', u, v, \lambda\right)$, that takes account of both the atmospheric MTF and the telescope OTF. Similarly, explicit expressions can be obtained from 7.44 for the two intensity terms that appear in the denominator of the above equation. Inserting these expressions into the above equation provides an explicit expression for $\mu\left(u', v', u, v, \lambda\right)$. However, because of its inordinate length, we do not attempt to show the expression here.

## 7.7 Average Intensity Envelopes for Point-Object Images

The intensity in a point-object image may be given in terms of the complex amplitude in the image (cf., 3.21) by

$$I(u, v, \lambda) = U(u, v, \lambda) \cdot U^*(u, v, \lambda). \quad (7.46)$$

The intensity may also be expressed in terms of angular coordinates,

$$I(\alpha, \beta, \lambda) = U(\alpha, \beta, \lambda) \cdot U^*(\alpha, \beta, \lambda). \quad (7.47)$$

The average intensity in the image, $\langle I(u, v, \lambda)\rangle$, may be considered as the degenerate case of function, $J\left(u', v', u, v, \lambda\right)$, in the limit, $u' = u$, $v' = v$. Equation 7.44 allows us to give an explicit expression for this function:

$$\langle I(u, v, \lambda)\rangle = \frac{1}{\lambda^2} \cdot \int_{-\infty}^{\infty} \int_{-\infty}^{\infty} M(\xi, \eta, \lambda)$$

$$\times \left\{ \int_{-\infty}^{\infty} \int_{-\infty}^{\infty} K(x+\xi, y+\eta, \lambda) \cdot K^*(x, y, \lambda) \cdot dx \cdot dy \right\}$$

$$\times \exp\left[\frac{-2\cdot\pi\cdot i}{\lambda\cdot f}\cdot(u\cdot\xi+v\cdot\eta)\right]\cdot d\xi\cdot d\eta \tag{7.48}$$

**Note on dimensions**: The right-hand side of the above equation is set out in a way that gives the average intensity as a dimensionless quantity. Many other similarly dimensionless expressions are scattered throughout the book, not only for the intensity, but also for the complex amplitude and other related functions such as the mutual intensity function. In instances where absolute, rather than relative, quantities are required, clearly it would be necessary to insert appropriately chosen multiplier constants to associate suitable dimensions. In this regard, the Electro-Optics Handbook (1974) may be found helpful.

The double integral in curly brackets in 7.48 is recognized as the autocovariance function of the telescope pupil function. The unit-normalized form of this function is the telescope OTF for incoherent illumination, $M_T(\xi, \eta, \lambda)$, (cf., 7.10). Replacing the double integral in 7.48 by this function gives the average image intensity, $\langle I(u, v, \lambda)\rangle$, as the Fourier transform of the product of the atmospheric MTF, $M(\xi, \eta, \lambda)$, and the telescope OTF for incoherent illumination, $M_T(\xi, \eta, \lambda)$, which may be written in the form,

$$\langle I(u, v, \lambda)\rangle = \frac{1}{\lambda^2}\cdot\int_{-\infty}^{\infty}\int_{-\infty}^{\infty} M(\xi, \eta, \lambda)\cdot M_T(\xi, \eta, \lambda)$$

$$\times \exp\left[\frac{-2\cdot\pi\cdot i}{\lambda\cdot f}\cdot(u\cdot\xi+v\cdot\eta)\right]\cdot d\xi\cdot d\eta \tag{7.49}$$

## 7.8 Statistics of the Complex Amplitude in Point-Object Images Formed by Large Telescopes

In cases where the width of the atmospheric MTF is significantly smaller than the telescope diameter, the image-forming waves in the telescope pupil plane may be considered (according to the Huygens–Fresnel principle) as constituted by a large number of elementary phasors with uncorrelated random phases. If the phase distribution of the light waves in the telescope pupil is also assumed to be approximately uniform in the primary interval, $-\pi$ to $\pi$, the statistical properties of the complex amplitude in the image will be those of a circular Gaussian complex random variable (Dainty, 1984). As discussed in Chap. 11, the real and imaginary parts of the complex amplitude in this case are uncorrelated, but both have the same variance and both have zero mean values. To ensure approximately uniform distribution of phase in the range, $-\pi$ to $\pi$, the rms OPD fluctuation of the waves in the pupil $\sigma$ (7.29) merely has to satisfy the condition,

$$\sigma > 0.4 \cdot \lambda. \tag{7.50}$$

### 7.8.1 *Reed's Theorem for Gaussian-Distributed Complex Random Variables*

When the complex amplitudes in the image at any arbitrary pair of wavelengths, $\lambda_1$ and $\lambda_2$, separately obey circular Gaussian statistics, an important theorem due to Reed (1962) provides the following statistical relationship between the two-point two-wavelength covariance functions of the intensities and the complex amplitudes in the image:

$$\begin{aligned}&\left\langle I\left(u', v', \lambda_1\right) \cdot I(u, v, \lambda_2)\right\rangle - \langle I\left(u', v', \lambda_1\right)\rangle \cdot \langle I(u, v, \lambda_2)\rangle \\ &= \left|\left\langle U\left(u', v', \lambda_1\right) \cdot U^*(u, v, \lambda_2)\right\rangle\right|^2.\end{aligned} \tag{7.51}$$

The above result can also be expressed in the form,

$$\begin{aligned}&\frac{\left\langle I\left(u', v', \lambda_1\right) \cdot I(u, v, \lambda_2)\right\rangle}{\langle I(u', v', \lambda_1)\rangle \cdot \langle I(u, v, \lambda_2)\rangle} - 1 \\ &= \frac{\left|\left\langle U\left(u', v', \lambda_1\right) \cdot U^*(u, v, \lambda_2)\right\rangle\right|^2}{\langle I(u', v', \lambda_1)\rangle \cdot \langle I(u, v, \lambda_2)\rangle}.\end{aligned} \tag{7.52}$$

The above result is of fundamental importance. The left-hand side of the equation has the form of a generalized two-point two-wavelength correlation function of the intensity in the image (cf., Sect. 3.6.3), while the right-hand side is seen as the squared modulus of the two-point two-wavelength correlation function of the corresponding complex amplitudes in the image.

### 7.8.2 *Unit-Normalized Two-Point Two-Wavelength Correlation Function of the Image Intensities*

The right-hand side of 7.52 is recognized as the squared modulus of the right-hand side of 7.36. Thus, we may write

$$\frac{\left\langle I\left(u', v', \lambda_1\right) \cdot I(u, v, \lambda_2)\right\rangle}{\langle I(u', v', \lambda_1)\rangle \cdot \langle I(u, v, \lambda_2)\rangle} - 1 = \left|\mu\left(u', v', u, v, \lambda_1, \lambda_2\right)\right|^2. \tag{7.53}$$

Recalling 7.43, the terms on both sides of the above equation take values in the range 0 to 1.

### 7.8.3 Two-Wavelength Correlation Function of the Intensity at a Single Point in the Image

When the two points coalesce (i.e., $u' = u$ and $v' = v$), 7.53 reduces to

$$\frac{\langle I(u, v, \lambda_1) \cdot I(u, v, \lambda_2) \rangle}{\langle I(u, v, \lambda_1) \rangle \cdot \langle I(u, v, \lambda_2) \rangle} - 1 = |\mu(u, v, u, v, \lambda_1, \lambda_2)|^2. \tag{7.54}$$

Both sides of this equation again take values in the range, 0 to 1.

### 7.8.4 Two-Wavelength Correlation Function of the Complex Amplitude at a Single Point in the Image

Suppose that the two-point two-wavelength correlation function of the complex amplitudes of the light waves in the telescope pupil, $S(\xi, \eta, \lambda_1, \lambda_2)$, falls to zero for $\xi$ and $\eta$ argument values significantly smaller than the dimensions of the telescope aperture. In this case, we may regard this function as though it were a Dirac delta function. This kind of approximation has been used previously by Beckmann and Spizzichino (1963), Goodman (1963), and Dainty (1984), and in spite of appearing coarse, it usually leads to quite accurate predictions. Using 6.45 as the starting point, $S(\xi, \eta, \lambda_1, \lambda_2)$ may be approximated in terms of the Dirac delta function, $\delta(\xi, \eta)$, by

$$S(\xi, \eta, \lambda_1, \lambda_2) = \delta(\xi, \eta) \cdot \exp\left[-2 \cdot \pi^2 \cdot \sigma^2 \cdot \left(\frac{1}{\lambda_1} - \frac{1}{\lambda_2}\right)^2\right]. \tag{7.55}$$

Combining this approximation for $S(\xi, \eta, \lambda_1, \lambda_2)$ with 7.33, and by allowing $u' \to u$ and $v' \to v$, we obtain

$$\begin{aligned} \langle U(u, v, \lambda_1) \cdot U^*(u, v, \lambda_2) \rangle &= \frac{1}{\lambda_1 \cdot \lambda_2} \cdot \exp\left[-2 \cdot \pi^2 \cdot \sigma^2 \cdot \left(\frac{1}{\lambda_1} - \frac{1}{\lambda_2}\right)^2\right] \\ &\times \int_{-\infty}^{\infty} \int_{-\infty}^{\infty} K(x, y, \lambda_1) \cdot K^*(x, y, \lambda_2) \cdot \exp\left[\frac{-2 \cdot \pi \cdot i}{f} \cdot (x \cdot u + y \cdot v) \cdot \left(\frac{1}{\lambda_1} - \frac{1}{\lambda_2}\right)\right] \cdot dx \cdot dy. \end{aligned} \tag{7.56}$$

Further, by unit-normalizing both sides of this equation (cf., 7.36), we obtain the result,

$$\mu(u, v, u, v, \lambda_1, \lambda_2) = \exp\left[-2 \cdot \pi^2 \cdot \sigma^2 \cdot \left(\frac{1}{\lambda_1} - \frac{1}{\lambda_2}\right)^2\right]$$
$$\times \frac{\int_{-\infty}^{\infty}\int_{-\infty}^{\infty} K(x, y, \lambda_1) \cdot K^*(x, y, \lambda_2) \cdot \exp\left[\frac{-2 \cdot \pi \cdot i}{f} \cdot (x \cdot u + y \cdot v) \cdot \left(\frac{1}{\lambda_1} - \frac{1}{\lambda_2}\right)\right] \cdot dx \cdot dy}{\left\{\int_{-\infty}^{\infty}\int_{-\infty}^{\infty} K(x, y, \lambda_1) \cdot K^*(x, y, \lambda_1) \cdot dx \cdot dy \cdot \int_{-\infty}^{\infty}\int_{-\infty}^{\infty} K(x, y, \lambda_2) \cdot K^*(x, y, \lambda_2) \cdot dx \cdot dy\right\}^{\frac{1}{2}}}. \quad (7.57)$$

#### 7.8.4.1 The Function at Image Center

The degree of correlation between the complex amplitudes at wavelengths $\lambda_1$ and $\lambda_2$ in the center of a point-object image may be obtained by setting $u = 0$ and $v = 0$ in 7.57:

$$\mu(0, 0, 0, 0, \lambda_1, \lambda_2) = \exp\left[-2 \cdot \pi^2 \cdot \sigma^2 \cdot \left(\frac{1}{\lambda_1} - \frac{1}{\lambda_2}\right)^2\right]$$
$$\times \frac{\int_{-\infty}^{\infty}\int_{-\infty}^{\infty} K(x, y, \lambda_1) \cdot K^*(x, y, \lambda_2) \cdot dx \cdot dy}{\left\{\int_{-\infty}^{\infty}\int_{-\infty}^{\infty} K(x, y, \lambda_1) \cdot K^*(x, y, \lambda_1) \cdot dx \cdot dy \cdot \int_{-\infty}^{\infty}\int_{-\infty}^{\infty} K(x, y, \lambda_2) \cdot K^*(x, y, \lambda_2) \cdot dx \cdot dy\right\}^{\frac{1}{2}}}. \quad (7.58)$$

Evaluations of $\mu(0, 0, 0, 0, \lambda_1, \lambda_2)$ from the above expression produce both real and complex values, with the function always taking values in the range 0 to 1.

#### 7.8.4.2 The Function at Image Center for Diffraction-Limited Telescopes

In the case where the telescope is diffraction-limited over a wavelength band that includes both $\lambda_1$ and $\lambda_2$, both pupil functions, $K(x, y, \lambda_1)$ and $K(x, y, \lambda_2)$, will be entirely real-valued. It follows that the telescope-related term in 7.58 must then always take the value unity.

$$\frac{\int_{-\infty}^{\infty}\int_{-\infty}^{\infty} K(x, y, \lambda_1) \cdot K^*(x, y, \lambda_2) \cdot dx \cdot dy}{\left\{\int_{-\infty}^{\infty}\int_{-\infty}^{\infty} K(x, y, \lambda_1) \cdot K^*(x, y, \lambda_1) \cdot dx \cdot dy \cdot \int_{-\infty}^{\infty}\int_{-\infty}^{\infty} K(x, y, \lambda_2) \cdot K^*(x, y, \lambda_2) \cdot dx \cdot dy\right\}^{\frac{1}{2}}} = 1. \quad (7.59)$$

The above result applies to all diffraction-limited telescopes regardless of the shape of their apertures. In particular, it applies to diffraction-limited circular aperture telescopes for which the pupil function was previously given by 7.7.

By combining 7.58 and 7.59, we obtain the desired expression for the degree of correlation between the complex amplitudes at $\lambda_1$ and $\lambda_2$ in the center of an image formed by any diffraction-limited telescope adjusted for correct focus:

$$\mu(0, 0, 0, 0, \lambda_1, \lambda_2) = \exp\left[-2 \cdot \pi^2 \cdot \sigma^2 \cdot \left(\frac{1}{\lambda_1} - \frac{1}{\lambda_2}\right)^2\right]. \quad (7.60)$$

Because the right-hand side of the above equation is entirely real, function $\mu(0, 0, 0, 0, \lambda_1, \lambda_2)$ is constrained to take real values in the range.

$0 \le \mu(0, 0, 0, 0, \lambda_1, \lambda_2) \le 1$. In Chap. 8 (Sect. 8.2), 7.60 will be used as the basis of a technique for measuring the rms OPD fluctuation, $\sigma$, of the image-forming light waves in the pupil of telescopes observing over atmospheric paths.

## 7.9 OTF for an Entire End-To-End Imaging Path

The product of the atmospheric MTF and the telescope OTF, which we denote by $M_E(\xi, \eta, \lambda)$, may be considered as the OTF corresponding to the entire end-to-end (i.e., object-to-image) propagation/imaging path:

$$M_E(\xi, \eta, \lambda) = M(\xi, \eta, \lambda) \cdot M_T(\xi, \eta, \lambda). \tag{7.61}$$

The atmospheric portion of the propagation path (function $M(\xi, \eta, \lambda)$ in the above equation) can be considered to include the effects of turbulence in the dome path portion. Similarly, for improperly focused telescopes, the effect of defocus may be considered included in the telescope OTF, $M_T(\xi, \eta, \lambda)$.

By taking the inverse Fourier transform of both sides of 7.49, $M_E(\xi, \eta, \lambda)$ may be expressed as the unit-normalized Fourier transform of the average intensity distribution in the image,

$$M_E(\xi, \eta, \lambda) = \frac{\int_{-\infty}^{\infty}\int_{-\infty}^{\infty} \langle I(u, v, \lambda)\rangle \cdot \exp\left[\frac{2\cdot\pi\cdot i}{\lambda\cdot f} \cdot (u \cdot \xi + v \cdot \eta)\right] \cdot du \cdot dv}{\int_{-\infty}^{\infty}\int_{-\infty}^{\infty} \langle I(u, v, \lambda)\rangle \cdot du \cdot dv}. \tag{7.62}$$

### *7.9.1 OTF for an Entire End-To-End Imaging Path for Space Telescopes*

For space-based telescopes, which of course lie entirely above Earth's atmosphere, and where we assume that they are looking out into space and not down into the atmosphere, the "atmospheric MTF" in the above expression, $M(\xi, \eta, \lambda)$, may be considered as taking the value unity for all argument values (i.e., $M(\xi, \eta, \lambda) = 1$). For this case, 7.61 simplifies to the form,

$$M_E(\xi, \eta, \lambda) = M_T(\xi, \eta, \lambda). \tag{7.63}$$

Because there is now no longer a random variable, the averaging brackets straddling the intensity function $I(u, v, \lambda)$ in 7.62 become redundant and the intensity envelope may then be regarded simply as the intensity point-spread function of the telescope, $PSF_I(u, v, \lambda)$. For this case, 7.62 reduces to the familiar Fourier transform relationship between the telescope OTF and the intensity point-spread function (Born & Wolf, 2003):

$$M_T(\xi, \eta, \lambda) = \frac{\int_{-\infty}^{\infty}\int_{-\infty}^{\infty} PSF_I(u, v, \lambda) \cdot \exp\left[\frac{2\cdot\pi\cdot i}{\lambda\cdot f} \cdot (u \cdot \xi + v \cdot \eta)\right] \cdot du \cdot dv}{\int_{-\infty}^{\infty}\int_{-\infty}^{\infty} PSF_I(u, v, \lambda) \cdot du \cdot dv}. \quad (7.64)$$

The normalization of this equation is, as required, such that $M_T(0, 0, \lambda) = 1$.

By taking the inverse Fourier transform of each side of 7.64, the unit-normalized form of the intensity point-spread function may be expressed in terms of the telescope OTF by

$$PSF_I(u, v, \lambda) = \frac{\int_{-\infty}^{\infty}\int_{-\infty}^{\infty} M_T(\xi, \eta, \lambda) \cdot \exp\left[\frac{-2\cdot\pi\cdot i}{\lambda\cdot f} \cdot (u \cdot \xi + v \cdot \eta)\right] \cdot d\xi \cdot d\eta}{\int_{-\infty}^{\infty}\int_{-\infty}^{\infty} M_T(\xi, \eta, \lambda) \cdot d\xi \cdot d\eta}. \quad (7.65)$$

### 7.9.2 OTF and Intensity PSF for a Diffraction-Limited Telescope with Circular Aperture

For this commonly occurring type of telescope, the OTF has circular symmetry as indicated previously by 7.12. Combining 7.12 and 7.65, the intensity PSF arises for this case Born and Wolf (2003) as the familiar Airy pattern for which the unit-normalized form was previously given by 4.29 in terms of the first-order Bessel function of the first kind, $J_1(\cdot)$ .

## 7.10 Mathematical Notation Used in This Chapter

The mathematical notation used in this chapter is indicated in Table 7.1.

**Table 7.1** Mathematical notation used in this chapter along with the SI dimensional units of the individual quantities

| Symbol | Quantity | Dimensions |
|---|---|---|
| λ | Wavelength | m |
| (x,y) | Cartesian coordinate system in telescope pupil | m |
| r | Radial coordinate in telescope pupil, $\left(= \sqrt{x^2 + y^2}\right)$ | m |
| (ξ,η) | Cartesian coordinate system in plane perpendicular to the direction of light travel | m |
| $\varepsilon$ | Radial coordinate in plane perpendicular to the direction of light travel, $\left(= \sqrt{\xi^2 + \eta^2}\right)$ | m |
| (u,v) | Cartesian coordinate system in telescope image plane | m |
| $q$ | Radial coordinate in telescope image plane, $(= \sqrt{u^2 + v^2})$ | m |
| $(\alpha, \beta)$ | Angular coordinate system in telescope image plane | "1" |
| $\vartheta$ | Radial angular coordinate in telescope image lane, $(= \sqrt{\alpha^2 + \beta^2})$ | "1" |
| H | Integrated OPD fluctuation over entire atmospheric path | m |
| $\sigma$ | rms of integrated OPD fluctuation, H | m |
| $\rho$ | Autocorrelation function of integrated OPD fluctuation, H | "1" |
| D | Telescope diameter | m |
| $d$ | Telescope central obstruction | m |
| $f$ | Telescope focal length | m |
| K | Telescope pupil function | "1" |
| $k$ | Wavenumber in air | m − 1 |
| B | Complex amplitude of wave arriving in telescope pupil | "1" |
| M | Atmospheric MTF | "1" |
| S | Two-point two-wavelength correlation function of complex amplitudes | "1" |
| $M_T$ | Telescope OTF | "1" |
| $\|M_T\|$ | Telescope MTF | "1" |
| $M_E$ | OTF for entire end-to-end (atmospheric and telescope) path | "1" |
| J | Mutual intensity function | "1" |
| U | Complex amplitude in image | "1" |
| I | Intensity in image | "1" |
| $\mu$ | Complex coherence factor in image | "1" |

Dimensionless quantities are indicated by "1"

## References

Beckmann, P., & Spizzichino, A. (1963). *The scattering of electromagnetic waves from rough surfaces*. Pergamon.

Born, M., & Wolf, E. (2003). *Principles of optics* (7th ed., revised). Cambridge University Press.

Dainty, J. C. (1984). Laser speckle and related phenomena. In J. C. Dainty (Ed.), *Topics in applied physics* (Vol. 9). Springer, Berlin.

Electro-Optics Handbook. (1974). *Electro-optics handbook*. Burle Industries, Inc.

Goodman, J. W. (1963). *Electronics laboratories technical report TR2303-1 (SEL-63–140)*. Stanford University.

Goodman, J. W. (2004). *Introduction to fourier optics* (3rd ed.). Roberts & Company Publishers.

Reed, I. S. (1962). On a moment theorem for complex Gaussian processes. *IRE Transaction on Information Theory, IT-8*, 194–195.

# Chapter 8
# Atmospheric Path Characterization

**Abstract** This chapter describes how atmospheric paths can be characterized by measuring certain key intensity properties of point-object (star) images formed by telescopes observing over these paths. The characterization is given in terms of the rms and the autocorrelation function of the integrated OPD fluctuation over the paths, $\sigma$ and $\rho(\xi, \eta)$. Once $\sigma$ and $\rho(\xi, \eta)$ have been established for a given path, so too are the two-point two-wavelength correlation function and the atmospheric MTF for that path. The wavefront structure function, the refractive index structure function, and the turbulence spectrum are also determined by $\sigma$ and $\rho(\xi, \eta)$. When the measures $\sigma$ and $\rho(\xi, \eta)$ are appropriately combined with the telescope pupil function, all meaningful statistical properties are determined for point-object images formed by that telescope. When the image intensity measurements are obtained using AO-equipped telescopes, the measured OPD properties are those of the AO-corrected image-forming waves, which we denote by $\sigma_{AO}$ and $\rho_{AO}(\xi, \eta)$. Naturally, for a properly functioning AO system, $\sigma_{AO} < \sigma$.

Measurement procedures are described in this chapter for obtaining measured estimates of both the variance, $\sigma^2$, and the autocorrelation function, $\rho(\xi, \eta)$, of the integrated optical path difference (OPD) fluctuation, $H(x, y)$, that accrues over extended atmospheric paths. These two key statistical measures quantify both the average strength and the average size distribution of the turbulence structure in the paths. They also determine most of the important statistical properties of images formed by telescopes observing over these paths. The measurement procedures are conducted in the telescope image plane; the images are always those of point-objects, such as unresolved stars, and the measured quantity is always intensity. Depending on which measurement procedure is employed, the intensity measurements are made in the images of point-object sources of either monochromatic or polychromatic illumination.

The measurement procedures enable accurate estimates of both $\sigma^2$ and $\rho(\xi, \eta)$. They also allow for the possibility of non-isotropic turbulence, i.e., where function, $\rho(\xi, \eta)$, is not circularly symmetric. In some instances, ultimate measurement accuracy may not be necessary; less accurate estimates of $\sigma^2$ and $\rho(\xi, \eta)$ may be sufficient,

 
T. S. McKechnie, *General Theory of Light Propagation and Imaging Through the Atmosphere*, Progress in Optical Science and Photonics 20,
https://doi.org/10.1007/978-3-030-98828-9_8

especially if they can be obtained quickly and easily. Techniques for making such measurements are described in Chaps. 13 and 18.

The measurement procedures described in this chapter and Chaps. 13 and 18 only allow OPD fluctuation estimates of $\sigma^2$ and $\rho(\xi, \eta)$ over the truncated wave portions actually collected by the telescope. For telescopes with circular apertures, these limited portions lie in the domain defined by

$$0 \leq \xi^2 + \eta^2 \leq \left(\frac{D}{2}\right)^2, \tag{8.1}$$

where $D$ is the telescope diameter.

To establish the behaviour of $\sigma^2$ and $\rho(\xi, \eta)$ beyond the argument range just indicated would require some sort of assumption about the turbulence structure function at scales larger than the telescope aperture. Such an assumption would instantly defeat one of the primary objectives of this book, namely a formulation free of speculative assumptions of this sort. It is also important to note that the $\sigma^2$ and $\rho(\xi, \eta)$ estimates obtained by any of the measurement procedures described in this chapter are sufficient to determine all of the key statistical properties of the complex amplitude and intensity in point-object images formed by telescopes of all aperture sizes, up to and including the aperture size of the measuring telescope. Except in instances where function $\rho(\xi, \eta)$ falls to zero comfortably within the argument limit indicated by (8.1), it would not be safe to use the measured estimates of $\rho(\xi, \eta)$ for apertures larger than that of the measuring instrument.

Astronomical seeing quality varies from night to night, and from hour to hour over the course of the night. Historically, to keep track of the variations, seeing logs have kept time records of metrics such as the 'full-width half-maximum (FWHM) angular size' of a reference star lying roughly in the observing direction. While such information has some value, it does not provide definitive seeing characterization. Thus, for example, a FWHM measurement made at visible wavelengths would not, by itself, enable prediction of the FWHM size of the star image at, say, IR wavelengths. Nor would it shed any light on whether the star images at these longer wavelengths would retain the seeing disc appearance or develop a core-and-halo appearance. The measurement procedures described in this chapter and Chaps. 13 and 18 enable $\sigma^2$ and $\rho(\xi, \eta)$ to be quantified, on site and as needed, to any required level of accuracy. Together these two measures enable accurate predictions of star image appearance (using the same telescope in the same seeing) for all visible and IR wavelengths.

Seeing logs based on measurements of $\sigma^2$ and $\rho(\xi, \eta)$ using the full aperture of the observing telescope could potentially interfere with the actual observation programs. However, there are ways of recording seeing log measurements of this type without causing such interference. Details are provided in Chap. 13 (Sect. 13.2.5).

Before the ~1989 advent of IR focal plane arrays (FPAs), it was not possible to record instantaneous star images at wavelengths longer than the 1-μm cut-off wavelength of (silicon) CCD imaging arrays. Consequently, until then it was not possible to exploit the unique properties of core and halo star images, which are routinely found at near-IR and longer wavelengths, and which hugely simplify the measurement procedures for quantifying $\sigma^2$ and $\rho(\xi, \eta)$.

Around 1975, star images produced by large ground-based telescopes invariably appeared in the form of seeing discs; nothing else and nothing better was expected. Consequently, prior to the advent of IR FPAs, there were few, if any, reliable methods for measuring the variance, $\sigma^2$, of the integrated OPD fluctuation over the wave portions collected by large ground-based telescopes.

However, around that time, significant advances were made in understanding both laser speckle and stellar speckle (Dainty, 1984). These opened up entirely new methods (Sect. 8.2) by which $\sigma^2$ and $\rho(\xi, \eta)$ could be obtained directly from seeing disc star images at visible wavelengths. The first method tried in the field required measuring the correlation coefficient between the speckle in the center of a star image at two different wavelengths. The previously developed two-point, two-wavelength correlation function, *S*, is then used to convert the measured coefficient into a value for $\sigma^2$. The value calculated by this method is entirely unaffected by telescope aberrations. This attribute greatly enhances the method's usefulness because details of telescope aberrations are rarely known with any certainty, and they tend to change anyway, from hour to hour as the ambient temperature changes.

## 8.1 Obtaining the Atmospheric MTF from Point-Object Images

An estimate of the atmospheric modulation transfer function (MTF), $M(\xi, \eta, \lambda)$ can be obtained by Fourier transforming the average intensity envelope of any unresolved star image formed by a large telescope. Denoting the measured average image intensity envelope by $\langle I(u, v, \lambda)\rangle$ 7.61 and 7.62 enable $M(\xi, \eta, \lambda)$ to be given in the required unit-normalized form by

$$M(\xi, \eta, \lambda) = \frac{\int_{-\infty}^{\infty}\int_{-\infty}^{\infty}\langle I(u, v, \lambda)\rangle \cdot \exp\left[\frac{2\cdot\pi\cdot i}{\lambda\cdot f}\cdot(u\cdot\xi + v\cdot\eta)\right]\cdot du\cdot dv}{M_T(\xi, \eta, \lambda)\cdot\int_{-\infty}^{\infty}\int_{-\infty}^{\infty}\langle I(u, v, \lambda)\rangle\cdot du\cdot dv} \tag{8.2}$$

where $M_T(\xi, \eta, \lambda)$ is the telescope OTF.

### *8.1.1 For Large Diffraction-Limited Telescopes*

For large telescopes built to diffraction-limited optical standards, the telescope OTF, $M_T(\xi, \eta, \lambda)$, at visible wavelengths is a considerably wider function than the atmospheric MTF, $M(\xi, \eta, \lambda)$. For such telescopes, it is often allowable to set the telescope OTF to unity over the spatial frequency range where the atmospheric MTF takes nonzero values. In such cases, 8.2 simplifies, providing the following approximate expression for the atmospheric MTF,

$$M(\xi, \eta, \lambda) = \frac{\int_{-\infty}^{\infty} \int_{-\infty}^{\infty} \langle I(u, v, \lambda) \rangle \cdot \exp\left[\frac{2 \cdot \pi \cdot i}{\lambda \cdot f} \cdot (u \cdot \xi + v \cdot \eta)\right] \cdot du \cdot dv}{\int_{-\infty}^{\infty} \int_{-\infty}^{\infty} \langle I(u, v, \lambda) \rangle \cdot du \cdot dv}. \quad (8.3)$$

#### 8.1.1.1 For the Case of Circular Symmetry

While we continue to assume large telescopes built to precise optical standards, we now restrict consideration to telescopes with circular apertures and, when appropriate, circular central obstructions. We also assume isotropic turbulence. In such cases, both the atmospheric MTF and the average image intensity envelope formed by the telescope are circularly symmetric. Both functions can then be fully described by their respective central sections, which are related by the Hankel transform (Abramowitz & Stegun, 1970),

$$M(\varepsilon, \lambda) = \frac{\int_0^{\infty} \langle I(q, \lambda) \rangle \cdot J_0\left[\frac{2 \cdot \pi \cdot q \cdot \varepsilon}{\lambda \cdot f}\right] \cdot q \cdot dq}{\int_0^{\infty} \langle I(q, \lambda) \rangle \cdot q \cdot dq}, \quad (8.4)$$

where the radial coordinate, $q$, in the image plane (Fig. 7.2), was previously defined by 7.3; $\varepsilon$ is the radial coordinate in the telescope pupil plane defined previously by 6.46; and $J_0(\cdot)$ is the zero-order Bessel function of the first kind.

By obtaining the inverse Hankel transform of both sides of 8.4, the average intensity envelope may be expressed in terms of the atmospheric MTF by

$$\langle I(q, \lambda) \rangle = \int_0^{\infty} M(\varepsilon, \lambda) \cdot J_0\left[\frac{2 \cdot \pi \cdot q \cdot \varepsilon}{\lambda \cdot f}\right] \cdot \varepsilon \cdot d\varepsilon. \quad (8.5)$$

### 8.1.2 *Long- and Short-Exposure Atmospheric MTFs*

As discussed in Sect. 7.1, long-exposure images refer to the images obtained by integrating over long time periods while making no attempt to correct image wander. The average short-exposure image can be obtained by integrating over similarly long time periods, but where image wander is continuously corrected during the exposure. We denote the long- and short-exposure versions of the atmospheric MTF by $M_L(\xi, \eta, \lambda)$ and $M_S(\xi, \eta, \lambda)$, respectively. The most general form of either of these atmospheric MTF versions can be obtained by inserting, as appropriate, the long- or short-exposure versions of $\langle I(u, v, \lambda)\rangle$ into 8.2. In other less general cases, either 8.3 or 8.4 may be used instead of 8.2.

### 8.1.3 Effective End-to-End OTF for a Telescope Equipped with Adaptive Optics

Adaptive optics (AO) refers to the technology area where an AO system is used to correct unwanted wavefront distortions that degrade the quality of images formed by telescopes and other optical systems. In the case of ground-based astronomical telescopes, the wavefront distortions are introduced by inhomogeneity in the atmospheric path. A brief description of AO and its implementation is given in Chap. 10 (Sect. 10.10).

The effect of adding an AO correction system to a telescope observing through the atmosphere amounts to adding yet one more phase screen to the multiple phase screen atmospheric path model described in Chap. 6 (cf., Fig. 6.2). In this instance, however, there is a crucial difference: the OPD fluctuation caused by this particular ‘phase screen’ now subtracts from, rather than adds to, the total integrated OPD fluctuation.

Residual uncorrected OPD error for an entire end-to-end imaging path (that is, the entire path from object to image) not only depends on the performance of the AO system in correcting the dynamic OPD fluctuation introduced by the atmosphere, it also depends on how well the AO system corrects the OPD contributions from other dynamic sources, such as telescope shake and dome turbulence (McKechnie, 2004). The fixed aberrations of the telescope also require correction, as does focus error. But here, rather than consider how the AO system corrects all of these different OPD error contributions separately, it is more convenient to deal in terms of a single OTF function that describes the residual uncorrected OPD fluctuation as a single entity. Such an OTF can be obtained as the Fourier transform of the average intensity envelope in AO-corrected images of unresolved stars.

We denote such an image envelope by $I_{AO}(u, v, \lambda)$ and note that this envelope can be measured in exactly the same way as before, except that now the AO system would be made to operate during the measurements. In its most general non-circularly symmetric form, the effective OTF of the entire AO-corrected imaging path, which we denote by $M_{AO}(\xi, \eta, \lambda)$, may be expressed in terms of $\langle I_{AO}(u, v, \lambda)\rangle$ by the Fourier transform relation,

$$M_{AO}(\xi, \eta, \lambda) = \frac{\int_{-\infty}^{\infty}\int_{-\infty}^{\infty}\langle I_{AO}(u, v, \lambda)\rangle \cdot \exp\left[\frac{2\cdot\pi\cdot i}{\lambda\cdot f}\cdot(u\cdot\xi + v\cdot\eta)\right]\cdot du\cdot dv}{\int_{-\infty}^{\infty}\int_{-\infty}^{\infty}\langle I_{AO}(u, v, \lambda)\rangle\cdot du\cdot dv}. \tag{8.6}$$

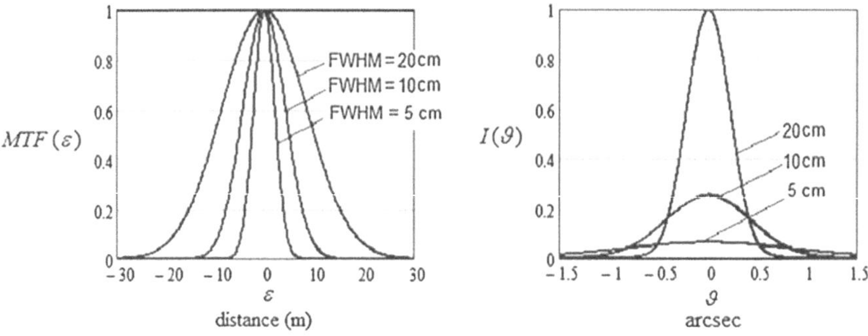


**Fig. 8.1** *Left* Atmospheric MTF plots for the visible wavelength, 0.55 μm, for typical FWHM values, 5, 10, and 20 cm. *Right* The corresponding average intensity envelopes

Because it is supposed here that the uncorrected fixed aberrations of the telescope are included in $\langle I_{AO}(u, v, \lambda)\rangle$, there is now no reason for including a telescope OTF term on the right-hand side of this equation (as was necessary in 8.2).

### *8.1.4 Atmospheric MTF Plots and Corresponding Intensity Envelopes*

Figure 8.1 (left) shows central sections through typical atmospheric MTFs for the visible wavelength 0.55 μm for the case of isotropic turbulence. The FWHM widths of the functions shown, 5 cm, 10 cm, and 20 cm, correspond roughly to poor, average, and good seeing conditions. Figure 8.1 (right) shows the corresponding average image intensity envelopes calculated from 8.5. Upon rotation, all of these intensity envelopes enclose identical volumes, i.e., identical amounts of light energy. This form of normalization is used frequently throughout the book and is particularly useful when side-by-side resolution comparisons are required.

## 8.2 Measurement of the rms OPD Fluctuation

### *8.2.1 Measurement for the Case $\sigma/\lambda \geq 0.4$ Using Two Narrowband Filters*

We saw in the last chapter (Sect. 7.8) that the image of a point-object formed by a large telescope at a wavelength short enough to fulfill the condition, $\sigma/\lambda \geq 0.4$, appears as a speckle pattern obeying Gaussian statistics. When such an image is formed by a diffraction-limited telescope, 7.60 provides the relationship between the correlation coefficient, $\mu(0, 0, 0, 0, \lambda_1, \lambda_2)$, of the complex amplitudes in the center of the image (i.e., where $u = 0$ and $v = 0$) at the two wavelengths, $\lambda_1$ and $\lambda_2$, in terms of the rms OPD fluctuation, σ, associated with the image-forming waves in the telescope pupil. Combining this relation with 7.54 gives

$$\mu_I(0, 0, 0, 0, \lambda_1, \lambda_2) = |\mu(0, 0, 0, 0, \lambda_1, \lambda_2)|^2 = \frac{\langle I(0, 0, \lambda_1) \cdot I(0, 0, \lambda_2)\rangle}{\langle I(0, 0, \lambda_1)\rangle \cdot \langle I(0, 0, \lambda_2)\rangle} - 1 = \exp\left[-4 \cdot \pi^2 \cdot \sigma^2 \cdot \left(\frac{1}{\lambda_1} - \frac{1}{\lambda_2}\right)^2\right], \tag{8.7}$$

where $\mu_I(0, 0, 0, 0, \lambda_1, \lambda_2)$ is the intensity correlation coefficient in the center of the image.

Solving the above equation for $\sigma$ gives

$$\sigma = \left|\frac{\lambda_1 \cdot \lambda_2}{2 \cdot \pi \cdot (\lambda_1 - \lambda_2)} \cdot \sqrt{-\ln\left[\frac{\langle I(0, 0, \lambda_1) \cdot I(0, 0, \lambda_2)\rangle}{\langle I(0, 0, \lambda_1)\rangle \cdot \langle I(0, 0, \lambda_2)\rangle} - 1\right]}\right|, \tag{8.8}$$

where $\ln(\cdot)$ is the Naperian logarithm to base $e$.

#### 8.2.1.1 Measurement Procedure

The schematic in Fig. 8.2 shows the speckle pattern image of a point-object—an unresolved star in this case—formed by a large ground-based telescope. The method of obtaining $\sigma$ from images of this type is based on the degree of correlation between the speckle patterns formed by light at two different wavelengths. The principle of the method was developed previously for measuring surface roughness (Elbaum et al., 1972; Sprague, 1972).

To obtain a measurement estimate of $\sigma$, the (speckle pattern) star image is first centered and focused on a small pinhole sampling aperture. A neutral density beam splitter separates the light portions that pass through the pinhole into two channels. Narrowband transmission filters located in each channel restrict the light reaching the two photomultiplier tube (PMT) detectors to narrow bands centered, respectively, on the two chosen wavelengths, $\lambda_1$ and $\lambda_2$. As indicated in the figure, the intensities simultaneously recorded by the detectors in the two channels are denoted, respectively, by $I(0, 0, \lambda_1)$ and $I(0, 0, \lambda_2)$.

The condition, $\sigma/\lambda \geq 0.4$, is generally fulfilled when large telescopes observe at visible wavelengths in mediocre or poor seeing conditions. Star images in this case appear as polychromatic speckle patterns, similar to the pattern shown in Fig. 8.2 (top left). The speckle pattern image in such cases is comprised of a very large number of partially correlated Gaussian speckle patterns; each individual pattern may be

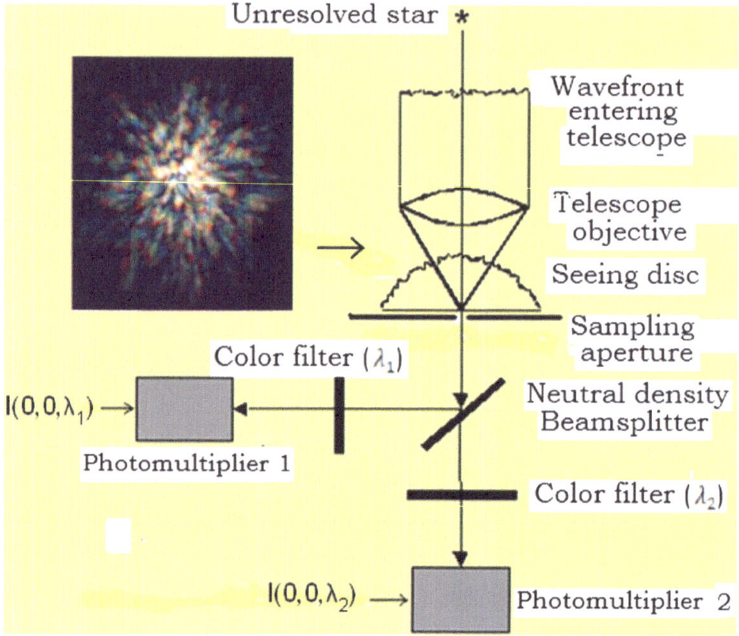


**Fig. 8.2** Experimental setup for obtaining a measured estimate of the rms OPD fluctuation, $\sigma$, of the image-forming light waves arriving in the pupil of the telescope. Large numbers of instantaneous intensity measurements are made in the center of a star image. The degree of correlation between the intensities obtained simultaneously in the two channels at two different wavelengths, $\lambda_1$ and $\lambda_2$, enables $\sigma$ to be calculated

considered to arise from just one of the many individual wavelengths that constitute the polychromatic light bundle (an infinite number in the case of a continuous spectrum).

To obtain an accurate estimate of the rms OPD fluctuation, $\sigma$, it is first necessary to obtain a large number of simultaneously measured intensity pairs, $I(0, 0, \lambda_1)$ and $I(0, 0, \lambda_2)$. The measured intensities are then used to obtain an estimate of the intensity correlation coefficient, $\mu_I(0, 0, 0, 0, \lambda_1, \lambda_2) = \frac{\langle I(0,0,\lambda_1)\cdot I(0,0,\lambda_2)\rangle}{\langle I(0,0,\lambda_1)\rangle\cdot\langle I(0,0,\lambda_2)\rangle} - 1$. The coefficient value is then inserted into 8.8 to provide the $\sigma$ estimate.

To minimize estimator bias due to "aperture averaging" (Dainty, 1984), the diameter of the pinhole aperture must be significantly smaller (say 10 times smaller) than the size of the average speckle; similarly, to minimize bias due to temporal averaging (Dainty, 1984), the detector integration time must be significantly shorter (say 10 times shorter) than the speckle relaxation time (which depends on atmospheric conditions). Typically, a 1-ms integration time might be used, short enough to, in effect, freeze the temporal fluctuations of the speckle. Such an integration period enables intensity pairs, $I(0, 0, \lambda_1)$ and $I(0, 0, \lambda_2)$, to be measured at a 1-kHz measurement rate. At such a rate, in a one-minute data collection run, about 60,000 intensity measurement pairs are obtained—a number large enough to provide an accurate $\sigma$ estimate. Actual $\sigma$ estimates based on astronomical telescope measurements using unresolved stars as "point-objects" are described in Sect. 8.2.3.

#### 8.2.1.2 Photon Noise

Even when the brightest stars are used to make the image intensity measurements, the combination of the tiny sampling aperture and the short (1-ms) integration period can result in significant photon noise artifacts. Unless these are compensated, both the calculated intensity correlation coefficient and the $\sigma$ estimate obtained from it (8.8) will exhibit significant bias.

Appendix F shows how appropriate bias correction factors, $a_1$ and $a_2$, can be obtained from the temporal autocorrelation functions of the measured intensity data sequences at each of the two wavelengths. It is also shown in that appendix how these factors are used to produce an unbiased $\sigma$ estimate, via the slightly modified version of 8.8 which can be written in the form,

$$\sigma = \left| \frac{\lambda_1 \cdot \lambda_2}{2 \cdot \pi \cdot (\lambda_1 - \lambda_2)} \cdot \sqrt{-\ln\left\{ \frac{1}{\sqrt{a_1 \cdot a_2}} \cdot \left[ \frac{\langle I(0,0,\lambda_1) \cdot I(0,0,\lambda_2)\rangle}{\langle I(0,0,\lambda_1)\rangle \cdot \langle I(0,0,\lambda_2)\rangle} - 1 \right] \right\}} \right|. \tag{8.9}$$

### 8.2.2 *Measurement for the Case σ/λ ≥ 0.4 Using a Broadband Filter*

When, $\sigma/\lambda \geq 0.4$, estimates of $\sigma$ can also be obtained from a single channel measurement arrangement of the sort depicted in Fig. 8.3. With this particular measurement method, a broadband spectral filter is used so that the intensity recorded by the detector is, in effect, the wavelength-integrated intensity that arises from a polychromatic speckle pattern. With this measurement method, image intensities are again obtained by sampling the speckle pattern images through a suitably small pinhole sampling aperture and using a suitably short integration period so as to minimize the biasing effects of both aperture averaging and temporal averaging. An accurate estimate of $\sigma$ can be calculated once a suitably large number of intensity measurements has been recorded (typically 60,000 individual measurements obtained at a 1-kHz rate over a one-minute data acquisition period). If the intensities are measured in the center of the star image (i.e., at the image location, $u = 0$ and $v = 0$), we can denote the individual measured (wavelength integrated) intensities by $I_I(0,0)$.

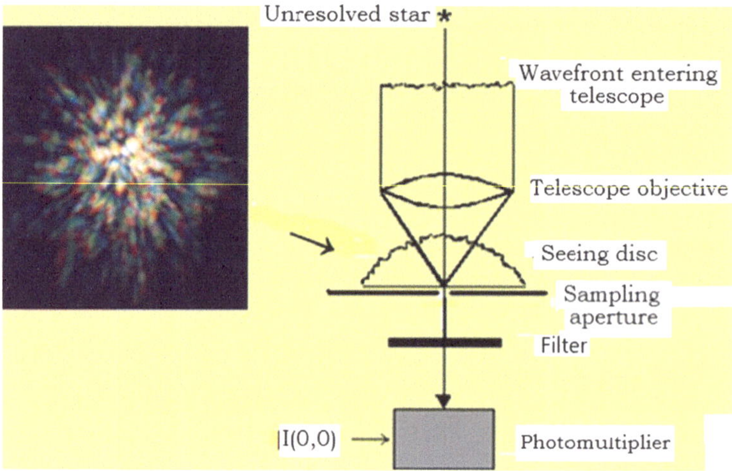


**Fig. 8.3** Single-channel arrangement for $\sigma$ measurement using a broad band-pass filter. The variance of the intensities measured in the center of the star image allows $\sigma$ to be calculated

A relatively simple algorithm can then be used to convert the random fluctuations of the measured $I_I(0,0)$ values into the σ estimate. We now derive that algorithm.[1]

In typical astronomical seeing conditions (~1-arcsec visible seeing), a suitable spectral filter might transmit almost the entire visible light spectrum, from 0.4 to 0.7 μm. Over such a broad spectral range, we would generally expect to find significant variations of the following: (1) the spectral output of the object star, (2) the spectral transmission of the telescope optics, and (3), the spectral response of the detector. For notational simplicity, we therefore define a single spectral function, which we denote by $G(\lambda)$ that captures (as the product) all of the various spectral dependencies just listed. The normalization of $G(\lambda)$ is unimportant for present purposes; only the relative spectral variation of this function matters. In terms of $G(\lambda)$, the integrated intensity measured by the detector in the center of the star image, $I_I(0,0)$,

$$I_I(0,0) = \int_0^\infty G(\lambda) \cdot I(0,0,\lambda) \cdot d\lambda \tag{8.10}$$

is given by

Because of the random temporal fluctuations of the speckle in the star image, the measured intensities, $I_I(0,0)$, also vary temporally. The contrast associated with the measured intensity fluctuation can be obtained from 8.10 in terms of the following proportionality,

[1] The effects of temporal averaging and aperture averaging are examined in detail in Sect. 11.6. The analysis given in that section allows for the possibility of larger sampling apertures and longer integration times as long as appropriate bias correction factors are used.

$$\frac{\left\langle I_I(0,0)^2\right\rangle - \langle I_I(0,0)\rangle^2}{\langle I_I(0,0)\rangle^2} \propto$$
$$\int_0^\infty\int_0^\infty G(\lambda)\cdot G(\lambda')\cdot\left[\frac{\langle I(0,0,\lambda)\cdot I(0,0,\lambda')\rangle - \langle I(0,0,\lambda)\rangle\cdot\langle I(0,0,\lambda')\rangle}{\langle I(0,0,\lambda)\rangle\cdot\langle I(0,0,\lambda')\rangle}\right]\cdot d\lambda\cdot d\lambda'. \tag{8.11}$$

Using Reed's theorem for Gaussian processes (Sect. 7.8.1), the above proportionality can be expressed in the form,

$$\frac{\left\langle I_I(0,0)^2\right\rangle - \langle I_I(0,0)\rangle^2}{\langle I_I(0,0)\rangle^2} \propto \int_0^\infty\int_0^\infty G(\lambda)\cdot G(\lambda')$$
$$\cdot\frac{\left|\langle U(0,0,\lambda)\cdot U^*(0,0,\lambda')\rangle\right|^2}{\langle I(0,0,\lambda)\rangle\cdot\langle I(0,0,\lambda')\rangle}\cdot d\lambda\cdot d\lambda', \tag{8.12}$$

from which the appropriate unit-normalized contrast may be written

$$\frac{\left\langle I_I(0,0)^2\right\rangle}{\langle I_I(0,0)\rangle^2} - 1 = \frac{\int_0^\infty\int_0^\infty G(\lambda)\cdot G(\lambda')\cdot\left\{\frac{\left|\langle U(0,0,\lambda)\cdot U^*(0,0,\lambda')\rangle\right|^2}{\langle I(0,0,\lambda)\rangle\cdot\langle I(0,0,\lambda')\rangle}\right\}\cdot d\lambda\cdot d\lambda'}{\left[\int_0^\infty G(\lambda)\cdot d\lambda\right]^2}, \tag{8.13}$$

where $U(0,0,\lambda)$ is the instantaneous complex amplitude of the sampled light in the center of the image at wavelength, $\lambda$, and $I(0,0,\lambda)$ is the corresponding instantaneous intensity.

Recalling 7.40, we see that the term in curly brackets in the numerator on the right-hand side of the above equation is simply the squared modulus of the unit-normalized spectral correlation function of the complex amplitudes in the center of the image, $\mu_I(0,0,0,0,\lambda_1,\lambda_2)$. The left-hand-side of 8.13 may be considered as the (unit-normalized) contrast of the integrated intensity, $\mu_I$, defined by

$$\mu_I = \frac{\left\langle I_I(0,0)^2\right\rangle}{\langle I_I(0,0)\rangle^2} - 1. \tag{8.14}$$

The quantity, $\mu_I$, takes values in the range,

$$0 \le \mu_I \le 1. \tag{8.15}$$

For quasi-monochromatic light [i.e., where a limitingly narrowband spectral filter is used), $G(\lambda)$ may be approximated by the Dirac delta function, $\delta(\lambda)$. The "integrated" intensity in this case then obeys Gaussian statistics and $\mu_I$ attains its maximum value, unity. For polychromatic light, smaller $\mu_I$ values arise; the images in this case each consist of a large number of partially correlated quasi-monochromatic Gaussian speckle patterns (an infinite number if the spectral content is continuous).

Using 7.60 to substitute for the term in curly brackets in 8.13, we obtain

$$\mu_I = \frac{\int_0^\infty \int_0^\infty G(\lambda) \cdot G(\lambda') \cdot \exp\left[-4 \cdot \pi^2 \cdot \sigma^2 \cdot \left(\frac{1}{\lambda} - \frac{1}{\lambda'}\right)^2\right] \cdot d\lambda \cdot d\lambda'}{\left[\int_0^\infty G(\lambda) \cdot d\lambda\right]^2}. \quad (8.16)$$

Assuming that a large enough number of wavelength integrated intensity measurements, $I_I(0, 0)$ , have been obtained, 8.14 then provides a suitably accurate $\mu_I$ estimate. The $\sigma$ estimate is obtained by inserting the $\mu_I$ estimate in the left-hand side of 8.16 and then finding the $\sigma$ value that satisfies the equation. Because of its integral form, 8.16 does not generally allow analytic solutions for $\sigma$; however, solutions can readily be obtained numerically.

#### 8.2.2.1 Behavior of $\mu_I$ for a Typical Band-Pass Filter

For illustration purposes here, we assume that the spectral function, $G(\lambda)$, is constant over the entire wavelength band transmitted by the broad band filter. Thus, this function may be represented by the rectangular function

$$\begin{aligned} G(x) &= 1 \quad \text{for} \quad \lambda_1 \leq \lambda \leq \lambda_2 \\ &= 0 \qquad \text{otherwise.} \end{aligned} \quad (8.17)$$

where $\lambda_1$ and $\lambda_2$ are the two bounding wavelengths.

Inserting the above form of $G$ into 8.16, gives the following expression for $\mu_I$:

$$\mu_I = \frac{\int_{\lambda_1}^{\lambda_2} \int_{\lambda_1}^{\lambda_2} \exp\left[-4 \cdot \pi^2 \cdot \sigma^2 \cdot \left(\frac{1}{\lambda} - \frac{1}{\lambda'}\right)^2\right] \cdot d\lambda \cdot d\lambda'}{(\lambda_2 - \lambda_1)^2}. \quad (8.18)$$

For the rectangular filter transmission function described by 8.17 above, the width of the broadband filter, $\Delta$, is given simply by

$$\Delta = \lambda_2 - \lambda_1. \quad (8.19)$$

Denoting the center wavelength of the filter by $\lambda_C$, we may express the lower and upper wavelength bounds of this filter, $\lambda_1$ and $\lambda_2$, as follows:

$$\lambda_1 = \lambda_C - \frac{\Delta}{2}, \quad (8.20)$$

and

Fig. 8.4 Contrast of the measured spectrally integrated intensities, $\mu_I$, plotted against filter bandwidth $\Delta$, for $\sigma = 0.3$, $\sigma = 0.6$, $\sigma = 0.9$ μm, where $\lambda_C = 0.55$ μm

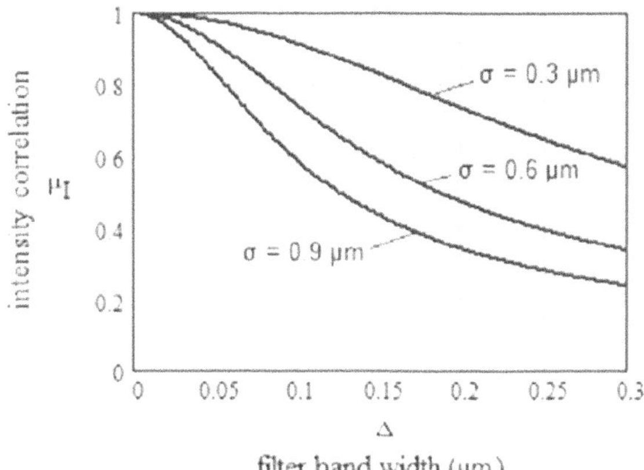


$$\lambda_2 = \lambda_C + \frac{\Delta}{2}. \tag{8.21}$$

Rewriting 8.18 in terms of $\Delta$ and $\lambda_C$, we obtain

$$\mu_I = \frac{1}{\Delta^2} \cdot \int_{\lambda_C - \frac{\Delta}{2}}^{\lambda_C + \frac{\Delta}{2}} \int_{\lambda_C - \frac{\Delta}{2}}^{\lambda_C + \frac{\Delta}{2}} \exp\left[-4 \cdot \pi^2 \cdot \sigma^2 \cdot \left(\frac{1}{\lambda} - \frac{1}{\lambda'}\right)^2\right] \cdot d\lambda \cdot d\lambda'. \tag{8.22}$$

Figure 8.4 shows the quantity $\mu_I$ (calculated from the above equation) plotted against filter bandwidth, $\Delta$, for three representative $\sigma$ values, where the center wavelength, $\lambda_C$, has been set at 0.55 μm.

### 8.2.3 Actual Field Measurements of $\sigma$

In 1975 and 1976, measurement expeditions were made on two occasions to the Royal Greenwich Observatory (RGO) by a research team from the Imperial College Physics Department.[2] At that time, the RGO telescopes were located on the grounds of Herstmonceux Castle (Fig. 8.5), thus avoiding the light pollution at the historical Greenwich site on the prime meridian.[3] The various team members had different objectives.

[2] Other postgraduate and postdoctoral Imperial College members attending the 1975/76 RGO trips included J.C. Dainty, G. Parry, R.J. Scadden, and J.G. Walker.

[3] The old Royal Observatory at Greenwich, the oldest part of which was designed by Sir Christopher Wren and built in 1675, is the site of the Airy Transit Circle which marks the Greenwich meridian. In 1884, the Greenwich meridian was adopted by international agreement as the world's prime meridian. Because of the deterioration of observing conditions at the old site, over a ten-year period beginning just after WWII, the library, Nautical Almanac Office and Chronometer Department, and finally the observing equipment was moved to a new site at Herstmonceux Castle in East

**Fig. 8.5** Herstmonceux Castle, East Sussex, UK, site of the Royal Greenwich Observatory (RGO) from 1948 to 1990. The castle was built at a cost of £3800 on land gifted by King Henry V (1386–1422) to Sir Roger Fiennes (1384–1449) for his loyalty and service, most notably at the Battle of Agincourt (1415). Photograph by courtesy of David Parker

The Author's primary objective was to obtain estimates of the rms OPD fluctuation introduced by the atmosphere, $\sigma$, using the two-wavelength intensity correlation technique described in Sect. 8.2.1.1 and illustrated schematically in Fig. 8.2. Bright stars were used as the "point-objects." The telescope used to make the observations—the Yapp 36-in. reflector—is pictured in Fig. 8.6. The bright star, Vega, which transits the local meridian close to the zenith, gave particularly consistent data.[4]

The measurement data discussed below were obtained, while Vega lay within 20° of the zenith. Thus, the observing paths through the atmosphere were no more than 7% longer than the shortest path through the zenith. With the detectors in the two channels set to simultaneously record the intensities, $I(0, 0, \lambda_1)$ and $I(0, 0, \lambda_2)$, at a 1-kHz collection rate, 60,000 measured intensity pairs were obtained in each of the various one-minute data collection runs.

Three narrowband filters centered at the visible wavelengths, 0.486, 0.55, and 0.656 μm, were used in all three filter-pair combinations, thus providing three different intensity-pair data sets. The bandwidths of all three filters were nominally the same, about 0.01 μm (10 nm).

---

Sussex. The new establishment was named the Royal Greenwich Observatory (RGO). Sometime later, in 1990, the decision was taken to move the observatory again (Mills, 2005), to a new site at Cambridge, adjacent to the University's Institute of Astronomy. While the Observatory itself was moved, the Equatorial Group of Telescopes, including the Yapp 36-in. reflector, was left behind at Herstmonceux Castle.

[4] At the Herstmonceux Castle observatory site (coordinates: 50.87° N, 0.35° W), the zeroth-magnitude star, Vega (RA = 18 h 37 min, DE = 38.68°), passes within 12° of the zenith. Vega's 3.3-mas disk is entirely unresolvable by the 36-in. Yapp reflector.

**Fig. 8.6** Top The Yapp 36-in. reflector telescope from inside the dome. The instrument was built in 1932 by Grubb Parsons of Newcastle upon Tyne, a gift from the American industrialist, William J. Yapp. Bottom The Yapp dome at its present location since 1958 in the 370-acre grounds of Herstmonceux Castle. Photographs by courtesy of the Observatory Science Centre

Table 8.1 shows the correlation coefficient of the intensity, $\mu_I$, calculated from the measured intensity data for each of the three filter-pair combinations. The corresponding estimates of the rms OPD fluctuation, $\sigma$, calculated using 8.9, are also shown in the table. The correlation coefficient values indicated are seen to vary over a wide range. As expected, the highest value arises for the most closely separated wavelength pair, 0.486 and 0.550 μm. However, irrespective of which intensity correlation coefficient value is inserted into 8.9, when the corresponding $\lambda_1$ and $\lambda_2$ values

**Table 8.1** Intensity correlation coefficients $\mu_I$ calculated from measured intensity data obtained using the 36-in

| WWavelength pair ($\lambda_1$ and $\lambda_2$) (μm) | Intensity correlation coefficient ($\mu_I$) | $\sigma$ estimate (μm) |
|---|---|---|
| 0.486 and 0.550 | 0.72 | 0.390 |
| 0.550 and 0.656 | 0.60 | 0.385 |
| 0.486 and 0.656 | 0.19 | 0.395 |

Yapp reflector at RGO for three different wavelength pairs using the setup shown in Fig. 8.2

are inserted, all three measured coefficients provide approximately the same value for the rms OPD fluctuation, σ.

The actual σ value obtained in all three cases was about 0.39 μm. It might be observed here that, if the three different intensity data sets had given substantially different σ values, the validity of 8.8 and 8.9 would have been in doubt. Reassuringly, good consistency was achieved.

### 8.2.4 *Telescope Aberrations Do not Affect the Measured σ Values*

One might suspect that the measured rms OPD fluctuation caused by the atmosphere, σ, (8.9) would be biased by the aberrations of the measuring telescope. However, for telescopes equipped with monolithic primary mirrors—as is the case for the Yapp instrument—the aberrations remain substantially constant over time intervals far longer than the typical, one-minute, data collection runs described in Sect. 8.2.2.

Because the OPD errors due to telescope aberrations are frozen during the data acquisition intervals, so too is the telescope intensity PSF. Telescope aberrations in this case have no capacity to bias the σ estimate. We can demonstrate this by example. Suppose that the telescope is severely aberrated, so that in the absence of atmosphere (or for a perfectly homogenous atmosphere where σ = 0 μm) the instrument forms a fixed speckle pattern image of the star. In this case, the light intensity passing through the pinhole remains constant for as long as the telescope aberrations remained fixed. Moreover, the light intensity passing through the pinhole retains the same fixed intensity distribution with respect to wavelength. (The actual distribution is of course identical to the distribution of wavelengths at the particular pinhole sampling point on the fixed speckle pattern intensity PSF. This condition results in perfect correlation between the light intensities at any pair of wavelengths, $\lambda_1$ and $\lambda_2$. Inserting $\mu_I(0, 0, 0, 0, \lambda_1, \lambda_2) = 1$ into 8.7 yields $\sigma = 0$ μm. Whatever the aberrations of the Yapp instrument might have been in 1975/ 76 is of no consequence. Equation 8.8, already provides an unbiased estimate of σ and needs no further modification. Equation 8.9 may also be expressed in the equivalent form,[5]

[5] Note that in the (2015) first edition of this book, 8.23 erroneously contained the extra term ${\sigma_T}^2$. The corrected form of 8.23 is shown on the next page.

$$\sigma = \left| \left[ \frac{-(\lambda_1 \cdot \lambda_2)^2}{4 \cdot \pi^2 \cdot (\lambda_1 - \lambda_2)^2} \cdot \ln[\mu_I(0, 0, 0, 0, \lambda_1, \lambda_2)] \right]^{\frac{1}{2}} \right|. \quad (8.23)$$

### 8.2.5 *Convergence of σ as* $\lambda_2 \rightarrow \lambda_1$

The two-wavelength technique for measuring $\sigma$ is based on the assumption (Chap. 5) that the two-point two-wavelength correlation function, $S\ (\xi, \eta, \lambda_1, \lambda_2)$ conserves perfectly during propagation in a homogeneous medium. This assumption is only strictly valid in the limit (5.35),

$$z \cdot (\lambda_2 - \lambda_1) \rightarrow 0. \quad (8.24)$$

Had the wavelengths chosen for making the measurements, $\lambda_1$ and $\lambda_2$, been too widely separated, the mere act of propagating over the atmospheric path would have led to further decorrelation of the complex amplitudes at the two wavelengths, in addition to the decorrelation that primarily interests us which is directly attributable to the turbulence structure in the path. In such cases, $\sigma$ estimates would show bias, with measured estimates generally found to be larger than the true (unbiased) values. However, estimate bias can be avoided by studying the convergence of the various $\sigma$ estimates for a number of different filter pairs corresponding to progressively smaller wavelength differences. Once convergence of the $\sigma$ estimates has been seen to occur, the resulting $\sigma$ value should provide an unbiased estimate.

As mentioned previously, since each of the intensity correlation coefficients shown in Table 8.1 is consistent with the same $\sigma$ value, $\sigma \approx 0.39\ \mu m$, it may be concluded that, for the near-vertical atmospheric paths measured, the required convergence had already occurred even for the most widely separated of the three wavelength pairs used in the measurements, 0.486 μm and 0.656 μm.

### 8.2.6 *Measurement of σ for the Case* $\sigma/\lambda < 0.4$

As discussed in Chap. 9 (Sect. 9.4.2), when the ratio $\sigma/\lambda$ takes values less than about 0.4, central cores begin to form in star images. Cores occur at visible wavelengths in good seeing conditions while, at near-IR and longer wavelengths, they occur in almost any seeing conditions. When cores are present, an approximate $\sigma$ estimate can be obtained by measuring the average Strehl intensity[6] in a star image. From the

[6] Strehl intensity may be obtained by dividing the measured irradiance in the center of the image by the irradiance in the center of the telescope's theoretical diffraction-limited intensity PSF, normalized to include the same total light energy.

measured SI value, the $\sigma$ estimate can be obtained using the formula (c.f., 4.31),

$$\sigma = \frac{\lambda}{2 \cdot \pi} \cdot [-\ln(SI)]^{\frac{1}{2}}, \tag{8.25}$$

where the telescope is assumed diffraction limited.

For an aberrated telescope where the rms OPD contribution of the telescope, $\sigma_T$, is known, an improved estimate of the atmospheric path OPD contribution, $\sigma$, can be obtained by using the modified expression,

$$\sigma = \left[ -\ln(SI) \cdot \frac{\lambda^2}{4 \cdot \pi^2} - \sigma_T^2 \right]^{\frac{1}{2}}, \tag{8.26}$$

where the assumption is again made that the OPD errors due to the telescope and the atmospheric path are statistically independent and therefore add in quadrature. Further discussion of how the average measured Strehl intensity may be used to generate $\sigma$ estimates is given in Chaps. 13 and 18.

### *8.2.7 Measurement of Residual OPD Fluctuation for an AO-Equipped Telescope*

The procedures described so far for measuring $\sigma$ using telescopes without AO capability can be applied just as readily to telescopes with AO capability. The only procedural difference is that the AO system would now be made to operate during the measurements. Assuming an adequately performing AO system, the measured residual uncorrected rms OPD fluctuation, denoted by $\sigma_{AO}$, would naturally be less than the $\sigma$ value obtained without the AO system operating. For $\sigma_{AO} \geq 0.4 \cdot \lambda$, a $\sigma_{AO}$ estimate can be obtained as previously using the two wavelength measurement technique (Sect. 8.2.1). The estimate is provided by the measured intensity data obtained from AO-corrected star images by using the relation (cf., 8.8),

$$\sigma_{AO} = \left| \frac{\lambda_1 \cdot \lambda_2}{2 \cdot \pi \cdot (\lambda_1 - \lambda_2)} \cdot \sqrt{-\ln\left\{\left[\frac{\langle I_{AO}(0,0,\lambda_1) \cdot I_{AO}(0,0,\lambda_2)\rangle}{\langle I_{AO}(0,0,\lambda_1)\rangle \cdot \langle I_{AO}(0,0,\lambda_2)\rangle} - 1\right]\right\}} \right| \tag{8.27}$$

To avoid estimate bias due to photon noise, a modified equation similar to that given by 8.9 can be used in place of the above equation.

If the alternative broad band-pass filter measurement technique (Sect. 8.2.2) were used instead, the $\sigma_{AO}$ estimate could again be obtained using numerical methods to establish the value that satisfies the following, slightly modified version of 8.16, where $\sigma$ has now been replaced by $\sigma_{AO}$,

$$\mu_I = \frac{\int_0^\infty \int_0^\infty G(\lambda) \cdot G(\lambda') \cdot \exp\left[-4 \cdot \pi^2 \cdot \sigma_{AO}{}^2 \cdot \left(\frac{1}{\lambda} - \frac{1}{\lambda'}\right)^2\right] \cdot d\lambda \cdot d\lambda'}{\left[\int_0^\infty G(\lambda) \cdot d\lambda\right]^2}, \quad (8.28)$$

where 8.14 would again be used to calculate $\mu_I$ from the measured intensity data. In cases where $\sigma_{AO} \leq 0.4 \cdot \lambda$ the $\sigma_{AO}$ estimate could again be obtained from the average measured Strehl intensity of the AO-corrected images. Denoting average Strehl intensity by $SI_{AO}$, the $\sigma_{AO}$ estimate can be obtained using the relation (cf., 8.25),

$$\sigma_{AO} = \frac{\lambda}{2 \cdot \pi} \cdot [-\ln(SI_{AO})]^{\frac{1}{2}}. \quad (8.29)$$

It may be noted here that the forms given by 8.28 and 8.29 both hinge on the assumption that $\sigma_{AO}$ accounts for all residual uncorrected OPD fluctuation and therefore includes uncorrected residual contributions from both the atmospheric path and the telescope optics.

### 8.2.8 *Dependence of $\sigma$ on Zenith Angle*

In general, the smallest rms OPD fluctuation value arises for the shortest possible (vertical) atmospheric path through the zenith. Denoting that value by $\sigma_Z$, for non-zenith paths, where the paths lengthen in proportion to $[1 - \cos(ZA)]$, where ZA is the zenith angle, it can readily be shown that $\sigma$ may be obtained in terms of $\sigma_Z$ using the formula

$$\sigma = \frac{\sigma_Z}{\sqrt{\cos(ZA)}}. \quad (8.30)$$

For zenith angles larger than 70°, this formula begins to lose accuracy because the path length increase factor, $1/\sqrt{\cos(ZA)}$, ignores the small effect of Earth's curvature.)

## 8.3 Obtaining the Autocorrelation Function of the OPD Fluctuation

The autocorrelation function, $\rho(\xi, \eta)$, of the integrated OPD fluctuation, $H(x, y)$, provides a measure of the average size of the turbulence structures in the propagation path. This function (previously defined by 6.30) may be obtained from the measured estimates of the atmospheric MTF, $M(\xi, \eta, \lambda)$, and the rms OPD fluctuation, $\sigma$, using the following relation (cf., 6.52),

$$\rho(\xi, \eta) = 1 + \frac{\lambda^2}{4 \cdot \pi^2 \cdot \sigma^2} \cdot \ln[M(\xi, \eta, \lambda)]. \tag{8.31}$$

For isotropic turbulence, where both $M(\xi, \eta, \lambda)$ and $\rho(\xi, \eta)$ are rotationally symmetric, the above relation reduces to

$$\rho(\varepsilon) = 1 + \frac{\lambda^2}{4 \cdot \pi^2 \cdot \sigma^2} \cdot \ln[M(\varepsilon, \lambda)], \tag{8.32}$$

where the radial coordinate $\varepsilon$ was previously defined by 6.46.

Figure 8.7 shows plots of $\rho(\varepsilon)$, for plot data calculated from 8.32 using the rms OPD fluctuation value, $\sigma = 0.39$ μm, obtained during the 1975/76 RGO measurement expeditions. Because the average seeing disk envelope, $\langle I(q, \lambda)\rangle$, was not precisely measured at the time the $\sigma$ measurements were made, the atmospheric MTF, $M(\varepsilon, \lambda)$, is not precisely known. (Otherwise, $M(\varepsilon, \lambda)$ could have been calculated from $\langle I(q, \lambda)\rangle$ using 8.4.) The two $\rho(\varepsilon)$ plots shown in the figure are therefore based on the measured estimate of $\sigma$ and estimated atmospheric MTFs calculated from 8.4 using intensity envelopes corresponding to 0.75-arcsec and 1.5-arcsec (FWHM) visible seeing conditions, these two seeing estimates deemed as the likely best and worst estimates of seeing at the time the $\sigma$ measurements were made. The two $\rho(\varepsilon)$ plots shown in Fig. 8.7 therefore indicate the narrowest and widest versions of the function at the time of the $\sigma$ measurements. The likeliest $\rho(\varepsilon)$ function therefore lies somewhere between the two plotted functions.

It is clear from the figure that most of the turbulence structure lies in the size range, 0–0.5 m. The average turbulence structure size indicated by these plots is therefore significantly smaller than that assumed in Kolmogorov theory where copious amounts of turbulence energy are expected in the size range >1 m. The issue of turbulence structure size is extremely important; average turbulence structure size hugely

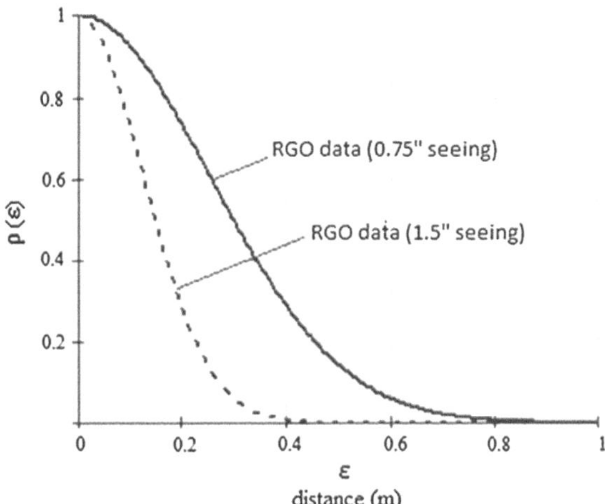


**Fig. 8.7** Plots of the autocorrelation function of the OPD fluctuation, $\rho(\varepsilon)$, based on the (RGO) measured value of the rms OPD fluctuation, $\sigma = 0.39$ μm. The plots are based on FWHM visible seeing estimated to lie in the range 0.75–1.5 arcsec. The plots indicate smaller average turbulence structure size than assumed in Kolmogorov theory

affects star image properties. Consequently, the issue of structure size is addressed at various junctures throughout the book. Chapter 12 is entirely devoted to the issue.

In Appendix C, where the function, $\rho(\varepsilon)$, is calculated from the turbulence measurements of Coulman et al. (cf., Fig. C.3), relatively small average size is again indicated. In fact, the calculated $\rho(\varepsilon)$ function in that appendix is found to be remarkably similar to the $\rho(\varepsilon)$ functions shown in Fig. 8.7.

### 8.3.1 Obtaining the Function for an AO-Equipped Telescope

In principle, Fourier decomposition allows the integrated OPD fluctuation $H(x, y)$ introduced by an atmospheric path to be separated into its various spatial frequency components. In general, AO systems demodulate different spatial frequencies by different amounts. Thus, the autocorrelation function of the OPD fluctuations obtained using a telescope with an operating AO system, which we denote by $\rho_{AO}(\xi, \eta)$, is likely to have a different shape than the function obtained when the AO system does not operate, $\rho(\xi, \eta)$. Function $\rho_{AO}(\xi, \eta)$ may be calculated (cf., 8.31).

from the AO-corrected atmospheric MTF, $M_{AO}(\xi, \eta, \lambda)$, using the relation,

$$\rho_{AO}(\xi, \eta) = 1 + \frac{\lambda^2}{4 \cdot \pi^2 \cdot \sigma_{AO}^2} \cdot \ln[M_{AO}(\xi, \eta, \lambda)], \tag{8.33}$$

where $\sigma_{AO}$ is the residual uncorrected rms OPD fluctuation measured with the AO system operating. Function $M_{AO}(\xi, \eta, \lambda)$ may be calculated from the average measured intensity envelope $\langle I_{AO}(u, v, \lambda)\rangle$ using the relation (cf., 7.62),

$$M_{AO}(\xi, \eta, \lambda) = \frac{\int_{-\infty}^{\infty} \int_{-\infty}^{\infty} \langle I_{AO}(u, v, \lambda)\rangle \cdot \exp\left[\frac{2 \cdot \pi \cdot i}{\lambda \cdot f} \cdot (u \cdot \xi + v \cdot \eta)\right] \cdot du \cdot dv}{\int_{-\infty}^{\infty} \int_{-\infty}^{\infty} \langle I_{AO}(u, v, \lambda)\rangle \cdot du \cdot dv}. \tag{8.34}$$

### 8.3.2 Significance of the Average Turbulence Structure Size

Figure 8.8 (left) shows two illustrative plots of the autocorrelation function, $\rho(\varepsilon)$, of the integrated OPD fluctuation; the plots were generated using the Gaussian functionality, $\rho(\varepsilon) = \exp(-(\varepsilon/w_0)^2)$, indicated previously by 6.50, where $w_0$ indicates the $1/e$ half-width of the function. Although both plots are consistent with nominally the same atmospheric MTF at visible wavelengths (shown at Right in the same figure) they correspond to two markedly different types of turbulence structure.

One plot corresponds to relatively weak turbulence with relatively small average structure size (described by $\sigma = 0.35$ µm and $w_0 = 0.4$ m); the other corresponds to

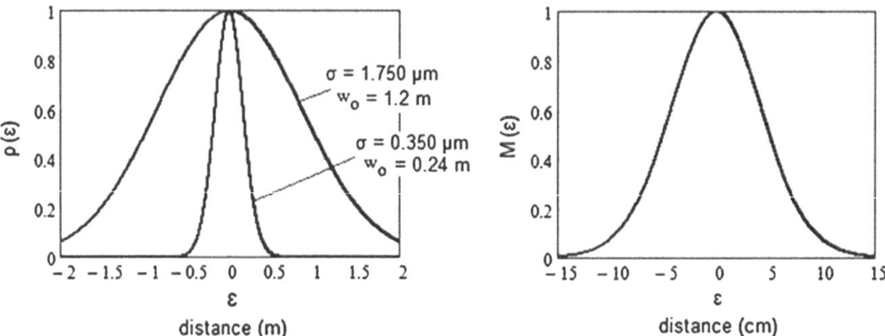


**Fig. 8.8** *Left* Autocorrelation functions of the integrated OPD fluctuation, $\rho(\varepsilon)$, for two distinct kinds of turbulence structure. Both are consistent with the atmospheric MTF shown *Right* for light at 0.55 μm. The plot corresponding to relatively small turbulence structure, $w_0 = 0.24$ m, is consistent with image cores at near-IR wavelengths. The plot, corresponding to 5× larger average turbulence structure size, $w_0 = 1.2$ m, is not consistent with near-IR image cores

five times stronger turbulence with five times larger average structure size (described by and $\sigma = 1.75$ μm and $w_0 = 2.0$ m). The fact that the atmospheric MTFs for these two very different kinds of turbulence are nearly identical indicates that, by itself, the atmospheric MTF at visible wavelengths is not a useful predictor of turbulence structure size. Thus, we identify a potentially serious problem: If the atmospheric MTF is the only measure used to characterize atmospheric paths, the opportunity exists for ambiguous and even entirely bogus notions to arise about turbulence structure size.

To generate atmospheric MTFs for the visible wavelength considered here, 0.55 μm, that have the same nominal width and shape, it is only necessary to choose values for the average turbulence size, $w_o$, and the integrated turbulence strength, $\sigma$, so that they bear the same correct $\sigma/w_0$ ratio. But, while the atmospheric MTFs are indeed seen to be closely similar in Fig. 8.8, the average structure sizes, as determined by the widths, $w_0$, of the two $\rho(\varepsilon)$ functions shown (left) in the same figure are seen to differ by a factor of five. Thus, we see that nominally the same atmospheric MTF can arise from atmospheric turbulence structure with widely different size scales.

The explanation for this apparently anomalous behavior can be seen by applying geometrical optics principles which, at short enough wavelengths, can provide reasonably correct descriptions of image size.[7] At visible wavelengths—which in the context of imaging through the atmosphere in average or poor seeing conditions may be considered suitably short—it is clear that, according to geometrical optics principles, the average wavefront gradient, and hence the average angular size of the image, must be proportional to the ratio $\sigma/w_o$. At visible wavelengths, therefore, even an incorrect choice of the average turbulence structure size, $w_o$, can still produce a

[7] In the limit of very short wavelengths, geometrical optics ultimately gives an accurate prediction of image appearance. When cores are absent from images, as is often the case with large telescopes at visible wavelengths, geometrical optics provides a reasonably adequate prediction of the average intensity envelope in a star image.

correct image size prediction just as long as a (proportionally) incorrect choice is made for the integrated turbulence strength, $\sigma$, so that the required $\sigma/w_o$ ratio is obtained.

However, image motion or, more precisely, the lack of it cannot be explained so easily. Image motions measured by Woolf et al. (1982) and by Mariotti and Di Benedetto (1984) were found to be several times smaller than anticipated by Kolmogorov theory; thus, the large-scale turbulence structures assumed to be present in that theory were evidently not present in the expected proportions. In spite of evidence of this sort directly contradicting Kolmogorov theory, for many years the issue of whether large turbulence structures really existed in the atmosphere remained largely moot. That status would change abruptly, however, when IR focal plane arrays first became available around 1990; these detectors immediately made it possible to record instantaneous short-exposure star images at IR wavelengths. When large telescopes equipped with these new detectors were turned to the heavens, all of a sudden, bright diffraction-limited image cores were routinely found dominating the images at near-IR wavelengths.

To correctly explain image cores at near-IR images formed by large astronomical telescopes (4- or 5-m class instruments being the largest in 1990), it is no longer enough to merely use the correct $\sigma/w_o$ ratio; it is now necessary to assign both $\sigma$ and $w_o$ with their correct absolute values. As discussed in Chap. 10 (Sect. 10.3.5), the routine appearance of image cores at near-IR wavelengths (without the assistance of AO other than image wander correction) infers that $\sigma$ must be significantly smaller than assumed by Kolmogorov theory. Consequently, to preserve $\sigma/w_o$ ratios consistent with the observed average image size at visible wavelengths, the average turbulence structure size $w_o$ must also be significantly smaller than assumed by Kolmogorov theory.

In average ($\sim$1 arcsec) visible seeing conditions, the plot of $\rho(\varepsilon)$ in Fig. 8.8 corresponding to the relatively weak and relatively small average turbulence structure ($\sigma = 0.35$ μm and $w_0 = 0.24$ m) readily explains the routine existence of image cores at near-IR wavelengths. On the other hand, the plot corresponding to the much larger average turbulence structure size with proportionately larger turbulence strength ($\sigma = 1.75$ μm and $w_0 = 1.2$ m) does not explain image cores at near-IR wavelengths. A more exhaustive, quantitative examination of how average turbulence structure size affects the properties of star images formed by large ground-based telescope is given in Chap. 12.

### 8.3.3 *Path Characterization During Daytime*

Star images formed by large telescopes can be seen just as easily during the day as they can at night. The atmospheric path characterization procedures described in Sects. 8.2 and 8.3 for obtaining estimates of $\sigma$ and $\rho(\xi, \eta)$ can therefore be used at any time of the day or night.

## 8.4 The Wavefront Structure Function

The unit-normalized wavefront structure function, $\sum(\xi, \eta)$, may be expressed in terms of the autocorrelation function of the integrated OPD fluctuation, $\rho(\xi, \eta)$, by the relation,

$$\Sigma(\xi, \eta) = 1 - \rho(\xi, \eta). \tag{8.35}$$

Combining the above equation with 8.31 allows the wavefront structure function to be given by

$$\Sigma(\xi, \eta) = -\frac{\lambda^2}{4 \cdot \pi^2 \cdot \sigma^2} \cdot \ln[M(\xi, \eta, \lambda)]. \tag{8.36}$$

### *8.4.1 Wavefront Structure Function for Isotropic Turbulence*

For isotropic turbulence, where both $\Sigma(\xi, \eta)$ and $M(\xi, \eta, \lambda)$ are rotationally symmetric, the above expression reduces to the rotationally symmetric form,

$$\Sigma(\varepsilon) = -\frac{\lambda^2}{4 \cdot \pi^2 \cdot \sigma^2} \cdot \ln[M(\varepsilon, \lambda)]. \tag{8.37}$$

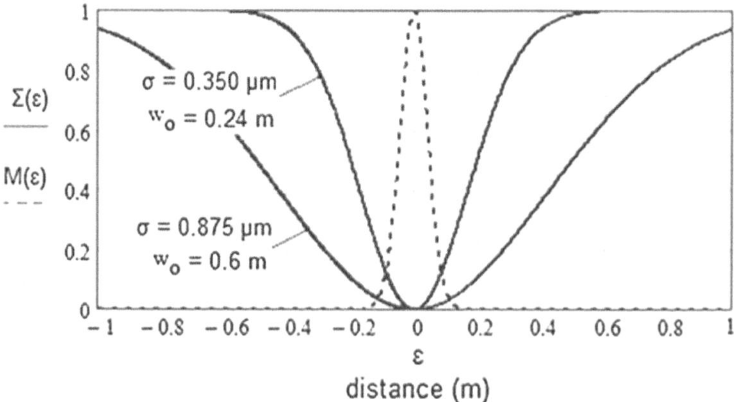


**Fig. 8.9** Wavefront structure functions (in *solid line*) for smaller and larger average turbulence structure sizes as prescribed by the $\sigma$ and $w_o$ values indicated. Both plots are consistent with the same atmospheric MTF ($FWHM = 10$ cm) at $\lambda = 0.55$ μm which is shown in *dashed line*

Figure 8.9 shows two wavefront structure functions, $\Sigma(\varepsilon)$, calculated from the above equation. Both functions shown correspond to approximately the same atmospheric MTF at $\lambda = 0.55\ \mu\text{m}$, a plot of which is shown in the same figure. However, plainly these two wavefront structure functions correspond to entirely different types of turbulence. To be consistent with the same atmospheric MTF at visible wavelengths, again it is only necessary to maintain the correct $\sigma/w_o$ ratio.[8]

Thus, just as was found to be the case for function, $\rho(\varepsilon)$, the atmospheric MTF at visible wavelengths does not necessarily provide either unique or definitive information about the unit-normalized wavefront structure function, $\Sigma(\varepsilon)$.

### 8.4.2 *Wavefront Structure Function for AO-Equipped Telescopes*

For telescopes equipped with AO, the wavefront structure function associated with the AO-corrected wavefronts may be obtained from the relation (cf., 8.37),

$$\Sigma_{AO}(\xi, \eta) = -\frac{\lambda^2}{4 \cdot \pi^2 \cdot \sigma_{AO}^2} \cdot \ln[M_{AO}(\xi, \eta, \lambda)] \tag{8.38}$$

## 8.5 Refractive Index Structure Function

The wavefront structure function, $\Sigma(\xi, \eta)$, represents a scalar field, whose values can be obtained from the appropriate path integrals over the (also scalar) refractive index field, $n(x, y, z)$. On the other hand, the average refractive index structure function over the propagation path, $SF(\xi, \eta)$, is a vector field determined by the vector gradient, grad $[\Sigma(\xi, \eta)] = \nabla[\Sigma(\xi, \eta)]$. Thus, we can write $SF(\xi, \eta)$ as follows:

$$SF(\xi, \eta) = \nabla[\Sigma(\xi, \eta)] = \left[\frac{\partial \Sigma(\xi, \eta)}{\partial \xi}, \frac{\partial \Sigma(\xi, \eta)}{\partial \eta}\right]. \tag{8.39}$$

Combining 8.36 with the above equation allows $SF(\xi, \eta)$ to be expressed in terms of the rms OPD fluctuation, $\sigma$, and the atmospheric MTF, $M(\xi, \eta, \lambda)$, as follows,

$$SF(\xi, \eta) = -\frac{\lambda^2}{4 \cdot \pi^2 \cdot \sigma^2 \cdot M(\xi, \eta, \lambda)} \left[\frac{\partial M(\xi, \eta, \lambda)}{\partial \xi}, \frac{\partial M(\xi, \eta, \lambda)}{\partial \eta}\right]. \tag{8.40}$$

[8] For Kolmogorov turbulence, the wavefront structure function obeys a 5/3-power law prior to flattening out at the outer scale limit, $L_0$. The refractive index structure function, SF, obeys a 2/3-power law before it also flattens out.

### 8.5.1 *Refractive Index Structure Function for Isotropic Turbulence*

Because of the rotational symmetry in this case, the refractive index structure function may be fully expressed in terms of the radial coordinate $\varepsilon$:

$$SF(\varepsilon) = -\frac{\lambda^2}{4 \cdot \pi^2 \cdot \sigma^2 \cdot M(\varepsilon, \lambda)} \left[ \frac{dM(\varepsilon, \lambda)}{d\varepsilon} \right]. \tag{8.41}$$

Plots of $SF(\varepsilon)$ calculated from the above equation are shown in Fig. 8.10. These plots correspond to the same atmospheric MTF and the same pairs of values of $\sigma$ and $w_o$ that were used previously to create the wavefront structure functions previously shown in Fig. 8.9.

## 8.6 Power Spectral Density of the Turbulence Structure

The power spectral density of the integrated OPD fluctuation over the atmospheric path, $\Phi(\mu, \upsilon)$, provides a measure of the distribution of turbulence energy as a function of spatial frequency. According to the Wiener–Khintchine theorem (Middleton, 1960; Thomas, 1971), the power spectral density of the integrated OPD fluctuation over an atmospheric path can be obtained as the Fourier transform of the autocorrelation function of the integrated OPD fluctuation over that path, $\rho(\xi, \eta)$ (cf., 6.30).

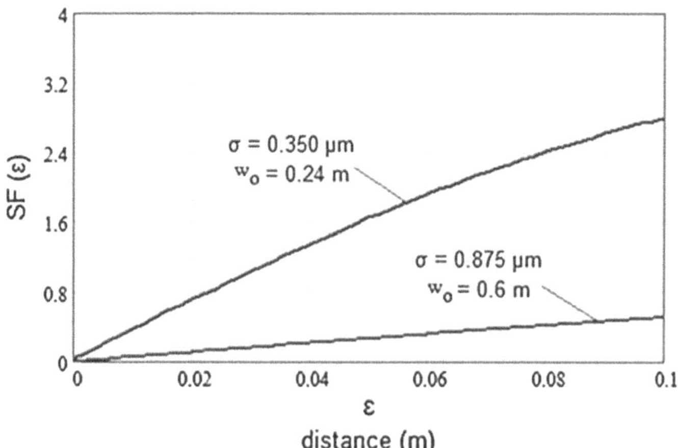


**Fig. 8.10** Refractive index structure functions for smaller and larger average turbulence structure sizes as prescribed by the $\sigma$ and $w_o$ values indicated. Both functions correspond to the same atmospheric MTF at $\lambda = 0.55\mu m$ (FWHM = 10 cm) depicted in Fig. 8.9

We now define spatial frequency arguments, $(\mu, \upsilon)$ in the telescope pupil plane (cf., Fig. 7.2):

$$\mu = \frac{1}{\xi}, \tag{8.42}$$

and

$$\upsilon = \frac{1}{\eta}. \tag{8.43}$$

In terms of these arguments, the power spectral density can be written in the form,

$$\Phi(\mu, \upsilon) = \sigma^2 \cdot \int_{-\infty}^{\infty} \int_{-\infty}^{\infty} \rho(\xi, \eta) \cdot \exp[-2 \cdot \pi \cdot i \cdot (\mu \cdot \xi + \upsilon \cdot \eta)] \cdot d\xi \cdot d\eta. \tag{8.44}$$

Using 8.31 to substitute for $\rho(\xi, \eta)$ in the above equation, the power spectral density can be expressed in terms of the atmospheric MTF, $M(\xi, \eta, \lambda)$, and $\sigma$ by the form,

$$\Phi(\mu, \upsilon) = \int_{-\infty}^{\infty} \int_{-\infty}^{\infty} \left\{\sigma^2 + \frac{\lambda^2}{4 \cdot \pi^2} \cdot \ln[M(\xi, \eta, \lambda)]\right\} \cdot \exp[-2 \cdot \pi \cdot i \cdot (\mu \cdot \xi + \upsilon \cdot \eta)] \cdot d\xi \cdot d\eta, \tag{8.45}$$

where $\ln(\cdot)$ denotes the Napierian logarithm.

### *8.6.1 Power Spectral Density for Isotropic Turbulence*

For isotropic turbulence, the power spectral density function is rotationally symmetric and may therefore be expressed in terms of the radial coordinate in spatial frequency space, $\omega$, defined by

$$\omega = \left(\mu^2 + \upsilon^2\right)^{\frac{1}{2}}. \tag{8.46}$$

Central sections through circularly symmetric power spectral density functions may be obtained by setting either $\mu$ or $\upsilon$ to zero in 8.44. Equally, they can be obtained from the Hankel transforms of the rotationally symmetric autocorrelation functions, as follows:

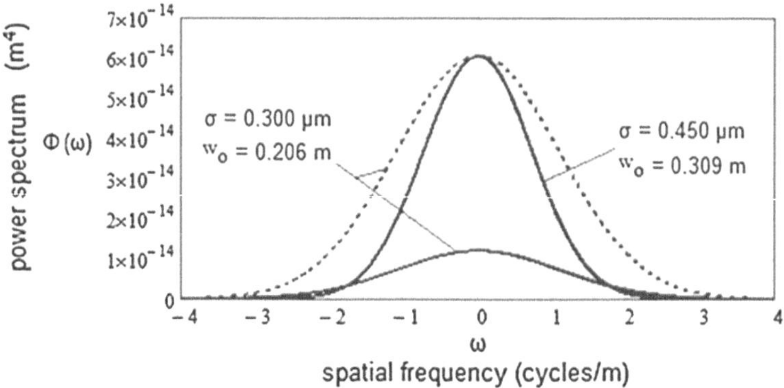


**Fig. 8.11** Power spectral density plots corresponding to the same atmospheric MTF for smaller and slightly larger average turbulence structure sizes as prescribed by the $\sigma$ and $w_0$ values indicated. The *dotted plot* is a renormalized version of the plot in firm line corresponding to $\sigma = 0.3$ μm and $w_0 = 0.206$ m, thus allowing direct comparison of the relative spatial frequency content of the two types of turbulence

$$\Phi(\omega) = 2 \cdot \pi \cdot \sigma^2 \cdot \int_0^{\infty} \rho(\varepsilon) \cdot J_0(2 \cdot \pi \cdot \omega \cdot \varepsilon) \cdot \varepsilon \cdot d\varepsilon, \tag{8.47}$$

where the radial coordinate, $\varepsilon$, was previously defined by 6.46.

Figure 8.11 shows two power spectral density plots calculated from the above equation. Although the two plots are plainly very different, they are nonetheless both consistent with the same atmospheric MTF ($FWHM = 10$ cm at $\lambda = 0.55$ μm) shown previously in Fig. 8.9. Thus, we conclude that the atmospheric MTF at visible wavelengths is not, by itself, a definitive predictor of the atmospheric turbulence power spectral density function.

### 8.6.2 *Power Spectrum of Residual OPD Fluctuation for AO-Equipped Telescopes*

In practice, some measure of residual OPD fluctuation always remains in AO-corrected image-forming wavefronts. The power spectrum of this residual fluctuation, which we denote by $\Phi_{AO}(\mu, \upsilon)$, may be expressed (cf., 8.45) in terms of the residual rms fluctuation, $\sigma_{AO}$, and the AO-corrected version of the atmospheric MTF, $M_{AO}(\xi, \eta, \lambda)$, by

$$\Phi_{AO}(\mu, \upsilon) = \int_{-\infty}^{\infty} \int_{-\infty}^{\infty} \left\{ \sigma_{AO}^2 + \frac{\lambda^2}{4 \cdot \pi^2} \cdot \ln[M_{AO}(\xi, \eta, \lambda)] \right\}$$

$$\cdot \exp[-2 \cdot \pi \cdot i \cdot (\mu \cdot \xi + \upsilon \cdot \eta)] \cdot d\xi \cdot d\eta. \tag{8.48}$$

The spatial frequency distribution of the residual OPD power indicated by $\Phi_{AO}(\mu, \upsilon)$ may be used to identify which spatial frequency ranges are well corrected by the AO system and which are not.

### 8.6.3 Volume Contained Under the Power Spectrum

The volume contained under the power spectrum, which we denote by $\Phi I$, provides a measure of the integrated OPD power over all spatial frequencies:

$$\Phi I = \int_{-\infty}^{\infty} \int_{-\infty}^{\infty} \Phi(\mu, \upsilon) \cdot d\mu \cdot d\upsilon. \tag{8.49}$$

By talking the inverse Fourier transform of both sides of 8.44, the right-hand side of the resulting equation readily simplifies, providing the relation,

$$\int_{-\infty}^{\infty} \int_{-\infty}^{\infty} \Phi(\mu, \upsilon) \cdot \exp[2 \cdot \pi \cdot i \cdot (\mu \cdot \xi + \upsilon \cdot \eta)] \cdot d\mu \cdot d\upsilon = \sigma^2 \cdot \rho(\xi, \eta). \tag{8.50}$$

By setting $\xi = \mu = 0$ in this equation, the left-hand-side reduces to $\Phi I$ (cf., 8.49), while the right-hand side reduces to $\sigma^2$. Thus, we have

$$\Phi I = \sigma^2, \tag{8.51}$$

which merely confirms the well-known result (Davenport & Root, 1958) that the integrated power over all spatial frequencies in a process, random or otherwise, is equal to the variance of the process itself.

**Table 8.2** Mathematical notation used in this chapter along with the SI dimensional units of the individual quantities

| Symbol | Quantity | Dimensions |
|---|---|---|
| $\lambda$ | Wavelength | $m$ |
| (x,y) | Cartesian coordinate system in telescope pupil | $m$ |
| $r$ | Radial coordinate in telescope pupil, ($r = \sqrt{x^2 + y^2}$) | $m$ |

(continued)

**Table 8.2** (continued)

| Symbol | Quantity | Dimensions |
|---|---|---|
| $(\xi, \eta)$ | Cartesian coordinate system in plane perpendicular to direction of light travel, Radial coordinate in telescope pupil, $(\varepsilon = \sqrt{\xi^2 + \eta^2})$ | $m$ |
| $\varepsilon$ | Radial coordinate in plane perpendicular to direction of light travel, | $m$ |
| $(u, v)$ | Cartesian coordinate system in telescope image plane | $m$ |
| $q$ | Radial coordinate in telescope image plane $(\mathrm{q} = \sqrt{u^2 + v^2})$ | $m$ |
| $(\alpha, \beta)$ | Angular coordinate system in telescope image plane | "1" |
| $\vartheta$ | Radial angular coordinate in telescope image plane, $(\vartheta = \sqrt{\alpha + \beta^2})$ | "1" |
| $(\mu, \upsilon)$ | Spatial frequency coordinate system in telescope image plane | $\mathrm{m}^{-1}$ |
| $\omega$ | Radial spatial frequency coordinate in telescope image plane, $(\omega = \sqrt{\mu^2 + \upsilon^2})$ | $\mathrm{m}^{-1}$ |
| $H$ | Integrated OPD fluctuation over entire atmospheric path | $m$ |
| $\sigma$ | rms of integrated OPD fluctuation, $H$ | $m$ |
| $\sigma_T$ | rms wavefront error of telescope | $m$ |
| $\rho$ | Autocorrelation function of the integrated OPD fluctuation, $H$ | "1" |
| $\sigma_{AO}$ | rms of AO-corrected image-forming waves | $m$ |
| $\rho_{AO}$ | Autocorrelation function of AO-corrected image-forming waves | "1" |
| $f$ | Telescope focal length | $m$ |
| $M$ | Atmospheric MTF | "1" |
| $M_{\mathrm{L}}$ | Long-exposure atmospheric MTF | "1" |
| $M_{\mathrm{S}}$ | Short-exposure atmospheric MTF | "1" |
| $M_{\mathrm{T}}$ | Telescope OTF | "1" |
| $I$ | Image intensity | "1" |
| $I_{\mathrm{I}}$ | Spectrally integrated image intensity | "1" |
| $I_{AO}$ | AO-corrected image intensity | "1" |
| $\mu$ | Complex coherence factor | "1" |
| $\mu_{AO}$ | Intensity contrast of AO-corrected image | "1" |
| $G$ | Relative spectral content of image intensity | "1" |
| $a_1$ and $a_2$ | Photon noise bias correction constants | "1" |
| $\lambda_C$ | Central wavelength of rectangular spectral distribution | $m$ |
| $\Delta\lambda$ | Full width of rectangular spectral distribution | $m$ |
| SI | Strehl intensity | "1" |
| $\mathrm{SI}_{\mathrm{AO}}$ | Strehl intensity of AO-corrected image | "1" |
| ZA | Zenith angle | "1" |

Dimensionless quantities are indicated by "1"

## References

Abramowitz, M., & Stegun, I. A. (1970). *Handbook of mathematical functions*. U.S. Government Printing Office.
Dainty, J. C. (1984). Laser speckle and related phenomena. In J. C. Dainty (Ed.), *Topics in applied physics* (Vol. 9). Springer.
Davenport, W. B., & Root, W. L. (1958). *An introduction to the theory of random signals and noise*. McGraw-Hill.
Elbaum, M., Greenbaum, M., & King, M. (1972). Wavelength diversity technique for reduction of speckle. *Optics Communication, 5*, 171–174.
Mariotti, M., & Di Benedetto, G. P. (1984). Pathlength stability of synthetic aperture telescopes. The case of the 25 cm Cerga interferometer. In H. Ulrich & K. Jaer (Eds.), *Proceedings of International Astronomical Union Colloquium 79* (pp. 257–265). European Southern Observatory.
McKechnie, T. S. (2004). Fundamental measurement procedures for establishing optical tolerances for ground-based telescopes. In S. C. Craig & M. J. Cullum (Eds.), *Proceedings of SPIE, Modeling and Systems Engineering for Astronomy*, Bellingham, WA (Vol. 5497_11, pp. 103–116).
Middleton, D. (1960). *Introduction to statistical communication theory*. McGraw Hill Book Co.
Mills, E. (2005). *The rise and fall of the Royal Greenwich Observatory*. Retrieved December 5, 2014, from http://www.millseyspages.com/astro_pages/rgo/rgo.html
Sprague, R. A. (1972). Surface roughness measurement using white light speckle. *Applied Optics., 11*, 2811–2816.
Thomas, J. B. (1971). *An introduction to applied probability and random process*. Wiley.
Woolf, N. J., McCarthy, D. W., & Angel, J. R. (1982). Performance of the MMT VII: image shrinkage in sub-arc second seeing at the MMT and 2.3 meter telescopes. In L. D. Barr & G. Burbridge (Eds.), *Advanced Technology Optical Telescopes, Proceedings of SPIE Engineering*, Bellingham, WA (Vol. 332, pp. 50–56).

# Chapter 9
# The Average Intensity Envelope of an Unresolved Star Image

**Abstract** This chapter examines the characteristics of the average intensity envelopes of star images formed by ground-based telescopes. These envelopes arise as the Fourier transform of the product of the atmospheric MTF and the telescope OTF. The seeing disc star images delivered by large (>1 m) telescopes typically measure about 1-arcsec across which, sadly, is about one or two orders of magnitude broader than would be delivered by diffraction-limited versions of the same telescopes observing in the absence of atmosphere. However, at the end of the chapter, we introduce the subject of image cores. In good seeing conditions, these features are sometimes seen in star images at visible wavelengths, but at IR wavelengths they appear routinely in almost any seeing conditions. If cores are present, the average intensity envelopes then comprise highly-resolved (diffraction-limited) core features surrounded by the much wider halo features. The intriguing properties of core and halo images are examined in detail in the next chapter.

As initially plane waves from distant stars propagate down through the turbulent atmosphere, they become increasingly disrupted. Nonetheless, once these waves arrive at an observer on the ground, they can still retain some semblance of smoothness—coherence—over wavefront portions measuring a few inches across. At visible wavelengths, this residual coherence allows small telescopes with good optics to deliver substantially diffraction-limited star images—Airy patterns for telescopes with good-quality optics and circular apertures (Fig. 4.7).

In contrast, star images formed by large telescopes are almost never diffraction-limited; the more severe random phase corrugations and amplitude scintillations that arise over the larger wavefront portions collected by these instruments fragment the images into speckle patterns, known as stellar speckle patterns (Fig. 4.9). The dynamical behavior of the atmosphere causes the individual speckle features in these images to continuously evolve over time, giving them an appearance similar to that of boiling water.

While instantaneous, or short-exposure, star images formed by large instruments are noisy speckle patterns, long-exposure images and even the average short-exposure image have smooth appearances due to the speckle noise averaging to zero.



T. S. McKechnie, *General Theory of Light Propagation and Imaging Through the Atmosphere*, Progress in Optical Science and Photonics 20,
https://doi.org/10.1007/978-3-030-98828-9_9

In this chapter, expressions are developed that describe the smooth (average) intensity envelopes that arise in long-exposure and average short-exposure images. A top-level examination is also made of the various properties and characteristics of these intensity envelopes.

The expressions developed apply to telescopes of all sizes, small and large. They also take account of all of the various controlling parameters, which include parameters describing the size and shape of the telescope aperture, telescope aberrations, central obstructions, the observation wavelength, and the atmospheric seeing quality. Explicit expressions are also given for the average intensity envelopes formed by the most frequently encountered type of telescope—telescopes with circular apertures.

From an initial examination of the expressions, it will be evident that at long-enough wavelengths—where the wavelengths become larger than the rms optical path difference (OPD) fluctuation of the image-forming waves in the telescope pupil—increasingly large fractions of the light energy in star images begin to concentrate into highly resolved features in the image centers. Such features are referred to as image cores, or simply cores.

The property that makes image cores so crucially important is that their angular widths do not depend on the atmospheric seeing[1]; instead they depend only on the imaging wavelength and the quality of the telescope optics. Whereas the angular width of the seeing disk image produced by a large (several meters in diameter) astronomical telescope might typically measure about 1 arcsec across, the angular width of an image core formed by the same instrument might typically be about an order of magnitude smaller.[2] Light not directed into the core remains in a surrounding speckle cloud, known as the halo. The angular size and shape of the halo closely relate to the angular size and shape of the seeing disk images typically formed by large telescopes at visible wavelengths in average or poor seeing conditions.

In better than average seeing conditions, cores can appear in star images at visible wavelengths. But even in poor seeing, when the core energy fraction at visible wavelengths is often negligible, strong cores can still be seen at near-IR and longer wavelengths. Core structure in star images is hugely important; to achieve the highest levels of resolution with any large ground-based telescope—whether or not the instrument is equipped with adaptive optics (AO)—it is essential to observe at wavelengths long enough to be consistent with the formation of star image cores.

The most general type of star image consists of both core and halo. In fact, images consisting of just a core or just a halo may be seen simply as degenerate cases of this type of image. While we merely introduce core and halo images in this chapter, the properties of this type of image are of such fundamental importance that they warrant a more thorough examination—one that will be given in the next chapter.

[1] Only the core energy fraction is determined by the seeing.

[2] For diffraction-limited telescopes, the angular radius of the core depends on the imaging wavelength and telescope diameter and is roughly given in radians by the formula, $1.22\,\lambda/D$.

## 9.1 Average Intensity Envelope for the Image of an Unresolved Star

An expression for the average intensity envelope, $\langle I(u, v, \lambda)\rangle$, of an unresolved star image formed by a telescope observing through the atmosphere was given previously by 7.49. For convenience, we reproduce that expression here:

$$\langle I(u, v, \lambda)\rangle = \frac{1}{\lambda^2} \cdot \int_{-\infty}^{\infty} \int_{-\infty}^{\infty} M(\xi, \eta, \lambda) \cdot M_T(\xi, \eta, \lambda) \times \exp\left[\frac{-2 \cdot \pi \cdot i}{\lambda \cdot f} \cdot (u \cdot \xi + v \cdot \eta)\right] \cdot d\xi \cdot d\eta, \tag{9.1}$$

where $(u, v)$ are the Cartesian coordinates in the telescope image plane (Fig. 7.2), $M(\xi, \eta, \lambda)$ is the atmospheric modulation transfer function (MTF), and $M_T(\xi, \eta, \lambda)$ is the telescope optical transfer function (OTF), an expression for which was given previously by 7.10 in terms of the telescope pupil function, $K(x, y, \lambda)$.

Using the atmospheric MTF expression given by 6.52, 9.1 may be written

$$\langle I(u, v, \lambda)\rangle = \frac{1}{\lambda^2} \cdot \int_{-\infty}^{\infty} \int_{-\infty}^{\infty} \exp\left\{\frac{-4 \cdot \pi^2 \cdot \sigma^2}{\lambda^2} \cdot [1 - \rho(\xi, \eta)]\right\} \cdot M_T(\xi, \eta, \lambda) \times \exp\left[\frac{-2 \cdot \pi \cdot i}{\lambda \cdot f} \cdot (u \cdot \xi + v \cdot \eta)\right] \cdot d\xi \cdot d\eta. \tag{9.2}$$

It may be observed here that 9.1 and 9.2 can be used for either long-exposure images or the average short-exposure image by merely inserting, as appropriate, the long- or short-exposure versions of $\sigma^2$, $\rho(\xi, \eta)$, $\langle I(u, v, \lambda)\rangle$, and $M(\xi, \eta, \lambda)$ into these equations. We recall from Chap. 8 (Sect. 8.1.2) that either set can be obtained depending on whether or not image wander is corrected during the star image measurements.

It might be noted that the intensity values provided by 9.1 and 9.2 (and by many other similar equations in this book) are given in dimensionless units. However, as previously noted in Sect. 7.7, appropriate intensity units may be assigned at any time by simply multiplying the dimensionless values that arise from these equations by appropriately dimensioned constants.

## 9.2 Average Intensity Envelope for Small Diffraction-Limited Telescopes with Circular Apertures

The intensity OTF of a telescope for incoherent light, $M_T(\xi, \eta, \lambda)$, was given previously as the (unit-normalized) autocorrelation of the pupil function (cf., 7.10). For telescopes with circular apertures, this function only takes nonzero values for argument values contained in the range

$$\xi^2 + \eta^2 = \varepsilon^2 \leq D^2. \tag{9.3}$$

An expression for the OTF of a diffraction-limited telescope with circular aperture was given previously by 7.12. By substituting that expression (with appropriately chosen arguments) into 9.2, the average intensity envelope formed by such a telescope may be written in the form

$$\begin{aligned}\langle I(u, v, \lambda)\rangle = \frac{1}{\lambda^2} \cdot \int_{-D}^{D}\int_{-D}^{D} \exp\left\{-\left(\frac{2\cdot\pi\cdot\sigma}{\lambda}\right)^2 \cdot [1-\rho(\xi,\eta)]\right\} \cdot \frac{2}{\pi} \\ \times \left\{\cos^{-1}\left[\frac{(\xi^2+\eta^2)^{\frac{1}{2}}}{D}\right] - \frac{(\xi^2+\eta^2)^{\frac{1}{2}}}{D} \cdot \sqrt{1-\frac{(\xi^2+\eta^2)}{D^2}}\right\} \\ \times \exp\left[\frac{-2\cdot\pi\cdot i}{\lambda\cdot f} \cdot (u\cdot\xi + v\cdot\eta)\right] \cdot d\xi \cdot d\eta.\end{aligned} \tag{9.4}$$

## 9.2.1 Case of Isotropic Turbulence

For isotropic turbulence, function $\rho(\xi, \eta)$ becomes circularly symmetric. A central section through the intensity envelope in this case can then be obtained from 9.4 by setting either $\xi$ or $\eta$ to zero. Equally, the central section could be obtained from the Hankel transform

$$\begin{aligned}\langle I(q, \lambda)\rangle = \frac{4}{\lambda^2} \cdot \int_0^D \exp\left\{-\left(\frac{2\cdot\pi\cdot\sigma}{\lambda}\right)^2 \cdot [1-\rho(\varepsilon)]\right\} \cdot \left[\cos^{-1}\left(\frac{\varepsilon}{D}\right) - \frac{\varepsilon}{D} \cdot \sqrt{1-\left(\frac{\varepsilon}{D}\right)^2}\right] \\ \times J_0\left(\frac{2\cdot\pi\cdot\varepsilon\cdot q}{\lambda\cdot f}\right) \cdot \varepsilon \cdot d\varepsilon,\end{aligned} \tag{9.5}$$

where $q$ is the radial coordinate in the image plane (Fig. 7.2).

An equivalent expression can also be given in terms of the radial angular coordinate $\vartheta = q/f$ (c.f., 7.4).

$$\begin{aligned}\langle I(\vartheta, \lambda)\rangle = \frac{4}{\lambda^2} \cdot \int_0^D \exp\left\{-\left(\frac{2\cdot\pi\cdot\sigma}{\lambda}\right)^2 \cdot [1-\rho(\varepsilon)]\right\} \cdot \left[\cos^{-1}\left(\frac{\varepsilon}{D}\right) - \frac{\varepsilon}{D} \cdot \sqrt{1-\left(\frac{\varepsilon}{D}\right)^2}\right] \\ \times J_0\left(\frac{2\cdot\pi\cdot\varepsilon\cdot\vartheta}{\lambda}\right) \cdot \varepsilon \cdot d\varepsilon.\end{aligned} \tag{9.6}$$

## 9.3 Average Envelopes for Small Diffraction-Limited Telescopes with Circular Apertures

In average nighttime seeing conditions, the instantaneous wave portions collected by small telescopes (with apertures smaller than about 4 in.) may be considered as being substantially planar. This condition may be represented setting $\sigma = 0$ in 9.5, so that this equation then reduces to the following:

$$\langle I(q,\lambda)\rangle = \frac{4}{\lambda^2}\cdot\int_0^D\left[\cos^{-1}\left(\frac{\varepsilon}{D}\right)-\frac{\varepsilon}{D}\cdot\sqrt{1-\left(\frac{\varepsilon}{D}\right)^2}\right]\cdot J_0\left(\frac{2\cdot\pi\cdot\varepsilon\cdot q}{\lambda\cdot f}\right)\cdot\varepsilon\cdot d\varepsilon. \tag{9.7}$$

An equivalent equation can also be obtained from 9.6 in terms of the radial angular coordinate, $\vartheta$,

$$\langle I(\vartheta,\lambda)\rangle = \frac{4}{\lambda^2}\cdot\int_0^D\left[\cos^{-1}\left(\frac{\varepsilon}{D}\right)-\frac{\varepsilon}{D}\cdot\sqrt{1-\left(\frac{\varepsilon}{D}\right)^2}\right]\cdot J_0\left(\frac{2\cdot\pi\cdot\varepsilon\cdot\vartheta}{\lambda}\right)\cdot\varepsilon\cdot d\varepsilon. \tag{9.8}$$

Equations 9.7 and 9.8 are both recognized as describing Airy patterns. Figure 9.1 shows such patterns calculated from 9.7 for a 4-in. (10 cm) diameter telescope for three different visible wavelengths. Random tilts in the incoming wavefronts can cause Airy pattern images like this to wander randomly in the image. If no attempt is made to correct such "image wander" while long-exposure images are being recorded, the Airy pattern image becomes smeared and hence degraded. The image degradation caused by such smearing (and the corresponding loss of resolution) can be seen in the intensity envelopes shown in Fig. 9.2. The plots in this figure were calculated from 9.5 using the $\sigma$ value and $\rho(\varepsilon)$ function obtained at RGO as

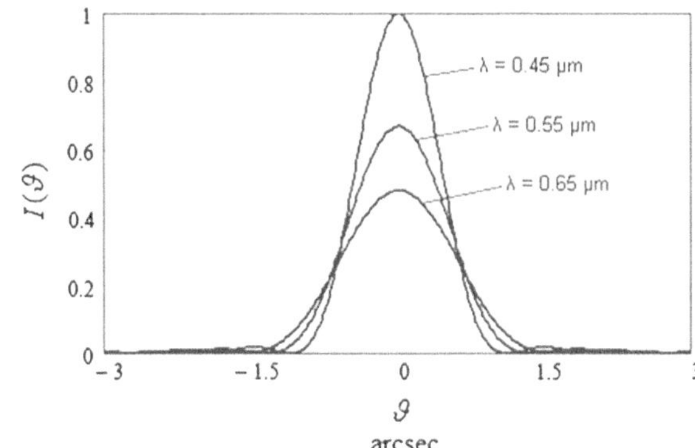


**Fig. 9.1** Instantaneous Airy patterns star images formed by a 4-in. diameter diffraction-limited telescope at blue, green, and red wavelengths. Upon rotation, all of these plots enclose the same volume (light energy)

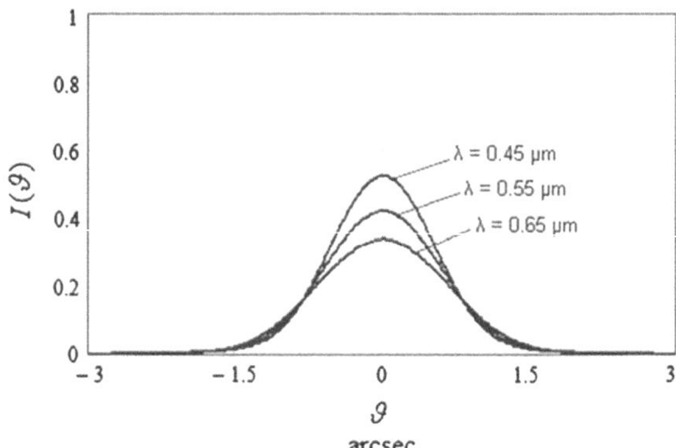


**Fig. 9.2** Long-exposure star images formed by the 4-in. diameter diffraction-limited telescope at blue, green, and red wavelengths. Image wander has significantly degraded the Airy pattern images seen in Fig. 9.1

described previously in Chap. 8. The use of the RGO measurement data is valid in this particular case because it was obtained using a 36-in. (0.915 m) telescope which entitles this data to be used for the 4-in. telescope considered here. (In fact it is valid for any telescope diameter up to 36 in.)

## 9.4 Seeing Disk Envelopes Formed by Large Telescopes

The average intensity envelope for a star image formed by a large ground-based telescope can take different forms depending on the ratio, $\sigma/\lambda$. Two cases are examined corresponding to (1) $\sigma/\lambda \geq 0.4$, and (2) $\sigma/\lambda < 0.4$.

For case (1), the instantaneous image comprises a speckle pattern halo with a relatively broad average intensity envelope, typically measuring about 1 arcsec across. For case (2), a highly resolved central core forms in the image (assuming that the optical quality of the telescope is good enough to adequately resolve this feature). This core is superposed on the much broader halo containing the remaining light fraction.

In average (~1 arcsec) visible seeing conditions, 4-m class telescopes with good optics, imaging in a wavelength band around 1 μm, should be able to deliver 0.05-arcsec image cores in the average short-exposure image (McKechnie, 1990, 1991, 1994). To produce such an image, it is only necessary to obtain a long-exposure image while correcting image wander during the exposure. If equipped with an AO system, a 4-m class instrument should be able to deliver image cores at even shorter wavelengths, resulting in even higher resolution levels.

Given an efficiently functioning AO system, in principle, image cores can be produced at any wavelength transmitted by the atmosphere, including the shortest wavelength, about 0.3 μm. Given a telescope with a large enough aperture equipped with such an AO system, there is no theoretical upper limit to telescope resolution.[3] Even though AO systems do not correct scintillation (which is invariably present to some extent in the image-forming wavefronts after propagation through the atmosphere), it will be shown later in Chap. 14 (Sect. 14.6.3) that even the strongest scintillation carries surprisingly little resolution penalty.

The illustrative intensity envelopes provided later in this chapter are based on isotropic turbulence where we use the autocorrelation functions of the OPD fluctuation, $\rho(\varepsilon)$, obtained from the RGO measurement data, functions that were shown previously in Fig. 8.7. From that figure, evidently $\rho(\varepsilon)$ may be approximated by the Gaussian function,[4]

$$\rho(\varepsilon) = \exp\left[-\left(\frac{\varepsilon}{w_o}\right)^2\right], \tag{9.9}$$

where the $1/e$ half-width, $w_o$, provides a measure of the average turbulence structure size. It might again be worth emphasizing here that the intensity envelopes based on the above $\rho(\varepsilon)$ functionality are for illustrative purposes only. Seeing measurements made at other sites and in other seeing conditions might result in other types of $\rho(\varepsilon)$ functionality.

### 9.4.1 Case (1): Speckle Images for the Case $\sigma/\lambda \geq 0.4$

The condition $\sigma/\lambda \geq 0.4$ frequently holds for large telescopes imaging at visible wavelengths in average or poor seeing conditions. In this case, the random phase angles of the imaging wavefronts are distributed in an approximately uniform manner over the primary phase angle range, $-\pi$ to $\pi$. As discussed in Chap. 11, for this type of phase distribution, the instantaneous image of a quasi-monochromatic point-object—such as an unresolved star viewed through a narrow-band filter—appears as a speckle pattern obeying Gaussian statistics. However, both the long-exposure intensity envelope and the average short-exposure intensity envelope for this type of image are entirely smooth due to speckle averaging.

By combining 9.5 and 9.9, the following expression arises for the average intensity envelope formed by a large diffraction-limited telescope with a circular aperture

$$\langle I(q,\lambda)\rangle = \frac{4}{\lambda^2}\cdot\int_0^D \exp\left\{-\left(\frac{2\cdot\pi\cdot\sigma}{\lambda}\right)^2\cdot\left[1-\exp\left[-\left(\frac{\varepsilon}{w_o}\right)^2\right]\right]\right\}$$

[3] This assumes that we can continue to build ever larger telescopes. Theoretically, a diffraction-limited 100-m telescope imaging at 0.3 μm should deliver angular 0.75-mas resolution. To further improve, resolution would require an even larger instrument.

[4] The wavefront structure function corresponding to a Gaussian representation of $\rho(\varepsilon)$ may be seen (Sect. 8.4) as similar to the function $\varepsilon^{5/3}$, rolled-off at an appropriate $L_0$ value.

$$\times \left[ \cos^{-1}\left(\frac{\varepsilon}{D}\right) - \frac{\varepsilon}{D} \cdot \sqrt{1 - \left(\frac{\varepsilon}{D}\right)^2} \right] \cdot J_0\left(\frac{2 \cdot \pi \cdot \varepsilon \cdot q}{\lambda \cdot f}\right) \cdot \varepsilon \cdot d\varepsilon. \tag{9.10}$$

This equation may be equivalently expressed in terms of the radial angular coordinate $\vartheta$ (7.4 and Fig. 7.2),

$$\langle I(\vartheta, \lambda) \rangle = \frac{4}{\lambda^2} \cdot \int_0^D \exp\left\{ -\left(\frac{2 \cdot \pi \cdot \sigma}{\lambda}\right)^2 \cdot \left[ 1 - \exp\left[ -\left(\frac{\varepsilon}{w_o}\right)^2 \right] \right] \right\}$$
$$\times \left[ \cos^{-1}\left(\frac{\varepsilon}{D}\right) - \frac{\varepsilon}{D} \cdot \sqrt{1 - \left(\frac{\varepsilon}{D}\right)^2} \right] \cdot J_0\left(\frac{2 \cdot \pi \cdot \varepsilon \cdot \vartheta}{\lambda}\right) \cdot \varepsilon \cdot d\varepsilon. \tag{9.11}$$

#### 9.4.1.1 Envelopes for a 4-m Diffraction-Limited Telescope Imaging at 0.55 μm

Figure 9.3 shows average intensity envelopes calculated from 9.11 as obtained by a 4-m diffraction-limited telescope imaging at the visible wavelength, $\lambda = 0.55\ \mu m$. All envelopes in the figure correspond to the same turbulence structure size distribution, as quantified by $w_o = 0.55$ m; differences between the various envelopes are due entirely to differences in turbulence strength, as set by the $\sigma$ values indicated in the figure.

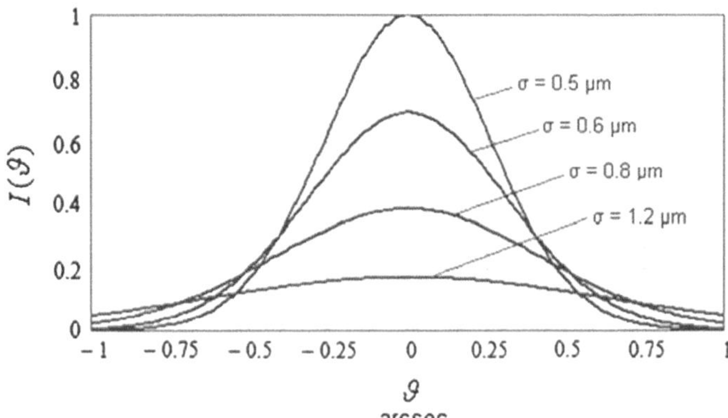


**Fig. 9.3** Star image intensity envelopes at $\lambda = 0.55$ μm for a 4-m diffraction-limited telescope for relatively large average turbulence structure size corresponding to $w_o = 0.55$ m. The FWHM angular widths of the envelopes 0.6, 0.75, 1.0, and 1.5 arcsec correspond to the various turbulence strengths indicated by the $\sigma$ values

The relationship between the full-width half-maximum (FWHM) angular width of a halo image, $\mathrm{FWHM}_{\mathrm{Halo}}$, and the seeing parameters, $\sigma$ and $w_o$, is developed later in Chap. 13 (cf., 13.46). However, it is convenient to anticipate this result for use in this section,

$$FWHM_{Halo} = 3.33 \cdot \frac{\sigma}{w_o}. \tag{9.12}$$

A significant result conveyed by this equation is that the FWHM angular size of a halo only, or seeing disk, star image is entirely determined by the ratio, $\sigma/w_o$. (In fact, the same is also true of other measures of the star's angular size, such as the star's angular width measured across the $1/e$ intensity levels.) The individual values attached to either $\sigma$ or $w_o$ do not, by themselves, control image size; only their ratio matters.

Inserting the appropriate $\sigma$ and $w_o$ values into the above relation gives the FWHM angular widths of the four envelopes in Fig. 9.3 as 0.6, 0.75, 1.0, and 1.5 arcsec (as may be confirmed by visual inspection of the actual envelopes). The various envelopes correspond to visible seeing quality ranging from good to poor; the normalization of the envelopes is such that, after rotation, they all enclose the same volume (i.e., total light energy).

The image envelops shown in Fig. 9.4 correspond to smaller turbulence structure as set by $w_o = 0.275$ m—a value exactly half of that used to generate the Fig. 9.3 envelopes. However, because the $\sigma$ values used, namely $\sigma = 0.25\,\mu m, \sigma = 0.30\,\mu m$, $\sigma = 0.40\,\mu m$, $\sigma = 0.25\,\mu m$, and $\sigma = 0.60\,\mu m$, are precisely half of the values used in Fig. 9.3, the angular widths of the intensity envelopes in both figures are approximately the same.

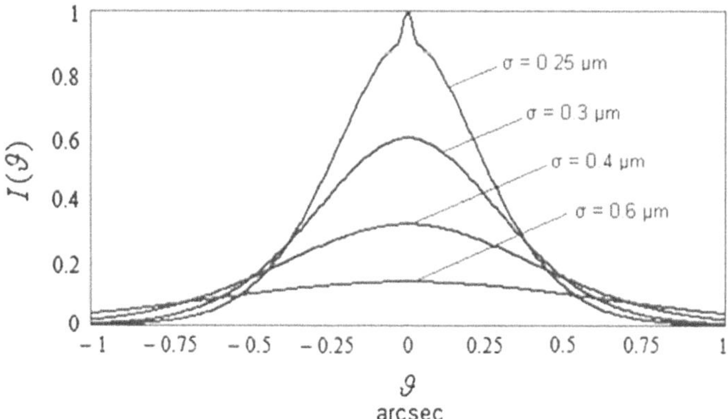


**Fig. 9.4** Star image intensity envelopes at $\lambda = 0.55\,\mu$m for a 4-m diffraction-limited telescope for relatively small average turbulence structure size corresponding to $w_o = 0.275$ m. The FWHM widths of the envelopes, 0.6, 0.75, 1.0, and 1.5 arcsec correspond to the various turbulence strengths indicated by the $\sigma$ values. The envelope corresponding to $\sigma = 0.25\,\mu m$ shows a central core emerging in the image

The plot in Fig. 9.4 corresponding to $\sigma = 0.25\,\mu m$ shows the emergence of a rudimentary image core; the image feature shown contains a mere 0.03% of the total light energy. We now discuss image cores containing much larger and more useful fractions of the total light energy in the image.

## 9.4.2 *Case (2): Core and Halo Images for the Case* $\sigma/\lambda < 0.4$

For $\sigma/\lambda < 0.4$, the phase angles of the wavefronts arriving in the telescope pupil are no longer uniformly distributed over the primary phase angle interval, $-\pi$ to $\pi$. Instead, they disproportionately cluster around the phase angle, zero—a condition that leads to central cores forming in the images. The presence of image cores enables significantly higher telescope resolution. As the ratio $\sigma/\lambda$ reduces, the core light energy fraction grows larger.

### 9.4.2.1 Satisfying the Condition $\sigma/\lambda < 0.4$ for Large Telescopes

For large telescopes, a number of circumstances can result in the condition $\sigma/\lambda < 0.4$ being satisfied, three of which we now identify:

(1) At visible (and longer) wavelengths in favorable seeing conditions when $\sigma$ is suitably small;
(2) At visible (and longer) wavelengths in less favorable seeing conditions where AO is used to, in effect, drive down $\sigma$ to a suitably small value;
(3) In less favorable seeing conditions, or where an imperfectly functioning AO system is used, we can simply observe at wavelengths long enough to satisfy the condition; often these wavelengths lie in the IR wavelength region.

When either circumstance (1) or (2) applies, the average image intensity envelopes obtained by a 4-m diffraction-limited telescope could look like those shown in Figs. 9.5, 9.6, and 9.7. The various envelopes shown in these figures corresponding to the visible wavelengths indicated.[5] The FWHM widths of the halos in these figures correspond to the sub-arcsec visible seeing values, 0.5, 0.35, and 0.25 arcsec. (These seeing values correspond, respectively, to $\sigma = 0.22$, 0.16, and 0.1 μm, with $w_o = 0.25$ m.) Seeing of this high quality is not uncommon at pristine, high-altitude sites, such as on Mauna Kea, HI (CFHT Site Characteristics, 2003). However, such "seeing" may also be created artificially by use of AO. Thus, the envelopes shown in these three figures may be considered as applying either to naturally occurring pristine seeing conditions or to equally "pristine seeing conditions" achieved with the assistance of AO. It might be noted that all of the image envelopes shown in these three figures enclose the same total amount of light energy.

[5] The visible wavelength, 0.78 μm, corresponds to the upper limit of scotopic vision.

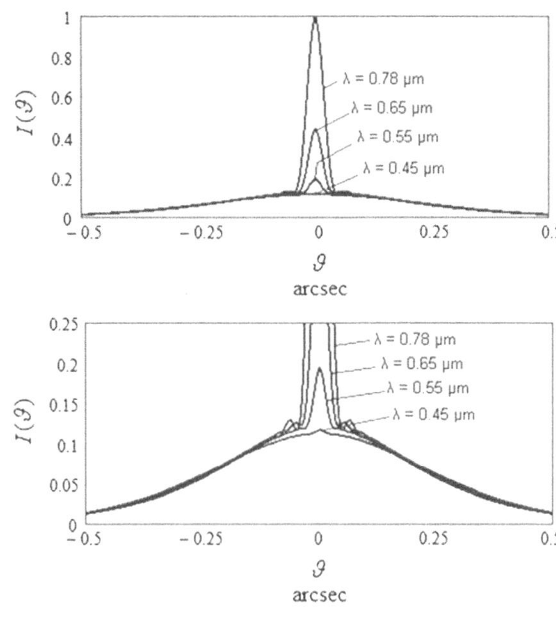


**Fig. 9.5** Star image intensity envelopes. *Top* Cores at various visible wavelengths formed by a 4-m diffraction-limited telescope in ~0.5-arcsec visible seeing conditions. *Bottom* Same plots with expanded vertical scale. Parameter values used are $\sigma = 0.22\ \mu$m, $w_o = 0.25$ m

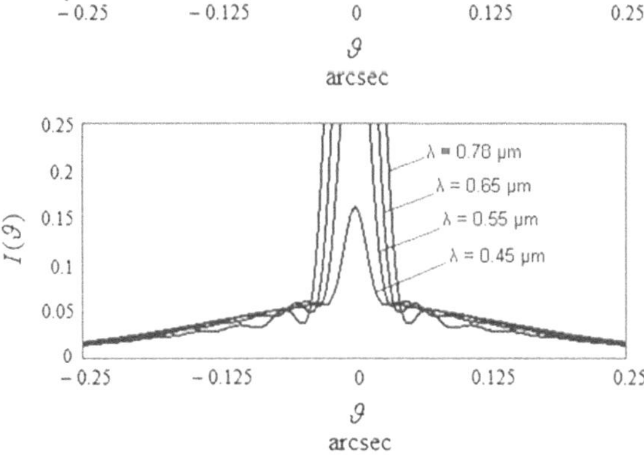


**Fig. 9.6** Star image intensity envelopes. *Top* Cores at various visible wavelengths formed by a 4-m diffraction-limited telescope in 0.35-arcsec visible seeing conditions. *Bottom* Same plots with expanded vertical scale. Parameter values used are $\sigma = 0.16\ \mu$m, $w_o = 0.25$ m

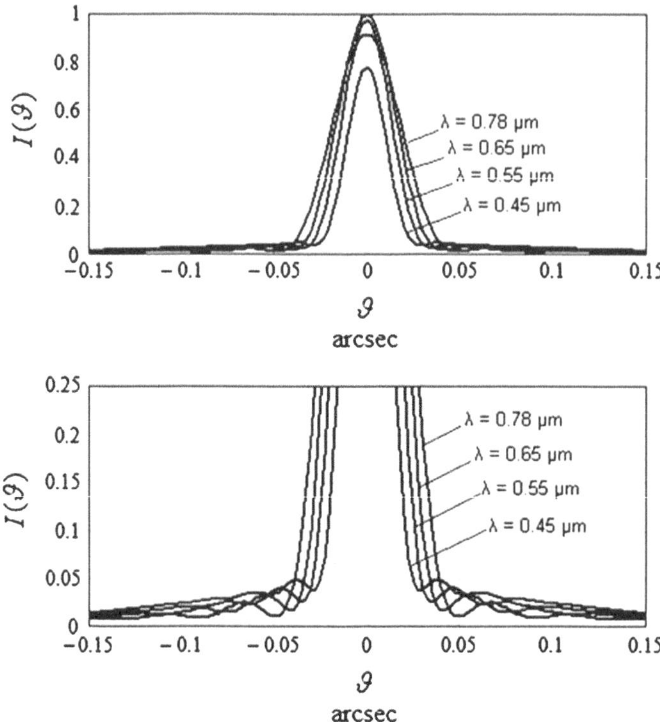


**Fig. 9.7** Star image intensity envelopes. *Top* Cores at the visible wavelengths indicated as formed by a 4-m diffraction-limited telescope in ~0.25-arcsec visible seeing conditions. *Bottom* Same plots with expanded vertical scale. Parameter values used: $\sigma = 0.1$ μm, $w_o = 0.25$ m

For significantly aberrated telescopes, or when telescopes observe in unfavorable seeing conditions, it may not be possible to enjoy the benefits of circumstances (1) and (2). Nonetheless, it is always possible to benefit from circumstance (3). The condition $\sigma/\lambda < 0.4$ can always be satisfied by simply imaging at a long-enough wavelengths where central cores must inevitably develop in the images. However, because the angular widths of image cores increase in proportion to the wavelength, the angular resolution obtained by fulfilling circumstance (3) is not generally as high as that obtained by fulfilling circumstances (1) or (2).

Halo-only images, which are often produced by large telescopes at visible wavelengths in average or poor seeing conditions, and core-only images, which are commonly produced by small telescopes, may both be considered degenerate cases of the core and halo image. Core and halo images are the most commonly-occurring type of star image that can be formed by any ground-based telescope; this type of image occurs irrespective of whether the telescope is equipped with AO. Such is the importance of core and halo images; the next chapter is entirely devoted to a detailed examination of the properties of this type of image.

## 9.5 Mathematical Notation Used in This Chapter

The mathematical notation used in this chapter is indicated in Table 9.1.

**Table 9.1** Mathematical notation used in this chapter along with the SI dimensional units of the individual quantities

| Symbol | Quantity | Dimensions |
|---|---|---|
| $\lambda$ | Wavelength | $m$ |
| $(\xi, \eta)$ | Cartesian coordinate system in plane perpendicular to direction of light travel | $m$ |
| $\varepsilon$ | Radial coordinate in plane perpendicular to direction of light travel $\left(= \sqrt{\xi^2 + \eta^2}\right)$ | $m$ |
| $(u, v)$ | Cartesian coordinate system in telescope image plane | $m$ |
| $q$ | Radial coordinate in telescope image plane $\left(= \sqrt{u^2 + v^2}\right)$ | $m$ |
| $(\alpha, \beta)$ | Angular coordinate system in telescope image plane | "1" |
| $\vartheta$ | Radial angular coordinate in telescope image plane $\left(= \sqrt{\alpha^2 + \beta^2}\right)$ | "1" |
| $I$ | Image intensity | "1" |
| $M$ | Atmospheric MTF | "1" |
| $M_T$ | Telescope OTF | "1" |
| $H$ | Integrated OPD fluctuation over entire atmospheric path | $m$ |
| $\Sigma$ | rms integrated OPD fluctuation, $H$ | $m$ |
| $\sigma$ | rms wavefront error of telescope | $m$ |
| $\rho$ | Autocorrelation function of $H$ | "1" |
| $w_0$ | $1/e$ half-width of Gaussian $\rho$ function | $m$ |
| $D$ | Telescope diameter | $m$ |
| $f$ | Telescope focal length | $m$ |

Dimensionless quantities are indicated by "1"

## References


CFHT Site Characteristics. (2003, January). *CFHT Observatory manual*, version 1.

McKechnie, T. S. (1990). Diffraction limited imaging using large ground-based telescopes. In *Proceedings of SPIE, V. 1236, Symposium on Astronomical Telescopes and Instrumentation for the 21st Century* (pp. 164–178). 11–17 February.

McKechnie, T. S. (1991). Light propagation through the atmosphere and the properties of images formed by large ground-based telescopes. *JOSA A, 8*, 346–365.

McKechnie, T. S. (1994). Another route to sharp images. *Sky and Telescope Magazine, 88*(2), 36–38.

# Chapter 10
# Core and Halo Star Images Formed by Large Telescopes

**Abstract** This chapter examines in detail the properties of cores and halos in star images. Strong cores emerge when $\sigma/\lambda < 0.4$. In average astronomical seeing conditions, cores dominate the images at near-IR and longer wavelengths. In the other regime, $\sigma/\lambda > 0.4$, halos instead dominate the images; in average seeing conditions, images at visible wavelengths usually appear as halos. Halo shape closely replicates the shape of seeing disc images often formed by large telescopes at visible wavelengths. Halo width is typically ~1 arcsec. In stark contrast, core shape is set by the telescope point-spread function. When cores dominate the images, extremely high resolution can be obtained. For large diffraction-limited instruments with circular apertures, resolution is given by $1.22 \cdot \lambda / D$; at near-IR wavelengths resolution is typically ~0.1 arcsec. Though image cores generally appear over a wide range of wavelengths, one particular wavelength—referred to as the optimum wavelength—gives rise to maximum irradiance at core center. This wavelength has obvious significance to HEL weapon systems.

Large ground-based telescopes can generally deliver higher resolution at near-IR wavelengths than at visible wavelengths. This matter has been discussed by several authors (Fried, 1966; Wolf, 1980). To obtain the highest levels of resolution from large ground-based telescopes irrespective of whether adaptive optics (AO) is used (cf., Sect. 10.10), it is crucially important to carry out the imaging at a wavelength, or more precisely over a wavelength band, consistent with significant fractions of the star image light energy residing in central cores. In this chapter, we examine in detail how the strength and shape of image cores depend on the various controlling parameters. These include parameters that describe the integrated strength and average structure size of the turbulence in the atmospheric path, the size and shape of the telescope aperture as well as the telescope aberrations, and the imaging wavelength, $\lambda$.

At longer wavelengths, where the quantity $\sigma/\lambda$ naturally takes smaller values, an increasingly large fraction of the light energy in a star image (cf., 10.3) concentrates into a central core; this of course causes an increase in the irradiance density ($W/m^2$) in the center of the image. However, because the angular size of the core grows in proportion to the wavelength, ultimately the use of ever-longer wavelengths leads to



T. S. McKechnie, *General Theory of Light Propagation and Imaging Through the Atmosphere*, Progress in Optical Science and Photonics 20,
https://doi.org/10.1007/978-3-030-98828-9_10

a reduction in the irradiance density in the image center. Consequently, as outlined in Sect. 10.4, at a certain intermediate wavelength, these two opposing effects combine in a way that maximizes core irradiance density. We refer to this wavelength as the optimum wavelength (McKechnie, 1991).

From energy conservation considerations, clearly the wavelength that maximizes the irradiance density in the center of a telescope image must correspond (at least roughly) to the wavelength that produces the image with the smallest angular footprint area. Such a wavelength must then roughly correspond to the wavelength at which the telescope delivers highest resolution. The optimum wavelength also has significance to high energy laser (HEL) weapon systems, but discussion of this subject is deferred to Chap. 16.

Telescope resolution can be further enhanced by adopting the core in a reference star image as the reference feature for both stabilizing the image and carrying out AO correction. As discussed In Sect. 10.7, the central core in star images is significantly less prone to angular jitter[1] than the light energy centroid. The reason is plainly evident (McKechnie, 1992). When a centroid tracking algorithm is used, the noise associated with the (continuously evolving) speckles in the halo causes significant random jitter in the location of the light energy centroid. On the other hand, when a core tracking algorithm is used, the jitter contribution caused by the halo speckle noise can be decimated by using an intensity thresholding algorithm. When such an algorithm is properly applied, the position estimates obtained from the image core—the only image feature that can consistently maintain intensity levels above the chosen threshold level—are significantly less affected by jitter than those of the light energy centroid. Thus, if the highest levels of precision are to be achieved for guiding, tracking, and image stabilization, generally central cores should be used as the reference feature, rather than the time-honored light energy centroid. It is therefore most important to track the reference star in a wavelength band at which an adequate fraction of the light energy in the star image resides in the core.

It is also shown in Sect. 10.7 that the instantaneous location of the core in a star image lies at a particularly significant location in the telescope image plane. It lies at the focus of the (imaginary) plane wave that least mean squares best fits the instantaneous wave in the telescope pupil that forms the star image at that instant. For telescopes with circular apertures, this location relates directly to the tip–tilt component of the Zernike polynomial expansion for image-forming waves in the telescope pupil.

As discussed later, in Chap. 17, when AO image corrections are carried out using Shack–Hartmann lenslet array devices to analyze the incoming wavefronts, once again it is important to choose a sensing wavelength band consistent with the formation of image cores in the spot images formed by the individual lenslets. The appropriate wavelength band depends both on the diameter of the individual Shack–Hartmann lenslets and on the initial (uncorrected) state of the image-forming waves in the telescope pupil.

---

[1] Angular jitter of the core refers to the random angular wanderings of these features in the image plane of the telescope.

## 10.1 Core and Halo Image Structure

The atmospheric modulation transfer function (MTF) (6.52) may be separated into a constant part and a variable part. In the equation below, these parts are depicted by the terms separately enclosed in curly brackets:

$$M(\xi,\eta,\lambda)=\left\{\exp\left[-\left(\frac{2\cdot\pi\cdot\sigma}{\lambda}\right)^2\right]\right\}$$
$$+\left\{\exp\left[-\left(\frac{2\cdot\pi\cdot\sigma}{\lambda}\right)^2\right]\cdot\left[\exp\left[\left(\frac{2\cdot\pi\cdot\sigma}{\lambda}\right)^2\cdot\rho(\xi,\eta)\right]-1\right]\right\}. \tag{10.1}$$

Core and halo image structure is illustrated in Fig. 10.1 (left). The average core envelope is shown in isolation (upper right); the average halo envelope is shown in isolation (lower right).

Assuming that the constant part, $\exp\left[-(2\cdot\pi\cdot\sigma/\lambda)^2\right]$, is non-negligible,[2] a central core should appear in the image. The average intensity envelope associated

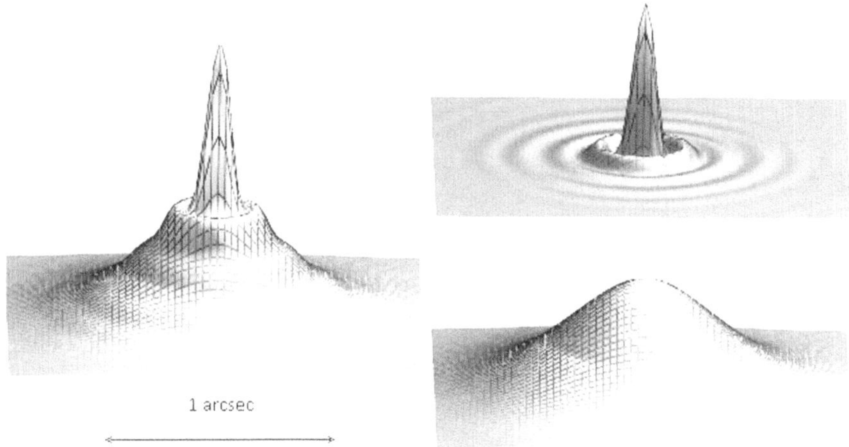


**Fig. 10.1** Left Star image exhibiting core and halo structure as obtained at the near-IR wavelength 2.2 μm by an aberrated 4-m diameter telescope in about 1.5-arcsec visible seeing conditions. Top right The core seen in isolation. Lower right The halo seen in isolation

[2] For many years, image cores were largely considered theoretical abstractions rather than practical vehicles for providing improved telescope resolution. Prior to about 1990, the lax optical specifications of large ground-based telescopes eliminated any possibility of obtaining diffraction-limited cores at visible wavelengths. Roger Griffin described the aberrated cores he saw directly by eye using the 200-in. Mt. Palomar and 100-in. Mt. Wilson telescopes (Appendix G) as significantly broader than the diffraction limit, a description consistent with the less than diffraction-limited optics used in these two revered instruments.

with this core, $\langle I_C(u, v, \lambda)\rangle$, may be obtained from 9.2 in the form,

$$\langle I_C(u, v, \lambda)\rangle = \frac{1}{\lambda^2} \cdot \exp\left[-\left(\frac{2 \cdot \pi \cdot \sigma}{\lambda}\right)^2\right] \cdot \int_{-\infty}^{\infty} \int_{-\infty}^{\infty} M_T(\xi, \eta, \lambda) \times \exp\left[-\frac{2 \cdot \pi \cdot i}{\lambda \cdot f} \cdot (u \cdot \xi + v \cdot \eta)\right] \cdot d\xi \cdot d\eta. \qquad (10.2)$$

## 10.1.1 Core Strength

Denoting the light energy fraction contained in the core by $E_C$, we can write

$$E_C = \exp\left[-\left(\frac{2 \cdot \pi \cdot \sigma}{\lambda}\right)^2\right], \qquad (10.3)$$

Figure 10.2 shows the core energy fraction, EC, plotted against the ratio, $\sigma/\lambda$. As $\sigma/\lambda \to 0$, evidently all of the light energy ultimately transfers into the core. However, when $\sigma/\lambda \geq 0.4$, for all practical purposes, the core energy fraction falls to zero; the image is then comprised solely of a halo.

## 10.1.2 Core Shape

The double integral term in 10.2 may be recognized as the Fourier transform of the telescope OTF which, as we saw previously (Sect. 7.10), is simply the intensity point-spread function (PSFI) of the telescope, i.e., the image delivered by the telescope in

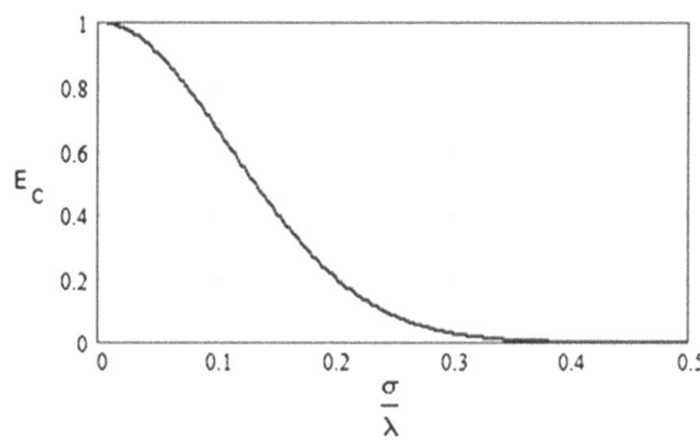


**Fig. 10.2** Core energy fraction, $E_C$, plotted against (wavelength normalized) rms optical path difference (OPD) fluctuation, $\sigma/\lambda$

the absence of atmosphere. The average intensity envelope associated with an image core is therefore given by

$$\langle I_C(u, v, \lambda) \rangle = \frac{1}{\lambda^2} \cdot \exp\left[ -\left( \frac{2 \cdot \pi \cdot \sigma}{\lambda} \right)^2 \right] \cdot PSF_I(u, v, \lambda). \quad (10.4)$$

The telescope point-spread function, $PSF_I(u, v, \lambda)$, can readily be obtained from the telescope pupil function (cf., 4.25). Thus, the average shape of the core is determined by the telescope optics rather than by the atmosphere. Better telescope optics result in more compact image cores which in turn provide sharper telescope resolution.

For a 4-m diffraction-limited telescope observing in average (1-arcsec) visible seeing conditions, image cores would likely form routinely at the near-IR wavelength 1.6 μm. The angular size of such cores (as given by $1.22 \cdot \lambda / D$) is about 0.1-arcsec. In ½-arcsec visible seeing conditions, image cores would likely form at wavelengths as short as 0.8 μm. Such cores would deliver resolution of about 0.05-arcsec (McKechnie, 1991, 1994), thus providing resolution comparable to that obtained by the Hubble Space Telescope at visible wavelengths.

### 10.1.3 Halo Strength

The halo strength, or halo light fraction, is determined by the light fraction not contained in the core. By referring back to 10.3, it may be seen that the halo light fraction, which we denote by $E_H$, is given by

$$E_H = 1 - E_C = 1 - \exp\left[ -\left( \frac{2 \cdot \pi \cdot \sigma}{\lambda} \right)^2 \right]. \quad (10.5)$$

### 10.1.4 Halo Shape

The light distribution in the halo portion of a (core and halo) star image formed by a large telescope is generally more widely distributed than the core light energy portion. Generally, the shape of the average halo envelope closely approximates that of a Gaussian function. For large telescopes with near diffraction-limited optics, the full-width half-maximum (FWHM) angular width of the halo is largely set by the atmospheric seeing at visible wavelengths which, in average astronomical seeing conditions, is about 1-arcsec. However, severe telescope aberrations can result in halo width growing significantly larger than the size set by seeing alone.

In an instantaneous image, the halo arises as a speckle pattern. The average halo intensity envelope, $\langle I_H(u, v, \lambda)\rangle$, may be obtained by subtracting the average intensity envelope of the core from the right-hand side of 9.2. Thus, we can write

$$\langle I_H(u,v,\lambda)\rangle = \frac{1}{\lambda^2}\cdot\int_{-\infty}^{\infty}\int_{-\infty}^{\infty}\exp\left[-\left(\frac{2\cdot\pi\cdot\sigma}{\lambda}\right)^2\right]\cdot\left\{\exp\left[\left(\frac{2\cdot\pi\cdot\sigma}{\lambda}\right)^2\cdot\rho(\xi,\eta)\right]-1\right\}\cdot M_T(\xi,\eta,\lambda)$$
$$\times\exp\left[-\frac{2\cdot\pi\cdot i}{\lambda\cdot f}\cdot(u\cdot\xi+v\cdot\eta)\right]\cdot d\xi\cdot d\eta. \quad (10.6)$$

Figure 10.1 (Lower right) previously showed the average halo intensity envelope calculated from this expression

## 10.1.5 Characteristics of Core and Halo Images

The average intensity envelope of a core and halo image can be expressed as the incoherent sum of the average envelopes of the core and the halo,

$$\langle I(u,v,\lambda)\rangle = \langle I_C(u,v,\lambda)\rangle + \langle I_H(u,v,\lambda)\rangle. \quad (10.7)$$

Due to the averaging involved in obtaining both the long-exposure image and the average short-exposure image, the core and the halo portions of both these image types have smooth, speckle-free envelopes. Figure 10.1 previously showed a typical core and halo average image envelope along with average envelopes shown separately for both the core and the halo.

In contrast to the smooth, speckle-free average short-exposure image, individual short-exposure images are immersed in speckle. The halo in an individual short-exposure image appears as a speckle pattern, with the core lying in the center. Since the core and halo light portions add coherently to form this image, the separate complex amplitude contributions of the core and the halo would first have to be summed together to obtain the combined complex amplitude distribution before calculating the final image intensity by obtaining the squared modulus of the result.

### 10.1.5.1 The Ubiquitous Core and Halo Image

As indicated in Chap. 9, the most general type of star image consists of both a core and a halo; the two other possible types of image, core-only and halo-only images, may be regarded simply as degenerate cases. Over the large optical wavelength range considered in this book—from the UV wavelength, 0.3 μm, to the far-IR wavelength, 1000 μm—irrespective of the magnitude of the rms optical path difference (OPD) fluctuation, $\sigma$, of the image-forming waves in the pupil of the observing telescope, over this huge range of wavelengths star images tend to exhibit the same family of appearances. For very short wavelengths, the image consists almost entirely of a halo. For very long wavelengths, the image consists almost entirely of a core. At intermediate wavelengths, the image consists of a hybrid mixture of both features.

If we neglect wavelength dispersion, $\sigma$ may be considered invariant with wavelength and the above image changes are as depicted in the generic plots shown in Fig. 10.3. (Eqs. 10.3 and 10.5 were used to calculate the respective core and halo energy fractions in this figure.)

In the somewhat arbitrary classification scheme used here, we consider that the image is essentially halo only when the halo contains more than 90% of the total light energy in the star image. Similarly, we consider the image to be core only when the core contains more than 90% of the light energy. For all other cases, where the core and halo have more equal divisions of the light energy, we consider the image as a core and halo image. The approximate wavelength transition values that separate the three different image types may be shown (cf., 10.3 and 10.5) to be the following:

$$\lambda > 20 \cdot \sigma \quad \text{core-only},$$

$$4 \cdot \sigma \leq \lambda \leq 20 \cdot \sigma \quad \text{core and halo},$$

$$\lambda < 4 \cdot \sigma \quad \text{halo-only}.$$

For telescopes equipped with AO, the same type of image classification scheme still applies. The only difference is that we would now replace the quantity $\sigma$ in the above scheme by the residual uncorrected rms OPD fluctuation, $\sigma_{AO}$. For an

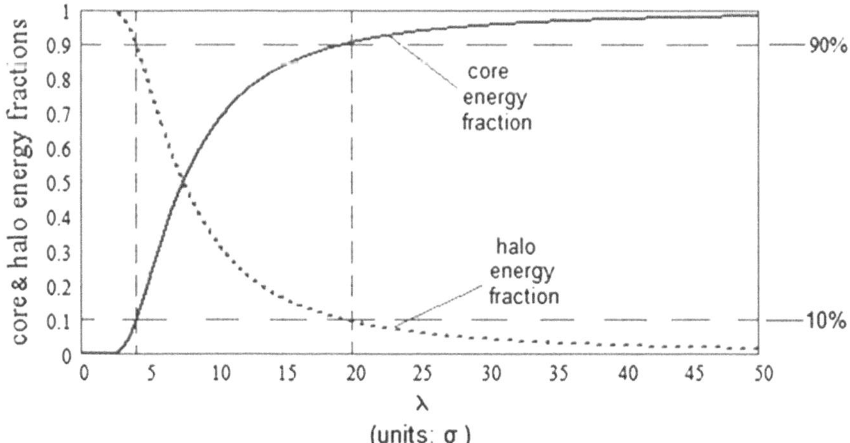


**Fig. 10.3** Generic appearance changes of star images as the imaging wavelength ranges over the extremes from very short to very long. As wavelength increases, the images morph from a halo-only stage through a core and halo stage until finally settling into a core-only stage

efficiently functioning AO system, we would naturally expect to find $\sigma_{AO} < \sigma$, and thus the wavelength transition points just indicated at $\lambda = 4 \cdot \sigma$ and $\lambda = 20 \cdot \sigma$ would occur at correspondingly shorter wavelengths given by $\lambda = 4 \cdot \sigma_{AO}$ and $\lambda = 20 \cdot \sigma_{AO}$. It should be emphasized that since $\sigma$ is an ensemble average quantity, the behavior depicted in Fig. 10.3 is also ensemble average behavior. Thus, image cores do not simply occur as a result of either "speckle imaging" or "lucky imaging." These two imaging techniques are discussed later in Sect. 10.11 and can provide additional resolution benefits. Just as long as the condition, $\lambda > 4 \cdot \sigma$, is fulfilled, central cores will appear in star images, most of the time on average.

## 10.2 Circularly Symmetric Core and Halo Image Envelopes Formed by Large Telescopes

For isotropic turbulence, large telescopes with circularly symmetric intensity PSFs[3] naturally produce circularly symmetric star images and, in general, these images display core and halo structure. In this section, a useful expression is given (10.8) that allows ready calculation of the average image intensity in central sections through this type of image.

The usefulness is twofold. First, the average intensity envelopes due to the core and halo arise as two separate terms in the expression, thus allowing these terms to be examined separately. Second, the equation is normalized so that, after rotation, the intensity envelopes always enclose the same volume (i.e., the same total light energy) regardless of the parameter value choices inserted into the equation. Consequently, intensity envelopes generated from 10.8 can be used to make side-by-side resolution comparisons between even the most widely asymmetrical telescope imaging arrangements. For example, there may be two entirely different telescopes, with different aperture sizes and shapes, different central obstruction sizes, and different aberrations. The two instruments could also be observing at different wavelengths and at different observing sites in entirely different seeing conditions.

The circular Gaussian function is again chosen here (cf., Sects. 9.4 and 9.9) to describe the autocorrelation function of the OPD fluctuation, $\rho(\varepsilon)$. (This function may be considered as the Gaussian function that best fits the actual measured function.) The simplifying assumption is also made that the variable part of the atmospheric MTF (10.1) falls off at a faster rate than the telescope optical transfer function (OTF). With these assumptions, it can be shown that the intensity envelope given previously by 9.2 can be expressed in the form (McKechnie, 1976, 1991),

$$\langle I(\vartheta, \lambda) \rangle = \frac{D^2 - d^2}{\lambda^2} \cdot \exp\left[ -\left( \frac{2 \cdot \pi \cdot \sigma}{\lambda} \right)^2 \right]$$

[3] By considering only circularly symmetric telescope PSFs, we effectively restrict ourselves here to telescopes with circular apertures, circular central obstructions, and circularly symmetric aberrations (which include defocus and spherical aberration).

$$\times \left\{ PSF_I(\vartheta, \lambda) + \frac{4 \cdot w_o^2}{D^2 - d^2} \cdot \sum_{n=1}^{\infty} \frac{\left(\frac{2 \cdot \pi \cdot \sigma}{\lambda}\right)^{2 \cdot n}}{n \cdot n!} \cdot \exp\left[ -\frac{(\pi \cdot w_o \cdot \vartheta)^2}{n \cdot \lambda^2} \right] \right\}. \quad (10.8)$$

where D is the telescope diameter, d is the central obstruction diameter, and λ is the imaging wavelength, with the atmospheric seeing described by the parameters, $\sigma$, and $w_0$ The appropriate form of the intensity PSF of the telescope, $PSF_I(\vartheta, \lambda)$, will be given in the next section. The average intensity envelopes for the core and halo features in the image are represented separately in the above equation by the two terms summed in the curly brackets. In practice, the upper limit for the summation, $\infty$, may be replaced with little loss of accuracy by a finite integer of the order 100. The numerical accuracy of any particular evaluation can be established by studying convergence while varying the summation limit.

### 10.2.1 Circularly Symmetric Telescope Point-Spread Functions

Equation 10.8 indicates that core shape is entirely determined by the PSF of the telescope, $PSF_I(\vartheta, \lambda)$. For telescopes with circularly symmetric apertures and circularly symmetric aberrations, the correct unit-normalized $PSF_I(\vartheta, \lambda)$ for use in 10.8 is given by

$$PSF_I(\vartheta, \lambda) = \left[ \frac{64}{(D^2 - d^2)^2} \right] \cdot \left| \int_{\frac{d}{2}}^{\frac{D}{2}} \exp\left[ -\frac{2 \cdot \pi \cdot i \cdot W(r)}{\lambda} \right] \cdot J_0\left[ \frac{2 \cdot \pi \cdot r \cdot \vartheta}{\lambda} \right] \cdot r \cdot dr \right|^2, \quad (10.9)$$

where W(r) denotes the circularly symmetric wavefront error introduced by the telescope and $J_0(\cdot)$ is the zero-order Bessel function of the first kind.

Wavefront error due to defocus increases as the square of the pupil radius; for first-order spherical aberration, wavefront error increases as the fourth power of the radius. Each of these wavefront aberration contributions can be specified in terms of their respective magnitudes at the edge of the pupil, which we denote here by $W_{defocus}$ and $W_{spherical}$, respectively. The combined wavefront error due to both contributions, W(r), may be expressed by the form,

$$W(r) = \frac{4 \cdot r^2 \cdot W_{defocus}}{D^2} + \frac{16 \cdot r^4 \cdot W_{spherical}}{D^4}. \quad (10.10)$$

If desired, 6th, 8th, and higher order spherical aberration terms could be added to the right-hand side of the above equation. The nth-order spherical aberration term can readily be seen to have the form, $2^n \cdot r^n \cdot w_n / D^n$.

For an aberration-free telescope, where $W_{defocus} = W_{spherical} = 0$, plainly

$$W(r)_{aberrnationfree} = 0 \quad \text{for} \quad \frac{d}{2} \leq r \leq \frac{D}{2}. \tag{10.11}$$

For telescopes built prior to about 1990, several waves of aberration were considered acceptable (Martin et al., 1991). Figure 10.4 shows how $2 \cdot \lambda$ (HeNe) peak to valley (P–V) wavefront error might look for a 4-m telescope without a central obstruction. To generate the plot shown, $W_{spherical}$ was set to $8 \cdot \lambda$ (HeNe) in 10.10, with best aberrational balance obtained by setting $W_{defocus}$ to the value, $-8 \cdot \lambda$ (HeNe).

#### 10.2.1.1 Intensity PSF for an Aberration-Free, Centrally Obstructed Telescope

For an aberration-free telescope with a circular central obstruction, the unit-normalized point-spread function given by 10.9 reduces to

$$PSF_I(\vartheta, \lambda) = 4 \cdot \left\{ \frac{1}{1 - \left(\frac{d}{D}\right)^2} \cdot \left[ \frac{J_1\left(\frac{\pi \cdot D \cdot \vartheta}{\lambda}\right)}{\frac{\pi \cdot D \cdot \vartheta}{\lambda}} - \left(\frac{d}{D}\right)^2 \cdot \frac{J_1\left(\frac{\pi \cdot d \cdot \vartheta}{\lambda}\right)}{\frac{\pi \cdot d \cdot \vartheta}{\lambda}} \right] \right\}^2, \tag{10.12}$$

where $J_1(\cdot)$ is the first-order Bessel function of the first kind.

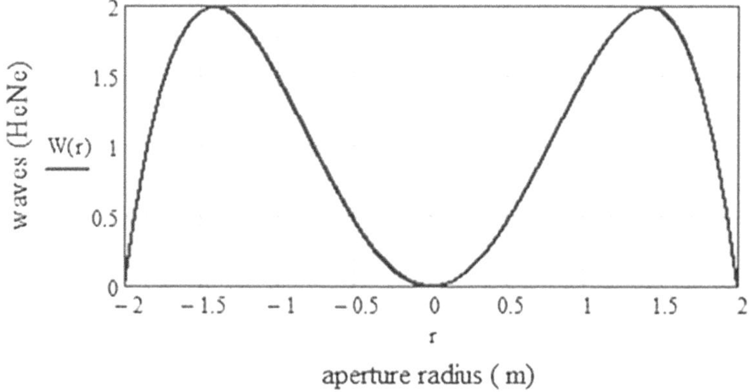


**Fig. 10.4** Wavefront error over a 4-m telescope aperture caused by 8-wave (HeNe) P–V first-order spherical aberration best balanced by −8 waves of defocus. The resulting 2-wave (HeNe) P–V wavefront aberration indicated might be considered typical of 4-m class telescopes constructed prior to about 1990

#### 10.2.1.2 Intensity PSF for an Aberration-Free, Unobstructed Telescope

For diffraction-limited telescopes without central obstructions (i.e., d = 0), 10.12 reduces to give the familiar unit-normalized Airy pattern,

$$PSF_I(\vartheta, \lambda) = 4 \cdot \left[ \frac{J_1\left(\frac{\pi \cdot D \cdot \vartheta}{\lambda}\right)}{\frac{\pi \cdot D \cdot \vartheta}{\lambda}} \right]^2. \tag{10.13}$$

Assuming that the image cores formed by the telescope contain a reasonable fraction of the total light energy, the angular resolution that can be obtained from these cores is then approximately given by the Rayleigh resolution limit, $1.22 \cdot \lambda/D$ (cf., Sect. 4.6).

### *10.2.2 Normalization of 10.8*

As mentioned earlier, the normalization of 10.8 is such that, after rotation, the total volume (i.e., light energy) enclosed under the envelope, $\langle I(\vartheta, \lambda)\rangle$, is independent of the values chosen for the parameters, D, d, $\lambda$, $\sigma$, $w_0$, and W(r). Thus, 10.8 can be used to make direct telescope resolution comparisons for widely different parameter value combinations. If required, appropriate units can be associated with the intensity envelopes by multiplying the dimensionless right-hand side of 10.8 by a suitably dimensioned constant (as discussed in "note on dimensions" in Sect. 7.7).

### *10.2.3 Numerical Accuracy of 10.8*

As may be confirmed by numerical evaluations, the intensity envelopes generated by 10.8 are approximately the same as those generated by the more exact expression given previously by 9.2. Such small numerical differences as might arise from these two equations stem from the additional assumption made in obtaining 10.8 that the variable part of the atmospheric MTF falls off at a much faster rate than the telescope OTF.

## 10.3 Theoretical and Observed Core and Halo Structure

This section examines theoretically the dependence of core and halo structure on factors such as telescope diameter, central obstruction diameter, telescope aberrations, the imaging wavelength, and also the size and strength of the average atmospheric turbulence structure. Intensity envelopes calculated from 10.8 are used to illustrate the various dependences. Actual star images depicting core and halo structure are shown later in this section. These images, obtained in 1990 using the 3.8-m Mayall telescope at 2.2 $\mu$m, bear a reasonable resemblance to theoretically predicted images; in particular, they show the huge resolution improvements that result from the presence of cores.

### *10.3.1 Core Dependence on Wavelength*

By observing at a long-enough wavelength, irrespective of the size of the rms OPD fluctuation, $\sigma$, cores containing large fractions of the light energy (10.3) must ultimately develop in the images. Figure 10.5 shows typical intensity envelopes for various near-IR wavelengths obtained by a 4-m diffraction-limited telescope in 1-arcsec visible seeing conditions.

Also shown in the figure for reference is the average halo image envelope for the visible wavelength, 0.55 $\mu$m. Image wander is assumed corrected in all cases. Thus, all of the images shown in this figure may be considered to be average short-exposure images.

The visible image envelope shows no sign of an image core; only the halo is evident in this image, measuring about 1-arcsec across, consistent with the assumed visible seeing conditions. As can be seen in the figure, a core is beginning to emerge at 1.0 $\mu$m, with cores ultimately dominating the images at longer wavelengths. The cores shown in Fig. 10.5 correspond to the fulfillment of condition (3) in Sect. 9.4.2.1.

### *10.3.2 Core Dependence on Seeing*

Figure 10.6 shows how the core energy fraction, $E_C$, depends on the seeing, the determining quantity being the rms OPD fluctuation, $\sigma$ (cf., 10.3). At visible wavelengths, significant core energy arises when $\sigma \leq 0.1$ $\mu$m. At the near-IR wavelength, 2.2 $\mu$m, significant energy requires $\sigma \leq 0.4$ $\mu$m. At long-enough wavelengths, cores ultimately dominate the images no matter how large the $\sigma$ value.

For a 4-m telescope observing in 1-arcsec visible seeing conditions where image wander is assumed corrected, $\sigma$ is typically about 0.35 $\mu$m. Such a value permits the formation of strong cores at near-IR and longer wavelengths. For diffraction-limited telescopes, the Strehl intensity in the image (cf., 4.30) is approximately given by the core energy fraction, $E_C$ (cf., 10.3). For a 4-m diffraction-limited telescope imaging in 1-arcsec visible seeing conditions at wavelengths > 5 $\mu$m, Strehl intensity can take values > 0.8 and thus exceed the threshold for substantially diffraction-limited imaging. At even longer, far-IR wavelengths, Strehl intensity can approach unity, indicating near-perfect image quality.

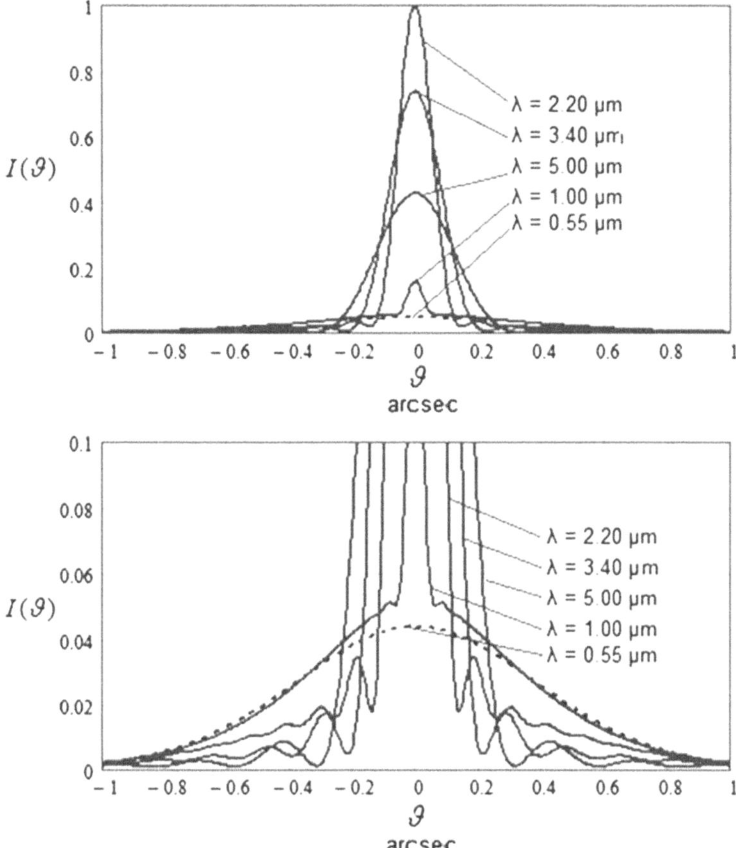


**Fig. 10.5** Top Intensity envelopes showing core and halo structure at the near-IR wavelengths, 1.0, 2.2, 3.4, and 5.0 μm formed by a diffraction-limited 4-m telescope in 1-arcsec visible seeing conditions. Bottom The same envelopes with a 10 × vertical scale change. Dotted line Halo image envelope for visible wavelength (0.55 μm) shown for reference (parameter values used: D = 4 m, d = 0 m, $\sigma = 0.35$ μm, $w_0 = 0.25$ m)

## 10.3.3 Core Dependence on Telescope Size

Figure 10.7 shows average star image intensity envelopes calculated from 10.8 and 10.9 for 2-m and 3-m diameter diffraction-limited telescopes in ~1-arcsec visible seeing conditions. The volumes contained under all of the envelopes in this figure are identical, and in fact, they are the same as those contained under the envelopes shown in Fig. 10.5 for a 4-m telescope.

For diffraction-limited telescopes, the angular size of the core reduces inversely as the telescope diameter, causing core footprint area to reduce as $1/D^2$. Because the total amount of light energy in the core also increases as $D^2$, the light irradiance

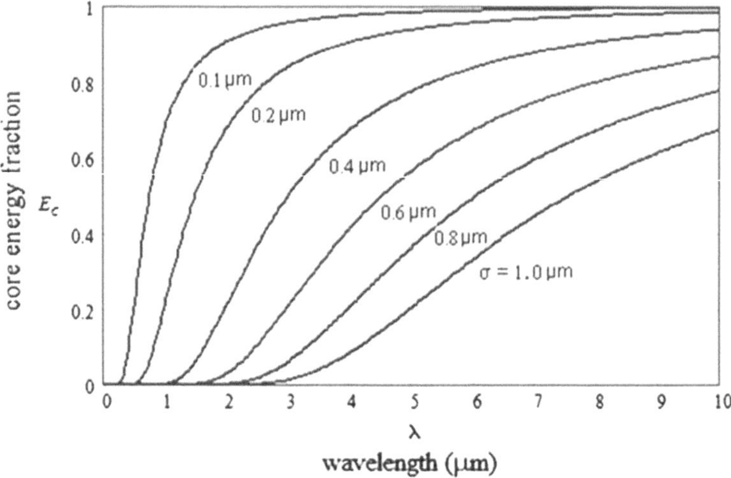


**Fig. 10.6** Wavelength dependence of core energy fraction, $E_C$, for various turbulence strengths prescribed by the σ values indicated. No matter how poor the seeing quality (i.e., no matter the size of σ), at long-enough wavelengths, ultimately $E_C \rightarrow 1$

density at core center grows as $D^4$. Thus, larger telescopes tend to produce more intense and better resolved cores.

The above (arguable) conclusion follows as a consequence of our choice of a relatively small average turbulence structure size, $w_0 = 0.25$ m (the same value used to produce the envelopes shown in Fig. 10.7). Had a significantly larger $w_0$ value been chosen, a somewhat different conclusion would have been reached. For larger $w_0$ values, as telescope aperture diameter increases, σ also tends to increase, resulting in core irradiance increasing as a rate slower than $D^4$. In fact, it is even possible that an overall decrease in core irradiance could occur. However, Roger Griffin seems to provide some support to the former conclusion by his observation (Griffin, 1973), "Moreover, the larger the telescope the less good is the seeing which seems to be necessary to allow the sharp bright core to form in the image."

### *10.3.4 Core Dependence on Telescope Aberrations*

The author had the privilege of using the 4-m Mayall telescope in 1991. At that time, P–V wavefront error for the instrument was estimated (Forbes, 1991) to be about 2-wave P–V (HeNe); aberrations of this magnitude were considered acceptable for telescopes constructed in the 1970s and 1980s. However, around 1990, after it became clear that the Mayall instrument could routinely deliver strong image cores at near-IR wavelengths even in quite average (~1-arcsec) seeing conditions, it also became clear that aberrations of this magnitude were significantly limiting the

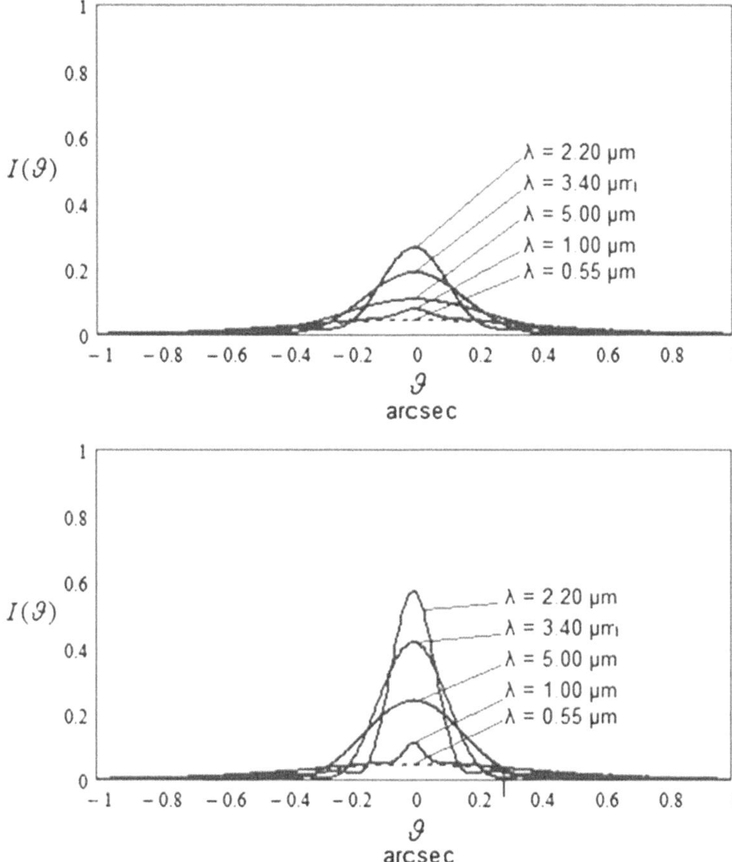


**Fig. 10.7** Intensity envelopes showing core and halo structure at the near-IR wavelengths, 1.0, 2.2, 3.4, and 5.0 μm, formed in 1-arcsec visible seeing. Top diffraction-limited 2-m telescope, Bottom diffraction-limited 3-m telescope. Parameter values used: D = 2 and 3 m, d = 0 m, $\sigma = 0.35$ μm, $w_0$ = 0.25 m. Dotted line Halo image envelope for visible wavelength (0.55 μm) shown for reference

resolution performance of the instrument. To illustrate the problem, we now consider two telescopes: (1) the actual Mayall telescope with 2-wave (HeNe) P–V wavefront aberration and (2) a hypothetical diffraction-limited version of the Mayall telescope.

It is assumed that both instruments are observing in the same average (~1-arcsec) visible seeing conditions, as prescribed by $\sigma = 0.35\mu m$ and $w_0 = 0.25$ m. The effective collection diameter of the Mayall telescope is 3.8 m; the central obstruction blocks a 1.65-m diameter central portion. For both telescopes, therefore, D = 3.8 m and d = 1.65 m. Wavefront error W(r) for the aberrated telescope version is set at about 2-wave (HeNe) P–V aberration, as shown in Fig. 10.8. (The actual wavefront error, W(r), depicted in this figure was calculated from 10.10 with $W_{spherical}$ set to

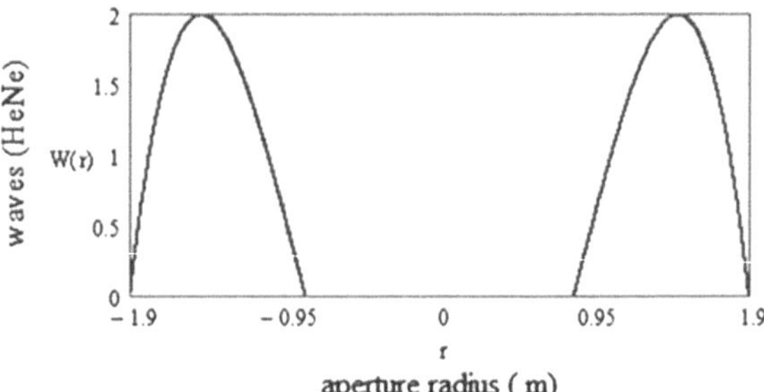


**Fig. 10.8** The wavefront aberration, 2 waves P–V (HeNe), used to calculate the intensity envelopes shown in Fig. 10.9 (bottom) for a hypothetically aberrated version of the Mayall telescope

12.1 waves (HeNe) and $W_{defocus}$ set to 14.4 waves.)[4] The average star image intensity envelopes obtained from the two telescopes (calculated using 10.9 with the above parameter values) are shown in Fig. 10.9.

At the visible wavelength, 0.55 μm, where the images formed by the two telescopes are halo only, the intensity envelopes are seen to be closely similar. Thus, telescope aberrations of the magnitude and type considered here do not appear to cause significant resolution degradation of visible wavelengths. However, at the near-IR wavelengths, 1.0, 2.2, and 3.4 μm, the same wavefront aberration significantly degrades the image cores formed at these wavelengths.

### *10.3.5 Image Cores Obtained in the Near-IR by the 4-m Mayall Telescope*

Figures 10.10 and 10.11 show previously published images (McKechnie, 1992; Pederson, 1990) obtained at 2.2 μm by the 3.8-m Mayall telescope. Figure 10.10 shows the single star, Zeta Ophiuchi, and Fig. 10.11 shows the 0.48-arcsec binary star, Eta Ophiuchi. The average, tip–tilt corrected short-exposure images are shown on the right. The conspicuous cores in these images contain about 40% of the total light energy. The corresponding long-exposure images (without tip–tilt correction) are shown on the left; cores are entirely absent from these images.

Core shape is determined by the telescope intensity PSF (10.2). The aberrations of the Mayall instrument, combined with the effect of the central obstruction, deplete light energy from the central disk, redistributing it into the surrounding rings, two of which are evident in Fig. 10.10 (Right). The redistribution caused by aberrations alone can be gauged by comparing the theoretical intensity envelopes shown in Fig. 10.9 for the two, aberrated and non-aberrated, telescope versions.

[4] A detailed study of the aberrations of the Mayall telescope, circa 1991, was conducted by Baldwin et al. (1992) prior to carrying out an optics upgrade program in 1992. Because the intensity envelopes shown in Fig. 10.8 for the aberrated telescope are based on a simplified set of aberration parameters, these envelopes only nominally describe the aberrated shapes of the image cores produced by the Mayall instrument at that time (shown at bottom in Fig. 10.9).

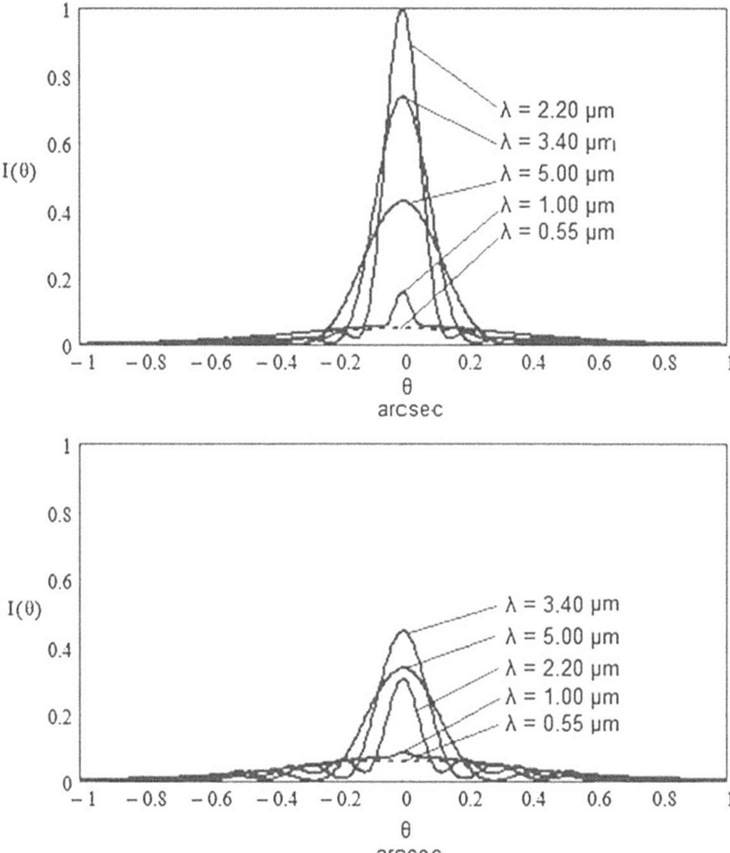


**Fig. 10.9** Intensity envelopes delivered by two hypothetical versions of the 3.8-m Mayall telescope with its 1.65 m central obstruction. Top Envelopes obtained from the diffraction-limited telescope version. Bottom Envelopes obtained from the version with 2-wave (HeNe) P–V aberration. (Parameter values used: D = 3.8 m, d = 1.65 m, $\sigma = 0.35$ m, $w_0 = 0.25$ m.) The halo image envelope dotted line for the visible wavelength, 0.55 μm, formed by the aberrated telescope is seen to be almost identical to the halo image for the aberration-free telescope. In contrast, the cores formed at IR wavelengths are significantly affected by aberrations

The ring features for the theoretical 2.2 μm average intensity envelope shown (Bottom) look quite similar to the ring systems in the actual 2.2 μm image obtained by the 4-m Mayall telescope shown in Fig. 10.10 (Right). The overall similarity between the actual image and the theoretical image sections shown in these figures may be taken as a measure of justification both for the Gaussian approximation for function $\rho(\xi, \eta)$ which is implicit in 10.8 and for the parameter value choices used for modeling the seeing, $\sigma = 0.35$ μm and $w_0 = 0.25$ m.

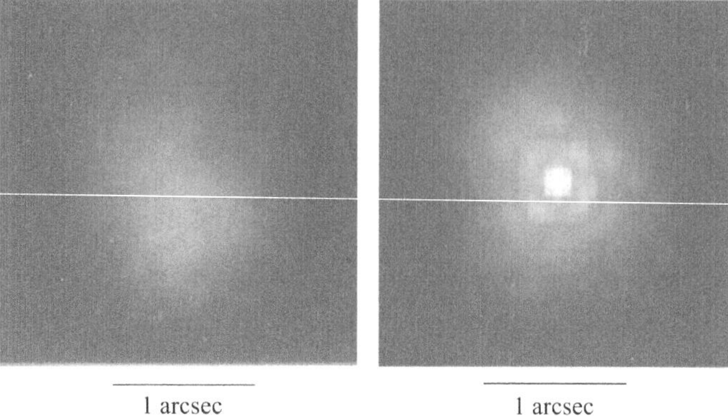


**Fig. 10.10** Left A long-exposure image of Zeta Ophiuchi at 2.2 μm obtained by the 3.8-m Mayall telescope in 1.25-arcsec visible seeing conditions. Right The average short-exposure image of the same star based on the addition of 500 short-exposure images using shift-and-add image wander correction. The angular size of the core is about 0.15 arcsec. Photographs by courtesy of J. Christou, National Optical Astronomy Observatory

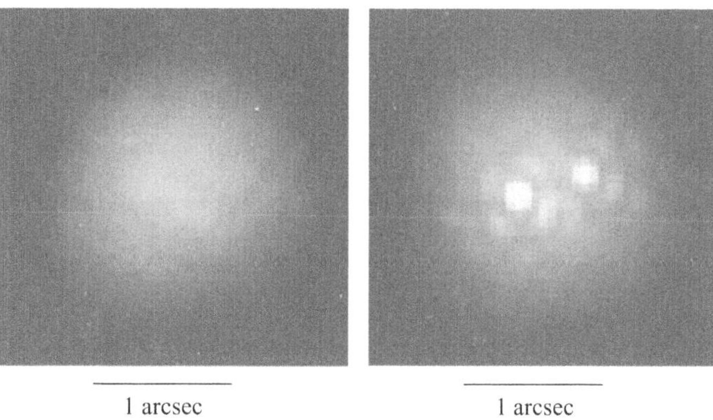


**Fig. 10.11** Left A long-exposure image of the 0.48-arcsec binary star, Eta Ophiuchi, obtained at 2.2 μm in 1.25-arcsec seeing conditions; the binary pairs are entirely unresolved. Right The average short-exposure image of the same binary star, now comfortably resolved, based on the addition of 500 short-exposure images using shift-and-add image wander correction. Photographs by courtesy of J. Christou, National Optical Astronomy Observatory

## 10.4 The Optimum Wavelength

For 4- or 5-m class telescopes observing in average (~1-arcsec) visible seeing conditions, star images at near-IR wavelengths are routinely dominated by cores. Assuming that the telescope optics are good enough (i.e., near diffraction-limited at near-IR wavelengths), as telescope diameter D increases, the image core becomes compressed into a smaller angular footprint. Assuming negligible growth in $\sigma$ as the diameter increases (i.e., assuming $w_0 \ll D$), irradiance ($W/m^2$) at core center should increase as $D^4$.[5] The irradiance in the center of the halo will also increase, but at a much slower rate, in proportion to $D^2$. Thus, as telescope diameter increases, the ratio of the core irradiance to that of the halo might be expected to grow (roughly) in proportion to $D^2$ (i.e., $D^4/D^2$).

As a consequence of $D^2$ dependence, for very large telescopes observing at near-IR and longer wavelengths (and at visible wavelengths in good seeing conditions), central cores must ultimately come to dominate the image centers. For images of this kind, the average irradiance in the center of the image is given by the following proportionality (c.f., 10.8):

$$\langle I(0,\lambda)\rangle \propto \frac{1}{\lambda^2}\cdot\exp\left[-\left(\frac{2\cdot\pi\cdot\sigma}{\lambda}\right)^2\right]\cdot PSF_I(0,\lambda). \tag{10.14}$$

If we now assume the telescope to be diffraction-limited at all wavelengths, we may set $PSF_I(0,\lambda) = 1$ (cf., 10.12 or 10.13). For such a telescope, maximum irradiance in the center of the image occurs at a wavelength given by solving the differential equation,

$$\frac{d\langle I(0,\lambda)\rangle}{d\lambda} = \exp\left[-\left(\frac{2\cdot\pi\cdot\sigma}{\lambda}\right)^2\right]\cdot\left[\frac{2\cdot\left(\lambda^2-4\cdot\pi^2\cdot\sigma^2\right)}{\lambda^5}\right] = 0. \tag{10.15}$$

The wavelength that satisfies the above equation is referred to here as the optimum wavelength (McKechnie, 1990, 1991). Denoting this wavelength by $\lambda_{opt}$, it is readily evident that

$$\lambda_{opt} = 2\cdot\pi\cdot\sigma. \tag{10.16}$$

The above expression for the optimum wavelength is only approximate because it does not take into account the small irradiance contribution of the halo in the center of the image. A more exact expression—one that does take account of the halo contribution—is given in Chap. 13 (Sect. 13.4). By combining 10.14 and 10.16, we see how the irradiance in the center of the image, $\langle I(0,\lambda_{opt})\rangle$, depends on $\lambda_{opt}$,

$$\langle I(0,\lambda_{opt})\rangle \propto \frac{1}{\lambda_{opt}^2}\cdot\frac{1}{e}, \tag{10.17}$$

where e is the Naperian logarithm base. From this proportionality, it is clear that shorter optimum wavelengths result in higher irradiance in the center of the image.

[5] The $D^4$ dependence assumes that $\sigma$ does not increase with the telescope diameter. Except for a limitingly small turbulence structure outer scale limit, at least some small $\sigma$ increase is inevitable, which would have the effect of core center irradiance growing at a rate slower than $D^4$.

For telescopes with circular apertures, circular central obstructions, and circularly symmetric wavefront aberrations, it follows from 10.9 that $PSF_I(0, \lambda)$ is given by

$$PSF_I(0,\lambda) = \frac{64}{\left(D^2 - d^2\right)^2} \cdot \left| \int_{\frac{d}{2}}^{\frac{D}{2}} \exp\left[ -\frac{2 \cdot \pi \cdot i \cdot W(r)}{\lambda} \right] \cdot r \cdot dr \right|^2. \quad (10.18)$$

Because the right-hand side of the above equation always takes the value unity for an aberration-free telescope, the normalization of this equation is such that the value taken by $PSF_I(0, \lambda)$ may be considered as the Strehl intensity of the telescope. For weakly aberrated telescopes, where the Strehl intensity—and hence the quantity $PSF_I(0, \lambda)$ —takes values in the approximate range, 0.25–1.0 (and where we also assume that $PSF_I(0, \lambda)$ varies only slowly with wavelength), the maximum irradiance in the center of the image can be given by the slightly modified version of 10.17,

$$\left\langle I\left(0, \lambda_{opt}\right)\right\rangle \approx \frac{1}{\lambda_{opt}^2} \cdot \frac{1}{e} \cdot PSF_I\left(0, \lambda_{opt}\right). \quad (10.19)$$

The irradiance levels indicated by 10.17 and 10.19 are seen to be less than those achieved at wavelength $\lambda_{opt}$ in the absence of atmosphere by the factor 1/e (~0.368). Even though the core that forms at the optimum wavelength is by far the largest contributor to the irradiance in the center of the image, we see that a larger overall fraction of the total light energy in this image (~0.632) actually resides in the halo.

## 10.5 Irradiance at Image Center as a Function of Wavelength

For telescopes with circular apertures, circular central obstructions, and circularly symmetric aberrations, the irradiance in the center of the average star image, $\langle I(0, \lambda)\rangle$, may be expressed (cf., 10.8) in the form,

$$\langle I(0,\lambda)\rangle = \frac{\left(D^2 - d^2\right)}{\lambda^2} \cdot \exp\left[ -\left( \frac{2 \cdot \pi \cdot \sigma}{\lambda} \right)^2 \right] \times \left\{ PSF_I(0,\lambda) + \frac{4 \cdot w_o^2}{\left(D^2 - d^2\right)} \cdot \sum_{n=1}^{\infty} \frac{\left(\frac{2 \cdot \pi \cdot \sigma}{\lambda}\right)^{2 \cdot n}}{n \cdot n!} \right\}, \quad (10.20)$$

where $PSF_I(0, \lambda)$ was given previously by 10.18.

For an unobstructed (i.e., d = 0) diffraction-limited telescope with a circular aperture, the irradiance $\langle I(0, \lambda)\rangle$ may be obtained from 9.5 in the form,

$$\langle I(0,\lambda)\rangle = \frac{4}{\lambda^2}\cdot\int_0^D \exp\left\{-\left(\frac{2\cdot\pi\cdot\sigma}{\lambda}\right)^2\cdot[1-\rho(\varepsilon)]\right\}$$
$$\times\left[\cos^{-1}\left(\frac{\varepsilon}{D}\right)-\frac{\varepsilon}{D}\cdot\sqrt{1-\left(\frac{\varepsilon}{D}\right)^2}\right]\cdot\varepsilon\cdot d\varepsilon, \qquad (10.21)$$

where we have used the Bessel function result, $J_0(0) = 1$.

Figure 10.12 shows the average irradiance in the center of the image, $\langle I(0, \lambda)\rangle$, plotted against wavelength for 1-, 2-, 4-, and 8-m diameter telescopes using a logarithmic vertical scale. The upper plots correspond to average (~1-arcsec) visible

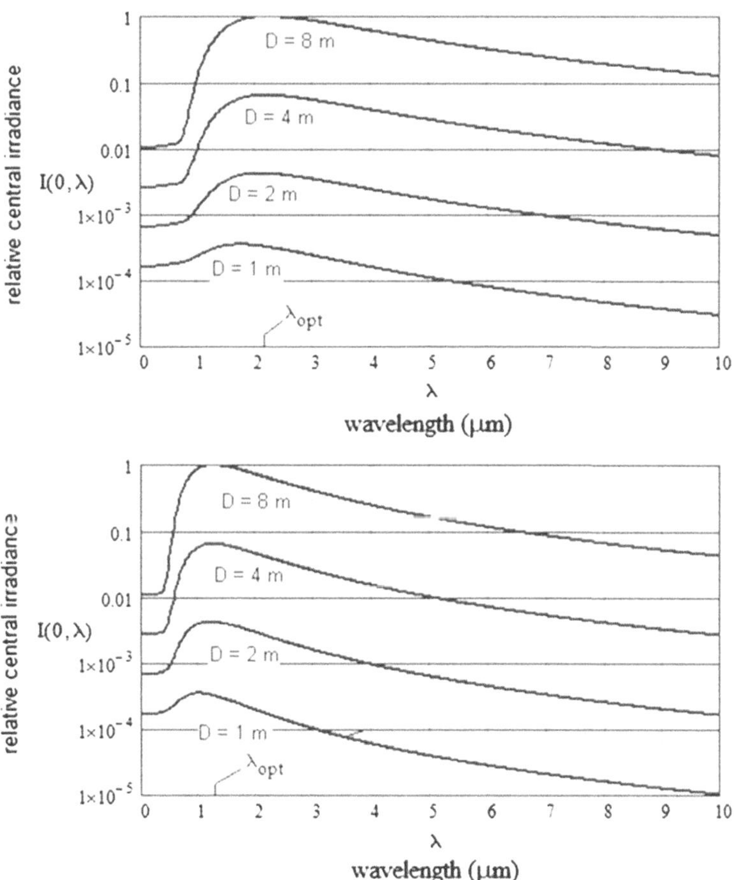


**Fig. 10.12** Typical irradiance variation in the center of a star image for 1-, 2-, 4-, and 8-m telescopes plotted against logarithmic vertical scales. Top for 1-arcsec visible seeing where $\sigma = 0.35$ μm and wo = 0.25 m. Bottom for 0.6-arcsec visible seeing where $\sigma = 0.20$ μm and $w_0 = 0.25$ m. The optimum wavelengths, $\lambda_{opt}$, are as indicated

seeing conditions. The lower plots correspond to slightly better seeing conditions (~0.6-arcsec).

Figure 10.13 shows the same set of plots using a linear vertical scale. Each of the plots has been normalized to unity to show more clearly the significantly higher irradiance levels that can be obtained by imaging at the optimum wavelength rather than at other non-optimum wavelengths. The irradiance improvement factors obtained at the optimum wavelength relative to those obtained at non-optimum wavelengths depend strongly on the diameter of the telescope. For the 1-m-diameter instrument, the improvement is by a factor of about 2. For the 8-m-diameter instrument, the improvement factor increases sharply to about 100.

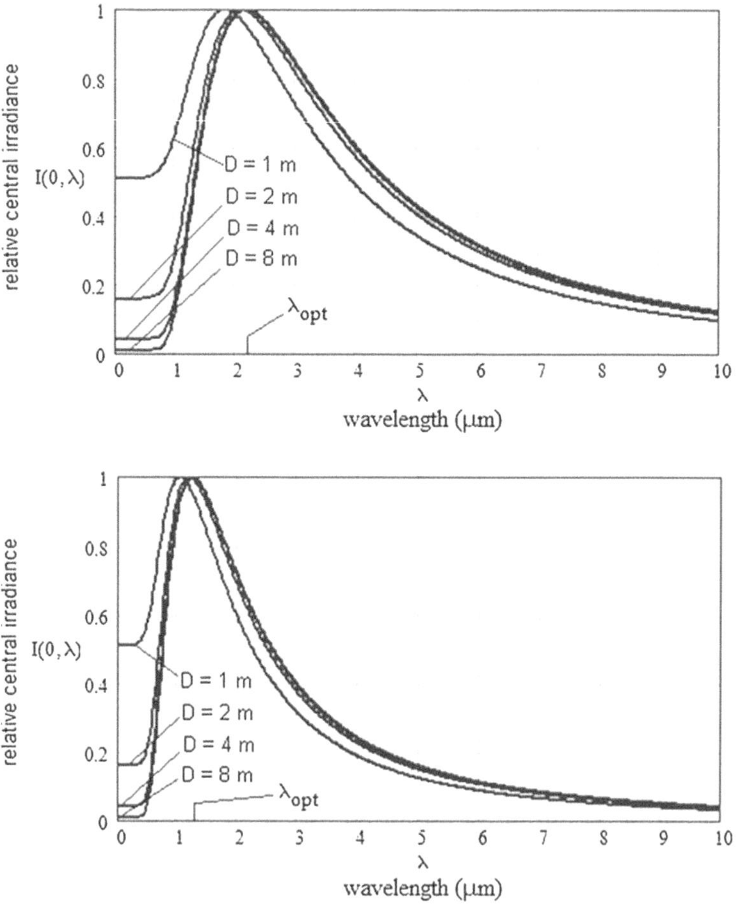


**Fig. 10.13** Plots against a linear scale of the irradiance attained in the center of a star image versus wavelength for 1-, 2-, 4-, and 8-m diffraction-limited telescopes. Top for 1-arcsec visible seeing where $\sigma = 0.35$ μm and $w_0 = 0.25$ m. Bottom for 0.6-arcsec visible seeing where $\sigma = 0.2$ μm and $w_0 = 0.25$ m. All plots are arbitrarily normalized to unity

For the $\sigma$ values used to calculate the plots in Figs. 10.12 and 10.13 (0.2 and 0.35 μm), 10.16 gives the approximate optimum wavelengths as $\lambda_{opt} = 1.25$ μm and. $\lambda_{opt} = 2.2$ μm. These two wavelengths are seen to be reasonably consistent with the irradiance plots in the figures where maximum irradiance is indeed found at about these two wavelengths. It is observed that the accuracy of the approximation, $\lambda_{opt} = 2 \cdot \pi \cdot \sigma$, steadily improves as the telescope diameter increases due to the increasing dominance of cores in the image centers.

## 10.6 Effect of Telescope Aberrations on Image Cores

### *10.6.1 The Effect on Irradiance at Core Center*

Whereas several waves of aberration in a large telescope cause only marginal degradation of the halo, the same aberrations can cause significant core degradation. Figure 10.14 shows center irradiance, $\langle I(0, \lambda)\rangle$, plotted against wavelength for an unobstructed (i.e., d = 0) diffraction-limited 4-m telescope; the figure also shows a plot of $\langle I(0, \lambda)\rangle$ for an aberrated version of the same telescope, where 2-wave (HeNe) P–V wavefront aberration is used, the amount shown previously in Fig. 10.4. Both plots in Fig. 10.14 are calculated using 10.20. For the diffraction-limited telescope version, 10.9 indicates that $PSF_I(0, \lambda) = 1$; for the aberrated version, 10.9 and 10.10 are used to calculate $PSF_I(0, \lambda)$. In the latter case, note that $PSF_I(0, \lambda)$ is wavelength dependent; for wavelengths longer than about 10 μm, $PSF_I(0, \lambda) \to 1$, which causes the two plots in Fig. 10.14 to converge at these longer wavelengths.

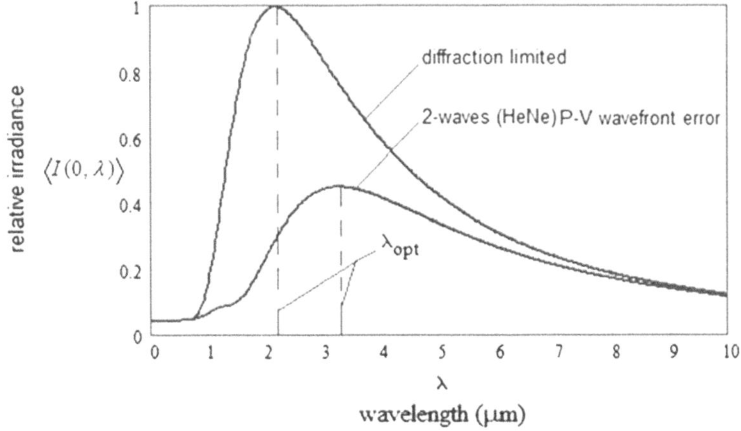


**Fig. 10.14** Average irradiance attained in the center of a star image versus wavelength for aberrated and unaberrated versions of 4-m telescopes observing in 1-arcsec visible seeing conditions (seeing described by $\sigma = 0.35$ μm and $w_0 = 0.25$ m). Telescope aberrations have the effect of shifting the optimum wavelength, $\lambda_{opt}$, to longer wavelengths

### *10.6.2 The Effect on the Optimum Wavelength*

Telescope aberrations have the effect of shifting the optimum wavelength toward longer wavelengths. For a telescope with rms wavefront aberration, $\sigma_T$, the optimum wavelength can again be calculated from 10.16, but only after replacing the rms quantity, $\sigma$, in that equation by the larger quantity, $\sqrt{\sigma^2 + \sigma_T^2}$, where the reasonable assumption is made that the wavefront aberrations of the telescope are statistically independent of the OPD fluctuations introduced by the atmospheric path. For the aberrated telescope case considered in Fig. 10.14, the optimum wavelength shifts from 2.2 to 3.3 μm. The irradiance pull-down of the core caused by aberrations occurs at all wavelengths, but is particularly evident for wavelengths in the vicinity of the optimum wavelength.

## 10.7 Instantaneous Core Location for Diffraction-Limited Telescopes

The important result is established in this section that the instantaneous location of the core lies at the focus of the plane wave that least mean squares best fits the instantaneous wavefront in the telescope pupil (Fig. 7.1). A proof of this result was given previously in a one-dimensional analysis (McKechnie, 1992). In this section, we give a more general proof in two dimensions.

With reference to the coordinate systems shown in Fig. 7.2, the complex amplitude in the image of a point-object, $U(\alpha, \beta, \lambda)$, is given by the Fourier transform of the complex amplitude of the image-forming waves in the telescope pupil. Using 7.6 as a starting point, for an aberration-free telescope where the pupil function, $K(x, y, \lambda)$, takes the value unity everywhere over the pupil, $U(\alpha, \beta, \lambda)$ may be expressed in unit-normalized form by

$$U(\alpha, \beta, \lambda) = \frac{\iint_{\mathrm{Pupil}} \exp\left[-\mathrm{i} \cdot \mathrm{k} \cdot \mathrm{H(x, y)}\right] \cdot \exp\left[\mathrm{i} \cdot \mathrm{k} \cdot (\alpha \cdot \mathrm{x} + \beta \cdot \mathrm{y})\right] \cdot \mathrm{dx} \cdot \mathrm{dy}}{\iint_{\mathrm{Pupil}} \mathrm{dx} \cdot \mathrm{dy}}, \tag{10.22}$$

where $k = 2 \cdot \pi / \lambda$ (cf., 3.33) and H(x, y) is the OPD fluctuation associated with the instantaneous wavefront in the telescope pupil (cf., Fig. 6.2). The above equation may be expressed more concisely in the form,

$$U(\alpha, \beta, \lambda) = \frac{\iint_{\mathrm{Pupil}} \exp\left\{-\mathrm{i} \cdot \mathrm{k} \cdot \left[\mathrm{H(x, y)} - (\alpha \cdot \mathrm{x} + \beta \cdot \mathrm{y})\right]\right\} \cdot \mathrm{dx} \cdot \mathrm{dy}}{\iint_{\mathrm{Pupil}} \mathrm{dx} \cdot \mathrm{dy}}. \tag{10.23}$$

It is observed here that the functions, H(x, y) and $(\alpha \cdot x + \beta \cdot y)$, both have the dimensions of length. Therefore, just as H(x,y) represents the instantaneous wavefront in the telescope pupil, function $(\alpha \cdot x + \beta \cdot y)$ also represents a wavefront—in this case, a plane wavefront tilted at vector angle $(\alpha, \beta)$ with respect to the z-axis. When a central core is present in the image, in the vicinity of this core the quantity $k \cdot [H(x, y) - (\alpha \cdot x + \beta \cdot y)]$ necessarily takes small values. In this vicinity, the right-hand side of 10.23 may be expanded by the Taylor series expansion,

$$U(\alpha, \beta, \lambda) = \frac{\iint_{\text{Pupil}} dx \cdot dy - \frac{k^2}{2} \cdot \iint_{\text{Pupil}} [H(x, y) - (\alpha \cdot x + \beta \cdot y)]^2 \cdot dx \cdot dy - i \cdot k \cdot \iint_{\text{Pupil}} [H(x, y) - (\alpha \cdot x + \beta \cdot y)] \cdot dx \cdot dy}{\iint_{\text{Pupil}} dx \cdot dy} \tag{10.24}$$

where $k^3 \cdot [H(x, y) - (\alpha \cdot x + \beta \cdot y)]^3$ and higher order terms have been neglected.

The intensity in the vicinity of the core, which we denote by $I(\alpha, \beta, \lambda)$, can be obtained as the squared modulus of $U(\alpha, \beta, \lambda)$ Thus, we may write

$$I(\alpha, \beta, \lambda) = 1 - k^2 \cdot \left\{ \frac{\iint_{\text{Pupil}} [H(x, y) - (\alpha \cdot x + \beta \cdot y)]^2 \cdot dx \cdot dy}{\iint_{\text{Pupil}} dx \cdot dy} - \left[ \frac{\iint_{\text{Pupil}} [H(x, y) - (\alpha \cdot x + \beta \cdot y)] \cdot dx \cdot dy}{\iint_{\text{Pupil}} dx \cdot dy} \right]^2 \right\}, \tag{10.25}$$

where again $k^3 \cdot [H(x, y) - (\alpha \cdot x + \beta \cdot y)]^3$ and higher order terms have been neglected.

The term in curly brackets in 10.25 is recognized as the mean square deformation, or variance, of the quantity $[H(x, y) - (\alpha \cdot x + \beta \cdot y)]$. This term therefore tells us that the intensity, or irradiance, at core center does not depend on the detailed characteristics of the OPD fluctuation H(x, y) per se; it only depends on the variance of that OPD fluctuation, measured with respect to the tilted (instantaneous best fit) plane. Moreover, 10.25 indicates that the intensity in the center of the core is less than the maximum possible value—unity—by an amount proportional to this variance.

It is clear from 10.25 that the angle vector $(\alpha, \beta)$ that minimizes the OPD variance corresponds to the angular location in the image at which the core attains maximum intensity, a location that we denote by $(\alpha_C, \beta_C)$. By the converse reasoning, it follows that the angular location in the image at which the core attains maximum intensity, $(\alpha_C, \beta_C)$, must lie at the focus of the plane wave that least mean square best fits the instantaneous wavefront in the telescope pupil; thus, the angle vector $(\alpha_C, \beta_C)$ determines the instantaneous location of the core.

Angle vector $(\alpha_C, \beta_C)$ may be established by solving the vector derivative equation,

$$\left(\frac{\partial}{\partial\alpha}, \frac{\partial}{\partial\beta}\right) I(\alpha, \beta, \lambda) = 0, \tag{10.26}$$

where $I(\alpha, \beta, \lambda)$ is given by 10.25.

It is observed here that, because 10.22 is unit-normalized, 10.25 is also unit-normalized. Thus, the quantity, $I(\alpha_C, \beta_C, \lambda)$, may be considered as providing a measure of the instantaneous Strehl intensity.

### *10.7.1 Angular Jitter of the Image Centroid*

According to a previous analysis (McKechnie, 1993), the light energy centroid in a star image generally undergoes larger random angular excursions in the image plane than those exhibited by the image core (assuming that the latter feature actually exists in the image). While the overall tilt component of the image-forming waves in the telescope pupil causes displacements of both the core and halo light portions in equal measures, the energy centroid of the halo undergoes additional displacements due to the random fluctuations of the instantaneous speckles that constitute this feature. The overall result is that the light energy centroid—the image entity traditionally used for guiding, tracking, and image stabilization—exhibits larger angular jitter excursions than the image core. As illustrated in Sect. 10.5 of the above-cited paper, more precise tracking and image stabilization can be obtained by using the central core as the reference feature, as opposed to the light energy centroid. For this reason, the remaining discussion in this chapter in relation to tracking and image stabilization emphasizes use of the core as the reference object, rather than the light energy centroid.

### *10.7.2 Instantaneous Core Location for Telescopes with Rectangular Apertures*

Again referring to the coordinate systems shown in Fig. 7.2, consider a telescope with a rectangular aperture whose side lengths in the x- and y-directions are denoted by a and b, respectively. By taking partial derivatives of the right-hand side of 10.25, the instantaneous location at which the core attains maximum irradiance, $(\alpha_C, \beta_C)$, may be written

$$[\alpha_c, \beta_c] = \left[ \frac{12}{a^3 \cdot b} \int_{-\frac{a}{2}}^{\frac{a}{2}} \int_{-\frac{b}{2}}^{\frac{b}{2}} x \cdot H(x, y) \cdot \mathrm{d}x \cdot \mathrm{d}y, \frac{12}{a \cdot b^3} \int_{-\frac{a}{2}}^{\frac{a}{2}} \int_{-\frac{b}{2}}^{\frac{b}{2}} y \cdot H(x, y) \cdot \mathrm{d}x \cdot \mathrm{d}y \right]. \tag{10.27}$$

### 10.7.3 *Instantaneous Core Location for Telescopes with Square Apertures*

For a telescope with a square aperture of side length, a, the instantaneous core location can be obtained by setting a = b in the above equation. Thus, we can write

$$[\alpha_c, \beta_c] = \left[ \frac{12}{a^4} \int_{-\frac{a}{2}}^{\frac{a}{2}} \int_{-\frac{a}{2}}^{\frac{a}{2}} x \cdot H(x, y) \cdot \mathrm{d}x \cdot \mathrm{d}y, \frac{12}{a^4} \int_{-\frac{a}{2}}^{\frac{a}{2}} \int_{-\frac{a}{2}}^{\frac{a}{2}} y \cdot H(x, y) \cdot \mathrm{d}x \cdot \mathrm{d}y \right]. \tag{10.28}$$

### 10.7.4 *Variance of Core Excursions for Rectangular Aperture Telescopes*

For the case of the rectangular aperture described in Sect. 10.7.2, the variance of the angular excursions of the core may be expressed by the vector form,

$$[\langle\alpha_c^2\rangle, \langle\beta_c^2\rangle] = \left[ \left\langle \left( \frac{12}{a^3 \cdot b} \cdot \int_{-\frac{a}{2}}^{\frac{a}{2}} \int_{-\frac{b}{2}}^{\frac{b}{2}} x \cdot H(x, y) \cdot \mathrm{d}x \cdot \mathrm{d}y \right)^2 \right\rangle, \left\langle \left( \frac{12}{a \cdot b^3} \cdot \int_{-\frac{a}{2}}^{\frac{a}{2}} \int_{-\frac{b}{2}}^{\frac{b}{2}} y \cdot H(x, y) \cdot \mathrm{d}x \cdot \mathrm{d}y \right)^2 \right\rangle \right]. \tag{10.29}$$

Because H(x,y) is a zero-mean function (6.43), the mean core location lies at the origin. Thus,

$$[\langle\alpha_c\rangle, \langle\beta_c\rangle] = [0, 0]. \tag{10.30}$$

The two vector components on right-hand side of 10.29 can both be expanded to give an expression in terms of quadruple integrals. Thus, we can write

$$\left[\left\langle\alpha_c^2\right\rangle, \left\langle\beta_c^2\right\rangle\right] =$$

$$\left[\frac{144}{a^6 \cdot b^2} \cdot \int_{-\frac{a}{2}}^{\frac{a}{2}} \int_{-\frac{b}{2}}^{\frac{b}{2}} \int_{-\frac{a}{2}}^{\frac{a}{2}} \int_{-\frac{b}{2}}^{\frac{b}{2}} x' \cdot x \cdot \left\langle H(x', y') \cdot H(x, y) \right\rangle \cdot dx' \cdot dy' \cdot \mathrm{d}x \cdot \mathrm{d}y, \right.$$

$$\left. \frac{144}{a^2 \cdot b^6} \cdot \int_{-\frac{a}{2}}^{\frac{a}{2}} \int_{-\frac{b}{2}}^{\frac{b}{2}} \int_{-\frac{a}{2}}^{\frac{a}{2}} \int_{-\frac{b}{2}}^{\frac{b}{2}} y' \cdot y \cdot \left\langle H(x', y') \cdot H(x, y) \right\rangle \cdot dx' \cdot dy' \cdot \mathrm{d}x \cdot \mathrm{d}y \right]. \tag{10.31}$$

where we observe that the averaging brackets have now been moved inward to enclose only those randomly varying parts of the integrands.

We now make the coordinate transformations,

$$\xi = x' - x \tag{10.32}$$

and

$$\eta = y' - y. \tag{10.33}$$

Noting that the Jacobian of this transformation is unity and making the appropriate modifications to the integration limits, 10.31 can be expressed in the form

$$\left[\left\langle \alpha_c^2 \right\rangle, \left\langle \beta_c^2 \right\rangle\right] =$$

$$\left[\frac{144}{a^6 \cdot b^2} \cdot \int_{-a}^{a} \int_{-b}^{b} \int_{-\frac{a}{2}}^{\frac{a}{2}} \int_{\frac{b}{2}}^{\frac{b}{2}} (x + \xi) \cdot x \cdot \langle H(x + \xi, y + \eta) \cdot H(x, y) \rangle \cdot \mathrm{d}\xi \cdot \mathrm{d}\eta \cdot \mathrm{d}x \cdot \mathrm{d}y, \right.$$

$$\left. \frac{144}{a^2 \cdot b^6} \cdot \int_{-a}^{a} \int_{-b}^{b} \int_{-\frac{a}{2}}^{\frac{a}{2}} \int_{-\frac{b}{2}}^{\frac{b}{2}} (y + \eta) \cdot y \cdot \langle H(x + \xi, y + \eta) \cdot H(x, y) \rangle \cdot \mathrm{d}\xi \cdot \mathrm{d}\eta \cdot \mathrm{d}x \cdot \mathrm{d}y \right]. \tag{10.34}$$

By using 6.31 to substitute for $\sigma^2 \cdot \rho(\xi, \eta)$ for $\langle H(x + \xi, y + \eta) \cdot H(x, y) \rangle$, the above equation may be written,

$$\left[\left\langle \alpha_c^2 \right\rangle, \left\langle \beta_c^2 \right\rangle\right] = \left[\frac{144}{a^6 \cdot b^2} \cdot \int_{-a}^{a} \int_{-b}^{b} \int_{-\frac{a}{2}}^{\frac{a}{2}} \int_{-\frac{b}{2}}^{\frac{b}{2}} (x + \xi) \cdot x \cdot \sigma^2 \cdot \rho(\xi, \eta) \cdot \mathrm{d}\xi \cdot \mathrm{d}\eta \cdot \mathrm{d}x \cdot \mathrm{d}y, \right.$$

$$\left. \frac{144}{a^2 \cdot b^6} \cdot \int_{-a}^{a} \int_{-b}^{b} \int_{-\frac{a}{2}}^{\frac{a}{2}} \int_{-\frac{b}{2}}^{\frac{b}{2}} (y + \eta) \cdot y \cdot \sigma^2 \cdot \rho(\xi, \eta) \cdot \mathrm{d}\xi \cdot \mathrm{d}\eta \cdot \mathrm{d}x \cdot \mathrm{d}y \right]. \tag{10.35}$$

### 10.7.5 *rms Core Excursions for Rectangular Aperture Telescopes*

The rms angular excursion of the core is given in vector form by $\left(\langle\alpha_c^2\rangle^{\frac{1}{2}}, \langle\beta_c^2\rangle^{\frac{1}{2}}\right)$. By using 10.35, this vector may be expressed as

$$\left[\left\langle\alpha_c^2\right\rangle^{\frac{1}{2}}, \left\langle\beta_c^2\right\rangle^{\frac{1}{2}}\right] = \left[\left(\frac{144}{a^6 \cdot b^2} \cdot \int_{-a}^{a}\int_{-b}^{b}\int_{-\frac{a}{2}}^{\frac{a}{2}}\int_{-\frac{b}{2}}^{\frac{b}{2}} (x+\xi)\cdot x\cdot \sigma^2 \cdot \rho(\xi,\eta)\cdot d\xi \cdot d\eta \cdot dx \cdot dy\right)^{\frac{1}{2}},\right.$$
$$\left.\left(\frac{144}{a^2 \cdot b^6} \cdot \int_{-a}^{a}\int_{-b}^{b}\int_{-\frac{a}{2}}^{\frac{a}{2}}\int_{-\frac{b}{2}}^{\frac{b}{2}} (y+\eta)\cdot y\cdot \sigma^2 \cdot \rho(\xi,\eta)\cdot d\xi \cdot d\eta \cdot dx \cdot dy\right)^{\frac{1}{2}}\right]. \tag{10.36}$$

#### 10.7.5.1 Analytic Expressions for the rms Core Excursion

Analytic expressions are developed in this section to show how rms core excursions depend on both the telescope aperture size and the strength and structure size of the turbulence in the imaging path. To proceed beyond the general form given by 10.36, it is necessary to assume some appropriate functionality for $\rho(\xi,\eta)$. We again choose the circular Gaussian function previously expressed by 9.9. In terms of that function, rms core excursions in the $\alpha$-direction, $\langle\alpha_c^2\rangle^{\frac{1}{2}}$, may be written as.

$$\left\langle\alpha_c^2\right\rangle^{\frac{1}{2}} = \left(\frac{144}{a^6 \cdot b^2} \cdot \int_{-a}^{a}\int_{-b}^{b}\int_{-\frac{a}{2}}^{\frac{a}{2}}\int_{-\frac{b}{2}}^{\frac{b}{2}} (x+\xi)\cdot x\cdot \sigma^2 \cdot \exp\left[-\left(\frac{\xi^2+\eta^2}{w_o^2}\right)\right]\cdot d\xi \cdot d\eta \cdot dx \cdot dy\right)^{\frac{1}{2}}. \tag{10.37}$$

A similar expression may be written for $\langle\beta_c^2\rangle^{1/2}$.

The following integral results allow simplification of 10.37 (as well as simplification of the corresponding equation for $\langle\beta_c^2\rangle^{1/2}$:

$$\int_{-\frac{b}{2}}^{\frac{b}{2}} dy = b, \tag{10.38}$$

$$\int_{-b}^{b} \exp\left[-\left(\frac{\eta}{w_o}\right)^2\right] \cdot d\eta = w_o \cdot \sqrt{\pi} \cdot erf\left(\frac{b}{w_o}\right), \tag{10.39}$$

$$\int_{-\frac{a}{2}}^{\frac{a}{2}} x \cdot (x + \xi) \cdot dx = \frac{a^3}{12}, \tag{10.40}$$

$$\int_{-a}^{a} \exp\left[-\left(\frac{\xi}{w_o}\right)^2\right] \cdot d\xi = w_o \cdot \sqrt{\pi} \cdot erf\left(\frac{a}{w_o}\right), \tag{10.41}$$

where erf(·) is the error function (Magnus et al., 1966). The above results enable the rms core excursions to be expressed in the vector form,

$$\left[\langle\alpha_c^2\rangle^{\frac{1}{2}}, \langle\beta_c^2\rangle^{\frac{1}{2}}\right] = \left[\left(\frac{12}{a^3 \cdot b} \cdot \pi \cdot \sigma^2 \cdot w_o^2 \cdot erf\left(\frac{a}{w_o}\right) \cdot erf\left(\frac{b}{w_o}\right)\right)^{\frac{1}{2}}, \left(\frac{12}{a \cdot b^3} \cdot \pi \cdot \sigma^2 \cdot w_o^2 \cdot erf\left(\frac{a}{w_o}\right) \cdot erf\left(\frac{b}{w_o}\right)\right)^{\frac{1}{2}}\right]. \tag{10.42}$$

The error function, erf(·), is plotted in Fig. 10.15. This function can be seen to asymptotically approach the value unity for large arguments, substantially reaching that value at the argument value, 2. Thus, for large rectangular aperture telescopes where we assume $a/w_0 > 2$ and $b/w_0 > 2$, the rms core excursion vector is given approximately by

$$\left[\langle\alpha_c^2\rangle^{\frac{1}{2}}, \langle\beta_c^2\rangle^{\frac{1}{2}}\right] = \left[\left(\frac{12 \cdot \pi \cdot \sigma^2 \cdot w_o^2}{a^3 \cdot b}\right)^{\frac{1}{2}}, \left(\frac{12 \cdot \pi \cdot \sigma^2 \cdot w_o^2}{a \cdot b^3}\right)^{\frac{1}{2}}\right]. \tag{10.43}$$

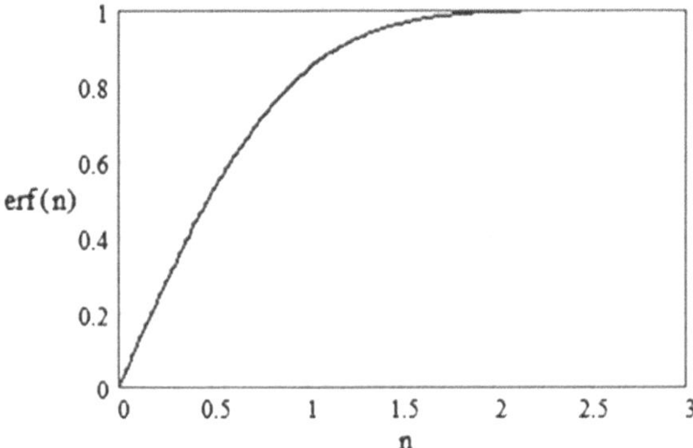


**Fig. 10.15** The error function, also known as the Gauss error function, plotted for positive arguments

#### 10.7.5.2 rms Core Excursions for Telescopes with Square Apertures

For a telescope with square aperture of side length, a, rms core excursions may be obtained by setting $b = a$ in 10.43. Thus, we can write

$$\left[\langle\alpha_c^2\rangle^{\frac{1}{2}}, \langle\beta_c^2\rangle^{\frac{1}{2}}\right] = \left[\frac{2 \cdot \sqrt{3 \cdot \pi} \cdot \sigma \cdot w_o}{a^2}, \frac{2 \cdot \sqrt{3 \cdot \pi} \cdot \sigma \cdot w_o}{a^2}\right]. \tag{10.44}$$

### *10.7.6 rms Core Excursions for Telescopes with Circular Apertures*

For telescopes with circular apertures where isotropic turbulence is assumed, an analysis similar to the one just given leads to the following expression for the rms core excursions in the radial direction (McKechnie, 1992),

$$\langle\vartheta_c^2\rangle^{\frac{1}{2}} = \frac{8 \cdot \sigma \cdot w_o}{D^2}. \tag{10.45}$$

#### 10.7.6.1 rms Core Excursion Relative to Core Size

For diffraction-limited telescopes with circular apertures, the angular size of the core is nominally given by $1.22 \cdot \lambda \cdot D$. Thus, we may express the rms angular excursions of the core relative to the angular size of the core itself, $R_{mw}$, as follows:

$$R_{mw} = \frac{8 \cdot \sigma \cdot w_o}{1.22 \cdot \lambda \cdot D} \approx \frac{6.56 \cdot \sigma \cdot w_o}{\lambda \cdot D}. \tag{10.46}$$

It is observed from the above relation that $R_{mw}$ increases in proportion to both the average turbulence strength (as described by $\sigma$) and the average structure size (as described by $w_o$), while it reduces in proportion to both the telescope diameter D and the imaging wavelength $\lambda$. For telescope apertures larger than 4 m and for typical values of $\sigma$, $w_o$, and $\lambda$, the above equation indicates that $R_{mw} \ll 1$. This tells us that core motions are generally much smaller, on average, than the angular size of the core itself. In the 1989 letter (Appendix G) where Roger Griffin discusses the image cores he saw at visible wavelengths, he states, "In my experience, angular excursions of the image due to quiver[6] were virtually nil."

[6] Image quiver is another name for image wander. In this book, the latter term is generally used.

## 10.8 Griffin's Naked-Eye Core Observations

Griffin (1973) reported routine, night-after-night, naked-eye observations of cores in star images while using the Mt. Wilson 100-in. and Mt. Palomar 200-in. telescopes. He describes the properties of these cores as follows:

(i) The formation of an image core of the type described above is a property of large telescopes, which possess it to an extent which increases with their aperture.
(ii) The size of the halo is the size corresponding to the angular deflection of rays through individual seeing elements, which would be seen in a small telescope (aperture ~10 in.). The size of the core is possibly limited by the optical imperfection of the telescope and may correspond to that of the image which may be obtained through a perfectly steady atmosphere.
(iii) The proportion of the total light in the image which is thrown into the core depends upon the size of the telescope and upon the perturbation, by the optics plus the seeing, of the wavefronts approaching the image.
(iv) In a large telescope, the effects of seeing diminish dramatically once they are below a certain level. Once the seeing perturbations fall below this level, a core starts to form in the image and rapidly increases in importance as the perturbations fall further.

In the same paper, Griffin makes the following assertions:

> I wish to assert, first of all as a matter of my own experience, that the image character is by no means the same in a large telescope as it is in a smaller one. Both telescopes give bad images in bad seeing (4 to 6 seconds of arc); but as the seeing improves, the characters of the images in the two telescopes diverge. In the small telescope (~ 10 inches) the image simply gets smaller as the seeing improves; in the large one (>60 inches) a core forms in the image, and more and more of the light is poured into the core. My impression is that the size of the outer halo of the image seen in the large telescope is that which would be considered as the image size in the small one; and that the core is something peculiar to the big telescope and may be very much smaller than the halo. Moreover, the larger the telescope the less good is the seeing which seems to be necessary to allow the sharp bright core to form in the image.
>
> According to this somewhat naive picture, therefore, seeing which arises on a spatial scale of 4 inches and is bad enough to cause the image to have an outer diameter of (say) 5 seconds of arc will not prevent much of the light from a 100-inch. aperture from being concentrated in a sharp bright core 0.2 seconds of arc across.

Although the relationship between core brightness and wavelength is not addressed in Griffin's, 1973 paper, in 1989 Griffin wrote (Griffin 1989, 2011) (cf., Appendix G), "I am interested to see the wavelength dependence of the images in your models ... the most amazing images I ever saw at the 100-in. (whose optics are better than those of the 200-in.) were when I undertook with my wife a limited program of infrared photographic spectroscopy and we were guiding through a filter that only transmitted the far-red and infrared. The image of Aldebaran (a very red first-magnitude star), in particular, seemed to be largely concentrated into a searing point of light, just as if one were looking straight at a laser."

The primary objectives of Griffin's observations were to obtain stellar spectra and measure stellar radial velocities. His visual observations of cores were made while manually guiding his telescopes, viewing the subject star directly by eye as he stabilized its image on the spectrograph slit. The eye has the natural ability to track and, in effect, compensate image wander. The images seen by Griffin may therefore be considered "short-exposure" images. His descriptions of cores at visible wavelengths are consistent with short-exposure rms OPD fluctuation, $\sigma = 0.15\mu m$. Had Griffin attempted to photograph the cores he saw without having any way of compensating image wander, the results might have resembled the disappointingly blurred, long-exposure images shown on the left of Figs. 10.10 and 10.11.

In a 1976 presentation at the Annual Meeting of the Optical Society of America, the Author compared Griffin's description of cores (as set out in items (i)–(iv) above) with predicted properties derived from 10.8 and 10.9, finding a close correspondence. The following passages are excerpted from the presentation abstract (McKechnie, 1976):

> Several astronomers claim to have seen central cores in star images when observing with large telescopes in conditions of good seeing ... we investigate this phenomenon and produce an expression for the brightness of the core in relation to that of the background seeing disk. The expression includes factors such as the aperture of the telescope, the aberrations of the telescope, the wavelength used to make the observation, and a factor due to the seeing. While all these factors are important, perhaps the most critical of them is the factor due to seeing ...the question arises as to whether, because of the seeing, the phenomenon can ever occur in practice ... in the Author's opinion, the observations of R.F. Griffin, made using the Hale 200-inch and the Mt. Wilson 100-inch, are in such close agreement with the predictions of the theory that it is almost certain that this observer has indeed seen central cores. By optimizing some of the factors that influence the brightness of the core, it may be possible to use this phenomenon as a tool for improving the resolution of large telescopes.

The only reasonable physical explanation for the occurrence of cores at visible wavelengths is that atmospheric turbulence structure size must be considerably smaller, on average, than assumed by Kolmogorov theory. Griffin remarked on this possibility in his 1973 paper as follows: "In telescopes of aperture larger than 4-in., the effect of seeing is more to increase the image diameter beyond the diffraction limit and less to move the image about bodily. From this observed decrease in image motion with increasing telescope aperture, we may deduce that the characteristic values of the [phase] delay gradients decrease on spatial scales larger than about 4 in. This, of course, is not a new result."

But not everyone at that time, circa 1973, was ready to accept such a radical possibility. After Griffin described cores in star images at a meeting of the Royal Astronomical Society (Griffin, 1990), one member "denounced" his descriptions as "a load of subjective rubbish." Goodman (1990) summarized his somewhat skeptical views about cores as follows: "It should be noted that the diffraction-limited 'core' in the average intensity PSF will very rapidly vanish with increasing phase variance and in many practical applications may be negligibly weak." For many years, Griffin's observations and claims were scorned and discounted. Yet the issue would not go away.

## 10.9 Cores at Near-IR Wavelengths

The 1975/76 rms OPD measurements made at Royal Greenwich Observatory (RGO) using the 36-in. Yapp telescope (described previously in Chap. 8, Sect. 8.2.3) consistently gave $\sigma$ values of about 0.39 $\mu$m[7]—values small enough to suggest that large telescopes located at more pristine viewing sites should routinely deliver image cores at near-IR and longer wavelengths and perhaps, in good seeing conditions, at visible wavelengths too. Around 1990, at about the time IR focal plane arrays were becoming available to astronomers, it would soon become clear that much sharper images were possible at near-IR wavelengths than previously supposed. After cophasing the six mirrors of the MMT, McCarthy et al. (1990) obtained image cores at 2.2 and 3.4 $\mu$m, finding the cores to persist for several minutes until thermal expansion effects degraded the alignment. Somewhat puzzled by their images, these authors concluded that the values of the coherence parameter, $r_0$, they were dealing with at these near-IR wavelengths, were significantly larger than the values extrapolated from the $r_0$ value at visible wavelengths using the standard Kolmogorov-based wavelength dependency, $r_0 \propto \lambda^{6/5}$. A ready explanation for the contradiction they had encountered is that both the average turbulence structure size and the rms OPD fluctuation were, in fact, much smaller than assumed by Kolmogorov theory. (A discussion of how a smaller outer scale limit can lead to anomalous, baseline-dependent measured estimates of the $C_n^2$ structure constants is given in Appendix I.) At about the same time, it was becoming clear to astronomers that telescope optical tolerances needed to be improved by almost an order of magnitude.

The 3.8-m United Kingdom Infrared Telescope (UKIRT) sited on Mauna Kea was constructed between 1975 and 1978. It saw first light in October 1979. The Infrared Flux Collector, as it was originally called, was conceived as an inexpensive "light bucket" telescope with relaxed specifications for use in the 1–30 $\mu$m wavelength region. The 1970 optical specification for the UKIRT instrument required that (in the absence of atmosphere) it should be able to concentrate most of the light in a star image into a 1-arcsec diameter patch—even if at least one member of the eight-person steering committee, R.F. Griffin, argued for a significantly tighter optical specification.

At a seminar presentation given in July 1990 at the Royal Observatory, Edinburgh (ROE), the administrative headquarters of the UKIRT instrument, the Author showed the highly resolved near-IR images that had been obtained by J. Christou just a few months earlier using the Mayall telescope. Taken aback by the superb 0.145-arcsec resolution obtained in the images, UKIRT astronomers were naturally eager to discover how they too might also obtain similar images. The answer was improbably simple[8]:

---

[7] The RGO site at Herstmonceux Castle is about 70 ft. above sea level. Astronomical images obtained at this site do not therefore benefit from the significantly better seeing conditions found at high-altitude sites.

[8] The answer was the same as that provided previously to D.W. McCarthy Jr. and K. Hege (cf., Sect. 14.8) at the September 26th 1989 seminar presentation (McKechnie, 1989).

(1) Record star images at the near-IR wavelength, 2.2 μm, where image cores were predicted to appear routinely even in modest 1.25-arcsec visible seeing.
(2) Increase telescope magnification by about 10 × so that the 'standard high resolution' 0.65-arcsec pixel spacing on the IR-FPA camera reduces to 0.05-arcsec – small enough to resolve the anticipated 0.1-arcsec image cores. (To resolve cores at 1-μm would have required a 0.025-arcec pixel spacing.)
(3) Capture short-exposure (0.001 s) images of bright stars to freeze any image motion that might otherwise smear out and obscure the image cores.
(4) Tighten telescope focus by ~ 10 × to sharpen, not only the ~ 1-arcsec halos but, more importantly, the 0.1-arcsec cores.

Over the next few months, ROE astronomers' attempts to obtain similar image cores using the UKIRT instrument ended in disappointment. Zero image cores were observed despite Mauna Kea seeing being considerably better, on average, than Kitt Peak seeing.

A program of UKIRT optics upgrades was soon begun by ROE,[9] and carried out between 1991 and 1998. The result was a huge improvement in image quality, leading to the subsequent claim (Hawarden, 1998) that UKIRT was now delivering the sharpest images of any ground-based telescope at that time. Hawarden also stated that "As a result of (the upgrades), the telescope's optical quality is good enough to provide images which, in sharpness, are close to the limit set by the wave nature of light (the diffraction limit) at wavelengths of 2 μm or longer (0.11-arcsec across the width of a point source image)...In fact at a wavelength 2 μm (the K band) the best UKIRT images thus far recorded rival those of the Hubble Space Telescope's NICMOS infrared imager for sharpness."

### Modeling Hawarden's UKIRT Images at 2 μm

Hawarden's descriptions of images obtained at 2 μm with the UKIRT instrument contain enough information to allow rough modelling. Denoting the average Strehl

---

[9] In 1896, the Royal Observatory, Edinburgh was moved two miles southwards from its old location in the center of the city on Calton Hill, to its present site on Blackford Hill (146 m altitude). The largest telescope at the observatory is a 36-in. Cassegrain reflector which was installed in 1930 but is now no longer operational. As a youth, the Author lived close to the observatory at 10 Observatory Road. As a teenager, he constructed a 6-in. Newtonian reflector telescope (see photograph on Page xxxv), grinding two ship's portholes together to make the primary mirror, faithfully following an early version of "How to make a telescope" by Texereau (1984). Disappointed by the blurry images, the author attributed the problem to figure errors of the paraboloidal primary mirror which, during fabrication, had been tested using a crude Foucault test, with the slit and knife-edge made from razor blades; the Foucault test was first described in 1858 by the French physicist, Jean Bernard Leon Foucault (1819–1868). However, in 1997—some thirty-five years after the instrument was constructed—after being urged to have it tested interferometrically by Mr. Garry Marions, a respected Optical Technician at the Developmental Optics Facility, US Air Force Research Laboratory (AFRL), Albuquerque, New Mexico, the instrument was finally tested this way. Somewhat taken aback, the author discovered that the instrument was in fact diffraction-limited, the interferogram indicating a P–V wavefront error of about 0.2 waves (HeNe). While researching for this book—but now fifty years after the construction of the instrument—the author learned that more than a century earlier, the seeing conditions in Edinburgh had been described as "appalling" (Footnote 8, in Chap. 4) by the second Astronomer Royal for Scotland, Charles Piazzi Smyth. The mystery of the blurry images was finally resolved.

intensity in Hawarden's images by $SI_E$ and denoting the cumulative rms OPD fluctuation of the image-forming waves caused by all sources of OPD fluctuation throughout the entire path from object to image by $\sigma_E$, by making the assumption that the total OPD fluctuation is Gaussian distributed, $SI_E$ may be given in terms of $\sigma_E$ by (cf., Sect. 8.2.6)

$$SI_E = \exp\left[-\left(\frac{2\cdot\pi\cdot\sigma_E}{\lambda}\right)^2\right]. \tag{10.47}$$

The primary contributions to $\sigma_E$ include contributions arising from atmospheric turbulence, telescope aberrations, and dome turbulence.

From 10.47, we can readily obtain the relation,

$$\sigma_E = \frac{\lambda\cdot\sqrt{-\ln(SI_E)}}{2\cdot\pi}. \tag{10.48}$$

Based on Hawarden's description that the images "are close to the limit set by the wave nature of light," we choose two possible candidate values for the Strehl intensity, $SI_E$: the value 0.67 and, more conservatively, the value 0.4. (Note that Strehl intensities $\geq 0.8$ indicate substantially diffraction-limited image quality.) For the two chosen $SI_E$ values, 10.48 gives the corresponding values for $\sigma_E$ as 0.2 and 0.3 μm.

Average visible seeing at Mauna Kea sites is generally considered to be about 0.7-arcsec FWHM. Surmising that the seeing may have been slightly better than average at the time Hawarden's superb images were recorded, we assume it to have been about 0.5-arcsec (FWHM). Using an equation derived later in Chap. 13 (13.46), the average turbulence structure sizes consistent with this seeing and the two $\sigma_E$ values just indicated (0.2 and 0.3 μm) correspond to $w_0 = 0.25$ m and $w_0 = 0.375$ m.

Central sections through image intensity envelopes based on the above $\sigma_E$ and $w_0$ values are shown in Fig. 10.16 that bear a reasonable correspondence to Hawarden's 2-μm images. Central sections are also shown for wavelengths, 0.55 and 1.0 μm. All envelopes in this figure were calculated using 10.8. (It might be recalled that all intensity envelopes calculated from this equation, upon rotation, enclose the same amount of light energy.) Because telescope aberrations are assumed to be already included in the chosen $\sigma_E$ values, the telescope intensity PSF that we insert into this equation may be considered that of a diffraction-limited instrument with the same aperture diameter (D = 3.8 m) and central obstruction (d = 0.35 m) as for the actual UKIRT instrument. Using these values for D and d, the appropriate intensity PSF can be obtained from 10.12.

According to the Rayleigh criterion, the angular resolution obtained from image cores at 2 μm delivered by the UKIRT instrument is about 0.13 arcsec. At 1 μm, resolution improves to about 0.065 arcsec. However, for the Strehl intensity values chosen, 0.67 and 0.4, the cores at this wavelength (as can be seen in Fig. 10.16) contain only a tiny fraction of the total light energy. Had we assumed the Strehl

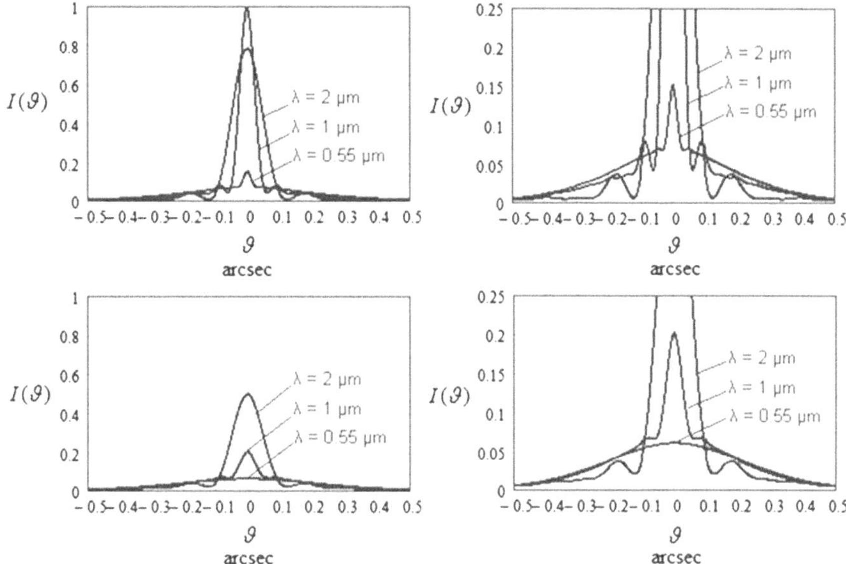


**Fig. 10.16** Surmised intensity envelopes for the images described by Hawarden using the UKIRT 3.8-m telescope. Top Images for $\sigma_E = 0.20\ \mu m$, $w_0 = 0.25$ m. Bottom Images for $\sigma_E = 0.30\ \mu m$, $w_0 = 0.375$ m. A 4 × vertical scale change has been applied to the images on the right. Parameter values used: D = 3.8 m, d = 0.35 m

intensity to be 0.8 in Hawarden's 2-μm images, 10.48 would give $\sigma_E \approx 0.15\mu m$. With the optimum wavelength then shortening to about 1 μm, dominant cores would have been found at that wavelength. But because Hawarden's account only discusses images at 2 μm, the 1-μm image envelopes shown in Fig. 10.16 are merely images predicted by the model; nothing in Hawarden's account allows us to check the validity of these images.

## 10.10 Adaptive Optics

Adaptive optics (AO) (Tyson, 2010) refers to the technological area where an AO system is used to correct unwanted wavefront distortions that degrade the quality of images formed by telescopes and other optical systems. The operating principle of an AO system first requires measurement of the phase distortions of the imaging waves using a sensor device, such as a Shack–Hartmann lenslet array (Platt & Shack, 2001). The measured distortions are then compensated using a device such as a deformable mirror (DM) or a segmented active mirror (Smithson et al., 1988). Currently, the most widely used DMs are microelectromechanical systems (MEMS) (Bifano et al., 2010; Gavel, 2009). However, deformable membrane mirrors (Brousseau et al., 2007) are also widely used, especially when large mirror surfaces are required. Deformable membrane mirrors are formed by stretching thin, flexible, conductive, and reflective membranes over a solid frame. Deformation is accomplished by applying voltages to actuators located either under or above the membranes.

Most AO systems initially use a large-stroke, fast response, two-axis, tip–tilt mirror since a significant fraction of the wavefront error generally arises as wavefront tilt—the lowest order distortion component with the exception of piston error. The tip–tilt mirror is followed by a DM device to correct higher order distortions.

In astronomical applications (Beckers, 1993; Hardy, 1998; Roddier, 1999, 2004) bright stars are normally used as the reference objects. It makes little difference whether resolved or unresolved stars are used for this purpose; planets can also be used as well as any other relatively compact object features; even sunspots can be used. A basic requirement of the reference object is that it should be suitably bright, consistent with providing signal-to-noise ratios that allow wavefront error to be measured to a suitably-high level of accuracy. (In the case of sunspots, suitably high contrast is required.) If natural stars are used as reference objects, to achieve suitably-high signal-to-noise ratios their visual magnitudes should not exceed about 12 (Chap. 17, Sects. 17.4–17.6).

AO only permits image correction within the isoplanatic patch that surrounds the reference object. By using the core in the reference star image rather than the light energy centroid as the reference object, it is postulated in Chap. 17 that significantly larger isoplanatic angles arise, which then enable image corrections over wider fields of view. The same ends can also be achieved by use of a technique known as multi-conjugate AO (Marchetti & Hubin, 2007). This method uses several DMs, where each DM is assigned the task of correcting the OPD fluctuation that arises from a specified depth portion of the atmospheric path lying in the neighborhood of a plane conjugate to the DM.

### 10.10.1 Tip–Tilt Correction

Usually, a large fraction of the total wavefront distortion is attributable to tip–tilt error. This error has various sources, but naturally includes a component due to atmospheric turbulence; other components also arise from telescope shake and telescope tracking and guiding errors. Irrespective of the source of the tip–tilt error, the error may be compensated either in real time by using a single fast response tip–tilt mirror or by subsequent post-processing of many short-exposure images using a computer-controlled shift-and-add algorithm.

Tip–tilt correction may be considered as a subset of AO correction. Tip–tilt correction is exactly the same as image wander correction. Thus, the average tip–tilt corrected image may simply be considered as the average short-exposure image. Remarkably high resolution can be obtained in a tip–tilt corrected image. As discussed earlier in Sect. 10.7, more precise tip–tilt correction can be achieved by stabilizing the image with respect to the core in a reference star image, rather than the light energy centroid.

### 10.10.2 Active Optics

Active optics may be considered as a subset of AO where much slower correction rates are used. Active optics is not intended to correct the rapidly occurring wavefront distortions introduced by the atmosphere. Its purpose is merely to correct the static wavefront errors of the telescope as well as any slowly varying (~1 Hz) wavefront errors caused by thermal expansion and gravitational loading changes of the telescope structure. Active optics technology was developed in the 1980s; its success paved the way for the 8- and 10-m Gemini and Keck telescopes. Instead of using thick/heavy monolithic primary mirrors such as found in the Palomar 200-in. instrument that naturally hold their shape as the pointing angle changes, active optics enables use of much thinner and lighter primary mirrors. An array of actuators control the mirror supports, enabling the mirror shape to be adjusted in real time, thus allowing the mirror to maintain correct figure in spite of its thin, light-weight design.

Active optics technology was first used on the Nordic Optical Telescope (NOT) sited at La Palma in the Canary Islands. This 2.6-m instrument saw first light in 1988 (Andersen et al., 1992). The ESO 3.6-m New Technology Telescope (NTT) saw first light soon afterward. Situated at La Silla in the Chilean Andes, this instrument produced exceptionally highly resolved images on its first night of use.

These two telescopes confirmed the viability of this type of primary mirror, paving the way for, not only the much larger astronomical reflector telescopes that would soon follow, including the Keck I and II instruments (first light, 1990 and 1996) and the Gemini North and South instruments (first light 1999 and 2000), but also the even larger extremely large telescopes (ELTs) for which first light is anticipated in the mid-2020s.

### 10.10.3 Laser Guide Stars

The limited number of natural stars in the sky bright enough to be used as AO correction reference objects significantly limits the areas of sky over which image corrections can be made. While use of the cores of reference star images may significantly increase sky coverage (Chap. 17), complete sky coverage is still not possible because of the uneven distribution of natural reference stars in the sky. To achieve complete sky coverage requires alternative strategies, such as the use of laser guide stars (LGSs).

LGSs are produced by laser beams projected from the observing telescope to create focused spots of light high up in the atmosphere. Light back-scattered, or re-emitted, from these high-altitude regions provides reference "point-objects" referred to as LGSs. While LGSs enable the correction of higher order wavefront distortions, natural reference stars are also required to stabilize the image by providing tip–tilt correction references.

Two types of LGSs are commonly used. Rayleigh guide stars rely on Rayleigh back-scattering from air molecules at altitudes of around 20 km. By using pulsed lasers, the effects of back-scattered light contributions from lower altitudes (which are out of focus and therefore unhelpful) can be eliminated by use of a time gate, or time window, that rejects all back-scatter that arises in the first several tenths of a millisecond (corresponding nominally to the out-and-back time of flight of the laser beam to ~20 km altitude). The window remains open for just a few microseconds, long enough to accept only the back-scattered light portions that arise from the targeted altitude range.

Sodium guide stars make use of lasers tuned to produce an output wavelength of about 589 nm. The beams are focused on a plane about 90 km overhead, where they excite naturally occurring sodium atoms found at that altitude. These atoms re-emit light at the characteristic sodium wavelength, thereby acting as LGSs (Wizinowich et al., 2006). In principle, LGSs at higher altitudes are preferred to those at lower altitudes. Because they are less affected by focus anisoplanatism (Fried, 1995), they allow more precise characterization of the phase errors introduced by the actual observing path of the telescope.

An informative overview of AO as applied to the very large telescope (VLT) and ELT can be found in the European Southern Observatory (ESO) booklet titled "The VTL adaptive Optics facility" by (Amico et al., n.d.). This booklet summarizes the many different component parts and technologies that combine together to provide a state-of-the-art AO system. Component parts include the wavefront sensor, the deformable secondary mirror, as used at the Four Laser Guide Star Facility (4LGSF).

## 10.11 Speckle Imaging

Speckle imaging is the name given to a number of related techniques for achieving diffraction-limited resolution from large ground-based telescopes in spite of the disruptive effect of the atmosphere. These techniques are based on large numbers of short-exposure images, where atmospheric motion is effectively frozen in each individual image. Exposure times are typically 10 ms for visible wavelengths and 100 ms for IR wavelengths. At visible wavelengths, the individual images are usually speckle patterns; at IR wavelengths, the images usually display core and halo structure.

In the most common application, the short-exposure images are combined, or stacked, together using a shift-and-add algorithm to recenter each image according to the brightest speckle (Baba et al., 1985). Since the average speckle size is determined by the diffraction-limited intensity PSF of the telescope, extremely high resolution can be obtained in the final composite image. Other speckle imaging techniques include lucky imaging and speckle interferometry. The statistical properties of speckle in star images, including core and halo images in which both speckles and cores are present, are described in some detail in Chap. 11.

### 10.11.1 Lucky Imaging

The imaging technique referred to as lucky imaging, or lucky exposure imaging, is a variant of speckle imaging in which resolution is enhanced further by creating the final image using only those images that are least affected by atmospheric turbulence. Typically 10%, or perhaps as few as 1%, of the images are selected; the remaining images are simply discarded.

Lucky imaging methods were first tried out in the 1950s and 1960s, but it was only with subsequent advancements in imaging technology (CCD cameras in particular) that the full benefits of the technique could be realized. The first numerical calculation of the probability of obtaining lucky exposures was developed by Fried (1978). Various methods have been tried for making the image selections. Baldwin et al. (2001) used Strehl intensity as the criterion.[10] In principle, images could also be selected on the basis of those formed by image-forming waves in the telescope pupil that display the least amount of rms OPD fluctuation. Use of such a criterion might then produce images similar to those obtained using AO; the final image obtained from either of these approaches is then based on image-forming waves with smaller average rms OPD fluctuation.

There are several drawbacks to speckle imaging and lucky imaging techniques. The relatively small isoplanatic angle of the atmosphere (Chap. 17) for speckle images (typically about 10-arcsec) restricts usable fields of view. Bright reference stars are also required within the field of view, and, due to the limited numbers of suitably bright stars, sky coverage is limited. The method is also unsuitable for use with complicated extended objects, for which imaging artifacts can arise that result in false image features. However, with relatively simple types of objects such as close binaries and other multiple star forms, remarkably good results can be obtained.

Bates and Cady (1980) showed that the result of shift-and-add image stacking of multiple short-exposure images of single stars using the brightest speckle as the reference feature produces star images closely resembling Airy patterns. This is the expected outcome for telescopes with circular apertures, even for telescopes that are far from diffraction-limited (cf., Sect. 11.4.10.1). In 2007, astronomers at Caltech and the University of Cambridge used a hybrid lucky imaging and an AO system to achieve the first diffraction-limited resolutions on the 200 in. Palomar instrument at visible wavelengths (Fienberg, 2008), achieving about 0.025 arcsec resolution in some binary star images.

By obtaining the individual images at wavelengths compatible with core and halo image structure while again using a bright star as the reference feature, the problems of field of view and sky coverage significantly ease; the isoplanatic angles associated with image cores are anticipated to be significantly larger than those of the individual

[10] Strehl intensity usually refers to the intensity attained in the center of the star image. The brightest speckle can arise anywhere within the extent of the seeing disc area so that this speckle does not necessarily lie close to the image center. Also, the brightest speckle generally appears and disappears in the image in a random discontinuous manner, making it difficult to rationalize the precise properties of the final shifted and stacked image.

speckles (cf., Sect. 17.4). There are also no restrictions to the type of object that can be imaged.

A disadvantage of image cores is that they can require longer imaging wavelengths; as wavelength increases, resolution reduces in inverse proportion. A second disadvantage of using image cores as reference features is that resolution is determined by the angular size of the core and, because this is determined by the telescope's optics, to achieve maximum benefit from the technique requires a diffraction-limited telescope. On the other hand, it could be argued that to achieve the best results with any speckle imaging technique requires the use of a diffraction-limited telescope anyway.

### *10.11.2 Speckle Interferometry*

The speckle imaging technique known as stellar speckle interferometry was devised by the French astronomer, Labeyrie (1970). This technique again enables diffraction-limited resolution from large telescopes, even if the instruments are significantly aberrated. Fourier analysis methods are used in this technique to analyze the high spatial frequency structure in the images; images can be reconstructed interferometrically from the Fourier plane information. The technique works best for simple types of objects such as binary stars. But because the technique does not retain phase information, full object reconstruction is not generally possible; image artifacts are likely to be found in the reconstructed images for more general types of extended object.

## 10.12 Mathematical Notation Used in This Chapter

The mathematical notation used in this chapter is indicated in Table 10.1.

**Table 10.1** Mathematical notation used in this chapter along with the SI dimensional units of the individual quantities

| Symbol | Quantity | Dimensions |
|---|---|---|
| $\lambda$ | Wavelength | m |
| $(x, y)$ | Cartesian coordinate system in telescope pupil | m |
| $r$ | Radial coordinate in telescope pupil, $\left(=\sqrt{x^2+y^2}\right)$ | m |
| $(\xi, \eta)$ | Cartesian coordinate system in plane perpendicular to the direction of light travel | m |
| $\varepsilon$ | Radial coordinate in plane perpendicular to the direction of light travel, $\left(=\sqrt{\xi^2+\eta^2}\right)$ | m |
| $(u, v)$ | Cartesian coordinate system in telescope image plane | m |
| $q$ | Radial coordinate in telescope image plane, $\left(=\sqrt{u^2+v^2}\right)$ | m |
| $(\alpha, \beta)$ | Angular coordinate system in telescope image plane | "1" |
| $\vartheta$ | Radial angular coordinate in telescope image plane $\left(=\sqrt{\alpha^2+\beta^2}\right)$ | "1" |
| $U$ | Image complex amplitude | "1" |
| $I$ | Image intensity | "1" |
| $I_C$ | Core intensity | "1" |
| $I_H$ | Halo intensity | "1" |
| $E_C$ | Core light energy fraction | "1" |
| $E_H$ | Halo light energy fraction | "1" |
| $M$ | Atmospheric MTF | "1" |
| $M_T$ | Telescope OTF | "1" |
| $H$ | Integrated OPD fluctuation over entire atmospheric path | m |
| $\sigma$ | rms integrated OPD fluctuation, H | m |
| $\sigma_T$ | rms wavefront error of telescope | m |
| $\rho$ | Autocorrelation function of H | "1" |
| $w_0$ | $1/e$ half-width of Gaussian $\rho$ function | m |
| $D$ | Telescope diameter | m |
| $d$ | Central obstruction diameter | m |
| $W$ | Telescope wavefront error | m |
| $a$ and $b$ | Side lengths of rectangular aperture | m |
| $f$ | Telescope focal length | m |
| $(\alpha_c, \beta_c)$ | rms angular core excursion vector | "1" |
| $SI$ | Telescope Strehl intensity | "1" |
| $PSF_I$ | Telescope intensity point-spread function | "1" |
| $\lambda_{opt}$ | Optimum wavelength | m |

(continued)

**Table 10.1** (continued)

| Symbol | Quantity | Dimensions |
|---|---|---|
| $R_{mw}$ | rms core excursion relative to core size | "1" |

Dimensionless quantities are indicated by "1"

# References


Amico, P., Arsenault, R., Bonaccini-Calia, D., Buzzoni, B., Comin, M., & Conzelmann, R. (n.d.). In E. La Penna (Preparation & Editing). The VTL adaptive optics facility booklet can be found at http://www.eso.org/sci/facilities/develop/ao/images/AOF_Booklet.pdf

Andersen, T., Larsen, O. B., Owner-Petersen, M., & Steenberg, K. (1992). Active optics on the Nordic optical telescope. In M. H. Ulrich (Ed.), *ESO conference and workshop proceedings. Progress in telescope and instrumentation technologies* (pp. 311–314).

Baba, N., Isobe, S., Norimoto, Y., & Noguchi, M. (1985). Stellar speckle image reconstruction by the shift-and-add method. *Applied Optics, 24*(10).

Baldwin, J., Gregory, B., Perez, G., & Elias, E. (1992, September 1). 4-m telescope optics upgrade project. CTIO, NOAO Newsletter No. 31.

Baldwin, J. E., Tubbs, R., Cox, G., Mackay, C., Wilson, R., & Andersen, M. (2001). Diffraction-limited 800 nm imaging with the 2.56 m Nordic optical telescope. *Astronomy and Astrophysics V*, 368(1), 1–4.

Bates, R., & Cady, F. (1980). Towards true imaging by wideband speckle interferometry. *Optics Communication, 32*, 365–369.

Beckers, J. M. (1993). Adaptive optics for astronomy, performance and applications. *Annual Review of Astronomy and Astrophysics V, 31*(1).

Bifano, T., Cornelissen, S., & Bierden, P. (2010). MEMS deformable mirrors in astronomical adaptive. In *1st AO4ELT Conference, 06003*. https://doi.org/10.1051/ao4elt/201006003. Article published by EDP sciences and available at http://ao4elt.edpsciences.org

Brousseau, D., Borra, E., & Thibault, S. (2007). Wave front correction with a 37-actuator ferrofluid deformable mirror. *Optics Express, 15*(26).

Fienberg, R. (2008). Sharpening the 200-inch (5100 mm). *Sky and Telescope Magazine, 07*(1).

Forbes, F. F. (1991). *Private communication*. NOAO.

Fried, D. L. (1966). Optical resolution through a randomly inhomogeneous medium for very long and very short exposures. *Journal of the Optical Society of America, 56*, 1372–1379.

Fried, D. (1978). Probability of getting a lucky short-exposure image through turbulence. *Journal of the Optical Society of America, 68*(12), 1651–1658.

Fried, D. (1995). Focus anisoplanatism in the limit of infinitely many artificial-guide-star reference spots. *Journal of the Optical Society of America a: Optics, Image Science, and Vision, 12*, 939–949.

Gavel, D. (2009). *SPIE MEMS Adaptive Optics III, 7209* (pp. 72090E-72095E). CA.

Goodman, J. W. (1990). *Statistical optics*. McMillan.

Griffin, R. F. (1973). On image structure, and the value and challenge of very large telescopes. *Observatory, 93*, 3–8.

Griffin, R. F. (1989). *Private letter*, 27 November.

Griffin, R. F. (1990). Giant telescopes, tiny images. *Sky & Telescope*, 469.

Griffin, R. F. (2011, January). Private email communication. R. F. Griffin: member of the 8-person UKIRT Steering Committee.

Hardy, J. (1998). *Adaptive optics for astronomical telescopes (Oxford series in optical and imaging sciences)*. Oxford University Press.

Hawarden, T. (1998, August 20). *The United Kingdom Infrared Telescope*. http://www.jach.hawaii.edu/UKIRT/public/tele-descrip.html

Labeyrie, A. (1970). Attainment of diffraction limited resolution in large telescopes by Fourier analyzing speckle patterns in star images. *Astronomy and Astrophysics, 6*, 85–87.

Magnus, W., Oberheffinger, F., & Soni, R. P. (1966). *Formulas and theorems for the special functions of mathematical physics*. Springer.

Marchetti, E., & Hubin, N. (2007). First ever multi-conjugate adaptive optics at the VLT achieves first light. http://www.eso.org/public/news/eso0719/

Martin, B., Hill, J. M., & Angel, R. (1991, March). The new ground-based optical telescopes. *Physics Today*, 22–30.

McCarthy, D. W., Jr., McLeod, B., & Barlow, D. J. (1990). Infrared array camera for interferometry with the cophased multiple mirror telescope. In *SPIE, V. 1236, Symposium on astronomical telescopes and instrumentation for the 21st century*, 11–17 February. Tucson, AZ.

McKechnie, T. S. (1976). Cores in star images. *Journal of the Optical Society of America, 66*(6), 635.

McKechnie, T. S. (1989, September 26). Seminar: Obtaining diffraction limited images at near infrared wavelengths using large ground based telescopes. (Attendees: D. W. McCarthy, Jr., K. Hege, Steward Observatory, Tucson, G. Loos, B. Venet, AFRL, R. Haddock, Lentec Corp.) Albuquerque, New Mexico, USA.

McKechnie, T. S. (1990). Diffraction limited imaging using large ground-based telescopes. In *Proceedings of SPIE, V. 1236, Symposium on astronomical telescopes and instrumentation for the 21st century* (pp. 164–178), 11–17 February.

McKechnie, T. S. (1991). Light propagation through the atmosphere and the properties of images formed by large ground-based telescopes. *Journal of the Optical Society of America A, 8*, 346–365.

McKechnie, T. S. (1992). Atmospheric turbulence and the resolution limits of large ground-based telescopes. *Journal of the Optical Society of America A, 9*, 1937–1954.

McKechnie, T. S. (1993). Atmospheric turbulence and the resolution limits of large ground-based telescopes: Reply to comment. *Journal of the Optical Society of America A, 10*(11), 2415–2417.

McKechnie, T. S. (1994, August). Another route to sharp images. *Sky & Telescope Magazine, 88*(2), 36–38.

Pederson, I. (1990, November). A window on turbulence. *Science News*, 335.

Platt, B., & Shack, R. (2001). History and principles of Shack-Hartmann wave front sensing. *Journal of Refractive Surgery, 17*, S573–S577.

Roddier, F. (1999). *Imaging through the atmosphere*. Cambridge University Press.

Roddier, F. (Ed.). (2004). *Adaptive optics in astronomy*. Cambridge University Press.

Smithson, R. C., Peri, M. L., & Benson, R. S. (1988). Quantitative simulation of image correction for astronomy with a segmented active mirror. *Applied Optics, 27*, 1615–1620.

Texereau, J. (1984). *How to make a telescope*. Willmann-Bell Inc.

Tyson, R. (2010). *Principles of adaptive optics* (3rd Ed.). Taylor & Francis.

Wizinowich, P. L., Le Mignant, D., Bouchez, A., Campbell, R., Chin, J., & Contos, A., et al. (2006). The W. M. Keck observatory laser guide star adaptive optics system: Overview. *Publications of the Astronomical Society of the Pacific, 118*(840), 297–309.

Wolf, E. (1980). *Progress in optics* (Vol. 19). North-Holland.

# Chapter 11
# Statistical Properties of Stellar Speckle Patterns

**Abstract** This chapter examines the statistical properties of speckle patterns—stellar speckle patterns—formed in short-exposure star images. Expressions are developed for the probability density functions of the amplitude, phase, complex amplitude, and intensity in these patterns. Expressions are also developed for the average size and shape of individual speckles; these depend on the imaging wavelength, the atmospheric seeing parameters, and the telescope pupil function. Modified expressions are provided when cores are present in the images. While cores are substantially fixed features in the images, the speckle continuously evolves over time. Average speckle size is determined by the intensity point spread function of a diffraction-limited version of the telescope; speckle size is not affected by aberrations. In polychromatic speckle, individual speckles are chromatically stretched into rainbow-like features by an amount proportional to the feature's radial distance from image center. Stellar speckle is a form of image noise; various speckle reduction strategies are analyzed.

Short-exposure images of stars obtained at visible wavelengths by large ground based telescopes generally arise in the form of speckle patterns, referred to as stellar speckle patterns. In this chapter, expressions are developed for the key statistical properties of the complex amplitude and intensity in this type of pattern. The statistical properties of laser speckle and related phenomena have been described in some detail by Dainty (1984) and more recently by Goodman (2006). In this chapter, we draw extensively from the former reference. It is noted, however, that many of the results given in that reference are in fact due to Goodman who was a contributing author.

Light waves arriving at the top of the atmosphere from distant unresolved stars can be regarded as nominally plane. However, as they propagate down through the turbulent atmosphere, they become increasingly disrupted. By the time they arrive at the entrance pupil of the observing telescope, they generally exhibit significant phase and amplitude fluctuations, the latter manifesting itself as intensity scintillation. The statistical properties of these light waves, in combination with the influence of the telescope optics, determine the statistical properties of stellar speckle patterns.

 
T. S. McKechnie, *General Theory of Light Propagation and Imaging Through the Atmosphere*, Progress in Optical Science and Photonics 20,
https://doi.org/10.1007/978-3-030-98828-9_11

Whereas it is prohibitively difficult to describe the statistical properties of the phase-perturbed and scintillated image-forming waves arriving at the telescope, as we conveniently saw in Chap. 6, the essential statistical properties of these waves are given by the statistical properties of the—considerably more manageable—integrated optical path difference (OPD) fluctuation over the atmospheric path, $H(x,y)$, the principal properties of which are the variance, $\sigma^2$, and autocorrelation function, $\rho(\xi, \eta)$.

Initially, we limit consideration to perfectly polarized light and assume that neither the atmosphere nor the telescope optics have any capacity to alter this state of polarization. Later, in Sect. 11.8, the analysis is expanded to include light of arbitrary state of polarization as well as the effect of the telescope optics which can cause either partial or complete depolarization.

Whereas the depolarizing effects caused by many commonly used types of telescope, e.g., Cassegrain and Ritchey–Chretien designs, are often negligible, other equally common telescope types, e.g., Newtonian telescopes and telescopes with Coude paths, can cause significant amounts of depolarization. The 45-degree mirror components in the latter telescope types treat light of different polarization states in different ways. Such behavior cannot be taken into account by scalar diffraction theory which does not take account of polarization; instead, it relies on small-angle approximations where behavior has no polarization dependence. Consequently, later in the chapter, when we examine the statistical properties of speckle formed by depolarizing telescopes, we are obliged, albeit temporarily, to abandon scalar diffraction theory in favor of vector diffraction theory.

## 11.1 Probability Density Function of the OPD Fluctuation

The integrated OPD fluctuation, $H(x,y)$, associated with the image-forming waves in the telescope pupil is assumed, as previously in Sect. 6.3, to be Gaussian distributed with zero mean. By also assuming that the statistics of $H(x,y)$ are stationary, we may omit the $(x,y)$ arguments and simply denote the OPD fluctuation by $H$. The probability density function (PDF) of $H$ may be expressed by Cramer (1954)

$$PDF(H) = \frac{1}{\sigma \cdot \sqrt{2 \cdot \pi}} \cdot \exp\left[-0.5 \cdot \left(\frac{H}{\sigma}\right)^2\right], \tag{11.1}$$

where $\sigma$ is the rms fluctuation of $H$. The normalization of this PDF is, as required, such that

$$\int_{-\infty}^{\infty} PDF(H) \cdot dH = 1. \tag{11.2}$$

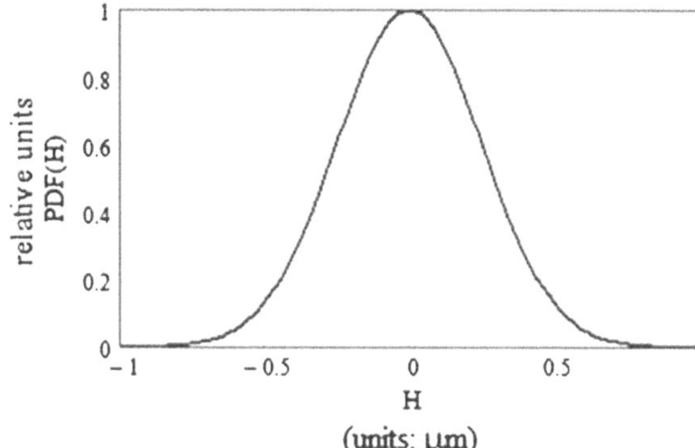


**Fig. 11.1** Typical Gaussian probability density function of the OPD fluctuation, $H$, associated with image-forming light waves in the pupil of a 4-m telescope. The rms fluctuation of the distribution shown: $\sigma = 0.35$ μm

Figure 11.1 shows a typical plot of the PDF of the integrated OPD fluctuation for the wavefront portions collected by a 4-m class telescope in 1-arcsec seeing after removal of tip–tilt contributions. The 0.35-μm σ value used to generate the plot is consistent with the near-IR cores shown in Figs. 10.10 and 10.11. In different seeing conditions, or when telescopes of other sizes are considered, smaller or larger σ values could arise just as readily.

## 11.2 Probability Density Function of the Phase

The phase angle, $\varphi(x, y)$, can be given in terms of the integrated OPD fluctuation, $H(x, y)$, by.

$$\varphi(x, y) = \frac{2 \cdot \pi \cdot H(x, y)}{\lambda}. \tag{11.3}$$

Because $H(x, y)$ is Gaussian distributed, so too is the phase and, because the statistics of H(x,y) are assumed stationary, so too are the statistics of $\varphi(x, y)$. Thus, again omitting the (x, y) arguments, we may write the PDF of $\varphi$ in the form

$$PDF(\varphi) = \frac{1}{\sigma \cdot \sqrt{2 \cdot \pi}} \cdot \frac{\lambda}{2 \cdot \pi} \cdot \exp\left[-\left(\frac{\varphi}{\frac{2 \cdot \pi \cdot \sigma}{\lambda}}\right)^2\right], \tag{11.4}$$

where again we observe that $\int_{-\infty}^{\infty} PDF(\varphi) \cdot d\varphi = 1$. The variance of the phase, $\langle \varphi^2 \rangle$, is given by

$$\langle \varphi^2 \rangle = \left(\frac{2 \cdot \pi}{\lambda}\right)^2 \cdot \sigma^2. \tag{11.5}$$

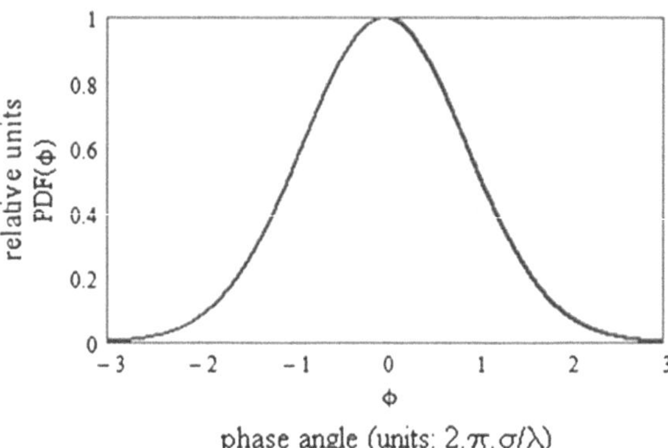


**Fig. 11.2** Generic Gaussian probability density function of the phase angle $\varphi$

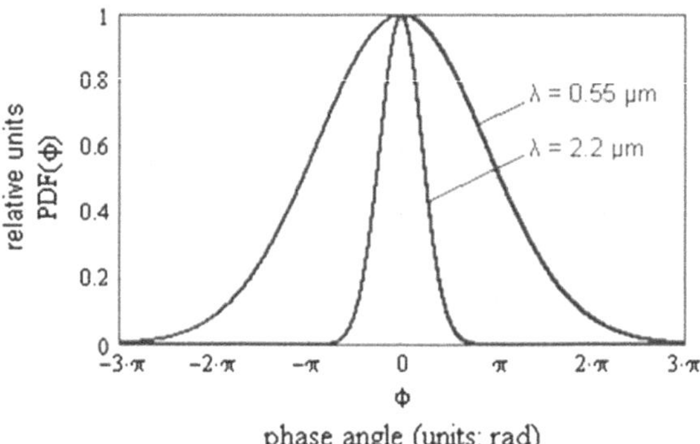


**Fig. 11.3** Plots of $PDF(\varphi)$ corresponding to the $PDF(H)$ plot shown in Fig. 11.1 for $\sigma = 0.35$ μm. Plots are shown for the visible and near-IR wavelengths indicated

Figure 11.2 shows a plot of $PDF(\varphi)$ generated from 11.4 where the generic phase angle units, $2 \cdot \pi \cdot \sigma/\lambda$, are used long the horizontal axis. For any given values of $\sigma$ and $\lambda$, the corresponding $PDF(\varphi)$ function can be generated from the PDF plot in this figure by simply using these values to appropriately scale the units along the horizontal axis. Figure 11.3 shows specific $PDF(\varphi)$ plots calculated using $\sigma = 0.35$ μm for both visible and near-IR wavelengths. The plot for the visible wavelength, 0.55 μm, is wider than the plot for the (longer) near-IR wavelength, 2.2 μm; the width differences reflect the inverse dependence of phase angle on wavelength.

### *11.2.1 PDF of Phase in the Primary Phase Range*

By adding or subtracting appropriate multiples of $2\pi$ to the phase angles, the resulting phase angles may be confined to the primary phase range,

$$-\pi \leq \varphi \leq \pi. \tag{11.6}$$

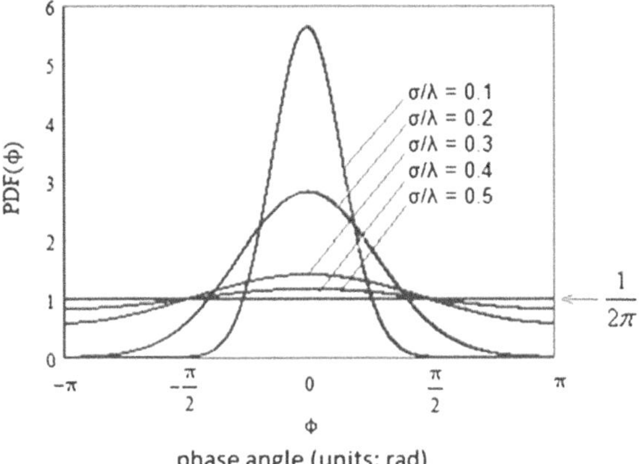


**Fig. 11.4** Probability density function of the phase angles $\varphi$ adjusted to lie in the primary range ($-\pi$ to $\pi$). Plots are shown for various $\sigma/\lambda$ ratios: 0.1, 0.2, 0.3, 0.4, 0.5, and 1.0

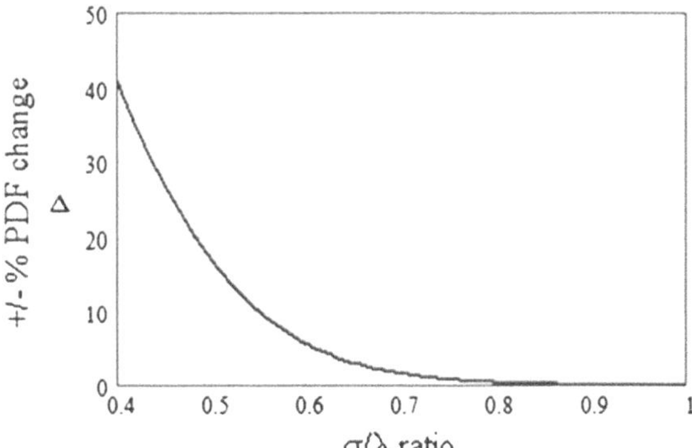


**Fig. 11.5** Maximum ($\pm$) departure of the function $PDF(\varphi)$ from the uniform phase distribution $1/2\pi$. For $\frac{\sigma}{\lambda} > 0.6$, the departure becomes negligibly small

Over this phase range, $PDF(\varphi)$ is no longer necessarily Gaussian distributed. As illustrated in Fig. 11.4, this function can now take a variety of forms depending on the turbulence strength and the wavelength, as controlled by the ratio $\sigma/\lambda$. Although the characteristics of the PDF plots in Figs. 11.3 and 11.4 now look entirely different, both PDF representations lead to the same mathematical descriptions of the statistical properties of the complex amplitudes and intensities in the images.

The maximum percentage departure of the $PDF(\varphi)$ plots shown in Fig. 11.4 from the uniform phase distribution value, $1/(2\pi)$, has been plotted in Fig. 11.5 against the quantity $\sigma/\lambda$. For $/\lambda \geq 0.5$, the departure is seen to be less than $\pm$ 17%; and for $\sigma/\lambda \geq 1$, the departure is less than $\pm$ 0.25%. For larger $\sigma/\lambda$ values, function $PDF(\varphi)$ asymptotically approaches the uniform distribution value $1/(2\pi)$. At the other extreme, that is, for $\sigma/\lambda \leq 0.2$, $PDF(\varphi)$ again becomes approximately Gaussian distributed, though the rms fluctuation of $\varphi(x, y)$ is now $\ll \pi$.

## 11.3 Star Image Characteristics Dependence on the Phase PDF

The various $PDF(\varphi)$ plots shown in Fig. 11.4 produce different kinds of image characteristics. Three principal types arise depending on the ratio, $\sigma/\lambda$. It might be noted here that we use slightly different transition values of $\sigma/\lambda$ than were used previously in Sect. 10.1.5.1. The reason for this is that even a core containing only 20% of the total light—as occurs when $\sigma/\lambda \approx 0.2$—is still the dominant feature in the center of a star image formed by a large telescope.

1. $\sigma/\lambda < 0.2$: **Dominant image cores**

In this regime, $PDF(\varphi)$ is found to be tightly clustered about the mean $\varphi$ value, zero. This type of PDF gives rise to a residual unscattered (specular) field component; telescopes image this component into a central core. Though some light energy still remains in the halo, the core dominates the center of the image, as discussed previously in Chap. 10. The statistical properties of the complex amplitude in this type of image may be approximately modeled by considering the core as a constant phasor of known amplitude which adds (coherently) to a Gaussian speckle pattern. Such modeling is carried out later, in Sect. 11.7.

2. $0.2 \leq \sigma/\lambda \leq 0.4$: **Core and halo images**

In this transition regime, the core contains proportionately less energy, the energy balance having transferred into the—now more prominent—halo. The statistical properties of the complex amplitude in this case are also modeled in Sect. 11.7, where again the core is considered as a constant phasor coherently added to a Gaussian speckle pattern.

3. $\sigma/\lambda \geq 0.4$: **Speckle images**

In this regime, the core all but vanishes. For $\sigma/\lambda = 0.4$, the energy fraction remaining in the core (11.3) is less than 0.002. For$/\lambda = 0.6$, the fraction collapses to $10^{-6}$ and, for most practical purposes, all of the light energy in the image can then be considered residing in the halo. In this limit, short-exposure images appear as random speckle patterns obeying circularly Gaussian statistics.

## 11.4 Circular Gaussian Speckle in Star Images

When fully coherent light is forward-scattered after transmission through a random diffuser screen,[1] a random diffraction pattern forms in the far field, referred to as a speckle pattern (Dainty, 1984). When the rms OPD fluctuation, $\sigma$, introduced by the scattering medium exceeds about 0.4 $\lambda$, so-called circular Gaussian speckle patterns form. Figure 11.6 shows the typical appearance of this type of speckle pattern. Such

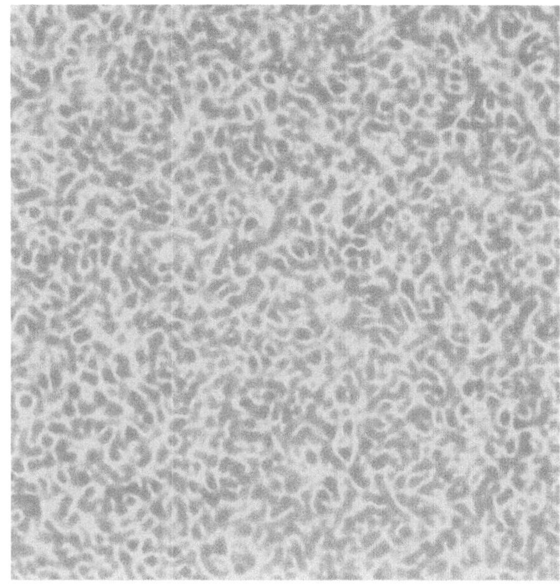

**Fig. 11.6** Coherent light speckle pattern obeying circular Gaussian statistics. This particular pattern was formed in the far field with respect to a uniformly illuminated random diffuser screen with a circular aperture. The size and shape of the "average" speckle in this case are determined by the size and shape of the Airy pattern formed in the far field by diffraction from the circular aperture

patterns frequently arise in star images formed by large telescopes, the "random diffuser screen" in this case being the atmosphere.[1]

According to Dainty, Gaussian speckle does not always necessarily provide a precise statistical description of speckle in star images. The predictive accuracy of the Gaussian model depends on how the properties of actual light waves (after disruption by the atmosphere) measure up against the assumed properties. Predictive accuracy also depends on the telescope aperture size, the larger the aperture the better the accuracy. Looking ahead to the next generation of extremely large telescopes (ELTs), with aperture diameters ranging from 25 to 39 m, the speckle properties laid out in this chapter may turn out to be more closely obeyed for these large instruments than they are in the images formed by smaller present-day instruments.

## *11.4.1 Stellar Speckle with Circular Gaussian Statistics*

We first suppose perfectly polarized plane light waves arriving at the top of the atmosphere from a distant unresolved star. Such waves may be considered perfectly spatially coherent. If we were to pass these waves through a narrowband filter, the emerging waves could then also be regarded as quasi-monochromatic and thus behave as though they were temporally coherent. (The entire spectral output of any star can be expressed in terms of a large number of quasi-monochromatic emissions, one for each wavelength represented in the output spectrum). Thus, quasi-monochromatic

[1] Speckle patterns arise equally when coherent light backscatters from a rough surface. However, scattering in transmission more closely resembles the atmospheric mechanism that causes stellar speckle patterns.

light waves arising from distant stars may be regarded as fully coherent, both spatially and temporally; the coherence properties of these waves closely approximate those of fully coherent light of the sort produced by a monochromatic laser.

Once these perfectly plane, perfectly coherent quasi-monochromatic light waves propagate down through the turbulent atmosphere and arrive in the pupil plane of the observing telescope, they generally exhibit significant phase and amplitude disruptions. We now suppose that the light waves arriving in the pupil plane have the following properties:

1. The magnitude of the phase disruption is large compared to the light wavelength, and the average lateral size of the disruption is assumed small compared to the size of the telescope aperture. With these assumptions, the image-forming wave portions collected by the telescope may be considered comprised of a very large number of elementary scattering areas—phasors—whose phases are all uncorrelated.
2. The individual phasors can be considered to act like sources of Huygens' secondary wavelets, the initial phases of which we assume to be uniformly distributed in the primary interval, $-\pi$ to $\pi$.

In typical astronomical seeing conditions, the width of the atmospheric modulation transfer function (MTF) at visible wavelengths is usually of the order of ten centimeters. Thus, the first of the above properties is generally realized by large telescopes with diameters larger than about one meter. The second property is approximately realized in cases where the integrated OPD fluctuation over the atmospheric path fulfills the condition, $\sigma/\lambda \geq 0.4$. As previously discussed in Sect. 11.2.1, this condition is generally fulfilled at visible wavelengths in average (~1-arcsec) seeing conditions.

When both properties are fulfilled simultaneously, short-exposure star images formed in the image plane of the telescope appear as random speckle patterns. If the light is also considered to be quasi-monochromatic, these speckle patterns appear as coherent speckle patterns obeying circular Gaussian statistics (Dainty, 1984) similar to the laser speckle pattern shown in Fig. 11.6. We now examine the properties of Gaussian speckle patterns.

### *11.4.2 First-Order Statistics of the Complex Amplitude*

As in Sect. 7.3, we again denote the complex amplitude in the image plane of the telescope by $U(\alpha, \beta, \lambda)$ and further denote the real and imaginary parts of this function, respectively, by $U^{(r)}$ and $U^{(i)}$. The first-order statistical properties of $U(\alpha, \beta, \lambda)$ tell us about the statistical properties at a single point in the image. Subject to the two properties assumed in the section above and also the quasi-monochromatic light assumption, the complex amplitude at any point in the image may be considered as the coherent addition (7.6) of a large number of elementary phasors that are uncorrelated in terms of both amplitude $|a_k|$ and phase $\varphi_k$.

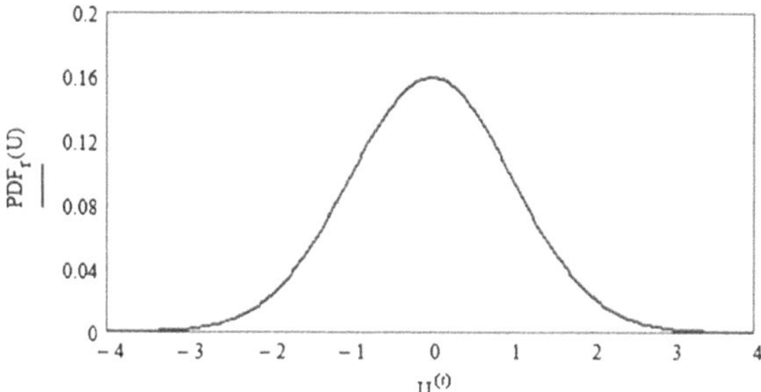


**Fig. 11.7** Probability density function for the real part of the complex amplitude, $U^{(r)}$ for Gaussian speckle. The PDF for the imaginary part, $U^{(i)}$, has the identical form

Thus, we may invoke the central limit theorem in the manner of Dainty (1984) and express the joint PDF of the real and imaginary parts of the complex amplitude in the form

$$PDF_{r,i}\left(U^{(r)}, U^{(i)}\right) = \frac{1}{2 \cdot \pi \cdot \sigma_a^2} \cdot \exp\left[-\frac{\left[U^{(r)}\right]^2 + \left[U^{(i)}\right]^2}{2 \cdot \sigma_a^2}\right], \tag{11.7}$$

where

$$U^{(r)} = \mathrm{Re}\{U\} = \frac{1}{\sqrt{N}} \cdot \sum_{k=1}^{N} |a_k| \cdot \cos(\varphi_k), \tag{11.8}$$

$$U^{(i)} = Im\{U\} = \frac{1}{\sqrt{N}} \cdot \sum_{k=1}^{N} |a_k| \cdot \sin(\varphi_k), \tag{11.9}$$

and

$$\sigma_a^2 = \frac{1}{N} \cdot \sum_{k=1}^{N} \frac{\langle |a_k| \rangle^2}{2}. \tag{11.10}$$

The joint PDF given by 11.7 may be referred to as a circular Gaussian PDF. Figure 11.7 shows the central section through the real axis of such a PDF. This section may be referred to as the marginal PDF, $PDF_r(U)$. The real and imaginary parts of the complex amplitude have zero-means, identical variances, and zero correlation. Speckle possessing these properties is commonly referred to as Gaussian speckle.

### 11.4.3 First-Order Statistics of the Intensity and Phase

Continuing to follow Dainty, we make the following random variable transformations:

$$I = \left[U^{(r)}\right]^2 + \left[U^{(i)}\right]^2 \tag{11.11}$$

and

$$\varphi = \tan^{-1}\left[\frac{U^{(i)}}{U^{(r)}}\right] \tag{11.12}$$

These transformations allow the joint PDF of the intensity and phase to be written in the form

$$\begin{aligned} PDF_{I,\varphi}(I,\varphi) &= \frac{1}{4\cdot\pi\cdot\sigma_a^2}\cdot\exp\left[-\frac{I}{2\cdot\sigma_a^2}\right] \text{ for } I \geq 0 \text{ and } -\pi \leq \varphi \leq \pi \\ &= 0 \text{ otherwise}. \end{aligned} \tag{11.13}$$

The marginal PDF of the intensity is obtained by integrating over phase,

$$\begin{aligned} PDF_I(I) &= \int_{-\pi}^{\pi} PDF_{I,\varphi}(I,\varphi)\cdot d\varphi = \frac{1}{2\cdot\sigma_a^2}\cdot\exp\left[-\frac{I}{2\cdot\sigma_a^2}\right] \quad \text{for } I \geq 0 \\ &= 0 \text{ otherwise}, \end{aligned} \tag{11.14}$$

which indicates that the intensity obeys negative exponential statistics. The marginal PDF of the phase is obtained by integrating over intensity,

$$\begin{aligned} PDF_\varphi(\varphi) &= \int_{0}^{\infty} PDF_{I,\varphi}(I,\varphi)\cdot dI = \frac{1}{2\cdot\pi} \quad \text{for } -\pi \leq \varphi \leq \pi \\ &= 0 \text{ otherwise}, \end{aligned} \tag{11.15}$$

indicating that the phases are uniformly distributed over the range, $-\pi$ to $\pi$. It might be noted that because the joint PDF is the product of the marginal PDFs, it follows that $I$ and $\varphi$ are statistically independent.

For Gaussian speckle, the mean intensity, $\langle I \rangle$, derives from 11.14 in the form

$$\langle I \rangle = 2\cdot\sigma_a^2 \tag{11.16}$$

The mean intensity, $\langle I \rangle$, may also be referred to as the first moment of intensity.

### *11.4.4 Moments of the Intensity*

The nth moment of the intensity $\langle I \rangle^n$ is given by

$$\langle I \rangle^n = n!\cdot\left(2\cdot\sigma_a^2\right)^n = n!\cdot\langle I \rangle^n \tag{11.17}$$

from which it follows that the second moment is given by

$$\langle I^2 \rangle = 2 \cdot \langle I \rangle^2 \tag{11.18}$$

This relation can be used to show that the variance of the intensity, $\sigma_I^2$, is equal to the square of the mean intensity,

$$\sigma_I^2 = \langle I^2 \rangle - \langle I \rangle^2 = \langle I \rangle^2 \tag{11.19}$$

From which we obtain

$$\sigma_I = \langle I \rangle \tag{11.20}$$

### 11.4.5 The Intensity PDF

Combining 11.14 and 11.16 allows us to write the intensity PDF in the negative exponential form (Dainty, 1984),

$$PDF_I(I) = \frac{1}{\langle I \rangle} \cdot \exp\left(-\frac{I}{\langle I \rangle}\right) \tag{11.21}$$

A PDF plot calculated from this equation is shown in Fig. 11.8. It may be observed that there is a higher probability of finding the intensity value zero in a Gaussian speckle pattern than of finding any other value. Gaussian speckle patterns are characterized by dark background fields, dotted with speckles of varying brightness, shapes, and sizes (cf., Fig. 11.6).

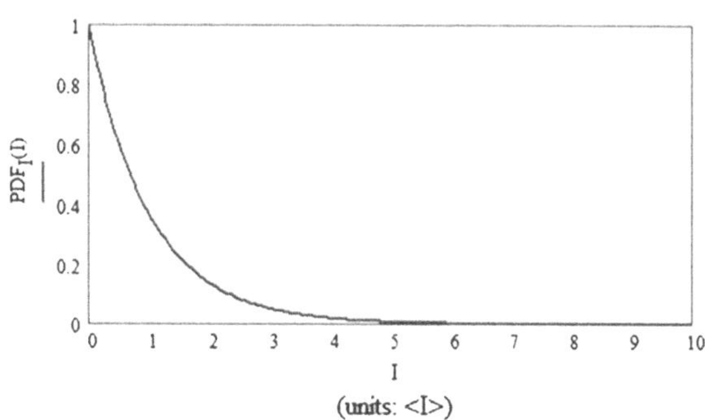


**Fig. 11.8** Negative exponential probability density function of the intensity in a Gaussian speckle pattern

## 11.4.6 Speckle Contrast

Speckle contrast refers to the modulation strength of the intensity fluctuations in speckle patterns and may be expressed in terms of the contrast ratio, $C$, a dimensionless quantity defined by

$$C = \frac{\sigma_I}{\langle I \rangle} \tag{11.22}$$

#### 11.4.6.1 Contrast for the Case of Gaussian Speckle

By combining 11.20 with 11.22, for Gaussian speckle we find that the contrast ratio takes the value unity,

$$C_{Gaus} = 1 \tag{11.23}$$

Gaussian speckle is notoriously noisy. The presence of Gaussian speckle in images can make it extremely difficult to recognize small objects. Speckle reduction techniques can be applied that have the effect of reducing the contrast ratio, $C$, ideally to zero. Speckle reduction methods are described in Sect. 11.6.

## 11.4.7 Signal-To-Noise Ratio

Signal-to-noise ratio, $S/N$, in a speckle pattern is defined as the reciprocal of contrast (Dainty, 1984):

$$\frac{S}{N} = \frac{1}{C} = \frac{\langle I \rangle}{\sigma_I} \tag{11.24}$$

#### 11.4.7.1 Signal-To-Noise Ratio for Gaussian Speckle

For Gaussian speckle (cf., 11.23), signal-to-noise ratio takes the value unity:

$$\left(\frac{S}{N}\right)_{Gaus} = \frac{1}{C_{Gaus}} = 1 \tag{11.25}$$

Application of speckle reduction procedures has the effect of increasing signal-to-noise ratio. Total speckle reduction occurs in the limit, $S/N \rightarrow \infty$.

### 11.4.8 Reduced Speckle as the Sum of Uncorrelated Gaussian Patterns

The square of the signal-to-noise ratio is an important quantity that we denote by $m$. This dimensionless quantity can be considered as the effective number of uncorrelated quasi-monochromatic Gaussian speckle patterns that constitute the final reduced speckle pattern:

$$m = \left(\frac{S}{N}\right)^2 = \frac{1}{C^2} \tag{11.26}$$

Reduced speckle occurs when a number (ideally very large) of partially correlated or uncorrelated speckle patterns add together incoherently. In the case where the individual patterns contributing to the sum all have the same average brightness and are all mutually uncorrelated, the $m$ value that results is given simply by the number of contributing patterns. If some, or even all, of the constituent patterns happen to be partially or fully correlated, $m$ is found to take a value less than the number of contributing patterns; the actual number may be found to be either fractional or integer. In general, for any reduced speckle pattern, $m$ lies somewhere in the range,

$$1 \leq m \leq \infty.$$

Large $m$ values indicate low contrast ratios ($C \ll 1$) and high $S/N$ ratios, and hence large amounts of speckle reduction. For Gaussian speckle (which may be considered as entirely unreduced), $m$ takes the value unity:

$$m_{Gaus} = 1 \tag{11.27}$$

### 11.4.9 Gaussian Speckle in Star Images Formed by Large Telescopes

Figure 11.9 shows a typical Gaussian speckle pattern image of an unresolved star formed by a 2-m diameter telescope in ~1-arcsec seeing. The same figure also shows the corresponding $\sim$ 1-arcsec wide average intensity envelope, $\langle I(\vartheta, \lambda)\rangle$, calculated from 9.11 using the parameter values, $D = 2$ m, $\lambda = 0.55$ μm, $\sigma = 0.35$ μm, and $w_0 = 0.25$ m. To generate the intensity PDF at any arbitrary chosen angular coordinate location, $\vartheta$, in the image, we must first establish the mean intensity, $\langle I(\vartheta, \lambda)\rangle$, at that location. This can be done by inserting the chosen $\vartheta$ value into previously given 9.11. The desired intensity PDF may then be obtained by inserting the said $\langle I(\vartheta, \lambda)\rangle$ value into 11.21.

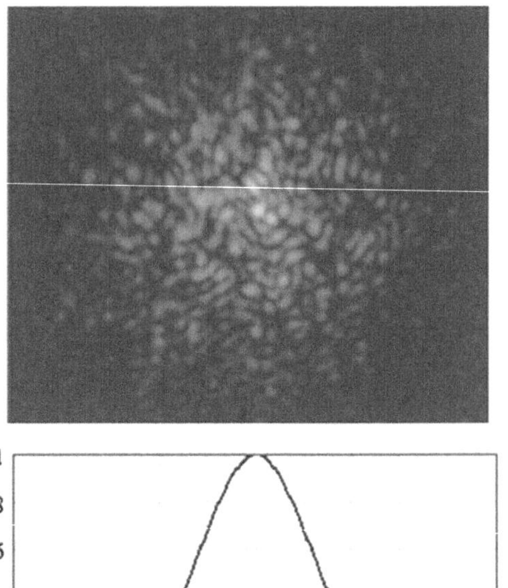

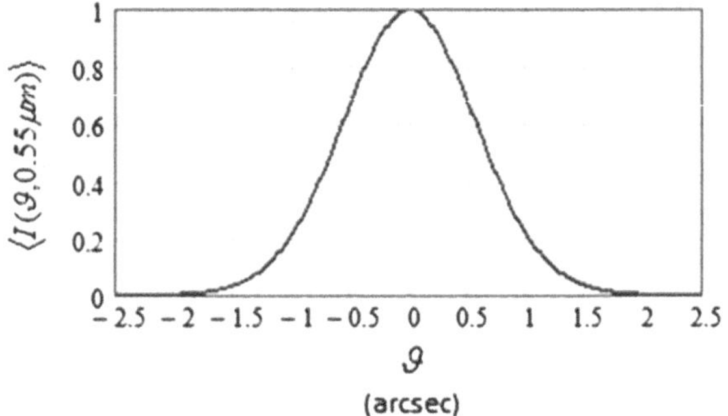


**Fig. 11.9** *Top* an unresolved star image at the visible wavelength, 0.55 μm, constituted by Gaussian speckle. *Bottom* the average intensity envelope for this speckle pattern plotted to the same angular scale

## *11.4.10 Second-Order Statistics of the Complex Amplitude*

The second-order statistical properties of speckle patterns tell us about the average size and shape of the individual speckles in these patterns. The joint PDF of the real and imaginary parts of the complex amplitude at two arbitrary locations in the image provides a complete description of the second-order statistical properties. For conciseness, we denote the complex amplitudes at any two arbitrary locations by $U_1$ and $U_2$, and denote their real and imaginary parts by $U_1^{(r)}$, $U_1^{(i)}$, $U_2^{(r)}$ and $U_2^{(i)}$. For speckle patterns obeying circular Gaussian statistics, the joint PDF may be given in the form Dainty (1984)

$$PDF\left(U_1^{(r)}, U_1^{(i)}, U_2^{(r)}, U_2^{(i)}\right) = \frac{\exp\left[-\frac{|U_1|^2+|U_2|^2-\mu\cdot U_1\cdot U_2^*-\mu^*\cdot U_1^*\cdot U_2}{2\cdot\sigma_a^2\cdot\left(1-|\mu|^2\right)}\right]}{4\cdot\pi^2\cdot\sigma_a^4\cdot\left(1-|\mu|^2\right)} \tag{11.28}$$

where the complex coherence factor, $\mu_U\left(u', v', u, v, \lambda\right)$, discussed previously in Sect. 3.8, has been abbreviated to $\mu$ in this equation. (In 11.28, the slightly different argument set, $\mu_U(u_1, v_1, u_2, u_2, \lambda)$, is used for $\mu$, but the meaning is essentially the same and should be clear.) Function $\mu$ describes the degree of correlation between the complex amplitudes at any two locations in the image, $(u', v')$ and $(u, v)$; this function is simply the autocorrelation function of the complex amplitude in the image, so that we may write

$$\mu(u', v', u, v, \lambda) = \frac{\langle U(u', v', \lambda) \cdot U^*(u, v, \lambda)\rangle}{\left[\langle |U(u', v', \lambda)|^2\rangle \cdot \langle |U(u, v, \lambda)|^2\rangle\right]^{\frac{1}{2}}} \quad (11.29)$$

The complex coherence factor, $\mu(u', v', u, v, \lambda)$ arises again later when we consider speckle reduction by aperture integration (Sect. 11.6.3). In fact, complex coherence factors arise when any speckle reduction mechanism is considered. To describe the statistical properties of reduced speckle, it is first necessary to establish the appropriate complex coherence factor in terms of the appropriate parameters that describe the speckle reduction mechanism.

#### 11.4.10.1 Size and Shape of the Average Speckle in a Stellar Speckle Pattern

The size and shape of the "average speckle" in a star image are both described by the autocorrelation function of the image intensity distribution. This function has the general form (cf., Sect. 3.6.3), $\frac{\langle [I(u', v', \lambda) - \langle I(u', v', \lambda)\rangle] \cdot [I(u, v, \lambda) - \langle I(u, v, \lambda)\rangle]\rangle}{\langle I(u', v', \lambda)\rangle \cdot \langle I(u, v, \lambda)\rangle}$. For the case of Gaussian speckle, we may use Reed's theorem (Sect. 7.8.1) to write

$$\frac{\langle [I(u', v', \lambda) - \langle I(u', v', \lambda)\rangle] \cdot [I(u, v, \lambda) - \langle I(u, v, \lambda)\rangle]\rangle}{\langle I(u', v', \lambda)\rangle \cdot \langle I(u, v, \lambda)\rangle} = \frac{|\langle U(u', v', \lambda) \cdot U^*(u, v, \lambda)\rangle|^2}{\langle |U(u', v', \lambda)|^2\rangle \cdot \langle |U(u, v, \lambda)|^2\rangle} \quad (11.30)$$

Since the right-hand side is simply the squared modulus of the unit-normalized mutual coherence function (cf., 3.26), we may use 7.44 to generate an explicit form for the autocorrelation function of the image intensity in terms of the parameters that describe the atmospheric path and the telescope optics.

When the width of the atmospheric MTF is significantly smaller than the size of the telescope aperture, the atmospheric MTF may be approximated by the two-dimensional Dirac delta function,

$$M(\xi, \eta, \lambda) = \delta(\xi, \eta) \quad (11.31)$$

and it is clear for this case that Gaussian speckle should form in star images.

Combining 11.31 and 7.44 enables the autocovariance function of the complex amplitude to be expressed in the form

$$\langle U(u', v', \lambda) \cdot U^*(u, v, \lambda)\rangle = \frac{1}{\lambda^2} \cdot \int_{-\infty}^{\infty} \int_{-\infty}^{\infty} K(x, y, \lambda) \cdot K^*(x, y, \lambda) \times \exp\left[\frac{-2 \cdot \pi \cdot i}{\lambda \cdot f} \cdot (x \cdot (u' - u) + y \cdot (v' - v)\right] \cdot dx \cdot dy. \tag{11.32}$$

This relation is analogous to the van Cittert-Zernike theorem in coherence theory (Born & Wolf, 2003; van Cittert, 1934; Zernike, 1938). The double integral is recognized as the Fourier transform of the intensity transmission function of the telescope. For a uniformly transmitting pupil (7.7), the double integral simplifies to a form functionally similar to the amplitude point-spread function of an aberration-free version of the telescope.

An explicit expression for the autocorrelation function of the complex amplitude in the speckle pattern image, $\mu(u', v', u, v, \lambda)$, may be obtained by combining 11.29 and 11.32. Thus,

$$\mu(u', v', u, v, \lambda) = \frac{\int_{-\infty}^{\infty} \int_{-\infty}^{\infty} K(x, y, \lambda) \cdot K^*(x, y, \lambda) \cdot \exp\left[\frac{-2 \cdot \pi \cdot i}{\lambda \cdot f} \cdot (x \cdot (u' - u) + y \cdot (v' - v))\right] \cdot dx \cdot dy}{\int_{-\infty}^{\infty} \int_{-\infty}^{\infty} K(x, y, \lambda) \cdot K^*(x, y, \lambda) \cdot dx \cdot dy} \tag{11.33}$$

For Gaussian speckle, where Reed's theorem applies, the autocorrelation function of the intensity in the speckle image can be obtained by combining 7.53 with the above equation to give

$$\frac{\langle I(u', v', \lambda) \cdot I(u, v, \lambda)\rangle}{\langle I(u', v', \lambda)\rangle \cdot \langle I(u, v, \lambda)\rangle} - 1 = |\mu(u', v', u, v, \lambda)|^2 = \left| \frac{\int_{-\infty}^{\infty} K(x, y, \lambda) \cdot K^*(x, y, \lambda) \cdot \exp\left[\frac{-2 \cdot \pi \cdot i}{\lambda \cdot f} \cdot (x \cdot (u' - u) + y \cdot (v' - v))\right] \cdot dx \cdot dy}{\int_{-\infty}^{\infty} \int_{-\infty}^{\infty} K(x, y, \lambda) \cdot K^*(x, y, \lambda) \cdot dx \cdot dy} \right|^2 \tag{11.34}$$

The following distance parameters are now defined in image plane space:

$$\Delta u = u' - u \tag{11.35}$$

$$\Delta v = v' - v \tag{11.36}$$

Using these relations to substitute for $u'$ and $v'$ allows us to write

$$\frac{\langle I(u+\Delta u, v+\Delta v, \lambda)\cdot I(u, v, \lambda)\rangle}{\langle I(u+\Delta u, v+\Delta v, \lambda)\rangle\cdot\langle I(u, v, \lambda)\rangle} - 1 = |\mu(u+\Delta u, v+\Delta v, u, v, \lambda)|^2 = \left|\frac{\int_{-\infty}^{\infty}\int_{-\infty}^{\infty} K(x, y, \lambda)\cdot K^*(x, y, \lambda)\cdot \exp\left[\frac{-2\cdot\pi\cdot i}{\lambda\cdot f}\cdot(x\cdot\Delta u + y\cdot\Delta v)\right]\cdot dx\cdot dy}{\int_{-\infty}^{\infty}\int_{-\infty}^{\infty} K(x, y, \lambda)\cdot K^*(x, y, \lambda)\cdot dx\cdot dy}\right|^2 \quad (11.37)$$

Assuming that the star image is reasonably well-focused, the mean intensity, $\langle I(u, v, \lambda)\rangle$, gradually diminishes as we move from the center of the image to the outer regions, as seen in the stellar speckle image in Fig. 11.9. However, for small-enough values of $\Delta u$ and $\Delta v$ (i.e., values ranging from zero to about the size of a typical speckle), the average intensity may be considered approximately constant, so that we may write

$$\langle I(u+\Delta u, v+\Delta v, \lambda)\rangle \approx \langle I(u, v, \lambda)\rangle. \quad (11.38)$$

To the extent that the above approximation holds true, the size and shape of the average speckle given by the right-hand side of 11.37 may be recognized as having the same functionality as the intensity point-spread function of an aberration-free version of the telescope (4.25). Thus, the size and shape of the average speckle are independent of, and thus unaffected by, telescope aberrations. Furthermore, irrespective of whether we deal with the center of the image or the outlying regions, the size and shape of the average speckle always remains the same; only the average brightness of the "average speckle" changes with image location, varying in proportion to $\langle I(u, v, \lambda)\rangle$.

#### 11.4.10.2 Speckle Size Dependence on Wavelength

The angular size of a diffraction-limited point-spread function (such as an Airy pattern) scales linearly with wavelength. The size of the "average speckle" therefore scales in the same way, as dictated by the $\lambda^{-1}$ factor in the Fourier kernel of 11.37.

#### 11.4.10.3 Speckle Size and Shape for Circular Aperture Telescopes

For telescopes with uniformly transmitting circular apertures, the intensity transmission function over the telescope pupil is given by

$$\begin{aligned} K(x, y, \lambda)\cdot K^*(x, y, \lambda) &= 1 \quad \text{for} \quad x^2+y^2 \le \frac{D^2}{4} \\ &= 0 \quad \text{otherwise}. \end{aligned} \quad (11.39)$$

By substituting this transmission function into 11.33, the autocorrelation function of the speckle intensity in a star image may be given by

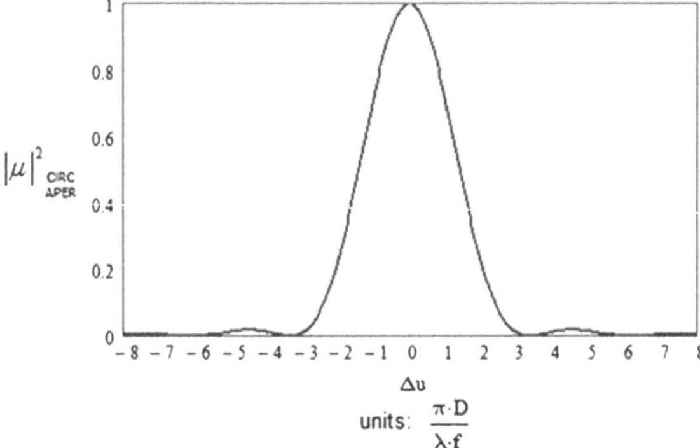


**Fig. 11.10** The central section, $\Delta v = 0$, through the autocorrelation function of the intensity in a stellar speckle pattern formed by a telescope with circular aperture for speckle obeying Gaussian statistics. This function describes the size and shape of the "average speckle" in the pattern. In this case, the function is identical to that of an Airy pattern. Telescope aberrations do not affect this outcome

$$\left[\frac{\langle I(u+\Delta u, v+\Delta v, \lambda)\cdot I(u, v, \lambda)\rangle}{\langle I(u+\Delta u, v+\Delta v, \lambda)\rangle\cdot\langle I(u, v, \lambda)\rangle}-1\right]_{\substack{Circ\\Aper}} =$$
$$|\mu(u+\Delta u, v+\Delta v, u, v, \lambda)|^2_{\substack{Circ\\Aper}} = \left[2\cdot\frac{J_1\left(\frac{\pi\cdot D\cdot\left(\Delta u^2+\Delta v^2\right)^{\frac{1}{2}}}{\lambda\cdot f}\right)}{\frac{\pi\cdot D\cdot\left(\Delta u^2+\Delta v^2\right)^{\frac{1}{2}}}{\lambda\cdot f}}\right]^2 \quad (11.40)$$

The familiar functionality seen on the right-hand side of the above equation (cf., 4.27) indicates that the size and shape of the "average speckle" formed by any telescope with a circular aperture is identical to the size and shape of the Airy pattern image formed by a diffraction-limited version of the same telescope. Figure 11.10 shows a ($\Delta v = 0$) central section through function, $|\mu(u+\Delta u, v+0, u, v, \lambda)|^2_{\substack{CIRC\\APER}}$, calculated from the above equation.

#### 11.4.10.4 Speckle Size and Telescope Aberrations

For all practical purposes, speckle size and shape are unaffected by telescope aberrations. This result may seem surprising, but it is a long-established result that has been confirmed by observations. However, in the case of extreme aberrations—beyond those found in telescopes that might be considered to be in any reasonable state of adjustment—speckle size and shape may ultimately show some aberration dependence. For example, suppose that the telescope was so grossly out of focus that the light received at some arbitrarily location in the image was now only drawn from a sub-portion of the image-forming wave in the telescope pupil. In such a case, the effective F/number of the image forming light pencil contributing light to that image location would now be larger than the F/number of the telescope itself. This would cause the average speckle at that image location to grow in size, in proportion to the F/number change.

## 11.5 Statistical Properties of Polychromatic Speckle Patterns

In this section, we examine the statistical properties of polychromatic speckle patterns. Since stars generally emit over a broad range of wavelengths, stellar speckle patterns are invariably polychromatic patterns.

Stellar emissions at any pair of wavelengths may be considered mutually incoherent. Therefore, the polychromatic speckle pattern in a star image may be considered as the incoherent sum of a number (infinite in the case of a star with a continuous spectral output) of monochromatic speckle patterns, where each constituent pattern corresponds to a specific wavelength present in the emission spectrum. In general, the individual monochromatic speckle patterns can be fully correlated, partially correlated, or entirely uncorrelated. In this section, we assume that the conditions set out in Sect. 11.4.1 are satisfied so that the individual quasi-monochromatic speckle patterns (that sum together to produce the polychromatic speckle pattern) all individually obey Gaussian statistics.

Typical detectors, such as the human eye or silicon detectors, respond to broad wavelength ranges. The wavelength-integrated intensity, $I_I(u, v)$, recorded by detectors of this type at some arbitrary location, $(u, v)$, in the image can then be expressed by

$$I_I(u, v) = \int_0^\infty I(u, v, \lambda) \cdot G(\lambda) \cdot d\lambda \tag{11.41}$$

where $G(\lambda)$ describes the spectral distribution of the light in the image. To simplify the analysis here, we consider that function $G(\lambda)$ takes account of all spectrally dependent elements in the imaging train (Sect. 8.2.2), including the spectral output of the source, the spectral sensitivity of the detector, and the spectral response of the telescope optics and optical coatings.

The statistical properties of the wavelength-integrated intensity, $I_I(u, v)$, depend on the degree of correlation between the individual Gaussian speckle patterns formed at each contributing wavelength. When all of the patterns are fully correlated, the statistical properties of the integrated intensity remain Gaussian and the contrast ratio remains at unity (Sect. 11.4.6). More generally, the constituent patterns are either partially correlated or entirely uncorrelated, in which case the contrast ratio for the integrated pattern takes values less than unity, which implies that the wavelength-integrated pattern is a reduced speckle pattern.

### 11.5.1 Autocovariance Function of the Integrated Polychromatic Intensity

The autocovariance function of the wavelength-integrated intensity derives from 11.41 as follows:

$$\langle I_I(u',v')\cdot I_I(u,v)\rangle - \langle I_I(u',v')\rangle\cdot\langle I_I(u,v)\rangle = \int_0^\infty\int_0^\infty G(\lambda_1)\cdot G(\lambda_2)\cdot\left[\langle I(u',v',\lambda_1)\cdot I(u,v,\lambda_2)\rangle - \langle I(u',v',\lambda_1)\rangle\cdot\langle I(u,v,\lambda_2)\rangle\right]\cdot d\lambda_1\cdot d\lambda_2 \tag{11.42}$$

Assuming that the individual constituent speckle patterns separately obey Gaussian statistics, Reed's theorem may be applied (7.51) to give

$$\langle I_I(u',v')\cdot I_I(u,v)\rangle - \langle I_I(u',v')\rangle\cdot\langle I_I(u,v)\rangle = \int_0^\infty\int_0^\infty G(\lambda_1)\cdot G(\lambda_2)\cdot\left|\langle U(u',v',\lambda_1)\cdot U^*(u,v,\lambda_2)\rangle\right|^2\cdot d\lambda_1\cdot d\lambda_2. \tag{11.43}$$

Equation 7.33 can be used to provide explicit expressions for both $U(u',v',\lambda_1)\cdot U^*(u,v,\lambda_2)$ and $\langle I_I(u',v')\cdot I_I(u,v)\rangle - \langle I_I(u',v')\rangle\cdot\langle I_I(u,v)\rangle$ in terms of the parameters describing the atmospheric path and the telescope optics. By substituting these expressions directly into 11.42, we could (in principle) obtain an explicit expression for the autocovariance function of the integrated intensity. Such an expression would provide measures of the size and shape of the "average speckle" in the wavelength-integrated speckle pattern; however, we do not attempt to reproduce that expression here because of its inordinate size.

For telescopes with circular apertures, the individual constituent Gaussian speckle patterns each exhibit a granular appearance of the kind seen in Fig. 11.9. Wavelength-integrated patterns, however, generally have more exotic appearances, as we now discuss.

#### 11.5.1.1 Radial Structure in Polychromatic Star Images

Radial structure in polychromatic speckle patterns has been discussed by Parry (Dainty, 1984). Radial structure is caused entirely by the telescope optics and it may be explained in broad terms by the fact that longer wavelengths diffract through larger angles than shorter wavelengths, thus causing chromatic stretching of individual speckle features in the image. The mechanism controlling the "stretching" may be seen by examining 7.34, which provides an expression for the function, $S_T(u',v',u,v,\xi,\eta,\lambda_1,\lambda_2)$, that describes the influence of the telescope optics on the two-point two-wavelength correlation function of the complex amplitudes in the image. For convenience here, we reproduce the numerator on the right-hand side of that equation:

$$\int_{-\infty}^{\infty}\int_{-\infty}^{\infty} K(x+\xi, y+\eta, \lambda_1)\cdot K^*(x, y, \lambda_2)$$
$$\times \exp\left\{\frac{-2\cdot\pi\cdot i}{f}\cdot\left[x\cdot\left(\frac{u'}{\lambda_1}-\frac{u}{\lambda_2}\right)+y\cdot\left(\frac{v'}{\lambda_1}-\frac{v}{\lambda_2}\right)\right]\right\}\cdot dx\cdot dy$$

Plainly, when the exponential term in the integrand takes its highest values (the modulus takes values in the range − 1 to 1), the entire double integral also tends to take high values, thus implying a high degree of correlation between the individual (Gaussian) speckle patterns in the image at the two arbitrary wavelengths considered, $\lambda_1$ and $\lambda_2$. The maximum value that can be attained by this term, *unity*, occurs when the term in curly brackets takes the value zero, an occurrence realized when the following two conditions are satisfied,[2]

$$\frac{u'}{\lambda_1}=\frac{u}{\lambda_2} \tag{11.44}$$

and

$$\frac{v'}{\lambda_1}=\frac{v}{\lambda_2} \tag{11.45}$$

Suppose now that we happened to notice a bright and, hence readily identifiable, speckle feature at wavelength $\lambda_1$ at a particular image location $(u', v')$. The above conditions imply that a similar (i.e., well correlated) speckle feature is likely to be observed at the other wavelength, $\lambda_2$, at a nearby image location, (u, v), given by

$$(u, v)=\left(\frac{\lambda_2}{\lambda_1}\cdot u', \frac{\lambda_2}{\lambda_1}\cdot v'\right) \tag{11.46}$$

The above relationship means that speckle features generally appear stretched in the radial direction in proportion to the quantity, $\lambda_2/\lambda_1\cdot\sqrt{u'^2+v'^2}$, the square root term indicating that the degree of stretching is proportional to the distance of the initial speckle feature from the center of the image. Such radially stretched speckle can be seen in the polychromatic speckle pattern shown in Fig. 4.9. Stretching may also be seen later in Chap. 12 where instantaneous star image realizations are shown in Figs. 12.7 and 12.8 for two closely separated visible wavelengths.

[2] It may be observed that the exponential term also takes the maximum value, *unity*, when $x\cdot((u')/\lambda_1-u/\lambda_2)+y\cdot((v')/\lambda_1-v/\lambda_2)=0$. But this only occurs for certain combinations of $x$ and $y$. In the case of interest, the exponential term takes the value unity for all $x$ and $y$.

## *11.5.2 The Spectral Correlation Function*

The unit-normalized single-point two-wavelength correlation function of the complex amplitudes in the image, which we refer to as the spectral correlation function, $\mu(u, v, u, v, \lambda_1, \lambda_2)$, was given previously by 7.40. A more explicit expression was also given by 7.57 which, for convenience, we now reproduce,

$$\mu(u, v, u, v, \lambda_1, \lambda_2) = \exp\left[-2 \cdot \pi^2 \cdot \sigma^2 \cdot \left(\frac{1}{\lambda_1} - \frac{1}{\lambda_2}\right)^2\right] \times \frac{\int_{-\infty}^{\infty}\int_{-\infty}^{\infty} K(x, y, \lambda_1) \cdot K^*(x, y, \lambda_2) \cdot \exp\left\{\frac{-2\cdot\pi\cdot i}{f} \cdot \left[(x \cdot u + y \cdot v) \cdot \left(\frac{1}{\lambda_1} - \frac{1}{\lambda_2}\right)\right]\right\} \cdot dx \cdot dy}{\left\{\int_{-\infty}^{\infty}\int_{-\infty}^{\infty} K(x, y, \lambda_1) \cdot K^*(x, y, \lambda_1) \cdot dx \cdot dy \cdot \int_{-\infty}^{\infty}\int_{-\infty}^{\infty} K(x, y, \lambda_2) \cdot K^*(x, y, \lambda_2) \cdot dx \cdot dy\right\}^{\frac{1}{2}}} \tag{11.47}$$

Evidently, the term $\exp\left[-2 \cdot \pi^2 \cdot \sigma^2 \cdot \left(\frac{1}{\lambda_1} - \frac{1}{\lambda_2}\right)^2\right]$ is entirely associated with the atmospheric path, while the term

$$\frac{\int_{-\infty}^{\infty}\int_{-\infty}^{\infty} K(x, y, \lambda_1) \cdot K^*(x, y, \lambda_2) \cdot \exp\left\{\frac{-2\cdot\pi\cdot i}{f} \cdot \left[(x \cdot u + y \cdot v) \cdot \left(\frac{1}{\lambda_1} - \frac{1}{\lambda_2}\right)\right]\right\} \cdot dx \cdot dy}{\left\{\int_{-\infty}^{\infty}\int_{-\infty}^{\infty} K(x, y, \lambda_1) \cdot K^*(x, y, \lambda_1) \cdot dx \cdot dy \cdot \int_{-\infty}^{\infty}\int_{-\infty}^{\infty} K(x, y, \lambda_2) \cdot K^*(x, y, \lambda_2) \cdot dx \cdot dy\right\}^{\frac{1}{2}}}$$

is entirely associated with the telescope optics. The latter term may be recognized as the Fourier transform of the product of the respective telescope pupil functions at the two wavelengths, $\lambda_1$ and $\lambda_2$, where the Fourier kernel can be seen to depend on these two wavelengths via the difference function, $(1/\lambda_1 - 1/\lambda_2)$.

#### 11.5.2.1 The Function for a Diffraction-Limited Telescope with Circular Aperture

For diffraction-limited telescopes with circular pupils (7.7), the (unit-normalized) spectral correlation function, $\mu(u, v, u, v, \lambda_1, \lambda_2)$, may be obtained from 11.47 in the form

$$\mu(u, v, u, v, \lambda_1, \lambda_2) = \frac{\langle U(u, v, \lambda_1) \cdot U^*(u, v, \lambda_2)\rangle}{[\langle I(u, v, \lambda_1)\rangle \cdot \langle I(u, v, \lambda_2)\rangle]^{\frac{1}{2}}} = \exp\left[-2 \cdot \pi^2 \cdot \sigma^2 \cdot \left(\frac{1}{\lambda_1} - \frac{1}{\lambda_2}\right)^2\right] \cdot \left[2 \cdot \frac{J_1\left(\frac{\pi\cdot D\cdot(u^2+v^2)^{\frac{1}{2}}}{f} \cdot \left(\frac{1}{\lambda_1} - \frac{1}{\lambda_2}\right)\right)}{\frac{\pi\cdot D\cdot(u^2+v^2)^{\frac{1}{2}}}{f} \cdot \left(\frac{1}{\lambda_1} - \frac{1}{\lambda_2}\right)}\right] \tag{11.48}$$

By invoking Reed's theorem (7.51), the spectral correlation function of the intensity may be obtained from the above expression in the form

$$\frac{\langle I(u, v, \lambda_1) \cdot I(u, v, \lambda_2)\rangle}{\langle I(u, v, \lambda_1)\rangle \cdot \langle I(u, v, \lambda_2)\rangle} - 1 = |\mu(u, v, u, v, \lambda_1, \lambda_2)|^2 =$$

$$\exp\left[-4 \cdot \pi^2 \cdot \sigma^2 \cdot \left(\frac{1}{\lambda_1} - \frac{1}{\lambda_2}\right)^2\right] \cdot \left[2 \cdot \frac{J_1\left(\frac{\pi \cdot D \cdot (u^2+v^2)^{\frac{1}{2}}}{f} \cdot \left(\frac{1}{\lambda_1} - \frac{1}{\lambda_2}\right)\right)}{\frac{\pi \cdot D \cdot (u^2+v^2)^{\frac{1}{2}}}{f} \cdot \left(\frac{1}{\lambda_1} - \frac{1}{\lambda_2}\right)}\right]^2 \tag{11.49}$$

This normalized expression takes its maximum value, unity, in the center of the image $u = v = 0$ when the two wavelengths, $\lambda_1$ and $\lambda_2$ coalesce. The Bessel function term indicates that the function oscillates as we go from the center of the image to outlying locations, the oscillations gradually dampening down as we go outwards. Because of the circular symmetry, the above equation may be more concisely expressed in terms of the radial coordinate $\vartheta$ in the image (defined previously by 7.4 and depicted in Fig. 7.2):

$$\frac{\langle I(\theta, \lambda_1) \cdot I(\theta, \lambda_2)\rangle}{\langle I(\theta, \lambda_1) \cdot I(\theta, \lambda_2)\rangle} - 1 = |\mu(\theta, \theta, \lambda_1, \lambda_2)|^2 =$$

$$\exp\left[-4 \cdot \pi^2 \cdot \sigma^2 \cdot \left(\frac{1}{\lambda_1} - \frac{1}{\lambda_2}\right)^2\right] \cdot \left[2 \cdot \frac{J_1\left(\pi \cdot D \cdot \theta \cdot \left(\frac{1}{\lambda_1} - \frac{1}{\lambda_2}\right)\right)}{\pi \cdot D \cdot \theta \cdot \left(\frac{1}{\lambda_1} - \frac{1}{\lambda_2}\right)}\right]^2 \tag{11.50}$$

Figure 11.11 shows plots of the spectral correlation function calculated from the above equation for a 4-m telescope. The two plot sets shown in this figure correspond to 0.5- and 1-arcsec visible seeing. Table 11.1 lists the wavelength pairs used to calculate the plot data. Each of the five pairs indicated in the table symmetrically straddles the same center wavelength, 0.55 $\mu m$.

For polychromatic speckle patterns, it can generally be assumed that the various constituent monochromatic speckle patterns are not fully correlated. Thus, the contrast ratio for the wavelength-integrated speckle pattern is generally less than unity (unity being the value associated with the individual constituent quasi-monochromatic Gaussian speckle patterns). Mechanisms that reduce speckle contrast in this way are referred to as speckle reduction mechanisms.

## 11.6 Speckle Reduction Applied to Stellar Speckle Patterns

Wavelength integration is just one of a number of mechanisms that can be used to effect speckle reduction. Though the light-sensitive areas of photocell detectors (e.g., the individual pixels in a focal plane array (FPA) detector) may be relatively small, they are not infinitesimally small. Thus, no physically realizable detector can be considered to be an ideal point detector. The nonzero detector dimensions cause

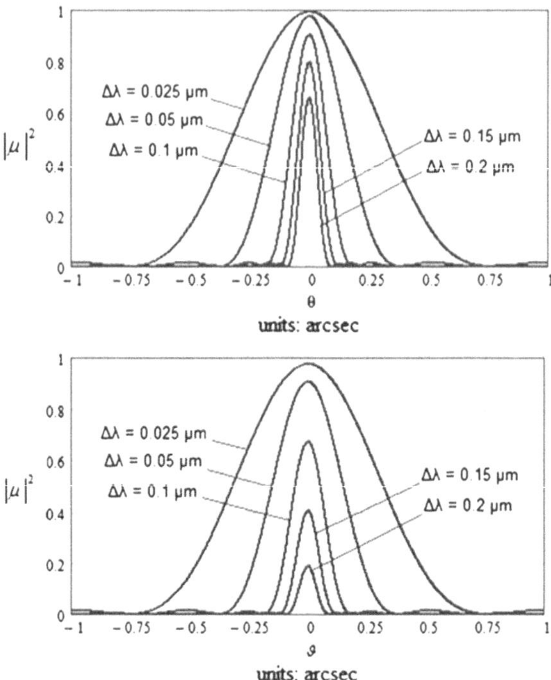


**Fig. 11.11** The behavior of the spectral correlation function of the intensity, $|\mu|^2$, as a function of image location, $\vartheta$, for the various visible wavelength pairs, $\lambda_1$ and $\lambda_2$, indicated in Table 11.1. The plots correspond to a 4-m diffraction-limited telescope observing in (*Bottom*) 1-arcsec seeing as set by $\sigma = 0.3\ \mu m$ and $w_0 = 0.25$m and (Top) 0.5-arcsec seeing as set by $\sigma = 0.15\ \mu m$ and $w_0 = 0.25$ m

**Table 11.1** The wavelength pairs, $\lambda_1$, and $\lambda_2$, used to generate the intensity correlation plots in Fig. 11.11

| $\lambda_1 (\mu m)$ | $\lambda_2 (\mu m)$ | $\Delta\lambda (\mu m)$ |
|---|---|---|
| 0.5375 | 0.5625 | 0.025 |
| 0.5250 | 0.5750 | 0.050 |
| 0.5000 | 0.6000 | 0.100 |
| 0.4750 | 0.6250 | 0.150 |
| 0.4500 | 0.6500 | 0.200 |

The various pairs all correspond to the same recipe, $0.55 \mu m \pm \Delta\lambda/2$, where the $\Delta\lambda$ are as indicated

speckle reduction by "aperture averaging."[3] Time averaging also causes speckle reduction, as witnessed by the absence of speckle in long-exposure star images.

While the speckle reduction mechanisms just mentioned might at first appear to operate in entirely different ways, in fact they all share the same fundamental operating principle: in every case, the reduced speckle pattern arises as the incoherent sum of a number (perhaps infinite) of partially correlated or entirely uncorrelated speckle patterns. Different methods of speckle reduction are distinguished simply by the different mechanisms employed to effect the decorrelation between the individual Gaussian speckle patterns that add together to produce the final reduced pattern.

[3] Aperture-averaged speckle can also be referred to as aperture-integrated speckle (Dainty 1984).

An account of speckle reduction as it relates to speckle formed by scattering from rough surfaces has been given elsewhere by the Author (Dainty, 1984). In this section, we develop expressions for the statistical properties of the intensity in reduced stellar speckle patterns, the scattering in this case caused by the turbulent atmosphere rather than by rough surfaces. For these reduced stellar speckle patterns, expressions are developed for the intensity PDF, the intensity variance, the contrast ratio, the signal-to-noise ratio, and the effective number of uncorrelated contributing Gaussian speckle patterns, $m$, that sum together to effect the reduction.

## 11.6.1 Speckle Reduction by Wavelength Integration

The variance of the intensity in a wavelength-integrated speckle pattern may be obtained from 11.41 as follows,

$$\left\langle I_I(u,v)^2\right\rangle - \langle I_I(u,v)\rangle^2 = \int_0^\infty\int_0^\infty G(\lambda_1)\cdot G(\lambda_2)\cdot[\langle I(u,v,\lambda_1)\cdot I(u,v,\lambda_2)\rangle - \langle I(u,v,\lambda_1)\rangle\cdot\langle I(u,v,\lambda_2)\rangle]\cdot d\lambda_1\cdot d\lambda_2 \tag{11.51}$$

By again assuming that the speckle statistics are circular Gaussian at each constituent wavelength, Reed's theorem (7.51) may be used to express the intensity variance in the form

$$\left\langle I_I(u,v)^2\right\rangle - \langle I_I(u,v)\rangle^2 = \int_0^\infty\int_0^\infty G(\lambda_1)\cdot G(\lambda_2)\cdot\left|\langle U(u,v,\lambda_1)\cdot U^*(u,v,\lambda_2)\rangle\right|^2\cdot d\lambda_1\cdot d\lambda_2 \tag{11.52}$$

#### 11.6.1.1 Contrast Ratio for Wavelength-Integrated Polychromatic Speckle

Contrast ratio was previously defined (11.22) as the rms intensity fluctuation divided by the mean intensity. For wavelength-integrated polychromatic speckle, we denote the contrast ratio by $C_{\Delta\lambda}$ and use 11.52 to express it in the form

$$C_{\Delta\lambda} = \frac{\left[\left\langle I_I(u,v)^2\right\rangle - \left\langle I_I(u,v)\right\rangle^2\right]^{\frac{1}{2}}}{I_I(u,v)} = \frac{\left[\int_0^\infty \int_0^\infty G(\lambda_1)\cdot G(\lambda_2)\cdot \left|\left\langle U(u,v,\lambda_1)\cdot U^*(u,v,\lambda_2)\right\rangle\right|^2 \cdot d\lambda_1 \cdot d\lambda_2\right]^{\frac{1}{2}}}{\int_0^\infty G(\lambda)\cdot \left\langle I(u,v,\lambda)\right\rangle \cdot d\lambda} \tag{11.53}$$

By incorporating any wavelength dependence of the quantity, $\langle I(u, v, \lambda)\rangle$, into function $G(\lambda)$ (as in Sect. 8.2.2), the above equation reduces to the form

$$C_{\Delta\lambda} = \frac{\left[\int_0^\infty \int_0^\infty G(\lambda_1)\cdot G(\lambda_2)\cdot \left|\mu(u,v,u,v,\lambda_1,\lambda_2)\right|^2 \cdot d\lambda_1 \cdot d\lambda_2\right]^{\frac{1}{2}}}{\int_0^\infty G(\lambda)\cdot d\lambda} \tag{11.54}$$

Because of the $(u, v)$ dependence of function, $\mu(u, v, u, v, \lambda_1, \lambda_2)$, the contrast ratio, $C_{\Delta\lambda}$, in a wavelength-integrated polychromatic stellar speckle pattern varies over the image. Highest contrast usually occurs in the center of the image, with lower contrast usually found in the outer regions. For diffraction-limited telescopes, where the pupil function, $K(x, y, \lambda)$, is entirely real-valued (cf., 11.47), function $\mu(u, v, u, v, \lambda_1, \lambda_2)$ tends to take higher values. Polychromatic stellar speckle patterns formed by diffraction-limited telescopes therefore tend to have higher contrast than those formed by aberrated instruments (where the same seeing conditions are assumed for both instruments).

#### Contrast Ratio for Diffraction Limited Telescopes with Circular Apertures

Equation 11.47 provides an expression for $\mu(u, v, u, v, \lambda_1, \lambda_2)$ in terms of the parameters that describe the atmospheric imaging path and the telescope optics. An explicit expression for the contrast ratio, $C_{\Delta\lambda}$, may be obtained by combining 11.47 with 11.54. For a diffraction-limited telescope with circular aperture, 11.48 may be combined with 11.54, allowing $C_{\Delta\lambda}$ to be written as

$$C_{\Delta\lambda} = \left\{ \frac{\int_0^\infty \int_0^\infty G(\lambda_1)\cdot G(\lambda_2)\cdot \exp\left[-4\cdot\pi^2\cdot\sigma^2\cdot\left(\frac{1}{\lambda_1}-\frac{1}{\lambda_2}\right)^2\right]}{\left[\int_0^\infty G(\lambda)\cdot d\lambda\right]^2} \right.$$
$$\left. \times \left[2\cdot\frac{J_1\left(\frac{\pi\cdot D\cdot\left(u^2+v^2\right)^{\frac{1}{2}}}{f}\cdot\left(\frac{1}{\lambda_1}-\frac{1}{\lambda_2}\right)\right)}{\frac{\pi\cdot D\cdot\left(u^2+v^2\right)^{\frac{1}{2}}}{f}\cdot\left(\frac{1}{\lambda_1}-\frac{1}{\lambda_2}\right)}\right]^2 \cdot d\lambda_1\cdot d\lambda_2 \right\}^{\frac{1}{2}} \tag{11.55}$$

## 11.6.2 Effective Number of Uncorrelated Speckle Patterns in the Integrated Pattern

The effective number of uncorrelated Gaussian speckle patterns, $m_{\Delta\lambda}$, comprising a wavelength-integrated speckle pattern may be given in terms of $C_{\Delta\lambda}$ (cf., 11.26) by

$$m_{\Delta\lambda} = \frac{1}{C_{\Delta\lambda}^2} \tag{11.56}$$

#### 11.6.2.1 *Range of Values Taken by* $m_{\Delta\lambda}$

In the limiting case of monochromatic light at wavelength $\lambda', G(\lambda)$ may be represented by the Dirac delta function,

$$G(\lambda) = \delta(\lambda - \lambda') \tag{11.57}$$

When this delta function form is substituted into 11.54, we find that $m_{\Delta\lambda} = 1$, which merely confirms that the "integrated" speckle in this case is Gaussian speckle. More generally, $m_{\Delta\lambda}$ takes values in the range,

$$1 \leq m_{\Delta\lambda} \leq \infty \tag{11.58}$$

## 11.6.3 Aperture-Averaged (or Pixel-Averaged) Speckle

Suppose that a Gaussian speckle pattern described by the intensity function, $I(u, v, \lambda)$, falls on an FPA detector consisting of rectangular-shaped pixels with side lengths, $a_P$ and $b_P$. The integrated intensity, $I_P(u, v, \lambda)$, collected by a single pixel centered at $(u, v)$ may be expressed by the integral

$$I_P(u, v, \lambda) = \int_{-\frac{a_P}{2}}^{\frac{a_P}{2}} \int_{-\frac{b_P}{2}}^{\frac{b_P}{2}} I(u + u_1, v + v_1, \lambda) \cdot du_1 \cdot dv_1 \tag{11.59}$$

Assuming that the average intensity in the speckle pattern remains approximately constant over distances comparable to the pixel dimensions, the variance of the collected intensity may be written as

$$\left\langle I_P(u,v,\lambda)^2\right\rangle - \langle I_P(u,v,\lambda)\rangle^2 = \int_{-\frac{a_P}{2}}^{\frac{a_P}{2}}\int_{-\frac{b_P}{2}}^{\frac{b_P}{2}}\int_{-\frac{a_P}{2}}^{\frac{a_P}{2}}\int_{-\frac{b_P}{2}}^{\frac{b_P}{2}} \left[\langle I(u+u_1,v+v_1,\lambda)\cdot I(u+u_2,v+v_2,\lambda)\rangle - \left\langle I(u,v,\lambda)^2\right\rangle\right] \times du_1\cdot dv_1\cdot du_2\cdot dv_2 \quad (11.60)$$

By again using Reed's theorem and by using an analysis similar to that which led to 11.53, the following expression can be obtained for the contrast ratio, $C_{\Delta A}$, for aperture-averaged speckle,

$$C_{\Delta A} = \frac{1}{a_P\cdot b_P}\cdot\left[\int_{-\frac{a_P}{2}}^{\frac{a_P}{2}}\int_{-\frac{b_P}{2}}^{\frac{b_P}{2}}\int_{-\frac{a_P}{2}}^{\frac{a_P}{2}}\int_{-\frac{b_P}{2}}^{\frac{b_P}{2}} |\mu(u_1,v_1,u_2,v_2,\lambda)|^2\cdot du_1\cdot dv_1\cdot du_2\cdot dv_2\right]^{\frac{1}{2}} \quad (11.61)$$

The effective number of uncorrelated Gaussian speckle patterns, $m_{\Delta A}$, comprising the aperture-averaged speckle pattern obtains directly from 11.61 in the form

$$m_{\Delta A} = \frac{a_P^2\cdot b_P^2}{\int_{-\frac{a_P}{2}}^{\frac{a_P}{2}}\int_{-\frac{b_P}{2}}^{\frac{b_P}{2}}\int_{-\frac{a_P}{2}}^{\frac{a_P}{2}}\int_{-\frac{b_P}{2}}^{\frac{b_P}{2}} |\mu(u_1,v_1,u_2,v_2,\lambda)|^2\cdot du_1\cdot dv_1\cdot du_2\cdot dv_2} \quad (11.62)$$

It may be observed that the above expression for $m_{\Delta A}$ is functionally similar to the one given previously for $m_{\Delta\lambda}$ by 11.54.

#### 11.6.3.1 Pixel Averaging for Telescopes with Circular Pupils

For detectors with square pixels (i.e., $a_p = b_p$), contrast ratio, $C_{\Delta A}$, and the effective number of uncorrelated speckle patterns, $m_{\Delta A}$, are given respectively by

$$C_{\Delta A} = \frac{\left[\int_{-\frac{a_P}{2}}^{\frac{a_P}{2}}\int_{-\frac{a_P}{2}}^{\frac{a_P}{2}}\int_{-\frac{a_P}{2}}^{\frac{a_P}{2}}\int_{-\frac{a_P}{2}}^{\frac{a_P}{2}} |\mu(u_1,v_1,u_2,v_2,\lambda)|^2\cdot du_1\cdot dv_1\cdot du_2\cdot dv_2\right]^{\frac{1}{2}}}{a_P^2} \quad (11.63)$$

and

$$m_{\Delta\lambda} = \frac{a_P^4}{\int_{-\frac{a_P}{2}}^{\frac{a_P}{2}}\int_{-\frac{a_P}{2}}^{\frac{a_P}{2}}\int_{-\frac{a_P}{2}}^{\frac{a_P}{2}}\int_{-\frac{a_P}{2}}^{\frac{a_P}{2}} |\mu(u_1,v_1,u_2,v_2,\lambda)|^2\cdot du_1\cdot dv_1\cdot du_2\cdot dv_2} \quad (11.64)$$

For telescopes with circular apertures, the relevant expression for $|\mu|^2$ was given previously by 11.40. Using the arguments appropriate to the present application, $|\mu|^2$ may be written as

$$|\mu(u_1, v_1, u_2, v_2, \lambda)|^2 = \left[ 2 \cdot \frac{J_1\left(\frac{\pi \cdot D \cdot \left((u_2-u_1)^2+(v_2-v_1)^2\right)^{\frac{1}{2}}}{\lambda \cdot f}\right)}{\frac{\pi \cdot D \cdot \left((u_2-u_1)^2+(v_2-v_1)^2\right)^{\frac{1}{2}}}{\lambda \cdot f}} \right]^2 \tag{11.65}$$

The degree of speckle reduction caused by pixel averaging depends on the size of the individual pixels relative to the average speckle size. It is convenient, therefore, to define the quantity, $R_{\Delta A}$, as follows:

$$R_{\Delta A} = \frac{a_P \cdot D}{1.22 \cdot \lambda \cdot f} \tag{11.66}$$

Defined in this way, $R_{\Delta A}$ takes values in the range, $0 - \infty$. Near-zero values correspond to pixel sizes much smaller than the size of an average speckle, so that individual pixels approximate "point detectors." For $R_{\Delta A} = 1$, pixel width is comparable to the size of the "average speckle." For $R_{\Delta A} > 1$, the pixels are larger than the average speckle size.

Figure 11.12 shows $m_{\Delta A}$ plotted against the parameter, $R_{\Delta A}$, with 11.64, 11.65, and 11.66 used to calculate the plot data. Figure 11.13 shows corresponding plots of $C_{\Delta A}$ against the parameter, $R_{\Delta A}$, with 11.63, 11.65, and 11.66 used to calculate the plot data.

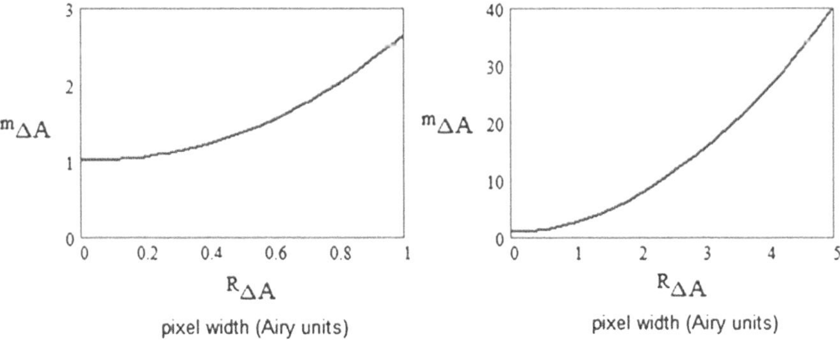


**Fig. 11.12** Speckle reduction as quantified by $m_{\Delta A}$ caused by pixel averaging, where a Gaussian speckle pattern formed by a telescope with a circular aperture falls on *square pixels*, as given by 11.66

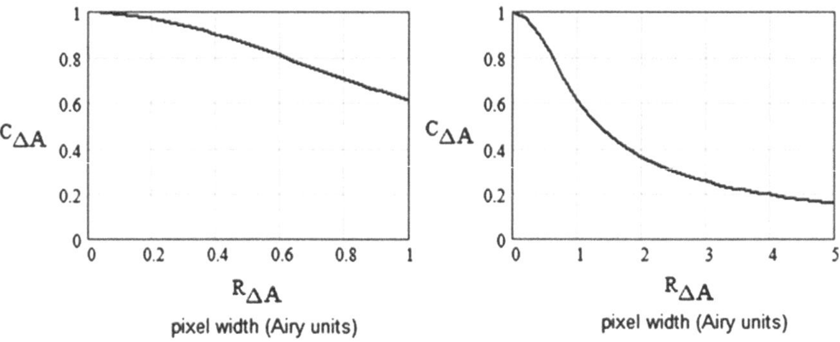


**Fig. 11.13** Speckle contrast reduction caused by pixel averaging where a Gaussian speckle pattern formed by a telescope with circular aperture falls on *square pixels*

## *11.6.4 Time-Averaged Speckle*

Turbulence churning and wind, both of which depend on altitude, cause temporal fluctuations of the speckle intensities in star images. Since all detectors have characteristic integration times (for the human eye, the integration time is about 1/25th second, and for CCD cameras integration time is often about 1/60th second), the speckle "observed" by any given detector is always, to some extent, time-averaged speckle.

A time-averaged, or time-integrated, speckle pattern may be regarded as the incoherent sum of many Gaussian speckle patterns that may be fully correlated, uncorrelated, or partially correlated. The degree of correlation between the various constituent patterns depends on the relaxation rate of the speckle intensity fluctuations. Denoting the intensity in the instantaneous speckle pattern formed in the image at location $(u,v)$, wavelength $\lambda$, and time $t$ by $I(u, v, \lambda, t)$, and by also assuming that the time-averaged intensity, $I_{\Delta t}(u, v, \lambda)$, is obtained using a single-point detector (to eliminate pixel-averaging effects), we may write

$$I_{\Delta t}(u, v, \lambda) = \int_{t}^{t+\Delta t} I(u, v, \lambda, t) \cdot dt \tag{11.67}$$

where $\Delta t$ is the integration time.

We may then express the intensity variance in the time-averaged intensity pattern by

$$\begin{aligned} &\left\langle I_{\Delta t}(u, v, \lambda)^2 \right\rangle - \langle I_{\Delta t}(u, v, \lambda) \rangle^2 = \\ &\int_{t}^{t+\Delta t}\int_{t}^{t+\Delta t} [\langle I(u, v, \lambda, t_1) \cdot I(u, v, \lambda, t_2) \rangle - \langle I(u, v, \lambda, t_1) \rangle \cdot \langle I(u, v, \lambda, t_2) \rangle] \cdot dt_1 \cdot dt_2 \end{aligned} \tag{11.68}$$

#### 11.6.4.1 Temporal Autocorrelation Function of the Complex Amplitude

The temporal autocorrelation function of the complex amplitude can be defined as follows:

$$\mu(u, v, \lambda, t_1, t_2) = \frac{\langle A(u, v, \lambda, t_1) \cdot A^*(u, v, \lambda, t_2)\rangle}{[\langle I(u, v, \lambda, t_1)\rangle \cdot \langle I(u, v, \lambda, t_2)\rangle]^{\frac{1}{2}}} \tag{11.69}$$

where function $\mu(u, v, \lambda, t_1, t_2)$ describes the (unit-normalized) degree of correlation of the complex amplitudes at wavelength $\lambda$ at image location $(u, v)$ at any two instants of time, $t_1$ and $t_2$. If we assume as previously that the speckle statistics are temporally stationary, then only the time difference, $(t_1 - t_2)$, carries any significance. Function $\mu$ may then be written in the form

$$\mu(u, v, \lambda, \Delta t) = \frac{\langle A(u, v, \lambda, t) \cdot A^*(u, v, \lambda, t + \Delta t)\rangle}{[\langle I(u, v, \lambda, t)\rangle \cdot \langle I(u, v, \lambda, t + \Delta t)\rangle]^{\frac{1}{2}}} \tag{11.70}$$

#### 11.6.4.2 Temporal Autocorrelation Function of the Intensity

By again using Reed's theorem, the unit-normalized temporal autocorrelation function of the intensity may be expressed in the form

$$|\mu(u, v, \lambda, \Delta t)|^2 = \frac{\langle I(u, v, \lambda, t) \cdot I(u, v, \lambda, t + \Delta t)\rangle}{\langle I(u, v, \lambda, t)\rangle \cdot \langle I(u, v, \lambda, t + \Delta t)\rangle} - 1 \tag{11.71}$$

#### 11.6.4.3 Contrast Ratio in Time-Averaged Speckle

The contrast in a time-averaged speckle pattern, which we denote by $C_{\Delta t}$, is given by

$$C_{\Delta t} = \frac{\left[\int_t^{t+\Delta t} \int_t^{t+\Delta t} |\mu(u, v, \lambda, t_1, t_2)|^2 \cdot dt_1 \cdot dt_2\right]^{\frac{1}{2}}}{\Delta t} \tag{11.72}$$

#### 11.6.4.4 Effective Number of Uncorrelated Speckle Patterns

The effective number of uncorrelated Gaussian speckle patterns, $m_{\Delta t}$, that incoherently sum together during the integration period to form the time-averaged pattern is given by the inverse square of $C_{\Delta t}$:

$$m_{\Delta t} = \frac{\Delta t^2}{\int_t^{t+\Delta t} \int_t^{t+\Delta t} |\mu(u, v, \lambda, t_1, t_2)|^2 \cdot dt_1 \cdot dt_2} \quad (11.73)$$

#### 11.6.4.5 Measurement of the Temporal Autocorrelation Function of the Intensity

At the arbitrary image location, $(u, v)$ and at wavelength, $\lambda$, function $|\mu(u, v, \lambda, t_1, t_2)|^2$ may be calculated using 11.71 from a time sequence of image intensity measurements, where we denote the individual intensity measurements by $I(u, v, \lambda, t_i)$ where $i = 1...n$ and where $n$ is a suitably large number. The time interval between measurements, $(t_{i+1} - t_i)$ must be short enough to adequately resolve the temporal fluctuations of the intensity.[4] In general, $|\mu(u, v, \lambda, t_1, t_2)|^2$ estimates so calculated tend to take smaller values at shorter wavelengths, indicating that intensity fluctuations in stellar speckle patterns occur at faster rates at these wavelengths. The decorrelation rate could also depend, at least in principle, on the $(u, v)$ location in the image.

#### 11.6.4.6 Candidate Temporal Correlation Functions of the Image Intensity

In windless conditions, the temporal autocorrelation function of the image intensity is largely determined by turbulence churning such that the function might be approximated by the Gaussian form,

$$|\mu(u, v, \lambda, \Delta t)|^2 = \exp\left[-\left(\frac{\Delta t}{t_o(u, v, \lambda)}\right)^2\right] \quad (11.74)$$

where $t_0(u, v, \lambda)$ denotes the average $1/e$ relaxation time of the intensity fluctuations. Figure 11.14 shows $m_{\Delta t}$ plotted against integration time, $\Delta t$, where time is expressed in units of $t_0$. Figure 11.15 shows corresponding plots of the contrast ratio, $C_{\Delta t}$.

In steady winds, the en bloc motion of the turbulence structure in the observing path also produces temporal fluctuations of the speckle in the image. If the winds are strong enough, the fluctuations caused by this mechanism might sometimes overwhelm the fluctuations caused by turbulence churning. In these conditions, the phase and amplitude structures present in the disrupted image-forming waves would tend

---

[4] As long as the sampling rate is at least twice the highest frequency present in the signal (for either temporally or spatially varying signals), the entire signal form can be perfectly reconstructed from the sampled measurements. Sampling at twice the highest frequency is called Nyquist rate sampling after Harry Theodor Nyquist (1889–1976), a Swedish Electronic Engineer who was an important contributor to communication theory. Although the Nyquist rate is sufficient to allow full signal reconstruction, for purely practical reasons, sampling rates are often chosen 10 times higher than the Nyquist rate.

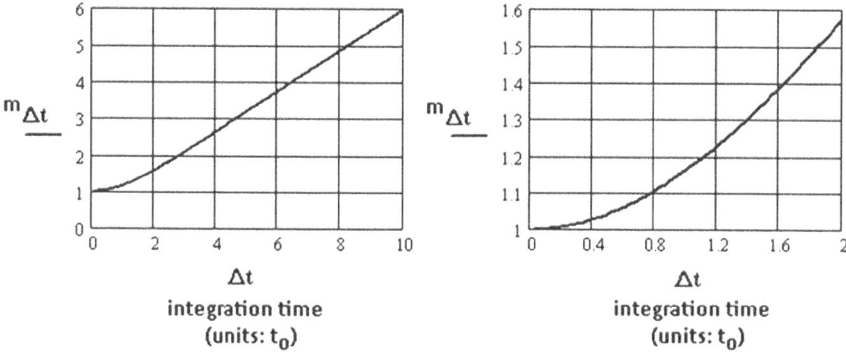


**Fig. 11.14** A Gaussian speckle pattern which continually evolves over time may be reduced by time-averaging. The amount of reduction as expressed by the quantity, $m_{\Delta t}$, increases as integration time increases

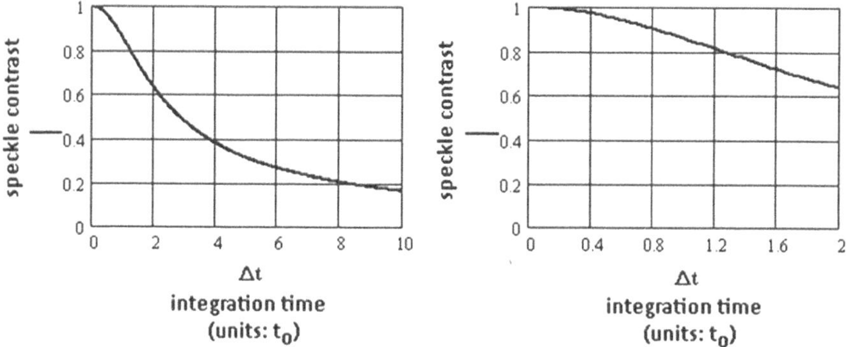


**Fig. 11.15** The contrast ratio, $C_{\Delta t}$, plotted against integration time, $\Delta t$, for a time-averaged continually evolving Gaussian speckle pattern

to move, en bloc, across the telescope aperture. (Similar behavior occurs in a slightly different context, discussed later in Sect. 15.6.3.2.)

For large telescopes with circular apertures, the temporal autocorrelation function of the image intensity in this case is approximately given by

$$|\mu(u, v, \lambda, \Delta t)|^2 = \left\{ \frac{2}{\pi} \cdot \left[ \cos^{-1}\left( \frac{v_W \cdot \Delta t}{D} \right) - \frac{v_W \cdot \Delta t}{D} \cdot \sqrt{1 - \left( \frac{v_W \cdot \Delta t}{D} \right)^2} \right] \right\}^2 \tag{11.75}$$

where $v_W$ is the wind speed component at right angles to the observing direction. The functionality seen in the above equation may be recognized as similar to that of the OTF for a diffraction-limited telescope with circular aperture (Sect. 7.4.1.1).

## 11.6.5 *Multiple Speckle Reduction Mechanisms Acting Simultaneously*

Higher levels of speckle reduction are usually achieved by arranging for multiple speckle reduction mechanisms to act simultaneously. In general, different speckle reduction mechanisms tend to interact in ways that complicate the analysis. However, in many cases, it is reasonable to assume that the different reduction mechanisms all act independently. As an example of two speckle reduction mechanisms acting simultaneously, suppose that the speckle reduction is carried out by combining the effects of pixel averaging by rectangular pixels (11.61 and 11.62) and time-averaging over an interval $\Delta t$ (11.72 and 11.73). Assuming that both mechanisms act independently, the contrast ratio in the reduced pattern, $C$, may be calculated as the product of the contrast ratio delivered by each mechanism separately

$$C = C_{\Delta A} \cdot C_{\Delta t} \quad (11.76)$$

where $C_{\Delta A}$ and $C_{\Delta t}$ can be obtained separately from 11.61 and 11.72, respectively.

The effective number of uncorrelated speckle patterns in the reduced pattern is then given by

$$m = m_{\Delta A} \cdot m_{\Delta t} \quad (11.77)$$

where $m_{\Delta A}$ and $m_{\Delta t}$ can be separately obtained using 11.62 and 11.73.

More generally, when $N$ independently-acting speckle reduction mechanisms act simultaneously, the corresponding values for $C$ and $m$ are given by

$$C = \prod_{j=1}^{N} C_j \quad (11.78)$$

and

$$m = \prod_{j=1}^{N} m_j \quad (11.79)$$

#### 11.6.5.1 Interaction Between Different Speckle Reduction Mechanisms

Because different speckle reduction mechanisms generally interact and thereby influence one another, the $C$ and $m$ values given by 11.78 and 11.79 are only approximations. If exact values are required, it becomes necessary to produce an expression for a single complex coherence factor that simultaneously takes account of all of the various parameters that control the various speckle reduction mechanisms and act together to effect the speckle reduction. For the three speckle reduction methods discussed thus far—wavelength integration, aperture averaging, and time-averaging—the required complex coherence factor would have the form, $\mu(u, u', v, v', \lambda_1, \lambda_2, \Delta t)$. As long as an explicit expression can be obtained for the appropriate function, precise values of $C$ and $m$ can be obtained by integrating the function, $|\mu(u, u', v, v', \lambda_1, \lambda_2, \Delta t)|^2$, over each of the (seven) parameter spaces corresponding to the seven arguments indicated.[5] However, the evaluation of a sevenfold integral is not a trivial matter. Generally, if more than three speckle reduction mechanisms act simultaneously, the computation soon becomes prohibitively complicated.

### *11.6.6 Intensity PDF for a Reduced Speckle Pattern*

Once the appropriate expression has been established for the complex coherence factor, $\mu$, the PDF of intensity in the reduced speckle pattern can be calculated exactly. The method of calculation makes use of a mathematical formulation known as the Karhunen–Loeve expansion. Since this method is both complicated and computationally intensive, we refer the reader elsewhere for details (Dainty, 1984; Davenport & Root, 1958). Here, we restrict consideration to a simpler form for the PDF known as the gamma PDF approximation.

#### 11.6.6.1 Gamma Approximation for the Intensity PDF

Only two parameters are needed to fully specify a gamma PDF of the intensity in a reduced speckle pattern. In principle, the two parameter values are chosen so that the gamma PDF best-approximates the exact PDF. A commonly used approach involves matching the first two moments of the gamma PDF to the exact first- and second-order moments of the intensity. Thus, we would choose the mean intensity in the reduced speckle pattern, $\langle I_R \rangle$, as one of the parameters and the effective number of uncorrelated speckle patterns comprising the reduced pattern, $m$, as the other (Dainty, 1984). The PDF may then be expressed in the form

$$
\begin{aligned}
PDF(I) &= \frac{\left(\frac{m}{\langle I_R \rangle}\right)^m \cdot I^{m-1} \cdot \exp\left(-m \cdot \frac{I}{\langle I_R \rangle}\right)}{\Gamma(m)} \quad \text{for } I > 0 \\
&= 0 \quad \text{otherwise,}
\end{aligned}
\tag{11.80}
$$

where $\Gamma(m)$ is the gamma function (Abramowitz & Stegun, 1970). Equation 11.80 gives the correct first and second moments of the integrated intensity, but does not necessarily give the correct higher-order moments. However, in the case where $m$

[5] An expression has been given elsewhere (McKechnie 1976) for the complex coherence factor for speckle in the image of a rough surface formed by an optical system with arbitrary aberrations when the illumination is both temporally and spatially partially coherent.

constituent Gaussian speckle patterns are all mutually uncorrelated (cf., Sect. 11.8.2), the PDF provided by this equation is exact and thus correctly provides all higher moments.

Figure 11.16 shows PDF plots calculated from 11.80 for various integer values of $m$. For an infinitely large value of $m$, the PDF takes the form of a Dirac delta function, located at the abscissa value, 1. Such a PDF would correspond to an ideal incoherent image, entirely free of speckle, the speckle replaced by a uniformly smooth image appearance with the constant intensity value, $\langle I_R \rangle$, applying throughout.

#### 11.6.6.2 Average Speckle Size in Reduced Speckle Patterns

The average speckle size in a reduced speckle pattern is not necessarily the same as that found in the individual constituent Gaussian speckle patterns. This result is obvious for integrated polychromatic speckle patterns (where the average speckle size for each wavelength represented in the integrated pattern is proportional to that wavelength), but less obvious for time-integrated and aperture-integrated speckle patterns. Suppose for example (cf., Sect. 11.6.4.6) that the atmosphere, driven by a steady wind, were to move en bloc so as to carry the phase and amplitude structures on the image-forming waves along with it. The average speckle size in time-averaged images formed by telescopes in this case would now be found to grow (relative to the average speckle size in the constituent instantaneous Gaussian speckle patterns) in the direction of the wind by a factor of about 1.2.[6] In the direction at right angles, the average speckle size also grows, but by an amount slightly less than 1.2.

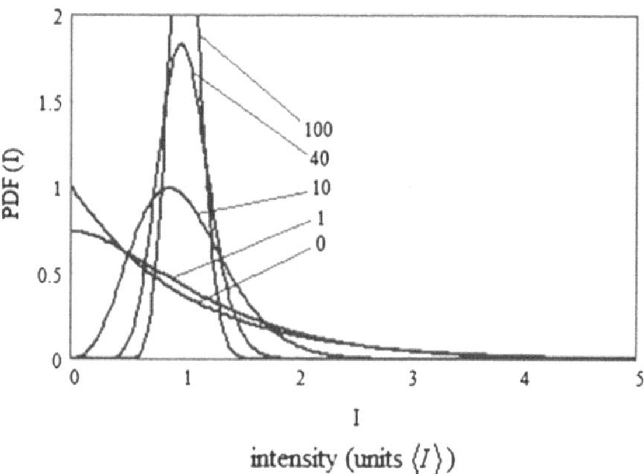


**Fig. 11.16** Gamma probability density functions for Gaussian speckle ($m = 1$) and reduced speckle patterns as described by the various other $m$ values indicated

[6] This case closely resembles a case studied previously by the Author (Dainty 1984). Theoretical and experimental results summarized in Fig. 4.8 of that reference show average speckle size for the time-averaged speckle pattern to be about 1.2 times larger than for each of the instantaneous constituent Gaussian speckle patterns.

In contrast to the above behavior, if the atmospheric turbulence were simply churning (with no overall en bloc motion), the average speckle size in the time-averaged image would then remain about the same as that of the individual constituent instantaneous Gaussian speckle patterns. For typical observing conditions, atmospheric churning and en bloc motion would likely occur together, resulting in some increase in average speckle size by a factor lying somewhere in the range, 1–1.2.

## 11.7 Stellar Speckle Statistics When Cores Are Present in the Image

As we saw in Sect. 11.2.1 and Fig. 11.4, when cores are present in the image (i.e., $\sigma \leq 0.4{\cdot}\lambda$), though the phase still remains Gaussian distributed over the range, $-\infty \leq \varphi \leq \infty$, it is now no longer uniformly distributed in the primary range, $-\pi \leq \varphi \leq \pi$. The new phase distribution continues to provide a circular Gaussian speckle pattern in the image—the halo—but an additional separate field component also arises in the image—the core. These two image features add together coherently to produce an overall light field that no longer obeys circular Gaussian statistics.

The size and shape of the core are determined by the intensity point-spread function of the telescope (10.4). For aberrated telescopes, although the core still contains exactly the same light energy fraction as before (10.3), the energy is now redistributed over a larger angular footprint; this results in a broader core feature with reduced central irradiance. At the same time, and yet still consistent with this core behavior, the size and shape of the average speckle in the halo are unaffected by telescope aberrations (cf., Sect. 10.1).

At every location in the core and halo images considered here, the complex amplitude may be considered as the coherent addition of a constant phasor term[7]—the core—and a Gaussian speckle pattern—the halo. The method of calculating the statistical properties of the complex amplitude and intensity for this type of field distribution has been given previously (Dainty, 1984). Before following Dainty's approach here, it is first necessary to establish certain relationships between the average intensity envelopes for the separate core and halo portions of the image.

[7] It is not strictly correct to consider the intensity variations of the core as zero. However, these variations are significantly less than those of the speckle found in the surrounding halo. While the intensity variations of the core progressively diminish with increasing telescope aperture size, the intensity variations in the surrounding speckle halo (as measured by the signal-to-noise ratio) remain more or less constant ($\sigma/\lambda \approx 1$). Core intensity variations further reduce at longer wavelengths as increasingly more and more of the total light energy concentrates into the core.

### 11.7.1 The Core and Halo Light Fractions in Star Images

The statistical properties of the complex amplitude at any point in a core and halo image may be obtained from the statistical properties of the complex sum illustrated in Fig. 11.17. An equation was given previously (9.2) for the average intensity envelope in a core and halo image, $\langle I(u, v, \lambda)\rangle$, in terms of the parameters that describe the atmospheric imaging path and the telescope optics. It is convenient to reproduce that equation here:

$$\langle I(u, v, \lambda)\rangle = \frac{1}{\lambda^2} \cdot \int_{-\infty}^{\infty} \int_{-\infty}^{\infty} \exp\left\{\frac{-4 \cdot \pi^2 \cdot \sigma^2}{\lambda^2} \cdot [1 - \rho(\xi, \eta)]\right\} \cdot M_T(\xi, \eta, \lambda) \times \exp\left[\frac{-2 \cdot \pi \cdot i}{\lambda \cdot f} \cdot (u \cdot \xi + v \cdot \eta)\right] \cdot d\xi \cdot d\eta \tag{11.81}$$

We also reproduce 10.7, which describes how the average intensity envelope may be decomposed into the respective core and halo portions:

$$\langle I(u, v, \lambda)\rangle = \langle I_C(u, v, \lambda)\rangle + \langle I_H(u, v, \lambda)\rangle \tag{11.82}$$

It is convenient to define the ratio of the average intensities in the core and halo portions of the image, $R_{CH}$(u,v,λ), as follows:

$$R_{CH}(u, v, \lambda) = \frac{\langle I_C(u, v, \lambda)\rangle}{\langle I_H(u, v, \lambda)\rangle} \tag{11.83}$$

The quantity, $R_{CH}$(u,v,λ), takes its highest values in the center of the image where the core dominates; it takes its lowest values in the outer parts where the halo dominates.

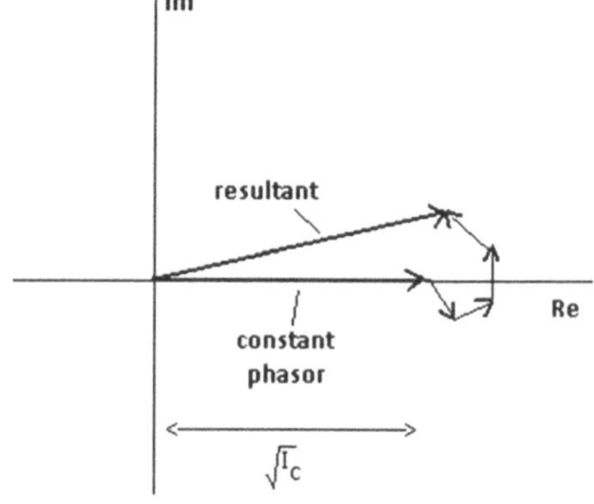


**Fig. 11.17** Constant phasor plus a random walk phasor in the complex plane

The following relationships readily derive from 10.82 and 10.83:

$$\langle I_C(u, v, \lambda)\rangle = \frac{\langle I(u, v, \lambda)\rangle \cdot R_{CH}(u, v, \lambda)}{1 + R_{CH}(u, v, \lambda)} \tag{11.84}$$

and

$$\langle I_H(u, v, \lambda)\rangle = \frac{\langle I(u, v, \lambda)\rangle}{1 + R_{CH}(u, v, \lambda)} \tag{11.85}$$

Substituting for $I_C(u, v, \lambda)$ and $I_H(u, v, \lambda)$ in 11.83 (using 11.4 and 11.6) allows $R_{CH}(u, v, \lambda)$ to be expressed in the form

$$\begin{aligned} R_{CH}(u, v, \lambda) &= \exp\left[\frac{-4 \cdot \pi^2 \cdot \sigma^2}{\lambda^2}\right] \cdot PSF_I(u, v, \lambda) \\ &\times \left\{\int_{-\infty}^{\infty}\int_{-\infty}^{\infty} \exp\left[\frac{-4 \cdot \pi^2 \cdot \sigma^2}{\lambda^2}\right] \cdot \left[\exp\left(\frac{-4 \cdot \pi^2 \cdot \sigma^2 \cdot \rho(\xi, \eta)}{\lambda^2}\right) - 1\right]\right. \\ &\left. \times M_T(\xi, \eta, \lambda) \cdot \exp\left[\frac{-2 \cdot \pi \cdot i}{\lambda \cdot f} \cdot (u \cdot \xi + v \cdot \eta)\right] \cdot d\xi \cdot d\eta\right\}^{-1}, \end{aligned} \tag{11.86}$$

where $PSF_I(u, v, \lambda)$ is the intensity PSF of the telescope, unit-normalized for the case of a diffraction-limited telescope (cf., 10.9).

Thus, 11.82 may be used to calculate $\langle I(u, v, \lambda)\rangle$ at any location in the image in terms of the parameters that describe the atmospheric path and the telescope optics.

Likewise, 11.86 may be used to calculate $R_{CH}(u, v, \lambda)$. Once $\langle I(u, v, \lambda)\rangle$ and $R_{CH}(u, v, \lambda)$ have been obtained in this way, 11.84 and 11.85 may be used to calculate the average core and halo intensity envelopes, $\langle I_C(u, v, \lambda)\rangle$ and $\langle I_H(u, v, \lambda)\rangle$. Once these two envelopes have been obtained, the problem is then sufficiently specified to allow calculation of the statistical properties of the complex amplitude and intensity anywhere in the image at any wavelength of interest.

### 11.7.2 Probability Density Function of the Complex Amplitude

As mentioned earlier, the real and imaginary parts of the complex amplitude in the halo portion of the image obey circular Gaussian statistics. The addition of a core to this pattern has the same effect as adding a constant phasor term, the magnitude of which varies in a known way over the image. Since the absolute phase reference may be chosen at our convenience, without loss of generality we associate the phase angle, zero, to this phasor. Recalling 11.7, the PDF of the real and imaginary parts of the total complex amplitude in such a speckle pattern may be written in the form Dainty (1984)

$$PDF_{r,i}\left(U^{(r)},U^{(i)}\right)=\frac{1}{2\cdot\pi\cdot\sigma_a^2}\cdot\exp\left[-\frac{\left[U^{(r)}-\sqrt{\langle I_C(u,v,\lambda)\rangle}\right]^2+\left[U^{(i)}\right]^2}{2\cdot\sigma_a^2}\right] \tag{11.87}$$

where $\sqrt{\langle I_C(u,v,\lambda)\rangle}$ is the average modulus of the amplitude in the core related light contribution. For Gaussian speckle, we found previously (11.16) that the quantity $2\cdot\sigma_a^2$ corresponds to the mean intensity. In the present case, this quantity refers only to the halo portion of the image. Thus,

$$2\cdot\sigma_a^2=\langle I_H(u,v,\lambda)\rangle \tag{11.88}$$

Combining 11.87 and 11.88 allows the PDF of the real and imaginary parts of the total complex amplitude in the image to be written as

$$PDF_{r,i}\left(U^{(r)},U^{(i)}\right)=\frac{1}{\pi\cdot\langle I_H(u,v,\lambda)\rangle}\cdot\exp\left\{-\frac{\left[U^{(r)}-\sqrt{\langle I_C(u,v,\lambda)\rangle}\right]^2+\left[U^{(i)}\right]^2}{\langle I_H(u,v,\lambda)\rangle}\right\} \tag{11.89}$$

## 11.7.3 Probability Density Function of the Intensity and Phase

By using the same variable transformations defined previously (Sect. 11.4.3), the joint PDF of intensity and phase may be obtained from 11.89 as follows:

$$\begin{aligned}&PDF_{I,\varphi}(I,\varphi)=\\&\frac{1}{2\cdot\pi\cdot\langle I_H(u,v,\lambda)\rangle}\cdot\exp\left[-\frac{I+\langle I_C(u,v,\lambda)\rangle-2\cdot\sqrt{I\cdot\langle I_C(u,v,\lambda)\rangle}\cdot\cos(\varphi)}{\langle I_H(u,v,\lambda)\rangle}\right]\text{for } I\geq 0\, and\, -\pi\leq\varphi\leq\pi\\&=0\quad\text{otherwise.}\end{aligned} \tag{11.90}$$

By integrating with respect to phase, we obtain the marginal PDF of intensity,

$$\begin{aligned}PDF_I(I)&=\frac{1}{\langle I_H(u,v,\lambda)\rangle}\cdot\exp\left[-\frac{I+\langle I_C(u,v,\lambda)\rangle}{\langle I_H(u,v,\lambda)\rangle}\right]\cdot I_0\left[\frac{2\cdot\sqrt{I\cdot\langle I_C(u,v,\lambda)\rangle}}{\langle I_H(u,v,\lambda)\rangle}\right]\text{for } I\geq 0\\&=0\quad\text{otherwise,}\end{aligned} \tag{11.91}$$

where $I_0(\cdot)$ is the modified zero-order Bessel function of the first kind (Abramowitz & Stegun, 1970).

Figure 11.18 shows plots of $\mathrm{PDF_I}(I)$ calculated from 11.91 for various

$R_{CH}(u, v, \lambda)$ values. When $R_{CH}(u, v, \lambda)$ takes the value zero (i.e., when the core energy fraction is zero), the PDF reduces to the negative exponential form for Gaussian speckle obtained previously (11.21). When $R_{CH}(u, v, \lambda)$ takes large values (i.e., when an overwhelming fraction of the light energy resides in the core), the PDF can be represented by a Dirac delta function, located at the mean intensity of the core at the $(u, v)$ image location considered.

Integrating the joint PDF (11.90) with respect to intensity provides the marginal PDF of the phase (Middleton, 1960)

$$\begin{aligned} PDF_{\vartheta}(\vartheta) &= \frac{\exp(-R_{CH})}{2 \cdot \pi} + \sqrt{\frac{R_{CH}}{\pi}} \cdot \cos(\varphi) \\ &\times \exp\left[-R_{CH} \cdot \sin^2(\varphi)\right] \cdot \Phi_0\left(\sqrt{2 \cdot R_{CH}} \cdot \cos(\varphi)\right) \text{ for } -\pi \le \varphi \le \pi \\ &= 0 \quad \text{otherwise,} \end{aligned} \tag{11.92}$$

where

$$\Phi_0(b) = \frac{1}{\sqrt{2 \cdot \pi}} \cdot \int_{-\infty}^{b} \exp\left(-\frac{y^2}{2}\right) \cdot dy \tag{11.93}$$

It might be noted here that the function $\Phi_0$ defined by 11.93 is not the same as, nor is it related to, the turbulence power spectral density function $\Phi(\cdot)$ that arose in Sect. 8.6.

Representative plots of $PDF_{\varphi}(\varphi)$ shown previously in Fig. 11.4 were calculated from 11.4 by resetting the phase angle to lie in the primary range, $-\pi$ to $\pi$. Equation 11.92 could equally have been used to provide the same set of plots by simply inserting the appropriate value for $R_{CH}(u, v, \lambda)$. When $R_{CH}(u, v, \lambda)$ takes the value

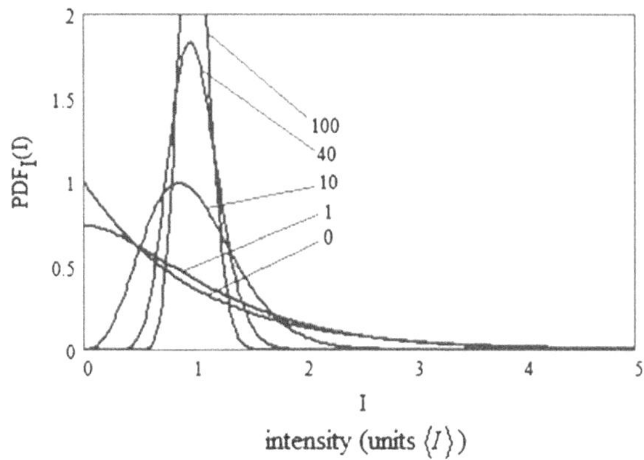


**Fig. 11.18** Probability density functions resulting from the coherent addition of a uniform intensity distribution to a Gaussian speckle pattern for various beam intensity ratios, $R_{CH}$, indicated

zero (i.e., the core contains zero light energy), 11.92 reduces to a uniform phase distribution corresponding to Gaussian speckle (11.15).

### 11.7.4 Second Moment of the Intensity and Variance

The second moment of intensity can be shown (Middleton, 1960) to be given by

$$\left\langle I(u,v,\lambda)^2\right\rangle = 2\cdot\langle I_H(u,v,\lambda)\rangle^2 + 4\cdot\langle I_H(u,v,\lambda)\rangle\cdot\langle I_C(u,v,\lambda)\rangle + \langle I_C(u,v,\lambda)\rangle^2. \tag{11.94}$$

From this equation (and by also using to 11.84 and 11.85), it follows that the variance may be written in the form

$$\sigma_I(u,v,\lambda)^2 = \left\langle I(u,v,\lambda)^2\right\rangle - \langle I(u,v,\lambda)\rangle^2 = \langle I_H(u,v,\lambda)\rangle^2\cdot(1+2\cdot R_{CH}(u,v,\lambda)) \tag{11.95}$$

### 11.7.5 Contrast and Signal-To-Noise Ratio

Contrast ratio, $C_{CH}(u,v,\lambda)$ is defined here in the usual way

$$C_{CH}(u,v,\lambda) = \frac{\sigma_I(u,v,\lambda)}{\langle I(u,v,\lambda)\rangle} \tag{11.96}$$

By using 11.85 and 11.95, $C_{CH}(u,v,\lambda)$ may be written in the alternative form

$$C_{CH}(u,v,\lambda) = \frac{\sqrt{1+2\cdot R_{CH}(u,v,\lambda)}}{1+R_{CH}(u,v,\lambda)} \tag{11.97}$$

Figure 11.19 shows contrast ratio plotted against beam ratio, $R_{CH}(u,v,\lambda)$. For small beam ratios (i.e.,$R_{CH}(u,v,\lambda) \to 0$), the contrast ratio, $C_{CH}(u,v,\lambda)$, tends to unity, corresponding to Gaussian speckle, a case that arises when the core contains negligible energy. It can also arise in the outer portions of the image where the halo dominates. At core center, where $R_{CH}(u,v,\lambda)$ takes its highest values, speckle contrast falls off, ultimately falling to zero as $R_{CH}(u,v,\lambda) \to \infty$. This limiting case corresponds to the case, $\sigma/\lambda \to 0$, where all of the light energy resides in the core.

Signal-to-noise ratio, which is simply the reciprocal of contrast, follows directly from 11.97:

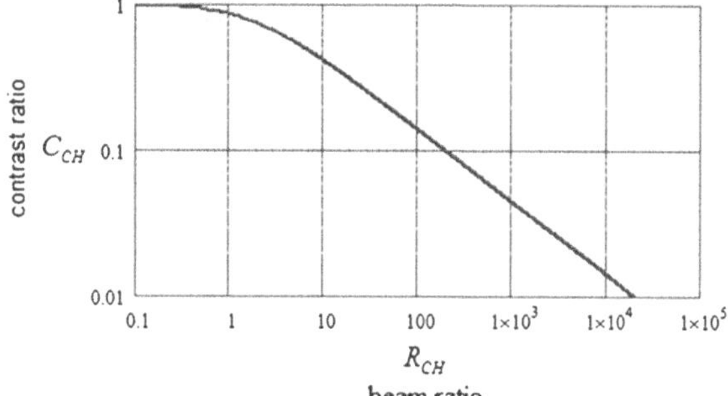


**Fig. 11.19** Variation of the speckle contrast ratio, $C_{\mathrm{CH}}$, in core and halo images as a function of the beam ratio, $R_{\mathrm{CH}}$

$$\left(\frac{S}{N}\right)_{CH} = \frac{\langle I(u, v, \lambda)\rangle}{\sigma_I(u, v, \lambda)} = \frac{1 + R_{CH}(u, v, \lambda)}{\sqrt{1 + 2 \cdot R_{CH}(u, v, \lambda)}} \tag{11.98}$$

## 11.8 Speckle Statistics for Light of Arbitrary State of Polarization

The statistical properties of stellar speckle developed thus far are based on the assumption (Sect. 11.4) that the light is initially perfectly polarized and, because of the small-angle assumption (originally made in Sect. 3.7 to justify the use of scalar diffraction theory), neither the atmospheric path nor the telescope optics were considered capable of significantly altering this initial state of polarization.[8] However, it may be observed that, though the various equations developed thus far to describe the statistical properties of speckle depend on the assumption of perfectly polarized light, nothing in these equations appears to discriminate between the many possible forms of "perfectly polarized" light, linearly polarized light in the vertical and horizontal directions being two of these forms. It may therefore be concluded that the initial polarization state is inconsequential. Just as long as we remain within the small-angle (scalar diffraction) regime, we may now relax the "perfectly polarized" requirement made in Sect. 11.4.1 and make the much broader claim that the statistical properties of speckle—given thus far—apply to light of any initial state of polarization, including light that is entirely unpolarized.[9]

To justify the application of scalar diffraction theory to telescope imaging (Sect. 3.7) requires one or other of two conditions to be fulfilled: (1) the telescope must be diffraction-limited for light in all possible states of polarization so that the telescope optics treat all polarizations equally and (2) for non-diffraction-limited

[8] We assume at this stage that components capable of altering the state of polarization (linear polarizers, wave plates, etc.) are not used. The telescope is also assumed to comprise only lenses and/or mirrors used in the small-angle (paraxial) limit.

[9] Another way of considering polarization in the small-angle limit is that the complex coherence factor for light in any two states of polarization is always unity.

telescopes, it must be assumed that they operate only in the small-angle (paraxial) region where polarization is not affected. The latter condition imposes a restriction on the telescope F/# as well as other restrictions discussed in the next section. For on-axis reflecting telescopes, such as Cassegrain instruments, where simple metallic mirror coatings are commonly used on the primary and secondary mirrors (aluminum or silver, with perhaps a protective layer of silicon dioxide), the small-angle requirement is reasonably satisfied as long as the telescope F/numbers are restricted to the approximate range, F/# $\geq 4$ (cf., Fig. 3.12).

### *11.8.1 Speckle Statistics for Depolarizing Telescopes*

For telescopes faster than F/4, the angles of incidence of some rays with respect to the optical surfaces can exceed 7°. (The highest angles arise for some of the marginal rays originating from off-axis object points.) Much larger angles of incidence can arise for Newtonian telescopes and telescopes with Coude paths; both employ 45-degree mirrors. Needless to say, the use of such steeply angled mirrors violates the small-angle assumption. When violations like this occur, telescope optics generally treat light differently depending on the state of polarization of the light; both the amplitude and phase of the light fields can be affected. In such instances, one has to resort to the formulations of vector diffraction theory, such as Jones matrix formulations as devised by Jones (1941a, 1941b, 1942). Other vector diffraction theory formulations can also be used as described by Born and Wolf (2003) and Wolf (1959). When large angles of incidence are encountered and vector diffraction theory is applied, the statistical properties of stellar speckle patterns are then found to depend on the initial state of polarization of the illumination. The general approach to calculating the speckle properties in the image of a point-object (whether this object is an unresolved star or some other type of point-object) is as follows: A large number of rays are generated from the point-object in a way that uniformly samples the image-forming light cone that originates from the object; it is assumed that all of the rays have the same initial state of polarization. Each of these rays is traced (by conventional skew ray tracing methods) from the object point all the way through the telescope optics to the final image. At every optical surface, a record is maintained of the state of polarization of each individual ray. Although the rays all have the same initial state of polarization, depending on their imaging path through the telescope optics, their final polarization states in the image are generally quite different.

Various schemes can be used to describe the initial polarization state. Jones vectors, for example, describe the polarization state in terms of a 2-D field vector where, in general, complex elements describe both the amplitudes and phases of the two orthogonal, linearly polarized components. Tracking the polarization state as the waves refract through, or reflect off, the individual optical surfaces in the telescope is accomplished by standard matrix multiplication methods, where each optical interface is represented by a $2 \times 2$ matrix.[10] In this way, the different amplitude and phase modifications that apply to different states of polarization can be tracked individually all the way to the final image plane. If additional polarizing elements (e.g., linear polarizers and wave plates) are included in the imaging path, appropriate $2 \times 2$ Jones matrices can be used as needed to describe the effects of these elements on the polarization state of the light as it passes through.

The state of polarization of the light field that finally arrives in the image plane can also be expressed in terms of a 2-D Jones vector. This vector is generated as the linear sum of the individual 2-D Jones vector components that arise in the image for each of the rays traced through the system. For a non-depolarizing telescope, the final light field vector would be identical to that of the initial state of polarization. For depolarizing telescopes, however, the final light field vector could be entirely different from the initial vector.

Even when the point-object emits perfectly polarized light, depolarizing telescope optics cause depolarization of the light, with either partial or complete depolarization of the light that finally arises in the image being possible. In general, the final speckle pattern formed in the image in such cases arises as the incoherent sum of two speckle patterns, one for each of the two orthogonal polarizations that may be used to represent the light field in the image. The two constituent patterns may be observed separately by viewing the image through a polarization analyzer.

Generally, the two patterns have different average intensities and, depending on the degree of depolarization caused by the telescope, they may be either partially correlated or totally uncorrelated.

By making the same assumptions that previously led to circular Gaussian speckle statistics (Sect. 11.4), it can be shown that the two orthogonal speckle patterns separately obey circular Gaussian statistics. The statistical properties of the intensity in the (incoherently) summed final image depend on the degree of correlation, $|\mu|^2$, between the intensities in the two orthogonal Gaussian speckle patterns, $\mu$ being the complex coherence factor between the complex amplitudes in the two patterns.

When the degree of intensity correlation is zero, i.e., $|\mu|^2 = 0$, the intensity distributions in the two orthogonal patterns would look entirely different in detail and their (incoherent) sum would then appear as a reduced speckle pattern. When the degree of correlation is unity, $|\mu|^2 = 1$, the intensity distributions for the two patterns would be identical (apart from a possible difference in average brightness); their incoherent sum would then, in effect, constitute a single Gaussian speckle pattern with contrast ratio, $\sigma/\langle I \rangle = 1$. More generally, $|\mu|^2$ takes values in the range, 0–1.

[10] The matrix elements convey essential information about both the refractive indices and thicknesses of the various coating layers on the optical substrate. Non-normal angles of incidence are dealt with by use of generalized refractive indices (Born and Wolf 2003).

### 11.8.2 Partially Polarized Speckle Fields and the Degree of Polarization

A partially polarized speckle field in the image can be represented as the sum of an entirely unpolarized field and a linearly polarized field (Wolf, 1959). Here, we denote the average intensity in the unpolarized field by $I_{UnPol}(u, v, \lambda)$ and the average intensity in the linearly polarized field by $I_{LinPol}(u, v, \lambda)$. It should be noted that, in general, the degree of polarization of the overall light field in the image depends on both the wavelength and the location in the image plane. Denoting the degree of polarization of the light at wavelength, $\lambda$, at the image point, $(u, v)$, by the function, $P(u, v, \lambda)$, we may express this function in terms of $I_{UnPol}(u, v, \lambda)$ and $I_{LinPol}(u, v, \lambda)$ as follows:

$$P(u, v, \lambda) = \frac{\langle I_{LinPol}(u, v, \lambda)\rangle}{\langle I_{UnPol}(u, v, \lambda)\rangle + \langle I_{LinPol}(u, v, \lambda)\rangle} \tag{11.99}$$

It is noted that the values taken by function $P(u, v, \lambda)$ lie in the range, $0 \leq P(u, v, \lambda) \leq 1$.

Following Dainty (1984), an equally valid and, in fact, more convenient way of representing a partially polarized speckle field in the image is to regard it as the sum of two uncorrelated speckle patterns, for which the average intensities, $I_1(u, v, \lambda)$ and $I_2(u, v, \lambda)$, can be expressed in terms of $P(u, v, \lambda)$ by

$$\langle I_1(u, v, \lambda)\rangle = \frac{\langle I(u, v, \lambda)\rangle}{2} \cdot [1 + P(u, v, \lambda)] \tag{11.100}$$

and

$$\langle I_2(u, v, \lambda)\rangle = \frac{\langle I(u, v, \lambda)\rangle}{2} \cdot [1 - P(u, v, \lambda)] \tag{11.101}$$

where plainly the average intensity in the combined speckle pattern, $I(u, v, \lambda)$, is given by

$$\langle I(u, v, \lambda)\rangle = \langle I_1(u, v, \lambda)\rangle + \langle I_2(u, v, \lambda)\rangle \tag{11.102}$$

For speckle patterns thus constituted, we may refer to Goodman's Table 2.1 (Dainty, 1984) and express the PDF of the intensity in the form

$$PDF(I) = \frac{1}{P(u, v, \lambda) \cdot \langle I(u, v, \lambda)\rangle} \times \left[\exp\left(\frac{-2 \cdot I}{[1 + P(u, v, \lambda)] \cdot \langle I(u, v, \lambda)\rangle}\right) - \exp\left(\frac{-2 \cdot I}{[1 - P(u, v, \lambda)] \cdot \langle I(u, v, \lambda)\rangle}\right)\right] \tag{11.103}$$

#### 11.8.2.1 Limiting Cases of Perfectly Polarized and Unpolarized Light

For perfectly polarized light, i.e., where $P(u, v, \lambda) = 1$, 11.103 reduces to the negative exponential PDF found previously for circular Gaussian speckle (11.21). For entirely unpolarized light, i.e., where $P(u, v, \lambda) = 0$, the light field in the image can be regarded as the incoherent sum of two uncorrelated Gaussian speckle patterns. The intensity PDF given by 11.103 then reduces exactly to the gamma PDF for the case $m = 2$ (11.80)

$$PDF(I) = \left(\frac{2}{\langle I(u, v, \lambda)\rangle}\right)^2 \cdot I \cdot \exp\left(\frac{-2 \cdot I}{\langle I(u, v, \lambda)\rangle}\right) \tag{11.104}$$

where it is noted that $\Gamma(2) = 1$.

### *11.8.3 Intensity PDFs for Depolarizing and Non-Depolarizing Telescopes*

Figure 11.20 shows three intensity PDF plots for speckle in the image of a point object. For all three, the object is assumed to emit the same type of perfectly polarized light. The differences observed between the plots arise from the different degrees of depolarization caused by the telescope optics. The plot marked $P = 1$ corresponds to speckle formed by a non-depolarizing telescope. For this case, the light field in the image remains perfectly polarized; the speckle then obeys circular Gaussian statistics

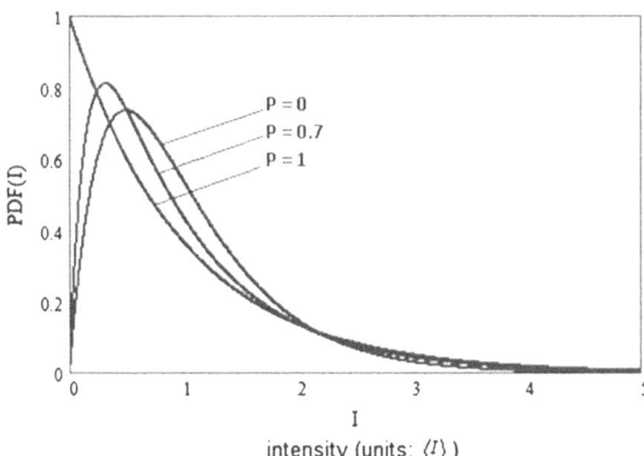


**Fig. 11.20** For point objects that emit perfectly linearly polarized light, the intensity probability density functions of the speckle in the image depend on the degree to which the telescope optics depolarize the light. For non-depolarizing telescopes, $P = 1$, and the intensity obeys a negative exponential distribution. For telescopes that cause complete depolarization, $P = 0$. For telescopes that partially depolarize, $0 < P < 1$

where the intensity PDF is a negative exponential distribution. Axially symmetric Cassegrain telescopes when used at prime focus (i.e., no Coude paths) would likely deliver stellar speckle patterns with this particular intensity PDF distribution for stars lying within a few degrees of the telescope's optical axis.

The plot marked $P = 0$ corresponds to speckle formed by a telescope that causes complete depolarization. Such a plot might arise for telescopes with Coude paths that include multiple 45-degree mirrors. In this case, the speckle can be considered as the incoherent sum of two entirely uncorrelated speckle patterns. The plot marked $P = 0.7$ corresponds to speckle formed by a telescope that causes only partial depolarization.

### 11.8.4 Moments of the Intensity for Partially Polarized Speckle

The first moment of intensity arises from the intensity PDF (11.103) simply as $I(u, v, \lambda)$. The second moment arises from the intensity PDF in the form

$$\left\langle I(u, v, \lambda)^2 \right\rangle = \langle I(u, v, \lambda) \rangle^2 \cdot \frac{\left[P(u, v, \lambda)^2 + 3\right]}{2} \tag{11.105}$$

### 11.8.5 Intensity Variance for Partially Polarized Speckle

The variance can be obtained directly from the first two intensity moments and may be written as

$$\sigma(u, v, \lambda)^2 = \left\langle I(u, v, \lambda)^2 \right\rangle - \langle I(u, v, \lambda) \rangle^2 = \langle I(u, v, \lambda) \rangle^2 \cdot \frac{\left[P(u, v, \lambda)^2 + 1\right]}{2} \tag{11.106}$$

### 11.8.6 Contrast and S/N Ratio for Partially Polarized Speckle

Contrast ratio, $C$, may be obtained directly from 11.106 in the form

$$C = \frac{\sigma(u, v, \lambda)}{\langle I(u, v, \lambda) \rangle} = \sqrt{\frac{P(u, v, \lambda)^2 + 1}{2}}. \tag{11.107}$$

Figure 11.21 shows speckle contrast ratio plotted against the degree of polari zation, $P$.

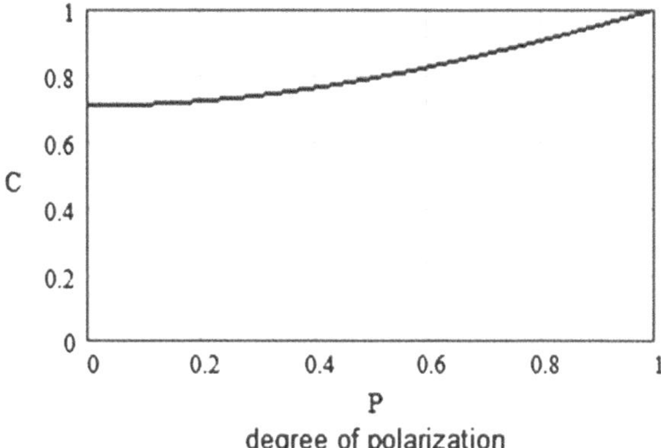


**Fig. 11.21** Contrast ratio, $C$, versus the degree of polarization, $P$, of the light in the image

Signal-to-noise ratio—the inverse of contrast ratio—follows directly from 11.107 and may be written in the form

$$\frac{S}{N}(u, v, \lambda) = \sqrt{\frac{2}{P(u, v, \lambda)^2 + 1}} \tag{11.108}$$

## *11.8.7 Summary of the Polarization Dependence of Speckle Statistics*

### 11.8.7.1 Non-depolarizing Telescopes

The statistical properties of speckle formed in the images of unresolved stars by non-depolarizing telescopes do not depend on the initial state of polarization of the light. For such telescopes, the speckle statistics are exactly the same irrespective of whether the light emitted by the star is perfectly polarized, partially polarized, or entirely unpolarized. The reason for this behavior is that the optics of non-depolarizing telescopes, whether or not they are diffraction-limited, treat the light in the same way regardless of its initial state of polarization.

Making the reasonable assumption that the atmosphere does not cause depolarization, the complex coherence factor for the light in the image is entirely determined by the telescope optics. For non-depolarizing telescopes, the state of polarization of this light is always the same as the initial state of polarization. Consequently, for such telescopes the complex coherence factor for light in the image originating from any pair of polarization states is always exactly unity. Because the speckle intensity distributions corresponding to each of the polarization states are then always perfectly correlated, when these distributions sum together to produce the final image, the final distribution is identical (apart from a brightness factor) to the intensity distributions that would have arisen separately from either of the two polarizations.

Non-depolarizing telescopes include almost all types of astronomical telescope where the small-angle approximation, $\cos(\vartheta \approx 1)$, is reasonably satisfied. Excluded from this category, however, are telescopes with optics faster than about $F/4$. Also excluded are Newtonian telescopes and telescopes with Coude paths where mirrors are used at high angles of incidence, usually 45°.

#### 11.8.7.2 Depolarizing Telescopes

For depolarizing telescopes, which may cause either partial or complete depolarization, the final speckle appearance depends on the initial state of polarization of the light and on the degree to which the telescope optics depolarizes that light.

In the case of light that is initially perfectly linearly polarized, the light in the final image may be either partially polarized or completely unpolarized depending on the degree of depolarization caused by the telescope. The statistical properties of speckle in the image then correspond to those of reduced speckle. For telescopes that only cause partial depolarization, the degree of polarization of the light in the image lies in the range, $0 < P(u, v, \lambda) < 1$, and the value taken by $m$ for the reduced speckle pattern in the image lies in the range, $1 < m < 2$. For telescopes that cause complete depolarization, the degree of polarization of the light in the image would be exactly zero, i.e., $P(u, v, \lambda) = 0$. In this case, the value taken by $m$ for the reduced speckle pattern in the image would be exactly 2.

For telescopes that cause partial depolarization and where the initial light is unpolarized, the properties of the reduced speckle pattern in the image are determined by the values taken by $P(u, v, \lambda)$ and $m$, where $0 < P(u, v, \lambda) < 1$ and $1 < m < 2$. The same two ranges also apply to the reduced speckle patterns produced by initially partially polarized light.

## 11.9 Mathematical Notation Used in This Chapter

The mathematical notation used in this chapter is indicated in Table 11.2.

**Table 11.2** Mathematical notation used in this chapter along with the SI dimensional units of the individual quantities

| Symbol | Quantity | Dimensions |
|---|---|---|
| $\lambda$ | Wavelength | m |
| $(x, y)$ | Cartesian coordinate system in telescope pupil | m |
| $r$ | Radial coordinate in telescope pupil,$(= \sqrt{x^2 + y^2})$ | m |
| $(\xi, \eta)$ | Cartesian coordinate system in plane perpendicular to direction of light travel | m |
| $\varepsilon$ | Coordinate in plane perpendicular to light travel direction, $(= \sqrt{\xi^2 + \eta^2})$ | m |
| $(u, v)$ | Cartesian coordinate system in telescope image plane | m |
| $q$ | Radial coordinate in telescope image plane,$\left(= \sqrt{u^2 + v^2}\right)$ | m |
| $(\alpha, \beta)$ | Angular coordinate system in telescope image plane | "1" |
| $\vartheta$ | Radial angular coordinate in telescope image plane,$\left(= \sqrt{\alpha^2 + \beta^2}\right)$ | "1" |
| $k$ | Wave number in air | $m^{-1}$ |
| $H$ | Integrated OPD fluctuation over entire atmospheric path | m |
| $\phi$ | Phase associated with OPD fluctuation, $H$ | "1" |
| $\sigma$ | rms of integrated OPD fluctuation, $H$ | m |
| $\rho$ | Autocorrelations function of integrated OPD fluctuation, $H$ | "1" |
| $M$ | Atmospheric MTF | "1" |
| $M_T$ | Telescope OTF | "1" |
| $D$ | Telescope diameter | m |
| $f$ | Telescope focal length | m |
| $K$ | Telescope pupil function | "1" |
| $PDF_I$ | Telescope intensity point-spread function | "1" |
| $U$ | Complex amplitude | "1" |
| I | Image intensity | "1" |
| $I_C$ | Core intensity portion | "1" |
| $I_H$ | Halo intensity portion | "1" |
| $\sigma_I$ | rms of intensity, $I$ | "1" |
| $U^{(r)}$ | Real part of complex amplitude | "1" |
| $U^{(i)}$ | Imaginary part of complex amplitude | "1" |
| $S/N$ | Intensity signal-to-noise ratio | "1" |
| $C$ | Speckle intensity contrast ratio | "1" |
| $m$ | Effective number of uncorrelated speckle patterns in reduced speckle pattern | "1" |
| $\mu$ | Complex coherence factor | "1" |
| $G$ | Relative spectral content in image | "1" |

(continued)

**Table 11.2** (continued)

| Symbol | Quantity | Dimensions |
|---|---|---|
| $R_A$ | Pixel size relative to average speckle size | "1" |
| $R_{CH}$ | Ratio of the intensity in the center of the average core to that in center of average halo | "1" |
| $P$ | Degree of polarization | "1" |

Dimensionless quantities are indicated by "1"

# References


Abramowitz, M., & Stegun, I. A. (1970). *Handbook of mathematical functions*. U.S. Government Printing Office.

Born, M., 7 Wolf, E. (2003). Principles of optics (7th ed., revised). Cambridge University Press.

Cramer, H. (1954). *Mathematical methods in statistics*. Princeton University Press.

Dainty, J. C. (1984). Laser speckle and related phenomena. In J.C. Dainty (Ed.), Topics in applied physics (Vol. 9). Springer

Davenport, W. B., & Root, W. L. (1958). *An introduction to the theory of random signals and noise*. McGraw-Hill.

Goodman, J. W. (2006). *Speckle phenomena: Theory and applications*. Roberts and Company Publishers.

Jones, R. (1941a). A new calculus for the treatment of optical systems. I. Description and discussion of the calculus. *Journal of the Optical Society of America, 31*(7), 488–493

Jones, R. (1941b). A new calculus for the treatment of optical systems. III The Sohncke theory of optical activity. *Journal of the Optical Society of America, 31*(7), 500–503.

Jones, R. (1942). A new calculus for the treatment of optical systems. IV. *Journal of the Optical Society of America A, 32*(8), 486–493.

McKechnie, T. S. (1976). Image plane speckle in partially coherent illumination. *Optical and Quantum Electronics, 8*, 61–67.

Middleton, D. (1960). *Introduction to statistical communication theory*. McGraw Hill Book Co.

van Cittert, P. (1934). Die wahrscheinliche schwingungsverteilung in einer von einer lichtquelle direkt oder mittels einer linse beleuchteten ebene. *Physica, 1*, 201–210.

Wolf, E. (1959). Coherence properties of partially polarized electromagnetic radiation. *Nuovo Cimento, 13*, 1165–1181.

Zernike, F. (1938). The concept of degree of coherence and its application to optical problems. *Physica, 5*, 785–795.

# Chapter 12
# Star Image Dependence on Turbulence Structure Size

**Abstract** This chapter describes how computer-generated, random instantaneous wavefront realizations can be created that conform to prescribed atmospheric turbulence structure functions. To illustrate the crucial importance of turbulence structure size on star image appearance, two wavefront realizations are generated, one where the outer scale is less than one meter, the other where the outer scale is three times larger. Instantaneous images are then calculated at both visible and near- IR wavelengths for the two wavefronts as formed by a large telescope (4 m in diameter for the case illustrated). The calculation involves Fourier transformation of the complex amplitudes associated with the wavefronts, followed by squaring the modulus to give the final image intensity distributions. By comparing the images thus created—particularly in regard to the formation of image cores—with short- exposure star images obtained by an actual 4-m telescope, insight is gained about the turbulence outer scale limit in the actual atmosphere.

In this chapter, we examine how structure size influences the properties of star images formed by large ground-based astronomical telescopes. Image properties at both visible and IR wavelengths are considered. Two hypothetical types of turbulence are examined: one where the average structure size is relatively small and the other where the average structure size is several times larger.

Both types of turbulence are able to "account" for the relatively large seeing disk, i.e., halo-only star images often delivered by large telescopes at visible wavelengths; the angular size of this type of image typically falls in the range 0.5–1.5 arcsec. However, at near-IR wavelengths, the images have markedly different appearances for the two turbulence structure sizes. For the larger size, halo-only seeing disk images again arise, with angular sizes only marginally smaller than that of the seeing disk images formed at visible wavelengths. For the smaller average size, diffraction-limited cores form in the images containing sizeable fractions of the total light energy. For 4- or 5-m class telescopes, the angular size of these cores is typically an order of magnitude smaller than that of the halo; for future extremely large telescopes (ELTs), core angular sizes are likely to be two orders of magnitude smaller than those of the halos. Light not directed into the core in a near-IR star image remains in the halo, the

 
T. S. McKechnie, *General Theory of Light Propagation and Imaging Through the Atmosphere*, Progress in Optical Science and Photonics 20,
https://doi.org/10.1007/978-3-030-98828-9_12

angular size of which is only marginally smaller than that of the seeing disk image at visible wavelengths.

In Sect. 12.3, computer-generated instantaneous star image realizations are created at both visible and near-IR wavelengths for the two, smaller and larger, types of turbulence. Comparison of the core and halo characteristics of these computer-generated images with actual star images obtained at near-IR wavelengths using the 3.8-m Mayall telescope (Kitt Peak, AZ) provides insight into which, if either, of the two turbulence structure size types was likely present in the atmospheric observing paths at the time these images were obtained.

## 12.1 The Autocorrelation Function of the OPD Fluctuation

As indicated previously (8.35), the autocorrelation function of the optical path difference (OPD) fluctuation associated with the image-forming waves in the telescope pupil, $\rho(\xi, \eta)$, is related to the unit-normalized wavefront structure function, $\Sigma(\xi, \eta)$, as follows,

$$\Sigma(\xi, \eta) = 1 - \rho(\xi, \eta) \tag{12.1}$$

For Kolmogorov turbulence, the wavefront structure function has a 5/3 power dependence on distance separation. In the vicinity of the outer scale limit, $L_0$, the function flattens off, thereafter remaining approximately constant. The unit-normalized central section ($\eta = 0$) for such a function may be expressed in the form,

$$\begin{aligned} \Sigma(\xi, 0)_{Kol} &= \left(\frac{\xi}{L_0}\right)^{\frac{5}{3}} && \text{for } 0 \le \xi \le L_0 \\ &= 1 && \text{for } L_0 \le \xi \le \infty \end{aligned} \tag{12.2}$$

The above function is plotted in Fig. 12.1. An abrupt gradient discontinuity can be seen at $\xi = L_0$. In addition, as may readily be shown by studying the second derivative of this function, an infinitesimally small radius of curvature occurs in the limit $\xi \to 0$. Both features are of course physically unrealistic. Therefore, also shown in the figure is a best-fit Gaussian-based structure function, $\Sigma(\xi, 0)_{KolBF}$, (cf., 8.35 and 9.9) that roughly mimics the $\xi^{5/3}$ Kolmogorov functionality yet displays more realistic physical behavior at the inner and outer scale limits,

$$\Sigma(\xi, 0)_{KolBF} = 1 - \exp\left(-\frac{\xi^2}{{w_o}^2}\right) \approx 1 - \exp\left(-\frac{\xi^2}{(0.8 \cdot L_0)^2}\right), \tag{12.3}$$

where we see that the reasonably close fit of the two functions plotted in the figure is the result of setting $w_o$ as follows,

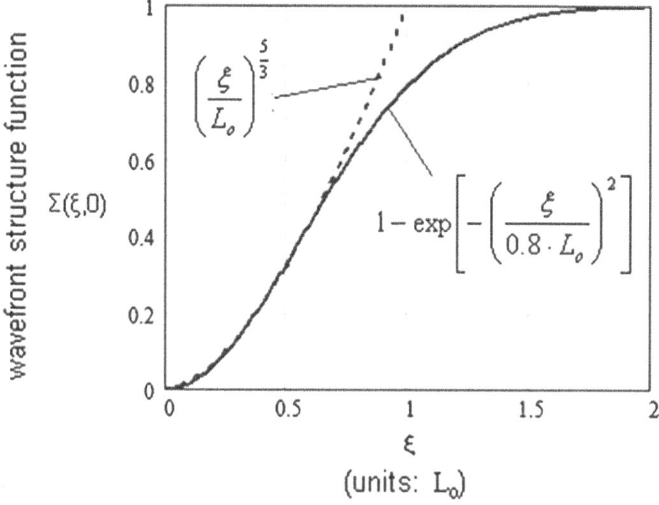


**Fig. 12.1** Kolmogorov wavefront structure function (*dashed line*) showing abrupt flattening at the outer scale limit, $L_0$, where 12.2 was used to calculate the plot. A more physically realistic structure function (*solid line*) showing gradual roll-off occurring around $L_0$

$$w_o = 0.8 \cdot L_0 \tag{12.4}$$

The autocorrelation function of the OPD fluctuation, $\rho(\xi, 0)$, corresponding to the wavefront structure function, $\Sigma(\xi, 0)_{KolBF}$, can be obtained from 12.1 and 12.3 in the form,

$$\rho(\xi, 0)_{KolBF} \approx \exp\left(-\frac{\xi^2}{(0.8 \cdot L_0)^2}\right) \tag{12.5}$$

It should again be emphasized that the Gaussian-based forms for $\Sigma(\xi, 0)_{KolBF}$ and $\rho(\xi, 0)_{KolBF}$ have been chosen here for illustration purposes only. Other functionalities and other forms of roll-off behavior could have been chosen that might have more closely modelled actual atmospheric turbulence. However, as becomes evident later, the "width" of these functions (as quantified by such measures as $L_0$, $w_o$, or FWHM) generally plays a larger role in determining image appearance than the detailed functionality itself. It might also be noted that the Gaussian-based function used here turns out to be remarkably consistent with the turbulence structure measurements made at various astronomical sites on three continents by Coulman et al. (1988). A detailed discussion of these measurements is given in Appendix C.

## 12.2 Generation of Instantaneous Wavefront Realizations

In this section, we describe how random instantaneous wavefront realizations may be generated whose statistical properties correspond to those of actual light waves arriving in the telescope pupil after having propagated over an extended atmospheric path. For simplicity, a one-dimensional analysis is given.

First, an intermediate "boxcar" wavefront is created by attaching computer-generated, quasi-random values of the OPD fluctuation, denoted by $H_I(x_i)$, at a large number, $N$, of points at equally spaced intervals, $\Delta x$. The $H_I(x_i)$ selections are drawn from a Gaussian-distributed ensemble with zero mean and variance $\sigma^2$ (cf., Sect. 6.2.3). The intermediate boxcar wavefront thus created extends over the baseline distance, $\Delta x \cdot (N-1)$, as illustrated in Fig. 12.2.

Apart from its Gaussian distribution, the other important properties of the random OPD selections, $H_I(x_i)$, may be summarized as follows,

$$\langle H_I(x_i)\rangle = 0 \quad \text{for all } i \text{ values, and} \tag{12.6}$$

$$\left\langle H_I(x_i)^2\right\rangle = \sigma^2 \quad \text{for all } i \text{ values.} \tag{12.7}$$

Also, since the $H_I(x_i)$ are all mutually uncorrelated, we may write as follows,

$$\begin{aligned}\langle H_I(x_i)\cdot H_I(x_k)\rangle &= \sigma^2 \quad \text{for } i = k,\\ &= 0 \quad \text{for } i \neq k.\end{aligned} \tag{12.8}$$

The intermediate boxcar wavefront generated in this way may be considered as having properties similar to those of white noise. This type of noise may be

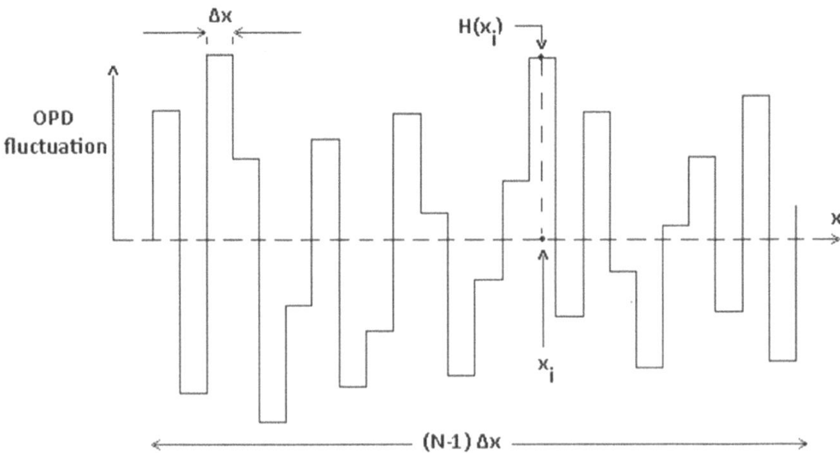


**Fig. 12.2** Boxcar intermediate wavefront

considered constituted by the sum of random spatial frequency components selected from a uniform probability distribution where the upper frequency limit is set by the Nyquist sampling frequency, which in this case is set by the sampling interval, $1/(2 \cdot \Delta x)$. An appropriately small interval, $\Delta x$, must be chosen, say $\Delta x \approx 1$ mm, to ensure that the highest spatial frequencies likely to be present in actual light waves arriving at the telescope are faithfully represented in the final wavefront realizations.

### 12.2.1 Smoothing the Intermediate Wavefront to Obtain the Final Wavefront

The final wavefront is created by smoothing the intermediate boxcar wavefront by an operation known as convolution (Davenport & Root, 1958). This operation may also be considered as spatial frequency filtering where the filtering is carried out using a low-pass filter, the purpose in this instance being to appropriately reduce, or eliminate, certain frequencies (or frequency bands) that are not consistent with the required wavefront structure function given by 12.3. The appropriate smoothing function in this case is simply the function whose self-convolution is identical to the desired autocorrelation function of the OPD fluctuation. For the autocorrelation function given by cf., 12.5, the appropriate smoothing function has the form,

$$\exp\left[-2 \cdot \left(\frac{\xi}{0.8 \cdot L_0}\right)^2\right]$$

To avoid end point artifacts commonly associated with convolution operations, we shall only attempt to generate the final wavefront realization over an inner subset of the $N$ sampled points, as set by the runner, $j$, where $j = N_S, \ldots, (N - N_S)$. A suitable choice of $N_S$ would necessarily have to fulfill the conditions $N_S \ll N$ and $N_S \gg 0.8 \cdot L_0/\Delta x$. Within the subrange, $N_S$ to $(N - N_S)$, the final wavefront realization, which we denote by $H(x_j)$ and anticipate as being free of end point artifacts, may be generated as follows,

$$H(x_j) = \frac{1}{K} \cdot \sum_{k=j-N_S}^{j+N_S} H_I(x_k) \cdot \exp\left[-2 \cdot \left(\frac{(k-j) \cdot \Delta x}{0.8 \cdot L_0}\right)^2\right], \quad \text{for } N_S < j < (N - N_S) \tag{12.9}$$

where the normalization constant, $K$, is given by

$$K = \left\{ \sum_{k=-N_S}^{N_S} \exp\left[ -4 \cdot \left( \frac{k \cdot \Delta x}{0.8 \cdot L_0} \right)^2 \right] \right\}^{\frac{1}{2}} \tag{12.10}$$

#### 12.2.1.1 Validity Check of the Final Wavefront Realization

By normalizing the final wavefront realization, $H(x_j)$, in the way just indicated, the statistical properties of this realization conform to the desired properties as we now demonstrate.

The autocovariance function of function $H(x_j)$ may be obtained directly from 12.9 in the form,

$$\begin{aligned} \langle H(x_{j+n}) \cdot H(x_j) \rangle &= \frac{1}{K^2} \\ &\times \sum_{k_1=j+n-N_S}^{j+n+N_S} \sum_{k_2=j-N_S}^{j+N_S} \langle H_I(x_{k_1}) \cdot H_I(x_{k_2}) \rangle \\ &\times \exp\left[ -2 \cdot \left( \frac{(k_1 - j - n) \cdot \Delta x}{0.8 \cdot L_0} \right)^2 \right] \cdot \exp\left[ -2 \cdot \left( \frac{(k_2 - j) \cdot \Delta x}{0.8 \cdot L_0} \right)^2 \right] \end{aligned} \tag{12.11}$$

By carrying out the summation with respect to $k_2$ and by making use of 12.8, the above equation simplifies to give

$$\langle H(x_{j+n}) \cdot H(x_j) \rangle = \frac{\sigma^2}{K^2} \cdot \exp\left[ -\left( \frac{n \cdot \Delta x}{0.8 \cdot L_0} \right)^2 \right] \cdot \sum_{k_1=j+n-N_S}^{j+n+N_S} \exp\left[ -4 \cdot \left( \frac{k_1 \cdot \Delta x}{0.8 \cdot L_0} \right)^2 \right] \tag{12.12}$$

By then carrying out the summation with respect to $k_1$ using 12.10 to substitute for the normalization constant, $K$, the above equation simplifies to the desired form,

$$\langle H(x_{j+n}) \cdot H(x_j) \rangle = \sigma^2 \cdot \exp\left[ -\left( \frac{n \cdot \Delta x}{0.8 \cdot L_0} \right)^2 \right] \tag{12.13}$$

The right-hand side of the above equation confirms that both the autocorrelation function of the OPD fluctuation (12.5) and the wavefront structure function associated with the fluctuation (12.3) do indeed have the desired functional forms. By setting $n$ to zero in the above equation, the variance of $H(x_j)$ can also be seen to take the desired value,

$$\langle H(x_j)^2 \rangle = \sigma^2 \tag{12.14}$$

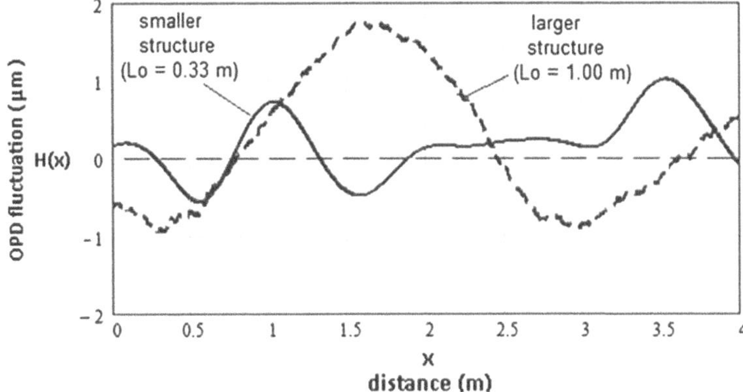


**Fig. 12.3** Wavefront realizations over a 4-m span. *Solid line* Small average turbulence structure size where $L_0 = 0.33$ m. *Dashed line* 3 times larger average turbulence structure size where $L_0 = 1.00$ m. The wavefront structure function for both of these wavefront realizations is given by 12.3 with the appropriate $L_0$ value inserted

## 12.3 Wavefront Realizations for Small and Large Turbulence Structure

Wavefront realizations obtained from 12.9 are shown in Fig. 12.3 for two turbulence structure scale sizes specified as follows: (1) relatively small average structure size corresponding to L = 0.33 m and (2) 3 times larger average structure size corresponding to $L_0 = 1.00$ m. The wavefront realizations are shown over a 4-m span so that they may be considered as instantaneous image-forming wavefronts in the pupil of a 4-m telescope.

As discussed previously in Chap. 7 (Sect. 7.3.2), by choosing the rms OPD values, $\sigma$, for each of the two wavefront realizations in proportion to the chosen $L_0$ values, in the short-wavelength limit, both wavefront realizations should produce images with about the same average angular size.[1] For the values actually used to produce the Fig. 12.3 wavefront realizations, $\sigma = 0.31$ μm and $\sigma = 0.93$ μm, visible wavelengths may be considered to lie within the "short-wavelength limit." It will be evident shortly that the actual image sizes corresponding to the two sets of values of $\sigma$ and $L_0$ are both about 1 arcsec, typical of star images delivered by large ground-based astronomical telescopes in average visible seeing conditions.

Figure 12.4 shows a second pair of wavefront realizations, in this case for (1) a relatively small average structure size corresponding to $L_0 = 0.33$ m and (2) 4 times larger average structure size corresponding to $L_0 = 1.33$ m. (12.3 was used as the unit-normalized wavefront structure function for these realizations with the appropriate $L_0$ value inserted.) In this figure, the instantaneous wavefront realizations

[1] In the short-wavelength limit, geometrical optics gives reasonably accurate image descriptions. The wavefront gradient distribution determines the angular size and shape of the average image.

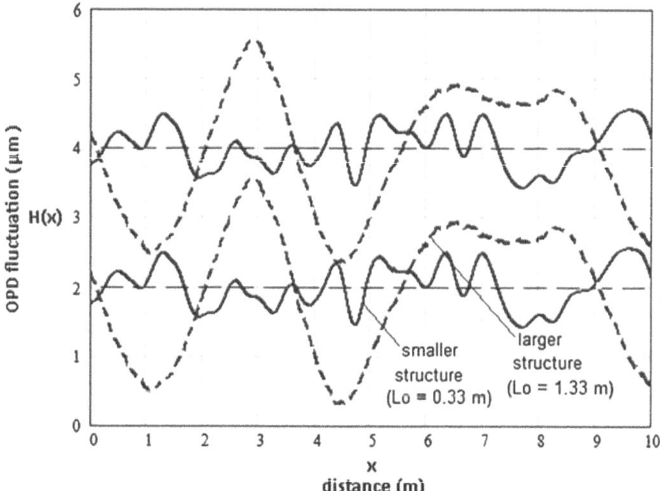


**Fig. 12.4** Wavefront realizations (all consistent with 1-arcsec seeing conditions) over a 10-m span at two closely separated instants of time. In the interval between the two instants, $6.67 \times 10^{-15}$ s, the wavefronts advance forward by $\sim 2\,\mu$m. *Solid line* Small average turbulence structure size ($\sigma = 0.31\,\mu$m, $L_0 = 0.33$ m). *Dashed line* 4 times larger average turbulence structure size ($\sigma = 1.24\,\mu$m, $L_0 = 1.33$ m)

for the two types of turbulence structure are depicted at two instants of time, separated by about $6.67 \times 10^{-15}$ s. During this extremely short time interval, the wavefronts advance forward by a distance of about 2 μm.

Because the rms OPD values used to calculate the Fig. 12.4 wavefronts, $\sigma = 0.31\,\mu$m and $\sigma = 1.24\,\mu$m, were again chosen in proportion to the $L_0$ value choices, all wavefront realizations in this figure produce images with approximately the same $\sim 1\, arcsec$ average angular size. These wavefront realizations may be regarded as typical of instantaneous image-forming wavefronts arriving in the pupil of a 10-m telescope for the two chosen (smaller and larger) average turbulence structure sizes.

By now considering that the actual wavelength of the light waves depicted in Fig. 12.4 also happens to be 2 μm (i.e., the same as the wavefront advancement that occurs over the time interval considered, $6.67 \times 10^{-15}$ s), the corresponding wavefront pairs at the two instants of time may then be considered as representing successive waves in the propagating wave trains. As is evident from the figure, in the case of the smaller average turbulence structure ($L_0 = 0.33$ m), the wavefront OPD disruptions appear relatively small compared to the wavelength interval; thus, the two corresponding wavefronts shown remain well separated from one another. Such a state of affairs is consistent with the formation of cores in the images formed from these wavefronts. It is also clear that, for the larger turbulence structure ($L_0 = 1.33$ m), where the wavefront OPD disruptions are proportionally larger, the two successive wavefronts encroach into regions that ideally should have been exclusive to only one or other of these wavefronts. Such a state of affairs is not consistent with the formation of image cores.

### 12.3.1 *Instantaneous Star Image Realizations at Visible Wavelengths*

Instantaneous star image realizations may readily be calculated from instantaneous wavefront realizations. To calculate the images, complex amplitudes of the form, $\exp\left[2 \cdot \pi \cdot i \cdot H(x_j)/\lambda\right]$, are first assigned to each of the sampled OPD values, $H(x_j)$, on the instantaneous image-forming wavefront in the telescope pupil (cf., 6.32). By Fourier transforming the discrete set of complex amplitudes thus generated, we can obtain the complex amplitudes at suitably sampled image locations. The final image intensity distribution can then be obtained by taking the squared modulus of these complex amplitudes. These two operations may be represented by the single equation,

$$I(\alpha, \lambda) = \left| \int_{Aperture} \exp\left[2 \cdot \pi \cdot i \cdot \frac{H(x)}{\lambda}\right] \cdot \exp\left[-2 \cdot \pi \cdot i \cdot \frac{\alpha \cdot x}{\lambda}\right] \cdot dx \right|^2 . \quad (12.15)$$

Figure 12.5 shows the actual instantaneous intensity distributions in the image at the visible wavelength, 0.55 μm, calculated from this equation for the two instantaneous wavefront realizations shown in Fig. 12.3, where we recall that these realizations correspond, respectively, to small ($L_0 = 0.33$ m) and 3 times larger ($L_0 = 1.00$ m) average turbulence structure size. The images shown in Fig. 12.5 may be considered as having been formed by a 4-m diffraction-limited telescope. Evidently, both images display Gaussian speckle characteristics. Also shown for reference in this figure are the corresponding average intensity envelopes for the two types of turbulence (calculated separately using 9.4); both envelopes indicate about the same 1-arcsec visible seeing quality.

From the similar overall appearances of the two instantaneous speckle pattern images shown in Fig. 12.5, evidently both small and large average turbulence structure sizes are equally capable of explaining image appearance at visible wavelengths in the (average) 1-arcsec seeing conditions considered.[2] Thus, we cannot draw any conclusions yet about which of the two types of turbulence might be the most representative of actual turbulence structure typically found in the atmosphere.

[2] The complex amplitudes associated with actual wavefronts arriving at a telescope generally comprise amplitude as well as phase variations. The image intensity distributions are shown in Fig. 12.5. These image intensity realizations may be considered as two possible image realizations drawn from a subset of the ensemble of all possible image apparitions, the entire ensemble inevitably including some images arising from wavefront realizations where there happens to be little scintillation over the finite (4-m) extent of the wavefront considered.

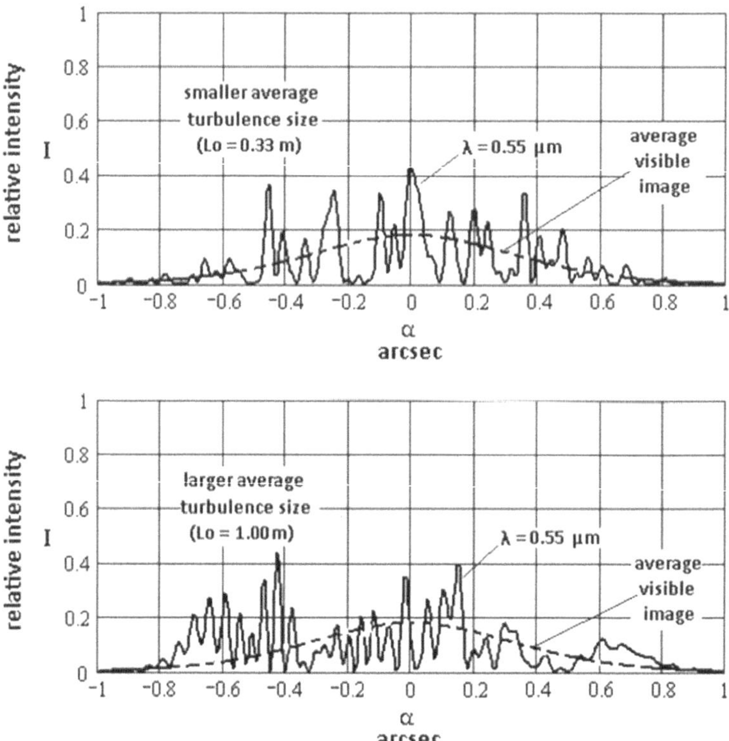


**Fig. 12.5** Instantaneous image intensity realizations calculated for a 4-m diffraction-limited telescope at the visible wavelength, 0.55 μm. *Top* Small average turbulence structure where $L_0 = 0.33$ m. *Bottom* 3 times larger average size where $L_0 = 1.00$ m. The average image envelopes (*dashed lines*) calculated from 9.6 are virtually identical for both types of turbulence, both of course smooth and speckle-free due to the averaging process

However, the account is not complete; we have yet to consider imaging behavior at IR wavelengths.

### 12.3.2 Instantaneous Star Image Realizations at Near-IR Wavelengths

Figure 12.6 shows the images obtained by the same 4-m diffraction-limited telescope at the near-IR wavelengths, 2.2 and 3.4 μm, for the wavefront realizations previously indicated in Fig. 12.3 (with 12.15 again used to calculate the image intensities). The images in Fig. 12.6 (Bottom) corresponding to the larger average turbulence structure size ($L_0 = 1.00$ m) again show speckle structure, not unlike that shown in Fig. 12.5 at visible wavelengths; the only obvious difference is that the average speckle size

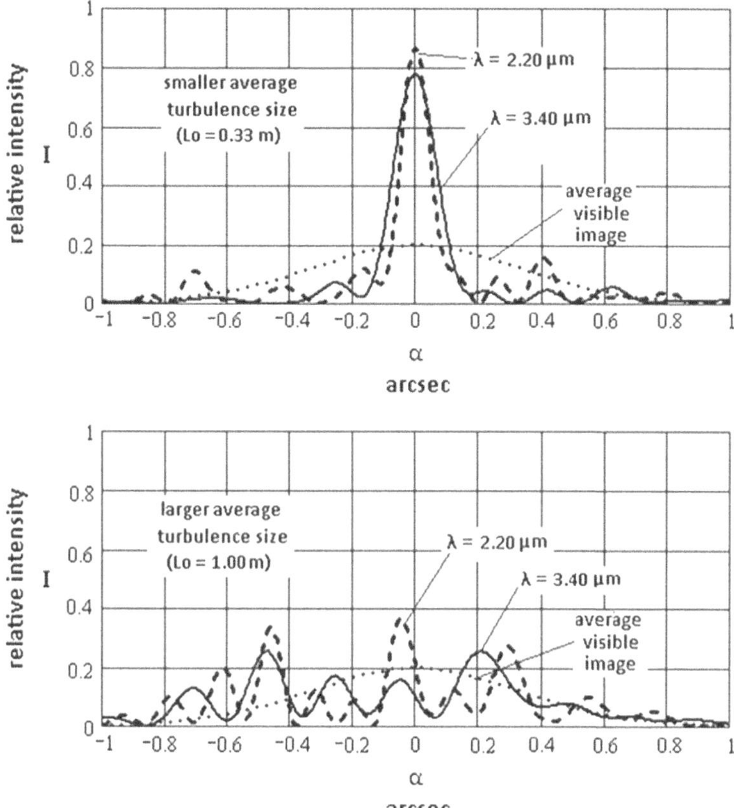


**Fig. 12.6** Instantaneous image intensity realizations calculated for a 4-m diffraction-limited telescope at the near-IR wavelengths 2.2 and 3.4 μm. *Top* Small average turbulence structure size ($L_0 = 0.33$ m). *Bottom* 3 times larger average turbulence structure size ($L_0 = 1.00$ m). *Dotted lines* The 1-arcsec average image envelopes at 0.55 μm (calculated from 9.6)

has now grown larger in proportion to the (longer) near-IR wavelengths considered. However, the images in Fig. 12.6 (Top) corresponding to the smaller average turbulence structure size ($L_0 = 0.33$ m) show entirely different characteristics. Strong central cores now appear in the images, superposed on background speckle halos.

### 12.3.3 Conclusions About the Average Turbulence Structure Size

The images in Fig. 12.5 show in a general way that, in the average $\sim 1 arcsec$ visible seeing conditions considered, both small and large average turbulence structure sizes are equally capable of explaining speckle pattern images at visible wavelengths.

During the period, 1964–1990, the outer scale limit, $L_0$, according to Kolmogorov theory, was generally supposed much larger than the aperture of even the largest telescope (4- and 5-m instruments being the largest at that time). Based on that supposition, resolution was expected to show only marginal improvement with increasing wavelength, as $\lambda^{-1/5}$.

According to this wavelength dependence, only a 30% resolution improvement is anticipated at 3.4 μm compared to resolution at visible wavelengths, while, at 2.2 μm, only a 25% improvement is anticipated. However, around 1990, when IR FPAs were first used on 4-m class telescopes, bright image cores were suddenly appearing routinely in short-exposure images formed at near-IR wave-lengths, even in quite average (~1-arcsec) visible seeing conditions. Images obtained by the Mayall telescope, shown previously in Figs. 10.10 and 10.11, were among the first to show IR image cores. It was now clear that resolution attainable from 4-m class ground-based telescopes at near-IR wavelengths is actually about 5- to 10-times higher than obtained at visible wavelengths.

By comparing the near-IR star image realizations shown in Figs. 12.5 and 12.6 for the (relatively) small and large average turbulence structure sizes with the actual star images shown in Figs. 10.10 and 10.11 obtained by the 3.8-m Mayall telescope at 2.2 μm, one cannot avoid the conclusion that the effective outer scale limit of turbulence structure size, $L_0$, in the typical astronomical seeing conditions prevailing when the images were obtained must have been significantly less than the very large values (>10 m) assumed in Kolmogorov theory.

### 12.3.4 Correlation Between Speckles at Closely Spaced Wavelengths

Figure 12.7 (Top) shows a wavefront realization (calculated over a 2-m aperture by the method described in Sect. 12.2) for 1-arcsec visible seeing conditions where the average turbulence structure size in the atmospheric path is assumed relatively small (corresponding to the parameter values $\sigma = 0.25\,\mu\text{m}$, $L_0 = 0.33\,\text{m}$). The corresponding image intensities obtained from a 2-m diffraction-limited telescope are shown in the same figure (Bottom) for the closely spaced wavelength pair, 0.50 and 0.60 μm. A high level of correlation is evident between the individual speckle features at these two closely spaced wavelengths. (Also shown for reference in the same figure is the average intensity envelope at 0.55 μm, calculated from 9.4, for the same $\sigma$ and $L_0$ values.)

Figure 12.8 (top) shows a wavefront realization over the same 2-m aperture for 3 times larger average turbulence structure size (corresponding to $\sigma = 0.75\,\mu m$, $L_0 = 1.00\,\text{m}$). The corresponding image intensities are shown (bottom) in the same figure for the same 2-m telescope and the same two wavelengths, 0.50 and 0.60 μm. The speckle patterns now appear much less correlated than before, and a larger amount

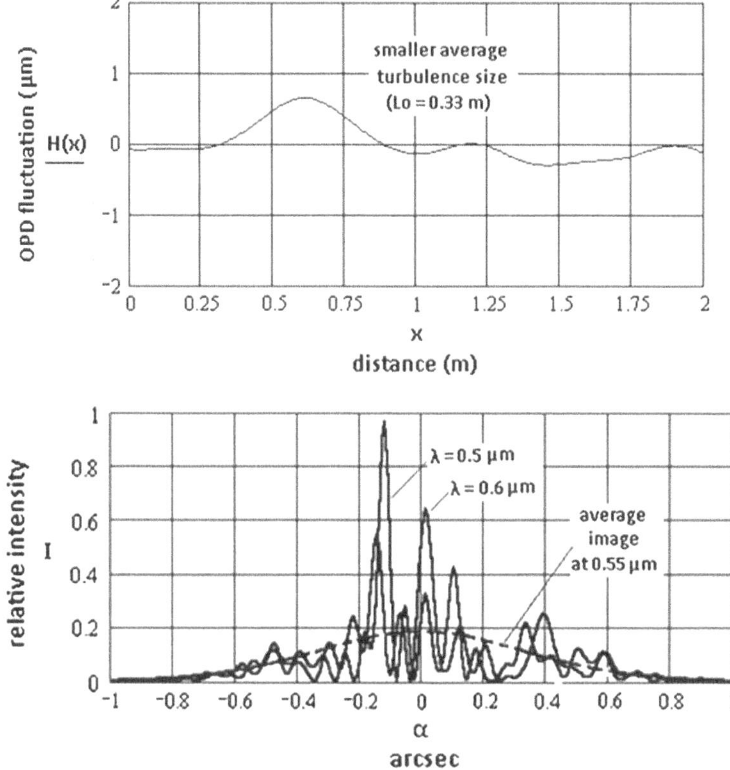


**Fig. 12.7** *Top* Random wavefront realization over a 2-m aperture for relatively small average turbulence structure size in 1-arcsec visible seeing. (Parameter value settings: $\sigma = 0.25\,\mu\text{m}$, $L_0 = 0.33\,\text{m}$.) *Bottom* The corresponding image realizations formed by a 2-m diffraction-limited telescope at wavelengths, 0.50 and 0.60 µm, showing a high degree of correlation (0.76) between the speckle intensities at the two wavelengths

of image wander is apparent.[3] (The average intensity envelope at 0.55 µm is also shown; this envelope is seen to be nearly identical to the envelope corresponding to smaller average turbulence structure size shown in Fig. 12.7.)

For a diffraction-limited telescope, the intensity correlation coefficient, $|\mu|^2$, for the speckle in the center of the star image at two arbitrary wavelengths, $\lambda_1$ and $\lambda_2$, was given previously (cf., 8.7) by $\exp\left[-4 \cdot \pi^2 \cdot \sigma^2 \cdot (1/\lambda_1 - 1/\lambda_2)^2\right]$. For the $\sigma$ values considered here, 0.25 and 0.75 µm, and the wavelength pair, 0.5 and 0.6 µm, the respective intensity correlation coefficients, $|\mu|^2$, take the values 0.76 and 0.08.

[3] Woolf et al. (1982) described images with the MMT and 2.3-m telescopes in which they saw less image motion than predicted by conventional theory, stating that "in data where the mean image size changed over a factor 2.5, no trend in image motion was discernible." However, according to Kolmogorov theory, image motion supposedly scales linearly with mean image size.

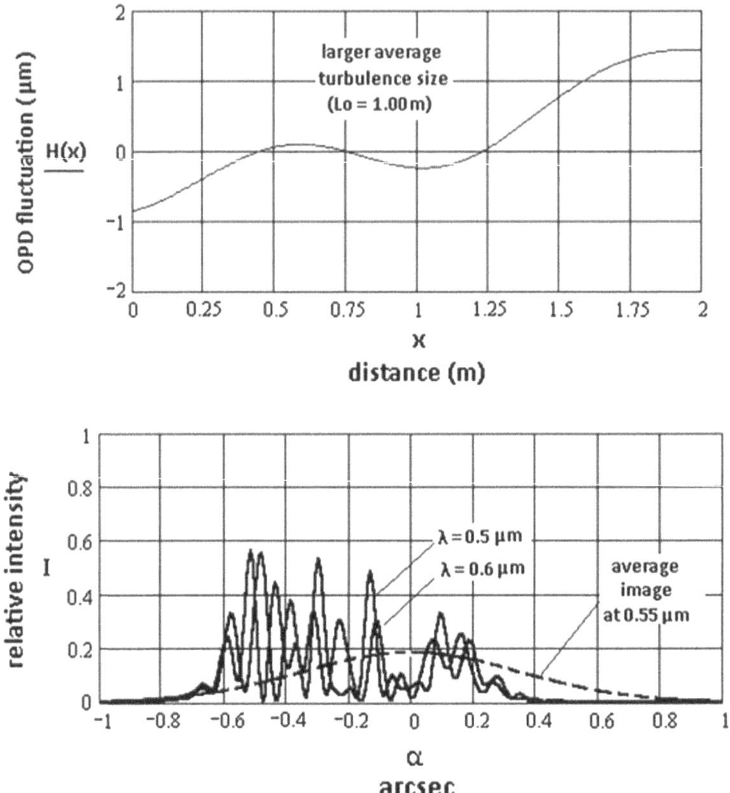


**Fig. 12.8** *Top* Random wavefront realization over a 2-m aperture for relatively large average turbulence structure size in 1-arcsec visible seeing. (Parameter values: where $\sigma = 0.75\,\mu m$.) *Bottom* The corresponding image realizations formed by the same 2-m diffraction-limited telescope at wavelengths 0.50 and 0.60 μm. The two speckle patterns are seen to have a lower degree of correlation (0.085) than those in Fig. 12.7 (cf., at abscissa value −0.24); a significantly larger amount of image wander is also evident

## 12.4 Atmospheric MTFs for Small and Large Turbulence

We saw in Sect. 12.3 that, in average (~1-arcsec) visible seeing conditions, visible image appearance can readily be explained in terms of either small or large average turbulence structure sizes. Evidently, if the "explanation" is based on an incorrect choice of the average size, the error can easily be compensated by making a proportionately incorrect choice for the rms OPD fluctuation, $\sigma$. (Note that for Kolmogorov turbulence, the variance of the OPD fluctuation, $\sigma^2$, is proportional to the integral over the atmospheric path, $\int C_n(z)^2 \cdot dz$. Thus, an incorrect choice of $\sigma$ would correspond to an incorrect choice of the value of this integral.)

Comparable behavior occurs with the atmospheric modulation transfer function (MTF), a function that can be obtained for present purposes by combining 6.52 and 12.5, which allows a $\eta = 0$ central section to be expressed in the form,

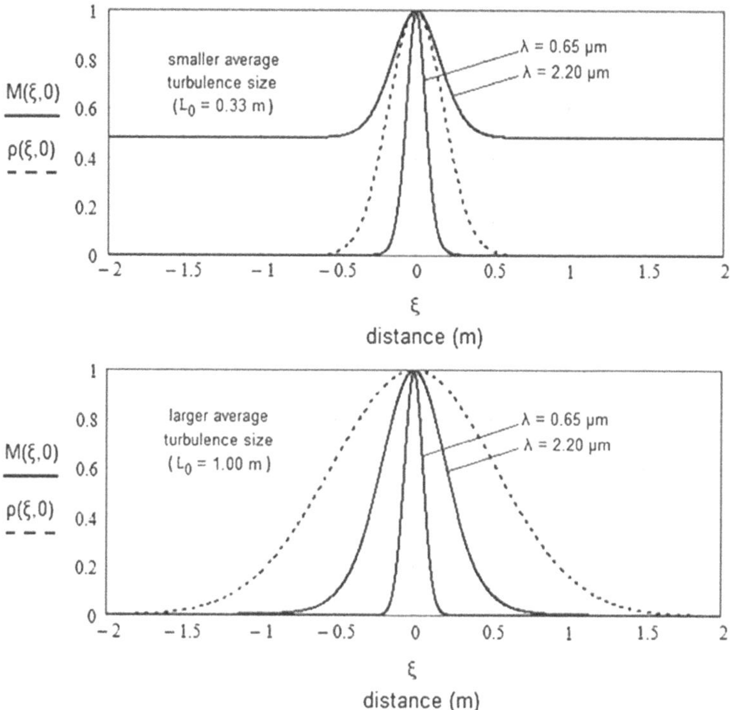


**Fig. 12.9** Atmospheric MTFs at visible and IR wavelengths for two different autocorrelation functions of the integrated OPD fluctuation. *Top* Relatively small average turbulence structure size. *Bottom* 3 times larger average turbulence structure size for visible and near-IR wavelengths. Plots calculated using 6.52 and 12.15 with parameter values (*top*) $\sigma = 0.30\,\mu\text{m}$, $L_0 = 0.33\,\text{m}$, and (*bottom*) $\sigma = 0.90\,\mu\text{m}$, $L_0 = 1.00\,\text{m}$. Both parameter sets correspond to about ~ 0.8-arcsec visible seeing

$$M(\xi,0) = \exp\left\{-\left(\frac{2\cdot\pi\cdot\sigma}{\lambda}\right)^2\cdot\left[1-\exp\left[-\left(\frac{\xi}{0.8\cdot L_0}\right)^2\right]\right]\right\} \tag{12.16}$$

As illustrated in Fig. 12.9, for the visible wavelength, $\lambda = 0.65\,\mu m$, evaluations of 12.16 for either the small or large average turbulence structure sizes—corresponding to the parameter value sets, ($\sigma = 0.30\,\mu\text{m}$, $L_0 = 0.33\,\text{m}$) and ($\sigma = 0.90\,\mu\text{m}$, $L_0 = 1.00\,\text{m}$)—result in two atmospheric MTFs having nominally the same 15-cm full-width half-maximum (FWHM). Thus, to produce the "correct" atmospheric MTF in this case, it is only necessary to choose the $\sigma$ values in proportion to the assumed average turbulence structure size (which is proportional to $L_0$). However, to explain the existence of image cores formed by a 4-m-diameter telescope at the near-IR wavelength, 2.2 μm, the atmospheric MTF at this wavelength must necessarily account for phase correlations over distance spans at least as large as the 4 m diameter of the telescope.

Evaluations of 12.16 at $\lambda = 2.2\,\mu\text{m}$ to obtain the atmospheric MTF sections, $M(\xi, 0)$, for the parameter value sets—($\sigma = 0.30\,\mu\text{m}$, $L_0 = 0.33\,\text{m}$) and ($\sigma = 0.90\,\mu\text{m}$, $L_0 = 1.00\,\text{m}$)—produce two quite different looking atmospheric MTFs. Both result in asymptotic values being attained within the 4-m extent of the telescope aperture. As can be seen from 12.16, the asymptotic values that arise in the limit of large $\xi$ are set by the quantity, $\exp\left[-(2 \cdot \pi \cdot \sigma/\lambda)^2\right]$. For the case of smaller average turbulence size, where $\sigma = 0.30\,\mu\text{m}$, the asymptotic value for $\lambda = 2.2\,\mu\text{m}$ is about 0.48; such a value indicates (cf., 10.3) that about 48% of the light energy in the image lies in a central core. For the case of larger average turbulence size, where $\sigma = 0.90\,\mu\text{m}$, the asymptotic value is a barely significant 0.001, indicating the absence of any significant core feature.

To explain the routine existence of cores in star images formed by 4-m class telescopes at the near-IR wavelength, 2.2 μm, examples of which were shown previously in Figs. 10.10 and 10.11, the distances over which the atmospheric MTF maintains significantly high values at this near-IR wavelength must be at least 4-m. As can be seen from the above analysis, to explain such behavior requires the average turbulence structure size to be smaller than 1 m.

## 12.5 Mathematical Notation Used in This Chapter

The mathematical notation used in this chapter is indicated in Table 12.1.

**Table 12.1** Mathematical notation used in *this* chapter along with the SI dimensional units of the individual quantities

| Symbol | Quantity | Dimensions |
|---|---|---|
| $\lambda$ | Wavelength | m |
| $(x, y)$ | Cartesian coordinate system in telescope pupil | m |
| $(\xi, \eta)$ | Cartesian coordinate system in plane perpendicular to the direction of light travel | m |
| $(\alpha, \beta)$ | Angular coordinate system in telescope image plane | "1" |
| $H$ | Integrated OPD fluctuation over entire atmospheric path | m |
| $\sigma$ | rms of integrated OPD fluctuation, $H$ | m |
| $\rho$ | Autocorrelation function of integrated OPD fluctuation, $H$ | "1" |
| $w_0$ | $1/e$ half-width for Gaussian $\rho$ function | m |
| $\Sigma$ | Wavefront structure function | "1" |
| $K$ | A constant | "1" |
| $M$ | Atmospheric MTF | "1" |
| $I$ | Image intensity | "1" |
| $L_0$ | Outer scale of Kolmogorov turbulence structure | m |
| $\mu$ | Complex coherence factor in image | "1" |

Dimensionless quantities are indicated by "1."

## References

Coulman, C. E., Vernin, J., Coqueugniot, Y., & Caccia, J. L. (1988). Outer scale of turbulence appropriate to modeling refractive index structure profiles. *Applied Optics, 27*, 155–160.

Davenport, W. B., & Root, W. L. (1958). *An introduction to the theory of random signals and noise*. McGraw-Hill.

Woolf, N. J., McCarthy, D. W., & Angel, J. R. (1982). Performance of the MMT VII: Image shrinkage in sub-arc second seeing at the MMT and 2.3 meter telescopes. In L. D. Barr & G. Burbridge (Eds.), *Advanced technology optical telescopes. Proceedings of SPIE Engineering* (Vol. 332, pp. 50–56). Bellingham.

# Chapter 13
# Approximation of Star Images Formed by Large Telescopes

**Abstract** So far, we have seen that unresolved star images can appear in core-only, core-and-halo, and halo-only guises. Since the Fourier transform integral equations that describe these images may not be transparent to all readers, in this chapter we simplify the mathematics by approximating star images by best-fitting Gaussian functions. Single Gaussian functions are used to approximate core-only or halo-only images; two summed Gaussian functions are used to approximate core-and-halo images. Because Fourier transformation of Gaussian functions can be performed analytically, the integral equations effectively vanish, leaving more transparent and relatively simple mathematics. To create the best-fit Gaussian approximation to an actual star image intensity envelop (obtained at some convenient wavelength) appropriate values must be set for three key parameters: the relative central intensities of the two Gaussian functions, $A_{CH}$, and their respective 1/e half-widths, $B_C$ and $B_H$. Once values have been obtained for these three quantities, it is a simple matter to calculate the atmospheric seeing parameters, $\sigma$ and $\rho(\xi, \eta)$, and the telescope Strehl intensity. If adaptive optics is used, $\sigma_{AO}$ and $\rho_{AO}(\xi, \eta)$ are obtained rather than $\sigma$ and $\rho(\xi, \eta)$. Once these measures have been obtained, star image intensity envelops—and all other star image properties—can be directly calculated for all other wavelengths in the near-UV to far-IR wavelength range.

In previous chapters, we saw that point-object images—unresolved star images in particular—can appear in a variety of forms depending on the telescope size, the telescope aberrations, the imaging wavelength, the seeing conditions and, not least, whether or not adaptive optics (AO) is used. While some images show just a core, and some just a halo, the most general type of star image displays both features.

Many of the equations used to describe the properties of these images appeared in the form of integral equations—a type of equation that has little transparency even for those familiar with this kind of mathematics. In this chapter, therefore, we develop a considerably simpler set of equations. In exchange for providing slightly less accurate image descriptions, the equation set provides greater physical transparency while significantly reducing the computational burden.



T. S. McKechnie, *General Theory of Light Propagation and Imaging Through the Atmosphere*, Progress in Optical Science and Photonics 20,
https://doi.org/10.1007/978-3-030-98828-9_13

The simplified equation set enables us to: (1) express the average image intensity envelopes for star images (in whatever core and halo guise they might take) in a relatively simple way in terms of the parameters that describe the telescope optics and the atmospheric seeing, and (2) develop an inverse set of equations that allow the telescope optics and atmospheric seeing parameters to be calculated from readily measurable properties of star image intensity envelopes.

The formulae developed capture the underlying physics just as before, yet reduce the to-and-fro calculations to a much simpler, back-of-the-envelop level. The simplifications are made possible by approximating the star image intensity envelopes for the most general type of core-and-halo star image as the sum of two Gaussian functions,[1] one of which is chosen to best fit the core and the other to best fit the halo.

For circularly symmetric image envelopes, the entire image can be fully specified by fixing just three independent parameters relating to the two Gaussian functions: One is the ratio of the central intensities in the individual core and halo portions of the image; the other two are the respective 1/e angular half-widths of the two image portions. Halo-only or core-only images correspond to the degenerate cases when the intensity ratio parameter takes one or other of the two limiting values, zero or infinity.

For non-circularly symmetric image envelopes, five independent parameters are required to specify the image intensity envelop; in this case, the core and halo image portions are each described in terms of elliptical Gaussian functions whose widths are specified in the two appropriate orthogonal directions. Irrespective of whether we consider telescopes with or without AO capability, star images always exhibit the same general type of core and halo characteristics. Thus, the back-of-the-envelop formulae developed in this chapter apply equally to telescopes with or without AO.

For some applications, the Gaussian approximations outlined in this chapter might be judged insufficiently accurate. For these applications, the option still remains of reverting back and using the more exact formulations given in Chaps. 8–10. A further option is provided in Chap. 18 where a more accurate and physically realistic three-Gaussian function model is developed, a model particularly well-suited for dealing with star image formation by the coming generation of Extremely Large Telescopes (ELTs).

[1] Functional forms other than Gaussian could have been used to approximate the intensity envelops. However, in Sect. 13.1, it is apparent that the use of Gaussian functions generally provides realistic approximations to actual average image intensity envelopes.

## 13.1 Gaussian Approximations for Unresolved Star Images

### 13.1.1 General Properties of Gaussian Functions

Suppose we approximate the average unit-normalized image intensity envelop of an unresolved star, $\langle I(\vartheta, \lambda)\rangle$, by the best-fit circularly symmetric Gaussian function,

$$\langle I(\vartheta, \lambda)\rangle = \exp\left[-\left(\frac{\vartheta}{\vartheta_0}\right)^2\right], \tag{13.1}$$

where $\vartheta_0$ denotes the $1/e$ angular half-width of the Gaussian envelop. Other equally valid, and directly related, measures of the angular widths of Gaussian intensity envelopes are the following:

The $1/e$ full-width, which we denote by $FW(1/e)$,

$$FW\left(\frac{1}{e}\right) = 2 \cdot \vartheta_o \tag{13.2}$$

the $1/e^2$ full-width, which we denote by $FW\left(1/e^2\right)$,

$$FW\left(\frac{1}{e^2}\right) = 2 \cdot \sqrt{2} \cdot \vartheta_o \approx 2.828 \cdot \vartheta_o \tag{13.3}$$

and the full-width half-maximum width which we denote by FWHM,

$$FWHM = 2 \cdot \sqrt{\ln(2)} \cdot \vartheta_o \approx 1.665 \cdot \vartheta_o \tag{13.4}$$

Any one of the above angular width measures may be used, as convenient, to provide a common yardstick for comparing the angular widths of images approximated by Gaussian forms as defined by 13.1.

The light energy contained under any circularly symmetric intensity envelop is proportional to the volume enclosed under that envelop. For the Gaussian envelop considered here, the enclosed volume is given by

$$2 \cdot \pi \cdot \int_0^\infty \exp\left[-\left(\frac{\vartheta}{\vartheta_o}\right)^2\right] \cdot \vartheta \cdot d\vartheta = \pi \cdot \vartheta_o^2 \tag{13.5}$$

An elliptical Gaussian function might sometimes provide a better approximation for non-circularly symmetric star image envelops. The volume enclosed under such an envelop is given by

$$\int_{-\infty}^{\infty}\int_{-\infty}^{\infty} \exp\left[-\left(\frac{\alpha^2}{\alpha_o^2}+\frac{\beta^2}{\beta_o^2}\right)\right]\cdot d\alpha\cdot d\beta = \pi\cdot\alpha_o\cdot\beta_o \tag{13.6}$$

It is noted for later use that the Fourier transform of the above elliptical Gaussian function can be written in the form,

$$\begin{aligned}
&\int_{-\infty}^{\infty}\int_{-\infty}^{\infty} \exp\left[-\left(\frac{\alpha^2}{\alpha_o^2}+\frac{\beta^2}{\beta_o^2}\right)\right] \\
&\times \exp\left[\frac{2\cdot\pi\cdot i\cdot(\alpha\cdot u+\beta\cdot v)}{\lambda}\right]\cdot d\alpha\cdot d\beta \\
&\quad = \pi\cdot\alpha_o\cdot\beta_o\cdot\exp\left[-\left(\frac{\pi^2\cdot(\alpha_o^2\cdot u^2+\beta_o^2\cdot v^2)}{\lambda^2}\right)\right]
\end{aligned} \tag{13.7}$$

The Fourier transform of the circular Gaussian function (13.1) can be readily obtained as a special case of the above equation; it can also be obtained from the Hankel transform,

$$\begin{aligned}
&2\cdot\pi\cdot\int_{0}^{\infty} \exp\left[-\left(\frac{\vartheta}{\vartheta_o}\right)^2\right]\cdot J_0\left(\frac{2\cdot\pi\cdot\vartheta\cdot r}{\lambda}\right)\cdot\vartheta\cdot d\vartheta \\
&\quad = \pi\cdot\alpha_o^2\cdot\exp\left[-\frac{\pi^2\cdot\alpha_o^2\cdot\left(u^2+v^2\right)}{\lambda^2}\right]
\end{aligned} \tag{13.8}$$

where $J_0(\cdot)$ is the zero-order Bessel function of the first kind.

### 13.1.2 Gaussian Approximations for the Telescope PSF and the Image Core

Because the shape of the average star image core formed by a telescope is solely determined by the telescope intensity point spread function (PSF) (Sect. 10.1), the Gaussian approximation for the intensity PSF of the telescope can also be used to approximate the image core. The only real difference between the core feature and the intensity PSF is that the latter only contains the fraction, $\exp\left[-(2\cdot\pi\cdot\sigma/\lambda)^2\right]$, of the total light energy in the image.[2]

We now develop the appropriate circular Gaussian functions that can be used to best fit both the telescope intensity PSF and the image core intensity envelop for the case of circularly symmetric images. Images of this kind are formed by refractor or

[2] The halo contains the remaining light energy fraction, $1-\exp\left[-(2\cdot\pi\cdot\sigma/\lambda)^2\right]$.

reflector telescopes with circular apertures, with or without circular central obstructions, when the instruments are either diffraction-limited or have circularly symmetric aberrations.

#### 13.1.2.1 Gaussian Approximation to an Airy Pattern

The unit-normalized intensity distribution in an Airy pattern image may be expressed by (cf. 4.29) in the form,

$$I_{Airy}(\vartheta,\lambda)=\left[2\cdot\frac{J_1\left(\frac{\pi\cdot D\cdot\vartheta}{\lambda}\right)}{\frac{\pi\cdot D\cdot\vartheta}{\lambda}}\right]^2 \tag{13.9}$$

where $\vartheta$ is the radial angular coordinate in image space (cf., Fig. 7.2), $D$ is the telescope diameter, and $\lambda$ is the wavelength. The first dark Airy ring occurs when the $J_1(\cdot)$ Bessel function argument takes the value 3.832. Thus, the first dark ring occurs at an angular coordinate given by the familiar formula,

$$\vartheta_{1st\,Dark\,Ring}=\frac{3.832\cdot\lambda}{\pi\cdot D}=\frac{1.22\cdot\lambda}{D} \tag{13.10}$$

Numerical evaluations of 13.9 show that the FWHM width of the Airy pattern is given by

$$FWHM_{Airy}=\frac{3.232\cdot\lambda}{\pi\cdot D}=\frac{1.03\cdot\lambda}{D} \tag{13.11}$$

We now wish to approximate the unit-normalized Airy pattern as defined by 13.9 by a circular Gaussian function that identically matches the Airy pattern in terms of both the unit-normalized central intensity and, upon rotation, the total enclosed volume. We represent this Gaussian function by

$$I_{GA}(\vartheta,\lambda)=\exp\left[-\left(\frac{\vartheta}{\vartheta_{GA}}\right)^2\right] \tag{13.12}$$

where the "GA" suffix used with the 1/e half-width of the function, $\vartheta_{GA}$, reflects the association of the Gaussian approximation to the Airy pattern. We now wish to establish the actual $\vartheta_{GA}$ value for which the Gaussian function, upon rotation, encloses the same volume (light energy) as enclosed under the Airy pattern. It can readily be shown that the volume under the Airy pattern intensity envelop described by 13.9 is as follows:

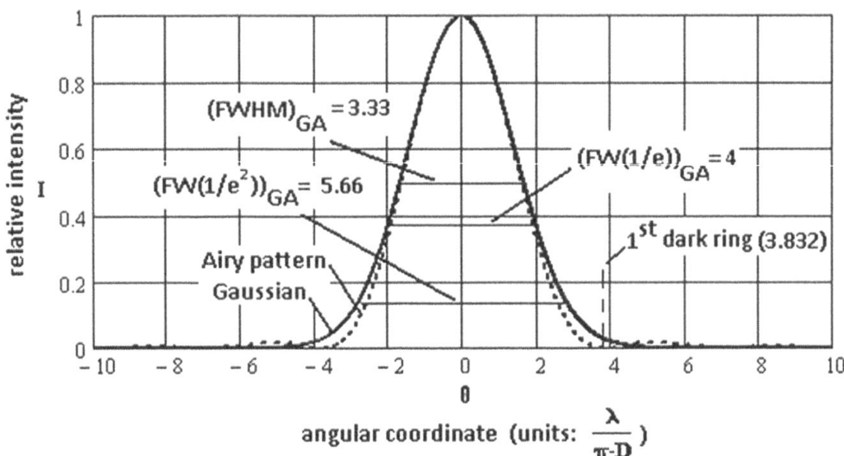


**Fig. 13.1** Unit-normalized Airy pattern and its Gaussian approximation. Upon rotation, both envelops enclose the same volume (light energy)

$$2 \cdot \pi \cdot \int_0^{\infty} \left[ 2 \cdot \frac{J_1\left(\frac{\pi \cdot D \cdot \vartheta}{\lambda}\right)}{\frac{\pi \cdot D \cdot \vartheta}{\lambda}} \right]^2 \cdot \vartheta \cdot d\vartheta = \frac{4 \cdot \lambda^2}{\pi \cdot D^2} \tag{13.13}$$

The volume enclosed under the Gaussian approximation to the Airy pattern given by 13.12 is given by $\pi \cdot \vartheta_{GA}^2$ (cf., 13.5). Thus, by choosing $\vartheta_{GA}$ as follows,

$$\vartheta_{GA} = \frac{2 \cdot \lambda}{\pi \cdot D} = \frac{0.637 \cdot \lambda}{D}, \tag{13.14}$$

the volumes contained under the Airy pattern and its Gaussian approximation are identical.

Summarizing here, with $\vartheta_{GA}$ set to the value given by 13.14, the Gaussian approximation to the Airy pattern (13.12) has exactly the same central intensity value as the Airy pattern, unity, and also encloses the same amount of light energy.

It might be noted that the angular quantity, $\vartheta_{GA}$, given by 13.14 is approximately half that of the Rayleigh angular resolution limit, $1.22 \cdot \lambda/D$. The Airy pattern described by 13.9 and its Gaussian approximation (defined by 13.12 and 13.14) are plotted in Fig. 13.1.

#### 13.1.2.2 Approximations for Aberrated Images

Various algorithms can be used to provide best-fit Gaussian approximations to the intensity PSFs formed by aberrated telescopes. Two algorithms are considered below, referred to as Algorithm (1) and Algorithm (2). The choice between the two for any particular application depends on the Strehl intensity of the telescope. Algorithm

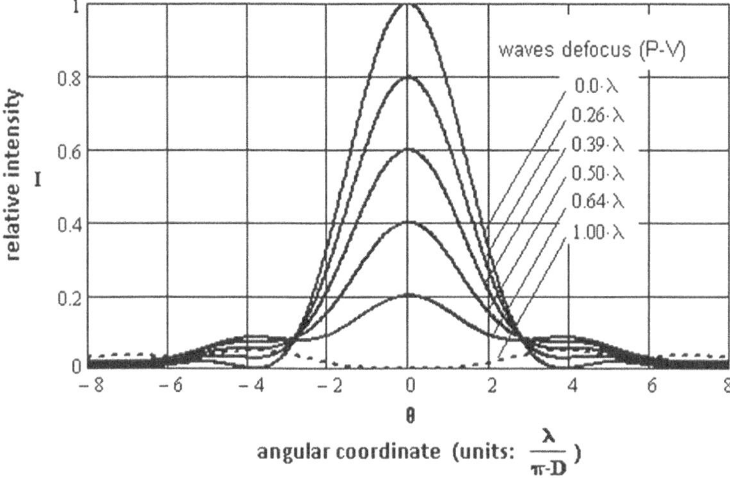


**Fig. 13.2** Degradation of an Airy pattern caused by defocus. P–V wavefront errors of less than $0.26\lambda$ are consistent with substantially diffraction-limited performance ($SI \geq 0.8$). However, one full wave of defocus causes catastrophic image degradation

(1) is generally the better choice for mildly aberrated images where Strehl intensity lies in the approximate range of $0.4 \leq SI(\lambda) \leq 1$; Algorithm (2) is generally better for more severely aberrated images where Strehl intensity lies in the range $0 \leq SI(\lambda) \leq 0.4$.

Both of these Strehl regimes are represented in Fig. 13.2 where point-object images are shown for a diffraction-limited telescope with circular aperture for various amounts of defocus. Evidently, one full wave (P-V) of defocus causes Strehl intensity to collapse towards zero.

Although it is not possible to build a perfectly diffraction-limited telescope,[3] if Strehl intensity can be held at, or above, the so-called Strehl limit, the telescope may be considered diffraction-limited for all practical purposes.

$$SI_{lim}(\lambda) = 0.8 \tag{13.15}$$

### Algorithm 1: Best-Fit Gaussian Approximations for Mild Telescope Aberrations

For Strehl intensities lying in the range of $0.4 \leq SI(\lambda) \leq 1$, the Gaussian approximation for the intensity PSF of a telescope may be expressed in the form

$$I_{G(PSF)}(\vartheta, \lambda) = SI(\lambda) \cdot \exp\left[-\left(\frac{\vartheta}{\vartheta_{G(PSF)}}\right)^2\right] \tag{13.16}$$

[3] The Strehl intensity delivered by the Hubble Space Telescope comfortably surpasses the value, 0.8, for all visible and IR wavelengths.

where we notice that this form maintains the correct Strehl intensity. By also choosing $\vartheta_{G(PSF)}$ thus,

$$\vartheta_{G(PSF)} = \frac{\vartheta_{GA}}{\sqrt{SI(\lambda)}} \tag{13.17}$$

where $\theta_{GA}$ was given by 13.14. The volume, or light energy, enclosed under the Gaussian approximation remains the same as that enclosed under the Airy pattern formed by an aberration-free version of the telescope (cf., 13.9). Combining 13.16 and 13.17 gives the Gaussian approximation for the intensity PSF in this Strehl intensity regime in the form

$$I_{G(PSF)}(\vartheta, \lambda) = SI(\lambda) \cdot \exp\left[-\left(\frac{\vartheta}{\left(\frac{\vartheta_{GA}}{\sqrt{SI(\lambda)}}\right)}\right)^2\right] \tag{13.18}$$

The average intensity envelop for a core formed by the same telescope follows directly from the above equation by simply attaching the appropriate multiplier factor (cf., 10.3) to account for the reduced energy fraction in the core,

$$\left\langle I_{G(Core)}(\vartheta, \lambda)\right\rangle = \exp\left[-\left(\frac{2 \cdot \pi \cdot \sigma}{\lambda}\right)^2\right] \cdot SI(\lambda) \cdot \exp\left[-\left(\frac{\vartheta}{\left(\frac{\vartheta_{GA}}{\sqrt{SI(\lambda)}}\right)}\right)^2\right] \tag{13.19}$$

**Algorithm 2: Best-Fit Gaussian Approximations for Badly Aberrated Telescopes**

For severe aberrations, telescope Strehl intensity typically takes values in the range, $0 \leq SI(\lambda) \leq 0.4$. As shown in Fig. 13.2, Strehl intensity can even go to zero if the aberrations are large enough. In such cases, because Strehl intensity no longer correlates with image width, it no longer provides a useful measure of telescope resolution. In this Strehl regime, Algorithm (2), which uses a least-mean-squares best-fit algorithm, provides a more sensible scheme for approximating the image.

With Algorithm (2), it is convenient to deal in terms of an effective Strehl intensity value which we denote by $SI_{Ef}(\lambda)$. This Strehl value relates to the intensity in the center of the best-fit Gaussian envelop, rather than to the intensity attained in the center of the actual image. The Gaussian approximation in this case may be expressed in the form,

$$I_{G(PSF)}(\vartheta, \lambda) = SI_{Ef}(\lambda) \cdot \exp\left[-\left(\frac{\vartheta}{\left(\frac{\vartheta_{GA}}{\sqrt{SI_{Ef}(\lambda)}}\right)}\right)^2\right], \tag{13.20}$$

where the normalization again preserves the enclosed volume, $\pi \cdot \theta_{GA}^2$. It might be observed here that the only difference between the above equation and 13.18 is that now the Strehl intensity, $SI(\lambda)$, has been replaced by the effective Strehl intensity, $SI_{Ef}(\lambda)$. The $SI_{Ef}(\lambda)$ value that produces the least-mean-square best-fit may be identified using numerical methods. Since the value so obtained directly correlates with image width, it provides a meaningful indication of image resolution.

The average intensity envelop for a core formed by the same telescope follows directly from the above equation by simply attaching the core energy multiplier factor,

$$\left\langle I_{G(Core)}(\vartheta, \lambda)\right\rangle = \exp\left[-\left(\frac{2 \cdot \pi \cdot \sigma}{\lambda}\right)^2\right] \cdot SI_{Ef}(\lambda) \cdot \exp\left[-\left(\left(\frac{\vartheta}{\left(\frac{\vartheta_{GA}}{\sqrt{SI_{Ef}(\lambda)}}\right)}\right)\right)^2\right] \tag{13.21}$$

Again, we observe that the only difference between the above equation and 13.19 is that the actual Strehl intensity, $SI(\lambda)$, has been replaced by an effective Strehl intensity, $SI_{Ef}(\lambda)$.

#### 13.1.2.3 Approximating the PSF of a Centrally Obstructed Telescopes

Most large astronomical telescopes are Cassegrain reflectors with circular apertures and circular central obstructions. For diffraction-limited instruments of this type, the intensity PSF was previously given by 4.45. For convenience, we reproduce that equation here,

$$I(\vartheta, \lambda) = 4 \cdot \left\{\left[\frac{J_1\left(\frac{\pi \cdot D \cdot \vartheta}{\lambda}\right)}{\frac{\pi \cdot D \cdot \vartheta}{\lambda}} - \left(\frac{d}{D}\right)^2 \cdot \frac{J_1\left(\frac{\pi \cdot d \cdot \vartheta}{\lambda}\right)}{\frac{\pi \cdot d \cdot \vartheta}{\lambda}}\right] \cdot \frac{1}{1 - \left(\frac{d}{D}\right)^2}\right\}^2 \tag{13.22}$$

where $D$ is the telescope diameter, and $d$ is the central obstruction diameter.

Figure 13.3 shows intensity PSFs for telescopes of this type for various obstruction ratios, $d/D$. The bright central disk evidently narrows as obstruction ratio increases. For simple types of object, such as two-point objects, central obstructions may actually improve telescope resolution. However, for extended objects, central obstructions tend to reduce image contrast, the cause evident in Fig. 13.4 where the light fraction contained in the central disk gradually decreases as obstruction ratio increases. Thus, for general applications, central obstructions should be kept as small as possible. By limiting the central obstruction diameter to the range

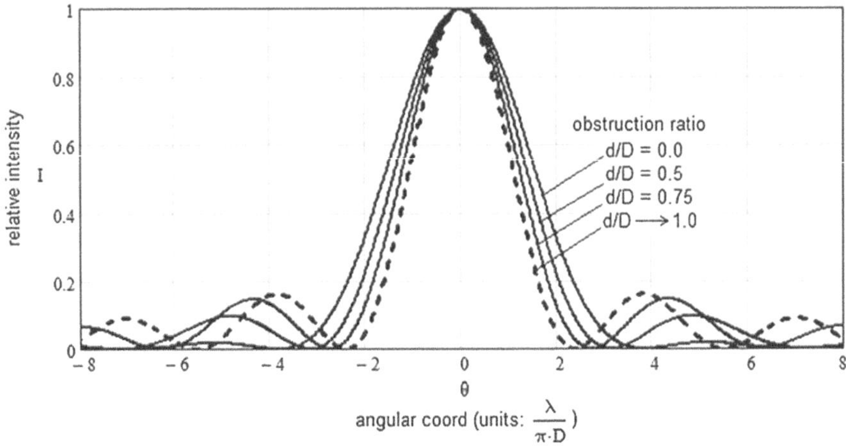


**Fig. 13.3** Points spread functions, all normalized to unity, for diffraction-limited telescopes with circular apertures and various central obstruction sizes, as set by the $d/D$ ratio values indicated. The *dotted line* shows the limiting case where $d/D \to 1$

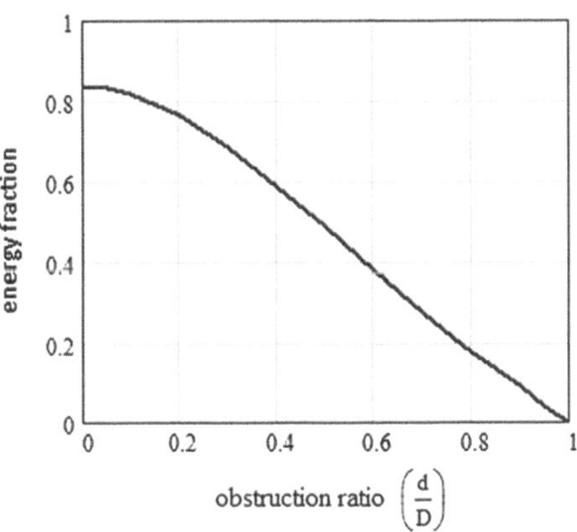


**Fig. 13.4** Light energy fraction in the central disk as a function of the central obstruction size for diffraction-limited telescopes with circular apertures

$d/D \leq 0.2$, two-point resolution and the contrast in extended object images can both be maintained at satisfactory high levels.[4]

To account for the light transfer caused by central obstructions from the central disk into the surrounding rings, the Gaussian approximation for the intensity PSF formed by a mildly aberrated (i.e., $0.4 \leq SI(\lambda) \leq 1$) centrally obstructed telescope, which we denote here by $I_{G(CO)}(\vartheta, \lambda)$ may be written

[4] The diameter of the central obstruction in the Mayall 3.8-m telescope is about 1.65 m, corresponding to an obstruction ratio of about 0.43. Even if the instrument were diffraction-limited, only 55% of the light energy in the PSF would lie in the central disk. Since the actual telescope has significant aberrations, the light fraction in the central disc would generally be less than 55%.

$$I_{G(CO)}(\vartheta,\lambda) = SI(\lambda)\cdot\left(1-(d/D)^2\right)\cdot\exp\left[-\left(\frac{\vartheta}{\left(\frac{\vartheta_{GA}}{\sqrt{SI(\lambda)\cdot(1-(d/D)^2)}}\right)}\right)^2\right] \tag{13.23}$$

where the normalization is again such that the enclosed volume is the same as that of an Airy pattern formed by a diffraction-limited version of the same telescope (without a central obstruction), $\pi\cdot\vartheta_{GA}^2$.

For a severely aberrated version of the same telescope (i.e., $0 \leq SI(\lambda) \leq 0.4$) SI(λ) can be replaced in the above equation by the effective Strehl intensity, $SI_{Ef}(\lambda)$.

Infinitesimally Narrow Annular Aperture

As the central obstruction diameter grows in size toward the limit $d \rightarrow D$, the total amount of light contained in the central disk is driven downward by the combined effects of two pull-down causes: (1) the light energy transmitted through an obstructed aperture diminishes as $(D^2-d^2)/D^2$ and, (2) the light energy fraction contained in the central disk (relative to the total light energy in the image) approximately decreases as $(D^2-d^2)/D^2$ the light energy balance having transferred into the surrounding ring system as shown in Fig. 13.6. Combining these two factors, the ratio of the amount of light contained in the central disk for an obstructed circular aperture to that for an unobstructed aperture, which we denote by $R_{\mathrm{obst}}$, is approximately given by

$$R_{obst} = \frac{\left(D^2-d^2\right)^2}{D^4} \tag{13.24}$$

The above functionality is plotted in Fig. 13.5 where we see that, in the limit $d/D \rightarrow 1$ the light fraction in the highly resolved central disk feature that could potentially deliver maximum two-point resolution has now collapsed to zero. We also saw previously (cf., Fig. 7.5) that in the same limit the telescope modulation transfer function (MTF) largely collapses to zero in the intermediate spatial frequency range. Thus, we see that there are now two very compelling reasons for avoiding large central obstructions.

In the limit, $d/D \rightarrow 1$, 13.22 reduces to the unit-normalized form (Born & Wolf, 2003),

$$I(\vartheta,\lambda) = \left[J_0\left(\frac{\pi\cdot D\cdot\vartheta}{\lambda}\right)\right]^2 \tag{13.25}$$

where $J_o(.)$ is the zero-order Bessel function of the first kind.

Because of the hugely reduced amount of light in the central disk, obviously we cannot afford to approach the limit, d → D, too closely. However, if we choose a central obstruction diameter marginally smaller than the telescope diameter and just short of this limit, say d $\approx$ 0.99 D, the resulting intensity PSF has the interesting, and possibly useful property that it does not substantially alter with defocus. Even with gross amounts of defocus—where there is a significant fractional increase in the distance between the pupil plane and image plane—the intensity PSF remains largely unchanged, other than being subject to a uniform angular rescaling in proportion to the change in the focal distance.

It must be acknowledged here that the two-Gaussian function approximation for the PSF of a centrally obstructed telescope becomes increasingly inadequate as d $\rightarrow$ D. Though the approximation is reasonably valid for d < 0.2 D, it is not valid for larger d values. The matter is addressed again in Chap. 18, where a three-Gaussian function star image model provides the necessary additional degrees of freedom to

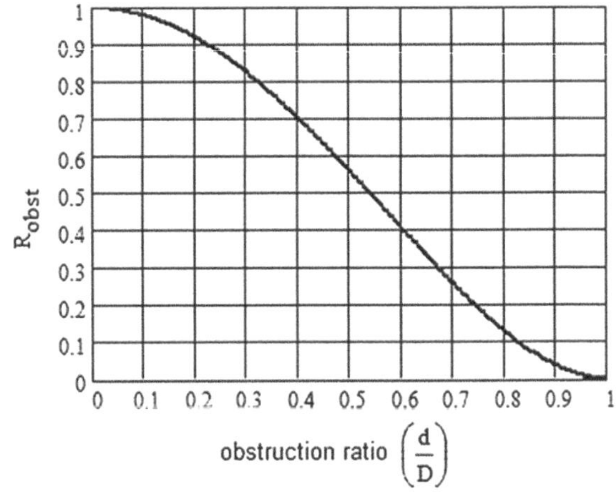


**Fig. 13.5** Effect of a circular central obstruction on the light fraction contained in the central disk of the point spread function formed by a diffraction-limited telescope as a function of the obstruction ratio, *d*/*D*

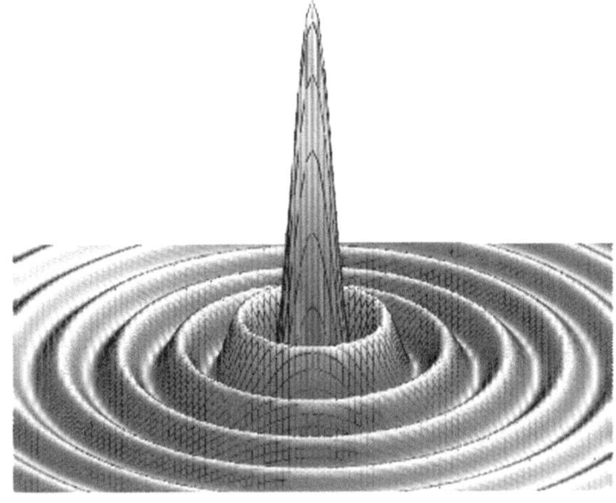

**Fig. 13.6** Diffraction pattern formed by a diffraction-limited telescope with an infinitesimally thin annular aperture (i.e., $d/D \rightarrow 1$). Equation 13.25 was used to calculate the pattern

give a comprehensive and precise accounting of the various issues associated with central obstructions for obstruction diameters, d, up to at least 0.5 D.

#### 13.1.2.4 Gaussian Approximation to Telescope PSF for Asymmetric Aberrations

For telescopes with non-rotationally symmetric aberrations, such as caused by astigmatism, the intensity PSF may be approximated by an elliptical Gaussian function of the form

$$I_{G(asym)}(\alpha,\beta,\lambda) = SI(\lambda)\cdot\left(1-(d/D)^2\right) \times \exp\left\{-\left[\left(\frac{\alpha}{\left(\frac{\vartheta_{GA}}{\sqrt{E_L}\cdot\sqrt{SI(\lambda)\cdot\left(1-(d/D)^2\right)}}\right)}\right)^2 + \left(\frac{\beta}{\left(\frac{\sqrt{E_L}\cdot\vartheta_{GA}}{\sqrt{SI(\lambda)\cdot\left(1-(d/D)^2\right)}}\right)}\right)^2\right]\right\} \quad (13.26)$$

were $E_L$ is the elongation of the ellipse, and the normalization of this equation again preserves the enclosed volume at $\pi.\vartheta_{GA}^2$. Elongation $E_L$ is defined as the ratio of the semi-minor and semi-major axis lengths.[5] The ratio takes values in the range, $0 \le E_L \le 1$ Without loss of generality, the long axis of the ellipse is assumed here to lie in the $\alpha-$ direction. When $E_L = 1$, the ellipse degenerates to a circle. When $E_L = 0$, the ellipse degenerates into one or other of two mutually perpendicular line images. In practice, these could correspond to the two line foci seen in point-object images formed by optical systems affected by astigmatism.

The average image core intensity envelop may be obtained from 13.26 in the same way as before by attaching the multiplier term, $\exp\left[-(2\cdot\pi\cdot\sigma/\lambda)^2\right]$, to account for the reduced light energy fraction.

#### 13.1.2.5 Angular Width of the Gaussian Approximation to the Core

For cores formed by either an aberrated or an aberration-free telescope, by denoting the $1/e$ half-width of the Gaussian approximation to such cores by $\vartheta_{Core}$, the various standard measures of angular width (Sect. 13.1.1) can be given in terms of $\vartheta_{Core}$ as follows:

[5] The (first) eccentricity of an ellipse, e, is given by $\left(1-b^2/a^2\right)^{1/2}$ where a and b are the semi-major and semi-minor axes lengths. The elongation, $E_L$, may be expressed in terms of the eccentricity $\left(1-e^2\right)^{1/2}$.

$$FW\left(\frac{1}{e}\right)_{Core} = 2 \cdot \vartheta_{Core} \quad (13.27)$$

$$FW\left(\frac{1}{e^2}\right)_{Core} = 2 \cdot \sqrt{2} \cdot \vartheta_{Core} \quad (13.28)$$

$$FWHM_{Core} = 2 \cdot \sqrt{\ln(2)} \cdot \vartheta_{Core} \quad (13.29)$$

### *13.1.3 Gaussian Approximations for Halo-Only Images*

As seen in Chap. 11 (Sect. 11.3), for $\sigma/\lambda \geq 0.4$ , the energy fraction in a star image core is less than 0.002, with the bulk of the light energy residing in the halo. Halo images are primarily associated with visible images obtained in unremarkable or poor seeing conditions when AO is not used. The long-exposure version of such an image is referred to as the seeing disk. In this section, we develop Gaussian approximations for the average image intensity envelop for this type of halo-only image.

We assume telescopes with circular apertures and diameters greater than about 1 m. We also assume telescopes with only mild wavefront aberrations where the wavefront error is significantly smaller than the optical path difference (OPD) fluctuation introduced by the atmosphere. Subject to these assumptions, the angular widths of the halo intensity envelops depend predominantly on the seeing conditions, as described by the rms OPD fluctuation, σ, and the turbulence structure size parameter $w_o$ (Sect. 10.3). In contrast, the angular widths of the cores are determined by the telescope intensity PSF (Sect. 13.1.2) via the wavelength, the telescope diameter, and the telescope aberrations. Generally, the angular widths of cores are considerably smaller than those of halos.

We initially suppose circularly symmetric telescope aberrations so that the telescope optical transfer function (OTF) can be represented by the circularly symmetric form, $M_T(\varepsilon \cdot \lambda)$. Further, we assume isotropic turbulence so that the atmospheric MTF is also circularly symmetric. Because of the assumption of mild telescope aberrations compared to the OPD fluctuation introduced by the atmospheric path, we can assume that $M_T(\varepsilon, \lambda)$ takes values close to unity in the ε-argument range where the atmospheric MTF, $M(\varepsilon, \lambda)$, takes non-zero values. Thus subject to these various assumptions the average intensity envelop of the halo is given by the proportionality (cf., 8.5),

$$\langle I(\vartheta, \lambda)\rangle \propto \int_0^\infty M(\varepsilon, \lambda) \cdot J_0\left(\frac{2 \cdot \pi \cdot \vartheta \cdot \varepsilon}{\lambda}\right) \cdot \varepsilon \cdot d\varepsilon \quad (13.30)$$

where the circular symmetry property has allowed the two-dimensional Fourier transform to be written as a one-dimensional Hankel transform. The radial coordinates in the telescope pupil and image planes, $\vartheta$ and $\varepsilon$, were depicted previously in Fig. 7.2. As previously, $J_o(.)$ is the zeroth-order Bessel function.

Using the atmospheric MTF form previously given by 6.55 and the autocorrelation function of the OPD fluctuation given by 9.9, the average intensity in the halo is given by the proportionality

$$\langle I(\vartheta,\lambda)\rangle \propto \int_0^{\infty}\left\{\exp\left[-\left(\frac{2\cdot\pi\cdot\sigma}{\lambda}\right)^2\cdot\left(1-\exp\left(\frac{\varepsilon^2}{w_o^2}\right)\right)\right]\right\} \times J_0\left(\frac{2\cdot\pi\cdot\vartheta\cdot\varepsilon}{\lambda}\right)\cdot\varepsilon\cdot d\varepsilon \tag{13.31}$$

For the halo-only images that concern us in this section where, in effect, we consider $\sigma/\lambda \geq 0.4$, numerical evaluations readily show that the atmospheric MTF described by the term in curly brackets in the above equation may be approximated by the following Gaussian function

$$M_G(\varepsilon,\lambda) = \exp\left[-\left(\frac{2\cdot\pi\cdot\sigma}{\lambda}\right)^2\cdot\left(1-\exp\left(\frac{\varepsilon^2}{w_o^2}\right)\right)\right] \approx \exp\left(-\frac{\varepsilon^2}{w_G^2}\right) \tag{13.32}$$

where it can readily be shown that $w_G$ is given by

$$w_G = \frac{\lambda\cdot w_o}{2\cdot\pi\cdot\sigma} \tag{13.33}$$

Figure 13.7 shows the exact and approximate atmospheric MTFs for the case $\sigma/\lambda = 0.66$. These plots allow us to gauge the accuracy of the above Gaussian approximation for the average intensity envelop for halo-only images. For larger

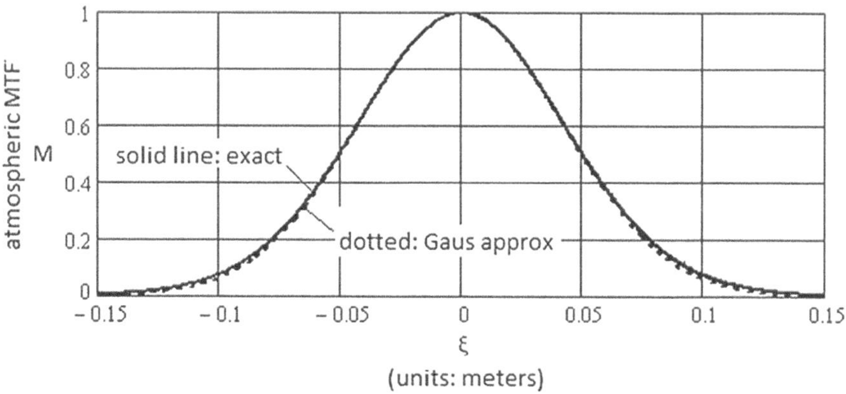


**Fig. 13.7** Typical atmospheric MTF, $M(\varepsilon,\lambda)$, and its Gaussian approximation, $M_G(\varepsilon,\lambda)$, in the regime $\sigma/\lambda \geq 0.4$ (parameter values used: $\lambda = 0.55$ μm, $\sigma = 0.365$ μm, $\sigma/\lambda = 0.66$)

$\sigma/\lambda$ ratio values, the exact and approximate intensity envelopes become practically indistinguishable.

By combining 13.30 and 13.32, the unit-normalized Gaussian approximation for the halo intensity envelop can be expressed in the form

$$\langle I_{Halo}(\vartheta,\lambda)\rangle = \frac{\int_0^{\infty}\exp\left[-\left(\frac{\varepsilon^2}{w_G^2}\right)\right]\cdot J_0\left(\frac{2\cdot\pi\cdot\vartheta\cdot\varepsilon}{\lambda}\right)\cdot\varepsilon\cdot d\varepsilon}{\int_0^{\infty}\exp\left[-\left(\frac{\varepsilon^2}{w_G^2}\right)\right]\cdot\varepsilon\cdot d\varepsilon} \tag{13.34}$$

Since the Hankel transform of a Gaussian function is itself a Gaussian function, the above approximation for the halo intensity envelop may be expressed in the form

$$\langle I_{Halo}(\vartheta,\lambda)\rangle = \exp\left[-\left(\frac{\vartheta}{\vartheta_{Halo}}\right)^2\right] \tag{13.35}$$

where $\vartheta_{Halo}$ denotes the 1/e half-width of the Gaussian halo approximation.

It can readily be shown from 13.33 to 13.35 that

$$\vartheta_{Halo} = \frac{\lambda}{\pi\cdot w_G} = \frac{2\cdot\sigma}{w_o} \tag{13.36}$$

By combining 13.35 and 13.36 and normalizing the result so that the intensity envelop (after rotation) encloses the same volume as the Airy pattern formed by a diffraction-limited version of the same telescope, $\pi\cdot\vartheta_{GA}^2$, we obtain

$$\langle I_{Halo}(\vartheta,\lambda)\rangle = \left(\frac{\lambda\cdot w_o}{\pi\cdot D\cdot\sigma}\right)^2\cdot\exp\left[-\left(\frac{\vartheta}{\left(\frac{2\cdot\sigma}{w_o}\right)}\right)^2\right] \tag{13.37}$$

The degradation of telescope resolution caused by the atmosphere may be gauged by comparing the intensity envelopes calculated from the above expression with Airy pattern intensity envelopes calculated from 13.9. For 4-m class telescopes, where the FWHM angular width of the halo is typically about 1-arscec, the FWHM angular width of the Airy pattern (1.22 λ/D) at visible wavelengths is of the order 0.035 arcsec, indicating resolution degradation by a factor of about thirty for a telescope of this size.

#### 13.1.3.1 Angular Widths of Gaussian Approximations to the Halo

The various standard measures of angular width (Sect. 13.1.1) may be given in terms of the 1/e half-width of the halo, $\vartheta_{Halo}$, as follows:

$$FW\left(\frac{1}{e}\right)_{Halo} = 2 \cdot \vartheta_{Halo} \tag{13.38}$$

$$FW\left(\frac{1}{e^2}\right)_{Halo} = 2 \cdot \sqrt{2} \cdot \vartheta_{Halo} \tag{13.39}$$

$$FWHM_{Halo} = 2 \cdot \sqrt{\ln(2)} \cdot \vartheta_{Halo} \tag{13.40}$$

Assuming that the width of the intensity PSF associated with the telescope optics is significantly less than the width of the halo associated with atmospheric turbulence, the various halo width measures indicated above may be expressed in terms of the atmospheric seeing parameters (cf., 13.37), σ and $w_o$, as follows:

$$FW\left(\frac{1}{e}\right)_{Halo} = \frac{4 \cdot \sigma}{w_o} \tag{13.41}$$

$$FW\left(\frac{1}{e^2}\right)_{Halo} = \frac{4 \cdot \sqrt{2} \cdot \sigma}{w_o} \tag{13.42}$$

$$FWHM_{Halo} = \frac{4 \cdot \sqrt{\ln(2)} \cdot \sigma}{w_o} \tag{13.43}$$

Figure 13.8 shows sections through exact and approximate forms of the halo intensity envelopes (calculated using 13.31 and 13.37) for the visible wavelength 0.55 μm. The other parameter values indicated in the figure caption are consistent with average (~1 arcsec) visible seeing conditions. In the regime $\sigma/\lambda \geq 1$, the exact

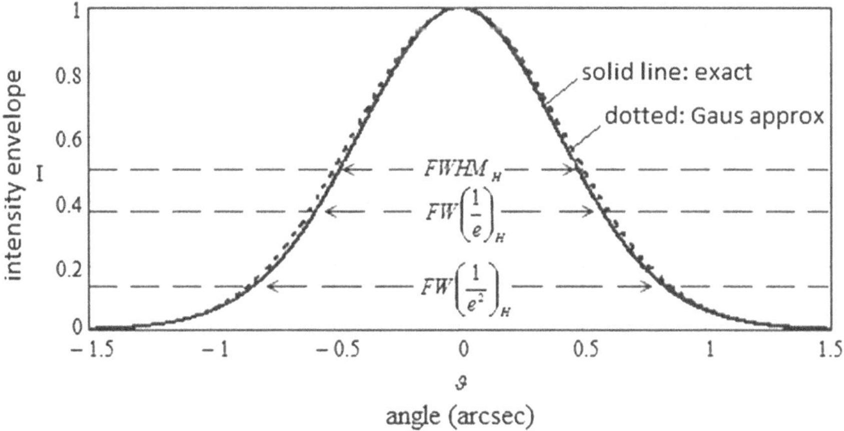


**Fig. 13.8** A halo image intensity envelop and the Gaussian approximation to this envelop; these envelopes correspond to the nearly identical atmospheric MTFs shown in Fig. 13.7 for the halo-dominated regime, $\sigma/\lambda \geq 0.4$ (parameter values used: $\lambda = 0.55$ μm, $\sigma = 0.365$ μm, $\sigma/\lambda = 0.66$, and $w_o = 0.25$ m)

and approximate envelopes exhibit even closer correspondence than that indicated in the figure.

**Expression for $\sigma/w_o$ in terms of the FWHM Angular Width of the Halo**

By rearranging 13.43, we obtain

$$\frac{\sigma}{w_o} = \frac{FWHM_{Halo}}{4 \cdot \sqrt{\ln(2)}} \tag{13.44}$$

where ln(·) is the Naperian logarithm. Since $4 \cdot \sqrt{\ln(2)} = 3.3302$, 13.44 may be re-expressed as follows:

$$\frac{\sigma}{w_o} = 0.3 \cdot FWHM_{Halo} \tag{13.45}$$

The above approximation may be rearranged to give

$$FWHM_{Halo} = 3.33 \cdot \frac{\sigma}{w_o} \tag{13.46}$$

**Example case for 1-arcsec FWHM visible seeing**

In the case of a halo-only image in 1-arcsec FWHM visible seeing conditions, 13.45 gives $\sigma/w_o = 0.3 \cdot$ arcsec ($1.454 \times 10^{-6}$ rad).

#### 13.1.3.2 Asymmetric Halos for Non-isotropic Turbulence

For non-isotropic atmospheric turbulence, function ρ(x,y) may be approximated (cf., 6.48) by the elliptical Gaussian function,

$$\rho(x, y) = \exp\left[-\left(\frac{x^2}{w_{ox}^2} + \frac{y^2}{w_{oy}^2}\right)\right] \tag{13.47}$$

where $w_{ox}$ and $w_{oy}$ are the $1/e$ half-widths of the average turbulence structure sizes in the $x$- and $y$-directions. By using an analysis similar to that given in Sect. 13.1.3, the average intensity envelop arising from such a ρ(x, y) function may be written in the form

$$I_{Halo}(\alpha, \beta, \lambda) = \left(\frac{\lambda}{\pi \cdot D \cdot \sigma}\right)^2 \cdot w_{ox} \cdot w_{oy} \cdot \exp\left\{-\left(\frac{\alpha^2}{\left(\frac{2\cdot\sigma}{w_{ox}}\right)^2} + \frac{\beta^2}{\left(\frac{2\cdot\sigma}{w_{oy}}\right)^2}\right)\right\} \tag{13.48}$$

where the $\alpha$- and $\beta$-axes are assumed aligned with the $x$- and $y$-axes, respectively, and where the equation is normalized so that the enclosed volume is again equal to the volume enclosed by the Airy pattern formed by a diffraction-limited version of the same telescope, $\pi \cdot \vartheta_{GA}^2$ (cf., 13.9). For isotropic turbulence, where $w_{ox} = w_{oy}$, the above equation reduces to the circularly symmetric form given previously by 13.37.

### *13.1.4 Gaussian Approximations for Core and Halo Images*

Having now seen how cores and halos can be individually approximated by Gaussian functions, approximate expressions can now be given for the most general image type—the core and halo image. By assuming circular symmetry for both core and halo, the average intensity, which we denote by $\langle I_{G(CH)}(\vartheta, \lambda)\rangle$, for this type of image may be expressed by the general form

$$\left\langle I_{G(CH)}(\vartheta, \lambda)\right\rangle \propto A_C \cdot \exp\left[-\left(\frac{\vartheta}{B_C}\right)^2\right] + A_H \cdot \exp\left[-\left(\frac{\vartheta}{B_H}\right)^2\right] \tag{13.49}$$

where $A_{\mathrm{C}}$ and $B_{\mathrm{C}}$, respectively, quantify the central intensity and $1/e$ half-width of the core, and $A_{\mathrm{H}}$ and $B_{\mathrm{H}}$ quantify the corresponding quantities for the halo.

Figure 13.9 shows in schematic form the typical appearance of a core and halo image. The figure also shows how this image decomposes into its separate core and halo portions. Later, in Sect. 13.1.7, we shall establish how the four parameters, $A_{\mathrm{C}}$, $B_{\mathrm{C}}$, $A_{\mathrm{H}}$, and $B_{\mathrm{H}}$, may be expressed in terms of the imaging wavelength, $\lambda$, the telescope diameter, $D$, the central obstruction diameter, $d$, the telescope Strehl intensity, $\mathrm{SI}(\lambda)$, and the seeing parameters, $\sigma$ and $w_o$.

#### 13.1.4.1 Gaussian Approximations for Non-circularly Symmetric Cores and Halos

Telescopes with non-circularly symmetric aberrations, such as astigmatism, give rise to asymmetrical core shapes. It is convenient to refer this type of aberration to the sagittal and tangential planes (Sect. 4.2.3) which define two orthogonal directions in the image plane.[6] Non-isotropic turbulence in the atmospheric path[7] could also cause asymmetry in $\rho(x, y)$ which in turn could lead to asymmetric halos. In general, the orthogonal axes for halos are different from those of cores.

As a first approximation, images comprised of asymmetric cores and halos can be described by elliptical Gaussian functions. Since the elongation directions of the

[6] The tangential plane includes both the object point considered and the axis of symmetry. The sagittal plane is orthogonal to the tangential plane; it contains the object point and intersects the optical axis in the telescope pupil. This plane contains the chief ray but not the optical axis; it is therefore a skew plane.

[7] Non-isotropic turbulence structure, which in principle could be the result of wind shear, requires a vector description rather than the scalar description used for isotropic turbulence.

core and halo generally lie in different directions, we denote the angular difference between these two directions by χ (Fig. 13.10). A general expression for the average intensity envelop for such an image may then be written in the form

$$\begin{aligned}\langle I_{G(CH)}(\alpha,\beta,\lambda)\rangle = A_C \cdot \exp\left[-\left(\frac{\alpha\cdot\cos(\chi)+\beta\cdot\cos(\chi)}{B_{C\alpha}}\right)^2\right] \\ \times \exp\left[-\left(\frac{\beta\cdot\cos(\chi)-\alpha\cdot\sin(\chi)}{B_{C\beta}}\right)^2\right] \\ + A_H\cdot\exp\left[-\left(\frac{\alpha}{B_{H\alpha}}\right)^2\right]\cdot\exp\left[-\left(\frac{\beta}{B_{H\beta}}\right)^2\right]\end{aligned} \quad (13.50)$$

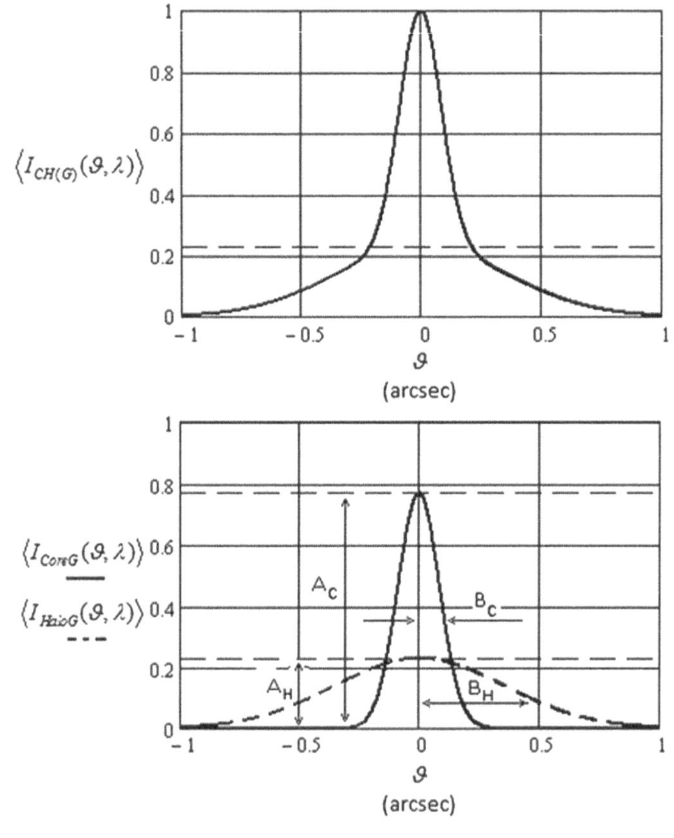

**Fig. 13.9** *Top* Gaussian approximation for a core and halo image. *Bottom* The same image decomposed into its separate (Gaussian) core and halo portions. The four—parameters that describe this type of image, $A_C$, $A_H$, $B_C$, and $B_H$, are as indicated

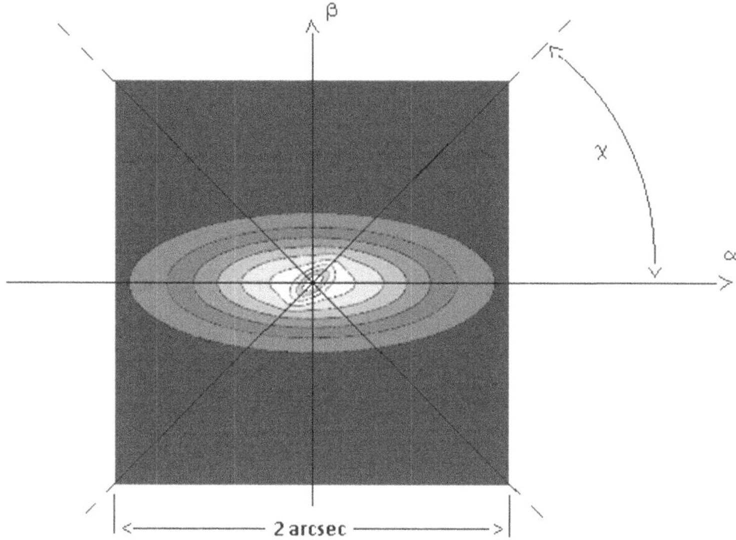


**Fig. 13.10** Image intensity contour plot for a core and halo image where both features are approximated by elliptical Gaussian functions and where there is an arbitrarily chosen 45° offset angle between the respective long-axes of the core and halo. Image calculated using 13.50 with parameter values $A_C = A_H = 1.0$, $B_{C\alpha} = 0.25$ arcsec, $B_{C\beta} = 0.1$ arcsec, $B_{H\alpha} = 1.35$ arcsec, $B_{H\beta} = 0.50$ arcsec, $\chi = 45°$

where $B_{C\alpha}$ and $B_{C\beta}$ are the $1/e$ angular half-widths of the core along the two orthogonal directions, and $B_{H\alpha}$ and $B_{H\beta}$ are the corresponding $1/e$ half-widths of the halo. Without loss of generality, it is assumed in 13.50 that the long axis of the halo lies in the $\alpha$-direction. The sine and cosine terms appearing in the above equation have their origins in the Cartesian coordinate rotation matrix,

$$\begin{bmatrix} \cos(\chi) & -\sin(\chi) \\ \sin(\chi) & \cos(\chi) \end{bmatrix}$$

For the degenerate case, $B_{C\alpha} = B_{C\beta} = B_C$ and $B_{H\alpha} = B_{H\beta} = B_H$, 13.50 reduces to the form given previously for the case of circular symmetry (13.49).

### 13.1.5 *Anticipated Accuracy of Gaussian Star Image Approximations*

Typical surface plots of the intensity envelop for a core and halo image are shown in Fig. 13.11. Actual and approximate envelopes are shown in the figure. The actual

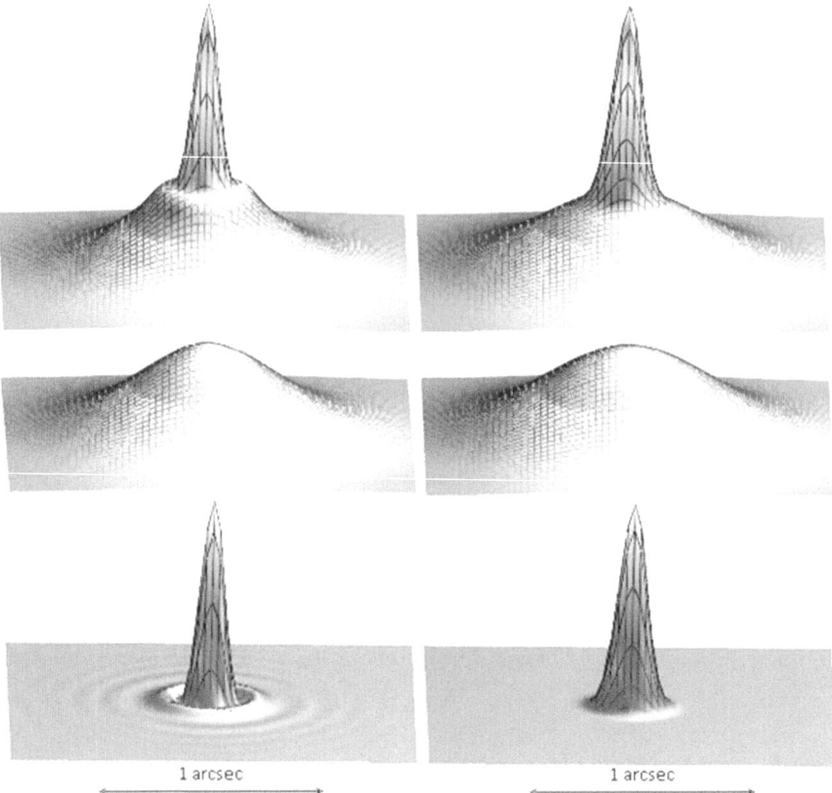


**Fig. 13.11** *Top left* Actual star image intensity envelop at wavelength 2.2 μm formed by a 3.8-m diameter telescope with one wave (HeNe) wavefront aberration. *Top right* The Gaussian approximation. Because the wavefront aberration of the telescope at 2.2 μm is approximately one quarter wave, Strehl intensity of the instrument at this wavelength is about 0.8 (i.e., substantially diffraction-limited). *Middle* The corresponding halo image portions are shown in isolation. *Bottom* The corresponding core image portions are shown in isolation

envelop is calculated using 10.8 and 13.9, and the Gaussian approximation is calculated using 13.49. The figure also shows actual and approximate surface plots of the cores and halos individually.

### *13.1.6 Normalization of Gaussian Star Image Approximations*

The average intensity in a core and halo star image exhibiting circular symmetry can be given in unit-normalized form by

$$\left\langle I_{G(CH)}(\vartheta,\lambda)\right\rangle = \frac{1}{A_C + A_H}\cdot\left\{A_C\cdot\exp\left[-\left(\frac{\vartheta}{B_C}\right)^2\right] + A_H\cdot\exp\left[-\left(\frac{\vartheta}{B_H}\right)^2\right]\right\} \tag{13.51}$$

We now define the quantity, $A_{CH}$, as the ratio of the central intensities in the core and halo portions in the image,

$$A_{CH} = \frac{A_C}{A_H} \tag{13.52}$$

By rewriting 13.51 in terms of $A_{CH}$, we obtain

$$\left\langle I_{G(CH)}(\vartheta,\lambda)\right\rangle = \frac{1}{(A_{CH}+1)}\cdot\left\{A_{CH}\cdot\exp\left[-\left(\frac{\vartheta}{B_C}\right)^2\right] + \exp\left[-\left(\frac{\vartheta}{B_H}\right)^2\right]\right\} \tag{13.53}$$

which is again unit-normalized.

Alternatively, the core and halo intensity envelop, $\left\langle I_{G(CH)}(\vartheta,\lambda)\right\rangle$, may be normalized so that unit volume is contained under the (rotated) envelop, in effect, normalizing the total enclosed light energy to unity:

$$\left\langle I_{G(CH)}(\vartheta,\lambda)\right\rangle = \frac{1}{\pi\cdot\left(A_C\cdot B_C^2 + A_H\cdot B_H^2\right)} \cdot\left\{A_C\cdot\exp\left[-\left(\frac{\vartheta}{B_C}\right)^2\right] + A_H\cdot\exp\left[-\left(\frac{\vartheta}{B_H}\right)^2\right]\right\} \tag{13.54}$$

The above equation can also be expressed in terms of the quantity $A_{CH}$,

$$\left\langle I_{G(CH)}(\vartheta,\lambda)\right\rangle = \frac{1}{\pi\cdot\left(A_{CH}\cdot B_C^2 + B_H^2\right)} \cdot\left\{A_{CH}\cdot\exp\left[-\left(\frac{\vartheta}{B_C}\right)^2\right] + \exp\left[-\left(\frac{\vartheta}{B_H}\right)^2\right]\right\} \tag{13.55}$$

A significant advantage of using the intensity envelop forms given by 13.53 and 13.55 is that since the ratio $A_{CH}$ is dimensionless, there is no need to concern with the actual dimensions of $A_C$ and $A_H$. Therefore, it does not matter whether these two quantities are expressed in radiometric measures, such as irradiance (W/m$^2$) or Watts per pixel, or whether they are simply vertical heights measured from an intensity section through the image. The essential information is entirely conveyed by the dimensionless ratio, $A_{CH}$, defined by 13.52.

#### 13.1.6.1 Degenerate Cases When the Image Is Either Core-Only or Halo-Only

These two degenerate cases correspond to $A_{\mathrm{CH}}$ taking the two limiting values, zero and infinity.

Core-Only Image

The core-only image corresponds to $A_{\mathrm{CH}} \rightarrow \infty$. For this case, 13.53 which is normalized to unity at $\vartheta = 0$ reduces to

$$\left\langle I_{G(Core)}(\vartheta, \lambda)\right\rangle = \exp\left[-\left(\frac{\vartheta}{B_C}\right)^2\right] \tag{13.56}$$

and 13.55 which is normalized to enclose unit volume reduces to

$$\left\langle I_{G(Core)}(\vartheta, \lambda)\right\rangle = \frac{1}{\pi \cdot B_C^2} \cdot \exp\left[-\left(\frac{\vartheta}{B_C}\right)^2\right] \tag{13.57}$$

Halo-Only Image

The halo-only image corresponds to $A_{\mathrm{CH}} = 0$. For this case, 13.53 reduces to the unit-normalized form

$$\left\langle I_{G(Halo)}(\vartheta, \lambda)\right\rangle = \exp\left[-\left(\frac{\vartheta}{B_H}\right)^2\right] \tag{13.58}$$

and 13.55 which is normalized to enclose unit volume reduces to

$$\left\langle I_{G(Halo)}(\vartheta, \lambda)\right\rangle = \frac{1}{\pi \cdot B_H^2} \cdot \exp\left[-\left(\frac{\vartheta}{B_H}\right)^2\right] \tag{13.59}$$

### 13.1.7 *Expressing $A_C$, $A_H$, $B_C$, and $B_H$ in Terms of the Telescope and Seeing Parameters*

For telescopes with circular apertures and circularly symmetric aberrations observing through isotropic turbulence, the following expressions for $A_{\mathrm{C}}$, $A_{\mathrm{H}}$, $B_{\mathrm{C}}$, $B_{\mathrm{H}}$, and $A_{\mathrm{CH}}$ can be derived from 13.14, 13.23, 13.35, and 13.36, giving these parameters in terms of the parameters that describe the telescope optics and the atmospheric seeing, $D$, $d$, $\lambda$, SI($\lambda$), $\sigma$, and $w_o$:

$$A_C \propto \frac{\pi \cdot SI(\lambda) \cdot \left(D^2 - d^2\right)}{4 \cdot \lambda^2} \cdot \exp\left[-\left(\frac{2 \cdot \pi \cdot \sigma}{\lambda}\right)^2\right] \tag{13.60}$$

$$B_C = \frac{2 \cdot \lambda}{\pi \cdot \sqrt{SI(\lambda) \cdot \left(D^2 - d^2\right)}} \tag{13.61}$$

$$A_H \propto \frac{w_o^2}{4 \cdot \pi \cdot \sigma^2} \cdot \left\{1 - \exp\left[-\left(\frac{2 \cdot \pi \cdot \sigma}{\lambda}\right)^2\right]\right\} \tag{13.62}$$

$$B_H = \frac{2 \cdot \sigma}{w_o} \tag{13.63}$$

It follows from 13.61 and 13.63 that $B_C$ and $B_H$ are related as follows:

$$B_C = \frac{\lambda \cdot w_o}{\pi \cdot \sigma \cdot \sqrt{SI(\lambda) \cdot \left(D^2 - d^2\right)}} \cdot B_H \tag{13.64}$$

The ratio $A_{CH}$ may be obtained from 13.60 to 13.62 in the form

$$\begin{aligned} A_{CH} &= \frac{A_C}{A_H} \\ &= \frac{\pi^2 \cdot \sigma^2 \cdot SI(\lambda) \cdot \left(D^2 - d^2\right)}{\lambda^2 \cdot w_o^2} \cdot \left\{\frac{\exp\left[-\left(\frac{2 \cdot \pi \cdot \sigma}{\lambda}\right)^2\right]}{1 - \exp\left[-\left(\frac{2 \cdot \pi \cdot \sigma}{\lambda}\right)^2\right]}\right\} \end{aligned} \tag{13.65}$$

Because of the inherent redundancy in the four-parameter ($A_C$, $A_H$, $B_C$, and $B_H$) description of the essentially three-parameter image, each of the four parameters may be expressed in terms of the other three:

$$A_C = \frac{1 - \pi \cdot A_H \cdot B_H^2}{\pi \cdot B_C^2} \tag{13.66}$$

$$A_H = \frac{1 - \pi \cdot A_C \cdot B_C^2}{\pi \cdot B_H^2} \tag{13.67}$$

$$B_C = \sqrt{\frac{1 - \pi \cdot A_H \cdot B_H^2}{\pi \cdot A_C}} \tag{13.68}$$

$$B_H = \sqrt{\frac{1 - \pi \cdot A_C \cdot B_C^2}{\pi \cdot A_H}} \tag{13.69}$$

## 13.2 Obtaining the Seeing Parameters and Telescope Strehl from Gaussian Image Approximations

Usually, $D$, $d$, and $\lambda$ are known at the outset, so there are usually only three unknown parameters, namely SI($\lambda$), $\sigma$, and $w_o$. In this section, relationships are given between the latter three parameters and the parameters, $A_{CH}$, $B_C$, and $B_H$, values for which can readily be obtained from appropriate measurements of the average intensity envelop of an unresolved star image. In practice, to obtain a sensible outcome, it is necessary to form the star image at a wavelength that delivers a roughly equal division of light energy between the core and the halo. In this way, we avoid the singularities where $A_{CH}$ takes the values zero or infinity which correspond to the degenerate cases where the image consists of either just a core or just a halo. A suitable wavelength range can be found by inspection of 10.3. The following approximate range would be suitable:

$$5 \cdot \sigma \leq \lambda \leq 13 \cdot \sigma$$

For this wavelength range, the proportion of light energy in the core lies in the range,

$$0.2 \leq E \leq 0.8$$

with the remaining light energy residing in the halo.

The desired expressions relating SI($\lambda$), $\sigma$, and $w_o$ to $A_{CH}$, $B_C$, and $B_H$ may be obtained from 13.61, 13.63, and 13.65 in the forms,

$$SI(\lambda) = \frac{4 \cdot \lambda^2}{\pi^2 \cdot B_C^2 \cdot \left(D^2 - d^2\right)} \tag{13.70}$$

$$\sigma = \frac{\lambda}{2 \cdot \pi} \cdot \sqrt{\ln\left[1 + \frac{B_H^2}{A_{CH} \cdot B_C^2}\right]} \tag{13.71}$$

$$w_o = \frac{2 \cdot \sigma}{B_H} = \frac{\lambda}{\pi \cdot B_H} \cdot \sqrt{\ln\left[1 + \frac{B_H^2}{A_{CH} \cdot B_C^2}\right]} \tag{13.72}$$

where ln(·) is the Naperian logarithm.

### 13.2.1 *Example Calculation of SI($\lambda$), $\sigma$, and $w_o$ from $A_{CH}$, $B_C$, and $B_H$*

Suppose that a core and halo image of the kind shown in Fig. 13.9 is obtained at wavelength, $\lambda = 2.2$ μm, using a telescope with a circular aperture of diameter, $D =$ 3.8 m, and central obstruction diameter, $d = 1.65$ m. Also suppose that measurements of the average image envelop yielded the following values for $A_{\mathrm{CH}}$, $B_{\mathrm{C}}$, and $B_{\mathrm{H}}$:

$$A_{\mathrm{CH}} = 3.33$$
$$B_{\mathrm{C}} = 0.122 \text{ arcsec},$$
$$\text{and } B_{\mathrm{H}} = 0.5 \text{ arcsec}.$$

Using 13.70–13.72, the following values arise for the telescope Strehl intensity, SI(λ), and the seeing parameters, σ, and $w_o$:

$$\mathrm{SI}(2.2\ \mu\mathrm{m}) = 0.48,$$
$$\sigma = 0.47\ \mu\mathrm{m},$$
$$\text{and } w_o = 0.39\ \mathrm{m}.$$

### 13.2.2 *Calculation of σ and $w_o$ for Reflector Telescopes*

It is assumed here that either (1) the wavefront error of the telescope $W(x, y, \lambda)$ is known so that the rms wavefront error, $\sigma_{\mathrm{T}}$, may be calculated, or (2) $\sigma_{\mathrm{T}}$ is simply known anyway. Telescope Strehl intensity may then be calculated from the expression,

$$SI(\lambda) = \exp\left[-\left(\frac{2 \cdot \pi \cdot \sigma_T}{\lambda}\right)^2\right] \tag{13.73}$$

where the tacit assumption has been made that the wavefront error is Gaussian distributed. With Strehl intensity thus quantified, we are no longer obliged to actually measure $B_{\mathrm{C}}$: a value for this parameter can be obtained directly from the relation (cf., 13.61),[8]

$$B_C = \frac{2 \cdot \lambda}{\pi \cdot \sqrt{SI(\lambda) \cdot \left(D^2 - d^2\right)}} \tag{13.74}$$

With $B_{\mathrm{C}}$ established in this way and by obtaining values of $A_{\mathrm{CH}}$ and $B_{\mathrm{H}}$ from image measurements, 13.71 and 13.72 can then be used to calculate the σ and $w_o$ values.

[8] However, in instances where we cannot be sure that the telescope is precisely focused, it might be safer to measure $B_{\mathrm{C}}$ directly from the image. Also, wherever opportunities arise, it might be good practice to check for consistency between measured and calculated values of the same parameter.

### 13.2.3 *Calculation of Residual Phase Error After AO Image Correction*

Even after AO wavefront corrections have been made, small residual amounts of phase fluctuation inevitably remain in the final image-forming wavefronts. These residuals can take two forms: (1) uncorrected portions of the fixed wavefront errors of the telescope, and (2) uncorrected portions of the dynamical phase errors caused by the atmosphere. Incorrect adjustment of the AO system can contribute to both error types, while sluggish AO temporal response can contribute to the dynamical phase residuals. These two quite different types of uncorrected wavefront error can be quantified separately by analyzing the average intensity envelop in a star image.

Generally, the appearance of AO-corrected images is not fundamentally different from image appearance when AO is not used. In general, both types of image show core and halo structure; any differences that might occur are limited to differences in degree only. AO-corrected images can differ from uncorrected images in the following ways: (1) The core likely contains a larger light energy fraction due to the reduced rms OPD error; (2) the core is likely to be more sharply defined due to at least partial correction of the fixed telescope errors; and (3) the reduced halo light portion is redistributed in a way determined by the spatial frequency response of the AO system to the spatial frequency content of the image-forming waves arriving at the telescope. We continue to denote the parameters of the uncorrected OPD fluctuation as before by $\sigma$ and $w_0$, but now also denote the corresponding parameters for the AO-corrected wavefronts by $\sigma_{AO}$ and $w_{0AO}$.

The primary purpose of an AO system is to reduce the rms OPD fluctuation of the image-forming wavefronts in the telescope pupil. Even if we only succeed in making $\sigma_{AO}$ slightly smaller than the uncorrected value, $\sigma$, perhaps by only as little as 25%, this can still lead to a huge increase in core brightness (cf., 10.3). The increase is of course at the expense of halo brightness. The shape of the halo is also likely to modify; it may turn out to be wider or narrower than before, depending on whether $w_{oAO}$ turns out to be smaller or larger than the initial value $w_o$.[9] Halo width is determined by the ratio $\sigma_{AO}/w_{oAO}$ (cf., 13.46). For a correctly working AO system where $\sigma_{AO} < \sigma$, the halo would likely appear more compact than before. However, such an outcome is not guaranteed. If $w_{oAO}$ happened to turn out small enough to fulfil the condition $\sigma_{AO}/w_{oAO} > \sigma/w$, 13.46 indicates that the AO-corrected halo could actually end up wider than before.

The most efficacious $w_{oAO}$ value depends on the imaging application. For purposes of imaging extended objects, it is generally better if the AO system produces the outcome, $w_{oAO} > w_o$; while, for simpler types of object such as binary stars, the outcome $w_{oAO} < w_o$ would generally be preferable. The effect of different $w_{oAO}$ values is illustrated in Fig. 13.12. Four average star image envelopes are shown in this figure, each obtained at $\lambda = 1.2$ μm using a supposed diffraction-limited 10-m telescope.

[9] $w_{oAO}$ can take values that are either smaller or larger than $w_o$, depending on the response of the AO system to different spatial frequencies.

The dashed line envelop (calculated using $\sigma = 0.6$ μm and $w_o = 0.4$ m) shows the image envelop formed by the telescope in about 1-arcsec seeing conditions when AO is not used. The other three solid line envelopes show different possible envelop outcomes when AO is used. All three correspond to $\sigma_{AO} = 0.45$ μm.

The observed differences between the three envelopes are therefore entirely attributable to the three different $w_{oAO}$ values considered, 0.2, 0.4, and 0.8 m. The AO-corrected images in the figure all show identical cores, the tiny (~0.02 arcsec) angular size of these cores offering about 50 times higher resolution than obtained in 1-arcsec visible seeing conditions without AO assistance. The solid line envelop corresponding to $w_{oAO} = w_o$ shows the anticipated outcome when the AO-corrected wavefronts have the same average lateral structure size as the uncorrected wavefronts. The solid line envelop corresponding to $w_{oAO} = 2 \cdot w_o$ shows a more highly resolved halo. For extended objects, such an envelop would likely produce sharper images. On the other hand, the solid line envelop for the wider halo, corresponding to $w_{oAO} = w_o/2$, might be better suited for resolving objects such as close binary pairs or exoplanets from their parent stars.

Even if it may not be obvious from Fig. 13.12, upon rotation, all four of the intensity envelopes in the figure enclose exactly the same amount of light energy. In some cases, a large portion of the light energy in some of the plots lies beyond the angular range plotted. Since all plots shown were calculated using 13.55, they must all enclose the same amount of light energy.

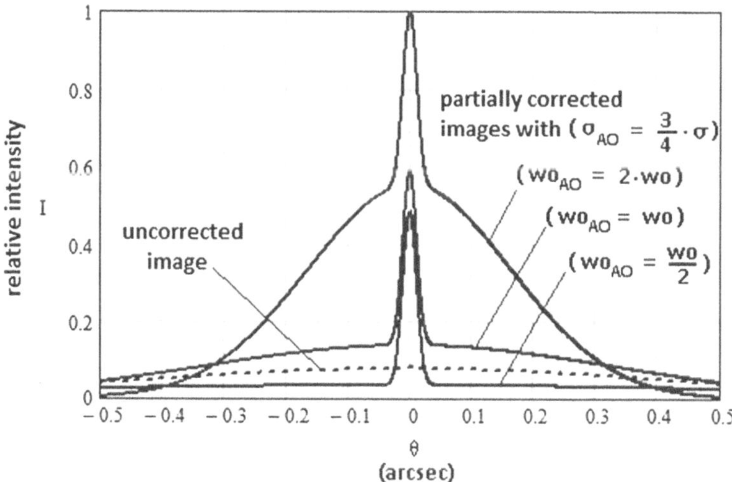


**Fig. 13.12** *Dashed line* An uncorrected star image envelop obtained at 1.2 μm by a diffraction-limited 10-m telescope in 1-arcsec visible seeing conditions. *Solid lines* Partially corrected images where $\sigma_{AO}$ is 25% less than the uncorrected $\sigma$ value. The AO-corrected envelopes have different appearances depending on the lateral structure size of the AO-corrected wavefronts as set by the quantity $w_{oAO}$

#### 13.2.3.1 Separating the Fixed from the Dynamical Wavefront Error Residuals

AO systems generally only provide partial correction of the fixed wavefront error of the telescope. The degree of correction can be sufficiently quantified for our purposes by establishing the AO-corrected Strehl intensity of the telescope, $SI(\lambda)_{AO}$. From a measurement of core $1/e$ half-width, $B_C$, $SI(\lambda)_{AO}$ may be calculated using the relation (cf., 13.74),

$$SI(\lambda)_{AO} = \frac{4 \cdot \lambda^2}{\pi^2 \cdot B_C^2 \cdot \left(D^2 - d^2\right)} \tag{13.75}$$

Once $SI(\lambda)_{AO}$ has been established in this way, the residual dynamical phase error remaining in the image-forming wavefronts after the AO correction, as expressed by $\sigma_{AO}$ and $w_{oAO}$, may be calculated by inserting the measured $B_C$, $B_H$, and $A_{CH}$ values into the following equations (cf., 13.71 and 13.72):

$$\sigma_{AO} = \frac{\lambda}{2 \cdot \pi} \cdot \sqrt{\ln\left[1 + \frac{B_H^2}{A_{CH} \cdot B_C^2}\right]} \tag{13.76}$$

$$\text{and} \quad w_{oAO} = \frac{2 \cdot \sigma_{AO}}{B_H} \tag{13.77}$$

Thus, from measurements of $B_C$, $B_H$, and $A_{CH}$, we can use the above three relations to quantify the performance of the AO system in terms of the extent to which it corrects the fixed wavefront errors of the telescope (via $SI(\lambda)_{AO}$) and the dynamical phase errors introduced by the atmosphere (via $\sigma_{AO}$ and $w_{oAO}$).

#### 13.2.3.2 Calculation of $\sigma_{AO}$ and $w_{oAO}$ After Correction of Fixed Telescope Errors

By fine-tuning the AO system, it might be possible to near perfectly correct the fixed wavefront errors of the telescope. In such cases, it would no longer be necessary to measure $B_C$. By substituting $SI(\lambda)_{AO} = 1$ into 13.75, we could obtain $B_C$ directly from

$$B_C = \frac{2 \cdot \lambda}{\pi \cdot \sqrt{D^2 - d^2}} \tag{13.78}$$

With $B_C$ thus established, the measured values of $A_{CH}$ and $B_H$ can then be used to calculate $\sigma_{AO}$ and $w_{oAO}$ using 13.76 and 13.77.

### 13.2.4 *Obtaining SI, $\sigma$, $w_{o\alpha}$, and $w_{o\beta}$ from Asymmetric Core–Halo Images*

To calculate SI($\lambda$), $\sigma$, $w_{o\alpha}$, and $w_{o\beta}$, it is first necessary to obtain measured values of $A_{CH}$, $B_{C\alpha}$, $B_{C\beta}$, $B_{H\alpha}$, and $B_{H\beta}$ from the average star image intensity envelop. The corresponding values for SI($\lambda$), $\sigma$, $w_{o\alpha}$, and $w_{o\beta}$ can be calculated using equations similar to those developed in Sect. 13.2:

$$SI(\lambda) = \frac{4 \cdot \lambda^2}{\pi^2 \cdot B_{C\alpha} \cdot B_{C\beta} \cdot \left(D^2 - d^2\right)} \tag{13.79}$$

$$\sigma = \frac{\lambda}{2 \cdot \pi} \cdot \sqrt{\ln\left[1 + \frac{B_{H\alpha} \cdot B_{H\beta}}{A_{CH} \cdot B_{C\alpha} \cdot B_{C\beta}}\right]} \tag{13.80}$$

$$w_{o\alpha} = \frac{2 \cdot \sigma}{B_{H\alpha}} = \frac{\lambda}{\pi \cdot B_{H\alpha}} \cdot \sqrt{\ln\left[1 + \frac{B_{H\alpha}^2}{A_{CH} \cdot B_{C\alpha} \cdot B_{C\beta}}\right]} \tag{13.81}$$

$$w_{o\beta} = \frac{2 \cdot \sigma}{B_{H\beta}} = \frac{\lambda}{\pi \cdot B_{H\beta}} \cdot \sqrt{\ln\left[1 + \frac{B_{H\beta}^2}{A_{CH} \cdot B_{C\alpha} \cdot B_{C\beta}}\right]} \tag{13.82}$$

#### 13.2.4.1 Degenerate Cases

Two interesting degenerate cases arise. For an asymmetric core and circularly symmetric halo (i.e., $w_{o\alpha} = w_{o\beta} = w_o$), the above equations reduce to

$$SI(\lambda) = \frac{4 \cdot \lambda^2}{\pi^2 \cdot B_{C\alpha} \cdot B_{C\beta} \cdot \left(D^2 - d^2\right)} \tag{13.83}$$

$$\sigma = \frac{\lambda}{2 \cdot \pi} \cdot \sqrt{\ln\left[1 + \frac{B_H^2}{A_{CH} \cdot B_{C\alpha} \cdot B_{C\beta}}\right]} \tag{13.84}$$

$$w_0 = \frac{2 \cdot \sigma}{B_H} = \frac{\lambda}{\pi \cdot B_H} \cdot \sqrt{\ln\left[1 + \frac{B_H^2}{A_{CH} \cdot B_{C\alpha} \cdot B_{C\beta}}\right]} \tag{13.85}$$

For an asymmetric halo and circularly symmetric core (i.e., $B_{C\alpha} = B_{C\beta} = B_C$), 13.79–13.82 reduce to

$$SI(\lambda) = \frac{4 \cdot \lambda^2}{\pi^2 \cdot B_C^2 \cdot \left(D^2 - d^2\right)} \tag{13.86}$$

$$\sigma = \frac{\lambda}{2 \cdot \pi} \cdot \sqrt{\ln\left[1 + \frac{B_{H\alpha} \cdot B_{H\beta}}{A_{CH} \cdot B_C^2}\right]} \tag{13.87}$$

$$w_{o\alpha} = \frac{2 \cdot \sigma}{B_{H\alpha}} = \frac{\lambda}{\pi \cdot B_{H\alpha}} \cdot \sqrt{\ln\left[1 + \frac{B_{H\alpha}^2}{A_{CH} \cdot B_C^2}\right]} \tag{13.88}$$

$$w_{o\beta} = \frac{2 \cdot \sigma}{B_{H\beta}} = \frac{\lambda}{\pi \cdot B_{H\beta}} \cdot \sqrt{\ln\left[1 + \frac{B_{H\beta}^2}{A_{CH} \cdot B_C^2}\right]} \tag{13.89}$$

### 13.2.5 Maintaining Detailed Seeing Logs While Observing

By periodically measuring the parameters, $A_{\mathrm{CH}}$, $B_{\mathrm{C}}$, and $B_{\mathrm{H}}$, from the average image of a suitable reference star (or by measuring $A_{\mathrm{CH}}$, $B_{\mathrm{C}\alpha}$, $B_{\mathrm{C}\beta}$, $B_{\mathrm{H}\alpha}$, and $B_{\mathrm{H}\beta}$ for the most general type of asymmetrical image) a detailed seeing log record can be kept. Such a log not only records the seeing characteristics of the observing path, but also the Strehl intensity of the telescope, SI(λ), which can vary as the temperature and gravitational-loading vectors change. The desired seeing parameters, $\sigma$ and $w_o$ (or $w_{o\alpha}$ and $w_{o\beta}$ for asymmetrical images) can be obtained from the measured data by using one or other of the three equations sets, 13.79–13.82, 13.83–13.85, or 13.86–13.89, as appropriate.

Maintaining a seeing log in this way requires monitoring a reference star in a wavelength band consistent with the formation of a core and halo image. As previously established in Sect. 13.2, such a band would lie somewhere in the approximate wavelength range given previously by $5 \cdot \sigma \leq \lambda \leq 13 \cdot \sigma$. For typical $\sigma$ values, the band would often lie in wavelength range of 1–5 μm. To maintain a seeing record of the type considered here without interfering with the actual astronomical imaging program itself, the record might be most conveniently based on the image of the reference star used to correct tracking errors; this image would be continually monitored anyway by the ancillary star tracking FPA described later in Chap. 17 (Sect. 17.4).

It might be observed that changes in the Strehl intensity of the telescope (which as just mentioned can be caused by thermal changes or primary mirror gravity loading changes) do not affect the seeing parameters recorded in the log. Such changes only affect the telescope Strehl intensity, values of which can be generated independently and in parallel with the seeing parameters.

## 13.3 Comparing Intensity Envelopes Formed by Two Different Telescopes

The expressions for the intensity envelop given by 13.55 (and by 13.57 and 13.58 for the degenerate core-only and halo-only cases) can be used to make direct side-by-side comparisons of the average star image intensity envelopes produced by asymmetrical telescope imaging arrangements (as discussed previously in Sect. 10.2). For example, comparisons can be made of images produced by telescopes for which the diameters, central obstructions, and aberrations are all different, and where one instrument might have AO capability while the other does not. The two instruments could also be located at two entirely different sites, in different seeing conditions, and observing at two different wavelengths.

Figure 13.13 shows an image intensity envelop (calculated from 13.57 using 13.61 for $B_C$) obtained at $\lambda = 0.55$ μm from a small telescope $D = 0.15$ m) where the telescope optics and the seeing are supposed good enough to deliver diffraction-limited images. Also shown in the figure are the core and halo images (calculated from 13.55) obtained at $\lambda = 1.65$ and 4.8 μm from a large diffraction-limited telescope (with the same diameter and central obstruction as the Mayall telescope). Upon rotation, all three envelopes enclose the same amount of light energy.

Figure 13.14 shows images (calculated from 13.55) obtained at $\lambda = 1.65$ μm for hypothetical 4-, 10- and 40-m diameter telescopes where it is assumed that AO is used in all three instruments and that the level of correction (as prescribed by the $\sigma_{AO}$ values indicated in the figure) steadily increases with the telescope aperture,

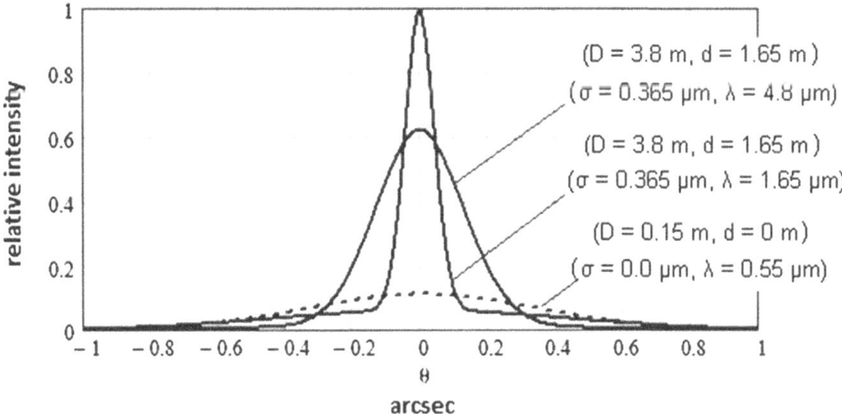


**Fig. 13.13** Typical average star image intensity envelopes formed by (*dotted line*) a small backyard telescope at the visible wavelength, 0.55 μm, and by (*solid lines*) a 3.8-m instrument at the IR wavelengths, 1.65 and 4.8 μm, in 1-arcsec visible seeing conditions. In all cases, image wander has been removed. Note that the value, $\sigma = 0$ μm, used for the small telescope is consistent with such an instrument delivering Airy pattern images, atmospheric turbulence having little disruptive effect on images formed by such a small instrument

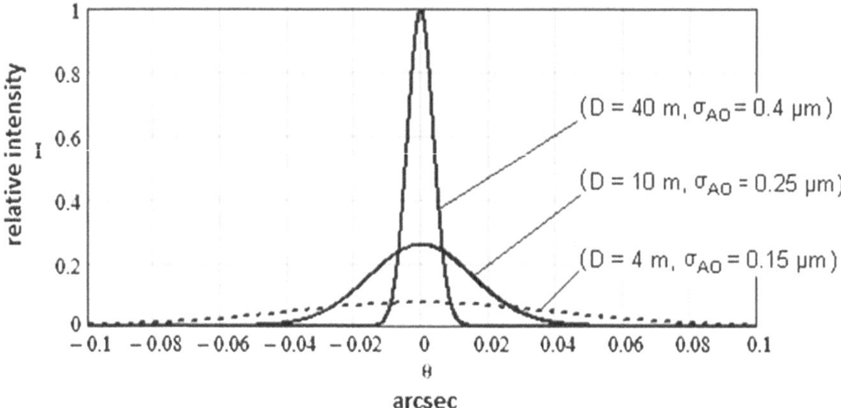


**Fig. 13.14** Average star image intensity envelopes formed at 1.65-μm by hypothetical 4-, 10-, and 40-m telescopes. Most of the halo light energy in these core and halo images actually lies beyond the ±0.1-arcsec angular range plotted

the increase attributable to the added complexity of controlling larger numbers of adaptive segments. In all three images shown, $w_{oAO}$ was set to the same value, 0.25 m.

The angular widths of the cores in Fig. 13.14 naturally reduce with increasing telescope diameter, the FWHM widths of the cores being 92 mas, 37 mas, and 9 mas, respectively. The light energy fractions in the cores (calculated from 10.3) also diminish with increasing telescope diameter, these fractions being 0.72, 0.40, and 0.1. However, despite the core energy fraction in the image produced by the 40-m telescope, 0.1, being considerably less than the core energy fractions in the images delivered by the 4 and 10-m telescopes, because this energy fraction is now concentrated into such a small (9 mas) angular patch, the irradiance level attained in the center of this core is far higher than the levels attained by the cores formed by the two smaller instruments. It might be noted here that this result is a consequence of $\sigma_{AO}$ growing with telescope size in the way specified. Different outcomes would of course arise if different growth rates were assumed.

## 13.4 Optimum Wavelength for Maximum Irradiance at Image Center

In Chap. 10, we found that the optimum wavelength was given approximately by $\lambda_{opt} = 2 \cdot \pi \cdot \sigma$, a result obtained by neglecting the halo contribution to the central intensity. In this section, we give a more exact analysis, one that now takes account of the halo contribution.

Equation 13.55 is used as our starting point. The light energy enclosed under this envelop is normalized to unity. Inserting the value $\vartheta = 0$ into the equation, the intensity in the center of an arbitrary core and halo image is given by

$$\langle I_{G(CH)}(0, \lambda) \rangle = \frac{A_{CH} + 1}{\pi \cdot (A_{CH} \cdot B_C^2 + B_H^2)} \quad (13.90)$$

Using 13.61, 13.63, and 13.65, to substitute for $A_{\mathrm{CH}}$, $B_{\mathrm{C}}$, and $B_{\mathrm{H}}$, the above expression may be written in the alternative form,

$$\langle I_{G(CH)}(0, \lambda) \rangle = \frac{\lambda^2 \cdot w_o^2 \left\{1 - \exp\left[-\left(\frac{2\cdot\pi\cdot\sigma}{\lambda}\right)^2\right]\right\} + (D^2 - d^2) \cdot \pi^2 \cdot \sigma^2 \cdot SI(\lambda) \cdot \exp\left[-\left(\frac{2\cdot\pi\cdot\sigma}{\lambda}\right)^2\right]}{4 \cdot \pi \cdot \sigma^2 \cdot \lambda^2} \quad (13.91)$$

We now assume that either the Strehl intensity remains invariant with the wavelength or the rate of change is slow enough to be neglected. Diffraction-limited reflector telescopes of course possess this property, with SI(λ)taking the value unity at all wavelengths. In such cases, we may denote the Strehl intensity by the constant value, SI, in terms of which the above equation can be written in the form,

$$\langle I_{CH(G)}(0, \lambda) \rangle = \frac{w_o^2}{4 \cdot \pi \cdot \sigma^2} \left\{1 - \exp\left[-\left(\frac{2 \cdot \pi \cdot \sigma}{\lambda}\right)^2\right]\right\} + \frac{(D^2 - d^2) \cdot \pi \cdot SI}{4 \cdot \lambda^2} \cdot \exp\left[-\left(\frac{2 \cdot \pi \cdot \sigma}{\lambda}\right)^2\right] \quad (13.92)$$

Differentiating the right-hand side of 13.92 with respect to wavelength and setting the result to zero allows us to calculate the wavelength at which maximum irradiance is attained in the center of the image. We refer to this wavelength as before (Sect. 10.4) as the optimum wavelength and can express it by

$$\lambda_{opt} = 2 \cdot \pi \cdot \sigma \cdot \frac{1}{\sqrt{1 + \frac{4 \cdot w_o^2}{SI \cdot (D^2 - d^2)}}} \quad (13.93)$$

For $\frac{4 \cdot w_o^2}{SI \cdot (D^2 - d^2)} \ll 1$, the above expression may be approximated by

$$\lambda_{opt} = 2 \cdot \pi \cdot \sigma \cdot \left[1 - \frac{2 \cdot w_o^2}{SI \cdot (D^2 - d^2)}\right] \quad (13.94)$$

The optimum wavelength given by this expression is proportional to $\sigma$, just as was found previously (cf., 10.16). The only essential difference is that the gradient

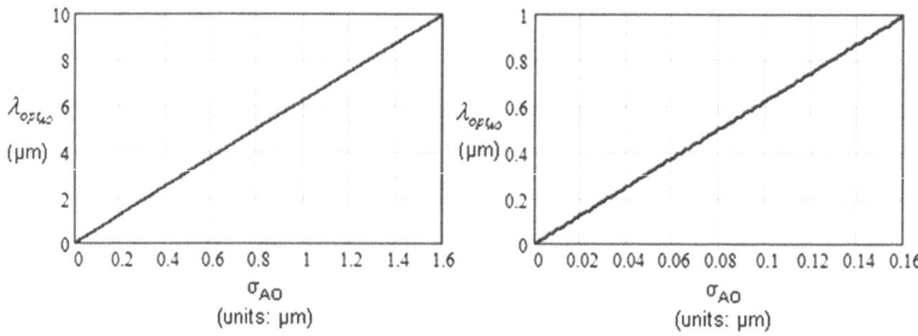


**Fig. 13.15** The optimum wavelength, $\lambda_{\text{optAO}}$, versus $\sigma_{AO}$ for extremely large telescopes where it is assumed that $D \gg w_{0AO}$

$\frac{d\lambda_{opt}}{d\sigma}$, previously given by $2 \cdot \pi$, is now given by the modified expression

$$\frac{d\lambda_{opt}}{d\sigma} = 2 \cdot \pi \cdot \left[1 - \frac{2 \cdot w_o^2}{SI \cdot \left(D^2 - d^2\right)}\right] \tag{13.95}$$

## 13.4.1 Optimum Wavelength for Extremely Large AO-Equipped Telescopes

It is assumed here that we deal with extremely large telescopes (ELTs) and that the fixed aberrations of these instruments are incorporated into $\sigma_{\text{AO}}$. Thus, we may set $\text{SI} = 1$ and, by also assuming that $D \gg w_{0\text{AO}}$ for these massive instruments, 13.94 reduces to the form,

$$\lambda_{opt_{AO}} = 2 \cdot \pi \cdot \sigma_{AO} \tag{13.96}$$

Evidently, for extremely large telescopes, the approximate expression for the optimum wavelength given previously by 10.16 becomes increasingly accurate. Figure 13.15 shows how, in the case of ELTs, the optimum wavelength likely varies as a function of the residual uncorrected rms OPD fluctuation, $\sigma_{\text{AO}}$.

## 13.4.2 Irradiance at Image Center for the Optimum Wavelength

The irradiance in the center of the image at the optimum wavelength may be obtained by substituting the expression for $\lambda_{\text{opt}}$ given by 13.94 into 13.92. Thus,

$$\langle I_{G(CH)}(0,\lambda)\rangle = \frac{4\cdot w_o^2 + SI\cdot(D^2-d^2)\cdot \exp\left[-\left(1+\frac{4\cdot w_o^2}{SI\cdot(D^2-d^2)}\right)\right]}{16\cdot\pi\cdot\sigma^2} \quad (13.97)$$

When SI = 1, as would be the case for diffraction-limited telescopes or for telescopes where AO has removed any fixed wavefront errors, the above equation marginally reduces to

$$\langle I_{G(CH)}(0,\lambda)\rangle = \frac{4\cdot w_o^2 + (D^2-d^2)\cdot \exp\left[-\left(1+\frac{4\cdot w_o^2}{(D^2-d^2)}\right)\right]}{16\cdot\pi\cdot\sigma^2} \quad (13.98)$$

#### 13.4.2.1 Obtaining Average Turbulence Size, $w_o$, from the Optimum Wavelength

In cases where $\sigma$, SI, and $\lambda_{opt}$ are known (the latter could be identified by examining images obtained at several different wavelengths in the neighborhood of the optimum wavelength), a useful expression for $w_o$ may be obtained by rearranging 13.93 to give

$$w_o = \frac{1}{2}\cdot\sqrt{SI\cdot(D^2-d^2)\cdot\left[\left(\frac{2\cdot\pi\cdot\sigma}{\lambda_{opt}}\right)^2-1\right]} \quad (13.99)$$

## 13.5 MTFs Corresponding to Gaussian Core and Halo Image Envelops

As indicated in Chap. 7, the product of the atmospheric MTF and the telescope OTF may be obtained by carrying out a 2-D Fourier transform of the average star image intensity envelop. For circularly symmetric images, we begin by using the two-Gaussian function approximation for the intensity envelop (cf., 13.55) which, for convenience, we reproduce here,

$$\langle I_{G(CH)}(\vartheta,\lambda)\rangle = \frac{1}{\pi\cdot(A_{CH}\cdot B_C^2 + B_H^2)} \cdot\left\{A_{CH}\cdot\exp\left[-\left(\frac{\vartheta}{B_C}\right)^2\right] + \exp\left[-\left(\frac{\vartheta}{B_H}\right)^2\right]\right\} \quad (13.100)$$

Circular symmetry allows the 2-D Fourier transform to be carried out as a 1-D Hankel transform. Thus, a central section through the function product, $M(\varepsilon,\lambda)\cdot M_T(\varepsilon,\lambda)$ is given by,

$$M(\varepsilon,\lambda)\cdot M_T(\varepsilon,\lambda)=\frac{2}{A_{CH}\cdot B_C^2+B_H^2}\\ \times\int_0^\infty\left\{A_{CH}\cdot\exp\left[-\left(\frac{\vartheta}{B_C}\right)^2\right]+\exp\left[-\left(\frac{\vartheta\cdot}{B_H}\right)^2\right]\right\}\\ \times J_0\left(\frac{2\cdot\pi\cdot\vartheta\cdot\varepsilon}{\lambda}\right)\cdot\vartheta\cdot d\vartheta \tag{13.101}$$

where we recall (Chap. 7) that $\varepsilon$ is the radial distance in the telescope pupil plane.

The normalization of 13.100 is such that, upon rotation, the volume enclosed under the envelop is always unity. The above Hankel transform therefore provides (the appropriately unit-normalized) product of the atmospheric MTF and the telescope OTF, (i.e., $M(0,\lambda)\cdot M_T(0,\lambda)=1$),

$$M(\varepsilon,\lambda)\cdot M_T(\varepsilon,\lambda)=\frac{1}{A_{CH}\cdot B_C^2+B_H^2}\\ \cdot\left\{A_{CH}\cdot B_C^2\cdot\exp\left[-\left(\frac{\pi\cdot B_C\cdot\varepsilon}{\lambda}\right)^2\right]\right.\\ \left.+B_H^2\cdot\exp\left[-\left(\frac{\pi\cdot B_H\cdot\varepsilon}{\lambda}\right)^2\right]\right\} \tag{13.102}$$

Equation 13.102 may equally be expressed as a function of spatial frequency in the image, $\nu$, defined previously by 7.13. The parameter $\nu$ may be expressed in terms of $\varepsilon$ and the telescope focal length, f, as follows:

$$\nu=\frac{\varepsilon}{\lambda\cdot f} \tag{13.103}$$

Thus, the atmospheric MTF and telescope OTF product given by 13.102 may be expressed in terms of $\nu$ in the form,

$$M(\nu,\lambda)\cdot M_T(\nu,\lambda)=\frac{1}{A_{CH}\cdot B_C^2+B_H^2}\\ \cdot\left\{A_{CH}\cdot B_C^2\cdot\exp\left[-(\pi\cdot B_C\cdot f\cdot\nu)^2\right]\right.\\ \left.+B_H^2\cdot\exp\left[-(\pi\cdot B_H\cdot f\cdot\nu)^2\right]\right\} \tag{13.104}$$

The cutoff frequency, $\nu_c$, which occurs at $\varepsilon=D$ is given by

$$\nu_c=\frac{D}{\lambda\cdot f} \tag{13.105}$$

The right-hand side of 13.104 is observed as the sum of two Gaussian functions, which we note has the same functionality as that of the corresponding average star image intensity envelop from which it arose (cf., 13.100).

This shared functionality is illustrated in Fig. 13.16 where the atmospheric MTF and telescope OTF products are shown (Bottom) for a large telescope. The corresponding average image intensity envelop is shown (Top). A characteristic of Fourier transformations is that they map narrow functions into wide functions, and vice versa. Thus, the broad halo portion of the image envelop maps into a feature containing only low spatial frequencies, while the narrow core portion maps into higher spatial frequencies.

The atmospheric MTF and telescope OTF products can also be expressed in terms of angular frequency in the image, which we denote by $\upsilon_\vartheta$. This parameter is given in terms of the spatial frequency $\nu$ by

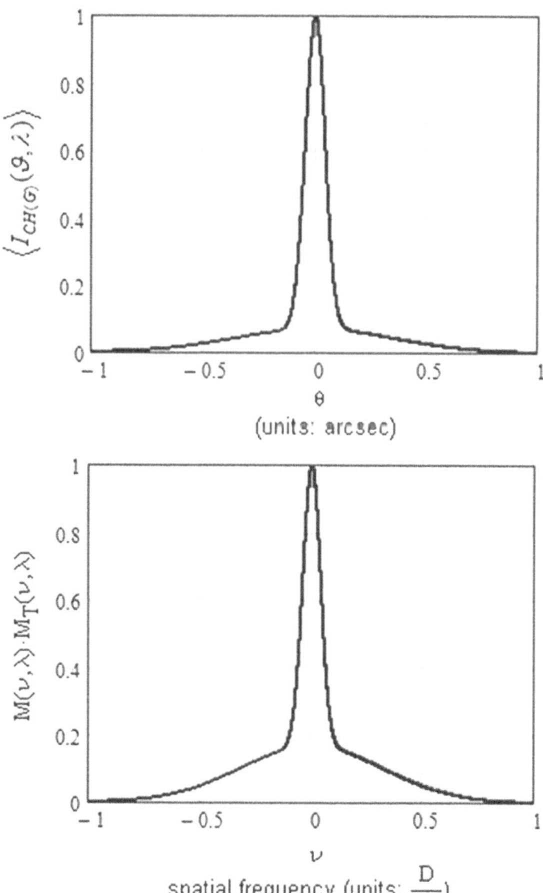


**Fig. 13.16** *Top* Typical core and halo average intensity envelop$\langle I_{G(CH)}(\vartheta,\lambda)\rangle$for a large telescope. *Bottom* The corresponding MTF product, $M(\nu,\lambda)\cdot M_T(\nu,\lambda)$. Parameter values used in calculations: $\lambda = 1.6\ \mu m$, $D = 3.8$ m, $d = 1.65$ m, $\sigma = 0.3405\ \mu m$, $w_o = 0.281$ m, $SI(1.6\ \mu m) = 1$, $A_{CH} = 13.3$, $B_C = 0.0614$ arcsec, $B_H = 0.5$ arcsec

$$\upsilon_\vartheta = f \cdot \nu \tag{13.106}$$

Substituting $\nu = \upsilon_\vartheta / f$ in 13.104 allows us to write

$$M(\upsilon_\vartheta, \lambda) \cdot M_T(\upsilon_\vartheta, \lambda) = \frac{1}{A_{CH} \cdot B_C^2 + B_H^2} \times \left\{ A_{CH} \cdot B_C^2 \cdot \exp\left[-(\pi \cdot B_C \cdot \upsilon_\vartheta)^2\right] + B_H^2 \cdot \exp\left[-(\pi \cdot B_H \cdot \upsilon_\vartheta)^2\right]\right\} \tag{13.107}$$

## 13.5.1 *The Cutoff Frequency for Gaussian MTF Approximations*

The cutoff for the product function $M(\varepsilon,\lambda) \cdot M_T(\varepsilon,\lambda)$ occurs at $\varepsilon_c = D$. The cutoff location may equally be expressed in terms of spatial frequency by $\nu_c = D / (\lambda \cdot f)$. In terms of the angular frequency parameter, $\upsilon_\vartheta$, the cutoff is given by $\upsilon_{\vartheta C} = D/\lambda$. Equations 13.102, 13.104, and 13.107, all of which are equivalent, may sometimes show small non-zero tails beyond the cutoff frequency. Response in this frequency region is of course not physically possible; it is merely an artifact of the Gaussian approximations.[10]

Residuals of this sort are invariably small, as can be seen from the $M(\varepsilon, \lambda) \cdot M_T(\varepsilon, \lambda)$ product function plotted in Fig. 13.16. An upper bounding limit for such residuals may be established by using 13.61, 13.63, and 13.65 to substitute for $A_{CH}$, $B_C$, $B_H$, and $\upsilon_{\vartheta C}$ in 13.107. Algebraic manipulation then leads to the following expression giving the apparent residual modulation at the cutoff frequency,

$$M(\nu_c, \lambda) \cdot M_T(\nu_c, \lambda) = \exp\left[-\left(\frac{2 \cdot \pi \cdot \sigma}{\lambda}\right)^2 - \frac{4}{SI}\right] + \left\{1 - \exp\left[-\left(\frac{2 \cdot \pi \cdot \sigma}{\lambda}\right)^2\right]\right\} \cdot \exp\left[-\left(\frac{2 \cdot \pi \cdot \sigma \cdot \sqrt{D^2 - d^2}}{\lambda \cdot w_o}\right)^2\right] \tag{13.108}$$

Evaluations of this equation show that the residuals are generally negligible. Beyond the cutoff frequency, $\nu_c$, the residuals diminish exponentially. For telescopes larger than 4 m observing in typical ~1-arcsec seeing conditions, even in the most stressful case where the telescopes are diffraction-limited (i.e., where SI(λ), the

[10] Gaussian functions decay exponentially but do not fall to absolute zero at any finite argument.

residual MTF value at $\nu_c$ is less than 0.02. For aberrated telescopes, where say $SI(\lambda) < 0:5$, the residual falls to less than 0.002. For larger aberrations, the residuals continue to diminish.

## 13.6 Mathematical Notation Used in This Chapter

The mathematical notation used in this chapter is indicated in Table 13.1.

**Table 13.1** Mathematical notation used in this chapter along with the SI dimensional units of the individual quantities

| Symbol | Quantity | Dimensions |
|---|---|---|
| $\lambda$ | Wavelength | $m$ |
| (x, y) | Cartesian coordinate system in telescope pupil | $m$ |
| $(\xi, \eta)$ | Cartesian coordinate system in plane perpendicular to light travel direction | $m$ |
| $\varepsilon$ | Radial angular coordinate in plane perpendicular to light travel direction $= \sqrt{\xi^2 + \eta^2}$ | $m$ |
| $(\alpha, \beta)$ | Angular coordinate system in telescope image plane | "1" |
| $\vartheta$ | Radial angular coordinate in telescope image plane $= \sqrt{\alpha^2 + \beta^2}$ | "1" |
| $H$ | Integrated OPD fluctuation over entire atmospheric path | $m$ |
| $\sigma$ | rms of integrated OPD fluctuation, $H$ | $m$ |
| $\rho$ | Autocorrelations function of integrated OPD fluctuation, $H$ | "1" |
| $w_0$ | $1/e$ half-width of Gaussian approximation to autocorrelation function of $H$ | $m$ |
| $\sigma_{AO}$ | rms of residual OPD fluctuation after AO correction | $m$ |
| $\rho_{AO}$ | Gaussian approximation to autocorrelation of residual OPD Fluctuation | "1" |
| $w_{0AO}$ | $1/e$ half-width of Gaussian autocorrelation function of residual OPD fluctuation after AO correction | $m$ |
| $D$ | Telescope diameter | $m$ |
| $d$ | Telescope central obstruction | $m$ |
| $f$ | Telescope focal length | $m$ |
| $SI$ | Telescope Strehl intensity | "1" |
| $M$ | Atmospheric MTF | "1" |
| $M_T$ | Telescope OTF | "1" |

(continued)

**Table 13.1** (continued)

| Symbol | Quantity | Dimensions |
|---|---|---|
| $I$ | Image intensity | "1" |
| $R_{\mathrm{obs}}$ | Ratio of light fraction in central disk to total light for a centrally obstructed telescope | "1" |
| $A_C$ | Intensity in the center of Gaussian core | "1" |
| $A_H$ | Intensity in the center of Gaussian | "1" |
| $A_{CH}$ | Ratio $A_{\mathrm{C}} = A_{\mathrm{H}}$ | "1" |
| $B_C$ | Angular 1/*e* half-width of Gaussian core | "1" |
| $B_H$ | Angular 1/*e* half-width of Gaussian halo | "1" |
| $\lambda_{opt}$ | Optimum wavelength | $m$ |
| $\nu$ and $\nu_{\mathrm{c}}$ | Spatial frequency and OTF cutoff spatial frequency | $\mathrm{m}^{-1}$ |
| $\chi$ | Angle between the elongation directions for an elliptical core and an elliptical halo | "1" |

Dimensionless quantities are indicated by "1"

## Reference

Born, M., & Wolf, E. (2003). *Principles of optics* (7th ed., revised). Cambridge University Press.

# Chapter 14
# Telescope Resolution and Optical Tolerance Specifications

**Abstract** The theoretical resolution limits of large ground-based telescopes are examined. Optical tolerance specifications are set out for large telescopes. Lax tolerances can be adopted if halo-only images are considered adequate; much tighter tolerances are required to exploit the much higher resolution levels provided by image cores. Telescope resolution is measured by the instrument's ability to resolve binary stars. The Rayleigh resolution criterion employed to meet the broad-ranging requirements of this chapter—where images can be influenced by atmospheric turbulence, telescope aberrations, central obstructions, AO, etc.—is the version where a binary object is considered just-resolved when the intensity in the center of the image is 26.5% less than the value attained in the peaks (i.e., 1.22 $\lambda$/D for diffraction-limited telescopes with circular apertures). Computer-generated and actual star images are shown for the Keck II instrument. Computer-generated binary star images are also shown for the future 39 m E-ELT instrument for various wavelengths in the range, 0.5–10 $\mu$m, with <10 mas resolution anticipated routinely in the optimum wavelength region.

The angular resolution of a telescope is usually described in terms of its ability to resolve a two-point object such as a binary star. The Rayleigh resolution criterion, devised by Lord Rayleigh (Strutt, 1880, 1902), relates narrowly to two-point objects imaged by diffraction-limited telescopes with circular apertures, where the two objects are equally bright and emit mutually incoherent light.[1] The single-point image formed by such a telescope is of course the Airy pattern. According to the Rayleigh criterion, a two-point object is just-resolved when the central disk of the Airy pattern image of one of the object points exactly overlies the first dark ring of

[1] Rayleigh's first resolution criterion, devised in 1879, referred to spectroscopes (Strutt, 1879). For the slit-based instruments considered in that paper, two wavelengths are just-resolved when the center of the (diffraction pattern) light distribution for one wavelength exactly overlies the first dark line of the light distribution for the other. In this case, the diffraction patterns are described by functions of the form, $(\sin(x)/x)^2$. The just-resolved interval, $x = \pi$, is therefore slightly different than that obtained for the Airy pattern form, $(2 \cdot J_1(x)/x)^2$, which applies to circular apertures where the first dark ring occurs at $x = 3.832$.



T. S. McKechnie, *General Theory of Light Propagation and Imaging Through the Atmosphere*, Progress in Optical Science and Photonics 20,
https://doi.org/10.1007/978-3-030-98828-9_14

the Airy pattern image of the other object point. According to Rayleigh, "this rule is convenient on account of its simplicity, and it is sufficiently accurate in view of the necessary uncertainty as to what exactly is meant by resolution."

The criterion just described is not suitable for the wider range of applications envisaged in this book. For example, it does not account for the effects of central obstructions, telescope aberrations, and atmospheric turbulence, all of which can modify the images so that they no longer resemble Airy patterns. Therefore, rather than using Rayleigh's criterion in the form stated above, we instead use an extended version of the criterion (Born & Wolf, 2003) according to which a two-point object is considered just-resolved when a 26.5% dip occurs in the image intensity midway between the geometrical images of the two points. Both criteria provide the same just-resolved separation for Airy pattern images.[2] The extended version, however, may be applied much more broadly—to images obtained in any given seeing conditions from almost any type of telescope, including telescopes equipped with adaptive optics (AO). Two other resolution criteria are also discussed—the Dawes criterion and the Sparrow criterion—but the discussion is brief, the purpose merely being to give a slightly broader background perspective to the subject of telescope resolution.

A "survey" of resolution given by den Dekker and van den Bos (1997) draws attention to a fundamental, though often forgotten, aspect of resolution: Since actual images are noisy (photon noise, electronic noise, etc.), they cannot be exactly replicated by images calculated using diffraction theory alone. As observed by the Italian physicist, Vasco Ronchi (1897–1988) (Ronchi, 1961), and reformulated by Goodman (1990), the ability to resolve a two-point object depends on the signal-to-noise ratio (SNR) of the detected image intensity. Because the various two-point resolution criteria discussed in this chapter by Rayleigh, Dawes, and Sparrow do not take noise into account, all of these criteria are subjective. One consequence is that the angular resolution limits set by these criteria are all different, although fortunately the differences are small enough not to create any real issue.

Telescope resolution is examined in detail for telescopes with circular apertures—the most frequently encountered type of telescope. Aberrated and diffraction-limited telescopes of this type are considered, including telescopes with central obstructions and telescopes with and without AO. The Gaussian point-object image approximations previously developed in Chap. 13 will be used to describe telescope resolution in terms of simple formulae, which allow telescope resolution to be quantified, not only in terms of the telescope parameters and the imaging wavelength, but also in terms of the atmospheric seeing parameters.

Optical tolerance specifications for large ground-based telescopes are also addressed in this chapter. Suitable specifications of course depend on the specific imaging application; they also depend on the imaging wavelength, the seeing quality at the observing site, and whether or not the telescope is equipped with AO.

For the Hubble Space Telescope, the total rms wavefront error allowance is about 0.025 μm (Allen et al., 1990) and indeed this astonishingly tight specification was

---

[2] The just-resolved separations given by these two versions of Rayleigh's criterion agree to within 0.01%.

finally achieved after the repair mission in 1993. This specification enables the Hubble instrument to deliver >0.9 Strehl intensity at visible wavelengths. Ground-based telescopes are generally built to considerably less stringent tolerances. For 4-m class instruments built in the 1970s and 1980s according to Kolmogorov prescriptions, allowable rms wavefront error was typically about 1 μm.[3] Such a prescription is adequate for delivering the coarse 0.5–1.5-arcsec halo-only images anticipated in that era. However, it is not adequate for delivering diffraction-limited image cores at near-IR wavelengths and is even less adequate for delivering image cores at visible wavelengths. Around 1990, tolerance specifications for ground-based telescopes of this class size abruptly tightened by almost an order of magnitude. The 3.58-m NTT telescope (first light in March, 1989) was one of the first to be constructed to higher optical specifications, albeit with the assistance of an active optic primary mirror.[4] On the very first night of use, it provided spectacularly sharp images (Griffin, 1990; Wilson, 2003).

We also consider in this chapter the resolution levels likely to be delivered by the extremely large telescopes (ELTs) that are expected to see first light within the next five years or so. It might be observed that there is no theoretical upper limit to telescope resolution; it is always possible to achieve ever-higher levels by simply building larger, AO-equipped telescopes (Gilmozzi et al.). The task ahead is simply one of overcoming practical engineering challenges. While these challenges are certainly not trivial, one might reasonably suppose that they will gradually succumb to future technological progress.

The following crucial issue is considered and analyzed in Sect. 14.6.3: To what extent does uncorrected scintillation undermine the ability of an otherwise perfectly functioning AO system to deliver diffraction-limited images? From the analysis, we discover that even the most severe scintillation that could possibly arise merely has the effect of reducing the Strehl intensity in star images by the factor $\pi/4 \approx 0.8$. Thus as long as the AO system can properly correct the phase corrugations of the image-forming waves in the telescope pupil, even the largest imaginable AO-equipped telescope should be capable of delivering substantially diffraction-limited images (i.e., Strehl intensity $\geq 0.8$) at any wave-length transmitted by the atmosphere, the shortest of which is about 0.3 μm.

The angular resolution delivered by a telescope—whether or not the instrument is equipped with AO—is ultimately limited by diffraction from the telescope aperture. According to the Rayleigh criterion, this limit is set by $1.22 \cdot \lambda/D$.

[3] The rms wavefront error of the Mayall 3.8-m telescope around 1990 was about 0.4 μm. The rms wavefront error of the UKIRT 3.8-m telescope, which at that time was incapable of delivering image cores at 3.4 μm, must have been in excess of 1.2 μm.

[4] Active optics correction of telescope optics may be considered as a subset of adaptive optics in which only slow corrections are made to the imaging wavefronts, with timescales of the order of minutes. Such slow corrections are capable of correcting the fixed wavefront errors of the telescope but are incapable of correcting the more rapidly varying atmospherically induced wavefront errors.

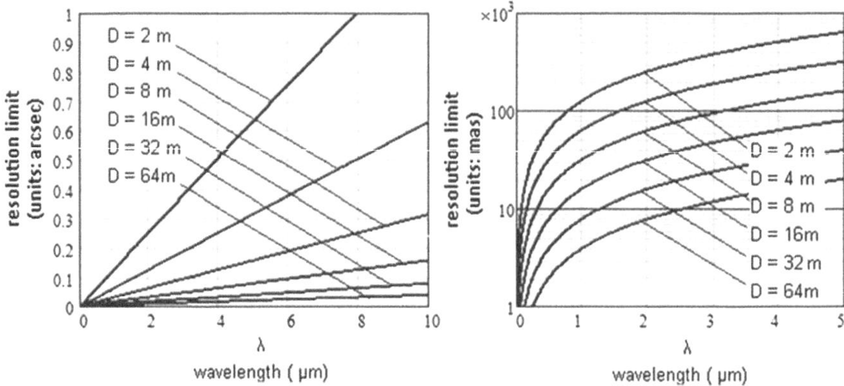


**Fig. 14.1** Rayleigh resolution limits versus wavelength for various telescope diameters. Left Linear vertical-axis scale in units of arcsec. Right Logarithmic vertical-axis scale in milli-arcsecond units (mas)

Table 14.1 shows the resolution limits generated from this formula at the visible wavelength, 0.55 μm, for various aperture diameters, while Fig. 14.1 shows how resolution varies with wavelength according to telescope aperture diameter. The plots in this figure apply equally to ground-based telescopes equipped with efficiently working AO systems and to diffraction-limited telescopes in space.

**Table 14.1** Limiting angular resolution at the visible wavelength 0.55 μm for various aperture sizes, including the aperture size of the unaided human eye under bright illumination

| Aperture diam (m) | Limiting resolution | |
|---|---|---|
| | arcsec | Mas |
| ~0.003 (eye)[a] | 50 | 50,000 |
| 0.1 | 1.5 | 1500 |
| 1 | 0.15 | 150 |
| 4 | 0.035 | 35 |
| 10 | 0.015 | 15 |
| 25 | 0.005 | 5 |
| 50 | 0.0025 | 2.5 |
| 100 | 0.0015 | 1.5 |

[a] The diameter of a fully dilated iris is about 7 or 8 mm. However, resolution in this case is severely degraded by spherical aberration. Under bright illumination, the iris diameter closes down to about 3 mm—the value indicated in the table. With the ~50-arcsec rod and cone spacing also taken into account, the resolution limit generally ascribed to the human eye is approximately 1-arcmin. Unsurprisingly, this value is similar to the measurement accuracy achieved by Tycho Brahe

## 14.1 Telescope Resolution Criteria

Historically, telescope resolving power has usually been assessed by the instrument's ability to resolve two-point objects, binary stars in particular. In this section, we review the various resolution criteria that can be used for this purpose. It is assumed that the two points are incoherently illuminated and of equal brightness. For diffraction-limited telescopes with circular apertures observing in the absence of atmosphere, the intensity variation in a central section through the image of two incoherently illuminated point-objects may be obtained from 4.44 by setting the complex coherence factor $\mu$ to zero. Thus, we can write

$$I(\vartheta, \lambda) = 4 \cdot \left\{ \left[ \frac{J_1\left(\frac{\pi \cdot D \cdot (\vartheta + \Delta\vartheta/2)}{\lambda}\right)}{\frac{\pi \cdot D \cdot (\vartheta + \Delta\vartheta/2)}{\lambda}} \right]^2 + \left[ \frac{J_1\left(\frac{\pi \cdot D \cdot (\vartheta - \Delta\vartheta/2)}{\lambda}\right)}{\frac{\pi \cdot D \cdot (\vartheta - \Delta\vartheta/2)}{\lambda}} \right]^2 \right\} \tag{14.1}$$

where $\Delta\vartheta$ is the angular separation of the-two point objects.

### 14.1.1 *Rayleigh Criterion*

The just-resolved separation at the Rayleigh limit is given by the celebrated expression,

$$\Delta\vartheta_{RayleighClassical} = \frac{3.831706 \cdot \lambda}{\pi \cdot D} = \frac{1.22 \cdot \lambda}{D} \tag{14.2}$$

where the first dark ring of the Airy pattern occurs at the Bessel function argument, 3.831706. As shown in Fig. 14.2, at the just-resolved separation, a 26.5% intensity dip arises midway between the geometrical images of the two points.

### 14.1.2 *Dawes Criterion*

The nineteenth-century English clergyman and astronomer, W. R. Dawes (1799–1868), nicknamed "eagle eye," based his resolution criterion on his own empirical naked-eye observations of double stars. According to his criterion, a binary star is just-resolved at the visible wavelength, 0.55 $\mu$m when a just-discernible intensity dip of about 5% occurs midway between the two image components (Fig. 14.2). At an arbitrarily chosen wavelength, $\lambda$, the Dawes limit may be interpreted as having the form,

$$\Delta\vartheta_{Dawes} = \frac{1.03 \cdot \lambda}{D} \tag{14.3}$$

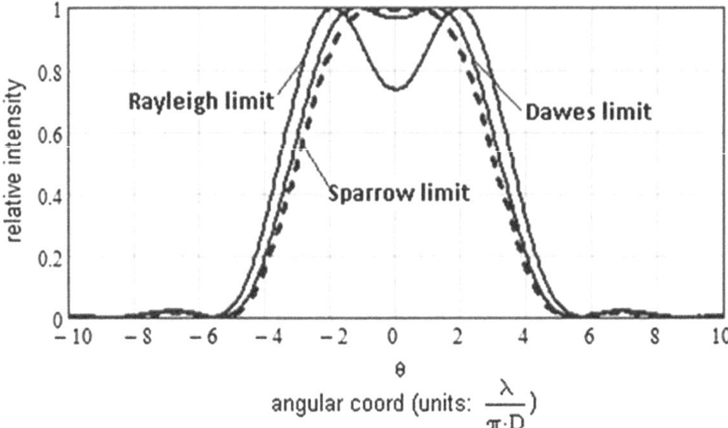


**Fig. 14.2** Intensity sections through a just-resolved two-point object at the Rayleigh, Dawes, and Sparrow limits. In the horizontal axis units used here, $\lambda/(\pi \cdot D)$, the just-resolved separations correspond, respectively, to the values, 3.83, 3.23, and 3.01

from which we see that the just-resolved separation for the Dawes limit is about 18% smaller than that of the Rayleigh limit.

### 14.1.3 Sparrow Criterion

According to the Sparrow criterion (Sparrow, 1916), a two-point object is just-resolved when the first two derivatives of the intensity midway between the two geometrical image points go to zero (Fig. 14.2). At this limiting separation, the intensity flattens out, entirely eliminating the central "dip." Angular resolution at the Sparrow limit is given approximately by

$$\Delta\vartheta_{Sparrow} = \frac{0.96 \cdot \lambda}{D} \tag{14.4}$$

The Sparrow limit separation is about 27% smaller than the separation at the Rayleigh limit. It has been argued that, rather than considering the Sparrow limit as a practical resolution limit, it might be better to consider it more as the limit at which resolution first becomes theoretically possible. All resolution criteria involve a certain level of eye/brain physiology and hence subjectivity. Although the resolution limits set by the above three criteria are not identical, it is reassuring to find that the actual differences are relatively small.

## 14.2 Effect of Central Obstructions on Resolution

For diffraction-limited telescopes with circular apertures and circular central obstructions, a central section of the intensity in the image of an incoherently illuminated two-point object, $I_{CO}(\vartheta, \lambda)$, may be obtained from 13.22 in the form:

$$I_{CO}(\vartheta, \lambda) = \frac{4}{\left[1-\left(\frac{d}{D}\right)^2\right]^2} \times \left\{ \left[ \frac{J_1\left(\frac{\pi \cdot D \cdot (\vartheta + \Delta\vartheta/2)}{\lambda}\right)}{\frac{\pi \cdot D \cdot (\vartheta + \Delta\vartheta/2)}{\lambda}} - \left(\frac{d^2}{D^2}\right) \cdot \frac{J_1\left(\frac{\pi \cdot d \cdot (\vartheta + \Delta\vartheta/2)}{\lambda}\right)}{\frac{\pi \cdot d \cdot (\vartheta + \Delta\vartheta/2)}{\lambda}} \right]^2 + \left[ \frac{J_1\left(\frac{\pi \cdot D \cdot (\vartheta - \Delta\vartheta/2)}{\lambda}\right)}{\frac{\pi \cdot D \cdot (\vartheta - \Delta\vartheta/2)}{\lambda}} - \left(\frac{d^2}{D^2}\right) \cdot \frac{J_1\left(\frac{\pi \cdot d \cdot (\vartheta - \Delta\vartheta/2)}{\lambda}\right)}{\frac{\pi \cdot d \cdot (\vartheta - \Delta\vartheta/2)}{\lambda}} \right]^2 \right\} \quad (14.5)$$

Figure 14.3 shows intensity sections calculated from the above expression for various central obstruction diameters, where the same point separation, $\Delta\vartheta = 1.22 \cdot \lambda/D$, is used in all cases. The points are of course just-resolved for the unobstructed aperture (i.e., d/D = 0), but evidently become more comfortably resolved as central obstruction diameter increases. Figure 14.4 shows how resolution steadily improves with increasing central obstruction diameter. In practice, the obstruction ratios for most large telescopes rarely exceed the value, 0.25, at which only about

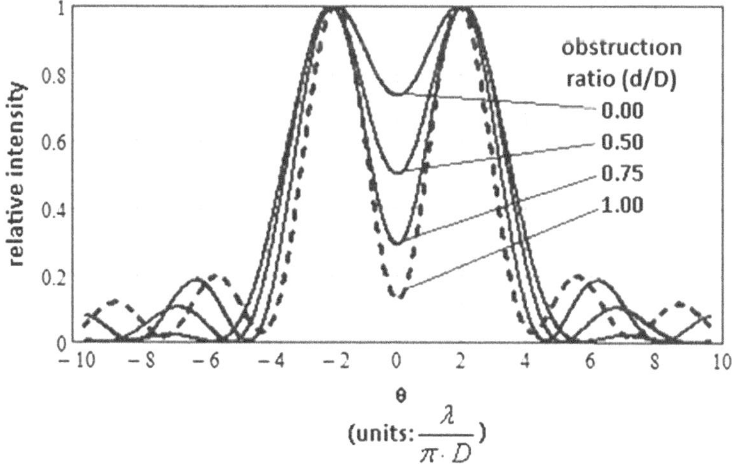


**Fig. 14.3** Intensity sections for a two-point object formed by diffraction-limited telescopes with various central obstruction diameters. All plots are normalized to unity. The dotted line plot corresponds to the limiting case, $d/D \rightarrow 1.0$

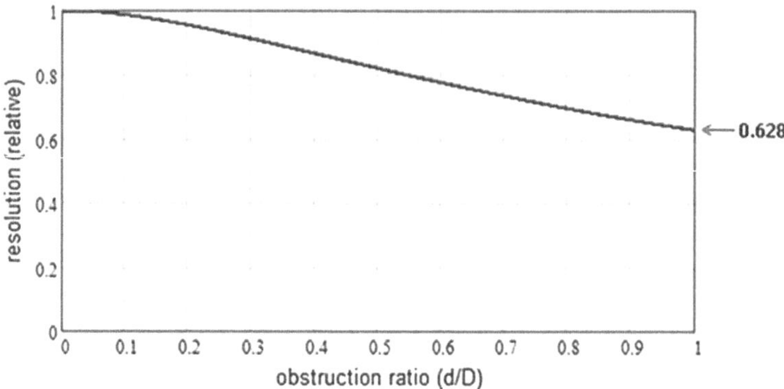


**Fig. 14.4** For diffraction-limited telescopes with circular apertures, the just-resolved separation according to the modified Rayleigh criterion slowly improves with increasing central obstruction diameter. The plot is normalized to unity for zero obstruction

a 7% resolution improvement arises.[5] Further discussion of the effect of central obstructions on resolution is given in Sect. 14.5.

## 14.3 Effect of Mild Aberrations on Resolution

In this section, we consider the effect of small amounts of aberration on telescope resolution. The aberration levels considered are consistent with Strehl intensities lying in the range, SI $\geq 0.8$. Figure 14.5 shows intensity sections through a two-point object for (1) an aberration-free telescope and (2) the same telescope with a quarter wave ($\lambda/4$) peak-to-valley (P-V) of defocus—an amount known to reduce Strehl intensity to about 0.8 (Born & Wolf, 2003). The same point separation, $1.22 \cdot \lambda/D$, was used to calculate both intensity sections shown.

As shown in the figure, both intensity sections show approximately the same, 26.5%, intensity dip in the center of the image. Thus, it may be concluded that defocus in the small amounts considered here does not significantly degrade resolution. The same conclusion also applies for the commonly occurring first-order Seidel aberrations, spherical aberration, coma, and astigmatism. Just as long as P-V wavefront aberration introduced by these aberration types does not exceed $\lambda/4$, Strehl intensity remains at or above the diffraction-limited threshold value, 0.8, and a two-point object can be resolved to nearly the same level as achieved by a diffraction-limited version of the same telescope.

[5] The Mayall telescope, where $d = 1.65m$ and $D = 3.8m$, has an exceptionally large central obstruction. Ignoring the instrument's aberrations, for binary stars with equal brightness, such a central obstruction offers a 15% resolution improvement.

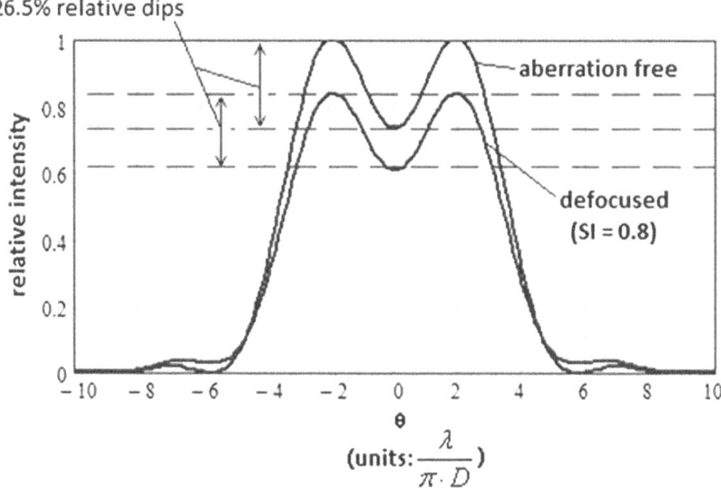


**Fig. 14.5** Image intensity sections for a two-point object separated at the just-resolved Rayleigh limit (3.832 units) for an aberration-free optical system (SI = 1) and an aberrated optical system where defocus has reduced the Strehl intensity to the limit for diffraction-limited imaging, SI = 0.8. In either case, the 26.5% relative intensity dip is maintained. Thus, as long as defocus does not exceed the amount set by the Strehl limit, the optical system retains its ability to resolve two-point objects to the same level as the system in perfect focus

## 14.4 Resolution Provided by Gaussian Approximation to the Airy Pattern

Equations 13.12 and 13.14 previously showed how an Airy pattern image may be approximated by a Gaussian function. In terms of that approximation, the central intensity section through an incoherently illuminated two-point object may be expressed in the form,

$$I_{Gaus}(\vartheta, \lambda) = \exp\left[-\left(\frac{\vartheta + \Delta\vartheta/2}{\vartheta_G}\right)^2\right] + \exp\left[-\left(\frac{\vartheta - \Delta\vartheta/2}{\vartheta_G}\right)^2\right] \tag{14.6}$$

where $\Delta\vartheta$ is the angular separation of the two-point object.

Figure 14.6 provides a comparison between the approximate intensity section calculated from the above expression and the exact intensity section that arises when both individual images are Airy patterns (cf., 14.1). Both plots show the same, 26.5%, intensity dip in the center of the image, so that both correspond to the just-resolved condition according to the Rayleigh criterion. Though the separation, $\Delta\vartheta = 1.22 \cdot \lambda/D$, fulfills this condition for the Airy pattern images, for the Gaussian image approximations, the required separation turns out to be slightly larger, specifically $\Delta\vartheta = 1.26 \cdot \lambda/D$. However, for most practical purposes, the small (<3%) difference

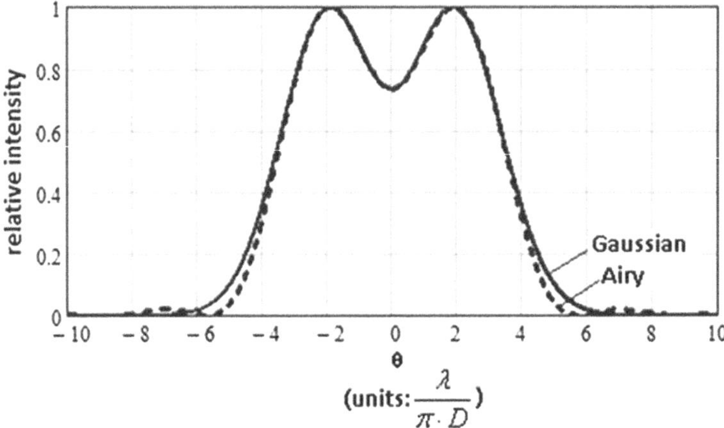


**Fig. 14.6** Intensity sections through the image of a two-point object separated at the Rayleigh limit showing the required 26.5% central dip. The sections show the case where the individual images are Airy patterns and the case where the individual Airy pattern images are approximated by Gaussian functions. To attain the required 26.5% central dip, slightly different point separations had to be used to generate the plots, 3.832 units for the Airy pattern images and 3.960 units for the Gaussian image envelope approximation

seen here is inconsequential.[6] Indeed, the fact that the difference is so small may be taken as a measure of validation for the use in this chapter of Gaussian image approximations for establishing telescope resolution.

## 14.5 Resolution for Images Displaying Core and Halo Structure

As discussed in Chap. 13, the most general type of point-object image usually displays core and halo structure, with core-only and halo-only images simply seen as special cases. We also saw in Chap. 13 how core and halo images can be approximated as the sum of two separate Gaussian functions. In this section, we provide corresponding approximations for the intensity in the images of two-point objects, where each of the individual (core and halo) images is approximated as the sum of two Gaussian functions. The illuminations arising from the two objects are assumed mutually incoherent, so that the analysis directly applies to binary star images.

To simplify the analysis further, we also assume that the individual point-object images have circular symmetry; thus, we restrict consideration to circular Gaussian functions. This assumption effectively restricts us here to telescopes with

[6] Compared to the 18% and 27% point separation differences for the Dawes or Sparrow criteria (14.3) and (14.4), the 3% difference indicated here is barely significant.

circular apertures, circular central obstructions, and circularly symmetric wavefront aberrations, if any.

In general, the individual components of a two-point object have unequal brightnesses. Brightness difference is indicated by the brightness ratio, BR. For binary stars, where we denote the apparent visual magnitudes of the two components by $m_1$ and $m_2$, the brightness ratio may be expressed in terms of the magnitude difference as follows,

$$BR = \left[100^{0.2}\right]^{(m_1 - m_2)} = 2.5119^{(m_1 - m_2)} \tag{14.7}$$

The quantity $100^{0.2} \approx 2.5119$ is known as Pogson's ratio, called after the English astronomer Norman R. Pogson (1829–1891).[7] Without loss of generality, it is assumed that the first of the two stars considered, Star1, is either as bright or brighter than the second, Star2. Thus $m_1 \leq m_2$ and BR takes values in the range (It is noted that the magnitude quantities used here, $m_1$ and $m_2$, are entirely unrelated to our use of the quantity, m, in Chap. 11 and other chapters in the context of reduced stellar speckle patterns.)

$$0 \leq BR \leq 1 \tag{14.8}$$

Using 13.55 to provide the Gaussian approximation for the average intensity envelope for the (core and halo) image of a single star, the corresponding intensity envelope for a binary star may be obtained by convolution of this Gaussian function approximation with two separated Dirac delta functions, the sifting property of which enables the average image intensity to be written in the form:

$$\begin{aligned} I_{TPOG}(\alpha, \beta, \Delta\alpha) = & \frac{1}{\pi \cdot \left(A_{CH} \cdot B_C^2 + B_H^2\right) \cdot (1 + BR)} \\ & \times \left\{ A_{CH} \cdot \exp\left[ -\left( \frac{(\alpha + \Delta\alpha/2)^2 + \beta^2}{B_C^2} \right) \right] \right. \\ & + \exp\left[ -\left( \frac{(\alpha + \Delta\alpha/2)^2 + \beta^2}{B_H^2} \right) \right] \\ & + BR \cdot A_{CH} \cdot \exp\left[ -\left( \frac{(\alpha - \Delta\alpha/2)^2 + \beta^2}{B_C^2} \right) \right] \\ & \left. + BR \cdot \exp\left[ -\left( \frac{(\alpha - \Delta\alpha/2) + \beta^2}{B_H^2} \right) \right] \right\} \end{aligned} \tag{14.9}$$

[7] In 1856, Pogson noted that in the stellar magnitude system introduced by the Greek astronomer Hipparchus, a first magnitude star is about 100 times brighter than a sixth magnitude star and suggested that this might be used as a standard. Thus, a first magnitude star is $100^{1/5}$ (or about 2.5119) times as bright as a second magnitude star.

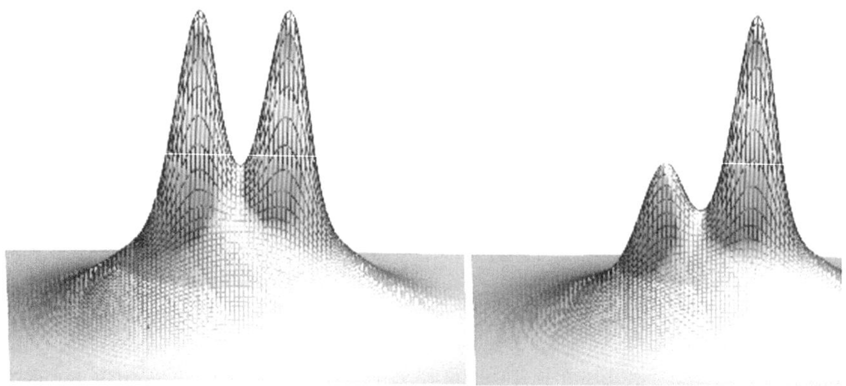

**Fig. 14.7** Images of binary stars separated by $\Delta\alpha = 0.4$ arcsec. Left Image for components of equal brightness. Right Image for a single-magnitude brightness difference. The individual star images display identical core and halo characteristic as prescribed by $A_{CH} = 2.7$, $B_C = 0.12$ arcsec, and $B_H = 0.5$ arcsec

where the angular separation of the binary star, $\Delta\alpha$ is assumed here, without loss of generality, to lie in the $\alpha$-direction. The normalization of the above equation is such that the total volume (light energy) enclosed under the 2-D intensity envelope is always unity.

The quantities $A_{CH}$, $B_C$, and $B_H$ may be obtained from 13.61, 13.63, and 13.65 by inserting the appropriate values for the parameters, D, d, λ, SI(λ), σ, and $w_0$. For AO-corrected images, the quantities $SI_{AO}(\lambda)$, $\sigma_{AO}$, and $w_{0AO}$ would be used instead of SI(λ), σ, and $w_0$.

Figure 14.7 shows the binary star images calculated from 14.9 for binary stars with equal and unequal brightnesses. The images in this figure are based on a 3.8-m telescope with a 1.65-m diameter central obstruction and 0.5 Strehl intensity at the near-IR observation wavelength, 2.2 μm, in 1-arcsec visible seeing conditions. (The parameter values used to create the surface plots in the figure are the following: D = 3.8 m, d = 1.65 m, σ = 0.495 μm and $w_0 = 0.413$ m, which equate to $A_{CH} = 2.7$, $B_C = 0.12$ arcsec, and $B_H = 0.5$ arcsec.)

### *14.5.1 Calculating the Just-Resolved Separation*

To apply Rayleigh's criterion, the binary star components must both have the same brightness. Thus, to examine the effect of core and halo structure on resolution, we must set BR = 1 in 14.9. The intensity envelope may then be written,

$$I_{TPOG}(\alpha, \beta, \Delta\alpha) = \frac{1}{2 \cdot \pi \cdot \left(A_{CH} \cdot B_C^2 + B_H^2\right)}$$

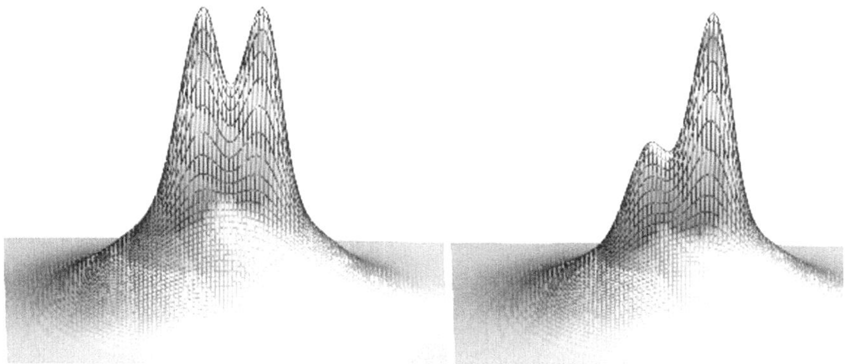

**Fig. 14.8** Left Binary star image at the just-resolved Rayleigh separation where a 26.5% intensity dip occurs. (Parameters used: $A_{CH} = 2.7$, $B_C = 0.12$ arcsec, $B_H = 0.5$ arcsec and $\Delta\alpha_{Rayleigh} = 0.2819$ arcsec). Right The image using the same parameters but with a single-magnitude brightness difference between the two components

$$\times \left\{ A_{CH} \cdot \exp\left[ -\left( \frac{(\alpha + \Delta\alpha/2)^2 + \beta^2}{B_C^2} \right) \right] + \exp\left[ -\left( \frac{(\alpha + \Delta\alpha/2)^2 + \beta^2}{B_H^2} \right) \right] + A_{CH} \cdot \exp\left[ -\left( \frac{(\alpha - \Delta\alpha/2)^2 + \beta^2}{B_C^2} \right) \right] + \exp\left[ -\left( \frac{(\alpha - \Delta\alpha/2)^2 + \beta^2}{B_H^2} \right) \right] \right\} \quad (14.10)$$

where it is assumed that $A_{CH}$, $B_C$, and $B_H$ have previously been established. Along the plane of symmetry between the twin image points (i.e., the plane defined by $\beta = 0$), the maximum intensity attained in the twin peaks, $I_{peak}$, and the intensity in the dip between these peaks, $I_{dip}$, can readily be obtained from 14.10 for any arbitrarily chosen $\Delta\alpha$ value. By numerically searching for the separation, $\Delta\alpha$, that satisfies the condition,[8]

$$\frac{I_{dip}}{I_{peak}} = 0.735 \quad (14.11)$$

we can establish the just-resolved angular separation according to the Rayleigh resolution criterion. We denote that separation here by $\Delta\alpha_{Rayleigh}$.

Figure 14.8 (left) shows the image of a just-resolved, equal brightness binary star where the angular separation is such that 14.11 is exactly satisfied. The other image (right) shows the image appearance for the same angular separation where there is

[8] While an analytic solution would have been more convenient, the Author was unable to find one.

now a single-magnitude brightness difference between the two components (i.e., BR = 0.398). Clearly, brightness differences significantly affect a telescope's ability to resolve binary stars.

### 14.5.2 Just-Resolved Separation for Core-Dominated Images

When all of the light in a binary star image resides in the twin image cores (corresponding to the case $\sigma/\lambda \to 0$), the Gaussian approximation for such an image can be obtained from 14.9 in the limit, $A_{CH} \to \infty$. Thus, we can write

$$\langle I_{TPOG,Core}(\alpha,\beta,\Delta\alpha)\rangle = \frac{1}{\pi \cdot B_C^2 \cdot (1+BR)} \times \left\{ \exp\left[-\left(\frac{(\alpha+\Delta\alpha/2)^2+\beta^2}{B_C^2}\right)\right] + BR \cdot \exp\left[-\left(\frac{(\alpha-\Delta\alpha/2)^2+\beta^2}{B_C^2}\right)\right]\right\} \quad (14.12)$$

where the normalization is again such that the intensity envelope encloses unit volume. Again, for an equal-brightness binary, numerical methods can be used with the above equation to find the just-resolved $\Delta\alpha_{Rayleigh}$ separation, for which the values, $I_{peak}$ and $I_{dip}$, satisfy 14.11.

### 14.5.3 Just-Resolved Separation for Halo-Dominated Images

When all of the light resides in the halos, with negligible amounts remaining in the cores (roughly corresponding to $\sigma/\lambda > 0.4$, the Gaussian approximation for the image of the binary star can be obtained from 14.9 in the limit, $A_{CH} \to 0$, thus

$$\langle I_{TPOG,Halo}(\alpha,\beta,\Delta\alpha)\rangle = \frac{1}{\pi \cdot B_H^2 \cdot (1+BR)} \times \left\{ \exp\left[-\left(\frac{(\alpha+\Delta\alpha/2)^2+\beta^2}{B_H^2}\right)\right] + BR \cdot \exp\left[-\left(\frac{(\alpha-\Delta\alpha/2)^2+\beta^2}{B_H^2}\right)\right]\right\} \quad (14.13)$$

Figure 14.9 shows image intensity envelopes, calculated from the above equation, for an equal brightness, 1-arcsec binary pair as observed by a 3.8-m telescope at the near-IR wavelength, 2.2 μm, when all of the light resides in the halos. Images are

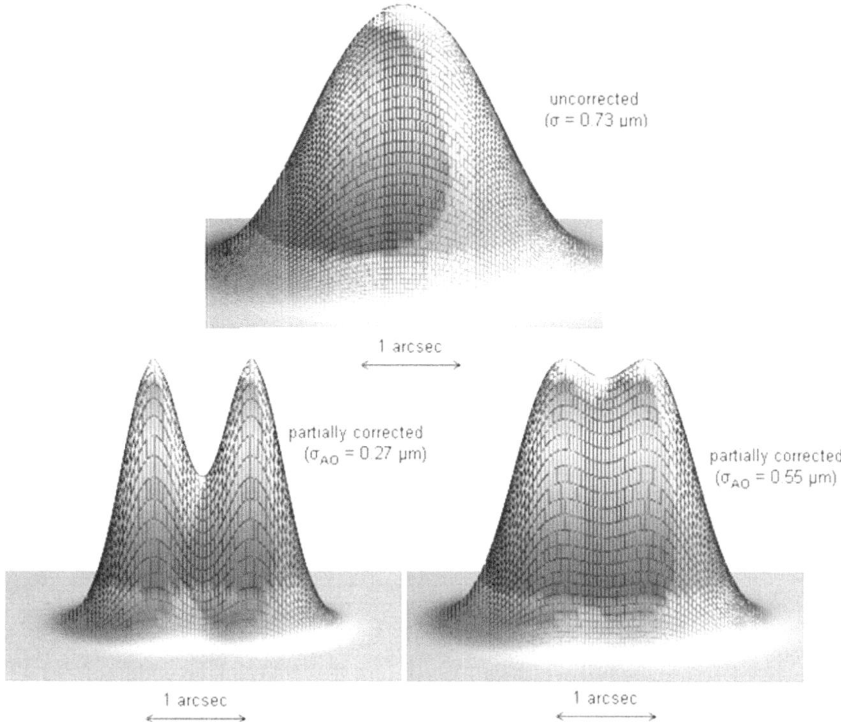


**Fig. 14.9** Top An uncorrected halo-only image of a 1-arcsec binary star obtained using a 4-m diffraction-limited telescope at 0.55 μm in 1.5-arcsec visible seeing conditions. Bottom Partially corrected images for two different levels of AO correction corresponding to the $\sigma_{AO}$ values indicated. Note that $\sigma/\lambda = 1$ for the top image envelope, while $\sigma_{AO}/\lambda = 0.655$ and $\sigma_{AO}/\lambda = 0.5$ for the bottom envelopes. Because the various ratios used here do not fulfill either of the conditions, $\sigma/\lambda < 0.4$ or $\sigma_{AO}/\lambda < 0.4$, cores are absent in all three images

shown for an uncorrected image and two partially AO-corrected images. For the case, BR = 1 numerical methods can be used with the above equation to find the just-resolved separation, $\Delta\alpha_{\text{Rayleigh}}$, for which the values, $I_{peak}$ and $I_{dip}$, satisfy 14.11.

### 14.5.4 Relative Angular Widths of Cores and Halos

With large telescopes, core-only images generally provide much higher resolution levels than halo-only images. The ratio, $B_C/B_H$, which we denote by $R_{RC}$, provides a convenient measure of the resolution improvement factor that occurs when cores are present in star images compared to resolution when only halos are present. Equations 13.61 and 13.63 allow $R_{CH}$ to be expressed in the form:

$$R_{CH} = \frac{B_C}{B_H} = \frac{\lambda \cdot w_0}{\pi \cdot \sigma \cdot \sqrt{SI(\lambda) \cdot \left(D^2 - d^2\right)}} \tag{14.14}$$

From the expression on the right, we see that $R_{RC}$ takes its smallest value when Strehl intensity takes its maximum value, $SI(\lambda) = 1$. Thus, maximum resolution benefit is obtained from image cores when the telescope is diffraction-limited. In the case of a diffraction-limited telescope with no central obstruction (i.e., $d = 0$), the above equation indicates that the smallest possible $R_{RC}$ value is given by

$$R_{CH_{MIN}} = \frac{\lambda \cdot w_0}{\pi \cdot \sigma \cdot D} \tag{14.15}$$

We may also use the full-width half-maximum (FWHM) image width (cf., 13.43) at visible wavelengths (where cores are assumed absent) to give the alternative expression for $R_{RCmin}$,

$$R_{CH_{MIN}} = \frac{4 \cdot \lambda \cdot \sqrt{\ln(2)}}{\pi \cdot D \cdot FWHM_{Halo}} \tag{14.16}$$

Evaluation of the above expression for a 3.8-m diffraction-limited telescope imaging at 2.2 μm in average visible seeing conditions ($FWHM_{Halo} = 1$ arcsec) gives $R_{RCmin} = 0.125$. Thus, when cores are present at this wavelength in such seeing conditions, they provide a resolution improvement factor of 8. For a 10-m diffraction-limited telescope imaging at wavelength 1.1 μm in the same seeing conditions, the above expression gives $R_{RCmin} = 0.025$, corresponding to a 40 times resolution improvement factor.

Equation 14.16 could give the (false) impression that, by simply imaging at short enough wavelengths, there is no limit to the resolution improvement obtainable from image cores. However, as wavelength decreases, the energy fraction invested in the core diminishes exponentially (cf., 10.3). Ultimately, at short enough wavelengths, cores disappear entirely from the image, at which point resolution reverts back to the lower level provided by the halo.

## 14.6 Resolution of AO-Equipped Telescopes

### 14.6.1 *Halo-Dominated Images*

For halo-dominated images, AO systems tend to reduce the angular width of the halo in proportion to $\sigma_{AO}/\sigma$ where, as before, $\sigma_{AO}$ is the rms of the residual optical path difference (OPD) fluctuation after AO correction and σ is the value before correction.[9] In instances where, even after the "correction," $\sigma_{AO}$ remains greater than $0.4 \cdot \lambda$, the AO-corrected images continue to be halo-dominated as previously illustrated in Fig. 14.9.

### 14.6.2 *Emergence of Cores in AO-Corrected Images*

If the AO system can create the condition, $\sigma_{AO} < 0.4 \cdot \lambda$, cores begin to appear in the images; the more the AO system succeeds in driving down the $\sigma_{AO}/\lambda$ ratio, the brighter and more dominant these cores become. This behavior is illustrated in Fig. 14.10 where an uncorrected image and two partially AO-corrected computer-generated images are shown for an equal brightness, 0.036-arcsec binary star as obtained at $\lambda = 0.55\,\mu\text{m}$ using a diffraction-limited 4-m telescope in 1-arcsec (FWHM) seeing conditions. (Eqs. 13.61, 13.63, and 13.65 were again used to calculate values for the parameters, $A_{CH}$, $B_C$, and $B_H$. The same parameter value set, $D = 4$ m, $d = 0$ m, $SI(0.55\,\mu\text{m}) = 1$, and $w_0 = 0.25$ m, is used to calculate all images in this figure.) For the uncorrected image, the value $\sigma = 0.36\,\mu\text{m}$ is used; for the partially corrected images, the values $\sigma_{AO} = 0.18\,\mu\text{m}$ and $\sigma_{AO} = 0.09\,\mu\text{m}$ are used. For the latter $\sigma_{AO}$ value, the core features dominate the image, with the 26.5% intensity dip indicating that the binary star is just-resolved according to the Rayleigh criterion.

### 14.6.3 *Strehl Intensity Limit Imposed by Uncorrected Scintillation*

In the absence of scintillation,[10] a large telescope equipped with a perfectly working AO system—that is, an AO system that perfectly corrects all wavefront phase errors—should in principle be able to deliver ideal, diffraction-limited star images with Strehl

[9] If $w_{0AO} \approx w_0$, resolution provided by the halo improves in proportion to $\sigma_{AO}/\sigma$. However, as discussed previously in Sect. 13.2.3, this relation does not hold generally since $w_{0AO}$ may take values other than $w_0$.

[10] Scintillation refers to the amplitude fluctuation of the waves arriving at the telescope. Unlike phase fluctuation, the amplitude fluctuation can be seen directly by eye as twinkling.

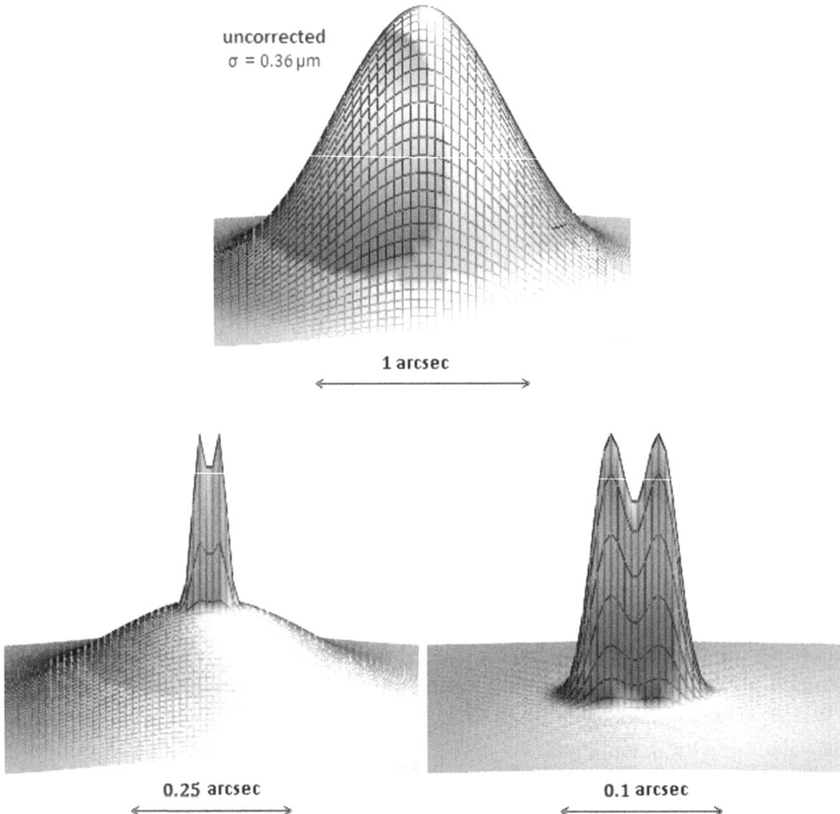


**Fig. 14.10** *Top* An uncorrected image of an extremely close, 0.036 arcsec (36-mas), binary star obtained using a 4-m diffraction-limited telescope at 0.55 μm in 1-arcsec visible seeing conditions. *Bottom* Partially AO-corrected images. Weak cores are evident (bottom left) for $\sigma_{AO} = 0.18$ μm, (note that $\sigma_{AO}/\lambda = 0.33$). For the case, $\sigma_{AO} = 0.09$ μm (*bottom right*), the cores dominate the image (note that $\sigma_{AO}/\lambda = 0.165$). Also note the different angular scales used in the three images

intensities attaining the maximum value, unity. However, when scintillation is present in the image-forming waves arriving at the telescope, and it is assumed that the otherwise perfectly working AO system has no capacity to correct this scintillation, it is then no longer possible to deliver such ideal images. In this section, we establish the theoretical maximum Strehl intensity value achievable by an AO system that perfectly corrects phase errors but has no capacity whatsoever to correct scintillation.

Scintillation is generally more severe for low-altitude observing paths and less severe for paths closer to the zenith. Scintillation is also more severe at visible wavelengths than at IR wavelengths. In this section, we analyze the worst possible case of scintillation: that is, the case where the atmospheric turbulence is supposed strong enough and the imaging path supposed long enough that the scintillation finally

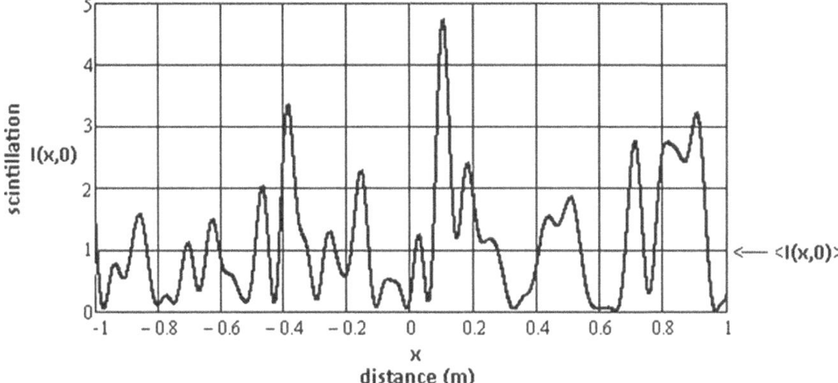


**Fig. 14.11** Fully developed (Gaussian) intensity scintillation in the section y = 0 over a 2-m telescope aperture. Random phase fluctuations (which are uncorrelated with the intensities) are also present. Whereas AO systems can correct phase errors, they have no ability to correct scintillation

developed in the image-forming waves arriving in the pupil plane of the telescope corresponds to fully developed Gaussian speckle at all visible and IR wavelengths.

The surprising outcome of the analysis is that it should still be possible to achieve the Strehl intensity value, $\pi/4 = 0.7854$, in AO-corrected star images. Since Strehl intensities $\geq 0.8$ are generally regarded as diffraction-limited, the analysis indicates that uncorrected scintillation—even in its most severe form—does not by itself exclude the possibility of AO-equipped telescopes delivering substantially diffraction-limited images, irrespective of the imaging wavelength.

Figure 14.11 shows intensity scintillation in the $y = 0$ section of a light wave arriving in the pupil of a large telescope where strong atmospheric turbulence (combined with a long enough atmospheric path) results in the development of Gaussian speckle. The average size of the scintillation structures shown in Fig. 14.11 has been chosen, somewhat arbitrarily,[11] to be about the same as that seen in the scintillation patterns shown previously in Fig. 3.11. However, just as long as the size of these structures is significantly less than the diameter of the observing telescope, the outcome of the analysis is the same irrespective of the size choice.

If we assume—as we have throughout the book—that the statistical properties of the intensity scintillation in the (x, y) pupil plane of the telescope (cf., Fig. 7.2) are spatially and temporally stationary over meaningfully large spatial and temporal intervals, it follows that the average intensity has the same constant value every-where and at all times in the pupil plane. Thus, we may write

$$\langle I(x, y) \rangle = I_{Ave} \tag{14.17}$$

where $I_{Ave}$ is constant.

[11] Scintillation structure size is inversely proportional to the scatter angle of the light arriving at the telescope.

Because the scintillation is assumed to obey Gaussian statistics, the PDF of the intensity fluctuation in the scintillation pattern can be expressed by the negative exponential form (cf., 11.21),

$$PDF_I(I) = \frac{1}{I_{Ave}} \cdot \exp\left(-\frac{I}{I_{Ave}}\right) \tag{14.18}$$

If the AO system possesses the ability to perfectly correct all phase errors, the phase of the AO-corrected waves over the entire telescope pupil may then be considered constant and, without loss of generality, we may consider the actual value to be zero (i.e., $\varphi(x, y) = 0$). It follows that the complex amplitude of the imaging waves, A(x, y), is then entirely real, with the actual value given by (cf., 3.22)

$$A(x, y) = \sqrt{I(x, y)} \tag{14.19}$$

where, without loss of generality, we choose the positive square root.

Denoting the complex amplitude of the star image in the telescope image plane (Fig. 7.2) by $A_{Im}(u, v)$, the complex amplitude in the center of the image, $A_{Im}(0, 0)$, can be obtained from the Fraunhofer diffraction integral (cf., 4.10) by directly summing the (real) complex amplitudes associated with the (assumed) large number of elementary phasors constituting the phase-corrected complex amplitude in the telescope pupil, A(x, y).[12] Thus, we can write

$$A_{Im}(0, 0) = \iint_{Pupil} A(x, y) \cdot dx \cdot dy = \iint_{Pupil} \sqrt{I(x, y)} \cdot dx \cdot dy \tag{14.20}$$

The corresponding intensity in the center of the image, $I_{Im}(0, 0)$, may then be expressed by (cf., 4.11)

$$I_m(0, 0) = \left(\iint_{Pupil} \sqrt{I(x, y)} \cdot dx \cdot dy\right)^2 \tag{14.21}$$

and thus, the average intensity in the center of the image may be written

$$\langle I_{Im}(0, 0)\rangle = \left(\iint_{Pupil} \langle\sqrt{I(x, y)}\rangle \cdot dx \cdot dy\right)^2 \tag{14.22}$$

Because I(x, y) obeys negative exponential statistics (14.18), the average quantity, $\sqrt{I(x, y)}$, may be obtained from the negative exponential PDF (cf., 14.17 and 14.18). However, because the statistics of I(x, y) are stationary with respect to x and y, $\sqrt{I(x, y)}$ must take the same value at all (x, y) locations. Thus, for present purposes, we may omit the (x, y) dependence and simply denote this quantity by $\sqrt{I}$. Thus,

[12] To justify this assumption, the average size of the scintillation structures must be at least several times smaller than the telescope diameter.

we may write

$$\langle\sqrt{I}\rangle = \int_0^\infty \frac{\sqrt{I}}{I_{Ave}} \cdot \exp\left(-\frac{I}{I_{Ave}}\right) \cdot dI \tag{14.23}$$

Combining 14.22 and 14.23 allows us to write the following expression for the average Strehl intensity, $\langle SI\rangle$

$$\langle SI\rangle = \frac{\left\{\iint_{Pupil}\left[\int_0^\infty \frac{\sqrt{I}}{I_{Ave}} \cdot \exp\left(-\frac{I}{I_{Ave}}\right) \cdot dI\right] \cdot dx \cdot dy\right\}^2}{\left[\iint_{Pupil} \sqrt{I_{Ave}} \cdot dx \cdot dy\right]^2} \tag{14.24}$$

Because of the absence of (x, y) dependence in the above equation, the integration operations with respect to x and y on both the numerator and the denominator become trivial. Equation 14.24 then reduces to the simpler form,

$$\langle SI\rangle = \frac{\left[\int_0^\infty \frac{\sqrt{I}}{I_{Ave}} \cdot \exp\left(-\frac{I}{I_{Ave}}\right) \cdot dI\right]^2}{I_{Ave}} \tag{14.25}$$

Using the result, $\int_0^\infty \sqrt{x} \cdot \exp(-x) \cdot dx = \sqrt{\pi}/2$, 14.25 simplifies to give the average Strehl intensity as

$$\langle SI\rangle = \frac{\pi}{4} = 0.7854 \tag{14.26}$$

Thus, even in the worst-case circumstance considered here, where scintillation has degenerated into fully developed Gaussian speckle, only a small fraction of the light energy, $(1 - \pi/4) = 0.2146$, scatters out of the core into the surrounding halo. The angular width of the halo varies inversely as the average scintillation structure size. If that size is significantly smaller than the telescope aperture size (as is the case for large telescopes), the halo then has a much larger angular size than the central core, thus enabling the core (containing 78.54% of the light energy) to dominate the center of the image.

#### 14.6.3.1 Strehl Intensity Achievable Under Typical Scintillation Conditions

For more typical astronomical imaging paths, the amount of scintillation in the image-forming waves from a distant star at the instant the waves arrive in the pupil plane of the telescope is generally significantly less than the limitingly large amount just considered. By approximating the probability density function of the intensities in these scintillation patterns by the Gamma distribution (Sect. 11.6.6.1), we may write the PDF of intensity in these patterns as follows:

$$PDF(I) = \frac{\left(\frac{m}{I_{Ave}}\right)^{m} \cdot I^{m-1} \cdot exp\left(-m \cdot \frac{I}{I_{Ave}}\right)}{\Gamma(m)} \tag{14.27}$$

where $\Gamma(\cdot)$ is the gamma function and the quantity, m, was defined previously (cf., 11.26) in terms of the contrast ratio of a speckle pattern which, in this case, is the scintillation pattern associated with the image-forming light waves arriving in the telescope pupil. Denoting the contrast ratio in this pattern by $C_{scin}$(cf., 11.22), and noting that $m = 1/C_{scin}^2$, we may rewrite 14.27 in the form,

$$\begin{aligned} PDF(I) &= \frac{1}{\Gamma\left(\frac{1}{C_{scin}^2}\right)} \cdot \left(\frac{1}{C_{scin}^2 \cdot I_{Ave}}\right)^{\left(\frac{1}{C_{scin}^2}\right)} \cdot I^{\left(\frac{1}{C_{scin}^2}-1\right)} \\ &\quad \cdot \exp\left(-\frac{I}{C_{scin}^2 \cdot I_{Ave}}\right) \quad \text{for } I \geq 0 \\ &= 0 \quad \text{otherwise} \end{aligned} \tag{14.28}$$

The average Strehl intensity achieved in a star image for this PDF is then given by

$$\langle SI \rangle = \frac{\left[\int_0^{\infty} \sqrt{I} \cdot \left(\frac{1}{C_{Scin}^2 \cdot I_{Ave}}\right)^{\left(\frac{1}{C_{Scin}^2}\right)} \cdot I^{\left(\frac{1}{C_{Scin}^2}-1\right)} \cdot \exp\left(-\frac{I}{C_{Scin}^2 \cdot I_{Ave}}\right) \cdot dI\right]^2}{\Gamma\left(\frac{1}{C_{scin}^2}\right) \cdot I_{Ave}} \tag{14.29}$$

A plot of $SI$ versus scintillation contrast ratio, $C_{scin}$, is shown in Fig. 14.12. The extreme case—where the scintillation pattern takes the form of Gaussian speckle and where the maximum achievable average Strehl intensity is π/4—corresponds to $C_{scin} = 1$. For astronomical telescopes observing within 45° of the zenith in 1.5 arcsec or better visible seeing conditions, where we might anticipate relatively weak scintillation, the figure indicates that the highest achievable Strehl intensity in star images is likely to be significantly higher than π/4.

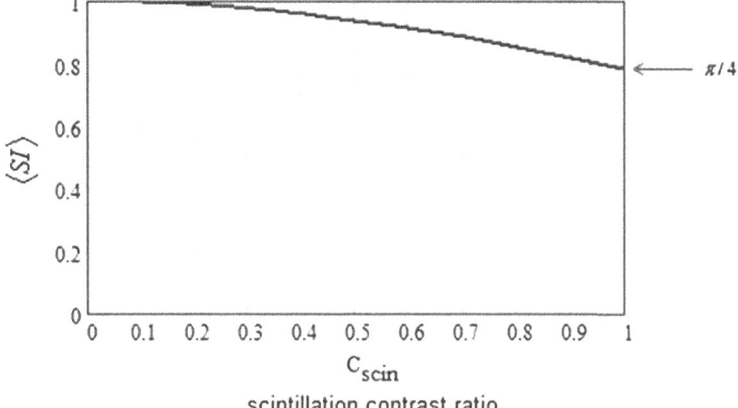


**Fig. 14.12** The maximum Strehl intensity that can be obtained from a large telescope equipped with an AO system (which is capable of correcting the phase errors but not the scintillation) plotted against the contrast ratio of the scintillation, $C_{scin}$. For mild scintillation, typical of viewing paths within 45° of the zenith where $C_{scin} \ll 1$, average Strehl intensity obtained would closely approach the ideal value, unity

#### 14.6.3.2 Equivalent Rms Phase Fluctuation to Obtain the Strehl Intensity, Π/4

The essential characteristic of a random phase screen is that it imprints phase fluctuations on traversing light waves without immediately introducing amplitude fluctuations, or scintillation, into the waves; scintillation only develops after the waves have propagated further beyond the phase screen. The complex amplitude "imprinted" on an initially plane wave by a random phase screen may be characterized by the amplitude and phase given by A(x, y) = constant and $\varphi(x, y)$ = random variable. This combination is seen to be exactly the opposite of the case dealt with in the previous section where we found that A(x, y) = random variable and $\varphi(x, y)$ = constant. Either of these combinations, of course, has the capacity to pull down Strehl intensity.

In this section, we investigate the relative Strehl intensity pull-down effects of phase and amplitude fluctuations on image-forming light waves. We do this by quantifying the rms OPD fluctuation that must be introduced into a plane image-forming wave by a random phase screen to pull down Strehl intensity in the image to $\pi/4$—which, as we have just seen, is the Strehl intensity that results in the image of a perfectly phase-corrected wave exhibiting uncorrected scintillation in its most severe (Gaussian speckle) form.

The approximate Strehl intensity in a star image formed by a diffraction-limited telescope from light waves scattered by a random phase screen was given previously by 10.3. From that equation, we can see that the following value of the $\sigma/\lambda$ ratio is required to pull down Strehl intensity to $\pi/4$:

$$\left(\frac{\sigma}{\lambda}\right)_{\frac{\pi}{4}} = \frac{\sqrt{\ln\left(\frac{\pi}{4}\right)}}{2 \cdot \pi} = 0.0782 \tag{14.30}$$

Figure 14.13 shows the average intensity envelopes for images of point-objects when the rms phase of the imaging wavefronts satisfies the above $\sigma/\lambda$ condition. The image envelopes are shown for the visible wavelength, $\lambda = 0.55\ \mu m$, for unobstructed, aberration-free, circular aperture telescopes of various diameters. For the visible wavelength considered here, we see that $\sigma$ must be set to $0.043\ \mu m$. The corresponding ideal Airy patterns are also shown (dashed lines) in the figure for reference. The angular size of an Airy pattern reduces inversely as the telescope diameter.

Consequently, the angular scales in Fig. 14.13 have been appropriately adjusted for each of the diameters considered in the figure. The angular size of the halos varies inversely as the average lateral size of the scintillation structures. The arbitrarily chosen size used for calculating the plots in this figure corresponds to that (~10 cm) shown in Fig. 14.11. Smaller scintillation structure sizes give rise to wider halos, and

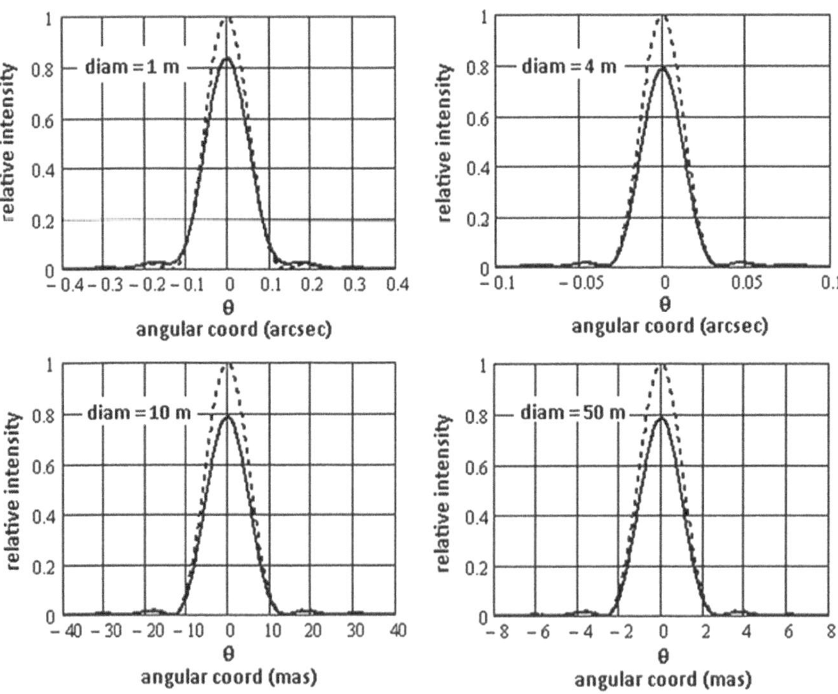


**Fig. 14.13** *Solid lines* The average image envelopes obtained at the visible wavelength, 0.55 μm, for AO-equipped telescopes of various sizes, where the AO system is assumed able to perfectly correct the phase but not the scintillation. *Dashed lines* ideal Airy patterns shown for reference. Note that the relative intensity for the 1-m telescope is marginally higher than for the larger instruments because the halo light fraction is less widely spread relative to the angular size of the core

vice versa. For the cases shown in Fig. 14.13, the 0.2146 halo light energy fractions are widely diffused, leaving the cores dominating the image centers.

## 14.7 Irradiance in Center of Star Images Formed by Large Telescopes

An expression for the intensity in the center of the image of an unresolved star formed by a large telescope with circular aperture, circular central obstruction, and Strehl intensity, SI(λ), was given previously by 13.91. By setting SI(λ) = 1 in that equation and by also replacing the quantity $\sigma$ by $\sigma_{AO}$ and the quantity $w_0$ by $w_{0AO}$, the following expression is obtained for the average intensity in the center of the image formed by an AO-equipped version of the telescope:

$$\langle I_{G(CH)}(0,\lambda)\rangle = \frac{1}{4\cdot\pi\cdot\sigma_{AO}^2\cdot\lambda^2} \times\left\{\lambda^2\cdot w_{0AO}^2\cdot\left\{1-\exp\left[-\left(\frac{2\cdot\pi\cdot\sigma_{AO}}{\lambda}\right)^2\right]\right\} +\pi^2\cdot(D^2-d^2)\cdot\sigma_{A0}^2\cdot\exp\left[-\left(\frac{2\cdot\pi\cdot\sigma_{AO}}{\lambda}\right)^2\right]\right\} \quad (14.31)$$

where the quantity, $\sigma_{AO}$, is assumed to include residual uncorrected telescope aberrations as well as uncorrected residuals of the dynamical OPD fluctuation introduced by the atmosphere. The quantity $w_{0AO}$ describes the 1/e half-width of the autocorrelation function of the uncorrected OPD fluctuation, where this function is assumed, as previously, to be a best-fit Gaussian function.

Figure 14.14 shows how the central intensity varies with wavelength for a 4-m diffraction-limited telescope (i.e., SI(λ) = 1) for several σ values. The plots in this figure were calculated using 13.91. (However, it is noted that by considering the σ and wo values indicated in the figure as simply the $\sigma_{AO}$ and $w_{0AO}$ for an AO-equipped telescope of the same size, an identical set of plots could have been generated from 14.31.)

### *14.7.1 Telescope Resolution and the Intensity in Center of a Star Image*

From energy conservation considerations, the angular width of a star image must be approximately related to the inverse square root of the central intensity. Thus, for large telescopes not equipped with AO, the angular width of the image may be given roughly by (cf., 13.91)

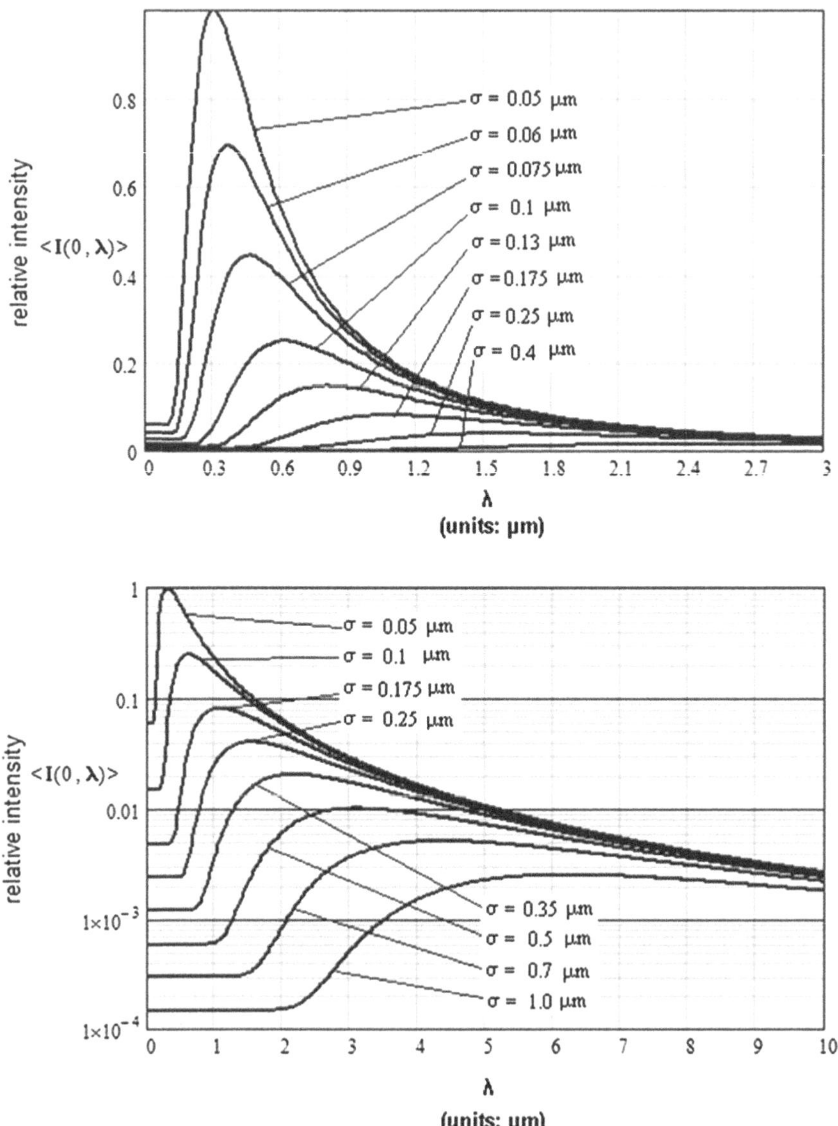


**Fig. 14.14** The average irradiance attained in the center of a star image formed by a 4-m diffraction-limited telescope plotted against wavelength for various amounts of uncorrected rms OPD fluctuation. Top Plots shown against a linear intensity scale. Bottom same plots for a logarithmic intensity scale. The same plots apply equally to 4-m telescopes equipped with AO by simply regarding the $\sigma$ and $w_0$ values indicated in the figure as the AO-corrected values, $\sigma_{AO}$ and $w_{0AO}$. For $w_0 = 0.25$ m, the value used to produce all of the plots, 1-arcsec (FWHM) visible seeing roughly corresponds to $\sigma = 0.4\,\mu$m; for other seeing conditions, $\sigma$ scales linearly with FWHM seeing (cf., 13.46)

$$AngularWidth \approx K \cdot \left\{ \frac{4 \cdot \pi \cdot \sigma^2 \cdot \lambda^2}{\lambda^2 \cdot w_0^2 \cdot \left[1 - \exp\left[-\left(\frac{2 \cdot \pi \cdot \sigma}{\lambda}\right)^2\right]\right] + \pi^2 \cdot (D^2 - d^2) \cdot SI(\lambda) \cdot \sigma^2 \cdot \exp\left[-\left(\frac{2 \cdot \pi \cdot \sigma}{\lambda}\right)^2\right]} \right\}^{\frac{1}{2}} \tag{14.32}$$

where K is a constant, a value for which will be established shortly. (Note that K as used here is unrelated to the telescope pupil function, K, which arose in Sect. 4.2.6.) Let us now consider a diffraction-limited telescope without central obstruction (i.e., where $SI(\lambda) = 1$ and $d = 0$). In particular, we consider the limiting case $\sigma/\lambda = 0$ where the image is simply an Airy pattern and for which the above equation reduces to

$$AngularWidth \approx K \cdot \frac{\lambda}{D} \cdot \sqrt{\frac{4}{\pi}} \tag{14.33}$$

If we now define Angular Width here as the radius of the first dark ring of the Airy pattern (i.e., $1.22 \cdot \lambda/D$) the constant of proportionality K must then be consistent with the relation,

$$K \cdot \frac{\lambda}{D} \cdot \sqrt{\frac{4}{\pi}} \approx 1.22 \cdot \frac{\lambda}{D} \tag{14.34}$$

from which it follows,

$$K \approx 1.08 \tag{14.35}$$

Using this value for K, the angular resolution according to the Rayleigh criterion for large telescopes not equipped with AO is then given approximately by

$$\Delta\vartheta_{Rayleigh} = 1.08 \times \left\{ \frac{4 \cdot \pi \cdot \sigma^2 \cdot \lambda^2}{\lambda^2 \cdot w_0^2 \cdot \left[1 - \exp\left[-\left(\frac{2 \cdot \pi \cdot \sigma}{\lambda}\right)^2\right]\right] + \pi^2 \cdot (D^2 - d^2) \cdot SI(\lambda) \cdot \sigma^2 \cdot \exp\left[-\left(\frac{2 \cdot \pi \cdot \sigma}{\lambda}\right)^2\right]} \right\}^{\frac{1}{2}} \tag{14.36}$$

For AO-equipped telescopes, a modified version of the above equation can be used which may be written the form,

$$\Delta\vartheta_{Rayleigh} = 1.08 \times \left\{ \frac{4 \cdot \pi \cdot \sigma_{AO}^2 \cdot \lambda^2}{\lambda^2 \cdot w_{0AO}^2 \cdot \left[1 - \exp\left[-\left(\frac{2 \cdot \pi \cdot \sigma_{AO}}{\lambda}\right)^2\right]\right] + \pi^2 \cdot (D^2 - d^2) \cdot \sigma_{AO}^2 \cdot \exp\left[-\left(\frac{2 \cdot \pi \cdot \sigma_{AO}}{\lambda}\right)^2\right]} \right\}^{\frac{1}{2}} \tag{14.37}$$

where the residual uncorrected fixed wavefront errors of the telescope are assumed to have been incorporated into $\sigma_{AO}$ which allows us to set $SI(\lambda)$ to unity.

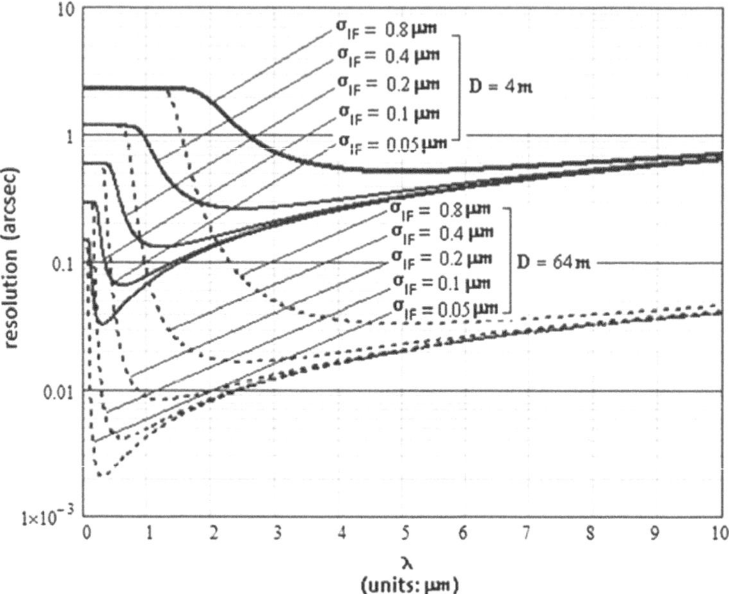


**Fig. 14.15** Approximate variation of telescope resolution plotted against wavelength for 4- and 64-m diameter telescopes where the effective width of a core and halo image is inferred (simplistically) from the irradiance level attained in the center of the image

Figure 14.15 shows how resolution (calculated from 14.36) varies with wavelength for 4- and 64-m telescopes. The plots shown apply equally to AO-equipped telescopes where the σ and $w_o$ values indicated are replaced by the appropriate $\sigma_{AO}$ and $w_{0AO}$ values. The choice of $w_0 = 0.25m$ used to create the plots in this figure was made simply because this value best explains the Mayall telescope images shown previously in Figs. 10.10 and 10.11.

Depending on its size, the parameter, $w_0$, greatly influences telescope resolution in the short-wavelength limit, $\lambda \rightarrow 0$. In this limit, star images are likely to be composed entirely of halos, the angular sizes of which are inversely related to the size of $w_o$. At longer wavelengths, in the range set by $\lambda \geq \lambda_{opt} \approx 2 \cdot \pi \cdot \sigma$ and where the images are dominated by cores, the $w_0$ value has less influence; for diffraction-limited telescopes, the angular size of the core scales inversely as the telescope diameter.

For AO-equipped telescopes, the $w_{oAO}$ values that arise depend on the spatial frequency response of the AO systems. If the response is uniform over all spatial frequencies represented in the image-forming waves, we would expect to find $w_{oAO} = w_o$. If the response is non-uniform, we would expect to find $w_{oAO} \neq w_o$.

In the two limiting core-only and halo-only cases, 14.36 and 14.37 accurately quantify resolution. However, in the general case where the image consists of both features, these equations merely provide weighted average resolution which cannot generally be considered trustworthy. The problem here is caused by our over-simplified approach; for the general case, it is simply not possible to precisely determine resolution in terms of a single parameter—in this case the intensity in the center of the image.[13] If a truly reliable assessment of resolution is needed for core and halo images, it is safer to avoid using 14.36 and 14.37 and instead use the more precise formulations of Sect. 14.5.

### *14.7.2 Diffraction-Limited Imaging and Imaging at the Optimum Wavelength*

In this section, we examine the properties of images formed by very large diffraction-limited telescopes (where D $\geq$ 4 m) at two uniquely interesting wave-lengths: (1) the optimum wavelength, $\lambda_{opt}$, at which maximum irradiance occurs in the center of a star image and (2) the wavelength $\lambda_{dl}$ at which the Strehl intensity in the image attains the value 0.8—i.e., the threshold of substantially diffraction-limited image quality. We initially consider telescopes without AO capability. However, the results obtained apply equally to telescopes with AO capability.

In the regime where Strehl intensity takes relatively high values, as it does for the two wavelengths considered, $\lambda_{opt}$ and $\lambda_{dl}$, Strehl intensity may be given approximately in terms of $\sigma$ and $\lambda$ by

$$SI(\lambda) = \exp\left[-\left(\frac{2 \cdot \pi \cdot \sigma}{\lambda}\right)^2\right] \tag{14.38}$$

where the contribution to the central intensity made by the halo (which is usually a broader feature than the core) has been assumed negligible. We saw previously (Sect. 10.4) that $\lambda_{opt} = 2 \cdot \pi \cdot \sigma$; we also saw that the Strehl intensity at the optimum wavelength is given by $SI\left(\lambda_{opt}\right) = 1/e = 0.368$.

It may readily be seen by evaluations of 14.38 that the wavelength at which Strehl intensity just attains the threshold value 0.8 is given by $\lambda_{dl} = 13.3 \cdot \sigma$. Thus, we can write,

$$\lambda_{dl} = \frac{13.3}{2 \cdot \pi} \cdot \lambda_{opt} \approx 2.12 \cdot \lambda_{opt} \tag{14.39}$$

The ratio of the average intensities attained in the center of the images at the two wavelengths, $\lambda_{opt}$ and $\lambda_{dl}$ is readily seen to be given by (cf., 10.14)

[13] The limitations are akin to those that apply to the Fried parameter, $r_0$, discussed in Appendix I. A single parameter like this cannot adequately describe the most general type of core and halo image.

**Table 14.2** $\lambda_{opt}$ and $\lambda_{dl}$ for some typical values of $\sigma$ or $\sigma_{AO}$

| $\sigma$ or $\sigma_{AO}$ (μm) | $\lambda_{opt}$ (μm) | $\lambda_{dl}$ (μm) |
|---|---|---|
| 0.05 | 0.315 | 0.665 |
| 0.1 | 0.63 | 1.33 |
| 0.2 | 1.26 | 2.66 |
| 0.4 | 2.51 | 5.32 |

$$\frac{\langle I(0,\lambda_{opt})\rangle}{\langle I(0,\lambda_{dl})\rangle}=\left(\frac{\lambda_{dl}}{\lambda_{opt}}\right)^2\cdot\frac{\exp\left[-\left(\frac{2\cdot\pi\cdot\sigma}{\lambda_{opt}}\right)^2\right]}{\exp\left[-\left(\frac{2\cdot\pi\cdot\sigma}{\lambda_{dl}}\right)^2\right]} \tag{14.40}$$

Inserting the wavelength values, $\lambda_{opt} = 2\cdot\pi\cdot\sigma$ and $\lambda_{dl} = 13.3\cdot\pi\cdot\sigma$, into this equation produces the result,

$$\frac{\langle I(0,\lambda_{opt})\rangle}{\langle I(0,\lambda_{dl})\rangle}\approx 2.06 \tag{14.41}$$

This result indicates that by choosing to image at the optimum wavelength, rather than the (2.12 times) longer wavelength at which diffraction-limited image quality is just achieved, the average central intensity in the resulting image increases by the factor, 2.06. It should also be clear that the same result also applies to large telescopes equipped with AO; the only difference is that the two wave-lengths, $\lambda_{opt}$ and $\lambda_{dl}$, would now be referred to the quantity $\sigma_{AO}$ rather than $\sigma$. Because $\sigma_{AO}$ is generally less than $\sigma$, both wavelengths in this case would be correspondingly shorter.

Whereas the wavelength choice, $\lambda_{dl}$, may or may not be better than $\lambda_{opt}$ for imaging applications (depending on the specifics of these applications), clearly the wavelength choice, $\lambda_{opt}$, would be the better choice for high energy laser (HEL) weapon system applications, as discussed further in Chap. 16.

Table 14.2 shows tabulated values of $\lambda_{opt}$ and $\lambda_{dl}$ or a number of $\sigma$ values. Figure 14.16 shows intensity envelopes for 10- and 50-m diameter diffraction-limited telescopes for the wavelengths, $\lambda_{opt}$ and $\lambda_{dl}$, for a number of different $\sigma$ values. As anticipated, the central intensities delivered at wavelength $\lambda_{opt}$ can be seen in the figure to be about 2 times larger than those delivered at $\lambda_{dl}$.

## 14.8 Optical Tolerances for Large Ground-Based Telescopes

From about 1965 to about 1990, the prevailing logic was that, since the atmosphere is going to severely degrade image quality anyway, there is no point in building ultra-precise ground-based astronomical telescopes. During that period, the resolution anticipated from large ground-based telescopes in average seeing conditions was

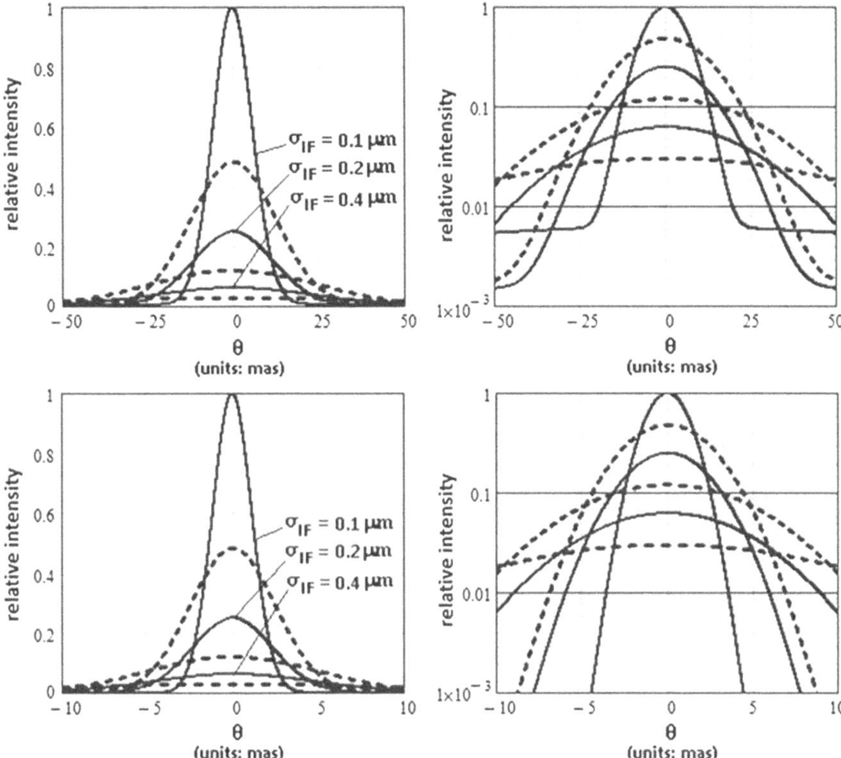


**Fig. 14.16** Average intensity envelopes in star images obtained by (Top) 10-m and (Bottom) 50-m diffraction-limited telescopes for several different $\sigma$ values. The plots in solid line correspond to the optimum wavelength $\lambda_{opt}$; the plots in dashed line correspond to wavelength $\lambda_{dl}$. The plots apply equally to AO-equipped telescopes where the $\sigma$ values indicated are simply considered as $\sigma_{AO}$ values. Left Linear vertical scale. Right Logarithmic vertical scale

about 1 arcsec, with 0.5 arcsec perhaps possible in more favorable seeing conditions. In accordance with this thinking, large astronomical telescopes constructed in the 1970s and 1980s were built to disappointingly coarse optical specifications.

The specification for the UKIRT 3.8-m telescope (1979 first light)—referred to at the time as a 'light-bucket' instrument—called for 80 % of the light to be concentrated into a 1-arcsec image patch diameter. The specification for the USAF 3.5-m telescope at the Starfire Optical Range in New Mexico (1994 first light) called for 80 % of the light to be contained in a 0.3-arcsec patch.[14] Energy fraction specifications of this type

[14] This method of specifying telescope optical performance has now fallen into disuse. In effect, it places an angular limit on the gradients associated with the wavefront errors of the telescope. However, because this specification method takes no account of the lateral distances over which the wavefront gradients are subtended, it provides no information about either the P-V or rms wavefront errors introduced by the telescope; such information is of course essential if diffraction-limited instruments are to be constructed.

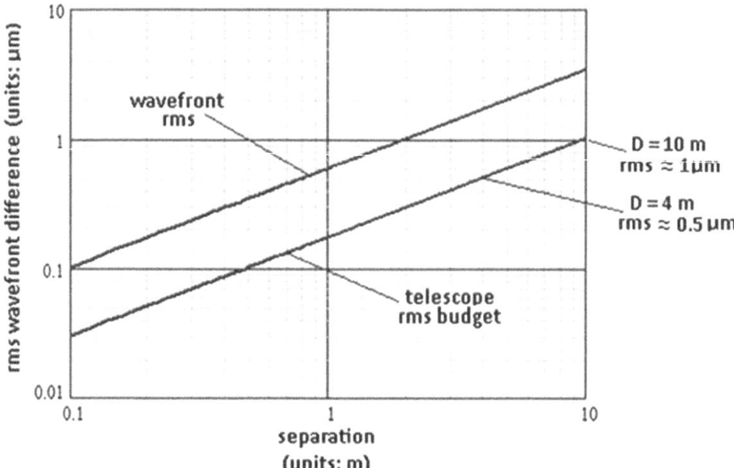


**Fig. 14.17** Typical pre-1990 optical tolerance specification for large ground-based astronomical telescopes. Beginning around 1990, tolerance specifications were tightened by about an order of magnitude

are essentially geometrical-optics-based specifications, justified at the time by the Kolmogorov-based notion that P-V wavefront disruptions caused by the atmosphere over large telescope apertures are significantly larger than the imaging wavelengths. With such large wavefront disruptions anticipated, star images were expected to appear as halo-only seeing disks.

Martin et al. (1991) gave wavefront error specifications based on the Kolmogorov understanding for large ground-based telescopes up to 10 m diameter. Ironically, their specifications (reproduced here in Fig. 14.17) coincided with a time when astronomers were becoming aware that much higher resolution was in fact attainable from ground-based telescopes. On its very first night of use in 1989, the NTT telescope (with its newly developed active optics primary mirror) delivered spectacularly good images (Wilson, 2003). At about the same time, IR imaging arrays were being installed on several large astronomical telescopes.

At the invitation of the US Air Force Research Laboratory (AFRL, Albuquerque, New Mexico), Steward Observatory astronomers, D.W. McCarthy Jr., and K. Hege, attended a Seminar[15] given by the Author (McKechnie, 1989) on the subject of obtaining diffraction-limited image cores at near-IR wavelengths from large ground-based astronomical telescopes. Some months earlier, the material presented at the seminar had been submitted for publication (McKechnie, 1991). Only weeks after the seminar, with the six primary mirrors of the MMT instrument on Mt. Hopkins accurately co-phased, McCarthy and Hege were able to deliver strong cores when imaging at 2.2 and 3.4 μm. These cores provided 0.1- and 0.15-arcsec resolution,

[15] Other attendees at the September 26, 1989, Seminar: G. Loos and B. Venet (AFRL), and B. Haddock (Lentec Corp.).

respectively, even though visible seeing at the telescope site at the time the images were recorded was no better than 1 arcsec (McCarthy Jr et al., 1990; McKechnie, 1990). Apart from the need to accurately co-phase the primary mirror elements, the only other requirement for obtaining these highly resolved image cores was the elimination of image wander— something that could readily be accomplished by simply capturing short-exposure images, effectively freezing image motion. (In long-exposure images, image wander, whatever the origin, smears the image cores causing them to lose their identity as image cores.

Images obtained shortly afterward by J. Christou using the 3.8-m Mayall telescope (shown previously in Figs. 10.10 and 10.11) also revealed prominent cores at 2.2 and 3.4 $\mu$m, even if they were significantly degraded by the instrument's aberrations, which at the time amounted to about 2-waves (HeNe) P-V (cf., Sect. 10.3.4). Calculations show that aberrations of this amount all but eliminate the possibility of obtaining useful cores at the H and J wavelength bands (1.65 and 1.25 $\mu$m). Had the optics of the Mayall instrument been constructed to diffraction-limited standards, the angular resolution delivered by cores at these two shorter wavelength bands would have been about 0.11 and 0.08 arcsec, respectively.

The optical performance of the Mayall instrument (circa 1990) can now be seen as typical of large ground-based instruments built in the 1970s and 1980s. Nowadays, most large telescopes built in that era have undergone optics upgrades, though they likely still fall short of diffraction-limited performance at visible wavelengths. An optics upgrade program was undertaken (circa 2010–2014) on yet another venerable instrument built in the same era, the 3-m NASA IRTF telescope on Mauna Kea (first light, 1979). The goal of the upgrade was to enable the instrument to deliver diffraction-limited images at both near-IR and visible wavelengths (Tollestrup and McKechnie, 2010; Tokunaga, 2010). The total P-V wavefront error budget allocated to the primary/secondary mirror combination had been set at $\lambda/4$ (HeNe). Had the upgrade been completed, a number of factors pointed to the NASA instrument delivering superbly resolved images. Average seeing on Mauna Kea is a superb 0.7 arcsec, about 40% better than at typical US mainland sites. The NASA instrument's thick monolithic primary mirror would have been more resistant to wind-induced flexing and oscillation to an extent not possible with the much thinner, truss-supported segmented mirrors used by more recent large telescopes, such as Keck I and II (Nelson et al., 1985). The final performance of the IRTF instrument had the potential to definitively established the true seeing limits imposed by the atmosphere above one of the world's most pristine high-altitude observing sites.

However, aware of the 1991–1998 performance upgrades made to the nearby UKIRT 3.8-m instrument, in 2014 NASA abandoned the 3-m IRTF upgrade program and began funding UKIRT under a scientific cooperation between Lockheed Martin, the University of Hawaii, and the U. S. Naval Observatory. The instrument is to be decommissioned after completion of the Thirty Meter Telescope.

### 14.8.1 *Optical Tolerances for Resolving Image Cores*

Ideally, to properly resolve image cores at a given wavelength, λ, the Strehl intensity of the telescope at that wavelength must be at least 0.8. Thus, a suitable Strehl intensity tolerance for the telescope may be expressed by

$$SI(\lambda) \geq 0.8 \tag{14.42}$$

If we assume that telescope wavefront error is approximately Gaussian distributed, the Strehl intensity may be expressed in terms of the telescope's rms wavefront error, $\sigma_T$, by

$$SI(\lambda) = \exp\left[-\left(\frac{2 \cdot \pi \cdot \sigma_T}{\lambda}\right)^2\right] \tag{14.43}$$

It may readily be shown from this relation that if the Strehl intensity is to remain at, or above, the diffraction-limited target value set by 14.42, the rms wavefront error introduced by the telescope optics, $\sigma_T$, must obey the condition

$$\sigma_T \leq \frac{\lambda}{13.3} \tag{14.44}$$

For common types of (low order) aberration, such as spherical aberration, coma, astigmatism, and defocus, Strehl intensity can be maintained at or above 0.8 just as long as the total P-V wavefront error of the telescope optics, $(\text{P-V})_\text{T}$, does not exceed λ/4 (Born & Wolf, 2003). Thus, an alternative and approximately equivalent way of specifying the tolerance indicated by 14.44 is given by

$$(P - V)_T \leq \frac{\lambda}{4} \tag{14.45}$$

**Budgeting the Wavefront Error**

The tolerance specifications just indicated must be budgeted over the various optical components and surfaces that comprise the entire optical train of the telescope. Though the primary mirror is only one of several components, its large size usually entitles it to a proportionately large share of the error budget. Typically, the primary mirror might be awarded half of the total wavefront error budget. For the total P–V allowance indicted by 14.45, the primary mirror allowance $(P\text{-}V)_{PM}$, might therefore be set as follows:

$$(P-V)_{PM} \leq \frac{\lambda}{8} \tag{14.46}$$

Because the rms wavefront error contributions from the various elements in the optical train tend to add in quadrature,[16] the primary mirror rms wavefront error allowance, $\sigma_{PM}$, consistent with 14.44 might be set at

$$\sigma_{PM} \leq \frac{\lambda}{13.3 \cdot \sqrt{2}} = \frac{\lambda}{18.8} \tag{14.47}$$

#### 14.8.1.1 Optical Tolerances for 4-m Class Telescopes

A suitable optical tolerance specification for any given telescope must take account of the shortest wavelength at which the instrument is required to deliver images. The seeing conditions at the site must also be taken into account as well as the instrument's diameter.

In average Kitt Peak visible seeing conditions (of just over 1 arcsec), a 3.8-m telescope, such as the Mayall instrument, should be able to routinely deliver star image cores at wavelengths down to about 1.25 μm. Inserting this wavelength into 14.44 and 14.45, sets the optical tolerance specifications for the Mayall instrument at $\sigma_T \leq 0.094\,\mu\text{m}$, and $(P-V)_T \leq 0.312\,\mu\text{m}$. The latter P–V specification is seen to be more than 4 times more stringent than the 2-waves HeNe (i.e., 1.25 μm) P–V wavefront error measured for the instrument around 1990 by Forbes (1991), National Optical Astronomy Observatory (NOAO).

At the more pristine seeing sites found in Hawaii and Chile, where average visible seeing is about 0.7 arcsec, a large diffraction-limited telescope in the 4-m size class might be expected to routinely deliver image cores at visible wavelengths (~0.55 μm). According to 14.44 and 14.45, to properly resolve these cores, tighter optical tolerance specifications are required, corresponding to $\sigma_T \leq 0.094\,\mu\text{m}$ and $(P-V)_T \leq 0.312\,\mu\text{m}$.

The rms wavefront error specifications required to adequately resolve image cores at the wavelengths, 1.25 and 0.55 μm, are shown in Fig. 14.18. Also shown in this figure is a typical rms wavefront error allowance for telescopes constructed prior to about 1990. For 4-m class telescopes, this figure indicates that to fully resolve image cores at 1.25 μm and 0.55 requires $\sigma_T$ values about 5× and 12× smaller, respectively, than were considered acceptable prior to 1990.

[16] To justify addition in quadrature in this kind of application merely requires that the wavefront error contributions made by the various optical components are either uncorrelated or statistically independent.

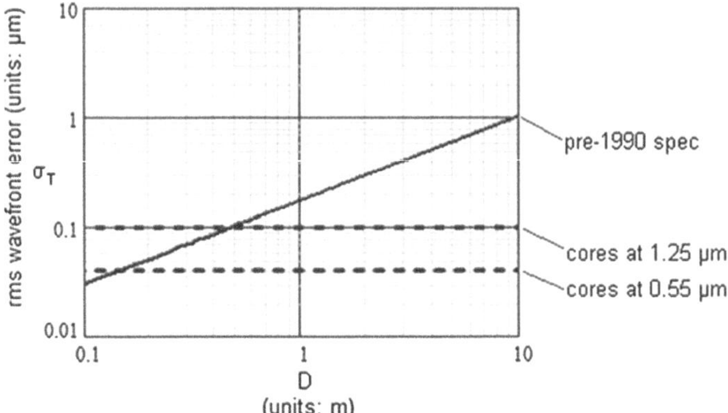


**Fig. 14.18** Optical specifications for large ground-based telescopes consistent with delivering fully resolved image cores at 0.55 and 1.25 μm. Also shown for comparison is the typical pre-1990 tolerance specification (Martin et al., 1991) indicated previously in Fig. 14.17

#### 14.8.1.2 Optical Tolerances for Next-Generation ELT Instruments

Thus far, the discussion has been about telescopes such as the 3.8-m Mayall telescope which use monolithic primary mirrors. However, with the successful implementation of segmented primary mirrors in the twin Keck 10-m instruments (Nelson et al., 1985),[17] the same fabrication principal has now been adopted for the primary mirrors of next-generation extremely large telescopes (ELTs) such as the 39-m E-ELT instrument. We now establish appropriate optical specifications for ELT telescopes constructed according to this segmented mirror principle.

We first assume that the ultimate objective with these new instruments is to achieve near-diffraction-limited image cores at visible wavelengths (from which it naturally follows that diffraction-limited performance will also be delivered at all longer wavelengths). To achieve such imaging performance, an appropriate tolerance for the rms OPD fluctuation of the final image-forming wavefronts produced by the telescope with the AO system operating, $\sigma_{AO}$, is given (cf., 14.44) by

$$\sigma_{AO} \leq 0.04\,\mu\text{m} \tag{14.48}$$

For telescopes only used at IR wavelengths, such as 1–5 μm, the above tolerance requirement might be relaxed by a factor of about 2.

The flow down requirements of the overall error budget indicated by 14.48 to the individual telescope components (such as individual primary mirror segments and the secondary mirror) naturally depend on the performance of the AO system; it may be assumed that a reasonable fraction of the wavefront error introduced by the various individual optical components can be compensated by the AO system.

[17] The primary mirrors of the two 10-m Keck telescopes each consist of eighteen hexagonal-shaped mirror segments, each measuring about 1.8 m across.

However, placing unnecessary burdens like this on AO systems might possibly result in lower overall AO performance. Thus, it might make sense to place relatively tight figure tolerances on the various individual optical components. Denoting the rms wavefront error allocation to individual components by $\sigma_{IC}$, an appropriate tolerance limit might be set by

$$\sigma_{IC} \leq 0.02\,\mu\text{m} \tag{14.49}$$

Assuming that the wavefront errors introduced by the various optical components are all uncorrelated, an estimate of the rms OPD fluctuation in the final image-forming wavefronts can be obtained by summing, in quadrature, the various individual contributions. Thus, it would require contributions of the magnitude indicated by 14.49 from four serial components in the optical train (summed together in quadrature) for the overall rms error allowance to add up to the rms budget limit set by 14.48.

**Phasing of Individual Mirror Segments**

The tolerance specification for co-phasing the individual primary mirror segments might also be set to the amount indicated by 14.49. But again, since the AO system can, at least in principle, compensate some of the residual phasing errors, a more complete examination of this matter would have to take account of the specifics of the AO system.

### *14.8.2 Stability of Multiple-Segment Primary Mirrors*

Large lightweight segmented primary mirrors supported by lightweight trusses may be susceptible to vibration, the primary excitation mechanisms being wind, air currents (Ulich, 1982, 1988), and background seismic activity (Ninneman et al. 1987).[18] As individual mirror segments deform and move relative to one another, they cause corresponding distortions and even discontinuities in the imaging wavefronts. Higher order vibration modes are particularly challenging; the fundamental mode can be readily compensated by the use of tip-tilt correction. If, after the AO correction has been made, the residual uncorrected rms wavefront error exceeds the amount indicated by 14.48, image quality delivered by the telescope will fall below the desired diffraction-limited level.

Although large lightweight mirror support trusses are no doubt exceptionally stiff in relation to their weight, in an absolute sense they may not be particularly stiff at all. The oscillations caused by (chaotic) wind currents pressing against huge mirror areas (areas extending over hundreds of square meters) may present the most critical problem facing future ELT instruments (Cho et al., 2001). The problem could be

[18] On Hawaii, seismic vibrations are caused by local volcanic activity. In California, at the 100-in. telescope on Mt. Wilson, seismic vibrations can be detected from waves breaking on the Pacific Ocean shore some 30 miles to the southwest.

eliminated by enclosing these instruments behind protective windows. The windows (possibly segmented) might enclose either just the telescope tube itself or, possibly, the entire telescope itself. In the latter case, to avoid unnecessary disruption of the air flow at the site, the plane of the windows might follow the local ground contour, which would effectively mean burying the instrument itself.

But even this relatively ambitious approach does not solve the entire problem. Wavefront errors caused by temperature differences between the window and the air on either side would have to be controlled. Otherwise, it might only be possible to produce high quality images during periods of temperature equilibrium. A helium-filled space behind the window would solve only half of the problem; the space in front of the window could not be similarly filled. Although the costs of such strategies might seem frighteningly high, they are likely dwarfed by the cost of space-borne telescopes of comparable size.

Enclosure windows would naturally have to be transmissive at the imaging wavelengths. For wavelengths up to about 3 μm, fused silica, with its extremely low coefficient of thermal expansion (CTE), might be an ideal substrate material. At longer wavelengths—where the AO system can more easily achieve diffraction-limited image quality—the windows might simply be rolled back. Wavefront error introduced by individual window segments would still have to conform to the high optical standards set by 14.49, though some relaxation might be possible if the AO system was capable of correcting excess amounts.

## 14.9 Defocus Tolerances for Large Telescopes

### *14.9.1 Allowance for Delivering Diffraction-Limited Image Cores*

If a perfectly diffraction-limited telescope is to properly resolve an image core at wavelength, $\lambda$, the maximum allowable wavefront error budget for defocus is about $\lambda/4$ (cf., 14.45). The actual defocus distance, or depth of focus, $\Delta f_{Core}$, corresponding to this amount of defocus can readily be shown to be (Born & Wolf, 2003)

$$\Delta f_{Core} = \pm 2 \cdot \lambda \cdot F^2 \qquad (14.50)$$

where the telescope F/number is defined in the usual way, F = f/D. If defocus exceeds the limit set by $2 \cdot \lambda \cdot F^2$, the Strehl intensity realized by the image core would rapidly fall below the target value for diffraction-limited imaging, 0.8.

For mildly aberrated telescopes, where the aberrations remain consistent with delivering image cores with Strehl intensities >0.8, the maximum allowable amount of defocus would naturally have to be less than that indicated by 14.50. However, calculation of the precise amount allowed would have to form part of a wider wavefront error budget analysis.

### *14.9.2 Allowance for Delivering Substantially Ideal Halo-Only Images*

To faithfully produce a seeing disk (i.e., halo-only) star image, a more generous defocus tolerance is allowed; the angular sizes of halos formed by large telescopes are usually many times greater than those of cores. A suitable defocus allowance may be obtained in this case by using a geometrical-optics approach.

In this approximate approach, the defocused image of a star can be approximately represented by a top-hat function whose angular width increases in proportion to the amount of defocus, $\Delta f_{Halo}$, according to $\Delta f_{Halo}/(F \cdot f)$, where F and f are, respectively, the F/number and focal length of the telescope. Denoting the FWHM angular width of the halo in a perfectly focused image by $FWHM_{Halo}$, the FWHM angular width of the same halo when degraded by defocus, $FWHM_{HaloDef}$, can be approximately obtained by convolution of the (top-hat) blur patch due to the defocus with the intensity distribution in the perfectly focused halo image. The FWHM width of the resulting defocused image is then given by the following summation in quadrature:

$$FWHM_{HaloDef} = \left[ FWHM_{Halo}^2 + \left( \frac{\Delta f_{Halo}}{F \cdot f} \right)^2 \right]^{\frac{1}{2}} \tag{14.51}$$

From energy conservation considerations, the ratio of the Strehl intensities achieved in the two halo envelopes, with and without defocus, is roughly proportional to the quantity,

$$\left[ \frac{FWHM_{HaloDef}}{FWHM_{Halo}} \right]^2$$

We now adopt (somewhat arbitrarily) the convention that a just-acceptable amount of defocus for a halo-only image is one that causes the Strehl intensity to reduce by the factor 0.8.[19] Thus, the just-acceptable Strehl intensity pull-down for a halo-only image can be expressed by

[19] The 0.8 factor used here should not be confused with the absolute Strehl intensity which, for substantially diffraction-limited images, is indeed 0.8. In the present context, the factor 0.8 represents an additional Strehl intensity pull-down factor. For example, for a perfectly focused halo image where the absolute Strehl intensity might typically be only about 0.01, after the 0.8 pull-down caused by the just-allowable amount of defocus, Strehl intensity would decrease to the absolute value 0.008.

$$\frac{FWHM_{Halo\,Def\,SI\,Limit}}{FWHM_{Halo}} = \frac{1}{\sqrt{0.8}} \tag{14.52}$$

Combining 14.51 and 14.52 indicates that the just-allowable amount of defocus $\Delta f_{Halo}$ for a halo-only image must satisfy the equation,

$$\frac{1}{\sqrt{0.8}} = \left[1 + \left(\frac{\Delta f_{Halo}}{F \cdot f \cdot FWHM_{Halo}}\right)^2\right]^{\frac{1}{2}} \tag{14.53}$$

Solving the above equation for $\Delta f_{Halo}$ gives

$$\Delta f_{Halo} = \pm 0.5 \cdot F \cdot f \cdot FWHM_{Halo} \tag{14.54}$$

By replacing f by F · D in the above equation, the defocus tolerance allowance can be expressed in the alternative form:

$$\Delta f_{Halo} = \pm 0.5 \cdot F^2 \cdot D \cdot FWHM_{Halo} \tag{14.55}$$

By comparing the depth of focus allowance given by the above equation with the corresponding depth of focus allowance for image cores (14.50), we see that both expressions have an $F^2$ dependence. However, while one shows a linear dependence on wavelength, λ, the other shows no wavelength dependence at all. Instead, it shows a linear dependence on both the telescope diameter D and the seeing quality as described by $FWHM_{Halo}$. These differences arise because of the fundamental difference between the core and halo image features.

### *14.9.3 Ratio of Depth of Focus Allowances for Resolving Cores and Halos*

The ratio of the defocus tolerance allowances for cores and halos may be obtained from 14.50 and 14.55 in the form:

$$\frac{\Delta f_{Halo}}{\Delta f_{Core}} = \frac{D \cdot FWHM_{Halo}}{4 \cdot \lambda} \tag{14.56}$$

This ratio does not depend on the telescope F/number, but it does depend on the seeing quality as described by $FWHM_{Halo}$, the telescope diameter D, and the imaging wavelength λ.

Table 14.3 shows evaluations of the ratio, $\Delta f_{Halo}/\Delta f_{Core}$, obtained from 14.56 for 2- to 50-m telescopes. It is assumed in these evaluations that the visible and near-IR images are all obtained in 1-arcsec visible seeing conditions (i.e., $FWHM_{Halo} \approx$ 1arcsec). In all cases shown in the table, it is observed that the ratio always takes

**Table 14.3** Ratio of the defocus tolerance allowances for halo-only images to that of core-only images for typical (FWHM = 1 arcsec) seeing at the visible wavelength 0.55 μm. All values shown are greater than unity, indicating that larger defocus tolerances are generally allowed for halo-only images

| λ (μm) | $D$ (m) | $\frac{Df_{Halo}}{Df_{Core}}$ |
|---|---|---|
| 0.55 | 2 | 4.4 |
| 0.55 | 4 | 8.8 |
| 0.55 | 10 | 22.0 |
| 0.55 | 25 | 55.0 |
| 0.55 | 50 | 110.0 |
| 1.1 | 2 | 2.2 |
| 1.1 | 4 | 4.4 |
| 1.1 | 10 | 11.0 |
| 1.1 | 25 | 27.5 |
| 1.1 | 50 | 55.0 |
| 2.2 | 2 | 1.1 |
| 2.2 | 4 | 2.2 |
| 2.2 | 10 | 5.5 |
| 2.2 | 25 | 13.8 |
| 2.2 | 50 | 27.5 |

values greater than unity, indicating as we might expect that halo-only images are less sensitive to defocus than images containing cores.

For the case of the 2-m telescope at the near-IR wavelength, 2.2 μm, the ratio is seen to approach unity. This behavior is somewhat misleading. At 2.2 μm in the 1-arcsec visible seeing conditions considered, a 2-m telescope would generally deliver strong image cores. Thus, it would no longer be appropriate to consider halo-only images. In this regime, the geometrical-optics approach used to treat halos is simply no longer valid.

## 14.10 Resolution Obtained by the Keck 10-m Telescopes

With the help of AO, the Keck instruments are able to produce diffraction-limited image quality (Strehl intensity >0.8) for wavelengths of 2.2 μm and above. At shorter wavelengths, the residual uncorrected rms OPD fluctuation produces significantly lower Strehl intensity values.

Figure 14.19 shows core and halo images of the 0.11-arcsec binary brown dwarf system CFBDSIR 1458 + 10 obtained by the Keck II telescope. The image on the left was obtained in the H band (1.5–1.8 μm). This image shows bright central cores, with speckle halos also evident containing a significant fraction of the total light energy. (The halos actually extend beyond the 0.5-arcsec field of view captured in the figure.) The right image in the same figure shows a false color image created by summing

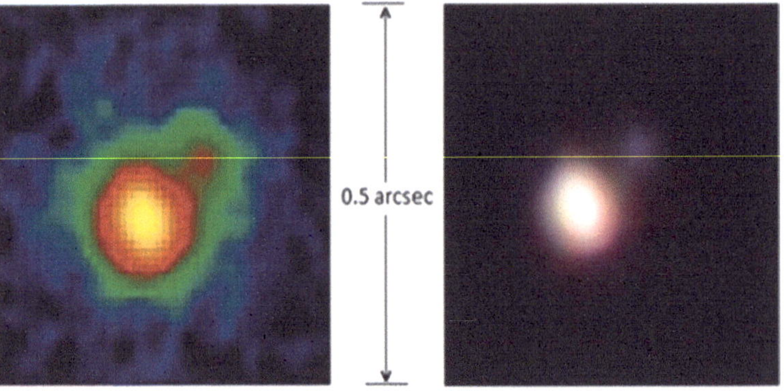


**Fig. 14.19** Core and halo images of the 0.11-arcsec binary brown dwarf CFBDSIR 1458 + 10 obtained by the Keck II telescope. Left Image obtained in the H-band (1.6 μm) using the instrument's laser guide star as the AO reference object. Right False color image created by combining images at four different IR wavelengths. The two components of the binary object differ in magnitude by 1.9 (i.e., BR ≈ 0.174) Images by Liu (2011), Institute for Astronomy, University of Hawaii, by courtesy of W. M. Keck Observatory

four separate images obtained at the IR bands, J (1.1–1.4 μm), H (1.5–1.8 μm), K (2.0–2.4 μm), and L (3.0–4.0 μm).

The average intensity envelope for a binary star image of the type shown in Fig. 14.19 can be modeled according to 14.10 by the general form:

$$\begin{aligned}\langle I(\alpha,\beta,\Delta,\chi)\rangle \propto\ & A_{CH}\exp\left[-\left(\frac{\left(\alpha+\frac{\Delta\cdot\cos(\chi)}{2}\right)^2+\left(\beta+\frac{\Delta\cdot\sin(\chi)}{2}\right)^2}{B_C^2}\right)\right] \\ & +\exp\left[-\left(\frac{\left(\alpha+\frac{\Delta\cdot\cos(\chi)}{2}\right)^2+\left(\beta+\frac{\Delta\cdot\sin(\chi)}{2}\right)^2}{B_H^2}\right)\right] \\ & +BR\cdot A_{CH}\cdot\exp\left[-\left(\frac{\left(\alpha-\frac{\Delta\cdot\cos(\chi)}{2}\right)^2+\left(\beta-\frac{\Delta\cdot\sin(\chi)}{2}\right)^2}{B_C^2}\right)\right] \\ & +BR\cdot\exp\left[-\left(\frac{\left(\alpha-\frac{\Delta\cdot\cos(\chi)}{2}\right)^2+\left(\beta-\frac{\Delta\cdot\sin(\chi)}{2}\right)^2}{B_H^2}\right)\right]\end{aligned} \tag{14.57}$$

where Δ is the angular separation of the binary star, χ is the inclination angle of the binary star axis (with respect to the α-axis), and BR is the brightness ratio of the two components.

Computer-generated images calculated from the above equation for the binary brown dwarf CFBDSIR 1458 + 10 are shown in Fig. 14.20 for a hypothetical Keck 10-m telescope for various visible and IR wavelengths. The images in this figure are all average short-exposure images. The image at the visible wavelength, 0.55 μm, indicates significantly degraded resolution; at this wavelength, the instrument delivers halo-only images.[20] The image at the IR wavelength, 4.8 μm, shows similarly low resolution, the cause in this case being increased diffraction spread angle at this longer wavelength. At wavelengths close to the optimum wavelength (~1.65 μm), much higher levels of resolution are obtained due to the presence of conspicuous and highly-resolved image cores. The image at 1.65 μm is reasonably consistent with the actual image obtained at 1.65-μm by the Keck II instrument shown on the left of Fig. 14.19.

The most striking feature of the images in Fig. 14.20 is the extremely high resolution levels achieved in what is nowadays called the sweet spot wavelength range—a range straddling the optimum wavelength. In this wavelength range, the 0.11-arcsec binary is comfortably resolved. Outside this range, at both shorter and longer wavelengths, resolution falls off dramatically, resulting in the binary object no longer being resolved.

The following parameter values were used to calculate the Fig. 14.20 images:

Telescope diam D = 10 m.

Central obstruction diam d = 1.8 m.

Binary brightness ratio BR = 0.174 (corresponding to a 1.9 magnitude difference)

Binary separation $\Delta$ = 0.11 arcsec (2.6 AU separation at distance 23 parsec), Binary axis tilt angle $\chi = 45°$

Uncorrected rms OPD fluctuation $\sigma_{AO} = 0.25$ μm (see Note 1 below) Telescope Strehl intensity $SI(\lambda) = 1$(residual uncorrected fixed telescope errors are assumed folded into the residual uncorrected $\sigma_{AO}$ value).

Average wavefront structure size $w_{0AO} = 1m$ (1/e half-width of the uncorrected wavefront structure. See Note 2 below).

Notes:

1. Strehl intensity at wavelength 2.2 μm is approximately given by $exp\left[-(2 \cdot \pi \cdot 0.25\,\mu\text{m}/2.2\,\mu\text{m})^2\right] \approx 0.8$, which indicates substantially diffraction-limited imaging at this wavelength. At wavelengths shorter than 2.2 μm, Strehl intensity falls below the value, 0.8. At these wavelengths, diffraction-limited imaging is not therefore achieved in the strict sense, even if the presence of image cores at these wavelengths still allows the binary pair to be comfortably resolved.
2. The structure function of the residual uncorrected wavefront errors (after the AO system has corrected wavefront phase to the best of its ability) depends on the response characteristics of the AO system as a function of the spatial frequency composition of the OPD fluctuation associated with the image-forming light

[20] Whereas the Keck instruments do not currently deliver diffraction-limited images at visible wavelengths, there is no reason why such images cannot be delivered at some time in the future, given improvements in AO technology.

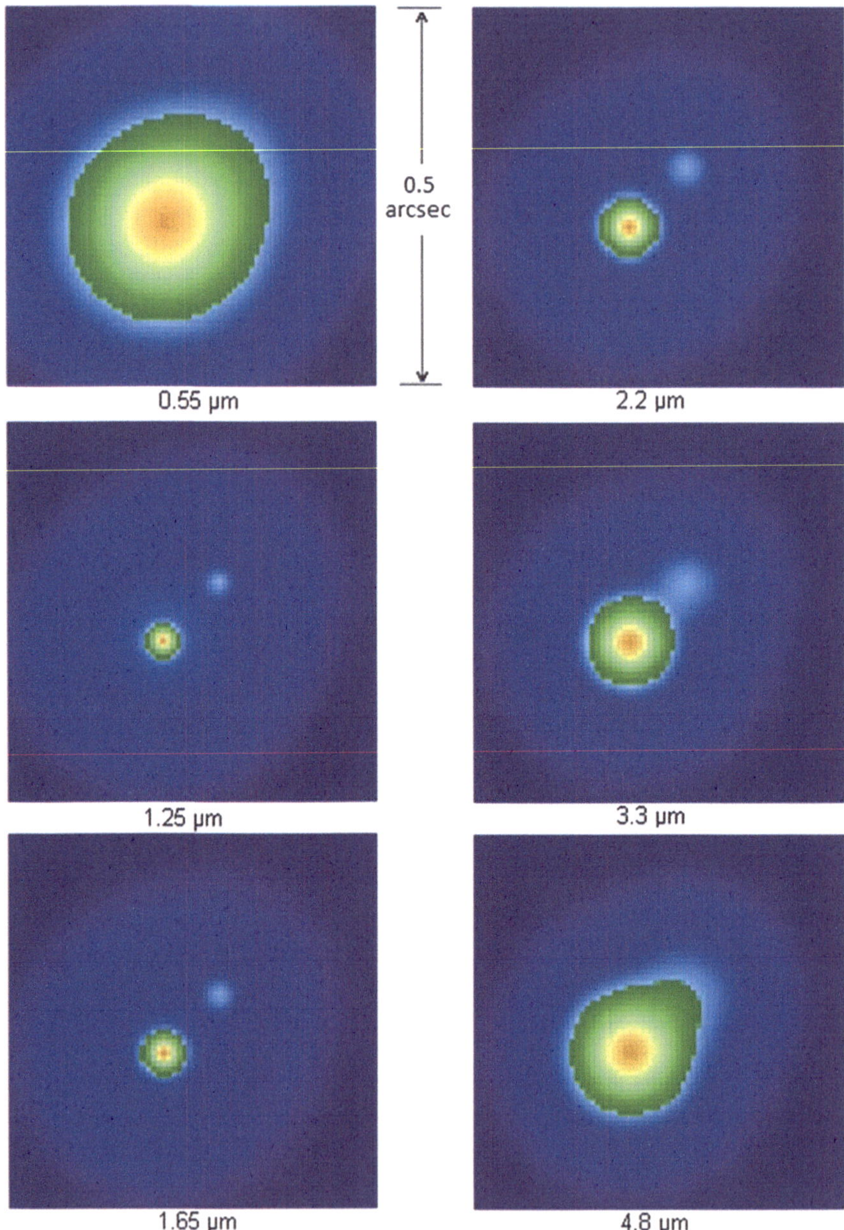


**Fig. 14.20** Computer-generated images of the binary brown dwarf CFBDSIR 1458 + 10 as this object might appear with the Keck II telescope at various visible and IR wavelengths. The angular scale is identical to that used in the actual Keck II shown in Fig. 14.19. (Parameter values used in calculations: $\sigma_{AO} = 0.25\,\mu\text{m}$, $w_{0AO} = 1m$)

waves in the telescope pupil. We assume here that the structure function has the approximate form, $exp\left[-(\varepsilon/w_{0AO})^2\right]$. The empirically chosen value, $w_{0AO} = 1m$, combines with $\sigma_{AO} = 0.25\ \mu\text{m}$ to give a FWHM halo width of about 0.17 arcsec (cf., 13.46), which roughly corresponds to the FWHM halo width as shown in Fig. 14.19 (left).

3. The appropriate values of $A_{CH}$, $B_C$, and $B_H$ used in 14.57 were determined by substituting the parameter values listed above and the wavelength of interest into 13.61, 13.63, and 13.65.
4. Based on the assumed value, $\sigma_{AO} = 0.25\ \mu\text{m}$, the optimum wavelength (cf., 10. 16) lies in the H-band region (1.5–1.8 μm). Though the image at this wavelength is not strictly diffraction-limited (SI (1.65 μm = 0.4), the binary pair nonetheless appears comfortably resolved. Though the halo light energy fraction (0. 63) is larger than that of the core, because this fraction is wildly scattered, cores dominate the star image centers.

The values assumed for $\sigma_{AO}$ and $w_{0AO}$ are based on estimated values consistent with the Keck II images shown in Fig. 14.19. Had full radiometric information been available for these images, more precise $\sigma_{AO}$ and $w_{0AO}$ values could have been established, possibly resulting in more precise image modeling.

### 14.10.1 Obtaining Diffraction-Limited Images at Visible Wavelengths

It is clear from Fig. 14.20 (Top left) that visible images currently delivered by the Keck instruments are far from diffraction-limited. We now establish a suitable optical specification for the AO-assisted Keck instruments that would allow these instruments to deliver diffraction-limited images at visible wavelengths. Figure 14.21 shows images (again calculated from 14.57) using the same above-listed parameter values, with the exception that the residual uncorrected rms OPD fluctuation $\sigma_{AO}$ has now been reduced from the original value, 0.25 μm, to significantly smaller values in the range 0.025–0.1 μm.

The Strehl intensities of the images in Fig. 14.21 are given approximately by $exp\left[-(2 \cdot \pi \cdot \sigma_{AO}/\lambda)^2\right]$, from which we see that diffraction-limited performance $SI(\lambda) \geq 0.8$ is not achieved at visible wavelengths unless residual uncorrected OPD fluctuation fulfills the condition, $\sigma_{AO} \leq 0.0375\ \mu\text{m}$.

To achieve diffraction-limited performance at the shortest visible wavelength, ~0.4 μm, an even smaller rms residual would be required, $\sigma_{AO} \approx 0.03\ \mu\text{m}$, a value about 8 times less than the value surmised as currently achieved by the Keck II AO system, $\sigma_{AO} \approx 0.25\ \mu\text{m}$.

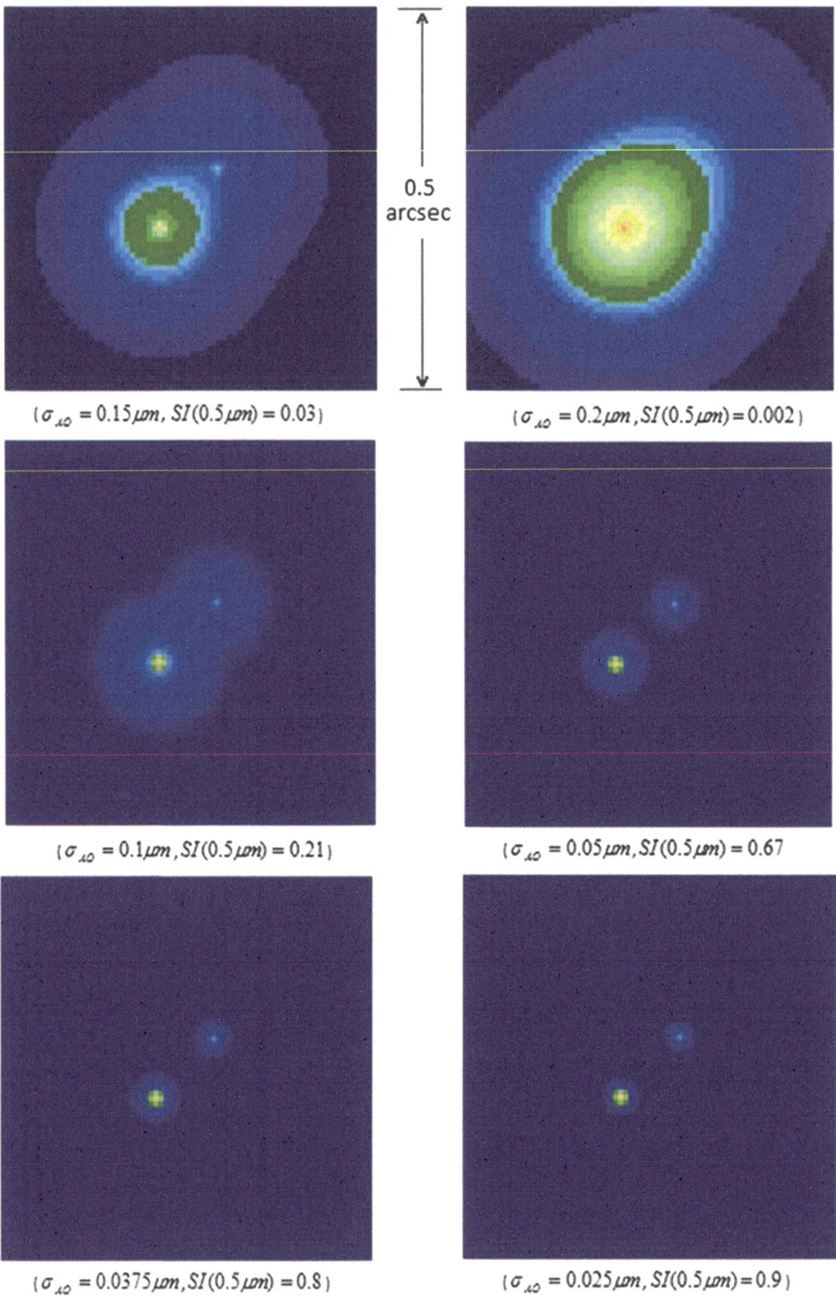


**Fig. 14.21** Computer-generated images of the binary brown dwarf CFBDSIR 1458 + 10 at the visible wavelength 0.5 μm as might be obtained from a hypothetical 10-m Keck telescope with improved AO performance. The residual uncorrected rms OPD fluctuation, $\sigma_{AO}$, and the corresponding Strehl intensities are indicated below each image. In all images, $w_{0AO}$ is set (somewhat arbitrarily) to the value, 1 m. Note that the angular scale used is identical to that used for the images shown in Figs. 14.19 and 14.20

## 14.11 Resolution Possibilities with Future ELT Instruments

As we have just seen, the AO-corrected images obtained by the Keck II at near-IR wavelengths (Fig. 14.19) are reasonably consistent with the residual uncorrected rms OPD fluctuation, $\sigma_{AO} \approx 0.25\ \mu m$. For the much larger ELT instruments currently under construction, the AO systems will have to control many more adaptive segments. This could result in residual uncorrected $\sigma_{AO}$ values greater than $0.25\ \mu m$ if we merely assume present day AO technology levels. However, by the time these massive instruments see first light, advances in AO technology might enable the uncorrected residual rms wavefront error to remain somewhere in the vicinity of $\sigma_{AO} \approx 0.25\ \mu m$; and in the longer term, one can anticipate improvements in AO performance that will gradually reduce the residual $\sigma_{AO}$ value to less than $0.25\ \mu m$. If the value,$\sigma_{AO} \approx 0.02\ \mu m$, is ultimately achieved, the ELT instruments would deliver diffraction-limited images (i.e., $SI(0.3\ \mu m) \geq 0.8$) at all wavelengths transmitted by the atmosphere, the shortest of these being the near-UV wavelength, $0.3\ \mu m$.

Because of the uncertainty about the actual $\sigma_{AO}$ values that will be achieved by the ELT instruments, the images shown in this section are somewhat speculative; at best, they provide only rough representations of the sort of images likely to be obtained by these instruments. Figures 14.22 and 14.23 show computer-generated images of the binary brown dwarf CFBDSIR 1458 + 10 (cf., Fig. 14.19) as they might appear with the 39-m European ELT at various visible and IR wavelengths.

(Eq. 14.9 was used to calculate the images shown in these figures.) The various images correspond to two residual rms OPD fluctuation values that may possibly be obtained at an early stage in the life of the 39-m instrument, $\sigma_{AO} = 0.25\ \mu m$ and $\sigma_{AO} = 0.4\ \mu m$.

Although there is a faint suggestion of a core in the Fig. 14.22 image at the visible wavelength, $0.55\ \mu m$, because of the barely significant (~0.0003) energy fraction in this core, resolution obtained at this wavelength remains disappointingly poor. In contrast, at wavelengths near the optimum wavelength, $1.57\ \mu m$, extremely high (~10 mas) resolution is obtained. At the wavelength, $10\ \mu m$, the core energy fraction is now very large, 0.94; however, increased diffraction spread at this longer wavelength results in larger core size with resolution falling back to about 65 mas. Figure 14.23 shows the corresponding set of images for $\sigma_{AO} = 0.4\ \mu m$. The optimum wavelength has now increased to about $2.5\ \mu m$; resolution achieved at this wavelength is about 16 mas. (As a general rule, the optimum wavelength is proportional to the $\sigma_{AO}$ value delivered by the AO system, with the just-resolved angular resolution growing larger in proportion.)

Figure 14.24 shows computer-generated intensity sections for the brown dwarf, CFBDSIR 1458 + 10, as obtained by a 39-m telescope corresponding to the $\sigma_{AO}$ values, 0.25 and $0.4\ \mu m$—the same values used to produce the images in Figs. 14.22 and 14.23. Figure 14.24 also shows image intensity sections arising for the case, $\sigma_{AO} = 0.15\ \mu m$. In favorable seeing/wind conditions, it is conceivable that the 39-m instrument might occasionally achieve this level of wavefront correction even

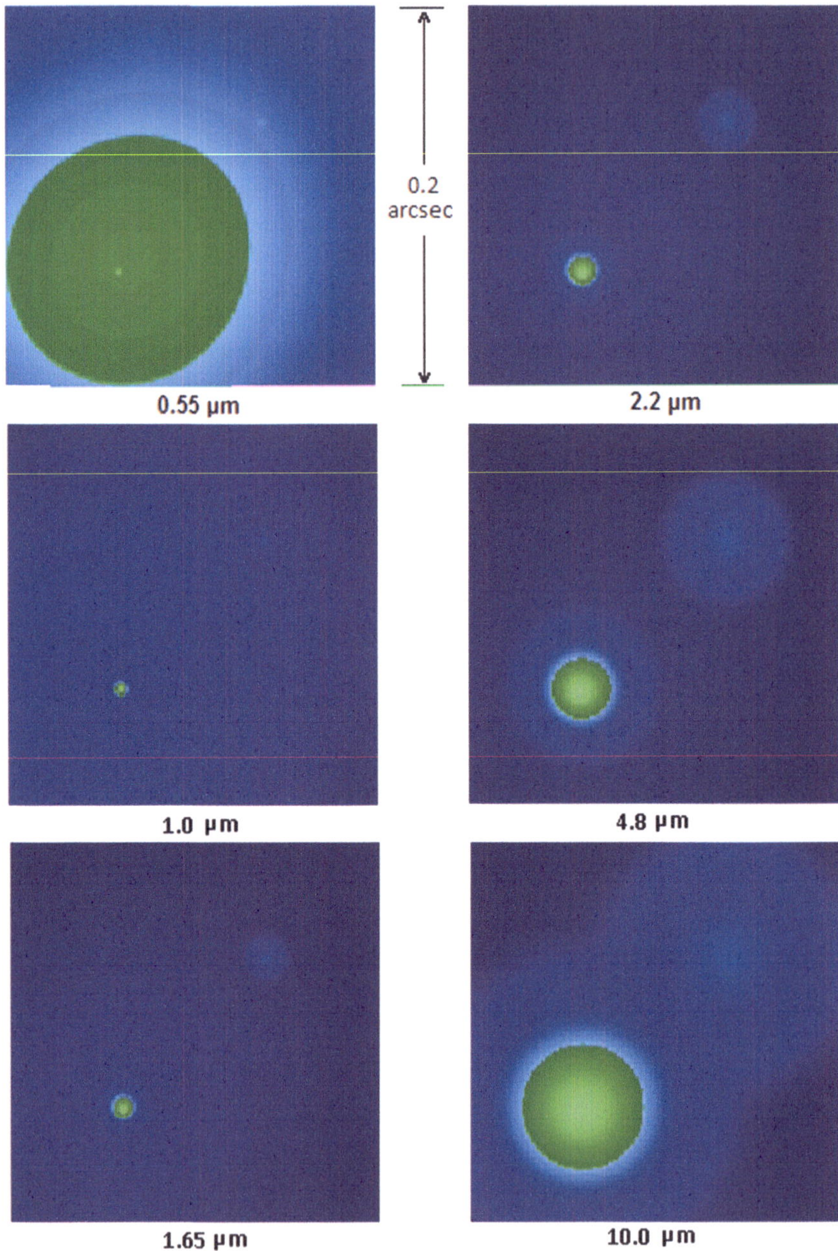


**Fig. 14.22** Computer-generated images showing how a future 39-m telescope might display the binary brown dwarf CFBDSIR 1458 + 10. For all images shown, $\sigma_{AO} = 0.25\,\mu$m and $w_{0AO} = 1m$. Note that the angular scale has been expanded by the factor 2.5 compared to that used for the actual Keck II images shown in earlier Figs. 14.19, 14.20, and 14.21

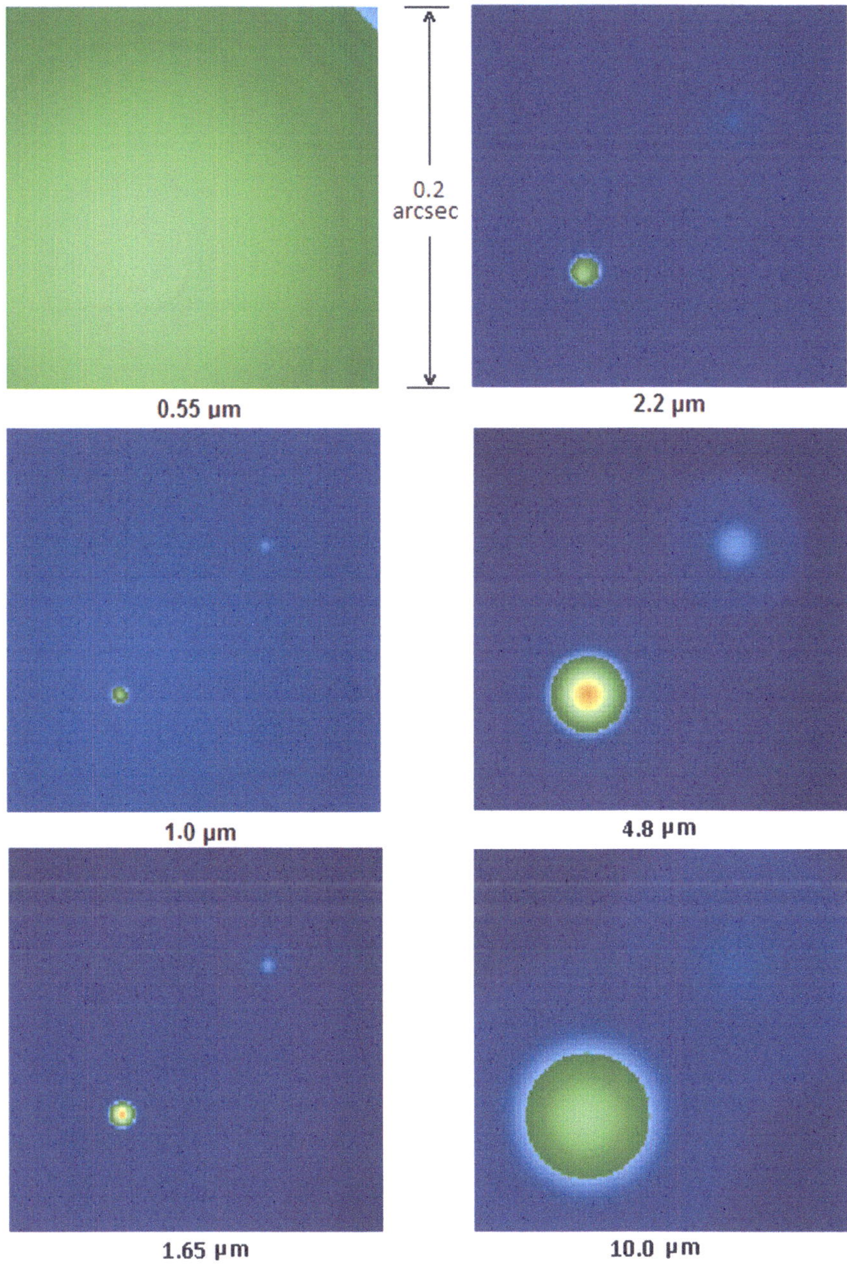


**Fig. 14.23** Computer-generated images showing how a 39-m telescope might display the binary brown dwarf CFBDSIR 1458 + 10. For all images shown, $\sigma_{AO} = 0.4\,\mu\text{m}$ and $w_{0AO} = 1m$. Note that the angular scale has been expanded by the factor 2.5 compared to the scale used for the images shown in Figs. 14.19, 14.20, and 14.21

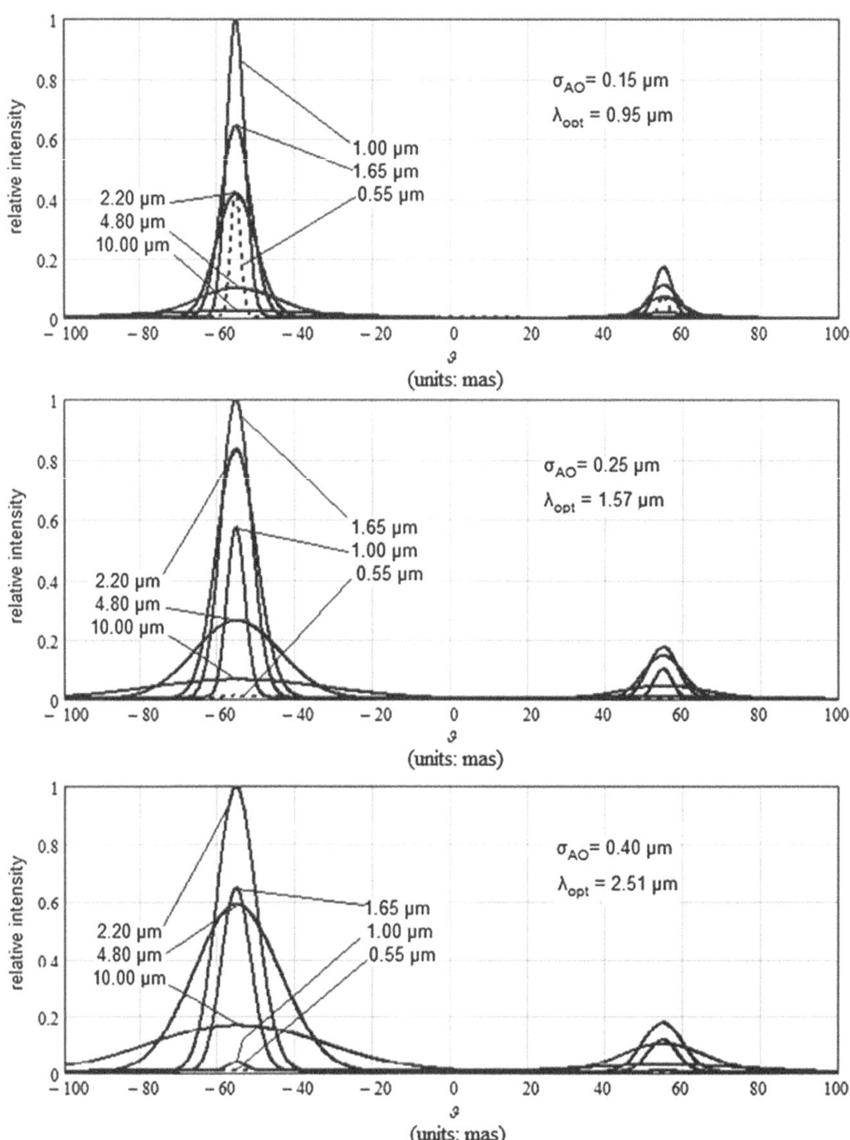


**Fig. 14.24** Computer-generated intensity sections through the binary brown dwarf CFBDSIR 1458 + 10 as they might be obtained by the future 39-m ELT instrument. Sections shown for the $\sigma_{AO}$ values indicated and $w_{0AO} = 1m$. The sections for $\sigma_{AO} = 0.25$ and 0.4 μm relate directly to the images in Figs. 14.22 and 14.23

at an early stage in its lifetime. This would result in visible images at 0.55 μm containing diffraction-limited cores. Though the energy fraction in these cores would still be relatively small—only about 0.05—these cores would nonetheless dominate the image centers, providing resolution down to about 3.5-mas. The optimum wavelength corresponding to $\sigma_{AO} = 0.15\ \mu\text{m}$ is about 0.95 μm; resolution obtained from the dominant cores at that wavelength would be about 6 mas.

To achieve diffraction-limited image quality (i.e., $SI(0.4\ \mu\text{m}) \geq 0.8$) at the shortest visible wavelength (about 0.4 μm) requires $\sigma_{AO} = 0.03\ \mu\text{m}$. Although this rms value has already been demonstrated by the Hubble Space Telescope, to achieve such a value in a large ground-based instrument represents a most formidable challenge.

## 14.12 Apparent Star Image Size and Its Dependence on Star Brightness

In Fig. 14.19, the brighter of the two components in the binary brown dwarf system, CFBDSIR 1458 + 10, appears significantly larger than the other component. Generally, size differences like this are apparent differences, unrelated to the actual angular subtenses of the stars themselves. In this section, we examine how the apparent size of a star image depends on the star's apparent brightness. In all cases, the stars are considered unresolved, so that they behave as ideal point-objects.

The relative sizes of stars of different apparent brightnesses are considered in two limiting cases: (1) where the star image primarily consists only of a core and (2) where the image primarily consists only of a halo. The more general case, where the image comprises both core and halo, can be dealt with by a hybrid analysis, details of which are left as an exercise to interested readers. For both cases considered here, the images are assumed circularly symmetric so that the average image intensity can be approximated (cf., Sect. 13.1) by

$$\langle I(\vartheta)\rangle = \exp\left[-\left(\frac{\vartheta}{B}\right)^2\right] \tag{14.58}$$

where B may be set to $B_C$ or $B_H$, as appropriate, for core-only or halo-only images. Figure 14.25 shows the intensity envelope (calculated from 14.58) for a star at the just-detectable visual magnitude, $m_T$. Also shown in this figure are the envelops for two other comfortably detected stars of magnitudes, $m_1$ and $m_2$, where $m_2 < m_1 < m_T$. The just-detectable intensity threshold, $I_T$, (indicated by the dotted line in the figure) may be considered as either the intensity level at which the star just becomes visible in the image or, alternatively, at the level where the star image attains some other prescribed intensity value; for example, in the case of color-coded contour images of the type shown in Fig. 14.19, a prescribed color might be chosen (somewhat arbitrarily) to represent the "just-detectable" intensity threshold.

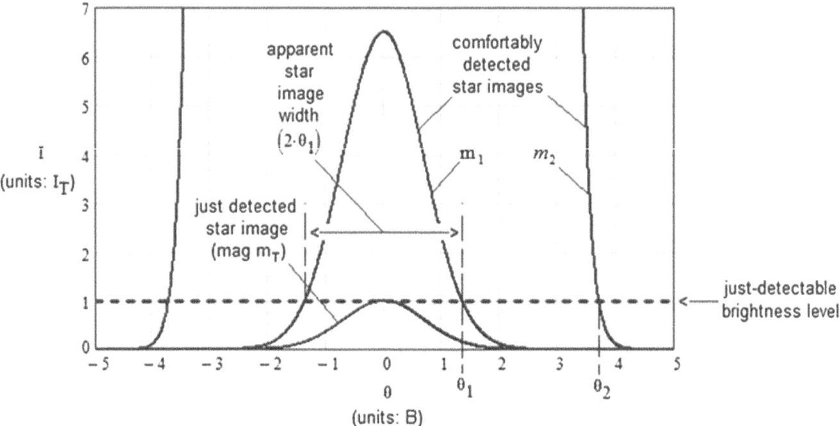


**Fig. 14.25** Intensity envelopes for a just-detectable star of visual magnitude $m_T$, and two comfortably detected brighter stars of magnitudes $m_1$ and $m_2$, where $m_2 < m_1 < m_T$

The intensity envelops for the three star images shown in the figure are given, respectively, by the expressions,

$$\langle I(m_T, \vartheta)\rangle = I_T \cdot \exp\left[-\left(\frac{\vartheta}{B}\right)^2\right] \tag{14.59}$$

$$\langle I(m_1, \vartheta)\rangle = I_T \cdot 100^{-0.2\cdot(m_1-m_T)} \cdot \exp\left[-\left(\frac{\vartheta}{B}\right)^2\right] \tag{14.60}$$

and

$$\langle I(m_2, \vartheta)\rangle = I_T \cdot 100^{-0.2\cdot(m_2-m_T)} \cdot \exp\left[-\left(\frac{\vartheta}{B}\right)^2\right] \tag{14.61}$$

where B is the 1/e half-width of these intensity envelops. It is noted that we have not yet specified whether these envelops refer to core or halo star images.

In the general case of a star of visual magnitude, m, the intensity envelop, $I(m, \vartheta)$, is given by

$$\langle I(m, \vartheta)\rangle = I_T \cdot 100^{-0.2\cdot(m_2-m_T)} \cdot \exp\left[-\left(\frac{\vartheta}{B}\right)^2\right] \tag{14.62}$$

The apparent size of a star is defined here as the angular full-width, $\vartheta_{Star}$, across the intensity envelop at the just-detectable threshold, $I_T$. (Since the image envelops are all assumed here to be circularly symmetric, we could equally use the word diameter rather than "width.") For the star at the just-detectable magnitude, $m_T$, the width is exactly zero. For the two brighter stars in Fig. 14.25, the 1/e full-widths are given by $2 \cdot \vartheta_1$ and $2 \cdot \vartheta_2$, respectively. In the general case of a star of magnitude, m, the apparent full-width, or diameter, of the star image is given by

$$\vartheta_{Star} = 2 \cdot B \cdot \sqrt{-\ln(100^{0.2 \cdot (m - m_T)})} = 2 \cdot B \cdot \sqrt{-\ln(10^{0.4 \cdot (m - m_T)})} \quad (14.63)$$

where it is assumed that $m < m_T$.

The apparent size of a star, $\vartheta_{Star}$, of arbitrary visual magnitude, m, could equally be obtained from the plots in Fig. 14.26 (calculated from the above equation) by using the following procedure:

First, the abscissa value, $m_T$, is identified on the horizontal axis; the point so identified is the Cartesian coordinate point $(m_T, 0)$. If we then identify the nearest plot that originates from that point and follow it until we reach the abscissa value, m,

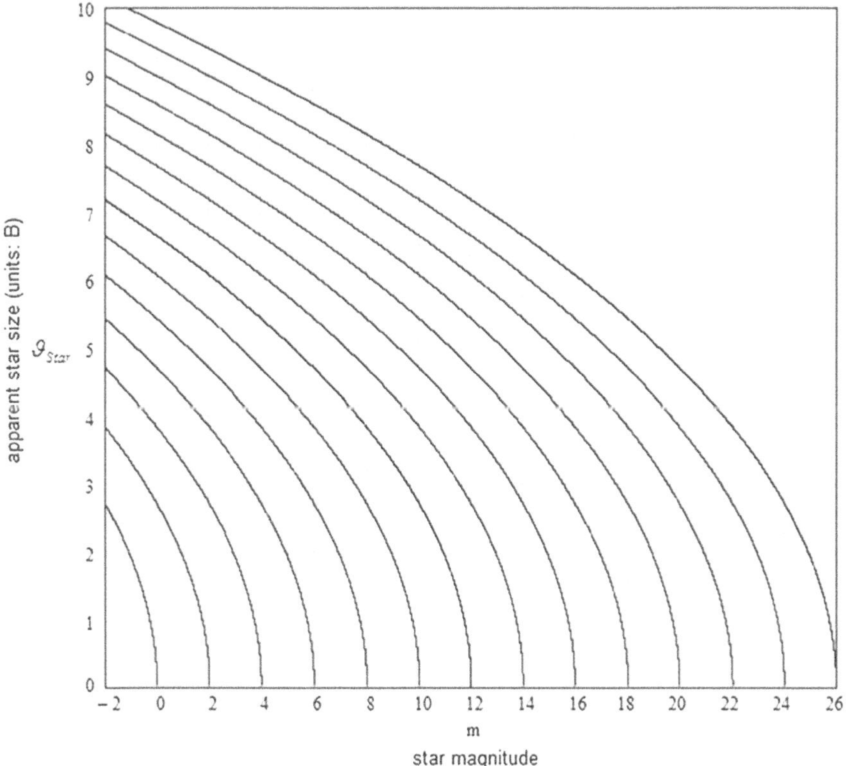


**Fig. 14.26** The apparent size of a star in the image depends on the magnitude of the star, m, and the just-detectable magnitude, $m_T$. To find the apparent size of a star of magnitude m from the figure, we first identify the plot that originates from the Cartesian coordinate point $(m_T, 0)$. We then "follow" that plot until the abscissa attains the value, m. The ordinate at that point gives the apparent full-width of the star (in units of B)

the ordinate corresponding to that abscissa then gives an approximate value for the apparent full-width, or diameter, of the star. (Interpolation can be used to provide a more exact full-width value in cases where $m_T$ is non-integer and therefore lies somewhere between the various plots shown in the figure.) As an example illustration of this procedure, for the case, $m_T = 20$, the apparent diameter of a star of visual magnitude, m = 18, arises as $\vartheta_{Star} = 2.715 \cdot B$.

Depending on whether we deal with core-only images or halo-only images, B would be replaced in the above equations by either $B_C$ or $B_H$, as appropriate. The two angular measures of core and halo 1/e half-widths, $B_C$ and $B_H$, were defined previously in Sect. 13.1.4.

The relative widths of the two comfortably detected stars in Fig. 14.25 is given by

$$\frac{\vartheta_{Star_2}}{\vartheta_{Star_1}} = \sqrt{\frac{-\ln\left(100^{0.2\cdot(m_2-m_T)}\right)}{-\ln\left(100^{0.2\cdot(m_1-m_T)}\right)}} \tag{14.64}$$

The result, $\ln\left(a^b\right) = b \cdot \ln(a)$, can be used to reduce the above equation to the simpler form,

$$\frac{\vartheta_{Star_2}}{\vartheta_{Star_1}} = \sqrt{\frac{m_2 - m_T}{m_1 - m_T}} \tag{14.65}$$

From the above relation, we see that the relative widths of any two star images depend on the just-detectable threshold magnitude, $m_T$. Since $m_T$ depends on several factors (including detector sensitivity, telescope size, and image exposure time) plainly, the relative size of stars of different magnitudes depends on more than just their relative brightness, as is now illustrated by the following example: The visual magnitudes used to produce the plots in Fig. 14.25 are the following: $m_T = 20$, $m_1 = 18$, and $m_2 = 5$. Inserting these values into 14.65 gives $\theta_{Star_2}/\theta_{Star_1} = 2.74$. However, if we were now to use a longer exposure to obtain the image, so that the limiting magnitude increased, say, to $m_T = 25$, the ratio $\theta_{Star_2}/\theta_{Star_1}$ would then reduce to 1.69.

### 14.12.1 Use of Binary Stars to Estimate Limiting Detectable Magnitude

If the components of a binary star have different brightnesses and the visual magnitudes are known, $m_1$ and $m_2$, the limiting detectable magnitude that can be obtained by the telescope (for the same set of imaging parameters, e.g., exposure time and imaging wavelength) may be obtained by first measuring the apparent sizes of the two stars, $\theta_{Star_2}$ and $\theta_{Star_1}$. The limiting magnitude, $m_T$, can then be obtained by rearranging 14.65 as follows:

$$m_T = \frac{m_1 \cdot \left(\frac{\vartheta_{Star_2}}{\vartheta_{Star_1}}\right)^2 - m_2}{\left(\frac{\vartheta_{Star_2}}{\vartheta_{Star_1}}\right)^2 - 1} \tag{14.66}$$

The magnitude difference between the two components of brown dwarf CFBDSIR 1458 + 10 at the H, J, and K near-IR wavelength bands is about 1.9; their combined magnitude is about 20. From this information, it can readily be shown that $m_1 = 20.17$ and $m_2 = 22.07$. By measuring the relative full-widths of the green first contour) platform around each binary component in Fig. 14.19 (right), we obtain $\theta_{Star_2}/\theta_{Star_1} = 1.5$. Inserting this value into the above equation, together with the values just established for $m_1$ and $m_2$, gives $m_T = 23.6$. Thus, the dimmer binary component is 1.53 magnitudes brighter than the just-detectable magnitude limit.

## 14.13 Resolution Obtained from Speckle Imaging

Speckle imaging techniques, discussed previously in Sect. 10.11, involve obtaining large numbers of short-exposure speckle pattern images of the object. Speckle images are not usually associated with high resolution. With some speckle imaging techniques where the various images are recentered according to the brightest speckle and then stacked together, a final composite image can be obtained that provides diffraction-limited resolution. However, the method only works well for certain simple types of object, such as binary and other multiple star types.

As discussed previously in Sect. 11.4, the average speckle shape is determined by the intensity PSF of an aberration-free version of the observing telescope. For a telescope with an unobstructed circular aperture, the PSF is an Airy pattern. For telescopes with central obstructions, the average speckle shape modifies according to 4.45.

For telescopes with unobstructed circular apertures, the angular resolution that can be obtained from speckle imaging techniques is the diffraction-limited value, $1.22 \cdot \lambda/D$, irrespective of whether an aberrated or aberration-free telescope is used to obtain the images.

Telescope optical quality and imaging performance may be assessed by a number of methods. Some of the more commonly used methods are now discussed.

### *14.13.1 The Star Test*

The star test has been described by Welford (1960, 1962) and by Malacara (1992). In this test, the optical system is set up to form the image of a point-object. In the laboratory, the point-object is typically a pinhole, perhaps illuminated by a laser beam. For astronomical telescopes, unresolved stars make ideal point-objects. In either case, the image—i.e., the intensity PSF of the telescope—is examined under high magnification. A skilled observer can deduce information about wavefront aberrations (such as coma and astigmatism) and can even make rough estimates of their magnitude; but, because this test does not provide definitive wavefront phase information, it is not well-suited for identifying and quantifying wavefront aberrations. On the other hand, star tests allow accurate measurement of Strehl intensity. For near-diffraction-limited telescopes where $SI(\lambda) \geq 0.8$, and also for lower quality telescopes that perhaps use AO to improve Strehl intensity toward this limit, the star test constitutes a reasonably definitive test.

### *14.13.2 Optical Transfer Function Tests*

The telescope optical transfer function (OTF) and modulation transfer function (MTF) were discussed previously in Sect. 7.4. These functions describe the modulation response of telescopes to sinusoidal object intensity patterns as a function of pattern spatial frequency.

The OTF of an optical system can be used to carry out object-to-image transformations for extended objects. To obtain the image of an astronomical object where the illumination is invariably incoherent, the intensity distribution over the object is transformed using the incoherent OTF of the telescope as the transformation kernel. (For coherently illuminated extended objects, the complex amplitude distribution over the object would be transformed using the amplitude transfer function of the telescope as the kernel.) The incoherent intensity PSF of a telescope may be obtained by Fourier transformation of the telescope OTF (cf., 7.49).

The quality of an astronomical telescope can be assessed by comparing the intensity OTF of the actual telescope with the intensity OTF of a diffraction-limited version of the same instrument. The latter OTF can readily be calculated from the shape geometry of the telescope aperture (see Sect. 7.4); for astronomical telescopes, aperture geometry often includes a central obstruction. Typical MTFs for telescopes with and without aberrations were shown previously in Fig. 7.4. The effect of central obstructions on the MTF of a diffraction-limited telescope with circular aperture was shown previously in Fig. 7.5.

Because the incoherent OTF provides complete amplitude and phase information, this function allows both exact and unique image intensity realizations to be obtained. In contrast, by not providing phase information, the incoherent MTF generally only provides partial information about the image.

### 14.13.3 Interferometric Tests

Unlike star tests, interferometric tests provide precise information about wavefront phase. This kind of test (Malacara, 1992) is particularly useful for testing the telescope optical components during fabrication; it is also useful for carrying out final telescope alignment. Astronomical telescopes may be tested and aligned, on-site, by common path interferometry using any suitably bright unresolved star as the reference point-object. An equal-path interferometer design, such as the one shown in Fig. 14.27, could be used to achieve suitably high fringe contrast despite the low temporal coherence properties of typical star light. In principle, the test can be carried out in any convenient wavelength band. For small telescopes, visible wavelengths might be used. For larger instruments, near-IR wavelengths would likely be more appropriate; an IR image core focused on the pinhole aperture shown in Fig. 14.27 should generally deliver a high-quality spherical reference wave. There is no requirement in this type of test for large ancillary test optics.

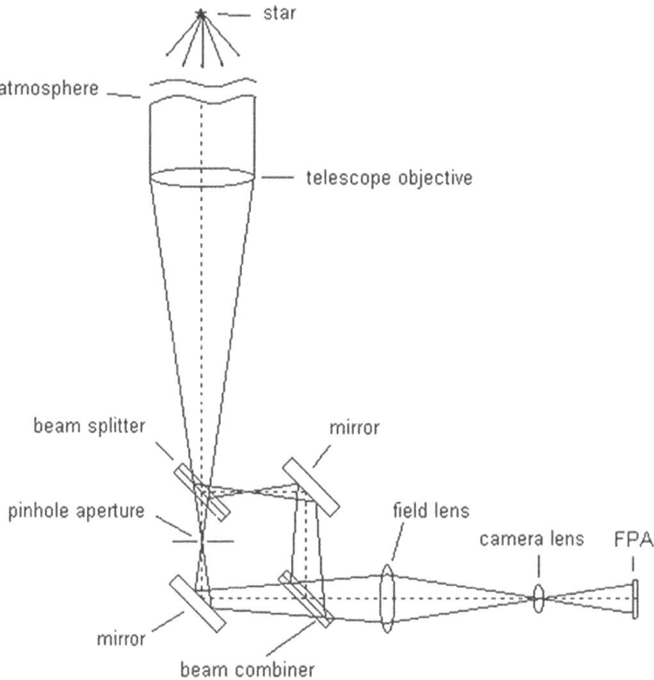


**Fig. 14.27** Common-path interferometer arrangement for on-site testing of large astronomical telescopes using a bright star as the "point source"

### *14.13.4 The Hartmann Wavefront Test*

This test (Malacara, 1992) can be performed, on-site, with any astronomical telescope using any suitably bright star as the illumination source.[21] A Hartmann screen placed in a (demagnified) secondary pupil plane enables wavefront gradient information to be obtained at a 2-D matrix of points over the telescope pupil. Information provided by this test is essentially the same as that provided by interferometry. Short-exposure Hartmann images allow quantification of the OPD fluctuation over the pupil caused by the combined effects of the atmosphere and the telescope optics. By using long-exposure Hartmann images, the effects of atmospheric turbulence average to zero, thereby allowing the fixed aberrations of the telescope to be identified. Tollestrup and Tokunaga (2010) used this technique successfully to quantify the aberration of the NASA 3-m telescope on Mauna Kea. This test may also be used to quantify residual uncorrected wavefront phase errors when AO is used. A detailed description of this type of test and its practical implementation has been given by Neal et al. (2002).

## 14.14 Mathematical Notation Used in This Chapter

The mathematical notation used in this chapter is indicated in Table 14.4.

[21] For this test, it makes little difference whether resolved or unresolved stars are used.

**Table 14.4** Mathematical notation used in this chapter along with the SI dimensional units of the individual quantities

| Symbol | Quantity | Dimensions |
|---|---|---|
| $\lambda$ | Wavelength | $m$ |
| (x, y) | Cartesian coordinate system in telescope pupil | $m$ |
| (u, v) | Cartesian coordinate system in telescope image plane | $m$ |
| $(\alpha, \beta)$ | Angular coordinate system in telescope image plane | "1" |
| $\theta$ | Radial angular coordinate in telescope image plane $= \sqrt{\alpha^2 + \beta^2}$ | "1' |
| $\sigma$ | rms of integrated OPD fluctuation, H | $m$ |
| $\rho$ | Autocorrelations function of integrated OPD fluctuation, H | "1" |
| $w_0$ | 1/e half-width of Gaussian approximation to autocorrelation function of H | $m$ |
| $\sigma_{AO}$ | rms of residual OPD fluctuation after AO correction | $m$ |
| $w_{0AO}$ | 1/e half-width of Gaussian autocorrelation function of residual OPD fluctuation | $m$ |
| $D$ | Telescope diameter | $m$ |
| $d$ | Telescope central obstruction | $m$ |
| $f$ | Telescope focal length | $m$ |
| $F$ | Telescope F/number | "1" |
| $\Delta f_C$ and $\Delta f_H$ | Focal depth tolerances for core-only and halo-only star images | $m$ |
| $m, m_1, m_2, m_T$ | Various star magnitudes as defined in text | "1" |
| BR | Linear brightness ratio for a binary star pair | "1" |
| SI | Telescope Strehl intensity | "1" |
| $I$ | Intensity in (x, y) pupil plane or $(\alpha, \beta)$ image plane, as appropriate | "1" |
| $A_C$ | Intensity in center of Gaussian approximation to core | "1" |
| $A_H$ | Intensity in center of Gaussian approximation to halo | "1" |
| $A_{CH}$ | Ratio $A_C/A_H$ | "1" |
| $B_C$ | Angular 1/e half-width of Gaussian core | "1" |
| $B_H$ | Angular 1/e half-width of Gaussian halo | "1" |
| $R_{CH}$ | Resolution improvement obtained from core compared to that obtained from a halo | "1" |
| K | A constant determined in the text | "1" |
| $C_{scin}$ | Intensity contrast in scintillation pattern | "1" |
| A | Amplitude | "1" |
| $\varphi$ | Phase | "1" |
| $\sigma_T$ | rms wavefront error of telescope | $m$ |
| P–V | Peak-to-valley wavefront error of telescope | $m$ |

(continued)

**Table 14.4** (continued)

| Symbol | Quantity | Dimensions |
|---|---|---|
| $\lambda_{opt}$ | Optimum wavelength | $m$ |
| $\lambda_{dl}$ | Shortest wavelength at which diffraction-limited imaging occurs | $m$ |
| $\Delta$ | Angular separation of binary star | "1" |
| $\vartheta_{star}$ | Apparent angular width of star due to the star's brightness | "1" |
| $\chi$ | Inclination angle of the binary star axis with respect to the $\alpha$-axis | "1" |

Dimensionless quantities are indicated by "1"

## References


Allen, L., Angel, R., Mangus, J. D., Rodney, G. A., Shannon, R. R., & Spoelhof, C. (NASA-TM-103443, 1990, November). *The Hubble Space Telescope optical systems failure report*. NASA.

Born, M., & Wolf, E. (2003). *Principles of optics* (7th ed., revised). Cambridge University Press.

Cho, M. K., Stepp, L., & Kim, S. (2001). Wind buffeting effects on the Gemini 8m primary mirrors. Gemini Preprint #75. 950 N. Gemini Observatory.

den Dekker, A., & van den Bos, A. (1997). Resolution: A survey. *Journal of the Optical Society of America A: Optics, Image Science, and Vision, 14*(3), 547–557.

Forbes, F. F. (1991). *Private communication*. NOAO.

Goodman, J. W. (1990). *Statistical optics*. McMillan.

Griffin, R. F. (1990, May). Giant telescopes, tiny images. *Sky & Telescope*, 469.

Liu, M. (2011, March 21). Brown Dwarf Duo: William Keck observatory picture gallery. Retrieved January 16, 2013, from http://keckobservatory.org/gallery/detail/milky_way/7.

Malacara, D. (1992). *Optical shop testing*. Wiley.

Martin, B., Hill, J. M., & Angel, R. (1991). The new ground-based optical telescopes. *Physics Today*, March, 22–30.

McCarthy, D. W., Jr., McLeod, B., & Barlow, D. J. (1990). Infrared array camera for interferometry with the cophased multiple mirror telescope. In *SPIE, Vol. 1236 Symposium on Astronomical Telescopes and Instrumentation for the 21st Century*, 11–17 February. Tucson, AZ.

McKechnie, T. S. (1989, September 26). Seminar: obtaining diffraction limited images at near infrared wavelengths using large ground based telescopes (Attendees: D. W. McCarthy, Jr., K. Hege, Steward Observatory, Tucson, G. Loos, B. Venet, AFRL, R. Haddock, & Lentec Corp.) Albuquerque, New Mexico, USA.

McKechnie, T. S. (1990). Diffraction limited imaging using large ground-based telescopes. In *Proceedings of SPIE, Vol. 1236, Symposium on Astronomical Telescopes and Instrumentation for the 21st Century*, 11–17 February (pp. 164–178).

McKechnie, T. S. (1991). Light propagation through the atmosphere and the properties of images formed by large ground-based telescopes. *JOSA A, 8*, 346–365.

McKechnie, T. S. (2010). Interferometric test method for testing convex aspheric mirror surfaces. In *Proceedings of SPIE, Symposium on Telescopes and Systems, June 27–July 2, SPIE* (Vol. 7739-31). San Diego, CA.

Neal, D. R., Copland, J., & Neal, D. A. (2002). Shack-Hartmann wave front sensor precision and accuracy. In *SPIE* (Vol. 4779). Bellingham, WA.

Nelson, J., Mast, T., & Faber, S. (1985). The design of the Keck observatory and telescope. In J. E. Nelson, T. S. Mast, & S. M. Faber (Eds.), *Keck observatory report 90*. Keck Observatory.

Ninneman, R. R., Sydney, P. F., Reamy, P. C., Lanier, T. V., Carter, S. E., & Olives, M. L. (1987). *Ambient vibration test results of proposed RME LSS site at the Air Force Maui Optical Station (AMOS)*. Air Force Weapons Laboratory, Kirtland AFB, N.M.

Ronchi, V. (1961). Resolving power of calculated and detected images. *Journal of the Optical Society of America, 51*, 458–460.

Sparrow, C. M. (1916). On spectroscopic resolving power. *Astrophysical Journal, 44*, 76–86.

Strutt, J. (1880). On the resolving power of telescopes. *Philosophical Magazine, X*, 116–119.

Strutt, J. (1902). *Collected papers* (Vol 3). Cambridge University Press.

Strutt, J. W. (1979). Investigations in optics, with special reference to the spectroscope. *Philosophical Magazine*, VIII.

Tollestrup, E. V., & Tokunaga, A. T. (2010). New phase compensating secondary mirrors for the NASA infrared telescope facility. In *Proceedings of SPIE, Vol. 7733, Ground-Based and Airborne Telescopes III, August 04*, Bellingham, WA.

Ulich, R. L. (1982). Performance of the multi-mirror telescope (MMT) V: Pointing and tracking of the MMT. In L. D. Barr & G. Burbidge (Eds.), *Proceedings of SPIE Vol. 332, Advanced Technology Optical Telescopes* (pp. 33–41).

Ulich, R. L. (1988). Overview of acquisition, tracking, and pointing system technologies. In J. E. Kimbrell (Ed.), *Proceedings of SPIE (Vol. 887), Acquisition, Tracking, And Pointing II* (pp. 40–63).

Welford, W. T. (1960). On the limiting sensitivity of the star test for optical instruments. *Journal of the Optical Society of America, 50*, 21.

Welford, W. T. (1962). *Geometrical optics*. North-Holland Publishing Co.

Wilson, R. N. (2003, September). The history and development of the ESO active optics system. *The Messenger*, 113. ESO.

# Chapter 15
# Laboratory Simulation of Images Formed by Large Telescopes

**Abstract** Scaling procedures are described for use in the design and construction of optical simulators that enable the creation of images in the laboratory with properties identical to those of actual images obtained by large telescopes in the field. The imaging optic used in the simulator—typically 5 mm in diameter—is designed to provide a suitably scaled replica of the telescope point-spread function. Scaling procedures are also described for designing the illumination optics so that coherence effects can be fully replicated in the images, including speckle displaying the appropriate statistical properties. The effect of the atmosphere can be simulated by a random phase screen (or, if isoplanatic effects have to be replicated, by several separated phase screens) located in the space in front of the imaging optic. Images can be simulated in either monochromatic or polychromatic light. The target objects are typically scaled models, but they could equally be actual target objects set up in the laboratory.

In this chapter, scaling procedures are described for use in the design and construction of scaled optical simulators that enable images to be created in the laboratory with properties identical to those of actual images obtained by large ground-based, or space-based, telescopes in the field. The effect of the atmosphere can be duplicated in the simulator either by a single random phase screen placed just in front of the imaging optic or, if atmospheric isoplanatic angle is also important, by multiple appropriately separated random phase screens, again placed just in front of the imaging optic. Simulated images can be produced using monochromatic or polychromatic light, as appropriate. Generally, the target object used in the simulator is a scaled model of the target object in the field, but it could also be a full-scale version of the actual target object. The imaging optic used in the simulator is designed to provide a suitably scaled replica of the telescope point-spread function (PSF).

For distant astronomical objects (such as single and multiple stars, star clusters, and galaxies), the object illumination can be considered incoherent. For less distant objects passively illuminated by the Sun, such as Earth satellites and other man-made objects in space, the object illumination, while no longer perfectly incoherent, can still be regarded as incoherent for all practical purposes. Stars can be simulated in



T. S. McKechnie, *General Theory of Light Propagation and Imaging Through the Atmosphere*, Progress in Optical Science and Photonics 20,
https://doi.org/10.1007/978-3-030-98828-9_15

the laboratory by rear-illuminating pinholes in metal foil. Extended objects such as galaxies can be simulated by scaled pictures or transparencies of actual galaxies, front- or rear-illuminated as convenient. Details are provided in the chapter of how suitably incoherent illumination can be created in the laboratory for all of the above applications.

For actively illuminated extended objects, such as satellites in low Earth orbit, the illumination generally has to be either partially coherent or fully coherent (temporally and/or spatially). These illumination types tend to produce additional speckle effects in the images, over and above the speckle effects caused by atmospheric turbulence. To faithfully reproduce all speckle phenomena in the simulated images, the spatial and temporal coherence properties of the illumination used in the optical simulator must be appropriately scaled from the corresponding properties of the target illumination in the field. Assuming that all necessary scaling procedures have been properly enacted in the designs of both the imaging optics and the illumination optics, the images created by the optical simulator can be made to faithfully replicate all meaningful properties of images formed by large telescopes in the field; for point-objects, such as stars, these image properties include speckle properties as well as core and halo properties. Images produced by telescopes equipped with adaptive optics (AO) can be simulated just as readily by using doctored phase screens that produce appropriately less optical path difference (OPD) fluctuation. A method of reducing the OPD fluctuation introduced by any given random phase screen by prescribed amounts is described in Sect. 15.2.2.2.

The optical image simulator described in Sect. 15.4 was constructed and used in early 1989 (McKechnie, 1990) at the time when NASA faced the problem of thermal insulation tiles detaching from the Space Shuttle's outer skin during the (high-vibration) launch phase. Because of the dangers posed by missing tiles during reentry, the agency was anxious to know the exact locations of any such tiles and thus had arranged for large ground-based telescopes to observe the craft as it orbited overhead. The purpose of the Space Shuttle optical simulator was to produce images of the spacecraft in the laboratory that faithfully replicated actual Space Shuttle images obtained in the field by a 4-m ground-based telescope. (At that time, only three telescopes were in existence larger than 4 m—the Palomar telescope, the MMT, and the Soviet BTA-6 instrument.) As demonstrated by the simulated images shown later in Sect. 15.4 (Fig. 15.6), while telescope resolution at visible wavelengths in that pre-AO era would generally not have been good enough to show missing tiles, there would have been a realistic chance of identifying them at the near-IR wavelength, 2.5 μm, where about a 10× resolution improvement was postulated.[1]

The diameter of the "objective" lens used in the Space Shuttle image simulator to represent the 4-m telescope objective was only 2.4 mm. Thus, the scale factor used for scaling the optics diameters was 1/1667. The length of the Space Shuttle target

[1] When the Space Shuttle image simulator was constructed in early 1989, it was mere postulation that a wavelength of about 2.5 μm would be close to the "optimum wavelength" where 0.15-arcsec resolution might be obtained from a 4-m class telescope observing in modest 1.5-arcsec visible seeing conditions. There had been no reports at that time by observational astronomers of unusually high levels of resolution at near-IR wavelengths.

model used in the simulator was 13.5 cm, the actual length of the craft itself being 39 m. The model scale factor was therefore about 1/288. These two scale factors can be chosen independently; there is no obligation to choose the same value for both.

For the Space Shuttle image simulations, the atmosphere was represented by a single random phase screen placed just in front of the imaging optic. Since only short-exposure images were simulated, the phase screen was held fixed. To simulate long-exposure images, the effects of the temporal evolution of atmospheric turbulence could have been replicated by appropriately translating, and/or rotating, the phase screen.

Wavelength scaling equations are also described in this chapter by which images produced in the field, say at a near-IR wavelength, can be faithfully simulated in the laboratory at a more convenient wavelength, most likely a visible wavelength.

While all meaningful image properties can be preserved when wavelength scaling is used (including speckle and core and halo properties), by carrying out the simulation at visible wavelengths, the images can be observed directly by eye or by CCD cameras, thus avoiding the need for expensive IR focal plane array (FPA) cameras.

Large ground-based astronomical telescopes not equipped with AO are rarely diffraction-limited. The Strehl intensities of these instruments at visible wavelengths typically lie in the range, 0.001–0.1. The Strehl intensity of the "objective" lens used in the optical simulator would, at the very least, have to approximately match that of the telescope in the field; but obviously, it would be even better if the wavefront error characteristics of the lens exactly replicated those of the telescope in the field.

## 15.1 Choice of Detector in the Optical Simulator

The field of view in the optical simulator would normally be chosen to match that of the actual telescope in the field. The detector used in the simulator would often be the same as the one used in the field. The only mandatory detector requirement is that it should be able to resolve image details to the same extent as the detector used in the field.

## 15.2 Choice of Scale Factors in the Optical Simulator

The subscript "S" is added to distance parameters and other parameters in the optical simulator setup to distinguish them from those used in the telescope setup in the field. Thus, for example, while the actual diameter of the telescope in the field is denoted by $D$, the diameter of the imaging lens in the simulator is denoted by $D_S$.

Three scaling factors are used to specify the imaging optics in the simulator. All three may be chosen independently. Referring to the schematics in Fig. 15.1, the target model scale factor, Scale$M$, is defined by the ratio of the width of the field of view in simulated target space, $\mathrm{Wid}_S$, to the width of the corresponding field of view

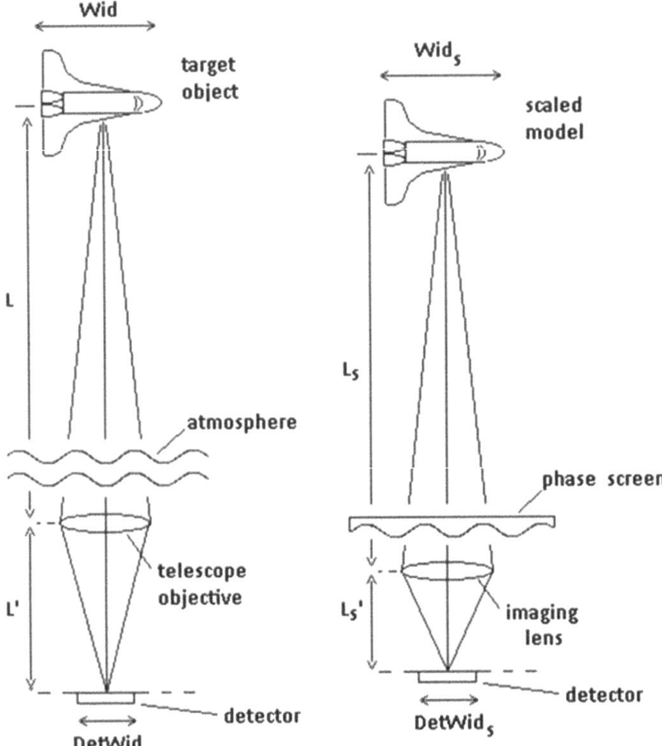


**Fig. 15.1** The telescope imaging geometry and the corresponding simulator imaging geometry (drawn to different scales). In the simulator, the atmosphere is represented by a phase screen

of actual image space in the field, Wid

$$ScaleM = \frac{Wid_S}{Wid}. \tag{15.1}$$

It may be noted here that the widths of the FPA detectors in actual and simulated image spaces are denoted by DetWid and DetWid$_S$ (Fig. 15.1). If exactly the same type of detector is used in actual and simulated image spaces, DetWid = DetWid$_S$.

The diameter of the simulator imaging objective is established by the scale factor, Scale$D$, defined by

$$ScaleD = \frac{D_S}{D} \tag{15.2}$$

The wavelength scale factor, Scale$\lambda$, is defined as the ratio of the wavelength at which the images are created in the optical simulator, $\lambda_S$, to the wavelength used to obtain the images in the field, $\lambda$. Thus,

$$Scale\lambda = \frac{\lambda_S}{\lambda} \tag{15.3}$$

When the two wavelengths are the same, as is often the case, we simply set $Scale\lambda = 1$.

### 15.2.1 Scaling Equations for Image Simulation at the Same Wavelength

In this section, the simplifying assumption is made that the wavelength used to create the simulated images is the same as that used to produce the actual telescope images in the field. The development of the scaling equations for the more general case where $Scale\lambda \neq 1$ is deferred until Sect. 15.2.2. Thus, in the present section, we set $\lambda_S = \lambda$, and hence, $Scale\lambda = 1$.

To simulate the atmospheric portion of the imaging path (Fig. 15.1), a random phase screen would be placed just in front of the imaging lens in the simulator. For the case, $\lambda_S = \lambda$, it is clear that the rms OPD fluctuation introduced by this phase screen, $\sigma_S$, would have to be the same as that introduced by the actual atmospheric path, $\sigma$. Thus, we can write

$$\sigma_S = \sigma \tag{15.4}$$

In principle, the scaling procedures described below can be applied to any type of telescope aperture geometry. However, to simplify the analysis here, we limit consideration to telescopes with circular apertures and, optionally, circular central obstructions.

#### 15.2.1.1 Scaling the Telescope Diameter and Maintaining Strehl Intensity

The diameter of the simulator imaging lens, $D_S$, is given in terms of the diameter of the actual telescope objective, $D$, by

$$D_S = ScaleD \cdot D \tag{15.5}$$

Though the scale factor, Scale$D$, can be chosen quite arbitrarily, the value chosen would generally be small. For example, to simulate a 10-m-diameter telescope, by choosing Scale$D$ to lie in the approximate range, 0.0001–0.001, the diameter of the simulator imaging optic would lie in the range, 1–10 mm; such a lens would be inexpensive and easily fabricated. The primary requirement of the simulator imaging lens is that the combination of this optic and the diffuser screen should faithfully reproduce an appropriately scaled version of the average intensity PSF produced by the telescope/atmosphere combination in the field. In general, the short-exposure version of this intensity PSF will display core and halo structures; as previously, core-only and halo-only images are regarded simply as degenerate cases.

The required scaling equations can all be derived from 15.1 to 15.5. Telescope aberrations are described in these equations in terms of Strehl intensity.[2] By a slight rearrangement of 13.70, Strehl intensity may be expressed in the form

$$SI(\lambda) = \frac{4 \cdot \lambda^2}{\pi^2 \cdot B_C^2 \cdot D^2 \cdot \left(1 - \left(\frac{d}{D}\right)^2\right)} \tag{15.6}$$

Since the Strehl intensity of the simulator optics must be the same as that of the actual telescope in the field, the following condition must be satisfied:

$$\frac{4 \cdot \lambda^2}{\pi^2 \cdot B_{CS}^2 \cdot D_S^2 \cdot \left(1 - \left(\frac{d_S}{D_S}\right)^2\right)} = \frac{4 \cdot \lambda^2}{\pi^2 \cdot B_C^2 \cdot D^2 \cdot \left(1 - \left(\frac{d}{D}\right)^2\right)} \tag{15.7}$$

It is clear from this condition that the diameter of the simulator imaging optic $D_S$ and the diameter of the central obstruction $d_S$ must obey the relation,

$$\frac{d_S}{D_S} = \frac{d}{D} \tag{15.8}$$

Recalling 15.5, the central obstruction diameter then scales as

$$d_S = ScaleD \cdot d \tag{15.9}$$

Thus, not unexpectedly, Scale$D$ determines both the diameter of the imaging optic, $D_S$, and the diameter of the central obstruction, $d_S$.

#### 15.2.1.2 Scaling Equations to Preserve the Core and Halo Angular Width Ratio

From 15.7, it is also clear that the product $B_{CS} \cdot D_S$ for the optical simulator must have the same value as the product $B_C \cdot D$ for the actual telescope. Thus, we require

$$B_{CS} \cdot D_S = B_C \cdot D \tag{15.10}$$

from which it follows

---

[2] If more detailed accounting of telescope aberrations was required than that offered by Strehl intensity alone, in principle, an imaging optic could be fabricated whose pupil function (Sect. 4.2.6) replicated that of the actual telescope, with due account taken of the diameter scaling factor, Scale$D$.

$$B_{CS} = \frac{B_C}{ScaleD} \tag{15.11}$$

To preserve the same relative widths of the core and halo, the $1/e$ angular half-width of the halo, $B_{HS}$, must scale in exactly the same way as $B_{CS}$. Thus, we can write

$$B_{HS} = \frac{B_H}{ScaleD} \tag{15.12}$$

#### 15.2.1.3 Scaling the Random Phase Screen Atmospheric Path Simulator

Combining 15.12 and 13.63 (where $B_H$ is seen to be inversely proportional to $w_o$), it is clear that the average lateral size of the scattering structure in the phase screen, $w_{oS}$, must scale linearly with the average turbulence structure size in the atmosphere, $w_o$. Thus, we require

$$\frac{w_{oS}}{D_S} = \frac{w_o}{D} \tag{15.13}$$

from which we obtain

$$w_{oS} = w_o \cdot ScaleD \tag{15.14}$$

#### 15.2.1.4 Maintaining the Same Relative Brightness of the Core and Halo

The ratio of the central intensities in the core and halo portions of the simulated image must be the same as that of actual images obtained in the field. Thus, we require $A_{CHS} = A_{CH}$. However, as the following equation shows, this result has in fact already been secured by the scaling relationships already in place:

$$A_{CHS} = \frac{\pi^2 \cdot \sigma_S^2 \cdot SI_S(\lambda) \cdot D_S^2\left(1-\left(\frac{d_S}{D_S}\right)^2\right)}{\lambda^2 \cdot w_{oS}^2} \cdot \left\{\frac{\exp\left[-\left(\frac{2\cdot\pi\cdot\sigma_S}{\lambda}\right)^2\right]}{1-\exp\left[-\left(\frac{2\cdot\pi\cdot\sigma_S}{\lambda}\right)^2\right]}\right\}$$

$$= \frac{\pi^2 \cdot \sigma^2 \cdot SI(\lambda) \cdot D^2\left(1-\left(\frac{d}{D}\right)^2\right)}{\lambda^2 \cdot w_o^2} \cdot \left\{\frac{\exp\left[-\left(\frac{2\cdot\pi\cdot\sigma}{\lambda}\right)^2\right]}{1-\exp\left[-\left(\frac{2\cdot\pi\cdot\sigma}{\lambda}\right)^2\right]}\right\} = A_{CH} \tag{15.15}$$

where we have used $\sigma_S = \sigma$, $SI_S(\lambda) = SI(\lambda)$, $d_S/D_S = d/D$, $d_S/w_{0S} = D/w_0$.

#### 15.2.1.5 Scaling the Optics and the Optical Layout of the Image Simulator

For an object of width, Wid, located at a finite distance, $L$ (Fig. 15.1), the angular field of view of the telescope needed to fully capture the object is given by

$$\vartheta_{FOV} = \frac{Wid}{L} \tag{15.16}$$

For astronomical objects, both Wid and $L$ may be extremely large. However, only their ratio (which is simply the angular field of view, $\vartheta_{\mathrm{FOV}}$) carries any significance. The corresponding angular field of view in simulated image space (Fig. 15.1) is then given by

$$\vartheta_{FOV\,S} = \frac{Wid_S}{L_S} \tag{15.17}$$

The just-resolved spatial interval in simulated object space must correspond exactly to the just-resolved interval delivered by the actual telescope in object space. Thus, depending on whether we consider the core or the halo as determining the "just-resolved" patch size, we may write

$$\frac{B_C}{\vartheta_{FOV}} = \frac{B_{CS}}{\vartheta_{FOV\,S}} \tag{15.18}$$

$$\text{or} \quad \frac{B_H}{\vartheta_{FOV}} = \frac{B_{HS}}{\vartheta_{FOV\,S}} \tag{15.19}$$

For our purposes, both expressions contain essentially the same information. By choosing the upper expression and combining it with 15.16 and 15.17, we obtain

$$L_S = \frac{B_{CS}}{B_C} \cdot \frac{Wid_S}{Wid} \cdot L \tag{15.20}$$

Recalling 15.1 and 15.11, we may now write

$$L_S = ScaleD \cdot ScaleM \cdot L \tag{15.21}$$

When the refractive index in both object and image spaces closely approximate unity (as they do for air and vacuum), the Lagrange invariant (Born & Wolf, 2003) simplifies in a way that allows us to write (Fig. 15.1),

$$L'_S = L_S \cdot \frac{DetWid_S}{Wid_S} \tag{15.22}$$

Thus, 15.21 and 15.22 allow calculation of the conjugate distances in the optical simulator, $L_S$ and $L'_S$, consistent with the scale factors, Scale$D$ and Scale$M$.

The focal length of the simulator imaging optic, $f_S$, may be obtained from $L_S$ and $L'_S$ by using the conjugate distance equation (Welford, 1962). When the imaging is performed in either air or vacuum, this equation can be written in the form

$$\frac{1}{L'_S} + \frac{1}{L_S} = \frac{1}{f_S} \tag{15.23}$$

In the above equation, the sign convention is consistent with that required for a positive imaging lens (i.e., $f_S$ positive) where the object and the image are both real (as opposed to virtual) and where $L_S$ and $L'_S$ are both positive (Fig. 15.1).

Solving the above equation for $f_S$ gives

$$f_S = \frac{L_S \cdot L'_S}{L_S + L'_S} \tag{15.24}$$

The $F$/number of the imaging optic is then given by

$$F_S = \frac{f_S}{D_S} \tag{15.25}$$

Because $L_S$ is inversely proportional to the scale factor product (cf., 15.21), Scale$D$ and Scale$M$, and $F_S$ is proportional to $L_S$, it might be prudent at this point to revisit the choices made for Scale$D$ and Scale$M$ to make sure that they make practical sense. Different choices can lead to smaller, or larger, $F_S$ values. If the choices lead to $F_S \geq 6$, the task of designing and fabricating the imaging optic so that it delivers diffraction-limited performance (if such is required) greatly simplifies. The choice $F_S \leq 15$ also prevents the simulator optical train length becoming unnecessarily long. Thus, a sensible (but certainly not mandatory) rule of thumb is that Scale$D$ and Scale$M$ should be chosen so that $6 \leq F_S \leq 15$.

Once suitable values for $D_S$, $L_S$, $L'_S$, $f_S$ and $F_S$ have been obtained from the above scaling equations, the simulator imaging optics are then fully specified, at least to first-order accuracy. If necessary, optical ray tracing methods can be used to create a more precise optical layout specification.

#### 15.2.1.6 Simulating Polychromatic Images

Assuming that the imaging lens is chromatically corrected, inspection of the scaling equations outlined thus far in Sect. 15.2.1 indicates that none of the parameters that specify the optical simulator layout ($D_S$, $L_S$, $L'_S$, $f_S$ and $F_S$) depend on the imaging wavelength, $\lambda$. For this straightforward case, an optical layout that adequately simulates monochromatic images can equally be used to simulate polychromatic images.

The only additional requirement here is that the spectral content of the illumination provided in the simulator must match that of the illumination provided in actual target space.

It might be observed that the angular size of the point-spread function of the imaging optic naturally scales with wavelength in the same way as it does with an actual telescope in the field. For example, if the intensity PSF happened to be an Airy pattern, its angular size would scale linearly with wavelength due to the $1/\lambda$ factor in the argument of the Bessel function describing the Airy pattern (cf., 4.27). A second observation is that any radial structure artifacts present in actual polychromatic telescope images (Fig. 4.9) will automatically appear in the simulated images.

### *15.2.2 Scaling When the Images Are Simulated at a Different Wavelength*

Sometimes it might be more convenient to create the simulated images at a wavelength other than the one at which the actual images were obtained in the field. For example, images obtained in the field at an IR wavelength might be more conveniently simulated in the laboratory at a visible wavelength. For such an image simulation, an IR imaging detector would no longer be needed; instead, much cheaper detector alternatives could be used, such as CCD imaging arrays. One could even dispense with detectors altogether and simply view the images directly by eye.

Thus far, the scale factor, Scale $\lambda$, has been set to unity. In this section, Scale $\lambda$ is now allowed to take any other convenient value. However, there are two significant consequences: (1) To maintain the same rms phase fluctuation of the image-forming waves in the simulator at the (now different) imaging wavelength requires a corresponding change in the rms OPD fluctuation, $\sigma_S$, introduced by the random phase screen, and (2) 15.21, which gives the object distance in the simulator, $L_S$, in terms of the actual object distance in the field, $L$, now has to be appropriately modified.

Suppose we wish to simulate telescope images obtained in the field at wavelength, $\lambda$, at an entirely different wavelength, $\lambda_S$, where $\lambda_S/\lambda = Scale\lambda$. Plainly, any image cores/halos present in both the simulated and actual images must contain the same light energy proportions. Thus, recalling 10.3, we may write

$$\exp\left[-\left(\frac{2\cdot\pi\cdot\sigma_S}{\lambda_S}\right)^2\right] = \exp\left[-\left(\frac{2\cdot\pi\cdot\sigma}{\lambda}\right)^2\right] \qquad (15.26)$$

from which it follows that the rms phase fluctuations introduced by the phase screen and the atmosphere, $\sigma_S$ and $\sigma$, are related by

$$\frac{\sigma_S}{\lambda_S} = \frac{\sigma}{\lambda}. \qquad (15.27)$$

It then follows that

$$\sigma_S = Scale\lambda \cdot \sigma \tag{15.28}$$

The modified version of 15.21 needed to take account of $Scale\lambda \neq 1$ can now be written in the more general form,

$$L_S = \frac{ScaleD \cdot ScaleM}{Scale\lambda} \cdot L \tag{15.29}$$

As required in the degenerate case where $Scale\lambda = 1$, the above equation reduces to the simpler form given by 15.21. Once $L_S$ has been calculated from the above equation, 15.5, 15.9, 15.22, 15.24, and 15.25 can be used as previously to calculate $D_S$, $d_S$, $L'_S$, $f_S$ and $F_S$, thus fully specifying the simulator optics' parameter values.

#### 15.2.2.1 Core and Halo Parameters for Actual and Simulated Images

Wavelength scaling leads to slightly modified forms of 15.1 and 15.12. These may be written

$$B_{CS} = \frac{Scale\lambda}{ScaleD} \cdot B_C \tag{15.30}$$

and

$$B_{HS} = \frac{Scale\lambda}{ScaleD} \cdot B_H \tag{15.31}$$

From these two equations, it follows that

$$\frac{B_{CS}}{B_{HS}} = \frac{B_C}{B_H} \tag{15.32}$$

This relationship simply states that the ratio of the angular widths of the core and halo in simulated images at wavelength $\lambda_S$ must be the same as the ratio of the corresponding quantities in actual images obtained in the field at wavelength $\lambda$.

Correct scaling further requires that the ratio of the central intensities in the core and halo portions of simulated images must bear the same ratio as the corresponding intensities in the images obtained in the field. However, as the reader may readily confirm by examination of the more general version of 15.15 that applies where $\lambda \neq \lambda_S$, that result has already been secured by the scaling procedures already in place. Thus, for the general case, $\lambda \neq \lambda_S$, we can formally write,

$$A_{CHS} = A_{CH} \tag{15.33}$$

#### 15.2.2.2 Modifying the rms OPD Fluctuation Introduced by the Phase Screen

Usually, the wavelength chosen for the image simulation, $\lambda_S$, would be shorter than the wavelength used to produce the actual images in the field, $\lambda$. Thus, usually Scale $\lambda < 1$, and hence, $\sigma_S < \sigma$. In such cases, commercially available[3] refractive index matching liquids can be used (as shown in Fig. 15.2) to adjust the rms OPD fluctuation introduced by the phase screen to the desired value, $\sigma_S$.

Denoting the refractive index of the phase screen substrate by $n$ and the rms surface height variation of the scattering surface by $\sigma_{\mathrm{H}}$, it may readily be seen that

$$\sigma_H = \frac{\sigma}{n-1}. \tag{15.34}$$

Suppose now that a containment window is used to sandwich refractive index liquid against the (rough) scattering surface of the phase screen, as shown in Fig. 15.2 (Right). To achieve the desired rms fluctuation, $\sigma_S$, the refractive index of the liquid, $n_L$, must satisfy the conditions (cf., 15.27),

$$\sigma_H \cdot (n - n_L) = \sigma_S = \sigma \cdot \frac{\lambda_S}{\lambda} \tag{15.35}$$

By using 15.34 to substitute for $\sigma_H$, $n_L$ may be expressed in the form,

$$n_L = n - \frac{\lambda_S}{\lambda} \cdot (n-1) = n - Scale\lambda \cdot (n-1) \tag{15.36}$$

In the degenerate case where $Scale\lambda = 1$, (i.e., $\lambda_S = \lambda$), the above relation gives $n_L = 1$, confirming what was already known that no index matching liquid

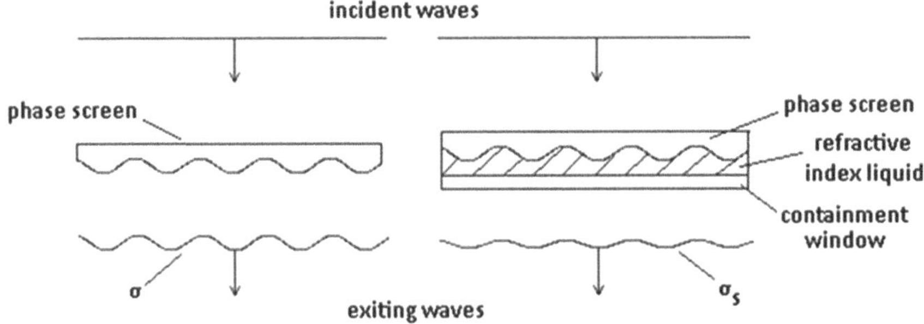


**Fig. 15.2** Index matching arrangement for reducing (by a prescribed factor) the rms OPD fluctuation introduced by the phase screen

[3] Suitable refractive index matching liquids can be obtained from Cargille Laboratories Inc., 55 Commerce Rd., Cedar Grove, NJ 07,009, USA.

is necessary for this case. For the less likely occurrences where $Scale\lambda > 1$, a refractive index liquid would be required with a refractive index greater than that of the phase screen substrate. If no such liquid were available, an entirely new phase screen would have to be fabricated to a prescription calling for a larger rms surface roughness value, $\sigma_H$. Assuming that the same substrate material is used, we would calculate $\sigma_H$ from the relation,

$$\sigma_H = \frac{\sigma_S}{n-1} = \frac{\sigma \cdot Scale\lambda}{n-1} \tag{15.37}$$

The average lateral size of the surface structure on the phase screen, $w_{0S}$, is not affected by wavelength scaling; nor is it affected by the rms surface height variation of the scattering surface of the phase screen, $\sigma_H$. Thus, $w_{0S}$ is again determined by the diameter scale factor, Scale$D$, according to 15.14.

## 15.3 Extended Incoherent Illumination and Image Simulation

To create faithful image simulations in the laboratory using a scaled target model, the model would be required to have approximately the same color and rms surface roughness as the actual target object. The model would also be required to have similar bidirectional reflectance distribution function (BRDF) scattering properties.[4] In addition, the temporal and spatial coherence properties of the illumination used in the simulator would also have to be appropriately scaled versions of the corresponding illumination properties in the field.

In general, passive illumination can be fully coherent, partially coherent, or incoherent; self-luminous objects may be considered as included in the last category. Strictly speaking, extended incoherent sources of finite angular extent, such as the Sun, produce partially coherent illumination. In many instances, however, especially for distant objects, the coherence patch size of sunlight may be significantly smaller than the dimensions of the telescope intensity PSF in object space, so that for all practical purposes, sunlight behaves as though it were incoherent. In the next section, we consider objects illuminated by partially coherent illumination and incoherent illumination. Discussion of coherent illumination is dealt with separately in Sect. 15.6.

[4] Sometimes, it might be convenient to use a photograph of the target rather than an actual target model. The photograph would have to contain image detail at least as fine as that resolved by the simulator imaging objective. Issues might arise, however, if the BRDF scattering properties of the photographic medium were different from those of the actual target.

## 15.3.1 *Incoherent and Partially Coherent Illumination*

In principle, suitable incoherent or partially coherent illumination for illuminating the target models in the simulator can be produced from any type of light source. Sources ranging from incandescent light bulbs to lasers can be used. Whatever the light source type, certain conditions must be fulfilled relating to how the light is manipulated and directed toward the target object.

Contrary to intuition, when light from an incandescent source is allowed to stream directly towards a target, it can produce highly spatially coherent illumination at the target. To produce incoherent or partially coherent illumination from such a source generally requires diffusing the light toward the target in a prescribed way. Appropriate diffusion characteristics can be created either by using a rough surface to back-scatter the light toward the target or by transmitting the light toward the target through a diffusing screen. In either case, the degree of spatial coherence of the illumination at the target is determined by the solid angle subtended by the illuminated area of the diffuser (as seen from the target) via the van Cittert-Zernike theorem (Born & Wolf, 2003).

To create appropriate incoherent or partially coherent illumination from a laser source requires a similar diffusion process. For monochromatic lasers as opposed to broadband lasers, it might sometimes be necessary to continuously move the diffuser (perhaps cyclically) to reduce the temporal coherence of the illumination. An appropriate motion rate for this purpose would be one that causes $>2\pi$ phase change in the scattered illumination in a time interval significantly shorter than the detector integration time. Techniques for creating incoherent or partially coherent illumination from laser light sources by use of moving diffusers have been described elsewhere (Dainty, 1984).

### 15.3.1.1 The Degree of Coherence of Illumination Depends on the Optical Setup

Before developing the scaling equations for the simulator illumination, it might first be useful to discuss in more detail the meaning of illumination-related terms such as "coherent," "partially coherent," and "incoherent." Light is a wave motion and, as with all waves, light waves interfere with one another. For coherent laser sources, the interference effects can be pronounced. Partially coherent light produced by extended incoherent light sources (such as the Sun) causes less pronounced effects; whether the effects can be seen at all depends critically on the imaging geometry. Partially coherent illumination in some imaging arrangements can produce images that, for all practical purposes, appear as though they were formed under incoherent illumination. In other arrangements, the same partially coherent illumination can give rise to images that resemble fully coherent images. Thus, terms such as coherent illumination and incoherent illumination are relative terms; one must therefore be careful to specify the "coherence properties" of the illumination in terms that take proper account of the imaging geometry.

Sunlight illumination is usually considered incoherent. In certain instances, however, it can behave as though it were partially coherent; in extreme instances, it can even behave as though it were almost perfectly coherent. The appearance of an object illuminated by sunlight depends on the imaging geometry and, in particular, on the spatial coherence of the illumination at the object. One measure of the spatial coherence patch size for direct sunlight can be obtained from the van Cittert-Zernike theorem in the form, $1.22 \cdot \lambda \cdot F_{I\,Sun}$, where $F_{I\,Sun}$ is the $F$/number of the illumination cone subtended by the Sun's (circular) disk at the target and uniform radiant emission strength is assumed over the disk.

On planet Earth, where the Sun's disk subtends about half of one degree, the illumination $F$/number of sunlight, is approximately $F_{I\,Sun} \approx 115$. The van Cittert-Zernike theorem indicates that the coherence patch size of sunlight at the visible wavelength, 0.55 μm, corresponding to this $F$/number measures about 0.05 mm across. For an adult looking down at the ground (~1.75 m below eye height), the just-resolved patch size (according to Rayleigh's criterion) is given by $1.22 \cdot \lambda \cdot 1.75\,\text{mm}/0.3\,\text{mm} \approx 0.4\,\text{mm}$. (In bright sunlight, the iris closes down to about 3 mm.) Looking down at the ground from heights greater than 1.75-m, the just-resolved patch size grows in proportion to the height. In either case, because the coherence patch size of sunlight illumination is significantly smaller than the just-resolved patch size of the eye at the target, under normal circumstances, speckle and other coherent light artifacts arising from sunlight are not usually observed. Sunlight illumination in these instances behaves as though it were incoherent.

But now consider a rough surface illuminated by sunlight when the surface is brought to within 25 cm of the eye. The patch size of the eye at this distance reduces to about 0.05 mm—a distance now comparable to the size of the sunlight coherence patch. By wearing strong reading glasses, the object could be brought even closer to the eye, say to within 10 cm, and still remain in good focus. Imaging geometries like this are illustrated in Fig. 15.3. At 10-cm viewing distance, the just-resolved patch size shrinks to only about 0.02 mm—a distance now significantly smaller than the 0.05-mm coherence patch of the illumination. For close-up imaging geometries like this, sunlight now behaves as though it were highly spatially coherent.

Figure 15.4 shows the typical polychromatic speckle pattern appearance of a rough surface illuminated by direct sunlight and viewed directly by eye in such a close-up imaging geometry. The high level of speckle contrast testifies to the high degree of coherence of the illumination.[5] If we now viewed this speckle pattern through a narrowband color filter (which would have the effect of increasing the temporal coherence of the illumination), the speckle pattern would increasingly begin to resemble a fully coherent laser speckle pattern.

If human beings ever set foot on the dwarf planet Pluto, they will be intrigued to discover that sunlight on this planet is more than an order of magnitude more spatially coherent than the sunlight that we see on Earth. On Pluto, the Sun subtends an angle 40 times less than it does on Earth. Consequently, compared to the 0.05-mm

[5] A description of how sunlight speckle patterns can best be observed has been given by Pedersen (2004).

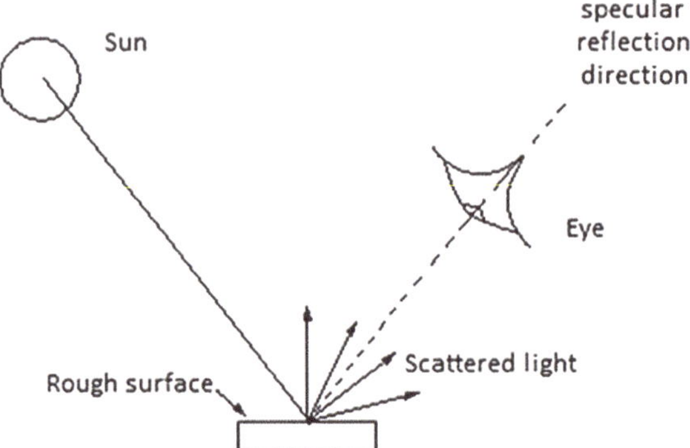


**Fig. 15.3** To maximize contrast in sunlight speckle—that is, polychromatic speckle seen under direct sunlight—the following setup conditions should be obeyed: (1) The surface should be rough, but not too rough. (2) The eye or eye/eyeglass combination should be as free of aberrations as far as possible. (3) The eye should be placed as close to the target as possible, consistent with retaining sharp focus. (4) The observer should view the rough surface in the specular direction with respect to the Sun (i.e., the direction in which the observer would see the Sun if the diffuser surface were replaced by a plane mirror)

**Fig. 15.4** Sunlight speckle obtained from an imaging geometry similar to that depicted in Fig. 15.3 where a color camera is used in place of the eye. By using a short focal length camera lens, the size of the intensity PSF of the lens at the rough surface can be made smaller than the ~0.05-mm spatial coherence patch of sunlight. Sunlight then behaves as though it were highly spatially coherent, resulting in high speckle contrast

coherence patch size on the Earth, the coherence patch size of sunlight on Pluto is about 2 mm. A person standing on Pluto looking down at the ground would routinely see speckle patterns similar to those shown in Fig. 15.4.

Curiously, the speckle would be limited to a circular patch measuring only a few degrees across, centered on the specular reflection direction with respect to the Sun. By stooping down closer to the ground, the angle subtended by the speckle patch would grow larger and the speckle contrast within the patch would grow stronger. If the visitor happened to notice a small Pluto insect, the closer he approached the creature, the more it would become immersed in speckle. While this might be somewhat annoying, by simply moving his/her head to one side or the other (thereby shifting the line of sight to the creature away from the specular direction with respect to the Sun), the visitor would see the speckle patch shift away from the creature, allowing it to be seen in relatively speckle-free form.

For its part, the tiny creature would have a far worse experience than the visitor. If its visual system were comparable to that of the human visual system—that is, a visual system comprising an imaging lens as opposed to the non-imaging compound eye system of the type usually associated with insects on Earth (consisting of thousands of individual photodetectors laid out on a convex surface)—the creature would find itself hideously plagued by speckle. If its eye level were only a few millimeters above the ground, the angular extent of the speckle patch in the specular direction would extend over tens of degrees and perhaps even cover the creature's entire field of view. (In fact, the situation would barely be any better for a comparable creature back on Earth.)[6] Nonetheless, the creature would not necessarily be doomed to a life perpetually confused by speckle. The mere act of moving, running or flying, would have the effect of reducing the speckle. To gain a substantial reduction, the time taken for the speckle to decorrelate (which is inversely related to the insect's speed of travel) would only have to be shorter than the creature's eye integration time.[7] Given a fast enough travel speed, or given a long enough eye integration time, the speckle could be made to vanish.

#### 15.3.1.2 Scaling Equations for the Illumination Optics

To produce either incoherent or partially coherent illumination in the object space of the simulator equivalent to that found in the object space of the actual telescope in the field, it is only necessary to preserve the ratio of the just-resolved image patch size to that of the coherence patch size in the actual and simulated object spaces.

For a diffraction-limited telescope, this relationship can be expressed in the form,

$$\frac{1.22 \cdot \frac{\lambda}{D} \cdot L}{1.22 \cdot \lambda \cdot F_I} = \frac{1.22 \cdot \frac{\lambda_S}{D_S} \cdot L_S}{1.22 \cdot \lambda_S \cdot F_{IS}} \tag{15.38}$$

[6] One could speculate that the problem of speckle in rough objects seen at close ranges might partly explain why insects and other small creatures evolved non-imaging compound eyes rather than direct imaging eye systems similar to those found in larger creatures such as humans.

[7] The same strategy could equally be adopted by the human visitor to Pluto. The happenings surmised here are based on an analysis (McKechnie 1976) of the statistical properties of speckle formed under illumination with arbitrary state of coherence (spatial and temporal). For poly-chromatic speckle patterns of this type, speckle contrast is found to decay rapidly as viewing angles depart from the specular direction. The decay rate can be quantified by using what might be considered a generalized form of the van Cittert-Zernike theorem given in the paper—one that takes account of extended polychromatic incoherent sources as opposed to extended quasi-monochromatic incoherent sources.

where $F_I$ and $F_{IS}$ are the $F$/numbers of the illumination cones in actual and simulated object spaces (Fig. 15.1). It may be noted that this expression is written in a way that permits wavelength scaling.

By combining the previously established scaling equations (15.2 and 15.29) with 15.38, it can readily be shown that

$$F_{IS} = \frac{F_I \cdot ScaleM}{Scale\lambda} \tag{15.39}$$

The quantities $L/D$ and $L_S/D_S$ may be considered as the $F$/numbers of the light collection cones that apply, respectively, to actual and simulated object spaces. Thus, we write

$$\frac{L}{D} = F_C \tag{15.40}$$

$$\frac{L_S}{D_S} = F_{CS} \tag{15.41}$$

The above expressions permit the 15.39 scaling requirement to be expressed in the alternative form (cf., Fig. 15.5),

$$\frac{F_{CS}}{F_{IS}} = \frac{F_C}{F_I} \tag{15.42}$$

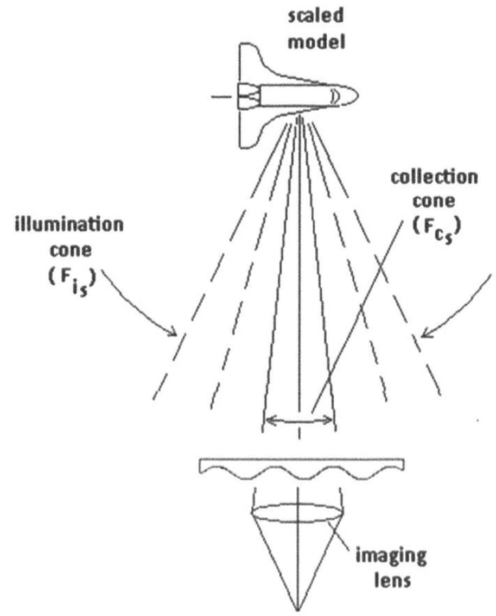


**Fig. 15.5** The desired spatial coherence characteristics of the target illumination in the optical simulator can be controlled by appropriately setting the $F$/number of the illumination cone. The scaling equations given in the text allow calculation of the $F$/numbers, $F_{IS}$ and $F_{CS}$, from the corresponding object space $F$/numbers, $F_I$ and $F_C$

Non-circular Extended Incoherent Sources

Whereas the scaling equation just given relates specifically to extended incoherent sources with circular apertures, e.g., the Sun, it is clear that extended sources of any other shape (square, rectangular, or other irregular shapes) scale just as readily by simply continuing to use the scaling multiplier given by 15.39, namely Scale$M$/Scale$\lambda$.

## 15.4 Space Shuttle Image Simulations

As mentioned at the beginning of the chapter, during the late 1980s and early 1990s, NASA was plagued by the problem of thermal insulation tiles detaching during the stressful, high-vibration Space Shuttle launch phase. In attempting to identify missing tiles, NASA arranged for ground-based telescopes to be trained on these craft as they orbited overhead using natural sunlight as the illumination. Atmospheric blurring, however, made it impossible to see missing tiles at visible wavelengths. In the 1.5-arcsec visible seeing conditions assumed for the simulations, the smallest just-resolved patch size on the Space Shuttle orbiting 160 km overhead measures about 1.20 m across.[8] (At viewing angles far from the zenith, the just-resolved patch size would be even larger.) Space Shuttle tiles typically measure about 15–20 cm across. Thus, at visible wavelengths (without the assistance of AO which, in 1989, was in its infancy and shrouded in military secrecy anyway), there was little prospect of seeing missing tiles.

In a November 1989 presentation at NASA headquarters in Washington, DC, the Author proposed that by imaging at the "optimum wavelength" significantly higher resolution would likely be obtained that might enable missing tiles to be identified. For a 4-m-diameter, diffraction-limited telescope observing in about 1.5-arcsec visible seeing conditions, this wavelength was surmised to be about 2.5 $\mu$m. For a 160 km target distance, the 0.15-arcsec angular size of the image cores anticipated at this wavelength translates to a just-resolved distance element at the spacecraft of about 13 cm[9]—small enough to offer reasonable prospect of identifying missing tiles. The image simulations in Fig. 15.6 show how the Space Shuttle, Challenger, might have appeared when viewed by a 4-m instrument as it passed overhead at an altitude of 160 km.[10]

The top image shows how the spacecraft would have appeared at the visible wavelength, 0.55 $\mu$m, in the absence of atmosphere, the phase screen having been removed from the imaging setup to obtain this image. The telescope intensity PSF in object space, which entirely determines the resolution obtained in this image,

[8] Space Shuttle orbital height varied from mission to mission. Average height was about 350 km.

[9] The telescope is assumed to be nearly diffraction-limited at 2.5 $\mu$m. Reasonable contrast difference is also assumed between tiled and untiled areas.

[10] The images shown in Fig. 15.6 were published previously (McKechnie 1990), but due to the disappointingly poor grayscale reproduction process, all three images appeared equally degraded. The fact that the near-IR wavelength actually did provide much better image resolution (as clearly shown in Fig. 15.6) was entirely lost.

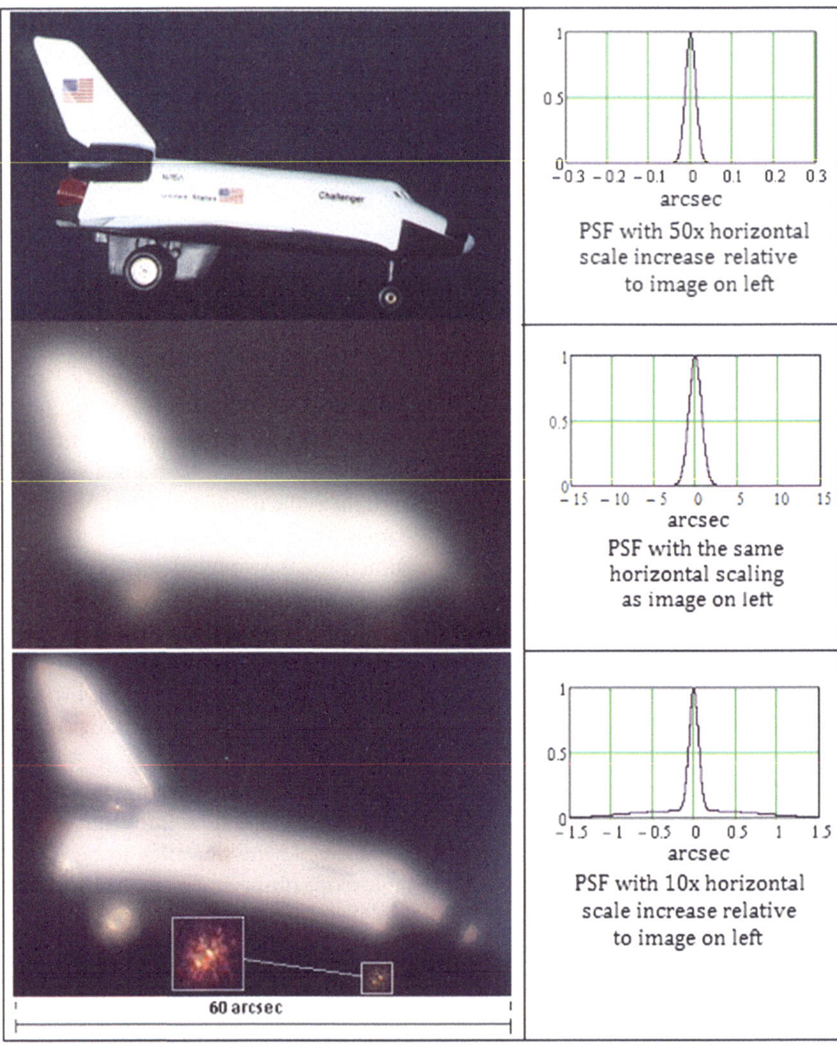


**Fig. 15.6** Laboratory image simulations of the Space Shuttle at 160-km altitude as seen by a 4-m diffraction-limited telescope. *Top* image obtained at 0.55 μm in the absence of atmosphere. *Center* image obtained at 0.55 μm as seen through the atmosphere in 1.5-arcsec visible seeing conditions. *Bottom* image obtained at 2.5 μm as seen through the same atmosphere. Core and halo structures can be seen in the twin wheel glints, the enlargement showing this structure more clearly. The applicable average intensity PSFs of the telescope/atmosphere combination are shown to the right of each image; but note the three significantly different angular scales used to display these PSFs

measures a mere 3 cm across; this PSF is shown (at right) in the figure. It is clear that missing tiles would have been readily identifiable in such a sharply resolved image.

The center image shows how the spacecraft might have appeared at the same visible wavelength, 0.55 μm, when viewed through the actual atmosphere in 1.5-arcsec visible seeing conditions. The geometry used to create this image was nominally the same as that used to obtain the top image, the only difference being the addition of a prescription random phase screen to represent the atmosphere. The phase screen was placed just in front of the imaging lens, in the near field with respect to this lens. The average intensity PSF of the telescope/atmosphere combination used to create the image is again shown to the right in the figure. As can be seen, almost all of the light in this PSF is contained in a 1.5-arcsec-wide halo. The severe blurring seen in this image makes it all but impossible to identify missing tiles.

The bottom image shows how the craft might have appeared when viewed through the same atmosphere at the near-IR wavelength 2.5 μm. Wavelength scaling was used to create this image. Thus, although the image was actually simulated at the visible wavelength, 0.55 μm, the image properties exactly replicate those of an actual Space Shuttle image obtained in the field at 2.5 μm. The core in the telescope/ atmosphere intensity PSF (shown to the right of the image) produces a sharply resolved underlying image, while the halo portion of the intensity PSF produces a background veiling glare that reduces overall image contrast. Nonetheless, it is clear that there is enough residual contrast in the image to allow missing tiles to be identified.

To carry out wavelength scaling in this instance, the rms OPD fluctuation introduced by the phase screen, 0.4 μm, had to be reduced in proportion to the wavelength scale factor, $Scale\lambda = 0.22$, to about 0.09 μm. Because the refractive index of the phase screen substrate was about 1.62, to effect this change required the index matching liquid to have the refractive index value, 1.484 (cf. 15.36).

### 15.4.1 Parameter Values Used for Simulating the Space Shuttle Images

To produce the Space Shuttle images shown in Fig. 15.6, two separate optical simulator setups were designed and constructed, one for the visible wavelength images and the other for the near-IR images (which in fact were also produced at visible wavelengths using wavelength scaling). Both setups used the same target model, a 1/288 scaled model about 13.5 cm in length; the length of the actual Space Shuttle is about 39 m. The parameter values specifying the telescope imaging arrangement in the field are listed in Table 15.1. The scale factors used for the two simulators—one set used for the 0.55-μm image simulations and the other for the 2.5-μm image simulations—are listed in Table 15.2. The parameter values specifying the two optical layouts are listed in Table 15.3.

**Table 15.1** Parameter values governing the visible and near-IR images obtained by the hypothetical 4-m diffraction-limited telescope in the field

| Telescope diameter | $D = 4$ m |
|---|---|
| Distance to Space Shuttle | $L = 160$ km |
| Space Shuttle length | 39 m |
| Telescope field of view | Wid = 39 m |
| Imaging wavelengths | $\lambda = 0.55$ μm and $\lambda = 2.5$ μm |
| Seeing conditions (visible) | 1.5 arcsec |
| Solar illumination cone | $F_{\mathrm{I}} = 115$ |

**Table 15.2** The various scale factors used for creating the Space Shuttle image simulations

| Model scale factor | Scale$M$ = 1/288 |
|---|---|
| Diameter scale factor | Scale$D$ = 1/1667 |
| Wavelength scale factors | Scale$\lambda$ = 1 and Scale$\lambda$ = 0.22 |

**Table 15.3** Parameter values used for creating the Space Shuttle image simulations shown in Fig. 15.6, consistent with the scale factor values indicated in Table 15.2

| | Top image | Center image | Bottom image |
|---|---|---|---|
| Actual telescope imaging wavelength ($\lambda$) (μm) | 0.55 | 0.55 | 2.5 |
| Wavelength used for image simulation ($\lambda_S$)(μm) | 0.55 | 0.55 | 0.55 |
| Target distance in simulator ($L_S$) (mm) | 335 | 335 | 1525 |
| Image conjugate distance in simulator ($L'_S$) (mm) | 87 | 87 | 395 |
| Space Shuttle model length (Wid$_{\mathrm{S}}$) (mm) | 135 | 135 | 135 |
| Detector width (DetWid$_{\mathrm{S}}$) (mm) | 35 | 35 | 35 |
| Imaging lens diameter ($D_{\mathrm{S}}$) (mm) | 2.4 | 2.4 | 2.4 |
| Imaging lens focal length ($f_{\mathrm{S}}$) (mm) | 69 | 69 | 314 |
| Imaging lens $F$/number ($F_{\mathrm{S}}$) | 29 | 29 | 131 |
| Illumination cone $F$/number ($F_{\mathrm{IS}}$) | 0.4 | 0.4[a] | 1.8 |

[a] See next section for more readily attainable illumination beam $F$/numbers still consistent with producing illumination at the target model with the required coherence properties

## 15.5 Practical Aspects of Illumination Used in Optical Simulators

From Table 15.3, it is evident that for the $\lambda = 0.55\mu m$ image simulations, the $F$/number of the illumination cone, $F_{IS} = 0.4$, is uncomfortably small; such an $F$/number would be difficult, though not impossible, to realize in practice. However, one can also envisage even more extreme imaging arrangements where the scale factor choices might require even smaller and more challenging F/number illumination cones (i.e., $F_{IS} \ll 1$).

In this section, we address the practical issues raised when the problem of unrealistically small illumination cone $F$/numbers arises.

From 15.39 and 15.42, it may be shown that the collection $F$/number of the imaging optic, $F_{\mathrm{CS}}$, and the required $F$/number of the illumination cone, $F_{\mathrm{IS}}$, are determined by the relations,

$$F_{CS} = F_C \cdot \left(\frac{ScaleM}{Scale\lambda}\right) \tag{15.43}$$

$$F_{IS} = F_I \cdot \frac{F_{CS}}{F_C} = F_I \cdot \left(\frac{ScaleM}{Scale\lambda}\right) \tag{15.44}$$

The main driver in these relations is the model scale factor, Scale$M$. (The wavelength scale factor, Scale$\lambda$, generally takes values of the order unity and therefore has a smaller relative effect.) As is readily verified, small Scale$M$ values lead to proportionately small values for both $F_{\mathrm{CS}}$ and $F_{\mathrm{IS}}$. For some choices of Scale$M$, the $F$/numbers, $F_{\mathrm{CS}}$ and $F_{\mathrm{IS}}$, might end up below practically achievable values.

Generally, because $F_{IS} < F_{CS}$, $F_{IS}$ is the more critical of the two $F$/numbers. For the most common type of passive illumination—sunlight on planet Earth—where $F_I \approx 115$, model scale factors less than 0.008 require simulator illumination beam cones in the range, $F_{CS} < 1$. The practical difficulty here can of course be avoided by simply choosing a model scale factor >0.008. However, if were to consider the even more stressful case where the actual target object is located somewhere in the atmosphere just below cloud level, the effective $F$/number of the diffused, rather than direct, sunlight illumination falling on the object will approximate the condition, $F_I \approx 1$. When we combine such an $F$/number with a model scale factor, $ScaleM \ll 1$, the $F$/number of the simulator illumination, $F_{\mathrm{IS}}$, becomes (cf., 15.44) prohibitively small. However, there is another way of looking at the problem that leads to more practicable $F$/numbers. To produce substantially incoherent illumination in the optical simulator merely requires the $F$/number of the illuminator beam, $F_{\mathrm{IS}}$, to be significantly smaller than the $F$/number of the simulator imaging optic, $F_{\mathrm{CS}}$. This alternative way of viewing the problem leads to a much more relaxed $F_{\mathrm{IS}}$ requirement than prescribed by 15.44.

If the spatial coherence patch size of the illumination falling on the target model is smaller than the just-resolved (intensity PSF) patch size of the imaging lens in the plane of the target model, the speckle pattern in the image will arise as a reduced speckle pattern. If the spatial coherence patch size is made even smaller, speckle contrast is ultimately driven below the contrast sensitivity of the detector, effectively making all speckle noise artifacts disappear. At that point, the illumination of the target model may be considered incoherent to the extent necessary to deliver satisfactory image simulations; further effort directed at making the illumination even more "incoherent" would then be pointless.

As previously indicated in Chap. 11, the speckle contrast in a reduced speckle pattern can be quantified by the ratio, $\sigma_I/\langle I\rangle$, where $\sigma_I$ is the rms fluctuation of the intensity in the speckle pattern and $\langle I\rangle$ is the mean intensity in the pattern. If the aperture of the simulator imaging optic and the illumination beam cone are both circular, it may readily be shown that the ratio $\sigma_I/\langle I\rangle$ is approximately given by the ratio of the coherence patch size to that of the just-resolved patch size,

$$\frac{\sigma_I}{\langle I\rangle} \approx \frac{1.22 \cdot \lambda_S \cdot F_{IS}}{1.22 \cdot \lambda_S \cdot F_{CS}} = \frac{F_{IS}}{F_{CS}} \tag{15.45}$$

In instances where the "detector" is simply the human eye, there would be little awareness of speckle noise once the intensity contrast, $\sigma_I/\langle I\rangle$, falls below about 0.05—the just-detectable contrast threshold of the eye. Therefore, for such applications, rather than attempting to set up the (impossibly small) illumination cone $F$/numbers that might be prescribed by 15.44, a sufficient level of incoherence can be established by simply meeting the requirement, $\sigma_I/\langle I\rangle \leq 0.05$. Combining this result with 15.45 leads to the much less stringent requirement for the $F$/number of the illumination beam cone, $F_{\text{IS}}$,

$$F_{IS} \leq 0.05 \cdot F_{CS} \tag{15.46}$$

For detectors other than the eye, different contrast sensitivity thresholds would of course apply. The constant, 0.05, used in the above inequality would then have to be appropriately modified.

In light of this new result, let us now revisit the $F$/number requirements (Table 15.3) for the illumination cones used in the two Space Shuttle simulators, where we recall that these were previously established as $F_{IS} = 0.4$ and $F_{IS} = 1.8$. If the detector was simply the human eye, it is apparent that almost identical images would be obtained by using the much larger and more practicable illumination cone $F$/numbers arising from 15.46, that is, $F_{IS} = 7$ and $F_{IS} = 32$.

## 15.6 Simulating Images of Actively Illuminated Targets

When target objects such as satellites are actively illuminated by laser beams, the illumination at the target can have a high degree of coherence. For targets with rough surfaces, telescope images formed under such illumination generally display additional speckle artifacts,[11] over and above those caused by atmospheric scattering. In this section, we describe scaling procedures that allow simulated images to be

[11] Smooth surfaces behave differently from rough surfaces; they reflect light specularly so that the target manifests itself only in the form of isolated glints and caustics arising from isolated locations on the surface that just happen to lie at right angles to the line of sight. It is not possible to form sensible images of such objects. As the target object moves and tumbles, the locations of the glints and caustics move in concert, sometimes continuously, sometimes discontinuously.

produced in the laboratory with speckle properties identical to those found in actual actively illuminated images obtained by telescopes in the field.

Speckle generally arises in its noisiest form when fully coherent illumination is used. Speckle noise degrades image quality and obscures image details that might otherwise have been distinguishable. However, a number of speckle reduction mechanisms can be used to quieten speckle noise. All of these mechanisms may be recreated in laboratory image simulators by implementing appropriate scaling procedures.

In addition to speckle noise, photon noise may further degrade images of faint objects obtained by large telescopes. Because the subject of photon noise is considered outside the scope of the book, other than noting that photon noise effects can readily be simulated in the laboratory using intensity filters to create light flux levels comparable to those actually attained in the field, no further discussion will be given to this particular noise mechanism.

### *15.6.1 Illumination and Imaging of Actively Illuminated Targets*

For simplicity, we limit the analysis here to telescopes with circular apertures and assume that the wavelength used to simulate the images is the same as that used in the field. Thus, $\lambda_S = \lambda$ (i.e., Scale $\lambda = 1$). We also assume monostatic illumination/imaging arrangements—that is, arrangements for which the axis of the laser illuminator beam and the telescope-viewing axis are coincident. We also loosely include in this monostatic category arrangements where the two axes are approximately coincident, such as when the laser beam is projected from a location lying only just outside the footprint of the telescope aperture, as shown schematically in Fig. 15.7.

As can be seen from that figure, the illumination portion back-scattered from a target (with a randomly rough surface) forms a random speckle pattern back in the plane of the telescope. (It is assumed that the telescope lies in the far field with respect to the target.) Consequently, images formed from these scattered waves by the telescope will be immersed in speckle. As the target moves, the speckle pattern in the telescope pupil plane moves in concert. For the arrangement shown in Fig. 15.7, where the target is moving at right angles to the line of sight, the speckle pattern moves at twice the speed of the target.[12] For coherently illuminated targets with rms surface roughness greater than about $\lambda/2$, the speckle pattern formed in the image is a Gaussian speckle pattern (Chap. 11).

---

[12] The speckle pattern formed back at the telescope moves en bloc as the target object moves; the "location" of the pattern in the plane of the telescope at any given instant is determined by the point of intersection of the specular reflection direction for light back-scattered from the target object in this plane. The doubling in pattern speed is due to the angle of the (specular) reflection being twice the angle of incidence, just as for light reflected from a mirror.

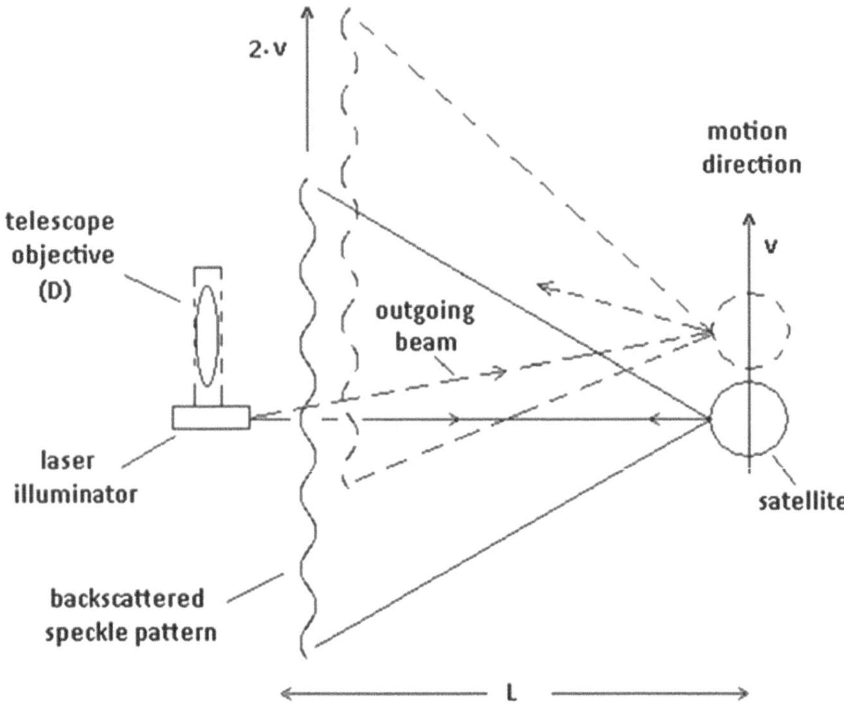


**Fig. 15.7** As the illuminator laser beam and the telescope track the target satellite, the backscattered light from the target forms a speckle pattern back in the neighborhood of the telescope. This pattern moves, en bloc, across the telescope aperture with a velocity twice that of the satellite. When the satellite also spins or tumbles, additional velocity components increase the relative speed of the speckle pattern with respect to the telescope. In time-integrated images, the combined effect of target velocity and target spin can produce large amounts of speckle reduction

For targets that neither spin nor tumble, after the target has moved through a distance, $D/2$ (where D is the telescope diameter), the speckle pattern in the plane of the telescope will have moved through the distance, $D$. At that instant, the speckle pattern portion that originally occupied the telescope aperture will have moved clear of the aperture, replaced now by an entirely new speckle pattern, one that is totally uncorrelated with the original pattern. The speckle in the target image evolves into a new speckle pattern, one that is again uncorrelated with the original pattern. The time taken for the new speckle pattern to develop in this way is given by

$$\Delta T = \frac{D}{2 \cdot v} \tag{15.47}$$

where the target speed $v$ is defined here as the speed component at right angles to the line of sight.

The time interval, $\Delta T$, may be considered as a measure of the relaxation time of the speckle in the telescope image. If the integration time of the detector is much shorter than $\Delta T$, the observed speckle pattern will be approximately Gaussian; if the integration time is much longer than $\Delta T$, a reduced speckle pattern will be seen instead. (It is assumed here that during the integration time, the telescope faithfully tracks the target, thus stabilizing the target image on the detector.) Thus, for actively illuminated moving targets, time integration offers a valuable speckle reduction mechanism. A more detailed examination of how this mechanism works (as well as the workings of various other speckle reduction mechanisms) is given in Sect. 15.6.3.

### *15.6.2 Additional Scaling Requirements When Active Illumination Is Used*

For target objects lying at great distances, the active illumination arriving at the target may be considered collimated. To produce comparable illumination in the simulator, the illuminator beam must first be expanded before ultimately being collimated toward the target model. The required width of the beam at the target model, $BW_S$, scales directly from the beam width at the actual target, BW, as follows:

$$BW_S = BW \cdot ScaleM \tag{15.48}$$

If the spectral emission function of the illuminator laser used in the field is assumed to be Gaussian, and the 1/*e* half-width of the Gaussian is denoted by $\Delta\lambda$, the temporal coherence length of the illumination, Coh*L*, may be approximately expressed by the well-known formula (Mandel & Wolf, 1995),

$$CohL = \frac{\lambda^2}{\Delta\lambda} \tag{15.49}$$

For targets surfaces slanted with respect to the line of sight, illumination beams with relatively short coherence lengths can cause additional amounts of speckle reduction. As we shall see in Sect. 15.6.3.3, to produce the same level of speckle reduction in simulated images as in actual images obtained in the field, the coherence length of the simulator illumination beam must be appropriately scaled relative to Coh*L*.

### *15.6.3 Simulation of Speckle Reduction Mechanisms*

Target motion and laser coherence length provide opportunistic mechanisms for reducing speckle, but several other speckle reduction mechanisms can also be used. In this section, we examine some of the more commonly used mechanisms. We also examine speckle reduction when two or more mechanisms are used simultaneously. To simplify the analysis, we assume that the same type of imaging FPA detector is used in both the optical simulator and the telescope in the field. Thus, we can set the pixel width and height in the image space of the simulator, $a_S$ and $b_S$, to the same values used in the telescope image space. Thus, $a_S = a$ and $b_S = b$. The signal-to-noise ratio, $S/N$, in a speckle pattern is defined here in the same way as in Chap. 11 (11.24) by

$$\frac{S}{N} = \frac{\langle I \rangle}{\sigma_I}. \tag{15.50}$$

For Gaussian speckle patterns, it may be recalled (11.25) that $S/N = 1$. Gaussian speckle patterns are notoriously noisy. When these speckle patterns arise in actively illuminated target images, they make it extremely difficult to identify fine target details. Added to this, when centroid or other types of algorithm are used to estimate target position, Gaussian speckle can introduce significant jitter into the position estimates. Speckle reduction techniques can be used to quieten this jitter.

In general, reduced speckle patterns are produced whenever a number (ideally large) of uncorrelated or partially correlated speckle patterns sum together incoherently. It is assumed here that all of the contributing patterns obey Gaussian statistics. To create a reduced speckle pattern with a suitably high signal-to-noise ratio, $S/N$, the effective number of uncorrelated Gaussian speckle patterns, $m$, that would have to sum together is given by

$$m = \left(\frac{S}{N}\right)^2 \tag{15.51}$$

(It is noted that the quantity, $m$, as used here is not related to its use in Chap. 14 where it denoted stellar magnitude. In the present chapter, $m$ is used as in Chap. 11 where it refers to reduced speckle patterns.)

If some, or all, of the contributing patterns happen to be partially correlated (as opposed to being totally uncorrelated), an even larger number of individual contributing patterns would be needed to deliver the same $S/N$ value. For reduced speckle patterns of the type considered here, the parameter $m$ takes values in the range, $1 \leq m \leq \infty$, and hence, $S/N$ takes values in the range, $1 \leq S/N \leq \infty$. It might also be observed that since $m$ indicates only the effective number of contributing Gaussian speckle patterns, this parameter can take non-integer as well as integer values.

A reasonable satisfactory level of speckle reduction might correspond to $S/N \geq 20$. However, as can be seen from 15.51, to achieve even this modest $S/N$ value requires summing together 400 uncorrelated speckle patterns. In practice, producing large numbers of uncorrelated speckle patterns can necessitate several different speckle reduction mechanisms acting simultaneously.

#### 15.6.3.1 Reduction by Pixel Integration

To faithfully record the intensity fluctuations in a speckle pattern, the detector must be capable of resolving the individual speckles; for an FPA imaging detector, the size of the individual pixels is obliged to be smaller than the average speckle size. (Recall that the size of the average speckle is set by the size of the intensity PSF delivered by an aberration-free version of the observing telescope.) However, because pixels cannot be infinitesimally small compared to the average speckle size, some degree of speckle averaging—pixel averaging—is inherent in images obtained by FPA detectors. Speckle patterns recorded by FPA detectors are therefore partially reduced speckle patterns. Unsurprisingly, the speckle reduction effect is identical to that obtained by aperture averaging, described previously in Chap. 11 (Sect. 11.6.3).

When a Gaussian speckle pattern falls on a rectangular-shaped pixel with dimensions, $a_S$ and $b_S$, the $S/N$ ratio of the observed pixel-averaged speckle pattern is given by (cf., 15.51, 11.62, and 11.65)

$$\frac{S}{N}=\frac{a_S\cdot b_S}{\left\{\int_{-\frac{a_S}{2}}^{\frac{a_S}{2}}\int_{-\frac{a_S}{2}}^{\frac{a_S}{2}}\int_{-\frac{b_S}{2}}^{\frac{b_S}{2}}\int_{-\frac{b_S}{2}}^{\frac{b_S}{2}}PSF_I(x_1,x_2,y_1,y_2)\cdot dx_1\cdot dx_2\cdot dy_1\cdot dy_2\right\}^{\frac{1}{2}}} \tag{15.52}$$

where $PSF_I(x_1, x_2, y_1, y_2)$ is the unit-normalized intensity PSF for a diffraction-limited version of the telescope (cf., Sect. 4.4). If we also assume that the intensity PSF remains invariant over suitably large isoplanatic regions, it may be expressed in terms of the argument differences, in the form $PSF_I(x_1 - x_2, y_1 - y_2)$.

If the telescope has a circular aperture of diameter, $D$, and a circular central obstruction of diameter, $d$, the intensity PSF, $PSF_I(x_1 - x_2, y_1 - y_2)$, can be expressed in the form (cf., 10.12),

$$\begin{aligned}\mathrm{PSF_I}(x_1-x_2,y_1-y_2)&=\frac{4}{1-\left(\frac{d}{D}\right)^2}\\&\times\left\{\left[\frac{J1\left(\frac{\pi\cdot D\cdot\left[(x_1-x_2)^2+(y_1-y_2)^2\right]^{\frac{1}{2}}}{\lambda\cdot f}\right)}{\frac{\pi\cdot D\cdot\left[(x_1-x_2)^2+(y_1-y_2)^2\right]^{\frac{1}{2}}}{\lambda\cdot f}}\right.\right.\\&\left.\left.-\left(\frac{d}{D}\right)^2\cdot\frac{J1\left(\frac{\pi\cdot d\cdot\left[(x_1-x_2)^2+(y_1-y_2)^2\right]^{\frac{1}{2}}}{\lambda\cdot f}\right)}{\frac{\pi\cdot d\cdot\left[(x_1-x_2)^2+(y_1-y_2)^2\right]^{\frac{1}{2}}}{\lambda\cdot f}}\right]^2\right\}\end{aligned} \tag{15.53}$$

For telescopes without central obstructions (i.e., $d = 0$) this expression reduces to

$$\mathrm{PSF_I}(x_1 - x_2, y_1 - y_2) = 4 \cdot \left[ \frac{J1\left(\frac{\pi \cdot D \cdot \left[(x_1 - x_2)^2 + (y_1 - y_2)^2\right]^{\frac{1}{2}}}{\lambda \cdot f}\right)}{\frac{\pi \cdot D \cdot \left[(x_1 - x_2)^2 + (y_1 - y_2)^2\right]^{\frac{1}{2}}}{\lambda \cdot f}} \right]^2 \tag{15.54}$$

Plots of the $S/N$ ratio for pixel-averaged Gaussian speckle (calculated from 15.52 and 15.54) are shown in Fig. 15.8 for the case of square pixels, $a_S = b_S$. The pixel dimension, $a_S$, has been expressed along the horizontal axis in terms of the Airy pattern unit—that is, the distance from the center of the Airy pattern to the first dark ring. In target space, this unit is defined as the distance, $1.22 \cdot \lambda \cdot L/D$; in the optical simulator target space, the unit is defined by the corresponding distance, $1.22 \cdot \lambda_S \cdot L_S/D_S$, where we observe that

$$1.22 \cdot \frac{\lambda_S \cdot L_S}{D_S} = 1.22 \cdot \frac{\lambda \cdot L}{D} \cdot ScaleM \tag{15.55}$$

When pixel width exactly equals one Airy unit, $S/N$ takes the value 1.585, indicating a small amount of speckle reduction. For $a_S > 5$ Airy units, $S/N$ increases approximately linearly with pixel size according to the relation,

$$\frac{S}{N} \approx 1.09 \cdot a_S \tag{15.56}$$

Although pixel averaging clearly offers some level of speckle reduction, a significant disadvantage of this method is that the speckle reduction is obtained at the

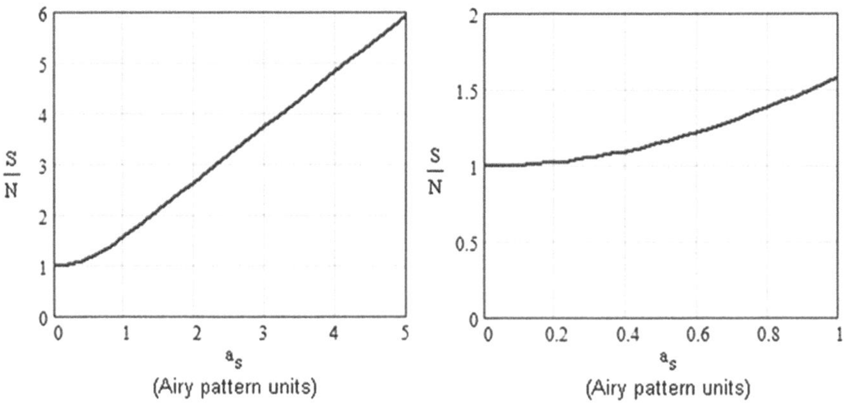


**Fig. 15.8** Signal-to-noise improvement due to pixel averaging of a Gaussian speckle pattern

expense of image resolution.[13] In subsequent sections, other methods of speckle reduction are examined that preserve resolution.

#### 15.6.3.2 Reduction by Exploitation of Target Velocity and Target Rotation

In Sect. 15.6.3.2, we saw that for moving targets, speckle reduction can be obtained without any loss of resolution by simply time averaging. Because target rotation (combined with time averaging) has a speckle reduction effect closely similar to that of target motion, these two speckle reduction mechanisms are quantified together.

It is assumed that the illuminator laser is fully coherent and that the target surface is rough compared to the illumination wavelength, so that instantaneous images delivered by the telescope are immersed in Gaussian speckle. As shown in Fig. 15.9, the target is assumed to move with translational velocity vector, $(\nu_X, \nu_Y, \nu_Z)$, while simultaneously rotating with angular velocity vector, $(\omega_X, \omega_Y, \omega_Z)$. As before, back-scattered illumination from the target forms a speckle pattern in the telescope pupil plane. As the target moves and rotates, this speckle pattern travels, en bloc, across the telescope pupil with (scalar) speed $V$ given by

$$V = 2 \cdot \sqrt{(v_X + L \cdot \omega_Y)^2 + (v_Y + L \cdot \omega_X)^2} \tag{15.57}$$

It may be observed that the above expression does not include terms stemming from either the velocity or the rotation components in the $z$-direction—that is, the

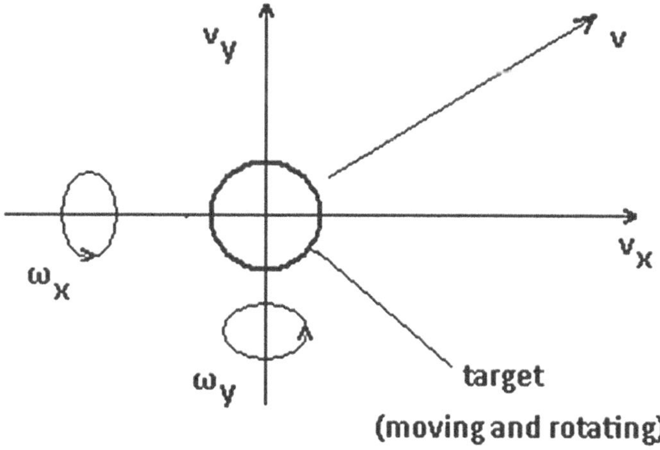


**Fig. 15.9** Schematic showing target moving with velocity, $(\nu_X, \nu_Y, \nu_Z)$, and rotating, or tumbling, with angular velocity, $(\omega_X, \omega_Y, \omega_Z)$. The $z$-axis lies along the line of sight

[13] In practice, rather than increasing the pixel size, the same result could equally be obtained by demagnifying the image on to the pixel array.

direction of the line of sight of the telescope. We ignore the effects of these two terms because any speckle pattern evolution that results from them occurs at rates several orders of magnitude slower than those caused by the other more dominant terms.

The degree of correlation between the instantaneous speckle patterns in the image at any two instants of time is governed by the degree of overlap of the respective instantaneous speckle patterns formed in the telescope pupil by the back-scattered light from the target (cf., Fig. 15.7). Denoting the detector integration time by $\Delta T$, the distance moved during this time interval by the speckle pattern in the telescope pupil, which we denote here by $\Delta\varepsilon$, is given by

$$\Delta\varepsilon = 2 \cdot \Delta T \cdot \sqrt{(v_X + L \cdot \omega_Y)^2 + (v_Y + L \cdot \omega_X)^2} \tag{15.58}$$

For telescopes with circular apertures, the degree of correlation, $\mu(\Delta\varepsilon)$, between the complex amplitudes arising in the image corresponding to two instantaneous speckle patterns in the telescope pupil plane separated by distance, $\Delta\varepsilon$, can be shown (Dainty, 1984) to be given by

$$\mu(\Delta\varepsilon) = \frac{2}{\pi} \cdot \left( \cos^{-1}\left(\frac{\Delta\varepsilon}{D}\right) - \frac{\Delta\varepsilon}{D} \cdot \sqrt{1 - \left(\frac{\Delta\varepsilon}{D}\right)^2} \right) \tag{15.59}$$

The $S/N$ ratio in the time-integrated speckle may then be calculated as follows:

$$\begin{aligned} \frac{S}{N} &= \left\{ \frac{\Delta\varepsilon}{\int_0^{\Delta\varepsilon} \left[ \frac{2}{\pi} \cdot \left( \cos^{-1}\left(\frac{x}{D}\right) - \frac{x}{D} \cdot \sqrt{1 - \left(\frac{x}{D}\right)^2} \right) \right]^2 \cdot dx} \right\}^{\frac{1}{2}} \quad \text{for } \Delta\varepsilon \le D \\ &= \left\{ \frac{\Delta\varepsilon}{\int_0^{D} \left[ \frac{2}{\pi} \cdot \left( a \cos^{-1}\left(\frac{x}{D}\right) - \frac{x}{D} \cdot \sqrt{1 - \left(\frac{x}{D}\right)^2} \right) \right]^2 \cdot dx} \right\}^{\frac{1}{2}} \quad \text{for } \Delta\varepsilon \ge D. \end{aligned} \tag{15.60}$$

In the regime, $\varepsilon \ge D$, the integral term on the denominator of the second of the two expressions always takes the fixed value 0.2724 $D$. In this case, $S/N$ is given by

$$\frac{S}{N} = 1.916 \cdot \sqrt{\frac{\Delta\varepsilon}{D}} \tag{15.61}$$

If the distance travelled by the speckle pattern in the telescope pupil during the integration time is significantly larger than the pupil diameter, $D$, significant speckle reduction occurs, as indicated in Fig. 15.10.

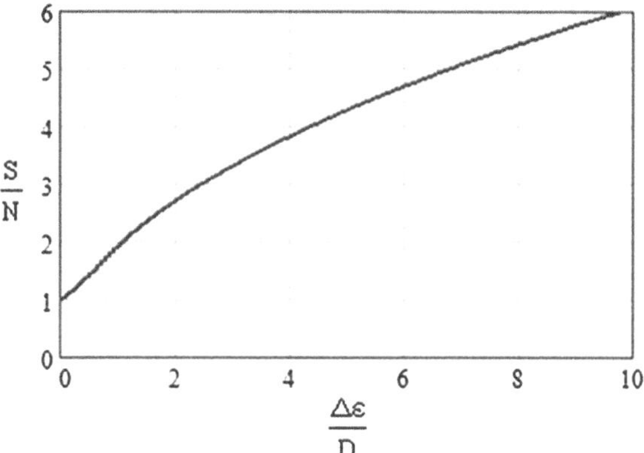


**Fig. 15.10** For moving and/or rotating targets, signal-to-noise ratio in time-averaged speckle images improves as the ratio, $\Delta\varepsilon/D$, where $\Delta\varepsilon$ is the distance moved by the back-scattered speckle pattern in the telescope pupil during the integration period, $D$ being the telescope diameter

By inserting typical parameter values into 15.61, it is clear that this mechanism can deliver extremely large amounts of speckle reduction. For example, consider a satellite passing overhead at altitude, $L = 160$ km, where orbital speed is about 8 km/s. Thus, we can set the velocity vector to (8 km/s, 0, 0). If the satellite is observed from the ground with a 4-m-diameter telescope using the camera exposure time, $\Delta T = \frac{1}{60}$ s, 15.58 and 15.61 yield $S/N \approx 15$. If the satellite also rotates at one revolution per second, with angular velocity vector, $(0, 2\pi \text{ rad/s}, 0)$, 15.58 and 15.61 yield $S/N \approx 175$.

In the above example, the satellite moves about 135 m during the (1/60th second) camera integration time. By ensuring that the telescope precisely tracks the target, this motion is effectively nulled, thus avoiding catastrophic image blurring. However, blurring caused by target rotation cannot be canceled so easily. The degree of blurring varies over the target depending on the distance between the target locality of interest and the axis of rotation of the target. For the case of a spherically shaped satellite 1 m in diameter, the blur patch size caused by rotation during the $1/60$ s integration time is about 5 cm in regions far (~0.5 m) from the rotation axis and near zero in regions close to the rotation axis. Differential blurring like this can of course be reduced by using shorter camera integration times, but only at the cost of reduced $S/N$ ratio.

### Scaling Target Velocity and Angular Rotation Rates to Replicate Reduced Speckle

In this section, scaling equations are set out for converting both the translational velocity and the angular velocity of an actual target in the field to the corresponding translational velocity and angular velocity of the target model in the simulator so that identical speckle decorrelation rates are achieved. By scaling according to these equations and by using the same detector integration time in the simulator as that used in the field, the $S/N$ ratio of the reduced speckle pattern obtained in the image simulator can be made identical to that of the reduced speckle pattern in the field. We denote the translational velocity and angular velocity vectors of the target in the field by $(\nu_X, \nu_Y, \nu_Z)$ and $(\omega_X, \omega_Y, \omega_Z)$ and the corresponding vectors for the target model in the simulator by $(\nu_{XS}, \nu_{YS}, \nu_{ZS})$ and $(\omega_{XS}, \omega_{YS}, \omega_{ZS})$.

To produce the same rate of speckle decorrelation in the simulated image as in the field for a target moving at translational velocity, $(v_X, v_Y)$, where we again ignore the small effect due to $v_Z$, it may be shown (by setting the rotation vector to zero in 15.58 and combining the result with 15.59) that the following relationship holds between the translational velocity vectors in actual target space and simulated target space,

$$\frac{(v_{XS}, v_{YS})}{D_S} = \frac{(v_X, v_Y)}{D} \tag{15.62}$$

Thus, recalling 15.2, the required translational velocity scaling equation may be written in the form,

$$(v_{XS}, v_{YS}) = ScaleD \cdot (v_X, v_Y) \tag{15.63}$$

Because Scale$D$ is usually $\ll 1$, according to 15.63, the required translational velocity of the target model in the simulator is usually much less than the translational velocity of the actual target object in the field.

The speckle decorrelation rate caused by rotation of the actual target in the field may be replicated in the simulator by arranging for a corresponding rotation of the target model in the simulator. The required angular velocity components for the target model, $v_{XS}$and$v_{YS}$, can readily be deduced from 15.58 and its counterpart equation for the optical simulator as follows:

$$(\omega_{XS}, \omega_{YS}) = \frac{(\omega_X, \omega_Y) \cdot D \cdot L_S}{D_S \cdot L} \tag{15.64}$$

Using 15.2 and 15.21, the above scaling equation may be written more simply,

$$(\omega_{XS}, \omega_{YS}) = \frac{(\omega_X, \omega_Y)}{ScaleM} \tag{15.65}$$

Because Scale$M$ appears in the denominator in the above equation and because this scale factor usually takes values $\ll 1$, the required angular velocity for the target model in the simulator is usually much greater than the angular velocity of the actual target object in the field. Thus, while the target model in the simulator generally has to move at a much smaller translational velocity than the actual target in the field, the opposite is true when it comes to angular rotation rate.

**Obtaining the Same Decorrelation Rate by Rotating Rather Than Moving the Target Model**

If the target model in the simulator were to actually move with translational velocity $(v_X, v_Y)$, the simulator imaging system would have to track and follow that motion, just as an actual telescope in the field must do to follow a moving target. However, there is a simpler way to proceed that avoids this inconvenience: The identical amount of speckle reduction can be obtained by holding the target model fixed while simply rotating the model, preferably about an axis of symmetry (e.g. the cylindrical axis of a missile) in the manner now prescribed.

The angular rotation rate of the target model required to produce the same speckle decorrelation rate as that caused by target model motion can be obtained by equating the translational velocity and angular velocity terms on the right-hand side of 15.58, thus giving

$$(\omega_{XS}, \omega_{YS}) = \frac{(v_{XS}, v_{YS})}{L_S} \tag{15.66}$$

#### 15.6.3.3 Reduction by Exploiting Laser Coherence Length

For illuminator laser beams with long coherence lengths (perhaps several meters) and targets with rms micro-roughness height variation, comparable to or greater than the laser wavelength, we should generally expect to find Gaussian speckle in instantaneous images (with $S/N = 1$). In contrast, for lasers with much shorter coherence lengths, we should expect to find reduced speckle (with $S/N > 1$). Thus, lack of coherence length in an illuminator laser beam is not necessarily detrimental; it offers yet another speckle reduction mechanism.

Coherence length is determined by the spectral content of the output beam, which we denote as previously by the function, $G(\lambda)$. (Coherence length varies approximately as the inverse of the width of this function.) For an argon-ion laser used without an etalon, $G(\lambda)$ could have the form of a Dirac comb, the individual comb members spread out over the entire visible spectrum. For many other laser types, the output beam comprises a single narrow spectral line whose energy content is approximately Gaussian-distributed. Denoting the central wavelength in the spectral distribution by $\lambda_C$ and the $1/e$ half-width of the distribution by $\sigma_\lambda$, the spectral output, which we denote by $G_G(\lambda)$ can then be expressed by

$$G_G(\lambda) = \exp\left[-\left(\frac{\lambda - \lambda_C}{\sigma_\lambda}\right)^2\right] \tag{15.67}$$

The coherence length of a laser with this spectral content, $CohL_G$, is given by

$$CohL_G = \frac{\lambda_C^2}{\sigma_\lambda} \tag{15.68}$$

Figure 15.11 shows the spectral content of a laser beam output at nominal wavelength, $\lambda_C = 0.55\ \mu m$, with 10-cm coherence length. The same figure also shows the spectral content of a laser beam at wavelength, $\lambda_C = 2.5\ \mu m$, with the same 10-cm coherence length.

The surface height variations over a target (measured in the direction of the line of sight of the telescope) are constituted by both micro-roughness height variations

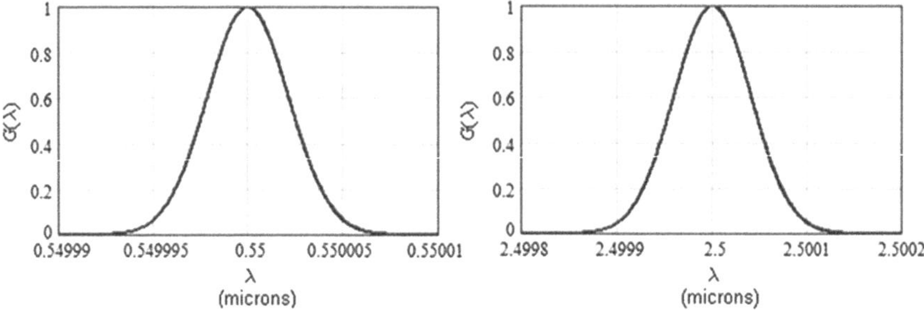


**Fig. 15.11** Spectral content of two-illuminator laser beams, both with the same 10-cm coherence length. *Left* laser beam with central wavelength $\lambda_C = 0.55\,\mu m$ and 1/*e* half-width, $\sigma_\lambda = 3 \times 10^{-6}\,\mu m$. *Right* laser beam with central wavelength $\lambda_C = 2.5\,\mu m$ and 1/*e* half-width $\sigma_\lambda = 6.25 \times 10^{-5}\,\mu m$

and macroscopic height variations, the latter associated with the structural shape and inclination of the target surfaces. Micro-roughness is a necessary element in our present discussion. Without it, there would be no speckle; the image would comprise just a few isolated glints and caustics. However, the amount of speckle reduction occurring in the image as a result of laser coherence length is, for most purposes, independent of the surface height variations associated with the micro-roughness structure. The reduction is overwhelmingly determined by the much larger height variations associated with the structural height variations over the target body. When the latter height variations are larger than the coherence length of the illuminator laser beam, significant speckle reduction can occur.

As is shown in Fig. 15.12, light arising at a given point in the image is drawn from a small target area centered on a point in the target plane conjugate to the image point of interest. The light contributions drawn from that object space area are weighted according to the local value of the telescope intensity PSF over the collection area. In general, the intensity PSF displays core and halo structure. To simplify the analysis here, we assume isoplanatic imaging behavior so that the intensity PSF may be considered invariant over the image. We also assume invariance of the intensity PSF with respect to wavelength. For visible and IR laser beams with coherence lengths greater than about 1 mm (i.e., $\sigma\lambda < 0.001\,\mu m$), the latter assumption would be reasonably valid.

Each of the individual quasi-monochromatic wavelengths that constitute G ($\lambda$) produces a Gaussian speckle pattern in the image. The Gaussian patterns formed at two different wavelengths, $\lambda_1$ and $\lambda_2$, generally show some degree of correlation, the actual amount depending on both the wavelength separation and the magnitude of the target body height variation over lateral spatial intervals set by the size of the telescope intensity PSF in the target plane. The (wavelength-integrated) final image may be considered as the incoherent sum of a very large number (perhaps infinite) of partially correlated Gaussian speckle patterns. If the individual contributing patterns are all highly correlated, the final wavelength-integrated speckle pattern

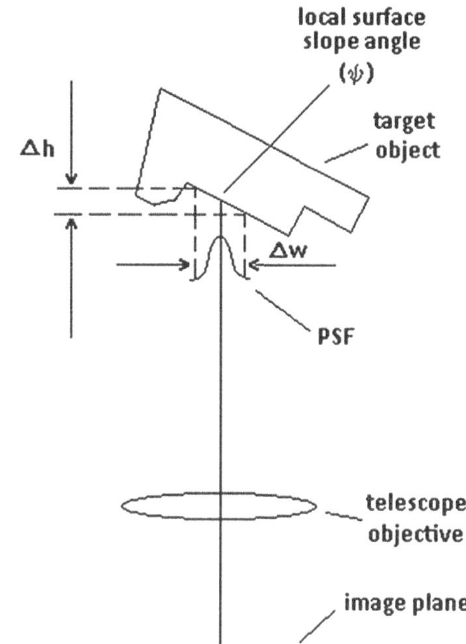


**Fig. 15.12** The effective target area that contributes light to an individual pixel is determined by the size of the telescope intensity PSF in target space and the pixel size in target space. Illuminator laser beams with short coherence lengths produce reduced speckle patterns in instantaneous images. The reduction amount varies locally over the target in a way governed by the height variation of the target surface, $\Delta h$, over the local light capture area of the intensity PSF

would remain approximately Gaussian; however, if they are poorly correlated, the final speckle pattern would show significant reduction.

**Reduction for the Most General (Core and Halo) Intensity PSF**

In this section, we develop an expression for the $S/N$ ratio for reduced speckle patterns formed in the image of a target object illuminated by a laser beam where the coherence length is defined by the spectral content of the beam, which we assume as before to be given by $G(\lambda)$. As is shown in Fig. 15.12, the surface height change $\Delta h$ that occurs over the extent of the intensity PSF, $\Delta w$, depends on the slope angle of the surface via the relation

$$\Delta h = \Delta w \cdot \tan(\psi) \tag{15.69}$$

where $\psi$ is defined as the angle between the surface normal and the line of sight of the telescope.

As discussed previously, the intensity PSF formed by the telescope/atmosphere combination generally displays core and halo structure. If circular symmetry is assumed, the unit-normalized intensity PSF may be expressed in the approximate form (cf., 13.51)

$$PSF(\vartheta) = \frac{1}{(A_C + A_H)} \cdot \left\{ A_C \cdot \exp\left[ -\left( \frac{\vartheta^2}{B_C^2} \right) \right] + A_H \cdot \exp\left[ -\left( \frac{\vartheta^2}{B_H^2} \right) \right] \right\} \tag{15.70}$$

where $\vartheta$ is the angular coordinate.

Let us now consider a target located at a distance $L$ from the telescope where the slope of the target surface with respect to the line of sight is assumed constant over distances comparable to the width of the intensity PSF in the target plane. The local height of the surface at any point in the intensity PSF—measured with respect to the height in the center of the intensity PSF—is then given by

$$h = L \cdot \vartheta_S \cdot \tan(\psi) \tag{15.71}$$

where $\theta_S$ is considered to be the surface slope angle in the direction of maximum slope. It follows that

$$\vartheta_S = \frac{h}{L \cdot \tan(\psi)} \tag{15.72}$$

By combining 15.70 and 15.72, it can be seen that the probability density function of surface heights, PDF($h$), over the extent of the intensity PSF (in the direction of the line of sight) may be expressed as a proportionality by

$$\begin{aligned} PDF(h) \propto \frac{1}{(A_C + A_H)} \cdot \Bigg\{ & A_C \cdot \exp\left[-\left(\frac{h}{B_C \cdot L \cdot \tan(\psi)}\right)^2\right] \\ & + A_H \cdot \exp\left[-\left(\frac{h}{B_H \cdot L \cdot \tan(\psi)}\right)^2\right]\Bigg\} \end{aligned} \tag{15.73}$$

The degree of correlation of the complex amplitudes at any given point in the image at an arbitrarily chosen pair of wavelengths, $\lambda_1$ and $\lambda_2$., can then be expressed (cf. 7.40) by

$$\begin{aligned} \frac{\langle U(\lambda_1) \cdot U^*(\lambda_2)\rangle}{[\langle I(\lambda_1)\rangle \cdot \langle I(\lambda_2)\rangle]^{\frac{1}{2}}} &= \mu(\lambda_1, \lambda_2) \\ &= \int_0^\infty PDF(h) \cdot \exp\left[-\left(2 \cdot \pi \cdot i \cdot (2 \cdot h) \cdot \left(\frac{1}{\lambda_1} - \frac{1}{\lambda_2}\right)\right)^2\right] \cdot dh \end{aligned} \tag{15.74}$$

where the left-hand side of the above equation is recognized as the spectral correlation function, $\mu(\lambda_1, \lambda_2)$ discussed previously in Sect. 7.5.3. The quantity, 2 $h$, in the equation accounts for the round-trip beam path associated with the surface height change, $h$.

Using 15.73 to substitute for PDF($h$) in 15.74 and evaluating what is, in effect, a Fourier transform integral (which may be considered as the characteristic function associated with the probability distribution of surface heights), we obtain

$$\mu(\lambda_1, \lambda_2) = \frac{1}{(A_C \cdot B_C + A_H \cdot B_H)}$$

$$\times \left\{ A_C \cdot B_C \cdot \exp\left[ -4 \cdot \pi^2 \cdot B_C^2 \cdot L^2 \cdot \tan(\psi)^2 \cdot \left( \frac{1}{\lambda_1} - \frac{1}{\lambda_2} \right)^2 \right] \right.$$
$$\left. + A_H \cdot B_H \cdot \exp\left[ -4 \cdot \pi^2 \cdot B_H^2 \cdot L^2 \cdot \tan(\psi)^2 \cdot \left( \frac{1}{\lambda_1} - \frac{1}{\lambda_2} \right)^2 \right] \right\} \quad (15.75)$$

By using the quantity $A_{CH}(= A_C/A_H)$, the above equation may be written in the alternative form

$$\mu(\lambda_1, \lambda_2) = \frac{1}{(A_{CH} \cdot B_C + B_H)}$$
$$\times \left\{ A_{CH} \cdot B_C \cdot \exp\left[ -4 \cdot \pi^2 \cdot B_C^2 \cdot L^2 \cdot \tan(\psi)^2 \cdot \left( \frac{1}{\lambda_1} - \frac{1}{\lambda_2} \right)^2 \right] \right.$$
$$\left. + B_H \cdot \exp\left[ -4 \cdot \pi^2 \cdot B_H^2 \cdot L^2 \cdot \tan(\psi)^2 \cdot \left( \frac{1}{\lambda_1} - \frac{1}{\lambda_2} \right)^2 \right] \right\} \quad (15.76)$$

The $S/N$ ratio of the wavelength-integrated speckle pattern in the final image can now be written in terms of $\mu(\lambda_1, \lambda_2)$ as follows (cf., 12.54):

$$\frac{S}{N} = \frac{\int_0^\infty G(\lambda).d\lambda}{\left[\int_0^\infty \int_0^\infty G(\lambda_1).G(\lambda_2) \cdot |\mu(\lambda_1, \lambda_2)|^2 \cdot d\lambda_1 \cdot d\lambda_2\right]^{\frac{1}{2}}} \quad (15.77)$$

Laser coherence length offers a useful speckle reduction mechanism because it does not degrade resolution. Unfortunately, however, we cannot rely solely on this mechanism. In image regions where the surface slope is zero (i.e., $\psi = 0$, and hence, $\mu(\lambda_1, \lambda_2) = 1$), the mechanism provides no speckle reduction whatsoever. In addition, only limited amounts of speckle reduction are possible for targets that lie close to the telescope (i.e., $L \rightarrow 0$). For such targets, 15.75 and 15.76 indicate that $\mu(\lambda_1, \lambda_2) \rightarrow 1$ which eliminates any possibility of obtaining speckle reduction from this mechanism.

The Space Shuttle images shown in Fig. 15.6 were obtained in sunlight illumination where the coherence length associated with the visible part of the Sun's broad output spectrum is extremely short—only a few microns. As a result, almost complete speckle reduction is oserved in these images—other than around the wheel glints where we note that because $\psi = 0$, coherence length offers no possibility of speckle reduction. In contrast, had the Space Shuttle been actively illuminated using fully coherent laser beams at the two wavelengths considered, 0.55 and 2.5 μm, significant speckle noise would have been observed in the images, as we now demonstrate.

Figure 5.14 shows how $S/N$ ratio varies over the surface of the Shuttle craft as a function of the local slope angle, for the case where the laser beams at 0.55 and 2.5 μm both have the same coherence length, 10 cm. The three plots shown were calculated using 15.67, 15.68, 15.76, and 15.77; the three intensity PSFs used in the calculations were those previously plotted on the right-hand side of Fig. 15.6. Table 15.4 lists the parameter values used to create the intensity PSFs.

From Fig. 15.13, it is evident that $S/N$ ratio increases as the angular size of the intensity PSF. It is also evident from 15.76 to 15.77 that the $S/N$ ratio also increases in proportion to the quantity, $L \cdot \tan(\psi)$, and inversely as the laser coherence length, $CohL_G$, the latter inversely related to the width of function $G(\lambda)$. In the limit of large intensity PSF widths, large object distances, and large coherence lengths, it may be

**Table 15.4** Parameter values used to calculate *S/N* plots shown in Fig. 15.13

| $\lambda_C$ | 0.55 μm (no atmosphere) | 0.55 μm (with atmosphere) | 2.5 μm (with atmosphere) |
|---|---|---|---|
| $L$ (km) | 160 | 160 | 160 |
| $\sigma\lambda$ (μm) | $3 \times 10^{-6}$ | $3 \times 10^{-6}$ | $6.25 \times 10^{-5}$ |
| Coh$L_G$ (cm) | 10 | 10 | 10 |
| $A_{CH}$ | $\infty$ ($10^{10}$) | 0 | 15 |
| $B_C$ (arcsec) | 0.017 | 0.017 | 0.077 |
| $B_H$ (arcsec) | 0.9 | 0.9 | 0.9 |

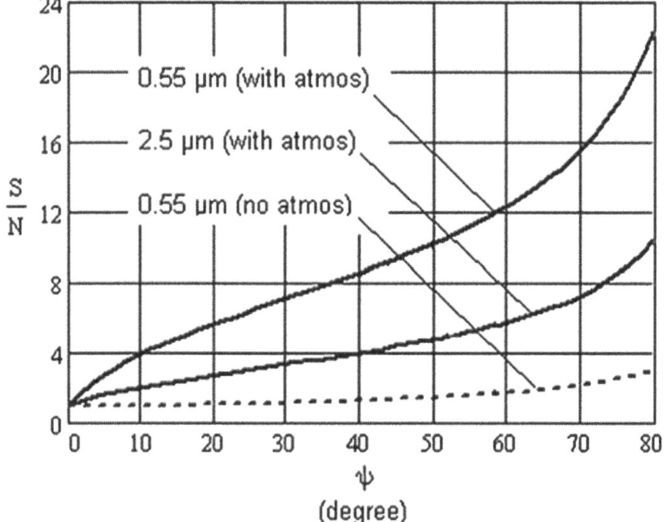


**Fig. 15.13** Plots showing how signal-to-noise in the instantaneous images of the Space Shuttle (previously shown in Fig. 15.6) might have varied over the body of the craft as a function of the local surface tilt angle, $\psi$, had the craft been actively illuminated by a laser rather than sunlight. The plots shown correspond to two different laser beams, both having the same 10-cm coherence length, one at wavelength 0.55 μm and the other at 2.5 μm. (The spectral content of these beams is shown in Fig. 15.11.) The *S/N* improvement depends on the size of the telescope intensity PSF in target space. The intensity PSFs used to create the plots are those depicted on the right of Fig. 15.6. More compact intensity PSFs provide less speckle reduction than broader intensity PSFs

shown that $S/N$ improves according to the proportionality,

$$\frac{S}{N} \propto \sqrt{\frac{PSF(width) \cdot L \cdot \tan(\psi)}{CohL_G}} \tag{15.78}$$

**Scaling the Coherence Length of the Illumination Laser Used in the Simulator** To achieve the same level of speckle reduction in a simulated image as in an actual image obtained in the field from the coherence length mechanism, it is clear that the ratio of the coherence length of the simulator laser, Coh$L_S$, to the height variation over the target model must be the same as the ratio of the coherence length of the laser used in the field to the height variation over the actual target. Since the height variations for the target model and the actual target scale in proportion to Scale$M$, it follows that

$$CohL_S = ScaleM \cdot CohL \tag{15.79}$$

Since Scale$M$ is usually less than unity, it follows that Coh$L_S$ is usually less than Coh$L$. Therefore, if the laser used in the simulator happened to be the same as that used in the field, it would be necessary to reduce its coherence length. Practical methods of carrying out such reductions have been described elsewhere (Dainty, 1984).

#### 15.6.3.4 Reduction by Deliberately Introducing Aberrations

In the previous section, we saw that reduced speckle occurs when angularly broad intensity PSFs combine with short coherence length illumination. Additional speckle reduction may be obtained by deliberately introducing telescope aberrations, such as defocus. As the intensity PSF grows in size, so does the amount of speckle reduction. However, a major disadvantage of this approach is that the speckle reduction is realized at the expense of target resolution.

#### 15.6.3.5 Reduction by Use of Multiple Offset Laser Beams

Figure 15.7 shows the illuminator laser located just below the telescope objective. The figure also shows the speckle pattern formed by illumination back-scattered from the target. If we were now to add a second illuminator laser, this time located just above the telescope objective, this laser would produce a nearly identical speckle pattern, except for the fact that the new pattern would be laterally offset with respect to the first pattern by an amount set by the lateral separation of the two lasers.

In Sect. 15.6.1, we saw that when speckle patterns are offset in this way by amounts that exceed the telescope diameter, they produce uncorrelated speckle patterns in the image. For the two-illuminator laser arrangement considered here, the signal-to-noise ratio of the reduced speckle pattern arising in the image is then given by $S/N =$

$\sqrt{2}$. If we were now to set up an array of $n$ identical lasers, with the lateral offset distances between all pair combinations always exceeding the telescope diameter, $D$, extrapolation of the same reasoning indicates that the signal-to-noise ratio for the reduced speckle pattern in the image is given by

$$\frac{S}{N} = \sqrt{n} \tag{15.80}$$

It might be noted that the speckle reduction obtained by this strategy is obtained without any loss of resolution.

If the two-illuminator lasers were laterally offset by an amount $\Delta\varepsilon < D$, the individual speckle patterns formed in the image by these two lasers would be partially correlated. The degree of correlation can be calculated from 15.59. The signal-to-noise ratio in the reduced speckle pattern formed by the summed illuminations from the two lasers would then be given by

$$\frac{S}{N} = \left(\frac{2}{1+\mu(\Delta\varepsilon)}\right)^{\frac{1}{2}}. \tag{15.81}$$

Because $\mu(\Delta\varepsilon) > 0$, when $\Delta\varepsilon < D$, the above relation indicates that $S/N$ ratio in the final image is now less than the maximum possible $S/N$ reduction factor, $\sqrt{2}$, offered by a two-laser arrangement.

For $n$ lasers acting together, similar reasoning tells us that to effect maximum $S/N$ speckle reduction factor, $\sqrt{n}$, all laser pairs must be separated by distances greater than or equal to the telescope diameter, $D$.

#### 15.6.3.6 Reduction by Use of Frame-to-Frame Averaging

Suppose that $n$ image frames were recorded using a short integration period, $\Delta T$, where the interval between the individual frames is longer than the relaxation time of the speckle in the image; the actual mechanism causing the speckle relaxation is not important. By summing all of these frames together, the final image comprises a reduced speckle pattern where, as indicated by 15.80, $S/N = \sqrt{n}$. Again, we see that this type of speckle reduction mechanism does not cause any loss of resolution.

#### 15.6.3.7 Reduction by Use of an Unpolarized Laser Beam

Generally, when a hard-bodied target such as a satellite is illuminated by an unpolarized laser beam, little or no speckle reduction occurs in the image. Although the final image is in effect constituted by two incoherently added Gaussian speckle patterns (one for each of the two orthogonal polarizations that constitute the unpolarized light), because the back-scattered light mostly results from a single-stage scattering process, the two constituent patterns tend to be highly correlated. Apart from a difference in brightness, the integrated speckle pattern recorded by the detector would likely be closely similar to the speckle patterns produced by a perfectly polarized beam.

To obtain significant speckle reduction from the mechanism of polarization, the scattering process at the target would have to cause significant depolarization of the incident illumination. If the "target" were either a cloud or an aerosol, for which multiple-scattering events would occur throughout the target depth, the resulting depolarization of the illuminator beam would indeed give rise to a final image comprised of the sum of two uncorrelated Gaussian speckle patterns. The signal-to-noise ratio in such a pattern would then be given by $S/N = \sqrt{2}$. It might be observed here that this outcome does not depend on the initial state of polarization of the outgoing illuminator beam.

#### 15.6.3.8 Reduction When Several Reduction Mechanisms Act Simultaneously

When some or perhaps all of the (seven) speckle reduction mechanisms described in Sect. 15.6.3 act together, in principle, it is possible to produce an exact expression for the S/N ratio of the reduced speckle pattern in the final image. An analysis of this type has been carried out (McKechnie, 1976) for the case of illumination that was both temporally and spatially partially coherent. The final expression however, turned out to be complicated. When more than two speckle reduction mechanisms act together, an exact analysis can soon become prohibitively complicated.

The analysis can be considerably simplified, however, by making the broad assumption that all of the speckle reduction mechanisms contributing to the reduced speckle pattern in the final image act independently.[14] With this assumption, the S/N ratio of the reduced speckle pattern in the final image is then given by the product of the signal-to-noise improvement factors delivered separately by each of the various contributing speckle reduction mechanisms, where each is considered to act in isolation. When n such mechanisms act simultaneously in this way, and we denote the signal-to-noise improvement factor caused by the $j$th mechanism by $S/N_j$, the approximate signal-to-noise ratio of the reduced speckle pattern in the final image can be calculated thus,

$$\frac{S}{N} = \prod_{j=1}^{n} \frac{S}{N}_j \tag{15.82}$$

Once the overall S/N ratio of the reduced speckle pattern has been so calculated, an approximate PDF for the intensity fluctuation can be obtained by using the gamma distribution given in Sect. 11.6.6.1 (cf., 11.80). The free parameter, m, for which a

[14] Some of the speckle reduction mechanisms discussed in Sect. 15.6.3 act independently; others do not. When any, or all, of the various reduction mechanisms interact with one another, the amount of speckle reduction indicated by 15.82 merely provides a rough upper-bound estimate of the $S/N$ improvement likely to be achieved.

value must be set in the equation, can be calculated thus,

$$m = \left(\frac{S}{N}\right)^2 = \left[\prod_{j=1}^{n} \frac{S}{N}_j\right]^2 \tag{15.83}$$

## 15.7 Mathematical Notation Used in This Chapter

The mathematical notation used in this chapter is indicated in Table 15.5.

**Table 15.5** Mathematical notation used in this chapter along with the SI dimensional units of the individual quantities

| Symbol | Quantity | Dimensions |
|---|---|---|
| $\lambda$ | Wavelength | $m$ |
| Scale$M$ | Target model scale factor | "1" |
| Scale$D$ | Scale factor relating diameter of simulator imaging optic to actual telescope diameter | "1" |
| Scale$\lambda$ | Scale factor relating wavelengths used for actual and simulated images | "1" |
| $D$ and $D_S$ | Diameters of telescope and simulator imaging optic | $m$ |
| $d$ and $d_S$ | Central obstruction diameters of telescope and simulator imaging optic | $m$ |
| Wid and Wids | Widths of detectors used to record actual and simulated images | $m$ |
| $L$ and $L'$ | Actual target distance and its conjugate image space distance | $m$ |
| $L_S$ and $L'_S$ | Target model distance and its conjugate distance in simulated image space | $m$ |
| $f$ and $f_S$ | Focal lengths of telescope and simulator imaging optic | $m$ |
| $F$ and $F_S$ | $F$/numbers of telescope and simulator imaging optic | "1" |
| $F_1$ and $F_{1s}$ | Illumination beam $F$/numbers in actual and simulated target spaces | "1" |
| $\sigma$ and $\sigma_S$ | rms of integrated OPD for actual and simulated atmospheric paths | $m$ |
| $w_0$ and $w_{0S}$ | $1/e$ half-widths of OPD autocorrelation functions for actual and simulated image paths | $m$ |
| $\sigma_{AO}$ | rms of residual OPD fluctuation after AO correction | $m$ |
| $\mathrm{wo}_{AP}$ | $1/e$ half-width of Gaussian autocorrelation function of residual OPD fluctuation | $m$ |

(continued)

**Table 15.5** (continued)

| Symbol | Quantity | Dimensions |
|---|---|---|
| $\vartheta$ and $\vartheta_S$ | Field of view in actual and simulated image spaces | "1" |
| $SI$ | Telescope Strehl intensity | "1" |
| $I$ | Intensity in (x, y) pupil plane or ($\alpha$,$\beta$) image plane as appropriate | "1" |
| $A_C$ | Intensity in center of Gaussian core | "1" |
| $A_H$ | Intensity in center of Gaussian | "1" |
| $A_{CH}$ | Ratio $A_C/A_H$ | "1" |
| $B_C$ | Angular 1/$e$ half-width of Gaussian core | "1" |
| $B_H$ | Angular 1/$e$ half-width of Gaussian halo | "1" |
| $\sigma_T$ | rms wavefront error of telescope | $m$ |
| $P-V$ | Peak-to-valley wavefront error of telescope | $m$ |
| $\psi$ | Angle between reference normal at target location and telescope line of sight | "1" |
| $\lambda_{opt}$ | Optimum wavelength | $m$ |
| $\lambda_C$ | Central wavelength produced by illuminator laser | $m$ |
| $\sigma\lambda$ | 1/$e$ half-width of illuminator laser beam output | $m$ |
| Coh$L$ and Coh$LS$ | Coherence lengths of illuminator lasers in actual and simulated target spaces | $m$ |

Dimensionless quantities are indicated by "1"

# References


Born, M., & Wolf, E. (2003). *Principles of optics* (7th ed. revised). Cambridge University Press.

Dainty, J. C. (1984). Laser speckle and related phenomena. In J. C. Dainty (Ed.), *Topics in applied physics* (Vol. 9). Springer.

Mandel, L., & Wolf, E. (1995). *Optical coherence and quantum optics*. Cambridge University Press.

McKechnie, T. S. (1976). Image plane speckle in partially coherent illumination. *Optical and Quantum Electronics, 8*, 61–67.

McKechnie, T. S. (1990). Diffraction limited imaging using large ground-based telescopes. In *Proceedings of SPIE, Symposium on Astronomical Telescopes and Instrumentation for the 21st Century* (Vol. 1236, pp. 164–178), 11–17 February.

Pederson, I. (2004, September 18). Sunlight speckle. *Science News Online, 166*(12). Retrieved from http://www.sciencenews.org/articles/20040918/mathtrek.asp

Welford, W. T. (1962). *Geometrical optics*. North-Holland Publishing Co.

# Chapter 16
# Laser Beam Propagation and Path Characterization

**Abstract** This chapter deals with laser beam propagation through the atmosphere. Applications include laser communication systems and High Energy Laser (HEL) weapon systems. Collimated beams are treated as well as beams that converge to a focus in the atmosphere. For terrestrial laser communication systems, laser beam quality is influenced by atmospheric turbulence and the aberrations of the beam-projection telescope. For HEL beams, other influences include thermal blooming and AO beam correction. For aircraft-mounted HEL systems, boundary layer turbulence arises, as does point-ahead angle for moving targets. Definitive beam path characterization techniques are described. Once values have been obtained for the key characterization parameters, an optimum wavelength region can be identified at which HEL beams (of the same beam power) concentrate maximum possible irradiance flux at target center, thereby maximizing system lethality and lethality range. Choice of outrider HEL wavelengths lying far from the optimum region can catastrophically reduce system performance, potentially dooming the system to failure from the outset.

Laser beam propagation through the atmosphere has been extensively studied, particularly in relation to Kolmogorov turbulence (Ishimaru, 1978; Strohbehn, 1985). In this chapter, we analyze the subject in terms of a generalized atmosphere. The analysis therefore applies to all possible types of atmospheric turbulence. It also applies to all types of laser beams, including communication laser beams and HEL beams. For the former, beam quality is degraded by atmospheric turbulence and the aberrations of the beam projection telescope. For the latter, thermal blooming further degrades beam quality. For moving targets, point-ahead angles arise and, for aircraft-mounted HEL systems, boundary-layer turbulence and air-frame vibrations cause yet more degradation (Perram et al., 2010).

For the applications just indicated, the objective is to control the size of the focused beam spot in a way that produces suitably high irradiance (Watts/m$^2$) at the target. For communication laser systems, higher irradiance leads to higher signal-to-noise ratios, which enable longer communication distances requiring fewer relay stations. For HEL laser systems, higher irradiance at the focus increases both beam lethality



T. S. McKechnie, *General Theory of Light Propagation and Imaging Through the Atmosphere*, Progress in Optical Science and Photonics 20,
https://doi.org/10.1007/978-3-030-98828-9_16

and lethality range, the latter enabling deployed systems[1] to provide wider theater coverage.

Laser beams projected from telescopes at relatively nearby targets follow conically convergent beam paths. Typical targets lie within the atmosphere, or perhaps just beyond it. The integrated optical path difference (OPD) fluctuations for convergent atmospheric beams are generally less than for collimated beams traveling over the same nominal paths.

Convergent HEL beam paths may be characterized either in the direction of the outgoing beam or in the reverse direction, as would be the case of an incoming beam originating from a point source in the target plane. The integrated OPD fluctuations are essentially the same for either path direction; minor differences discussed in Sect. 16.8 can generally be ignored. The beam path characterization procedures for convergent beam paths (Sect. 16.7) closely follow those described in Chaps. 8, 13 and 18 for collimated beam paths.

For HEL beams, path characterization must be carried out with the HEL beam operating so that thermal blooming effects are included in the full end-to-end beam path characterization. AO pre-corrections are generally applied to the outward projected HEL wave fronts to minimize the rms variation of the integrated OPD fluctuation over the entire end-to-end beam path. Once the minimum rms value has been established, we invoke 10.16 to establish the center of the optimum HEL wavelength region. Wavelengths falling within this region enable maximum theoretically possible irradiance concentration at the HEL beam focus (for the same output beam power).

Naturally, the chosen wavelength would also have to fall within an atmospheric transmission window. With reference to Fig. 3.9, there are many favorable windows spanning the visible to far-IR wavelength range (0.5 μm to >> 50 μm). For specific applications, most of these windows are off limits due to the overriding requirement for a near-optimum HEL wavelength choice. Other desirable attributes of the chosen HEL wavelength are discussed in Sect. 16.9.2.

Use of HEL wavelengths far removed from the optimum region—particularly those lying on the shorter wavelength side—can lead to hugely under-performing HEL systems (McKechnie, 1991b) with irradiance flux concentration at the target typically one or two orders of magnitude less than the theoretical maximum. Discussion of this crucially important matter is given in Sect. 16.9 in relation to the Airborne Laser (ABL) weapon system.

## 16.1 OPD Line Integrals for Convergent Beam Paths

Consider the straight line in Fig. 16.1 connecting the point (0, 0, 0) in the target plane to an arbitrary location in the entrance pupil of the telescope, (x, y, L). Denoting an

[1] To provide coverage of a given theater area, the required number of deployed weapon platforms reduces approximately as the inverse square of the lethality range.

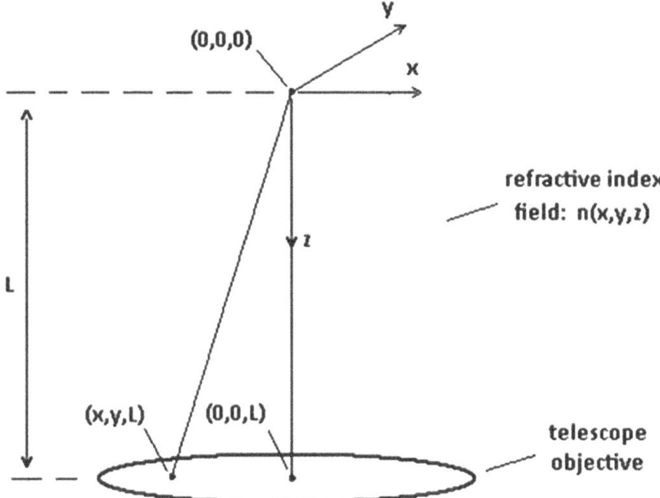


**Fig. 16.1** Geometry for calculating the OPD line integrals for the various rays in a convergent beam pencil from the target point (0, 0, 0) to an arbitrary point in the telescope pupil, $(x, y, L)$

arbitrary point on this line by $(x', y', z')$, the equations describing this line are as follows

$$x' = \frac{x}{L} \cdot z', \tag{16.1}$$

$$y' = \frac{y}{L} \cdot z'. \tag{16.2}$$

Denoting the refractive index field as before by $n(x, y, z)$, the OPD integral along the line may be expressed by

$$\mathrm{OPD}(x, y) = \int_0^L n\left(\frac{x \cdot z'}{L}, \frac{y \cdot z'}{L}, z'\right) \cdot \frac{1}{\cos(\chi)} \cdot \mathrm{d}z', \tag{16.3}$$

where the factor, $1/\cos(\chi)$, accounts for the increased length of the path due to beam convergence and $\chi$ is the angle made by the line with respect to the $z$-axis,

$$\chi = \tan^{-1}\left(\frac{\sqrt{(x^2 + y^2)}}{L}\right). \tag{16.4}$$

[Angle $\chi$ as used here should not be confused with our previous use of $\chi$ in Chap. 13 (Sect. 13.1.4.1)].

Suppose that an initially spherical light wave is projected from the telescope objective shown in Fig. 16.1 and directed toward the point (0, 0, 0) in the target plane. If the propagation space is homogeneous, the beam comes to a diffraction-limited focus. However, for a non-homogeneous space—as determined by the random fluctuations of function n(x, y, z)—the beam focuses to a wider, and correspondingly less intense, intensity distribution. The statistical properties of the complex amplitude and intensity in the focused spot are determined (just as in Chap. 6) by the statistical properties of the OPD line integrals from points in the telescope pupil plane to the point (0, 0, 0) in the target plane. The zero-mean fluctuating part of the OPD line integrals, which we denote by $H_{CBP}(x, y)$ and where the subscript, CBP, denotes convergent beam path, is given by

$$H_{\mathrm{CBP}}(x, y) = \mathrm{OPD}(x, y) - \langle \mathrm{OPD}(x, y) \rangle, \tag{16.5}$$

where OPD($x$, $y$) was indicated previously by 16.3.

The variance of $H_{\mathrm{CBP}}(x, y)$ is given by

$$\sigma_{\mathrm{CBP}}^2 = \left\langle H_{\mathrm{CBP}}(x, y)^2 \right\rangle, \tag{16.6}$$

where the averaging brackets $\langle\cdot\rangle$ again denote the ensemble average. The autocorrelation function of $H_{\mathrm{CBP}}(x, y)$ may be written,

$$\rho_{\mathrm{CBP}}(\xi, \eta) = \frac{\langle H_{\mathrm{CBP}}(x+\xi, y+\eta) \cdot H_{\mathrm{CBP}}(x, y) \rangle}{\left\langle H_{\mathrm{CPB}}(x, y)^2 \right\rangle}. \tag{16.7}$$

Combining 16.6 and 16.7, the autocovariance function of $H_{\mathrm{CBP}}(x, y)$ may be written

$$\langle H_{\mathrm{CBP}}(x+\xi, y+\eta) \cdot H_{\mathrm{CBP}}(x, y) \rangle = \sigma_{\mathrm{CBP}}^2 \cdot \rho_{\mathrm{CBP}}(\xi, \eta). \tag{16.8}$$

### 16.1.1 Phase Screen Stack Representation of Convergent Beam Paths

Although OPD line integrals of the type given by 16.3 contain all necessary information for describing the statistical properties of the beam path, it is again conceptually helpful to model the beam path in terms of a large number, $n_{\mathrm{L}}$, of random phase screens stacked together as shown in Fig. 16.2. (If $n_{\mathrm{L}}$ is chosen large enough, the random phase screen stack can model the optical path to any required degree of precision.) Further, suppose that we now construct a coherency matrix $J'$ for all of these phase screens, as was done previously in Sect. 6.2.1 (6.19). By using the appropriate unitary transformation matrix, we may diagonalize the coherency matrix, $J'$, to give a second equivalent coherency matrix, $J$, comprised of the eigenvalue set $\lambda_1$, $\lambda_2$, ..., $\lambda_n$. These eigenvalues may be considered to represent a set of $n$ uncorrelated random

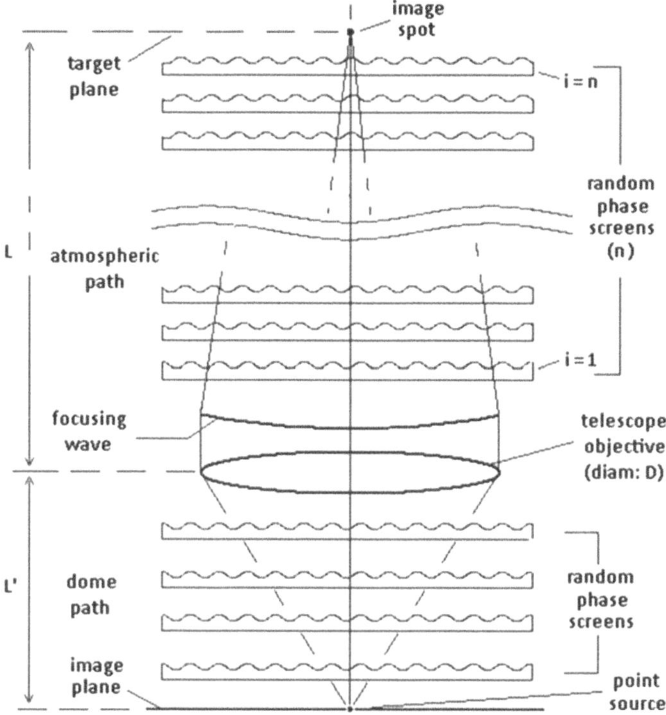


**Fig. 16.2** Schematic depiction of $n$ uncorrelated random phase screens representing the atmospheric path for a convergent beam in target space. The figure also shows a second similar stack of uncorrelated random phase screens representing the inhomogeneous air path inside the beam projection telescope

phase screens (where $n \ll n_L$) whose light-scattering effect is statistically the same as that of the much larger number of correlated phase screens first supposed.

Figure 16.2 shows the atmospheric portion of the path represented in terms of the $n$ uncorrelated random phase screens. (For later reference, the figure also shows a second set of phase screens representing the path inside the telescope. Since the imaging pencils in the latter path are also convergent, this portion of the path may be treated in a similar way to the atmospheric path portion.) By adopting the convention that the first phase screen lies close to the telescope and the $n$th screen lies close to the focus at $z = 0$ and by denoting the OPD fluctuation introduced by the $i$th phase screen by $h_{\mathrm{CBP}_i}(x, y)$, the integrated OPD fluctuation introduced by the entire phase screen stack, $H_{\mathrm{CBP}}(x, y)$, may be expressed in terms of all the $h_{\mathrm{CBP}_i}(x, y)$ by

$$H_{\mathrm{CBP}}(x, y) = \sum_{i=1}^{n} h_{\mathrm{CBP}_i}\left(x \cdot \frac{(L - z_i)}{L}, y \cdot \frac{(L - z_i)}{L}\right), \qquad (16.9)$$

where the arguments on the right-hand side of the above equation are scaled in a way that accounts for beam convergence (cf., 16.3).

## 16.2 Autocorrelation Function of the Integrated OPD Fluctuation for Convergent Paths

The diameter of the nominally spherical image-forming wave portion as it passes through the *i*th phase screen may be given in terms of the telescope diameter, *D*, by $(1 - i/n)\,D$. Thus, as the beam passes through the entire phase screen stack, beam diameter varies over the range, 0–*D*. The variance of the OPD fluctuation introduced by a particular phase screen depends on the location of that screen with respect to the focused spot at (0, 0, 0). Denoting the variance introduced by the *i*th phase screen by $\sigma^2_{CBP_i}$, we may express this quantity in terms of the zero-mean OPD fluctuation, introduced by the *i*th phase screen as follows,

$$\sigma^2_{\mathrm{CBP}_i} = \left\langle h_{\mathrm{CBP}_i}\left(x \cdot \frac{L - z_i}{L}, y \cdot \frac{L - z_i}{L}\right)^2\right\rangle \quad \text{for } i = 1, \ldots, n, \tag{16.10}$$

where the average indicated by the brackets for the *i*th phase screen refers to the average obtained over the circular footprint area of the beam as it passes through the phase screen, the diameter of this footprint area being $(1 - i/n)\cdot D$.

The autocorrelation function of the OPD fluctuation introduced by *i*th phase screen may be written in a similar way,

$$\rho_{\mathrm{CBPi}}(\xi, \eta) = \frac{\left\langle h_{\mathrm{CBPi}}\left((x + \xi) \cdot \frac{L-z_i}{L}, (y + \eta) \cdot \frac{L-z_i}{L}\right) \cdot h_{\mathrm{CBPi}}\left(x \cdot \frac{L-z_i}{L}, y \cdot \frac{L-z_i}{L}\right)\right\rangle}{\sigma^2_{\mathrm{CBPi}}} \quad \text{for } i = 1, \ldots, n, \tag{16.11}$$

where the $(\xi, \eta)$ arguments represent distances referred to the $(x, y)$ telescope pupil plane.

By combining (16.6)–(16.9), we obtain the following expression for the autocorrelation function of the integrated OPD fluctuation arising from the entire stack of phase screens,

$$\rho_{\mathrm{CBP}}(\xi, \eta) = \frac{\sum_{i=1}^{n} \sigma^2_{\mathrm{CBP}i} \cdot \rho_{\mathrm{CBP}i}\left(\xi \cdot \frac{L-Z_i}{L}, \eta \cdot \frac{L-Z_i}{L}\right)}{\sum_{i=1}^{n} \sigma^2_{\mathrm{CBP}i}}. \tag{16.12}$$

Apart from the scaled arguments, this result is similar to the one given previously in Chap. 6 (6.60) for collimated beams. Thus, the rms quantity, $\sigma_{CBP}$, and the function, $\rho_{CBP}(\xi, \eta)$, may be obtained from image intensity measurements by the same procedures previously described in Chaps. 8 and 13. Once $\sigma_{CBP}$ and $\rho_{CBP}(\xi, \eta)$ have been established, all significant statistical properties of the complex amplitude and intensity in the focused spot image can be determined.

### 16.2.1 OPD Autocorrelation Function Width for Convergent Beam Paths

Figure 16.3 shows typical plots of the autocorrelation function of the integrated OPD fluctuation as they might arise for a convergent beam. To generate these illustrative plots, isotropic turbulence was assumed and the same circular Gaussian autocorrelation function (cf., 5.23) was assumed to apply over the entire beam path, the $1/e$ half-width of this function denoted, as previously, by $w_o$. Because of the beam convergence, the autocorrelation function of the OPD fluctuation introduced by the ith phase screen may be expressed by the scaled version of 5.23),

$$\rho_{\mathrm{CBP}i}(\varepsilon) = \exp\left[-\left(\frac{\varepsilon}{w}\cdot\frac{L - z_i}{L}\right)^2\right] \quad \text{for } i = 1, \ldots, n, \qquad (16.13)$$

where $\varepsilon$ may be considered as the radial coordinate in the telescope pupil.

As the beam converges toward focus, the diameter of the circular intercept area on individual phase screens steadily diminishes. Consequently, the rms OPD fluctuation introduced by phase screens located near to the focus tends to be less than that

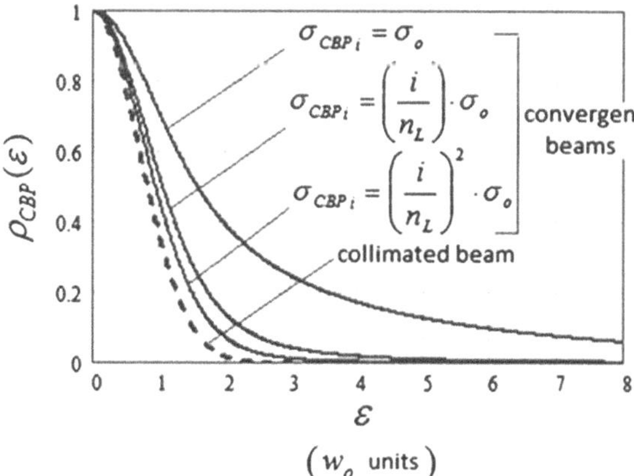


**Fig. 16.3** The width of the autocorrelation function of the integrated OPD fluctuation, $\rho_{\mathrm{CBP}}(\varepsilon)$, depends on how the rms OPD fluctuation, $\sigma_{\mathrm{CBP}i}$, introduced by each individual phase screen modifies as a function of the distance of that screen from the focus. Function, $\rho_{\mathrm{CBP}}(\varepsilon)$, is shown for the three candidate modification schemes indicated. The three functions are necessarily wider than the corresponding function for a collimated beam (*dotted line*)

introduced by those located further away. For the plots shown in the figure, it is supposed that rms OPD quantity, $\sigma_{CBPi}$, reduces as the distance of the $i$th phase screen from the focus according to the following two candidate schemes:

$$(1) \qquad \sigma_{\mathrm{CBP}i} = \left(\frac{L - z_i}{L}\right) \cdot \sigma_o \quad \text{for } i = 1, \ldots, n, \tag{16.14}$$

$$(2) \qquad \sigma_{\mathrm{CBP}i} = \left(\frac{L - z_i}{L}\right)^2 \cdot \sigma_o \quad \text{for } i = 1, \ldots, n, \tag{16.15}$$

where $\sigma_o$ is an arbitrarily chosen constant rms value. Equation 16.14 indicates a linear reduction in $\sigma_{CBPi}$ as the beam approaches focus, while 16.15 indicates a quadratic reduction. Also shown for reference is a plot of $\rho_{CBP}(\varepsilon)$ for the case where $\sigma_{CBPi}$ takes the same constant value for all of the individual phase screens, even for those limitingly close to the focus:

$$(3) \qquad \sigma_{\mathrm{CBP}i} = \sigma_o \quad \text{for } i = 1, \ldots, n, \tag{16.16}$$

The behavior indicated by 16.16 corresponds, in effect, to turbulence structure obeying a white noise size distribution which, though interesting as a limiting case, is physically unrealistic.

Irrespective of which of the above three candidate schemes might be most representative of typical atmospheric turbulence, the corresponding autocorrelation functions, $\rho_{\mathrm{CBP}}(\varepsilon)$, shown in Fig. 16.3, all share the same common property: Their widths are all greater than that of the autocorrelation function that would have arisen for a collimated beam traversing the same nominal path. The plot for this case is shown (dotted line) in the figure.

## 16.3 OPD Autocorrelation Function for Paths Inside the Telescope

Because the beam path inside the telescope tube is also convergent (or divergent for beams traveling in the opposite direction), the effect of turbulence inside the telescope may be analyzed in a manner similar to that just described. The path can again be represented by a very large number of, in general, partially correlated random phase screens. A coherency matrix can again be set up for this phase screen stack. As before, by diagonalizing this matrix, an equivalent phase screen stack can be established comprised of a smaller number of uncorrelated random phase screens.

The OPD fluctuation introduced by turbulence in this convergent path is, as before, given by the sum of the OPD contributions from each of the phase screens in the stack, where scaled arguments are again used, similar to those used to obtain (16.9). The total OPD fluctuation introduced by the telescope tube path, $H_{\mathrm{Tube}}(\xi, \eta)$, may then be expressed in the form

$$H_{\text{Tube}}(\xi,\eta) = \sum_{i=1}^{n_T} h_i\left(x \cdot \frac{L - z_i}{L}, y \cdot \frac{L - z_i}{L}\right) \tag{16.17}$$

where $n_{\text{T}}$ indicates the number of uncorrelated phase screens constituting the path inside the telescope and where we note that the $(x, y)$ coordinate system again refers to the telescope pupil plane and the scaling factor, $(L - z_i)/L$, accounts for the beam convergence.

The variance and autocorrelation function of $H_{Tube}(\xi, \eta)$ may be written (cf., 16.6 and 16.7) as follows:

$$\sigma_{\text{Tube}}{}^2 = \langle H_{\text{Tube}}(x)^2 \rangle \tag{16.18}$$

and

$$\rho_{\text{Tube}}(\xi,\eta) = \frac{\langle H_{\text{Tube}}(x+\xi, y+\eta) \cdot H_{\text{Tube}}(x,y) \rangle}{\left\langle H_{\text{Tube}}(x,y)^2 \right\rangle} \tag{16.19}$$

## 16.4 OPD Autocorrelation Function for Telescope Coude Paths

Because the beam pencils in Coude beam paths are collimated, the integrated OPD fluctuation arising from these path portions can be calculated in the same way as for the collimated atmospheric beam paths considered in Chap. 6. The only difference now is that because the beam diameter in the Coude path, $D_{\text{Coude}}$, is less than the telescope diameter, $D$, we must use scaled coordinates to express the integrated OPD fluctuation, $H_{\text{Coude}}(x, y)$, over this path. Thus, we write,

$$H_{\text{Coude}}(x,y) = \sum_{i=1}^{n_C} h_i\left(x \cdot \frac{D_{\text{Coude}}}{D}, y \cdot \frac{D_{\text{Coude}}}{D}\right), \tag{16.20}$$

where $n_{\text{C}}$ indicates the number of uncorrelated phase screens representing the Coude path and where the $(x, y)$ coordinate system again refers to the telescope pupil plane.

The variance and the autocorrelation function of $H_{\text{Coude}}(x, y)$ can be written as follows:

$$\sigma_{\text{Coude}}^2 = \left\langle H_{\text{Coude}}(x,y)^2 \right\rangle \tag{16.21}$$

and

$$\rho_{\text{Coude}}(\xi,\eta) = \frac{\langle H_{\text{Coude}}(x+\xi, y+\eta) \cdot H_{\text{Coude}}(x,y) \rangle}{\left\langle H_{\text{Coude}}(x,y)^2 \right\rangle} \tag{16.22}$$

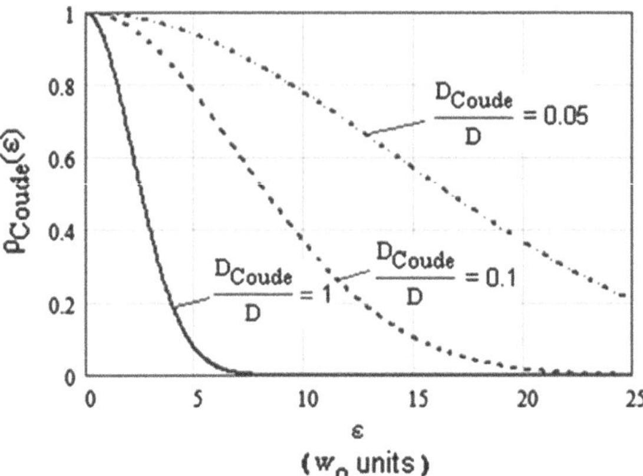


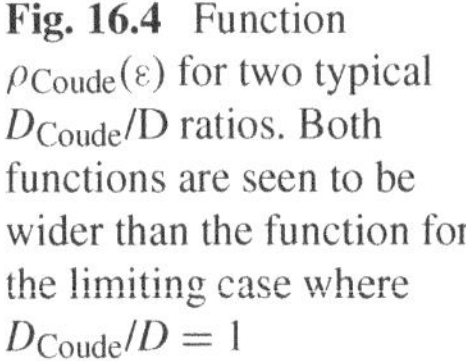
**Fig. 16.4** Function $\rho_{\text{Coude}}(\varepsilon)$ for two typical $D_{\text{Coude}}$/D ratios. Both functions are seen to be wider than the function for the limiting case where $D_{\text{Coude}}/D = 1$

Figure 16.4 shows $\rho_{\text{Coude}}(\xi, \eta)$ plots for two possible ratios, $D_{\text{Coude}}/D = 0.1$ and $D_{\text{Coude}}/D = 0.05$. It may be observed that smaller $D_{\text{Coude}}/D$ ratios result in wider autocorrelation functions. As a comparison reference, the autocorrelation function shown (solid line) corresponds to the case where the same integrated turbulence strength is contributed by a random phase screen located in the telescope pupil plane, rather than in the Coude path; this plot effectively corresponds to the case $D_{\text{Coude}}/D = 1.0$.

## 16.5 Integrated OPD Fluctuation for Entire End-To-End Beam Paths

The quality of the spot image formed by the beam projected by the telescope depicted in Fig. 16.2 is determined by the total integrated OPD fluctuation accruing from all portions of the beam path. We have already seen how to calculate the OPD contributions due to turbulence in convergent beams in target space and inside the telescope, $H_{\text{CBP}}(x,y)$ and $H_{\text{Tube}}(x,y)$, as well as for the contribution from the collimated Coude beam path, $H_{\text{Coude}}(x,y)$. OPD contributions also arise from telescope aberrations which we denote here by $H_{\text{Aberr}}(x, y)$. Other OPD fluctuation sources can also be envisaged. For aircraft-mounted systems, an OPD term for boundary-layer turbulence must be added to the sum, as would contributions arising from thermal blooming and tilt anisoplanatism.

The integrated OPD fluctuation accruing over the entire end-to-end beam path (McKechnie, 2004), which we denote by $H_{\text{INT}}(x,y)$, is simply the sum of all the various contributions,

$$H_{\text{INT}}(x, y) = H_{\text{CBP}}(x, y) + H_{\text{Aberr}}(x, y) + H_{\text{Tube}}(x, y) + H_{\text{Coude}}(x, y) + \cdots . \tag{16.23}$$

where we recall that each of the various OPD contributions represented by the different terms on the right-hand side is defined as a zero-mean contribution. The variance and autocorrelation function of $H_{\mathrm{INT}}(x, y)$ may therefore be expressed in the usual way by

$$\sigma_{\mathrm{INT}}^2 = \left\langle H_{\mathrm{INT}}(x, y)^2 \right\rangle, \tag{16.24}$$

and

$$\rho_{\mathrm{INT}} = \frac{\langle H_{\mathrm{INT}}(x + \xi, y + \eta) \cdot H_{\mathrm{INT}}(x, y) \rangle}{\left\langle H_{\mathrm{INT}}(x, y)^2 \right\rangle}. \tag{16.25}$$

## 16.6 Reducing End-to-End Integrated OPD Fluctuation by Use of AO

For AO-equipped beam projection telescopes, the effect of AO corrections may be considered as simply adding yet one more OPD contribution to the right-hand side of 16.23. But, in this case, unlike all other contributions this contribution is a negative contribution. Denoting the residual OPD fluctuation after AO correction by $H_{\mathrm{INTAO}}(x, y)$, it may be expressed as follows:

$$\begin{aligned} H_{\mathrm{INTAO}}(x, y) = {} & H_{\mathrm{CBP}}(x, y) + H_{\mathrm{Aberr}}(x, y) + H_{\mathrm{Tube}}(x, y) \\ & + H_{\mathrm{Coude}}(x, y) + \ldots + H_{\mathrm{AO}}(x, y). \end{aligned} \tag{16.26}$$

The OPD 'contribution' of the term, $H_{\mathrm{AO}}(x, y)$, may be regarded as negatively correlated with the other OPD terms on the right-hand side of 16.26. We can obtain estimates of the variance and autocorrelation function of $H_{\mathrm{INTAO}}(x, y)$ by operating the AO system while carrying out the same image measurement procedures described previously in Chaps. 8 and 13. As discussed shortly (Sect. 16.8), if the target plane is cooperative/accessible, we may choose to characterize the beam path either in the direction of the outgoing projected beam or in the direction of an incoming beam from a point source in the target plane.

## 16.7 Integrated OPD Fluctuation for an End-to-End Beam Path

When AO correction is used, the variance and autocorrelation function of the residual integrated OPD fluctuation as expressed by 16.26 can be written in the usual manner,

$$\sigma_{\mathrm{INTAO}}{}^{2} = \langle H_{\mathrm{INTAO}}(x, y)^{2} \rangle \tag{16.27}$$

and

$$\rho_{\mathrm{INTAO}}(\xi, \eta) = \frac{\langle H_{\mathrm{INTAO}}(x + \xi, y + \eta) \cdot H_{\mathrm{INTAO}}\mathrm{AO}(x, y) \rangle}{\langle H_{\mathrm{INTAO}}(x, y)^{2} \rangle}. \tag{16.28}$$

Once $\sigma_{\mathrm{INTAO}}$ and $\rho_{\mathrm{INTAO}}(\xi, \eta)$ have been obtained from image intensity measurements, expressions can immediately be set up for the two-point two-wavelength correlation function, $S(\xi, \eta, \lambda_1, \lambda_2)$, and the atmospheric Modulation Transfer Function (MTF), $M(\xi, \eta, \lambda)$, by simply inserting the measured $\sigma_{\mathrm{INTAO}}$ value and the function, $\rho_{\mathrm{INTAO}}(\xi, \eta)$, into equations of the forms given previously by 6.52 and 6.45. Once $S(\xi, \eta, \lambda_1, \lambda_2)$ and $M(\xi, \eta, \lambda)$ have been established in this way, we can then calculate all significant monochromatic and polychromatic properties of the AO-corrected focused spot in the target plane.

## 16.8 Reversibility of Light and Path Characterization Options

The notion of reversibility of light—that is, the notion that light follows exactly the same path in the reverse travel direction—is a consequence of the wave theory of light (Welford, 1962) but only as a geometrical optics approximation; diffraction effects are not fully reversible. Although "reversibility of light" cannot strictly be considered as a rigorous general principle, in many applications it is a valid approximation. For example, in interferometry, we tacitly assume the reversibility principle when the beam direction through the component under test is considered optional and chosen at our convenience.

For the optical arrangement shown in Fig. 16.2, the beam path could be characterized either by (1) locating a point source (as shown in the figure) in the image plane of the beam projection telescope and then measuring the appropriate spot image properties at the target plane or (2) by locating the point source in the target plane and measuring the corresponding properties in the telescope image plane. Either way, the majority of the beam path—that is, that path portion where the beam is expanded—is explored equally. OPD differences between the two paths primarily arise at the ends of the path where the beam cone narrows. Over the tiny beam diameters in these regions, air refractive index is effectively invariant. Such OPD differences as might arise in these regions can therefore be considered small enough to be ignored.

If the beam path shown in Fig. 16.2 is measured with the beam traveling in the direction indicated in the figure—that is, from a point source in the image plane of the beam projection telescope to the distant target plane—the large magnification involved would require use of a large target-board detector to characterize the focused spot. Such a "detector" might comprise a 2-D matrix array of individual "point" detectors, typically spread out over an area covering several square meters. However, by characterizing the path in the other direction—that is, from a point source in the target plane to the image plane of the beam projection telescope—we can use standard off-the-shelf FPA detectors. Path characterization can then be carried out using the same measurement procedures described previously in Chaps. 8 and 13.

## 16.9 Characterizing High Energy Laser (HEL) Beam Paths

While terrestrial laser communication beam paths merely have to contend with atmospheric turbulence, for HEL beam paths additional mechanisms contribute to the total integrated OPD fluctuation. For HEL air-to-air engagements, where the beam projection telescope and the target platform are in relative motion, the task of characterizing HEL beam paths is particularly challenging.

In this type of engagement, OPD temporal fluctuation frequencies are much higher than before due to the aircraft's air speed, necessitating higher bandwidth AO correction rates which result in reduced AO correction accuracy. Aircraft buffeting and vibration cause corresponding vibrations of the telescope and laser optics which then act as further sources of OPD fluctuation. Boundary-layer turbulence around the aircraft fuselage further adds to the OPD total and, for HEL beams, thermal blooming adds yet more, the part-chaotic nature of this heating and convection phenomenon making only partial correction possible.

Despite best-effort use of AO to reduce the total integrated OPD fluctuation from all sources, a significant residual remains uncorrected. This residual ultimately determines how precisely the HEL wavefronts can be brought to focus on the target plane and thus determines the irradiance potency that can be brought to bear on the target.

Consider the air-to-air engagement scenario depicted schematically in Fig. 16.5 where a HEL weapon system, mounted on an aircraft flying at a representative altitude and speed, projects a HEL beam at a cooperative test target mounted on a second aircraft, or test drone, flying independently at a representative altitude, speed, and engagement distance. A probe beam is used to illuminate the target. A corner cube reflector (or some similar device) mounted at an appropriate location in the target plane is used to provide an AO reference feature. It is also supposed that the outgoing pre-corrected HEL beam is directed in a way that takes account of point ahead angle. Our objective now is to characterize the irradiance flux distribution of the focused HEL beam in the target plane in a way that allows calculation of the residual uncorrected OPD fluctuation in the HEL beam—in terms of $\sigma_{INTAO}$ and $\rho_{INTAO}(\xi, \eta)$—as the beam converges towards focus.

Ideally, we would like to achieve the condition, $\sigma_{INTAO} \approx 0$, so that the HEL beam comes to a perfect diffraction-limited focus. In practice $\sigma_{INTAO}$ will be non-zero, so that lower irradiance flux levels are in fact realized. The actual levels are determined by the combination of the measured value of $\sigma_{INTAO}$ and the shape of the measured $\rho_{INTAO}(\xi, \eta)$ function.

In principle, any of the path characterization techniques described in Chaps. 8, 13 and 18 could be used to characterize the residual uncorrected OPD fluctuation in

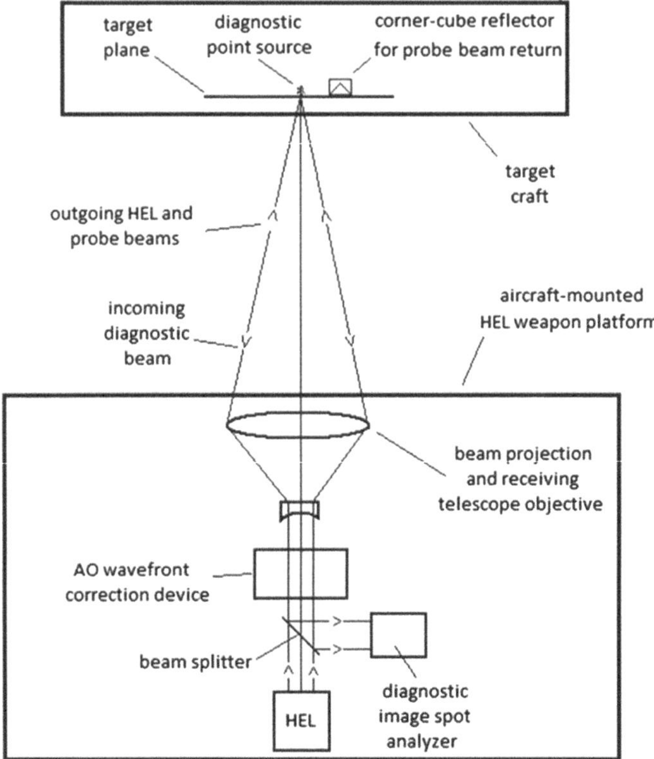


**Fig. 16.5** Schematic showing an optical test setup for characterizing the HEL beam path in an air-to-air engagement scenario. Both the outgoing HEL laser beam and the incoming diagnostic beam explore the same nominal beam path, albeit in opposite directions. Both beams access the same AO wavefront correction device so that they undergo identical AO corrections

the focusing HEL beam. However, in view of the extreme difficulty of this particular characterization task, we elect to use the relatively simple path characterization procedure described in Chaps. 13 and 18, and instead of measuring the residual uncorrected OPD fluctuations of the HEL beam itself, we choose to measure the residual uncorrected OPD fluctuation in a diagnostic beam that traverses the same nominal path in the opposite direction, as per the above "reversibility of light" discussion.

The diagnostic beam originates from a 'point source' located in the target plane at the nominal focus of the HEL beam (Fig. 16.5). Suitable optics would be used to project this beam back toward the HEL weapon platform. The divergence angle of the beam would be chosen so that the beam fully fills the aperture of the HEL beam projection telescope (which may now be regarded as a beam projection/receiving telescope).

The diagnostic beam must provide core and halo images in the diagnostic spot analyzer. Since core and halo characteristics depend on both the diagnostic beam wavelength and the final measured $\sigma_{INTAO}$ value, preliminary experimentation would be needed to establish a suitable wavelength. As indicated in Sect. 13.2, the wavelength selected will lie in the approximate range, $5 \cdot \sigma_{INTAO} \leq \lambda \leq 13 \cdot \sigma_{INTAO}$.

The diagnostic wavelength might typically lie in the wavelength range from near- to long-wavelength IR. Because of the wide range of possible wavelength choices, it would be essential to choose reflective components throughout the test setup, particularly the projection/receiving telescope optics, thus eliminating unnecessary refractive index dispersion effects.

Whatever the diagnostic beam wavelength choice, it would be necessary to limit the wavelength bandwidth, $\Delta\lambda$, to about $\Delta\lambda \approx \lambda/20$, where $\lambda$ is the nominal center wavelength. For broadband sources, band-pass filters could be used to set the desired bandwidth.

The diagnostic beam portion collected by the projection/receiving telescope would be routed as shown in Fig. 16.5 via the same AO wavefront correction device that shapes the outgoing HEL wave fronts. A beam splitter mirror placed as shown just in front of the HEL redirects the diagnostic beam into the 'diagnostic image spot analyzer.' Because the diagnostic beam in this arrangement nominally travels the same path as the HEL beam, the residual uncorrected OPD fluctuation in this beam closely replicates the residual uncorrected OPD fluctuation in the HEL beam. To include the effects of thermal blooming in the path characterization, it would be necessary to operate the HEL beam during characterization.

The core and halo images formed in the diagnostic image spot analyzer can be analyzed in the manner previously described in Chap. 13 (Sect. 13.2.3) by best-fitting the intensity envelopes of these images as the sum of two Gaussian functions—one representing the core, the other the halo. The core and halo image would be characterized as before by obtaining the ratio of the central intensities of the core and halo image portions, $A_{CH}$, and the 1/e half-widths of the respective core and halo Gaussian functions, $B_C$ and $B_H$.

The $A_{CH}$, $B_C$, and $B_H$ values so obtained enable us to quantify the residual fixed aberrations of the HEL beam projection telescope—that is, those telescope aberration residuals that the AO system failed to correct. Denoting the diagnostic wavelength by $\lambda_{Diag}$, the effect of these residuals can be expressed in terms of the Strehl intensity achieved by the AO-corrected telescope optics at the diagnostic wavelength (cf., 13.75) as follows:

$$SI_{AO}(\lambda_{Diag}) = \frac{4 \cdot \lambda_{Diag}^2}{\pi^2 \cdot B_C^2 \cdot (D^2 - d^2)} \tag{16.29}$$

If the telescope Strehl intensity so obtained, $SI_{AO}(\lambda_{Diag})$ were found to be substantially less than unity, it would enhance final system performance if the AO system were fine-tuned at this point to maximize the Strehl value. The residual dynamical OPD fluctuation of the image-forming wavefronts may then be calculated in terms of the residual rms OPD fluctuation, $\sigma_{INTAO}$, and the 1/e half-width of the Gaussian best-fit approximation to the autocorrelation function of the OPD fluctuation, $w_{0AO}$. The appropriate $\sigma_{INTAO}$ value can be calculated (cf., 13.76) using the relation,

$$\sigma_{INTAO} = \frac{\lambda_{Diag}}{2 \cdot \pi} \cdot \sqrt{\ln\left(1 + \frac{B_H^2}{A_{CH} \cdot B_C^2}\right)}, \tag{16.30}$$

and the appropriate $w_{0AO}$ value can be calculated (cf., 13.77) using the relation,

$$w_{oAO} = \frac{2 \cdot \sigma_{INTAO}}{B_H}, \tag{16.31}$$

where the autocorrelation function of the residual OPD fluctuation is approximately given by

$$\rho_{INTAO}(\varepsilon) = \exp\left[-\left(\frac{\varepsilon}{w_{oAO}}\right)^2\right]. \tag{16.32}$$

While circular symmetry has been tacitly assumed here, in instances where the image falling on the diagnostic spot analyzer was not circularly symmetric, a more general analysis could be given in terms of elliptical Gaussian functions as laid out in Sect. 13.2.4.

### *16.9.1 Optimum Wavelength for Maximum HEL Irradiance at the Target*

It is important to choose a HEL wavelength that lies close to the optimum wavelength so as to maximize irradiance flux at the target. Once the diagnostic measurements have produced an estimate of the residual uncorrected OPD fluctuation, $\sigma_{INTAO}$, the optimum HEL wavelength for the measured path may be obtained from of the approximate formula (c.f., 10.16),

$$\lambda_{OptHEL} = 2 \cdot \pi \cdot \sigma_{INTAO}. \tag{16.33}$$

A more precise value for the optimum wavelength could be calculated using a formula similar to that given previously by 13.93. But it should be noted that whereas the Strehl intensity value calculated by from 16.29 applies to the diagnostic wavelength denoted here by $\lambda_{\text{Diag}}$, the Strehl intensity value actually required in the calculation is the value at the optimum HEL wavelength, $\lambda_{\text{OptHEL}}$. A mapping equation is therefore needed to relate the Strehl intensities at these two wavelengths.

If we assume that the projection/receiving telescope optics are entirely reflective, the residual uncorrected wavefront error of the telescope, and hence the rms error variation will be independent of wavelength. If we also assume that the residual rms error is small—as expected for an adequately performing AO system—10.25 can be used to express the Strehl intensity at the two wavelengths, $SI_{AO}(\lambda_{Diag})$ and $SI_{AO}(\lambda_{OptHEL})$, by $(1 - 4 \cdot \pi^2 \cdot \sigma_{INTAO}^2/\sigma_{Diag}^2)$ and $(1 - 4 \cdot \pi^2 \cdot \sigma_{INTAO}^2/\sigma_{OptHEL}^2)$, respectively. It is then a simple matter to show that these two Strehl intensities are approximately related by

$$SI_{AO}(\lambda_{OptHEL}) = SI_{AO}(\lambda_{Diag})^{\left(\frac{\lambda_{Diag}}{\lambda_{OptHEL}}\right)^2}. \tag{16.34}$$

This relation now allows us to replace 16.33 with the more precise expression for the optimum wavelength,

$$\lambda_{OptHEL} = 2 \cdot \pi \cdot \sigma_{INTAO} \cdot \frac{1}{\sqrt{1 + \frac{4 \cdot w_{\varrho AO}^2}{SI_{AO}(\lambda_{Diag})^{\left(\frac{\lambda_{Diag}}{\lambda_{OptHEL}}\right)^2}}}}. \tag{16.35}$$

A solution to this transcendental equation for $\lambda_{OptHEL}$ can readily be obtained numerically. The use of a near-optimum HEL wavelength enables the delivery of maximum irradiance flux at the target and thus maximizes HEL system lethality. Near-optimum HEL wavelengths also maximize lethality range, which translates to wider theater coverage for the same number of deployed HEL weapon systems.

An optimally designed system should generally use a wavelength consistent with the $\sigma_{INTAO}$ value measured over the longest envisaged target path. Although this wavelength would be less than optimum for shorter ranges, the irradiances delivered at these shorter ranges would nonetheless be greater than that delivered at the maximum range. It might be observed here that, given the luxury of an adjustable-wavelength laser, such as a free electron laser (FEL), the HEL wavelength could be adjusted as needed to deliver maximum possible irradiance flux at the target for any given target distance.

#### 16.9.1.1 HEL Wavelength Choice for the Airborne Laser (ABL) Program

The objective of the flagship Airborne Laser (ABL) program was to develop and build an advanced HEL weapon system mounted aboard a Boeing 747 aircraft (Kasser & Sen, 2013) capable of destroying theater ballistic missiles (TBMs) from standoff distances of several hundred kilometers. The HEL beam strike would be made during the missile's boost phase when the missile is most vulnerable. (TBMs typically travel at speeds of 2 km/s and can strike targets at distances ranging from 300 to 3500 km.)

The ABL program ran for sixteen years, from 1996 to 2012, initially under the control of the US Air Force (USAF) until 2001 when control transferred to the Missile Defense Agency (MDA).

The chemical oxygen iodide laser (COIL) with its 1.315-$\mu$m output wavelength was selected as the HEL for the ABL program. COIL was invented in 1977 and developed throughout the 1980s at the US Air Force Research Laboratory (AFRL), Albuquerque, New Mexico. Whatever the scientific reasoning behind the COIL laser choice, the limited understanding of the physics of light propagation through the atmosphere at the time the choice was made hid the fact that 1.315-μm was particularly ill-suited to the ABL application. Ironically, instead of maximizing ABL lethality range and beam potency at the target—as was no doubt the intention—this wavelength choice all but minimized these two crucial attributes, casting a dark cloud over the ABL program from the outset.

The above assessment of the ABL wavelength choice owes little to hindsight. It dates from 1989[2] following the derivation of the 'optimum wavelength' equation, $\lambda_{Opt} = 2\,\pi\,\sigma$, from which it was instantly clear that 1.315 μm was disturbingly far from optimum for the ABL application.

The Author's original (1989) argument against use of the COIL wavelength was based on a comparison study between the rms residual OPD variation for typical AO-corrected ABL beam paths and the rms residual OPD variation for typical Mt. Palomar 200-inch observing paths, that instrument being the foremost telescope in the world at that time (McKechnie, 1991a, Sect. 10 and Eq. 10.77). In this section, we update the argument, using Keck II telescope observing paths as the comparison reference, thus taking account of the current state of AO technology.

Suppose for the sake of argument that 1.315-μm had indeed been the optimum HEL wavelength choice for the ABL application and thus would enable delivery of maximum theoretically possible irradiance flux at the target. To be consistent with that supposition, 16.33 indicates that $\sigma_{INTAO} \approx 0.2\mu m$ (where we recall that $\sigma_{INTAO}$ is the rms variation of the residual uncorrected OPD fluctuation in the outwardly projected HEL beam after best-effort AO pre-corrections are applied). Whether or not 1.315-μm was in fact an appropriate wavelength choice for the ABL program can now be judged by whether or not there was any realistic expectation of achieving such a miniscule $\sigma_{INTAO}$ value (0.2 μm) over the envisaged, hundreds of kilometer ABL beam paths.

Perched on Mauna Kea, Hawaii, 4200 m above sea level and above 40% of the atmospheric mass, the Keck II instrument enjoys some of the world's finest seeing conditions. With the benefit of AO, the sharpest star images currently formed by this instrument (Sect. 14.10) are obtained at a wavelength of about 2 μm, indicating that this near-IR wavelength is close to optimum for Keck II observing paths (in assumed quiescent observing conditions). Equation 16.33 then indicates that the residual rms OPD fluctuation, $\sigma_{INTAO}$, for the AO-corrected image-forming waves produced by the Keck II instrument must be about 0.32-μm.

Since this rms value is already 60% larger than the 0.2-μm value required to justify/rationalize the 1.315-μm ABL wavelength choice, clearly we have uncovered a problem. Less clear is the full enormity of the problem as now explained.

Aircraft-mounted HEL weapon systems must contend with many more sources of OPD fluctuation than the Keck II instrument. Moreover, the OPD contributions from some of these sources are likely to be far larger than any encountered by Keck II. Thus, the rms residual OPD fluctuation in the HEL light waves projected towards

[2] The equation was developed in 1989 while the Author was providing optics consultancy support to the Falcon Nuclear Laser program at Sandia National Laboratories, Albuquerque, New Mexico.

the target must be considerably larger than 0.2-μm. One then asks: what might be a more realistic rms value and how will this affect ABL system performance?

The HEL wave fronts initially emerging from the hot turbulent gas-filled COIL laser cavity naturally exhibit strong rapidly varying OPD fluctuations. Airframe induced vibrations of the telescope and laser optics, boundary layer turbulence, and thermal blooming over the atmospheric path add yet more OPD fluctuation. A portion of the OPD fluctuation arising from tilt-anisoplanatism (a consequence of point ahead angle) is unmeasurable and therefore uncorrectable, which further adds to the overall OPD burden.

Shallow slant angle (~1 degree) ABL beam paths hundreds of kilometers in length enclose more than an order of magnitude more air mass—and thus introduce correspondingly more OPD fluctuation—than near-vertical Keck II observing paths.[3] Moreover, for daytime ABL engagements, direct heat from the Sun more than doubles the atmospheric turbulence OPD contribution.

The several-hundred mph speed of the ABL aircraft together with the target missile speed greatly increase the temporal frequency rates of the OPD fluctuations, necessitating higher AO-correction rates and faster computational speeds. This calls for shorter OPD sampling intervals, resulting in increased photon shot noise effects and reduced overall AO-system S/N noise ratio. By having to contend with all these additional OPD sources, plainly the task of AO pre-correcting the outgoing ABL HEL wave fronts is orders of magnitude more challenging than the corresponding wavefront correction task faced by the Keck II instrument.

Suppose we make the—conservative—estimate that the residual uncorrected rms OPD fluctuation for a typical ABL beam path was in fact about 1.6 μm — a mere five-times greater than the 0.32 μm value typical of AO-corrected Keck II imaging paths. Equation 16.33 tells us that the optimum HEL wavelength is then about 10 μm. While there happens to be an atmospheric window at 10 μm and even a suitable HEL laser the $CO_2$ laser with its 10.6-μm output wavelength—the overall ABL prognosis is now disturbingly bleak. The average irradiance flux in the center of the focused HEL beam spot, $I(0, \lambda_{HEL})$, is given in terms of the HEL wavelength, $\lambda_{HEL}$, and the rms residual uncorrected OPD fluctuation, $\sigma_{INTAO}$, by (c.f., 10.4)

$$\langle I(0, \lambda_{HEL})\rangle = \frac{1}{\lambda_{HEL}^{2}} \cdot \exp\left[-\left(\frac{2 \cdot \pi \cdot \sigma_{INTAO}}{\lambda_{HEL}}\right)^{2}\right]. \quad (16.36)$$

Figure 16.6 shows how, according to 16.36, irradiance flux at beam focus varies with HEL wavelength. The sharp falloff at wavelengths shorter than the optimum wavelength indicates that, if the chosen HEL wavelength lies in that wavelength region, ABL system performance is instantly decimated. The same is true of wavelength choices longer than the optimum wavelength, even if the fall-off rate is less severe.

[3] Because extinction losses increase exponentially with air mass, this further adds to the irradiance flux short-fall at the target.

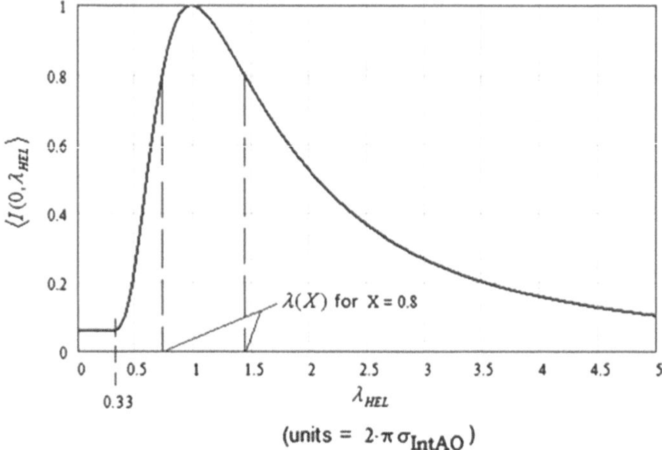


**Fig. 16.6** Plot shows how irradiance delivered in the center of a focused HEL beam varies as a function of HEL wavelength, $\lambda_{HEL}$. Maximum irradiance is attained at the optimum wavelength (corresponding to the value, unity, in the dimensionless wavelength units along the horizontal axis). Suitable HEL wavelengths lie in the approximate range, $0.75 \leq \lambda_{HEL} \leq 1.45$, indicated by the two vertical dashed lines which correspond to the abscissa value 0.8. According to the reasoning in Sect. 16.9.1.1, the actual ABL wavelength choice, 1.315 μm, lies to the left of abscissa value, 0.33

According to 10.17, the irradiance flux at the optimum wavelength, $\lambda_{Opt}$, diminishes in proportion to $1/\lambda_{Opt}^2$. Consequently, if 10.6 μm were indeed the optimum HEL wavelength for the ABL application, maximum theoretically possible irradiance flux at the target would now be some sixty-five times $[(10.6/1.315)^2]$ less than anticipated under the original (1980s) supposition that 1.315-μm was near-ideal for ABL applications. The effect of such a pull-down factor on ABL performance is painfully obvious; yet there is one further pull-down factor to be included.

If the optimum ABL wavelength was indeed 10.6 μm, Fig. 16.6 indicates that the use of any wavelength more than 3-times shorter than 10.6 μm (which includes 1.315 μm) would be subject to a further ~ 10-times irradiance pull-down. Combining both pull-down factors, 10 and 65, the overall irradiance flux actually delivered by the COIL 1.315-μm beam would be a catastrophic 650-times less than anticipated under the original—but now plainly mistaken—notion that 1.315 μm was the ideal choice for the ABL program.

The irradiance pull-down factor, PD, for the general case where the actual optimum ABL wavelength is significantly longer than 1.315 μm is given approximately by,

$$PD \approx 10 \cdot \left( \frac{\lambda_{opt}}{1.315 \cdot \mu m} \right)^2. \tag{16.37}$$

Recalling that the assumed rms uncorrected OPD fluctuation, $\sigma_{INTAO} = 1.6 \mu m$, is considered a conservative estimate, the actual optimum wavelength may likely be considerably longer than 10.6 μm. If it were 20 μm, the above equation would indicate 2300-times less irradiance flux at the target than originally anticipated. Plainly, such enormous target irradiance short-falls spelled disaster for the ABL program.

In theory, the ABL system could still achieve the desired irradiance flux levels by simply shortening the stand-off distance requirements. But in this case, the ABL lethality range would have to shrink by a factor of about $\sqrt{650}$ ($\approx 25$) and perhaps by even larger factors ... The originally envisaged hundreds of kilometer ABL standoff distances would now have to shrink to, at most, a few tens of kilometers, rendering the ABL end product wholly unfit for purpose.

### Just Acceptable HEL Wavelength Range for the ABL System

The two wavelengths lying on either side of the optimum wavelength in Fig. 16.6, where the irradiance flux falls off by the arbitrary factor X (where $0 \leq X \leq 1$) may be identified by solving the quadratic equation,

$$\left(\frac{\lambda_{OptHEL}}{\lambda(X)}\right)^2 \cdot \exp\left[-\left(\frac{\lambda_{OptHEL}}{\lambda(X)}\right)^2\right] = X. \tag{16.38}$$

Solutions may be obtained by numerical methods, or by analytical means using the appropriate Lambert W-function (Corless et al., 1996). If we now consider (somewhat arbitrarily) that the HEL wavelength sufficiently approximates the optimum wavelength as long as $X \geq 0.8$, 16.38 can be used to show that a suitable ABL HEL wavelength, $\lambda_{HEL}$, would have to lie somewhere in the range,

$$0.75 \cdot \lambda_{OptHEL} \leq \lambda_{HEL} \leq 1.45 \cdot \lambda_{OptHEL}$$

The two extremes of the "just-acceptable" HEL wavelength range are depicted by the dotted vertical lines in Fig. 16.6. If the actual optimum wavelength for the ABL system had indeed been 10 μm, the just acceptable wavelength range in that case would lie between 7.5 and 14.5 μm. If the optimum wavelength had been as long as 20 μm the just acceptable wavelength range would lie between 15 and 29 μm. Regrettably, the actual HEL wavelength chosen for the ABL system, 1.315 μm, falls well outside either of these ranges.

While it would be critically important that the HEL wavelength choice lies within the just-acceptable optimum wavelength range, the final wavelength selection would have to take account of a wider set of considerations. These include the availability of a functioning HEL laser with sufficiently high output power at a wavelength with acceptably low atmospheric absorption losses over representative ABL beam paths. Perram et al. (2010) have discussed some of these other considerations. In most instances, the generously broad, just-acceptable HEL wavelength ranges just indicated would usually allow simultaneous capture of all these other requirements.

### Final ABL System Tests and Program Closure

The ABL performance number projections used to promote and ultimately launch the ABL program in the 1980s and 1990s must have been vastly different from the performance numbers just indicated. In early-1990, the Author began a warning campaign to alert the relevant management and technical communities of the dire consequences of the 1.315-μm HEL wavelength choice. The campaign included more than a dozen oral presentations given between 1990 and 2003 to optics- and ABL-related audiences at AFRL, the Pentagon, and the Missile Defense Agency (MDA).

The talks were not well received. At best, the message was seen as inconvenient. There was no willingness by anyone in the audiences to engage in substantial scientific\mathematical dialogue aimed at establishing why the ABL community and the Author had reached such diametrically different ABL performance assessments. Crucial new (1989) advancements in understanding, such as the existence of optimum HEL wavelengths for maximizing beam potency at the target, were simply shrugged off.

The final campaign talk in 2003 was given at MDA headquarters. Confronted by a united wall of opposition, it was clear that ABL technical and management leadership had closed ranks and wished no further discussion of the wavelength issue.

The ABL program carried on for nine more years despite mounting problems and increasing scrutiny from a skeptical US Congress. In late 2011, final field testing of the completed ABL system found its destructive potency to be only a tiny fraction of required levels. Soon afterwards, in early 2012, Congress axed the program and decommissioned the 747 aircraft, leaving little to show the for the $6-billion expenditure.

Today, some ten years after program closure, there has still been no formal acknowledgment of the HEL wavelength issue. On the contrary, evidence from June 2016 indicates coordinated effort to suppress discussion of the issue (Appendix J). Those responsible continue to obstruct the free emergence of scientific truth, part of a historical pattern of deception and cover-up that unconscionably misled Congress, preventing it from fulfilling its responsibility of effective oversight of the US taxpayer-funded ABL program. The 45th vice president of the United States, Albert A. Gore, Jr., has described this – not uncommon – behavioral response in his best-selling book, An inconvenient truth (2006):

> "It's difficult to get a man to understand something if his salary depends on him not understanding it"

### 16.9.2 *Top-Level Feasibility Analysis of HEL Weapon Systems*

In this section, we provide a simplified top-level analysis procedure that provides rough initial guidance as to whether or not a proposed HEL weapon system has a realistic chance of destroying a specified target over a prescribed horizontal atmospheric path.

It is assumed that path characterization experiments have been carried out as described in Sect. 16.9 over a representative path using a projection telescope similar, or identical, to the one envisaged for the HEL weapon system. It is also assumed that a reasonably representative value has been obtained for the residual uncorrected rms OPD fluctuation, $\sigma_{INTAO}$, one that accounts for all uncorrected OPD residuals.

Inserting this $\sigma_{INTAO}$ value into 16.35, establishes the optimum HEL wavelength, $\lambda_{OptHEL}$, for the path. It is assumed that a HEL is available with an output wavelength lying in the just-acceptable wavelength range with the required output beam power, $P_{HEL}$ (Watts). It is further assumed that ablation experiments have been conducted with this laser to establish the minimum (Watt/m$^2$) irradiance flux, $W_{Targ}$, required to destroy the target within the prescribed beam dwell time. Once the various system parameter values have all been fixed, we may then assess whether the HEL system will in fact be capable of destroying the target in the envisaged engagement scenario.

Ignoring the small effect of the telescope central obstructions, the irradiance flux distribution in the core of the focused spot in the target plane will be that of an Airy pattern formed by a diffraction-limited version of the projection telescope, but of course now containing a reduced portion of the total light energy. The total light energy contained under the Airy pattern core formed at wavelength, λ, by a telescope of diameter, D, in a target plane at distance, L, can be calculated as follows:

$$2 \cdot \pi \cdot \int_0^\infty \left[ 2 \cdot \frac{J_1\left(\frac{\pi \cdot D \cdot r}{\lambda \cdot L}\right)}{\left(\frac{\pi \cdot D \cdot r}{\lambda \cdot L}\right)} \right]^2 \cdot r \cdot dr = 4 \cdot \pi \cdot \left( \frac{\lambda \cdot L}{\pi \cdot D} \right)^2 \tag{16.39}$$

where $J_1(\cdot)$ is the first-order Bessel function of the first kind.

At the optimum wavelength, $\lambda_{OptHEL}$, the core only contains the fraction $1/e$ (cf., 10.19) of the total light energy, where "$e$" is the Naperian logarithm base. The remaining light fraction resides in the surrounding halo. Denoting the total transmission pull-down factor due to HEL beam losses through the telescope optics and absorption losses over the atmospheric path by $T_o$, it follows from 16.39 that the irradiance flux in the center of the Airy pattern core at beam focus, $W_{max}$, is given by

$$W_{max} = \frac{\pi \cdot D^2 \cdot T_o \cdot P_{HEL}}{4 \cdot e \cdot L^2 \cdot \lambda^2_{OptHEL}} \cdot \frac{W}{m^2} \tag{16.40}$$

Elsewhere in the Airy pattern light distribution at the arbitrary distance, $r$, from the center of the Airy pattern, the irradiance delivered, $W(r)$, is given by

$$W(r) = \left[2 \cdot J_1\left(\frac{\pi \cdot D \cdot r}{\lambda_{OptHEL} \cdot L}\right) / \left(\frac{\pi \cdot D \cdot r}{\lambda_{OptHEL} \cdot L}\right)\right]^2 \cdot W_{max}. \quad (16.41)$$

(Note that the function $W(r)$ used here is unrelated to its previous use in Chap. 10 where it denoted telescope wavefront error.)

By using 16.33, we may substitute for $\lambda_{OptHEL}$ in 16.40 to obtain an alternative expression for the maximum irradiance at the target,

$$W_{max} = \frac{D^2 \cdot T_o \cdot P_{HEL}}{16 \cdot \pi \cdot e \cdot L^2 \cdot \lambda_{OptHEL}^2} \frac{W}{m^2}. \quad (16.42)$$

Whether or not the proposed HEL weapon system can meet its performance objectives depends on fulfilling the condition, $W_{max} \geq W_{Targ}$. Thus, we require,

$$\frac{D^2 \cdot T_o \cdot P_{HEL}}{16 \cdot \pi \cdot e \cdot L^2 \cdot \sigma_{INTAO}^2} \geq W_{Targ}. \quad (16.43)$$

If the above condition is not fulfilled, it would be concluded that the HEL system does not meet its performance objective, in which case design changes would be necessary to increase performance to the desired level. Such changes could include increasing the HEL beam power, $P_{HEL}$, increasing the optical transmission factor, $T_o$, increasing the telescope diameter, D, or improving AO system performance to reduce the uncorrected rms OPD fluctuation, $\sigma_{INTAO}$. With the last-mentioned approach, the maximum acceptable value for rms uncorrected OPD fluctuation, $\sigma_{MA}$, may be calculated using (cf., 16.43).

$$\sigma_{MA} = \frac{D}{4 \cdot L} \cdot \sqrt{\frac{P_{HEL} \cdot T_o}{\pi \cdot \varepsilon \cdot W_{Targ}}}. \quad (16.44)$$

With this new $\sigma_{MA}$ value, 16.33 tells that the HEL system would now have to operate at a new (shorter) wavelength given by $2 \cdot \pi \cdot \sigma_{MA}$.

### 16.9.3 Final Recourse When Design Changes Fail to Deliver Performance

If, after implementing some, or all, of the above design modifications, the HEL system is still unable to meet performance requirements, the only recourse at that point would be to relax the target standoff distance requirement and accept a smaller lethality range. Though disappointing, such recourse may be the only real option, a reflection of the fact that, given the practical and technological limitations in this development area, there are certain practical limits to HEL weapon system performance.

### *16.9.4 Example Calculation for Hypothetical HEL Weapon*

Description of proposed system: In this example calculation, it is supposed that a representative 40-km HEL beam path has been characterized using the same 1.5-m diameter telescope that will ultimately project the HEL beam at the target.

| | |
|---|---|
| HEL wavelength: | $\lambda_{\mathrm{HEL}} = 4.1\ \mu\mathrm{m}$, |
| HEL power output: | $P_{HEL} = 10^6$ W |
| Transmission through optics and atmospheric path: | $T_o = 0.5$ |
| Irradiance required at spot center | $I_{Targ} = 10^7\ \mathrm{W/m^2}$ |
| Projection telescope diameter: | $D = 1.5$ m |
| Target distance | $L = 40$ km |

Given the above parameter values, the question set in this example calculation is the following: Is there a realistic possibility that the HEL weapon system will meet its objective and be able of destroy the type of target considered at the required standoff distance? Inserting the values just listed for D, $T_o$, L, $P_{HEL}$, and $\lambda_{OptHEL}$ into 16.40 gives the delivered irradiance flux at the target, $W_{max} = 1.2 \cdot 10^7 \cdot \mathrm{Watts/m^2}$. Since this value just exceeds the minimum irradiance requirement, $W_{Targ} = 10^7 \cdot \mathrm{Watts/m^2}$, it may be concluded that the HEL weapon system in this instance is capable of meeting its performance objective.

## 16.10 Optimum Wavelengths for Laser Communication Systems

For laser communication systems, use of wavelengths close to the optimum wavelength enable maximum signal to be delivered into the receiver aperture. Also, by appropriately choosing the receiver aperture size, it can be arranged that the majority of the collected light energy arises from the central core portion of the beam spot. By thus excluding most of the light in the surrounding halo—and hence most of the associated speckle noise—higher overall S/N ratio is achieved, permitting higher data rates or, alternatively, longer communication paths spanned by fewer relay stations.

## 16.11 Mathematical Notation Used in This Chapter

The mathematical notation used in this chapter is indicated in Table 16.1.

**Table 16.1** Mathematical notation used in this along with the SI dimensional units of the individual quantities

| Symbol | Quantity | Dimensions |
|---|---|---|
| $\lambda_{HEL}$ | HEL wavelength | $m$ |
| $\lambda_{Diag}$ | Wavelength of diagnostic beam used to characterize path | $m$ |
| $\lambda_{OptHEL}$ | Optimum HEL wavelength | $m$ |
| $(x, y)$ | Cartesian coordinate system in plane perpendicular to optical axis | $m$ |
| $n$ | Refractive index distribution over atmospheric path | "1" |
| $L$ | Distance from projection telescope to beam focus | $m$ |
| $\chi$ | Inclination angle of ray in focusing beam to optical axis | "1" |
| $H$ | Integrated OPD fluctuation over beam path | $m$ |
| $\sigma$ | rms of integrated OPD fluctuation, $H$ | $m$ |
| $\rho$ | Autocorrelation function of integrated OPD fluctuation, $H$ | "1" |
| $w_0$ | $1/e$ half-width of Gaussian approximation to autocorrelation function of $H$ | $m$ |
| $\sigma_{AO}$ | rms of residual OPD fluctuation after AO correction | $m$ |
| $\rho_{AO}$ | Autocorrelation function of residual OPD fluctuation after AO correction | "1" |
| $w_{0AP}$ | $1/e$ half-width of Gaussian autocorrelation function of residual OPD fluctuation | $m$ |
| $D$ | Telescope diameter | $m$ |
| $d$ | Telescope central obstruction diameter | $m$ |
| $D_{\text{Coude}}$ | Beam diameter in Coude path (if applicable) | $m$ |
| $f$ | Telescope focal length | $m$ |
| $I$ | Intensity distribution at beam focus | "1" |
| $SI$ | Telescope Strehl intensity | "1" |
| $A_C$ | Intensity in the center of Gaussian core at beam focus | "1" |
| $A_H$ | Intensity in the center of Gaussian halo at beam focus | "1" |
| $A_{CH}$ | Ratio $A_C/A_H$ | "1" |
| $B_C$ | Angular $1/e$ half-width of Gaussian core | "1" |
| $B_H$ | Angular $1/e$ half-width of Gaussian halo | "1" |
| $To$ | Total beam loss factor over atmospheric path and through telescope optics | "1" |

(continued)

**Table 16.1** (continued)

| Symbol | Quantity | Dimensions |
|---|---|---|
| $W$ | Irradiance flux at HEL beam focus | kg $s^{-3}$ (W/$m^2$) |
| $W$Targ | Irradiance flux at HEL beam focus required to destroy target | kg $s^{-3}$ (W/$m^2$) |
| PD | Irradiance flux pull-down at target | "1" |

Dimensionless quantities are indicated by "1"

# References

Corless, R. M., Gonnet, G. H., Hare, D. E., Jeffrey, D. J., & Knuth, D. E. (1996). On the Lambert W function. *Advances in Computational Mathematics, 5*, 329–359.

Ishimaru, A. (1978). *Wave propagation and scattering in random media* (Vol. 2). Academic.

Kasser, J., & Sen, S. (2013). *The United States airborne laser test bed program; a case study*. Retrieved December 18, 2014, from http://www.academia.edu/4098242/The_United_States_Airborne_Laser_Test_Bed_program_A_case_study

McKechnie, T. S. (1991a). Light propagation through the atmosphere and the properties of images formed by large ground-based telescopes. *Journal of Optical Society of America A, 8*, 346–365.

McKechnie, T. S. (1991b). Focusing infrared laser beams on targets in space without using adaptive optics. In *Proceedings of SPIE, Propagation of high energy laser beams through Earth's atmosphere* (Vol. 1408, pp. 119–135).

McKechnie, T. S. (2004). Fundamental measurement procedures for establishing optical tolerances for ground-based telescopes. In S. C. Craig & M. J. Cullum (Eds.), *Proceedings of SPIE modeling and systems engineering for astronomy, Bellingham, WA* (Vol. 5497_11, pp. 103–116).

Perram, G. P., Cusumano, S. J., Hengehold, R. L., & Fiorino, S. T. (2010). *Introduction to laser weapon systems* (J. S. Accetta (Ed.)). The Directed Energy Professional Society; Managing.

Strohbehn, J. W. (1985). *Laser beam propagation in the atmosphere* (Vol. 25). *Topics in applied physics*. Springer.

Welford, W. T. (1962). *Geometrical optics*. North-Holland Publishing Co.

# Chapter 17
# Atmospheric Isoplanatic Angle

**Abstract** In this chapter an expression is developed for calculating the isoplanatic angles for a generalized atmosphere. The expression is evaluated for vertical atmospheric paths at visible and IR wavelengths as well as for two types of turbulence: Kolmogorov turbulence and the much smaller turbulence measured in 1988 by Coulman et al. For the former type, isoplanatic angles are disappointingly small (~10 arcsec). For the latter, the angles are similarly small for star images comprised entirely of speckle. However, for star images containing central cores—which are routinely anticipated at near-IR and longer wavelengths for this smaller type of turbulence—the analysis predicts that the twin cores in binary star images could remain angularly locked together (i.e., isoplanatic) for angles as large as 10 arcmin. If confirmed, such angles would greatly lessen the need for expensive laser guide star systems. By using star image cores as reference features for image stabilization and AO corrections (as opposed to the less stable light energy centroids) the sky coverage fraction provided by the limited number of suitably bright natural reference stars in the sky could expand by a factor in excess of one thousand. The chapter concludes with a quantitative examination of the age-old enigma: "Why do stars twinkle but not planets? "Of the wide variety of materials contained in this book, the contents of this chapter are perhaps the most speculative; at the same time, they are some of the more exciting. As we shall see, they offer the prospect of highly resolved, AO corrected images with high sky coverage fractions using only the limited number of suitably bright natural stars as reference objects.

The concept of isoplanaticity was previously discussed in Chap. 4 (Sect. 4.7) where we saw that it relates particularly to the imaging of extended objects. An isoplanatic patch refers to an area in either image space or object space over which both the amplitude and intensity point-spread functions (PSFs) of a telescope, or other optical system, remain substantially invariant.

For ground-based astronomical telescopes, the overall isoplanatic patch size depends on the isoplanatic characteristics of both the atmospheric path and the telescope optics. In this chapter, we suppose that both the amplitude and intensity PSFs associated with the telescope itself remain invariant over the entire image plane

 
T. S. McKechnie, *General Theory of Light Propagation and Imaging Through the Atmosphere*, Progress in Optical Science and Photonics 20,
https://doi.org/10.1007/978-3-030-98828-9_17

so that we can examine, in isolation, the isoplanatic restrictions imposed by the atmosphere. The examination is based on the simplest of all extended objects: the two-point object.

Light emitted by astronomical objects can usually be considered incoherent. Therefore, for any extended astronomical object fully contained within the isoplanatic patch, the image arises as the convolution of the object intensity distribution with the intensity PSF of the telescope/atmosphere combination. If natural stars are used as reference objects either for carrying out image stabilization or for making AO image corrections, and the goal is to produce diffraction-limited images, it is essential that the chosen reference star lies within the same isoplanatic patch as the object. Because isoplanatic patch size angles deriving from Kolmogorov assumptions are relatively small, typically measuring less than 10 arcsec across, and because there are only a limited number of stars in the sky bright enough to be used as reference stars, sky coverage was thought to be restricted to a few-hundred tiny, isolated island patches in the sky.

From about 1964 to 1990, star image cores at visible and near-IR wavelengths were widely considered illusory simply because they were squarely at odds with Kolmogorov theory. Consequently, their significance to atmospheric isoplanatic angle went unrecognized. However, star image cores obtained 1989/90 (Sect. 14.8) quickly changed perceptions.

By confirming the Author's long-held contention that star image cores should appear routinely at near-IR and longer wavelengths (and even at visible wavelengths in 0.5-arcsec or better seeing conditions) there was now a concrete basis for the broader prediction that the isoplanatic angles associated with star image cores should be orders of magnitude larger than the several arcseconds anticipated by Kolmogorov theory. There was now a real prospect of huge sky coverage increases by using the cores in reference star images (rather than star image centroids) as reference features for stabilizing images and making AO corrections.

Isoplanatic angle is examined in this chapter for two fundamentally different types of turbulence: (1) Kolmogorov turbulence, and (2) the much smaller turbulence variety measured by Coulman et al. (1988). It is noted that the turbulence structure size characteristics measured by these authors was remarkably similar to the structure sizes deduced from the star image measurements made by the Author at RGO in 1975/76 (Chap. 8).

Kolmogorov turbulence assumptions lead to the prediction of relatively small isoplanatic angles, typically ranging from about 3 arcsec at visible wavelengths to about 25 arcsec at 3.4 μm. Meanwhile, the smaller turbulence structure measured by Coulman et al. is not consistent with Kolmogorov turbulence assumptions and consequently leads to the expectation of much larger associated isoplanatic angles. As discussed previously (McKechnie, 1992) image core isoplanatic angles are likely to be very large, perhaps of the order 600 arcsec (10 arcmin). Such angles are one or two orders of magnitude larger than those anticipated by Kolmogorov theory, and thus offer the prospect of solid angle sky coverage enlargement factors of the order $10^3$ or $10^4$ using only natural guide stars as reference objects.

In 1991, experiments sponsored by Sandia National Laboratories[1] were carried out using the 3.8-m Mayall telescope (Kitt Peak, AZ) to test the validity of the large isoplanatic angle postulation for star image cores. Short-exposure images of binary stars were obtained, and evidence was indeed found of isoplanatism holding for binary separations up to 2 arcmin; the limited field of view of the Mayall instrument prevented exploration of larger angles. Funding cuts to the supporting program[2] prevented follow-up testing, leaving the matter tantalizingly unresolved.

In Sect. 17.2, an experimental setup is described, similar to the one used in 1991, that permits core isoplanaticity measurements at angles up to 10 arcmin. If such large angles are confirmed experimentally, they could significantly influence the way astronomers go about obtaining images with existing 10-m class and future 20- to 39-m class telescopes. The prospect beckons of these huge instruments delivering diffraction-limited images—with near-complete sky coverage—using only natural stars as reference objects. If this prospect were actually realized, there would be less need for laser guide stars. The much larger isoplanatic angles would require more sophisticated secondary imaging optics to deliver diffraction-limited images over the larger fields of view; however, the cost of such optics would be considerably less than the cost of implementing laser guide star systems.

## 17.1 Isoplanatic Angle Background

For binary stars with separations of just a few arcsec, the twin speckle images at visible wavelengths formed by large telescopes would generally look closely similar, even at the level of individual speckle features. However, with increasing binary star separation, the speckle pattern images would look increasingly different.

The isoplanatic angle for any given viewing path can be established by measuring the degree of correlation between the intensities at any two corresponding locations in the twin speckle pattern images. The isoplanatic angle may be considered as the angle at which the degree of correlation falls to some suitably chosen threshold value, such as 0.5. For most practical purposes, an isoplanatic patch can be considered as a circular patch area whose angular radius is set by the isoplanatic angle. This angle is mostly determined by atmospheric turbulence lying at intermediate and high altitudes. As light waves from well-separated binary star components travel towards the telescope in the altitude range between, say, 1 and 20 km, the image-forming light waves from each of the stars pass through a mixture of commonly shared and separate turbulence structures. The resulting differences in the OPD fluctuations introduced by the two partially overlapping light paths cause decorrelation of the

[1] The experiment was supported by Sandia National Laboratories. Others who attended the three-night experiment included, D. R. Neal, M. Kaufmann, and R. Michie. National Optical Astronomy Observatory (NOAO) kindly provided the use of the telescope and the assistance of technical personnel, F. F. Forbes, R. G. Probst, and R. Kraus.

[2] The isoplanatic measurement activity was carried out as part of the Falcon Nuclear Laser development program.

complex amplitudes (and hence the intensities) in the twin speckle images, effectively reducing the isoplanatic angle. When the same light waves travel through lower altitude regions, they tend to pass through essentially the same turbulence structures so as to barely affect the isoplanatic angle.

Dainty (1984) discusses isoplanatic angle measurements made at visible wavelengths by various researchers. Measurements by Lohmann and Weigelt (1979) and Weigelt (1979) indicated some level of correlation for stars as far apart as 22 arcsec; however, their measurements mostly showed that correlation had substantially fallen to zero for angles in the range 2–5 arcsec. Such small angles severely limit the sky coverage offered by the limited number of stars bright enough to be used as reference stars.

Nowadays, to gain more complete sky coverage, some large telescopes (e.g., the 8-m Gemini instruments) are equipped with multiple laser probe beams. The individual beams are focused on an upper atmospheric layer, chosen for its ability to return a useful portion of the light energy back towards the telescope. In this way, a constellation of artificial guide stars is created in an upper atmospheric layer, the guide star matrix extending over the same area as the image-forming wave portions ultimately collected and imaged by the telescope.

Once the individual sub-portions of the image-forming wavefronts have been corrected, computer-controlled real-time stitching algorithms are used to create a composite, phase-corrected image-forming wave spanning the entire telescope collection aperture. Plainly, the number of laser guide stars required for full wavefront correction increases in proportion to the square of the telescope aperture. As telescopes continue to grow larger, and with the generation of ELTs fast approaching, the laser guide star approach becomes increasingly challenging. Adding to the challenge, final image sharpness remains fundamentally limited by focal anisoplanatism[3] and the fact that laser guide stars are far from ideal point-objects. During the upward ascent of the laser beams, atmospheric turbulence blurs and broadens the guide stars.

If isoplanatic angles associated with image cores do indeed turn out to be as large as ~10-arcmin, the corresponding increase in sky coverage reduces the need for elaborate laser guide star systems. However, the need does not vanish entirely; the non-uniform distribution of suitably bright natural stars makes it unrealistic to expect full sky coverage.

[3] Focal anisoplanatism is a phenomenon caused by the conically convergent projected beam that forms a laser guide star in the upper atmosphere exploring a slightly different atmospheric path than the collimated image-forming light beams from distant astronomical objects. The resulting non-common path regions lead to residual uncorrected OPD fluctuation which ultimately limits the sharpness of the AO-corrected images.

## 17.2 Calculating Isoplanatic Angle

To simplify matters here, we give a 1-D analysis. However, if isotropic turbulence is assumed, the results obtained equally apply to the 2-D case. Initially, we assume vertical atmospheric paths; non-vertical paths are considered later, in Sect. 17.2.1.

Figure 17.1 shows a binary star object, where the two components are separated by angle $\vartheta$; also shown are the corresponding twin images formed by the telescope. To calculate the degree of correlation between the complex amplitudes at corresponding locations in the twin images—that is, at any pair of points separated by angle $\vartheta$ along the binary star axis—we take a similar approach to the one previously developed in Chap. 6. A similar atmospheric model is used comprising $n$ uncorrelated random phase screens, where the $j^{\text{th}}$ phase screen introduces OPD fluctuation, $h_j(x)$. This model allows the line integrals for each of the two beam paths (solid lines) to be replaced by the sum of $n$ discrete OPD contributions. (To avoid unnecessary clutter in the figure, only two atmospheric phase screens are shown.)

If the turbulence statistics are assumed temporally and spatially stationary (so that the statistics associated with the $h_j(x)$ are also temporally and spatially stationary), the covariance of the complex amplitudes arising at any two corresponding locations in the twin images may be expressed by

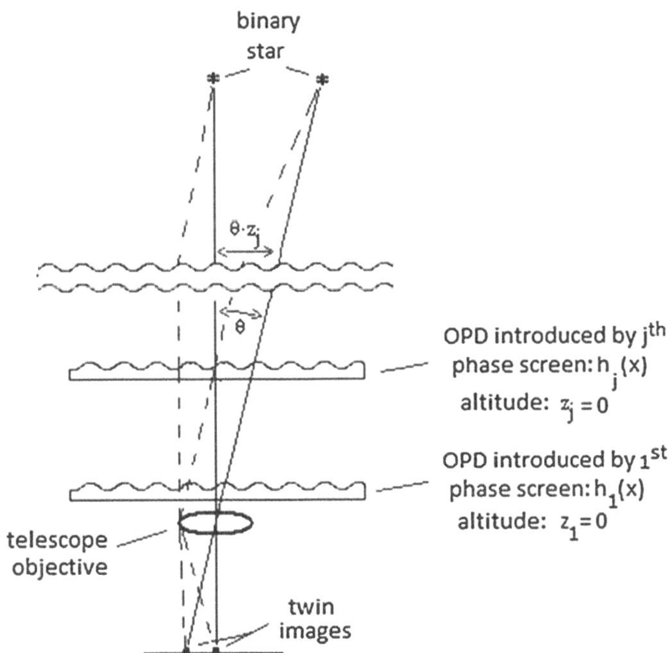


**Fig. 17.1** Geometry for calculating isoplanatic angle using a random phase screen stack atmospheric model

$$\langle U(0,\lambda)\cdot U^*(\vartheta,\lambda)\rangle =$$
$$\left\langle \exp\left[\frac{-2\cdot\pi\cdot i}{\lambda}\cdot\sum_{j=1}^{n}h_j(x)\right]\cdot\exp\left[\frac{2\cdot\pi\cdot i}{\lambda}\cdot\sum_{j=1}^{n}h_j\left(x+\vartheta\cdot z_j\right)\right]\right\rangle \tag{17.1}$$

Whereas the locations of the individual phase screens in the Chap. 6 application did not have to be carefully specified (they merely had to lie somewhere in the near-field region with respect to the telescope), in the present analysis the phase screens must be distributed with respect to altitude in the same way as the actual turbulence strength distribution in the atmosphere. For most practical purposes, a suitable distribution scheme could be obtained by uniformly distributing a reasonably large number of random phase screens (say about 20) over the atmospheric path where the individual scattering strengths of the phase screens, as represented by $\sigma_j^2$, could be obtained by scaling the overall variance of the OPD fluctuation introduced by the entire atmospheric path, $\sigma^2$, according to the $C_n^2$ distribution over the path.

When all the terms in the exponent of 17.1 are cross-multiplied, $n^2$ individual terms arise. However, as a consequence of the $n$ phase screens all being mutually uncorrelated, once the averages have been obtained, only $n$ nonzero terms remain, the cross-multiplied (uncorrelated) terms averaging to zero. Thus, 17.1 simplifies to

$$\langle U(0,\lambda)\cdot U^*(\vartheta,\lambda)\rangle = \left\langle \exp\left[\frac{-2\cdot\pi\cdot i}{\lambda}\cdot\sum_{j=1}^{n}\left(h_j(x)-h_j\left(x+\vartheta\cdot z_j\right)\right)\right]\right\rangle \tag{17.2}$$

By moving the averaging brackets inboard to capture only the randomly varying quantities, the above equation may be expressed in terms of the product operator, $\Pi$,

$$\langle U(0,\lambda)\cdot U^*(\vartheta,\lambda)\rangle = \prod_{j=0}^{n}\exp\left[\frac{-2\cdot\pi\cdot i}{\lambda}\cdot\left\langle h_j(x)-h_j\left(x+\vartheta\cdot z_j\right)\right\rangle\right] \tag{17.3}$$

The above equation is closely related to previous 6.34 and 6.41. By now assuming that the OPD fluctuations introduced by each phase screen, $h_j(x)$, are all individually Gaussian distributed, 17.3 may be written in the form similar to previous 6.52,

$$\langle U(0,\lambda)\cdot U^*(\vartheta,\lambda)\rangle = \prod_{j=1}^{n}\exp\left[\frac{-4\cdot\pi^2\cdot\sigma_j^2}{\lambda^2}\cdot\left[1-\rho_j\left(\vartheta\cdot z_j\right)\right]\right] \tag{17.4}$$

Function $U(0,\lambda)\cdot U^*(\vartheta,\lambda)$ has properties akin to those of the mutual intensity function (cf., 3.26). The unit-normalized form of this function, which we denote by $\mu(\vartheta,\lambda)$, is simply the complex coherence factor for the complex amplitudes at any pair of image locations separated by the previously indicated angle $\vartheta$. Function $\mu(\vartheta,\lambda)$ may now be expressed in the form,

$$\begin{aligned}\mu(\vartheta,\lambda) &= \frac{\langle U(0,\lambda)\cdot U^*(\vartheta,\lambda)\rangle}{\left[\langle |U(0,\lambda)|^2\rangle\cdot\langle |U(\vartheta,\lambda)|^2\rangle\right]^{\frac{1}{2}}}\\ &= \frac{\prod_{j=1}^{n}\exp\left[\frac{-4\cdot\pi^2\cdot\sigma_j^2}{\lambda^2}\cdot\left[1-\rho_j(\vartheta\cdot z_j)\right]\right]}{\prod_{j=1}^{n}\exp\left[\frac{-4\cdot\pi^2\cdot\sigma_j^2}{\lambda^2}\right]}\\ &= \frac{\prod_{j=1}^{n}\exp\left[\frac{-4\cdot\pi^2\cdot\sigma_j^2}{\lambda^2}\cdot\left[1-\rho_j(\vartheta\cdot z_j)\right]\right]}{\exp\left[\frac{-4\cdot\pi^2\cdot\sigma^2}{\lambda^2}\right]}\end{aligned} \tag{17.5}$$

where we have used the result (cf., 6.59) $\sigma^2 = \sum_{j=1}^{n}\sigma_j^2$.

The above expression for $\mu(\vartheta,\lambda)$ is quite general; it does not assume any particular type of turbulence structure characteristics, Kolmogorov or other. In Sects. 17.2.2 and 17.2.3, explicit expressions are obtained from this expression for two types of turbulence: (1) Kolmogorov turbulence with a large outer scale limit and (2) non-Kolmogorov turbulence with a smaller average turbulence size consistent with that measured by Coulman et al. Numerical evaluations of the two expressions show that these two different kinds of turbulence lead to entirely different atmospheric isoplanatic angle characteristics.

### *17.2.1 Effect of Zenith Angle*

The isoplanatic behavior at zenith angle, ZA, may be obtained from the 17.5 above by replacing $z_j$ by $z_j/\cos(ZA)$. Thus, we may write

$$\begin{aligned}\mu(\vartheta,\lambda,ZA) &= \frac{\langle U(0,\lambda,ZA)\cdot U^*(\vartheta,\lambda,ZA)\rangle}{\left[\langle |U(0,\lambda,ZA)|^2\rangle\cdot\langle |U(\vartheta,\lambda,ZA)|^2\rangle\right]^{\frac{1}{2}}}\\ &= \frac{\prod_{j=1}^{n} exp\left[\frac{-4\cdot\pi^2\cdot\sigma_j^2}{\lambda^2}\cdot\left[1-\rho_j\left(\frac{\vartheta\cdot z_j}{\cos(ZA)}\right)\right]\right]}{\prod_{j=1}^{n}\exp\left[\frac{-4\cdot\pi^2\cdot\sigma_j^2}{\lambda^2}\right]}\\ &= \frac{\prod_{j=1}^{n} exp\left[\frac{-4\cdot\pi^2\cdot\sigma_j^2}{\lambda^2}\cdot\left[1-\rho_j\left(\frac{\vartheta\cdot z_j}{\cos(ZA)}\right)\right]\right]}{\exp\left[\frac{-4\cdot\pi^2\cdot\sigma^2}{\lambda^2}\right]}\end{aligned} \tag{17.6}$$

where we note that the earlier assumption of isotropic turbulence is implicit in this expression. It may be observed that the zenith angle, ZA, and angle, $\vartheta$, always appear in 17.6 paired together in the quotient form $\vartheta/[\cos(\text{ZA})]$. This indicates that as zenith angle increases, the isoplanatic angle diminishes in proportion to cos(ZA).

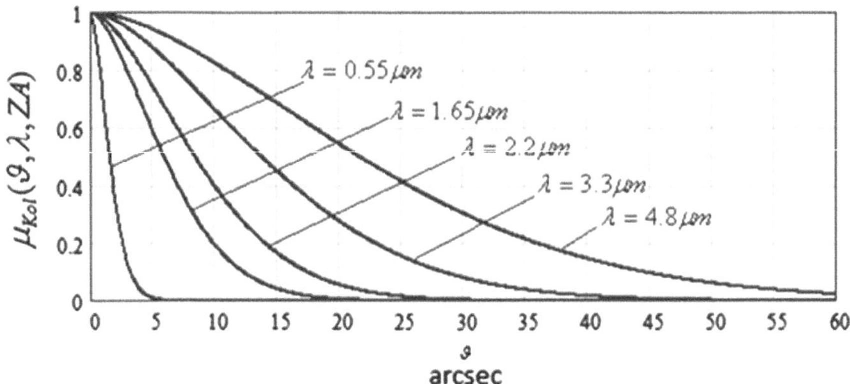


**Fig. 17.2** Isoplanatic angle for a vertical atmospheric path based on Kolmogorov turbulence and the SLC night model for $C_n^2$ where a large outer scale limit is assumed ($L_0 \gg 10m$). The plots shown are consistent with isoplanatic angle increasing as the 6/5 power of wavelength

### 17.2.2 Isoplanatic Angles for Kolmogorov Turbulence

For Kolmogorov turbulence where a large outer scale limit is assumed, 17.6 reduces to the form

$$\mu_{Kol}(\vartheta, \lambda, ZA) = \prod_{j=1}^{n} \exp\left[\frac{-2 \cdot \pi^2 \cdot \Delta z \cdot C_n^2(z_i) \cdot 2.91}{\lambda^2} \cdot \left(\frac{\vartheta \cdot z_j}{\cos(ZA)}\right)^{\frac{5}{3}}\right] \quad (17.7)$$

where $\Delta z$ is the separation interval between the $n$ uniformly distributed random phase screens in the atmospheric path model. This interval is given by

$$\Delta z = z_{j+1} - z_j \quad (17.8)$$

Plots of 17.7 are shown in Fig. 17.2 for vertical paths through zenith (i.e., ZA = 0) for a number of wavelengths using the $C_n^2$ values obtained from the submarine laser communication (SLC) night model. A plot of $C_n^2$ versus altitude for the SLC model can be found in Appendix C.

To produce the $\mu_{Kol}(\vartheta, \lambda, ZA)$ plots in Fig. 17.2, $L_0$ was set to the large value, 10 m. Thus, even for the largest angle plotted, $\vartheta = 60$ arcsec, for all $z_j$ values in the 0–20 km altitude range in which significant amounts of atmospheric turbulence are present, all of the various $\vartheta \cdot z_j$ products used in the evaluation of 17.7 are consistent with the condition.

$$\vartheta \cdot z_j < L_0 \quad (17.9)$$

At the visible wavelength, 0.55 μm, Fig. 17.2 shows $\mu_{Kol}(\vartheta, \lambda, ZA)$ falling to 50% after only about 1.6 arcsec. At 4.8 μm, the 50% value is attained at about 22 arcsec. The plots in Fig. 17.2 reflect the characteristic property for Kolmogorov turbulence that isoplanatic angle grows as the 6/5 power of wavelength. If we now, somewhat arbitrarily, define the isoplanatic angle $\vartheta_I$ as the angle at which $\mu_{Kol}(\vartheta, \lambda, ZA)$ falls to the value 0.5, the plots in Fig. 17.2 may be seen as consistent with the following approximate formula for isoplanatic angle:

$$\vartheta_{I,Kol}(\lambda) \approx 1.62 \cdot \left(\frac{\lambda}{0.55\mu\text{m}}\right)^{\frac{6}{5}} \cdot \cos(ZA)\ \text{arcsec}. \tag{17.10}$$

According to this formula, for vertical paths, the isoplanatic angle at 3.4 μm is about 14 arcsec. Thus, over the visible and near-IR wavelength range (in the nighttime seeing conditions described by the SLC model for $C_n^2$), isoplanatic angle typically ranges from 1.6 to 14 arcsec. At high-altitude observatory sites, these angles naturally tend to increase. For telescopes sited on Mauna Kea (4000-m altitude), the angular range just indicated might roughly double, perhaps to the range, 3–30 arcsec.

### 17.2.3 Isoplanatic Angles for Non-Kolmogorov Turbulence

Coulman et al. (1988) made extensive measurements of the outer scale limit of turbulence structure size, $L_0$, at sites in France, the USA, and Chile; they found that the measured values varied significantly with altitude. Averaged over the entire atmospheric depth, their measurements show that most of the turbulence structure lies in the size range, 0–0.5 m, which indicates an effective outer scale limit considerably smaller than the tens-of-meters limits assumed in Kolmogorov theory. Details of the $L_0$ measurements made by Coulman et al. are provided in Appendix C.

It is assumed in Appendix C that the atmospheric path can be represented (as previously justified in Chap. 6, Sect. 6.2.1) by a random phase screen model consisting of $n$ uncorrelated phase screens. It is also assumed in Appendix C that the wavefront structure function associated with each individual phase screen obeys the 5/3 power law,[4] prior to flattening out at the outer scale limit (corresponding to the altitude of that phase screen) measured by Coulman et al. The autocorrelation function of the OPD fluctuation introduced by the $j$th phase screens $\rho_j(\varepsilon)$ resulting from these two assumptions, is given in Appendix C by C.6.

The $\sigma_j$ values for each phase screen may be scaled according to the $C_n^2$ distribution with altitude using the relation

$$\sigma_j = \sigma \cdot \sqrt{\frac{C_{n_j}^2}{\sum_{k=1}^{n} C_{n_k}^2}} \tag{17.11}$$

[4] At the time Coulman et al. made their measurements, the 5/3-power law was commonly assumed. The crucial difference noticed by these authors is that the measured outer scale limits, rather than being larger than the largest telescope apertures, were significantly smaller than the apertures.

where, as previously, $\sigma$ is the rms of the integrated OPD fluctuation over the entire atmospheric path.

To calculate the isoplanatic angle plots shown later in this section, we again adopt the SLC $C_n^2$ night model (Appendix C); this model is consistent with about 5-arcsec full-width half-maximum (FWHM) seeing at visible wavelengths. In other seeing conditions, appropriate $C_n^2$ values can be generated by scaling the SLC model numbers in proportion to the square of the observed FWHM visible seeing angle. Thus, for example, if seeing worsened so that FWHM seeing over a given atmospheric path doubled in size, the corresponding $C_n^2$ values for this degraded seeing could be obtained by multiplying the original values by the factor four. It is noted that multipliers such as these do not affect the $\sigma_j$ values calculated from 17.11; these multipliers automatically cancel because they occur in both the numerator and the denominator. The seeing conditions can be factored into 17.11 simply by inserting appropriate values of $\sigma$.

By substituting the expressions for $\sigma_j$ and $\rho_j(\varepsilon)$ given by 17.11 and C.6 into 17.6, the degree of correlation for the non-Kolmogorov turbulence measured by Coulman et al. may be expressed in the form

$$\mu_{Coul}(\vartheta, \lambda, ZA) = \frac{1}{\exp\left[-\left(\frac{2\cdot\pi\cdot\sigma}{\lambda}\right)^2\right]} \cdot \prod_{j=1}^{n} \exp\left[\frac{-4\cdot\pi^2\cdot\sigma^2}{\lambda^2} \cdot \frac{C_{n_j}^2}{\sum_{k=1}^{n} C_{n_k}^2} \cdot \left(\frac{\vartheta\cdot z_j}{\cos(ZA)\cdot L_0(z_j)}\right)^{\frac{5}{3}}\right] \tag{17.12}$$

where we note that for large $\vartheta$ values satisfying the condition, $\vartheta \cdot z_j \geq \cos(ZA) \cdot L_0(z_j)$, the term $\vartheta \cdot z_j / (\cos(ZA) \cdot L_0(z_j))$ always takes the value unity (cf., C.6).

Figure 17.3 shows plots of $\mu_{Coul}(\vartheta, \lambda, ZA)$ calculated from 17.12. The various plots in this figure correspond to different FWHM visible seeing values in the range 0.3–1.25 arcsec, with the rms OPD fluctuation values, $\sigma$, calculated from the FWHM seeing values using the relation (cf., 13.43),

$$\sigma = FWHM_{Halo} \cdot \frac{w_o}{4\cdot\sqrt{\ln(2)}} \tag{17.13}$$

To be consistent with the wavefront structure function, $\sum(\varepsilon)$ that best fits the turbulence measured by Coulman et al. (cf., C.11 and Fig. C.4), it is necessary to assign a value of about 0.32 m to the parameter, $w_0$.[5] Thus, using this empirically chosen $w_0$ value, the $\sigma$ values corresponding to the chosen FWHM seeing values are as follows:

[5] The $w_o$ value chosen here is also consistent with the 4-m telescope images at 2.2 μm shown in Chap. 10 (Figs. 10.10 and 10.11).

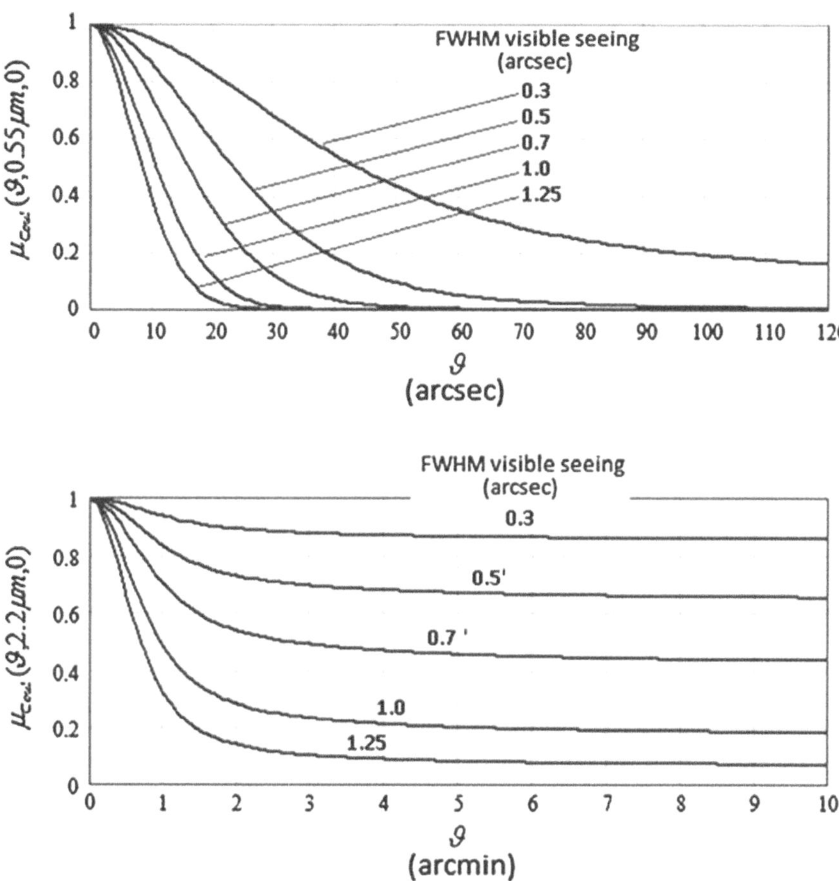


**Fig. 17.3** Isoplanatic angle characteristics of the atmosphere for turbulence structure of the kind measured by Coulman et al. The isoplanatic behavior is described by the complex coherence factor, $\mu_{Coul}(\vartheta, \lambda, 0)$, for (*Top*) $\lambda = 0{:}55$ μm and (*Bottom*) $\lambda = 2{:}2$ μm; the asymptotic flattening of the latter set of plots is due to the strong cores at 2.2 μm. Note that the horizontal scales for the two plots differ by a factor of five

$$FWHM_{Halo} = \begin{pmatrix} 0.3 \\ 0.5 \\ 0.7 \\ 1 \\ 1.25 \end{pmatrix} \text{arcsec}, \; \sigma = \begin{pmatrix} 0.14 \\ 0.23 \\ 0.33 \\ 0.47 \\ 0.58 \end{pmatrix} \mu\text{m}.$$

#### 17.2.3.1 Isoplanatic Angles at Visible Wavelengths

The visible wavelength plots of $\mu_{Coul}(\vartheta, \lambda, 0)$ in Fig. 17.3 (Top) are similar to those shown for Kolmogorov turbulence in Fig. 17.2. The plot corresponding to 1-arcsec visible seeing indicates that significant decorrelation occurs at $\vartheta = 5$ arcsec, even if some residual correlation is still evident at $\vartheta = 20$ arcsec; such behavior is consistent with the measured isoplanatic angle data cited by Dainty (1984) and described earlier in Sect. 17.1.

In exceptionally good seeing conditions, isoplanatic behavior can change dramatically. The plot corresponding to 0.3-arcsec seeing, instead of falling steadily to zero, now falls asymptotically to a nonzero value. This behavior is consistent with the presence of cores in star images at visible wavelengths.

#### 17.2.3.2 Isoplanatic Angles at Near-IR and Longer Wavelengths

At the near-IR wavelength 2.2 μm, the plots in Fig. 17.3 (Bottom) all tend to nonzero asymptotic values, behavior directly linked to the presence of cores at near-IR and longer wavelengths even in quite average (~1 arcsec) visible seeing conditions. It might be noted that the horizontal axis has now been extended in this figure to 10 arcmin. The asymptotic limiting value finally attained by function $\mu(\vartheta, \lambda, ZA)$ is set by the core energy fraction, $exp\left[-(2 \cdot \pi \cdot \sigma/\lambda)^2\right]$.

The isoplanatic behavior at near-IR and longer wavelengths can be explained by the core energy fractions remaining highly correlated out to very large angles, while the halo speckle fractions decorrelate at much smaller angles. The angle at which the core energy fractions finally lose correlation (as they ultimately must) is determined by turbulence energy at intermediate scales, say between 4 and 50 m. The measurements of Coulman et al. indicate negligible turbulence energy in this size range, while Kolmogorov theory anticipates much larger amounts. The turbulence energy fraction in this intermediate size range may vary with the seeing conditions, with smaller fractions arising in good seeing conditions and larger amounts in more turbulent conditions.

As previously discussed in Sect. 15.4, even a relatively small core energy fraction permits diffraction-limited resolution of simple objects such as binary stars. The larger light energy fraction remaining in the halo is usually much more widely distributed, leaving the cores dominating the star image centers. For extended objects, the halo light energy fraction reduces overall image contrast, but does not otherwise prevent image cores from producing a superposed highly resolved diffraction-limited image.

#### 17.2.3.3 Absence of Large Turbulence Structure in Coulman et al.'s Measurements

The turbulence structure size measurements of Coulman et al. indicate zero structure larger than 4 m at any altitude. In one sense, this finding is not unexpected if one assumes that the measurements were made in seeing conditions suitable for astronomical observations; it is certainly consistent with the cores at 2.2 μm obtained by the 3.8-m Mayall telescope in 1.25-arcsec visible seeing shown in Figs. 10.10 and 10.11. However, it is nonetheless inconceivable that there is absolutely zero turbulence energy at structure sizes larger than 4-m. One could speculate here that the turbulence energy fraction at scales greater than 4-m is simply small enough to fall below the detection threshold of the measurement equipment used by Coulman et al.

Turbulence structure at intermediate scales, say between 4 and 50 m, is critical to determining whether the isoplanatic angle limit for image cores can indeed be as large as the postulated 10-arcmin. The actual amounts present will determine the extent to which present-day 10-m class instruments and next-generation ELTs can benefit from these much larger isoplanatic angles. With the ELT instruments, the effect of turbulence structure at scales larger than 50-m will be largely limited to en block image movement which will have little influence on isoplanatic angle. By carrying out image measurements (of the sort described in Chaps. 8 and 13) using the largest existing (10-m class) telescopes, a better understanding could be established of the turbulence structure size distribution out to scales comparable to the 10-m apertures of these instruments. Such measurements would eliminate speculation about turbulence structure in the 4–10 size range and allow a better understanding of image core isoplanaticity and how it affects next-generation ELT instruments.

## 17.3 Why Stars Twinkle but not Planets?

Scintillation patterns were shown previously in Chap. 3 (Fig. 3.11)[6] formed by light from Sirius falling directly on the primary mirror of the Author's 8-in. Newtonian reflector telescope. The strong intensity fluctuations seen in the figure are attributable to the large (70°) zenith angle of Sirius when the images were recorded. The angular subtense of Sirius is a mere 0.006 arcsec (Table 17.1), which is much smaller than typical atmospheric isoplanatic angles at visible wavelengths. The same is true for any other star, the 'largest' subtending a mere 0.06-arcsec angle. Planets, on the other hand, can subtend angles as large 60-arcsec, which is significantly larger than atmospheric isoplanatic angles at visible wavelengths.

[6] The scintillation patterns shown in Fig. 3.11 were obtained by the author in the backyard of his home in Albuquerque, New Mexico. Partly due to the relatively high, 1850-m, altitude of the site, stars lying near the zenith rarely exhibit twinkling and likewise neither do planets. However, it is still possible to see strong twinkling (as in Fig. 3.11) by viewing stars at large zenith angles.

**Table 17.1** Approximate visual magnitudes and the angular subtenses of stars, planets, and the dwarf planet Pluto

| Object | Angular subtense (arcsec) | Magnitude | Comments |
|---|---|---|---|
| Betelgeuse | 0.055 | 0.2–1.2 | Variable |
| Sirius | 0.006 | −1.4 | – |
| Mercury | 7 | −0.3 | At max elongation (50% illum) |
| Venus | 25 | −4.1 | At max elongation (50% illum) |
| Mars | 25 | −2.9 | At opposition |
| Jupiter | 50 | −2.9 | At opposition |
| Saturn | 20 | −0.5 | At opposition |
| Uranus | 3 | 5.3 | At opposition |
| Neptune | 2 | 7.8 | At opposition |
| Pluto | 0.1 | 13.5 | At opposition |

If Sirius had been replaced by any of the planets that subtend the largest angles – Venus, Mars, Jupiter, and Saturn – the same imaging arrangement would have produced demodulated scintillation patterns, markedly different from those shown in Fig. 3.11.

There would have been far less intensity fluctuation and greatly reduced color contrast, the rich colors seen in the figure replaced by near-white pastel shades.

Scintillation patterns from planets are, in effect, reduced scintillation patterns. Just as with reduced speckle patterns, they arise as the integrated sum of many partially correlated scintillation patterns. In this case, each contributing pattern arises from sunlight reflected or back-scattered towards Earth from a single unique location on the planet's extended disc.

The intensity fluctuation in such reduced scintillation patterns is determined by the effective number of uncorrelated patterns that sum together to produce the final pattern; the actual number is proportional to the solid angle subtended by the planet as seen from Earth. The square root of that number, which is proportional to the angular diameter of the planet, provides a measure of the demodulation of the intensity fluctuation in the reduced scintillation pattern that would replace the pattern seen in Fig. 3.11 on the telescope's primary mirror.

For planets subtending the largest angles, the demodulation factor is often large enough to reduce the intensity fluctuation to a level below the response threshold of the eye (c.f., Sect. 15.5). The 'scintillation pattern' falling on the primary mirror would then appear as a uniformly smooth, nominally white light distribution. Since the unaided human eye is unable to resolve planetary discs, it simply focuses the steady white light received from the planet into a single point on the retina, thus forming the final image: a fixed point of light in the sky, shining with a steady light and showing no trace of the twinkling behavior exhibited by neighboring stars.

To simplify the 'twinkling' analysis given below, we assume all planetary disks to be circular. We also assume that the disks are fully illuminated by sunlight, and further assume that they always subtend angles less than the (1 arcmin) angular resolution limit of the unaided eye. Thus, both planets and stars appear as unresolved point-objects. We also assume for both stars and planets that the eye responds to the total received light over the entire visible spectrum.

The demodulation factor corresponding to the reduced intensity scintillation perceived by an observer looking directly at the planet is proportional to the ratio of the angular subtense of the planet to the angular width of the correlation function, $\mu(\vartheta, vis\lambda, ZA)$. Referring the Fig. 17.2 plots it is clear that, at visible wavelengths in unexceptional seeing conditions, that this function is approximately Gaussian and may be expressed in the form

$$\mu(\vartheta, vis\lambda, ZA) = \exp\left[-\left(\frac{\vartheta}{\vartheta_{Ie}}\right)^2\right] \quad (17.14)$$

where $\vartheta_{\mathrm{Ie}}$ is the $1/e$ half-width of the function. Near zenith (i.e., ZA $\approx 0°$) $\vartheta_{\mathrm{Ie}}$ typically takes values of the order of several arcsec. At larger zenith angles [where we recall that the isoplanatic angle reduces in proportion to cos(ZA)], $\vartheta_{\mathrm{Ie}}$ may take values less than 1 arcsec.

Using an approach similar to that used to treat speckle reduction in Chap. 11 (Sect. 11.6), the effective number of uncorrelated scintillation patterns, $N_{\mathrm{P}}$, that constitute the integrated scintillation pattern seen by a naked-eye observer viewing a “twinkling” planet is given by

$$N_P = \frac{\pi \cdot \left(\frac{\vartheta_P}{2}\right)^2}{\int_0^{\vartheta_P} 2 \cdot \pi \cdot \vartheta \cdot \left\{\frac{2}{\pi} \cdot \left[\cos^{-1}\left(\frac{\vartheta}{\vartheta_P}\right) - \left(\frac{\vartheta}{\vartheta_P}\right) \cdot \sqrt{1 - \left(\frac{\vartheta}{\vartheta_P}\right)^2}\right]\right\} \cdot \exp\left[-2 \cdot \left(\frac{\vartheta}{\vartheta_{Ie}}\right)^2\right] \cdot d\vartheta} \quad (17.15)$$

where $\vartheta_{\mathrm{P}}$ is the angle subtended by the diameter of the planetary disc, and $\vartheta_{\mathrm{Ie}}$ is a measure of the isoplanatic angle for the atmospheric path. The numerator in the above equation is simply the solid angle subtended by the planetary disc, while the term in curly brackets in the denominator is the (unit-normalized) autocorrelation function of the planet's circular disk (cf., 7.12).

The Gaussian term in the denominator of the above equation arises from 17.14, with the additional factor of two included because this term refers to the correlation function for the intensities rather than the complex amplitudes. Since the integrated OPD fluctuation over the path may be assumed as previously to be Gaussian distributed, it is assumed here that we can again make use of Reed's theorem for Gaussian processes (Sect. 7.8.1). The effective solid angle of each of the uncorrelated intensity patches distributed over the planetary disk is given by $\int_0^{\infty} 2 \cdot \pi \cdot \vartheta \cdot exp\left[-2 \cdot (\vartheta/\vartheta_{Ie})^2\right] d\vartheta$; this integral simplifies analytically to give the solid angle subtended by such a patch as $\pi \cdot \vartheta_{Ie}^2/2$.

The magnitude of the observed intensity fluctuations of a twinkling star—which we quantify here in terms of the twinkling contrast, $C_{\mathrm{S}}$—may be expressed (cf.,

11.22) in terms of the star's normalized moment-to-moment intensity fluctuation, $I_S(t)$, by

$$C_S = \left[\frac{\langle I_S(t)^2\rangle - \langle I_S(t)\rangle^2}{\langle I_S(t)\rangle^2}\right]^{\frac{1}{2}} \tag{17.16}$$

Depending on the seeing quality, the site altitude, and the zenith angle, $C_S$ likely takes values in the range $0 \leq C_S \leq 1$. Steady light corresponds to $C_S = 0$, while fully saturated scintillation (which approximates Gaussian speckle) corresponds to $C_S = 1$.

The twinkling contrast of a planet, which we denote by $C_P$, can be defined in a similar manner,

$$C_P = \left[\frac{\langle I_P(t)^2\rangle - \langle I_P(t)\rangle^2}{\langle I_P(t)\rangle^2}\right]^{\frac{1}{2}} \tag{17.17}$$

$C_P$ may be considered as a demodulated version of $C_S$. The demodulation factor, which we denote here by DF, is defined as the ratio of the twinkling contrast of the planet to that of a nearby star,

$$DF = \frac{C_P}{C_S} \tag{17.18}$$

The demodulation factor, DF, may be considered to act in the same way as the speckle contrast reduction factor discussed in Chap. 11, which we recall was given by the square root of the effective number of uncorrelated speckle patterns in the reduced speckle pattern. In the present application, where the effective number of uncorrelated scintillation patterns is given by $N_P$ (17.15), DF is given by

$$DF = \frac{1}{\sqrt{N_P}} \tag{17.19}$$

By using 17.15, the demodulation factor, DF, can be written in the expanded form

$$DF = \frac{2}{\sqrt{\pi}} \cdot \frac{1}{\vartheta_P} \times \left\{ \int_0^{\vartheta_P} 2 \cdot \pi \cdot \vartheta \cdot \left\{ \frac{2}{\pi} \cdot \left[ \cos^{-1}\left(\frac{\vartheta}{\vartheta_P}\right) - \left(\frac{\vartheta}{\vartheta_P}\right) \cdot \sqrt{1 - \left(\frac{\vartheta}{\vartheta_P}\right)^2} \right] \right\} \right.$$
$$\left. \times \exp\left[-2 \cdot \left(\frac{\vartheta}{\vartheta_{Ie}}\right)^2\right] \cdot d\vartheta \right\}^{\frac{1}{2}} \tag{17.20}$$

Figure 17.4 shows how DF varies with the angular subtense of the planet, $\vartheta_P$, for various discrete atmospheric isoplanatic angles, $\vartheta_{Ie}$.

For $\vartheta_P \gg \vartheta_{Ie}$, which is often the case for the larger planets, the right-hand side of the above expression for DF reduces to the approximate form $\sqrt{2} \cdot \vartheta_{Ie}/\vartheta_P$. Combining this approximation with 17.18 provides the following simple relation- ship between the twinkling contrasts of stars and planets,

$$C_P = \frac{\sqrt{2} \cdot \vartheta_{Ie}}{\vartheta_P} \cdot C_S \tag{17.21}$$

By measuring or otherwise obtaining values of $C_S$, $C_P$, and $\vartheta_{Ie}$, the angular subtense of the planet, $\vartheta_P$, can be obtained from the relation,

$$\vartheta_P = \sqrt{2} \cdot \vartheta_{Ie} \cdot \frac{C_S}{C_P} \tag{17.22}$$

### *17.3.1 Eye Sensitivity to Twinkling*

The eye has little sensitivity to intensity fluctuations once contrast falls below about 0.05. Thus, in conditions where stars might show strong twinkling (i.e., $C_S \gg 0.05$), for small-enough DF values, planets might not appear to twinkle at all (i.e., $C_P < 0.05$). For the high contrast scintillation patterns previously shown in Fig. 3.11 for the bright star Sirius, it is estimated that the contrast, $C_S$, may have been about 0.5. (This estimate is only rough because no radiometric measurements were made.) As an example illustration, we now calculate the twinkling contrast of the scintillation patterns that might have arisen had Jupiter been the target object rather than Sirius,

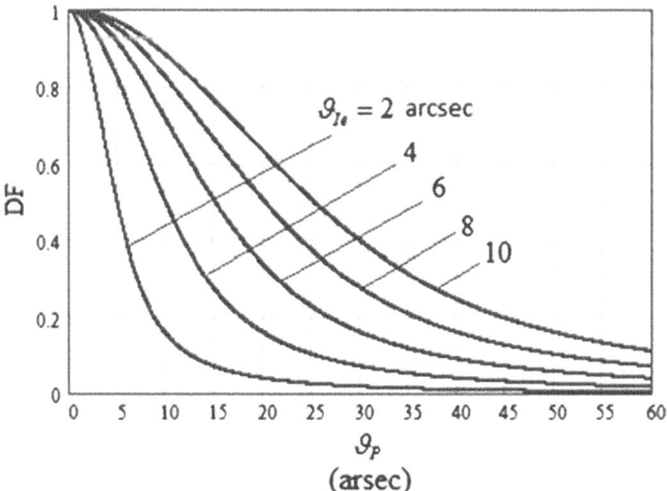


**Fig. 17.4** Plots of the demodulation factor, DF, against planetary diameter, $\vartheta_P$. The quantity, DF, gives the ratio of the twinkling contrast of planets to that of nearby stars. The plots are shown for a number of typical atmospheric isoplanatic angles, $\vartheta_{Ie}$

where we assume that all other factors, such as the seeing conditions and the zenith angle, remain the same.

The parameter values used in the calculation are as follows:

| | |
|---|---|
| $ZA \approx 70°$ | (Sirius zenith angle when the Fig. 3.11 scintillation patterns were recorded). |
| $C_S \approx 0.5$ | (estimated contrast value for the Fig. 3.11 Sirius scintillation patterns). |
| $\vartheta_{Ie} = 1.37$arcsec | (isoplanatic angle at ZA 70° corresponds to $\vartheta_{Ie} = 4$ arcsec at $ZA = 0°$). |
| $\vartheta_P = 50$ arcsec | (approximate average angular subtense of Jupiter at opposition). |

Inserting the above parameter values into 17.21 gives the twinkling contrast value for Jupiter, $C_P \approx 0.02$. Since this value is below the eye sensitivity threshold ($\sim 0.05$), it is concluded that, in the conditions used to view Sirius at the time the Fig. 3.11 scintillation patterns were recorded, Jupiter would have shone with a steady light.

### *17.3.2 Minimum Angular Size for Planets to Cease Twinkling*

In given viewing conditions (as set by the $C_S$ and $\vartheta_{Ie}$ values), ultimately any planet whose angular subtense exceeds a certain critical size will no longer appear to twinkle. The actual diameter limit, which we denote by $\vartheta_{P,Lim}$, may be obtained from 17.22 by setting $C_P$ to the eye threshold contrast value, 0.05. Thus, we obtain

$$\vartheta_{P,Lim} = \sqrt{2} \cdot \vartheta_{Ie} \cdot \frac{C_S}{0.05} \tag{17.23}$$

For the scintillation patterns shown in Fig. 3.11 (where $C_S \approx 0.5$ and $\vartheta_{Ie} = 1.37$ arcsec), the above expression gives the critical angle $\vartheta_{P,Lim} \approx 20$ arcsec. According to Table 17.1, in certain phases, Venus, Mars, and Jupiter can all subtend angles larger than 20 arcsec. Thus, in these phases and in the viewing conditions considered, these planets would not appear to twinkle. Saturn with its 20-arcsec subtense lies right on the cusp but, if we also take its ring system into account, it is probably safe to say that it would not twinkle either. Mercury and Uranus both fall well short of the required angle, and thus, both would twinkle vigorously in the conditions considered; the twinkling of the latter would rival that of stars. The twinkling of Neptune and Pluto would also rival that of stars—if only these objects were bright enough to be visible to the unaided eye in the first place!

### *17.3.3 Planetary Twinkling and Estimating Atmospheric Isoplanatic Angle*

If the angular diameter of a planet is known, $\vartheta_P$,[7] the isoplanatic angle, $\vartheta_{Ie}$, of the atmospheric path in the direction of that planet can be obtained from simultaneous measurements of the twinkling contrast of both the planet and a nearby star, $C_P$ and $C_S$. The isoplanatic angle, $\vartheta_{Ie}$, may be obtained in this case by rearranging 17.21, to give

$$\vartheta_{Ie} = \frac{C_P}{C_S} \cdot \frac{\vartheta_P}{\sqrt{2}} \tag{17.24}$$

If a more accurate expression is required—one that could also be used in cases where $\vartheta_P$ was not significantly larger than $\vartheta_{Ie}$—such an equation can be obtained by combining 17.18 and 17.20 to yield

$$\frac{C_P}{C_S} = \frac{2}{\sqrt{\pi}} \cdot \frac{1}{\vartheta_P} \times \left\{ \int_0^{\vartheta_P} 2 \cdot \pi \cdot \vartheta \cdot \left\{ \frac{2}{\pi} \cdot \left[ \cos^{-1}\left(\frac{\vartheta}{\vartheta_P}\right) - \left(\frac{\vartheta}{\vartheta_P}\right) \cdot \sqrt{1 - \left(\frac{\vartheta}{\vartheta_P}\right)^2} \right] \right\} \cdot \exp\left[ -2 \cdot \left(\frac{\vartheta}{\vartheta_{Ie}}\right)^2 \right] \cdot d\vartheta \right\}^{\frac{1}{2}} \tag{17.25}$$

This equation can readily be solved for $\vartheta_P$ using numerical methods. But plainly, to obtain an accurate result from this method, a detector with higher contrast sensitivity than the human eye would be required to obtain suitably accurate measurements of $C_P$ and $C_S$, particularly for the former where relatively small values (~0.01) often arise. To avoid aperture averaging (which further reduces measured contrast), the collection aperture of the measuring telescope would have to be significantly smaller than the lateral structure size of the scintillation structures (cf., Fig. 3.11). For planets riding high in the sky near the zenith, a 25-mm telescope aperture might be appropriate. For planets at lower elevation angles, smaller apertures would be necessary. A rough measure of the required size can be obtained by scaling down the above-suggested 25-mm aperture in proportion to cos(ZA), where ZA is the zenith angle.

At visible wavelengths, a good quality 25-mm telescope resolves down to about 5 arcsec and thus can easily resolve the larger planets. To measure the total integrated amount of light received from a given planet, the measurement setup would have to

---

[7] The planetary disc is assumed here, as previously, to be approximately circular as well as being uniformly illuminated by sunlight. The superior planets, Mars, Jupiter, and Uranus, may be considered suitably circular for present purposes. A modified analysis would be required to deal with the inferior planets, Mercury and Venus, which show phase-like behavior when not at opposition. Saturn would also require a modified analysis because of its ring system.

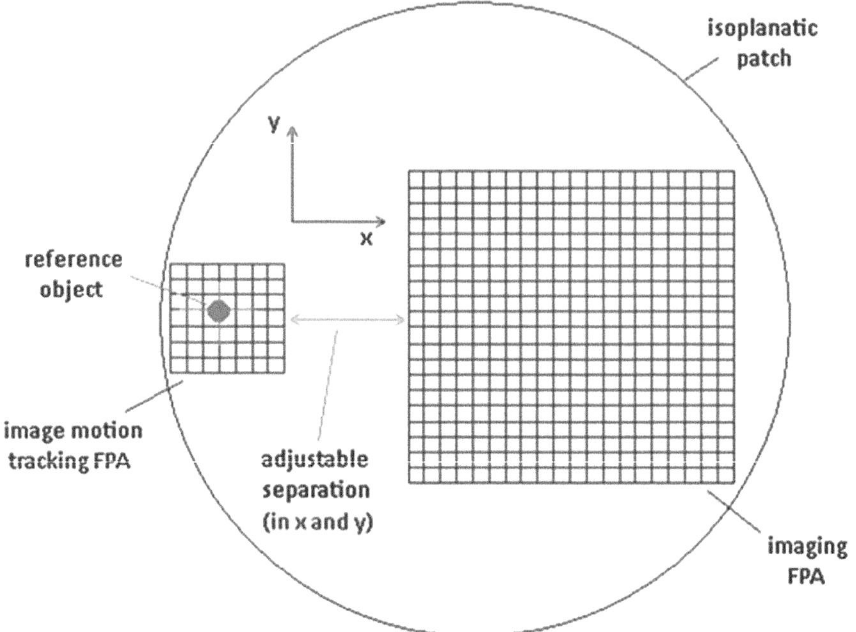


**Fig. 17.5** Image stabilization scheme using a second offset FPA array (at *left*) for imaging a nearby reference star at a wavelength consistent with the reference star image having a central core. The isoplanatic properties of image cores enable the highest levels of image stabilization to be achieved on the nearby imaging FPA (*right*)

ensure that the image of the entire planetary disk fell within the sensitive area of the detector.

By recording the planet's zenith angle during the measurements, ZA, an approximate value for the isoplanatic angle in the direction of the zenith could also be obtained by multiplying the angle given by either 17.24 or 17.25 by the factor 1/cos(ZA) (cf., Sect. 17.2.1).

## 17.4 Use of Natural Stars for Stabilizing Images in Large Telescopes

The large isoplanatic angles associated with image cores can be exploited for image stabilization purposes using a two-channel imaging arrangement such as shown in Fig. 17.5. Two imaging focal plane arrays (FPAs) can be seen in this figure. The larger array is used to record the actual image; the smaller one (perhaps just a 250 × 250 pixel array) is used to track motions of a nearby reference object. (The latter tracking array could also be used to generate the measurement data needed to maintain a definitive seeing log of the sort discussed previously in Chap. 13, Sect. 13.2.5.) By stabilizing the core of the reference object on the tracking FPA using a fast-steering mirror (FSM) located upstream in the telescope optical train, the image falling on the imaging FPA would be similarly stabilized (since the light beams forming this image also reflect off the same FSM).

The location of the reference object tracking FPA relative to the larger FPA used to record the actual image would be adjusted as needed in the telescope image plane so as to accommodate reference stars lying anywhere with respect to the imaging FPA within about a 10 arcmin radius, thus allowing exploitation of the largest isoplanatic angles likely to be associated with image cores. Once the tracking FPA has been correctly adjusted to receive the image of the chosen reference object, it would then be locked in position, ready for image acquisition.

As discussed in Chap. 10, more precise image stabilization can be achieved by using the core in a reference star image, rather than the light energy centroid. To ensure that cores are actually present in the reference star images, the wavelength band chosen for the tracking channel would have to lie at, or near, the optimum wavelength (Sect. 10.4); at the optimum wavelength, the core contains about 37% of the total light energy. Whereas large telescopes, say 4-m in diameter, with good optics typically image the core light energy fraction into a diffraction-limited feature perhaps 0.1 arcsec across, ELTs will likely deliver even sharper image cores, perhaps measuring only 0.01 arcsec across. Though image cores can often contain less total light energy than the surrounding halos, because the halo light energy fractions are spread over much wider angular areas, typically 1 arcsec across, image cores nonetheless dominate the star image centers.

Depending on telescope size and telescope aberrations, the irradiance level attained in the center of an image core can exceed the irradiance level in the background halo by a factor typically lying in the range, 10–1000.

### *17.4.1 Estimating the Location Where the Core Attains Maximum Intensity*

As previously discussed (Sect. 10.7) for telescopes with circular apertures, the image location where the core in a reference star image attains maximum intensity provides the most precise estimate of the instantaneous tip–tilt component in the Zernike polynomial for the OPD fluctuation over the instantaneous image-forming wave in the telescope pupil. This location can be most precisely determined by restricting consideration to only those light energy portions that fall on a small area of the tracking FPA straddling the core (cf., Fig. 17.6).

For the one-dimensional arrangement shown in the figure, the location at which the core attains peak intensity may be established to sub-pixel accuracy by using the algorithm,

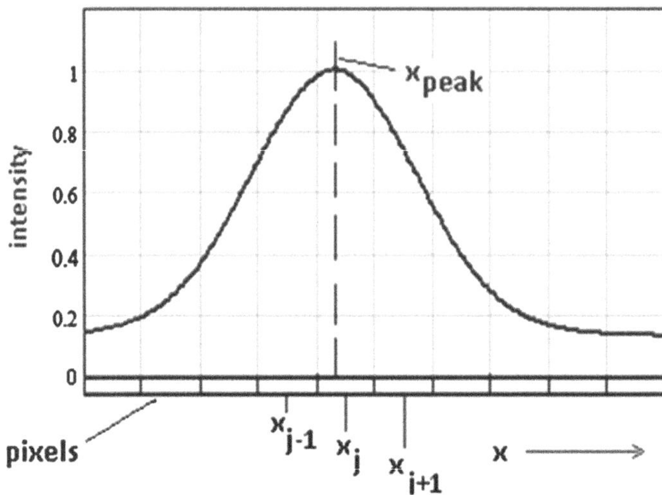


**Fig. 17.6** The location at which the core attains maximum intensity provides the most precise estimate of the Zernike tip–tilt component of the instantaneous wavefront in the telescope pupil. Core location can be most accurately established by restricting consideration to the signal received by the small pixel array subset that captures the light energy in the core, while also excluding most of the halo light energy. The speckle noise associated with the latter would otherwise introduce spurious jitter into the position estimates

$$x_{peak} = \frac{\sum_{j=nx_1}^{nx_2} \left(x_j \cdot I_j\right)}{\sum_{j=nx_1}^{nx_2} I_j} \tag{17.26}$$

where $x_j$ indicates the location of the $j$th pixel, and $I_j$ is the light intensity falling on that pixel. To minimize position estimate jitter (caused mainly by speckle in the background halo), the summation limits would be chosen so that only those pixels in the vicinity of the core center are included in the summation. Suitable limits to the left and right of the core peak, $nx_1$ and $nx_2$, may be obtained by choosing values roughly consistent with the conditions,

$$\frac{I_{nx_1}}{I_{peak}} \approx 0.5 \tag{17.27}$$

$$\frac{I_{nx_2}}{I_{peak}} \approx 0.5 \tag{17.28}$$

In two dimensions, the X- and Y-locations of the peak may be calculated using the vector relationship

$$\left(x_{peak}, y_{peak}\right) = \left(\frac{\sum_{j=nx_1}^{nx_2} \sum_{k=ny_1}^{ny_2} \left(x_j \cdot I_{j,k}\right)}{\sum_{j=nx_1}^{nx_2} \sum_{k=ny_1}^{ny_2} I_{j,k}}, \frac{\sum_{j=nx_1}^{nx_2} \sum_{k=ny_1}^{ny_2} \left(y_k \cdot I_{j,k}\right)}{\sum_{j=nx_1}^{nx_2} \sum_{k=ny_1}^{ny_2} I_{j,k}}\right) \tag{17.29}$$

where again the summation limits $nx_1$, $nx_2$, $ny_1$, and $ny_2$ are chosen to only gather signal from the core and thus exclude most of the (noisy) halo light energy. More sophisticated algorithms could of course be used to give more precise location estimates. Curve-fitting methods could be used, with ($x_{\text{peak}}$, $y_{\text{peak}}$) determined by the locations at which the first two partial derivatives (with respect to $x$ and $y$) of the best- fit curve go to zero.

## 17.4.2 Radiometry of Reference Star Cores

The light energy fraction contained in the core of a reference star image may be expressed in terms of the rms OPD fluctuation of the image-forming wavefronts. If the entire telescope aperture is considered, the light energy fraction in the core is given by (cf., 10.3) $exp\left[-(2\cdot\pi\cdot\sigma/\lambda)^2\right]$. When sub-aperture imaging is considered, such as when the core is formed by an individual lenslet member of a Shack–Hartmann array, σ would generally be replaced by a smaller value determined by the size of the individual sub-apertures. In a typical Shack–Hartmann wavefront sampling arrangement, the appropriate σ value would be chosen consistent with perhaps a 0.8 light energy fraction residing in the core. (The final energy fraction choice would obviously have to best accommodate instant-to-instant seeing variations.)

Suppose that the entire image core of a reference star of visual magnitude, $m_{\text{v}}$, falls on the tracking FPA, and we ignore the small energy losses due to dead space between pixels. The total number of photoelectrons produced by the core, $\text{PE}_{\text{core}}$, can be obtained as shown in Appendix E from Planck's blackbody emissions formula (cf., E8) by incorporating the core energy fraction, $E_{\text{C}}$, into that equation to give

$$PE_{Core} = 2.51189^{-m_v}\cdot\left(\frac{1}{\lambda}\right)^4\cdot\frac{\exp\left(\frac{h\cdot c}{0.55\,\mu\text{m}\cdot k\cdot T}\right)-1}{\exp\left(\frac{h\cdot c}{\lambda\cdot k\cdot T}\right)-1}\cdot E_C$$
$$\times\, QE(\lambda)\cdot\Delta t\cdot\Delta\lambda\cdot\left(D^2-d^2\right)\cdot T_o(\lambda)\cdot T_A(\lambda)^{\left(\frac{1}{\cos(ZA)}\right)}$$
$$\times\left(7.4675\cdot 10^{-9}\frac{\text{m}}{\text{s}}\right) \qquad (17.30)$$

where $D$ is the telescope diameter, $d$ is the central obstruction diameter, $T$ is the effective surface temperature of the reference star, λ is the mean imaging wavelength, Δλ is the imaging bandwidth, $\Delta t$ is the detector integration time, $T_0(\lambda)$ is the telescope optics transmission factor, $T_{\text{A}}(\lambda)$ is the atmospheric transmission factor for a vertical path (see Fig. 3.7), ZA is the zenith angle of the reference star, QE(λ) is detector quantum efficiency, and $E_{\text{C}}$ denotes the core energy fraction (cf., 10.3).

For telescopes equipped with AO, σ would be replaced in this formula by $\sigma_{\text{AO}}$. The term in 17.30 directly attributable to wavelength, which we denote by $R$(0.55 μm, λ) and which controls the number of photoelectrons liberated in the wavelength band centered on wavelength, λ, relative to the number collected over the same wavelength

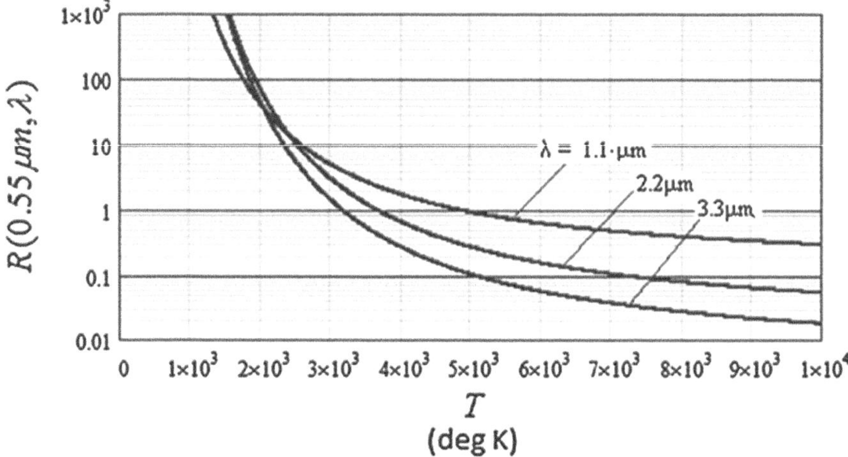


**Fig. 17.7** Relative radiant output of a star at the three near-IR wavelengths, 1.1, 2.2, and 3.4 μm, compared to the output at the visible wavelength, 0.55 μm, as a function of the effective surface temperature of the star, $T$

bandwidth centered on 0:55 μm (i.e., the wavelength at which the visual magnitude of the reference star, $m_V$, applies) may be extracted from 17.30 in the form,

$$R(0.55\,\mu\text{m}, \lambda) = \left(\frac{0.55\,\mu\text{m}}{\lambda}\right)^4 \cdot \frac{\exp\left(\frac{h \cdot c}{0.55\,\mu\text{m} \cdot k \cdot T}\right) - 1}{\exp\left(\frac{h \cdot c}{\lambda \cdot k \cdot T}\right) - 1}. \tag{17.31}$$

Figure 17.7 shows plots of $R(0.55\ \mu\text{m}, \lambda)$ versus effective surface temperatures, $T$, for three near-IR wavelengths. For stars with effective surface temperatures of about 4000 K, the function takes values close to unity. For stars with other effective surface temperatures, the function takes higher or lower values, possibly ranging over several orders of magnitude.

#### 17.4.2.1 Number of Photo Electrons Liberated by the Brightest Pixel

To accurately establish core location, the pixel spacing must be chosen so as to comfortably resolve the core. Assuming an Airy pattern core shape, the required width is roughly given by the radius of the first dark ring. Thus, if the core nominally occupies an $n \times n$ pixel area (where $n$ is a number typically in the range 3–5) the core light energy fraction falling on the pixel located closest to the center of the core, $\text{EF}_{\text{brightest pixel}}$, is approximately given by

$$EF_{brightest\ pixel} \approx \frac{1}{n^2} \tag{17.32}$$

**Table 17.2** Parameter values used in radiometric Example 1

| Parameter | Value |
|---|---|
| Telescope diam, $D$ | 10 m |
| Central obstruction diam, $d$ | 0.0 m |
| Reference star visual mag, $m_V$ | 10 |
| Reference star effective temp, $T$ | 5780 K |
| Nominal central wavelength, $\lambda$ | 2.2 μm |
| Bandwidth, $\Delta\lambda$ | 0.2 μm |
| Integration time, $\Delta t$ | 0.01 s |
| Detector quantum efficiency, QE($\lambda$) | 0.6 |
| Telescope optics trans, $T_O(\lambda)$ | 0.5 |
| Vertical path atmos trans, $T_A(\lambda)$ | 0.78 |
| Reference star zenith angle, ZA | 0.0° |
| No. of pixels across Airy 1st dark ring, $n$ | 4 |
| Core energy fraction, $E_C$ | 0.368 |

Combining 17.30 and 17.32 gives the approximate number of photoelectrons produced by this, the most brightly illuminated pixel,

$$PE_{brightest\ pixel} = 2.51189^{-m_v} \cdot \left(\frac{1}{\lambda}\right)^4 \cdot \frac{\exp\left(\frac{h \cdot c}{0.55\,\mu\text{m} \cdot k \cdot T}\right) - 1}{\exp\left(\frac{h \cdot c}{\lambda \cdot k \cdot T}\right) - 1} \cdot E_C \cdot \frac{1}{n^2}$$
$$\times QE(\lambda) \cdot \Delta t \cdot \Delta\lambda \cdot \left(D^2 - d^2\right) \cdot T_o(\lambda) \cdot T_A(\lambda)^{\left(\frac{1}{\cos(ZA)}\right)} \cdot \left(3.492 \cdot 10^{-8} \frac{\text{m}}{\text{s}}\right) \tag{17.33}$$

## Radiometric Calculation, Example 1
## Image stabilization using the image core of a bright star as the reference feature

This example relates to image stabilization with a 10-m telescope using the central core in a reference star image as the reference feature. Our aim is to establish whether there is enough signal to allow accurate estimates of core location; in particular, we would like to know the number of photoelectrons produced by the pixel most brightly illuminated by the core (with 17.33 used to make the calculation). To ensure that a core actually exists in the reference star image, we choose to form the image at the near-IR wavelength, 2.2 μm, and assume here that this wavelength lies reasonably close to the optimum wavelength so that we can set the core energy fraction, $E_C$, to the approximate value, $1/e$ ($\approx$0.368). Table 17.2 provides a list of the parameter values used in this example calculation.

For the values in the Table, 17.33 indicates that the brightest pixel generates about 7000 photoelectrons. This number is large enough to ensure a reasonably robust signal-to-noise ratio and thereby allow reasonably accurate estimates of core location; yet the number falls comfortably short of saturation levels that might unduly

**Table 17.3** Parameter values used in radiometric Example 2

| Parameter | Value |
|---|---|
| Shack–Hartmann lenslet diam, $D_S$ | 0.12 m |
| Reference star visual mag, $m_v$ | 3.0 |
| Reference star effective temp, $T$ | 5780 K |
| Nominal central wavelength, $\lambda$ | 0.55 μm |
| Bandwidth, $\Delta\lambda$ | 0.375 μm |
| Integration time, $\Delta t$ | 0.01 s |
| Detector quantum efficiency, QE($\lambda$) | 0.6 |
| Telescope optics trans, $T_o(\lambda)$ | 0.5 |
| Vertical path atmos trans, $T_A(\lambda)$ | 0.78 |
| Reference star zenith angle, ZA | 0.0° |
| No. of pixels across Airy 1st dark ring, $n$ | 4 |
| Core energy fraction, $E_C$ | 0.8 |

bias these estimates. (For typical FPA detectors, well depth is typically of the order 200,000 photoelectrons.) Core location estimates may be used in real time to drive a tip/tilt mirror to stabilize the image. Alternatively, these estimates could be used later to carry out computer-controlled shift-and-add procedures. Either way, with image motion nulled, the end result is a star image with near diffraction-limited sharpness, one that may be considered to be the average short-exposure image.

The number of photoelectrons calculated in the above example represents the outcome at a single point in parameter space. However, the chosen "point" plainly corresponds to a region of parameter space likely to produce reasonably good image stabilization. Other parameter value combinations that might deliver even more optimal stabilization could be established by using an optimization algorithm to search for combinations that minimize bias and jitter in the core location estimates.

Such an optimization would have to take account of all controlling factors, including seeing conditions (described by both the temporal and spatial fluctuations of the OPD fluctuation of the imaging waves) and the performance characteristics of the AO system. The optimization would also have to take into account the noise characteristics of the detector. However, since detectors and detector noise are not treated in any substantive way in this book, an analysis of this kind goes beyond the scope of the book. These general comments equally apply to the two other radio-metric example calculations, Example 2 and Example 3, given later, in Sect. 17.5.1.

Note on signal-to-noise ratio: Signal-to-noise ratio delivered by an individual pixel depends on the photon noise, or shot noise, associated with the signal[8] as well as the readout and electronic noise arising from the detector itself. At IR wavelengths (where a significant portion of detector noise is caused by thermal emissions), detectors are often cooled to cryogenic temperatures; visible wavelength detectors such as

[8] For the Poisson noise dealt with here, in the case where the average number of photons falling on a detector during a prescribed integration period is large (say $\gg 10$), the rms variation of the observed number is given approximately by the square root of that number.

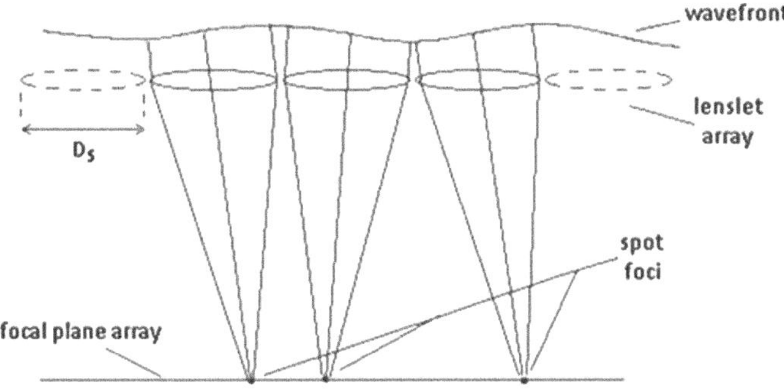


**Fig. 17.8** Schematic of a Shack–Hartmann lenslet array for sensing wavefront gradient

CCD arrays can also benefit from cooling, usually thermoelectric cooling. For more detailed information about detectors and detector noise, the reader is referred to the literature and, in particular, to the book "Detection of Light" by Rieke (2003).

## 17.5 Use of Natural Stars as Reference Objects for AO Image Correction

Figure 17.8 shows a typical wavefront sensing arrangement based on a Shack– Hartmann lenslet array. Each lenslet captures a small portion of the incoming wave from the chosen reference object and focuses it on to an FPA detector. The local wavefront gradients can be calculated from the observed displacements of the image spots relative to their long-term average locations. By integrating these gradients, a phase map can be generated for the instantaneous image-forming wave over the entire telescope pupil. The phase data so obtained can be used to drive an AO corrector device that reduces the wavefront phase fluctuation, ideally to zero.

The amount of light in each individual focused spot is proportional to the square of the diameter of the individual lenslets. It is now convenient to describe lenslet diameters in terms of the diameters of their conjugate images in the telescope pupil plane. We denote the latter diameters by $D_S$. Thus, for a telescope of diameter $D$, approximately $(D/D_S)^2$ individual lenslets are needed to capture the entire image-forming wave that arrives in the telescope pupil.

As in Sect. 17.4, to obtain the most accurate estimates of wavelength gradient, it is essential that cores exist in each individual spot image, typically containing about 37% of the light energy. In typical ($\sim$1 arcsec) seeing conditions where the lenslet diameter $D_S$ might be of the order of 20 cm, such cores might often arise at visible wavelengths. For lenslet diameters $\gg$ 20 cm, longer wavelengths might be needed to produce sufficiently bright cores. The methods outlined in Sect. 17.4.1 can again be used to establish to sub-pixel precision the locations on the FPA where the core in each individual focused spot attains maximum intensity.

### 17.5.1 *Radiometry of Natural Stars Used as AO Reference Objects*

Just as for the case dealt with in Sect. 17.4 where the objective was to stabilize the reference star image, in the present instance where the objective is to AO-correct the reference star image, we must appropriately scale each of the Shack–Hartmann array image spots on the FPA. The scaling can be accomplished by choosing the lenslet F/number in the Shack–Hartmann array, denoted here by $F_L$, so that the core in each lenslet image occupies only a small $n \times n$ pixel subset on the tracking FPA. Such a choice is roughly consistent with the condition,

$$F_L \approx \frac{n \cdot a_P}{\lambda} \tag{17.34}$$

where $a_P$ is the pixel width and square pixels are assumed. With this particular scaling arrangement and the assumption that each spot image approximates an Airy pattern, the distance spanned by the $n$ adjacent pixels approximately matches the radius of the first dark ring of the Airy pattern. Again, $n$ would typically be chosen to have a value in the range 3–5.

The total number of photoelectrons produced by the most brightly pixel can be obtained from 17.33 by replacing the telescope diameter $D$ by the scaled Shack–Hartmann lenslet diameter, $D_S$. Since individual Shack–Hartmann lenslets are not generally centrally obstructed, the expression may be simplified by omitting the $d^2$ dependence to give,

$$PE_{brightest\ pixel} = 2.51189^{-m_v} \cdot \left(\frac{1}{\lambda}\right)^4 \cdot \frac{\exp\left(\frac{h \cdot c}{0.55\,\mu\text{m} \cdot k \cdot T}\right) - 1}{\exp\left(\frac{h \cdot c}{\lambda \cdot k \cdot T}\right) - 1} \cdot E_C \cdot \frac{1}{n^2}$$
$$\times QE(\lambda) \cdot \Delta t \cdot \Delta\lambda \cdot D_S^2 \cdot T_o(\lambda) \cdot T_A(\lambda)^{\left(\frac{1}{\cos(ZA)}\right)} \cdot \left(3.492 \cdot 10^{-8} \frac{\text{m}}{\text{s}}\right) \tag{17.35}$$

To achieve diffraction-limited AO correction at visible wavelengths, it would generally be necessary to use a small $D_S$ value of the order 10 cm. However, because $D_S$ is usually much smaller than the telescope diameter $D$, inevitably the photoelectron numbers arising from the above expression are far smaller than those obtained in the Example 1 evaluation of 17.33. To ensure suitably high signal-to-noise ratios in this case, the only real option is to use significantly brighter reference stars. However, as we shall see in later Sect. 17.6, this results in a substantial sky coverage penalty.

**Radiometric Calculation, Example 2**
**AO image correction at 0.55 μm for telescopes of arbitrary size using natural stars as reference objects**

In this example, we use 17.35 to calculate the number of photoelectrons generated by the brightest pixel at the spot focus formed by a Shack–Hartmann lenslet of diameter $D_S$ =0.12 m. The diameter choice here matches that used by the AO systems of the Keck telescopes (Chanan et al., 1998). We assume that the lenslet images are obtained at the visible wavelength, 0.55 μm; we also assume that, because of the relatively small $D_S$ value, the Strehl intensities of the lenslet images are consistent with a relatively high $E_C$ value. Table 17.3 lists the parameter values used in the calculation.

According to 17.35, the brightest pixel for the chosen parameter values again generates about 7000 photoelectrons. As in radiometric Example 1, this number provides a signal-to-noise ratio consistent with precise core location estimates. However, to realize this relatively generous signal, requires a reference star of visual magnitude, $m_v = 3.0$. Since there are only about 300 stars in the sky of this brightness, or brighter, clearly we have an acute sky coverage problem. This problem is examined further in Sect. 17.6.

**Radiometric Calculation, Example 3**
**AO image correction at 2.2 μm for telescopes of arbitrary size using natural stars as reference objects**

Via this example, we find that the sky coverage problem significantly eases when, rather than attempting to carry out the wavefront sensing at visible wavelengths, sensing is carried out at IR wavelengths, in particular, at the near-IR wavelength, 2.2 μm. 17.35 is again used to calculate the number of photoelectrons generated by the brightest pixel. Visible seeing is assumed to be about 0.7 arcsec—about average for the Mauna Kea site where the Keck telescopes are located and hence representative of likely seeing at the nearby site proposed for the 30-m Telescope (TMT). In such seeing, it might reasonably be assumed that instantaneous reference star images formed at 2.2 μm by each of the individual 1.8-m mirror segments are nearly diffraction-limited.[9] Thus, we set $D_S = 1.8$ m and $E_C = 0.7$. It is also assumed that the temporal fluctuations of the image intensities at this wavelength occur at a slower rate than for the visible wavelength used in Example 2. (At longer wavelengths, larger OPD fluctuations are required to effect the same phase change.) In this example, therefore, we choose a longer integration time in proportion to the increased wavelength. Table 17.4 summarizes the parameter values used in the calculation.

For these values, 17.35 indicates that the brightest pixel again generates about 7000 photoelectrons. However, to achieve this healthy signal level, the visual magnitude of the reference star has now increased to $m_v = 8.53$, thus permitting much greater sky coverage. A more detailed account of sky coverage is given in the next section.

[9] The 3.8-m Mayall telescope despite its significant aberrations (circa 1990) could routinely produce well-defined cores at 2.2 μm even in average, 1.25-arcse, Kitt Peak visible seeing. On Mauna Kea where average seeing is about 0.7 arcsec, it may reasonably be assumed that a 1.8-m-wide diffraction-limited telescope primary mirror segment would produce image cores at 2.2 μm with Strehl intensities approaching 0.8.

**Table 17.4** Parameter values used in radiometric Example 3

| Parameter | Value |
|---|---|
| Shack–Hartmann lenslet diam, $D_S$ | 1.8 m |
| Reference star visual mag, $m_V$ | 8.53 |
| Reference star effective temp, $T$ | 5780 K |
| Nominal central wavelength, $\lambda$ | 2.2 μm |
| Bandwidth, $\Delta\lambda$ | 0.4 μm |
| Integration time, $\Delta t$ | 0.04 s |
| Detector quantum efficiency, $QE(\lambda)$ | 0.6 |
| Telescope optics trans, $T_O(\lambda)$ | 0.5 |
| Vertical path atmos trans, $T_A(\lambda)$ | 0.78 |
| Reference star zenith angle, ZA | 0.0° |
| No. of pixels across Airy 1st dark ring, $n$ | 4 |
| Core energy fraction, $E_C$ | 0.7 |

## 17.6 Sky Coverage When Natural Stars Are Used as Reference Objects

In the absence of laser guide stars, AO image correction of an astronomical object requires a suitably bright reference star lying within the same isoplanatic patch. Because of the random, but non-uniform star distributed in the sky, sometimes suitably bright reference stars are not available. To establish with certainty whether or not a suitable star actually exists, we could refer to a star atlas, such as the Millennium Star Atlas (Sinnott & Perryman, 1997). Short of doing this, we might make a rough estimate of the probability of finding a suitable star by making the simplifying, if crude, assumption that the stars are uniformly distributed throughout the sky.[10]

Figure 17.9 shows how the cumulative number of stars in the celestial sphere ($4\pi$ steradian) increases as a function of the limiting visual magnitude, $m_V$. The data used to create the plot were taken from the aforementioned Millennium Star Atlas, which provides star numbers to just beyond the 10th magnitude. The observed behavior between magnitudes 6 and 10 indicates that cumulative star numbers increase by a factor 2.91 for each unit increase in magnitude. For magnitudes greater than 10, it is assumed here that the star numbers continue to increase in the same way. Thus, the star numbers indicated in Fig. 17.9 can be reproduced using the approximate formula,

$$Number\ of\ stars \approx 2.91^{(m_v+2)} \tag{17.36}$$

Over the magnitude range, 3–21, the above formula may be considered accurate to within ±25%. Over the narrower range, 6–10, the formula is accurate to within

[10] In fact, a disproportionate number of stars lie in the direction of the galactic center. However, an equally disproportionate number of interesting objects may also lie in the same direction.

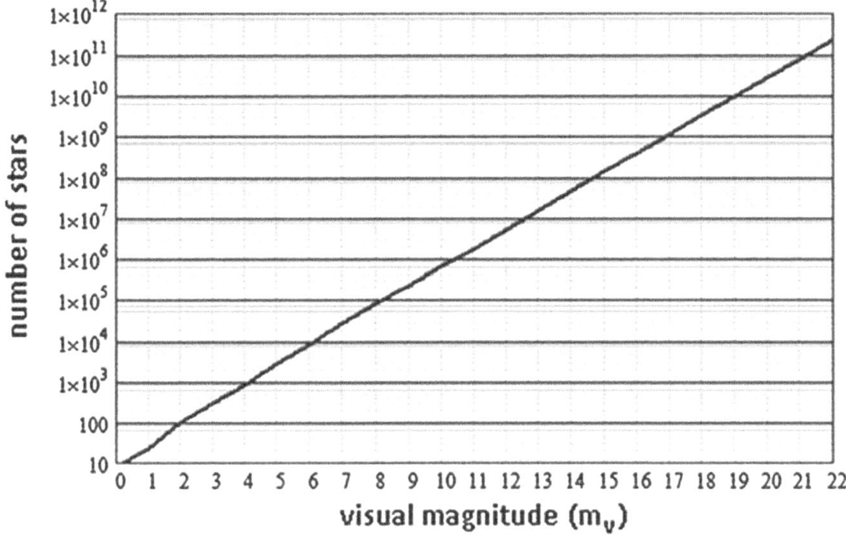


**Fig. 17.9** Cumulative numbers of stars in the sky brighter than apparent visual magnitude, $m_V$

±4%. As may readily be shown, the number of isoplanatic patches $N_{IP}$ required to fill the entire (4π steradian) celestial sphere can be given in terms of the isoplanatic angle, $\vartheta_I$, by the approximate formula

$$N_{IP} \approx \frac{4}{\vartheta_I^2} \tag{17.37}$$

The probability of finding a star within the isoplanatic patch as bright as, or brighter than magnitude $m_V$ gives a rough indication of the sky coverage fraction provided by stars up to that magnitude. Using 17.36 and 17.37, the sky coverage fraction, SC, arising from this coarse reasoning is given by

$$\begin{aligned} SC &\approx \frac{\vartheta_I^2 \cdot 2.91^{(m_v+2)}}{4} \quad \text{for} \quad \vartheta_I \leq 2 \cdot 2.91^{-\left(\frac{m_v+2}{2}\right)} \\ &\approx 1 \qquad\qquad\qquad \text{otherwise} \end{aligned} \tag{17.38}$$

where $0 \leq \mathrm{SC} \leq 1$.

Figure 17.10 shows sky coverage fractions plotted against isoplanatic angle for various limiting visual magnitudes.

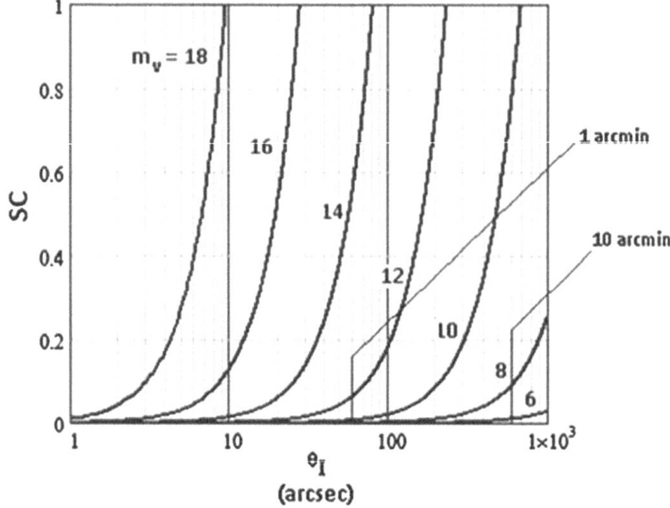


**Fig. 17.10** Approximate sky coverage fractions, SC, using only natural stars as AO reference objects up to the various limiting magnitudes indicated, plotted as a function of atmospheric isoplanatic angle, $\vartheta_I$. (All plots shown are based on the cumulative star numbers given in Fig. 17.9)

### *17.6.1 Sky Coverage for Image Stabilization at Near-IR Wavelengths*

We saw in the earlier radiometric calculation (Example 1, Sect. 17.4.2.1) that when image stabilization is carried out at the near-IR wavelength, 2.2 μm, reasonably robust S/N ratios can be obtained for 10-m class telescopes if we use natural star reference objects not exceeding visual magnitude $m_V = 10$. The sky coverage fraction offered by natural stars in this application of course also depends on the size of the isoplanatic angle.

If the isoplanatic angle associated with image cores does indeed extend out to 10 arcmin, as postulated on the basis of the turbulence structure characteristics measured by Coulman et al. (Sect. 17.2.3 and Appendix C), inserting $\vartheta_I \approx 10$ arcmin and $m_V = 10$ into 17.38 gives the sky coverage fraction, SC = 0.78. Though 17.38 is based on the coarse assumption of a uniform distribution of stars in the sky, this sky coverage fraction is nonetheless encouragingly high; it indicates that reasonably good sky coverage may be available for large telescopes imaging at near-IR wavelengths, but not equipped with laser guide star systems. (The cautionary reminder may be given here that, for Kolmogorov turbulence, 17.10 gives $\vartheta_I \approx 10$ arcsec at 2.2 μm. Sky coverage in this instance is then limited to the barely useful fraction, SC = 0.0002.)

### 17.6.2 Sky Coverage for Image Stabilization at Visible Wavelengths

In radiometric Example 2 (Sect. 17.5.1), we found that to provide an adequate light intensity level on the individual Shack–Hartmann lenslet spot foci at visible wavelengths required use of reference stars up to magnitude $m_v = 3$.Yet, even for the relatively favorable turbulence structure of the kind measured by Coulman et al., isoplanatic angles at these wavelengths are still only of the order 10 arcsec (Fig. 17.3, Top). Combining $\vartheta_\mathrm{I} \approx 10$ arcsec and $m_v = 3$ into 17.38 gives the sky coverage fraction, $\mathrm{SC} = 1.2 \times 10^{-7}$. This tiny, barely useful fraction underscores the critical need for reference objects other than natural stars for high resolution AO imaging at visible wavelengths.

### 17.6.3 Coverage for AO Correction at Near-IR Wavelengths

In radiometric Example 3 (Sect. 17.5.1), where AO correction was carried out at the near-IR wavelength, 2.2 μm, we found that reasonably robust signals were obtained from natural reference stars up to visual magnitude $m_\mathrm{v} = 8{:}53$. The relatively large magnitude indicated here is mostly attributable to the much larger Shack–Hartmann lenslet diameters that can be used at this longer near-IR wavelength ($D_\mathrm{S} = 1.8$ m). Again, assuming that core isoplanaticity does indeed extend to 10 arcmin, inserting $\vartheta_\mathrm{I} = 10$ arcmin and $m_\mathrm{v} = 8.53$ into 17.38 gives the sky coverage fraction, $\mathrm{SC} = 0.16$.

Although this sky coverage fraction is low, it is nonetheless useful, and, by making trade-offs between the various parameter values, it could be inflated to higher, more useful values. For example, by accepting a 4-times signal reduction (i.e., now only 1750 photoelectrons are generated by the brightest pixel as opposed to 7000), we could then use natural stars of visual magnitude up to $m_\mathrm{v} = 10$; in this case, 17.38 indicates that we obtain almost total sky coverage, with $\mathrm{SC} \approx 0.78$.

For Kolmogorov turbulence, where smaller isoplanatic angles are anticipated at 2.2 μm ($\vartheta_I \approx 10$ arcsec), even the use of reference stars up to magnitude $m_\mathrm{v} = 10$ provides only the barely useful sky coverage fraction, $\mathrm{SC} \approx 0.0002$. For the current generation of large telescopes, and also for the next generation of ELT instruments, the vastly different sky coverage fractions that arise for Kolmogorov turbulence and the non-Kolmogorov turbulence of the type measured by Coulman et al. underscore the crucial importance of a more precise understanding of the size characteristics of atmospheric turbulence.

### 17.6.4 Coverage for AO Correction at Visible Wavelengths

When $D_S$ is set to the relatively small value, 0.12 m, required for making AO corrections at visible wavelengths, the irradiance levels at the Shack-Hartmann lenslet foci are stubbornly low. As we saw in radiometric Example 2 (Sect. 17.5.1), to obtain a reasonably high signal level at the visible wavelength 0.55 μm (7000 photoelectrons for the brightest lit pixel) requires using reference stars brighter than magnitude 3. Even with the optimistic assumption of a 10 arcsec isoplanatic angle at 0.55 μm, the sky coverage fraction given by Eq. 17.38 is an extremely disappointing and barely useful, SC $= 1.2 \times 10^{-7}$.

### 17.6.5 Coverage Using Natural Reference Stars with ELT Instruments

Following the success of the segmented primary mirror approach used with the Keck 10-m instruments, the same principle is being adopted by the forthcoming ELT instruments. As previously mentioned, the primary mirror segment size used in the Keck instruments, 1.8 m, has been chosen as the segment size for the TMT instrument. In this section, we examine the sky coverage fractions that might be obtained when natural stars are used as the AO reference objects at near-IR and longer wavelengths with the various ELT instruments, the TMT instrument in particular.

The examination is based on a Shack–Hartmann wavefront sensing arrangement where the lenslet diameter, $D_S$, is chosen to be 1.8 m, so that each of the adaptive elements uniquely identifies with a specific primary mirror segment. It is assumed that the individual segments are figured to diffraction-limited standards, with the objective of obtaining AO-corrected images that closely approach diffraction-limited image quality at near-IR and longer wavelengths. Two types of turbulence are considered: Kolmogorov turbulence with an outer scale limit > 40-m and the non-Kolmogorov turbulence of the type measured by Coulman et al.

In the pristine seeing conditions found at the Keck site on Mauna Kea—which may also be representative of the nearby site proposed for the TMT instrument—it might be expected that, in average or better than average seeing conditions, each individual 1.8-m (diffraction-limited) mirror segment if used in isolation will deliver near-diffraction-limited images at near-IR and longer wavelengths. For a telescope composed of many such segments, to obtain AO-corrected images of diffraction-limited quality from the image-forming wavefronts over the entire telescope aperture, in principle it should only be necessary to make fast tip–tilt and piston adjustments to the wavefront portions collected by each individual mirror segment.

Again 17.36 can be used to calculate the total number of photoelectrons liberated by the most brightly illuminated pixel in the reference star image. Using the parameter values set out in Table 17.4, this equation indicates the most brightly illuminated pixel will again liberate about 7000 photoelectrons. We now consider how such a signal level translates to sky coverage for both Kolmogorov turbulence and the turbulence measured by Coulman et al.

#### 17.6.5.1 Kolmogorov Turbulence

For this type of turbulence, a typical isoplanatic angle for the near-IR wavelength, 2.2 μm, is about 10 arcsec. Inserting $\vartheta_{\mathrm{I}} \approx 10$ arcsec and $m_{\mathrm{v}} = 8.53$ into 17.38 gives the barely useful sky coverage fraction, SC $\approx 0.00005$. Thus, for Kolmogorov turbulence, the use of natural stars as AO reference objects for near-IR imaging offers little prospect of useful sky coverage fractions.

#### 17.6.5.2 Non-Kolmogorov Turbulence Measured by Coulman et al.

For this type of turbulence, where the possibility exists of 10-arcmin isoplanatic angles via image cores, sky coverage prospects greatly improve. Inserting $\vartheta_{\mathrm{L}} \approx 10$ arcmin and $m_{\mathrm{v}} = 8.53$ in 17.38 gives the sky coverage fraction, SC $\approx 0.16$. Although this fraction represents an improvement, it still falls well short of complete sky coverage. But, by settling for less signal (say 3000 photoelectrons), reference stars up to magnitude $m_{\mathrm{v}} = 9{:}5$ could be used, providing the sky coverage fraction, SC $\approx 0.5$. By making other trade-offs, such as increasing integration time, $\Delta t$, near-complete sky coverage comes within reach.

However, recalling that the sky coverage estimates given here depend on the—as yet only partially proven—postulation of 10-arcmin isoplanatic angles associated with image cores (as well as the fact that these estimates are also based on the coarse assumption of the stars being uniformly distributed in the sky), caution is needed before staking too much on these estimates. As mentioned at the beginning of the chapter, some of the content of this chapter is speculative. The only way to resolve this speculation is to carry out careful isoplanatic angle measurement experiments at very large, 10-arcmin, angles. The outcomes of these experiments may vindicate the speculation, or perhaps point to some middle ground. But, with evidence already reported of star images displaying significantly less image motion than predicted by Kolmogorov theory, and other evidence of binary star images exhibiting only tiny relative motions, it is inevitable that these experiments will indeed lead to much higher resolution and much larger sky coverage from ground-based telescopes, even if the choice is made to forego expensive laser guide star systems in favor of the much less expensive option of using natural stars as reference objects.

## 17.7 Mathematical Notation Used in This Chapter

The mathematical notation used in this chapter is indicated in Table 17.5.

**Table 17.5** Mathematical notation used in this chapter along with the SI dimensional units of the individual quantities

| Symbol | Quantity | Dimensions |
|---|---|---|
| $h_j$ | OPD fluctuation introduced by *j*th phase screen | $m$ |
| $\sigma_j$ | rms OPD fluctuation introduced by *j*th phase screen | $m$ |
| $\rho j$ | Autocorrelation function of OPD fluctuation introduced by *j*th random phase screen | "1" |
| $H$ | Integrated OPD fluctuation over entire atmospheric path | $m$ |
| $\sigma$ | rms of integrated OPD fluctuation | $m$ |
| $\rho$ | Autocorrelation function of integrated OPD fluctuation | "1" |
| $U$ | Complex amplitude | "1" |
| $\vartheta$ | Angular separation of two-point object | "1" |
| $\vartheta_{\mathrm{I}}$ | Atmospheric isoplanatic angle | "1" |
| $\vartheta_P$ | Angle subtended by planet's disk (assumed circular) | "1" |
| $I_S$ and $I_P$ | Intensity in twinkling images of stars and planets | "1" |
| $N_P$ | Effective number of uncorrelated scintillation patterns due to planet's angular size | "1" |
| $C_S$ and $C_P$ | Contrast ratios of intensity fluctuation in images of twinkling stars and planets | "1" |
| $DF$ | Demodulation of intensity fluctuation in twinkling planet image relative to that of a star | "1" |
| $ZA$ | Zenith angle | "1" |
| $L_0$ | Outer scale turbulence limit | $m$ |
| $C_n$ | Atmospheric refractive index structure constants | $\mathrm{m}^{-1/3}$ |
| $D$ | Telescope diameter | $m$ |
| $d$ | Telescope central obstruction | $m$ |
| $D_S$ | Effective diameter of Shack–Hartmann lenslets | $m$ |
| $T$ | Effective surface temperature of star | K |
| $m_v$ | Visual magnitude of star | "1" |
| $f$ | Telescope focal length | $m$ |
| $F_L$ | F/number of Shack–Hartmann lenslets | "1" |
| $a_p$ | Pixel width (assumed square) | $m$ |
| PE | Number of liberated photoelectrons | "1" |
| $R$ | Ratio of star irradiance emission at wavelength k to that at wavelength 0.55 μm | "1" |

(continued)

**Table 17.5** (continued)

| Symbol | Quantity | Dimensions |
|---|---|---|
| $QE$ | Detector quantum efficiency | "1" |
| $\Delta t$ | Image integration time | $s$ |
| $\lambda$ and $\Delta\lambda$ | Center imaging wavelength and bandwidth | $m$ |
| $T_O$ and $T_A$ | Light transmission fractions through telescope optics and atmosphere | "1" |
| $E_C$ | Core light energy fraction | "1" |
| $N_{IP}$ | Number of isoplanatic patches in $4\pi$ steradian | "1" |
| $SC$ | Sky coverage fraction | "1" |

Dimensionless quantities are indicated by "1"

# References


Chanan, G., Troy, M., Dekens, F., Michaels, S., Nelson, J., & Mast, T. (1998, January). Phasing the mirror segments of the Keck telescopes; the broadband phasing algorithm. *Applied Optics, 37*(1), 140–155.

Coulman, C. E., Vernin, J., Coqueugniot, Y., & Caccia, J. L. (1988). Outer scale of turbulence appropriate to modeling refractive index structure profiles. *Applied Optics, 27*, 155–160.

Dainty, J. C. (1984). Laser speckle and related phenomena. In J. C. Dainty (Ed.), *Topics in applied physics* (Vol. 9). Springer.

Lohmann, A. W., & Weigelt, G. P. (1979). Atmospheric turbulence and solar diameter measurement. *Optik, 53*, 167.

McKechnie, T. S. (1992). Atmospheric turbulence and the resolution limits of large ground-based telescopes. *JOSA A, 9*, 1937–1954.

Rieke, C. H. (2003). *Detection of light: From the ultraviolet to the submillimeter* (2nd ed.). Cambridge University Press.

Sinnott, R., & Perryman, M. (1997). *Millennium star atlas* (Vol. 3). Sky Publishing Corporation.

Weigelt, G. P. (1979). High resolution astrophotography: New isoplanaticity measurements and speckle holography applications. *Optica Acta, 26*, 1351–1357.

# Chapter 18
# Extremely Large Telescopes (ELTs): Imaging Performance and Imaging Path Characterization

**Abstract** A self-contained analytical equation tool set is developed in this chapter using an approach similar to that used in Chap. 13 where star images formed by large ground-based telescopes, with or without AO capability, were approximated by the best-fit sum of two Gaussian functions. In this chapter, the star images are approximated by the best-fit sum of three Gaussian functions. This advancement enables more faithful representation of the underlying physics of star image formation and more accurate modeling of star image intensity envelops. With statistical accuracy likely to improve with increasing telescope diameter, the equation tool set is particularly well-suited for characterizing and exploring star image formation and star image appearance with AO-equipped Extremely Large Telescopes (ELTs). The principal purpose of the equation tool set is to make the new general theory accessible and useful to the broad majority of ELT and other large telescope users, some of whom may have little or no background in statistical optics, physical optics Fourier optics, etc. The principles and disciplines of these various branches of physics (and the associated mathematics) are smoothly and painlessly captured by the tool set equations, allowing virtually anyone using these equations to characterize, analyze, predict, and optimize ELT star images formed over the entire range of visible and IR wavelengths. ELT imaging quality and resolution are quantified in terms of the average short-exposure intensity Point Spread Function (PSF) which may be obtained, as needed, by recording a long-exposure (core-and-halo) image of an unresolved star while making real-time image wander corrections. Once such an image has been modeled in terms of the three-Gaussian function approximation, other tool set equations enable calculation of the key properties of the residual (ideally zero) uncorrected Optical Path Difference (OPD) fluctuations of the image-forming light waves in the telescope pupil. Once these OPD properties have been established, sweet-spot wavelength regions can be identified and star image intensity PSFs can be calculated at all visible and IR wavelengths in the extended optical range, 0.3–1000 μm. Star images and star image intensity sections calculated from the equation set illustrate the incredible variety of star image shapes and sizes anticipated from the various ELT instruments.



T. S. McKechnie, *General Theory of Light Propagation and Imaging Through the Atmosphere*, Progress in Optical Science and Photonics 20,
https://doi.org/10.1007/978-3-030-98828-9_18

A self-contained analytical equation package is developed in this chapter with a full menu of applications to imaging with large ground-based reflector telescopes.[1] The equations are particularly well-suited for characterizing and exploring the statistical properties of star images formed by AO-equipped Extremely Large Telescopes (ELTs). The immense size of the light wave portions collected by these instruments naturally leads to more rapid emergence of the ensemble average star image properties. This in turn leads to more rapid emergence of the corresponding properties of the residual OPD fluctuation in the AO-corrected image forming light waves. Depending on the imaging wavelength, the atmospheric seeing conditions, and the efficacy of the AO systems, star images delivered by these huge instruments show an enormous range of variation. Some of the material contained in this chapter was presented in 2016 at the SPIE Conference on Large Telescopes, held in the Author's home town, Edinburgh (McKechnie, 2016)

ELT imaging performance is quantified in this chapter in terms of a single fundamental function: the average short-exposure intensity Point Spread Function (PSF). This function ultimately determines the highest level of imaging performance and resolution that can be obtained from each of the ELT instruments. The PSF properties are directly related to the properties of the residual OPD fluctuation in the final AO-corrected image-forming light waves.

As with other large ground-based telescopes, instantaneous star images obtained by ELTs will generally exhibit a fixed central core surrounded by a much broader halo, the latter comprising a time-frozen speckle pattern. The average short- exposure image looks distinctly different; it has a smooth, softer appearance, the time-averaging process having removed all trace of the continuously evolving speckle. "Long-exposure image PSFs" in the traditional sense—where no attempt is made to correct image motion—are not considered here as they generally provide unduly pessimistic measures of telescope imaging performance.

Image wander is caused by wave front tip/tilt variations. Since these can include significant contributions from telescope shake, they are generally the largest contributors to the total OPD fluctuation. Image wander correction is therefore an essential capability found in all AO correction systems.[2] Long-exposure images of unresolved stars obtained by AO-equipped ELT instruments therefore naturally provide the average short-exposure intensity PSF. For brevity in this chapter, we shall refer to this crucial function as simply the 'intensity PSF.'

The intensity PSF is strongly wavelength dependent. In addition to describing the intensity distribution in unresolved star images, this function also enables calculation of the intensity distribution in images of extended astronomical objects. Such objects usually behave as though incoherently illuminated. Their appearance may

[1] The formulation cannot be used with refractor telescopes because of their more complicated chromatic aberration characteristics.

[2] In Chap 10, Sect. 10.7, it was established that the most precise way of removing image motion is by stabilizing the image core.

This operation is directly equivalent to removing the tip/tilt component from the Zernike polynomial expansion describing the OPD fluctuation over the image-forming wavefronts for circular telescope pupils.

therefore be calculated by convolution of the intensity PSF with the object intensity illumination function.

To meaningfully characterize ELT imaging performance in terms of the intensity PSF, the reference star must lie close to the observing direction and, to ensure a fully definitive intensity PSF, the star image must be obtained at a wavelength consistent with the most general type of star image: the core and halo image.

## 18.1 Modelling the Intensity PSF as the Best-Fit Sum of Three Gaussian Functions

Once an image of an unresolved star has been obtained in the above manner with any of the AO-equipped ELT instruments, the next step is to approximate the image—the intensity PSF—by the best- fit sum of three Gaussian functions. Because of the additional degrees of freedom afforded by a three- rather than two-Gaussian function model (Chap. 13) modeling accuracy naturally improves. The model also more faithfully represents the underlying physics of image formation in large telescopes. Almost any least mean squares algorithm can be used to establish best-fit values of the six parameters that determine the three Gaussian function model: namely, the three central heights and the three 1/e half-widths of these functions. Levenberg (1944) developed the first least mean squares curve fitting algorithm suitable for use with electronic computers. It was subsequently re-discovered by Marquardt (1963) and became known as the Levenberg-Marquardt Algorithm (LMA). The LMA algorithm and derivative versions found ready application in lens design (Wynne, 1959) and have been incorporated into widely used optical design programs such as Zemax. The LMA algorithm is particularly well-suited for carrying out the three-Gaussian function curve fitting task. The task statement is set out in Appendix K in the form of six equations that must be satisfied simultaneously.

## 18.2 Residual OPD Fluctuation in AO-Corrected ELT Imaging Wave Fronts

Once the AO-system has best-effort corrected the phase-disrupted image-forming wave fronts collected by the ELT from a distant unresolved star, the residual uncorrected OPD fluctuation determines the final appearance of the star image. The essential properties of the residual OPD fluctuation are its root mean square (rms) variation and its unit-normalized autocorrelation function (c.f., Sect. 3.6.3).[3] In general, the residual OPD fluctuation comprises two separate components:

[3] They also enable calculation of the average short-exposure intensity Modulation Transfer Function (MTF) and the more general Two-point, Two-wavelength Correlation Function of the complex amplitudes.

1. Dynamical OPD fluctuation, caused by wind-driven motions and churning of the turbulence structure in the atmospheric path. In the case of ELT instruments, chaotic wind-induced undulations of the light-weight segmented primary mirrors could further contribute to dynamical OPD fluctuation.[4]
2. Fixed OPD fluctuation, caused by gravity sag and figure errors in the primary and secondary mirror surfaces (and possibly tertiary mirrors and Coude path mirror surfaces).

Due to their entirely different origins, the fixed and dynamical OPD contributions are statistically uncorrelated and necessarily retain that property even after the AO system has reduced their rms values and modified their autocorrelation functions. Recall Sect. 6.3, where functions of uncorrelated variables are necessarily themselves uncorrelated.[5] A consequence of the uncorrelated property is that the fixed and dynamical OPD residuals can be quantified separately.

## 18.3 Focusing ELT Instruments

As with all other telescopes, star images formed by ELT instruments can be considered best focused when the rms variation of the residual OPD fluctuation takes its minimum value. Since this condition corresponds to maximum intensity in the center of star images (Sect. 10.7) best focus corresponds to maximum Strehl intensity.[6]

## 18.4 Characterizing End-To-End Imaging Paths for ELT Instruments

By approximating the intensity PSF as the sum of Gaussian functions, the Fourier transform relationship between this intensity PSF and the properties of the residual OPD fluctuation can be expressed analytically. The Fourier transform operation naturally separates the residual OPD fluctuation into its fixed and dynamical components, expressing each in terms of its respective rms variation and autocorrelation function.

[4] Laminar air flow passing on either side of thin flexible membranes—which could possibly include ELT primary mirrors—causes chaotic undulations of the membrane itself. This phenomenon is responsible for the random flapping behavior of flags, even in light wind conditions and the random flexing behavior of suspension bridges. If such a problem were to arise with any of the ELTs, suitably designed winglets distributed over the primary mirror surface might help quieten the phenomenon.

[5] The 'function' referred to here is the temporal/spatial response function of the AO system which, in general, is neither linear nor separable with respect to space and time.

[6] In a departure from Chap. 13, we no longer express the residual fixed aberrations of the telescope in terms of Strehl intensity, $SI(\lambda)$.

Instead we use the rms OPD parameter associated with the telescope optics which equally specifies telescope Strehl intensity.

## 18.5 Calculating the Intensity PSFs at Other Visible and IR Wavelengths

Once the residual OPD properties have been established in the above manner, all necessary information is then available to calculate the intensity PSFs (formed by the same instrument observing over the same atmospheric path) at any other wavelength, or wavelengths, in the extended optical region, 0.3–1000 μm. The optimum wavelength and the surrounding sweet-spot wavelength region can also be identified. Depending on atmospheric conditions and the temporal/spatial frequency response characteristics of the AO system, star image intensity PSFs and sweet-spot wavelength regions can vary from night to night and even from hour to hour.

## 18.6 Applications of the Analytic Equation Package to AO-Equipped ELTs

By recording the intensity PSF with an AO-equipped ELT in the manner described before each observing run, the intensity PSF of the ELT can be accurately calculated for all other proposed imaging wavelengths or wavelength bands. This information allows an informed decision about whether ELT imaging performance (in the prevailing atmospheric conditions) is adequate for the proposed imaging task.[7]

By also recording the intensity PSF at regular time intervals during observing runs, definitive seeing logs can be maintained. The information provided by such logs can be used in a variety of ways, including issuance of real time prompts when imaging performance falls below required levels and identifying alternative imaging wavelength bands that might better accomplish the imaging task. The information could also assist image post-processing procedures and thus help extract maximum information from the images. It could also help determine whether ambiguous image details were actual image features or merely imaging artifacts.

Because ELT resolution is so strongly wavelength dependent, it would generally be beneficial to match the imaging task to a best-suited imaging wavelength band (assuming we have a choice in the matter). In Sect. 18.9.1, we look at the problem of identifying faint exoplanets nested in the glare of much brighter parent stars. For such applications, success or failure sharply depends on the imaging wavelength choice.

[7] As with UKIRT and the Hubble Space Telescope, and in fact most other large telescopes, there is no guarantee that any of the ELT instruments will quickly deliver its full imaging potential. It took draconian upgrades and two further decades after first light before UKIRT achieved anything like its true potential and a Space Shuttle repair visit before the Hubble instrument finally began delivering its full imaging potential.

## 18.7 Analysis and Characterization of AO-Equipped ELT Imaging Performance

It is assumed hereafter that the various ELT instruments all have nominally circular apertures and nominally circular central obstructions. Isotropic turbulence is also assumed so that the intensity PSFs formed by these instruments are nominally circularly symmetric and can therefore be fully described by central sections.

Though we have no detailed knowledge in advance of the characteristics of the residual OPD fluctuation for any of the forthcoming AO-equipped ELT instruments, we may nonetheless proceed by assuming that the autocorrelation functions of the residual fixed and dynamical OPD fluctuations may each be reasonably approximated by best-fitting Gaussian functions. We denote the 1/e half-widths of these functions by $w_{oF}$ and $w_{oD}$, respectively. By assuming (without loss of generality) that the dynamical and fixed OPD contributions, $OPD_D(x)$ and $OPD_F(x)$, are zero-mean functions, the unit-normalized best-fitting Gaussian function approximations to these autocorrelation functions can be written,

$$\rho_F(\xi) = \frac{\langle OPD_F(x+\xi) \cdot OPD_F(x) \rangle}{\langle OPD_F(x)^2 \rangle} \approx \exp\left[-\left(\frac{\xi}{w_{oF}}\right)^2\right] \tag{18.1}$$

and

$$\rho_D(\xi) = \frac{\langle OPD_D(x+\xi) \cdot OPD_D(x) \rangle}{\langle OPD_D(x)^2 \rangle} \approx \exp\left[-\left(\frac{\xi}{w_{oD}}\right)^2\right], \tag{18.2}$$

where spatial stationarity has been assumed (Sect. 3.6.1). Generic Gaussian autocorrelation functions, $\rho_F(\xi)$ and $\rho_D(\xi)$, are shown in Fig. 18.1.

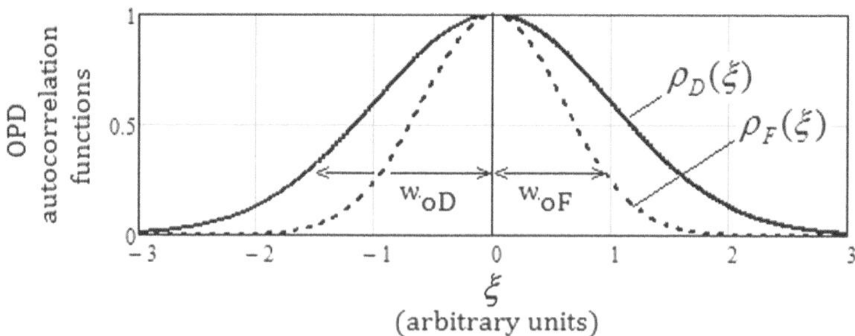


**Fig. 18.1** Gaussian approximations for the autocorrelation functions of the residual fixed and dynamical OPD fluctuation, $\rho_F(\xi)$ and $\rho_D(\xi)$. The 1/e half-widths of these functions are denoted by $w_{oF}$ and $w_{oD}$, where *e* is the base of Naperian logarithms

The rms variation of the fixed and dynamical residual OPD fluctuations, $\sigma_F$ and $\sigma_D$, are given as follows:

$$\sigma_F = \sqrt{\langle OPD_F(x)^2 \rangle} \tag{18.3}$$

and

$$\sigma_D = \sqrt{\langle OPD_D(x)^2 \rangle}. \tag{18.4}$$

For AO-equipped ELTs, the division of light energy between core and halo in star images is solely determined by the rms residual dynamical OPD fluctuation, $\sigma_D$. The rms variation of the residual fixed OPD, $\sigma_F$, does not influence the amount of light energy in the core; it merely determines how this light energy is distributed, and hence determines the angular size and shape of the core. Making the standard assumption that the residual dynamical OPD fluctuation is Gaussian distributed, the light fraction in the core, $E_C$, is given in terms of $\sigma_D$ and the imaging wavelength, $\lambda$, by (c.f., 10.3)

$$E_C = \exp\left[-(2 \cdot \pi \cdot \sigma_D/\lambda)^2\right]. \tag{18.5}$$

The light energy fraction scattered into the halo, $E_H$ is then given by

$$E_H = 1 - \exp\left[-(2 \cdot \pi \cdot \sigma_D/\lambda)^2\right]. \tag{18.6}$$

In Fig. 10.3, we saw how light energy in star images is distributed between the core and halo as a function of wavelength. For convenience, we reproduce that figure below (Fig. 18.2). At short wavelengths consistent with $\lambda/\sigma_D < 4$, star images are primarily halo-only images—i.e., seeing disc images. Over the relatively broad intermediate wavelength range, $4 < \lambda/\sigma_D < 20$, star images comprise both core and halo; maximum resolution is generally attained in this wavelength range. At longer wavelengths where $\lambda/\sigma_D > 20$, star images are largely core-only images. AO corrections have the effect of reducing $\sigma_D$, allowing cores to appear at shorter wavelengths. AO corrections also reduce $\sigma_F$, resulting in better-defined cores with larger light fractions retained in the central discs.

The cross-over wavelength, $\lambda_{XO}$, where the core and halo (Fig. 18.2) each contain 50% of the total light energy (i.e., $E_C = E_H$) occurs at a wavelength determined by $\sigma_D$ as follows:

$$\lambda_{XO} = \frac{2 \cdot \pi \cdot \sigma_D}{\sqrt{\ln(2)}} = 7.5427 \cdot \sigma_D. \tag{18.7}$$

where we note that $\sqrt{\ln(2)} = 0.832555$.

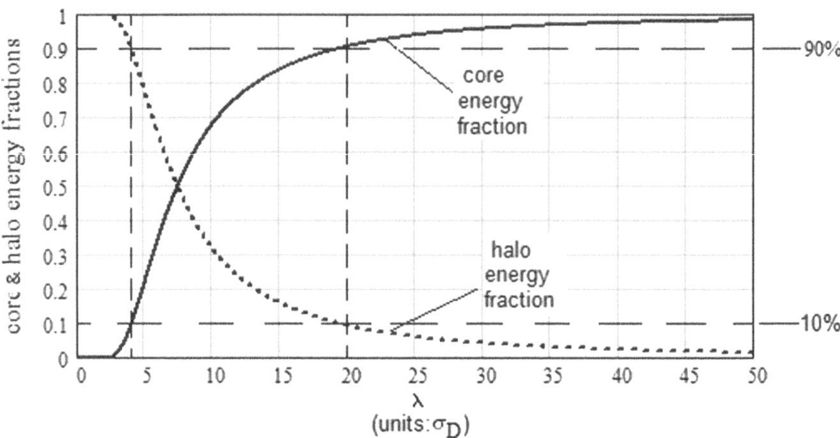


**Fig. 18.2** Star image appearance in ELTs depends on the imaging wavelength. As $\lambda/\sigma_D$ increases, star images morph from a halo-only stage to a core and halo stage until ultimately becoming core-only

## 18.7.1 The Ubiquitous Core and Halo Star Image

A central section through the intensity PSF for the most general type of star image formed by an ELT instrument is shown schematically in Fig. 18.3. The section shows a central core surrounded by a significantly broader halo. The core can generally be subdivided into a central disc (as in an Airy disc) and a pedestal, the latter comprised of light scattered out of the central disc by the residual fixed OPD fluctuation. The energy fraction remaining in the central disc portion of the core, $E_{CD}$, is given by (c.f., 18.5)

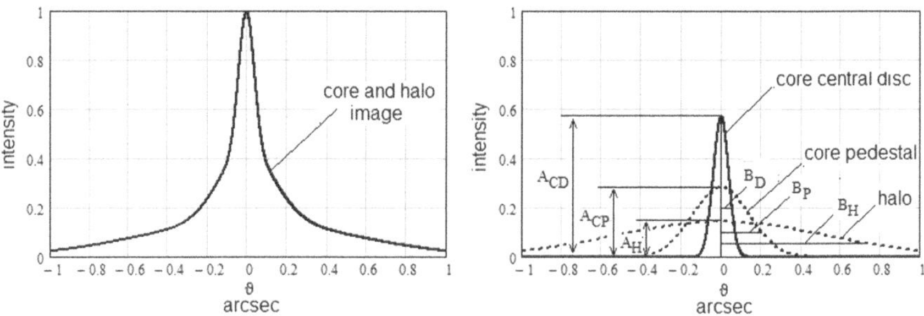


**Fig. 18.3** (Left) A section through the most general type of average short-exposure intensity PSF consisting of a core and halo. The core can generally be decomposed into a central disc sitting atop a pedestal, the latter comprised of light scattered out of the central disc by the residual fixed telescope aberrations. (Right) The intensity PSF separated into its three constituent parts: the core central disc, the core pedestal, and the halo

$$E_{CD} = \exp\left[-\left(\frac{2 \cdot \pi \cdot \sigma_D}{\lambda}\right)^2\right] \cdot \exp\left[-\left(\frac{2 \cdot \pi \cdot \sigma_F}{\lambda}\right)^2\right], \quad (18.8)$$

while the energy fraction in the core pedestal, $E_{CP}$, is given by

$$E_{CP} = \exp\left[-\left(\frac{2 \cdot \pi \cdot \sigma_D}{\lambda}\right)^2\right] \times \left\{1 - \exp\left[-\left(\frac{2 \cdot \pi \cdot \sigma_F}{\lambda}\right)^2\right]\right\}. \quad (18.9)$$

Irrespective of the residual fixed aberrations of an ELT, the light energy fraction remaining in the core's central disc, $E_{CD}$, is always distributed in the same way as the light energy in the central disc in star images formed by a diffraction-limited version of the same instrument. We will ignore for the moment the effect of the central obstruction. While residual telescope aberrations reduce the light energy fraction in the central disc, $E_{CD}$, they do not alter the shape distribution of that light energy, which remains that of the distribution in the central disc of an Airy pattern.

The following light energy fraction combinations (c.f., 18.5, 18.6, 18.8 and 18.9) necessarily sum to unity:

$$E_C + E_H = E_{CD} + E_{CP} + E_H = 1. \quad (18.10)$$

## *18.7.2 Normalized Forms of the Three-Gaussian Function Star Image Approximation*

In Fig. 18.3 (Right) $A_{CD}$, $A_{CP}$ and $A_H$, denote the central intensities in each of the three best-fit Gaussian functions (whose sum approximates the star image intensity PSF). The 1/e angular half-widths of these functions are denoted by $B_D$, $B_P$ and $B_H$. Denoting a central section through the intensity PSF by $I(\theta)$, this function may be expressed in unit-normalized form,

$$\langle I(\theta)\rangle = \frac{1}{(A_{CD} + A_{CP} + A_H)} \times \left\{A_{CD} \cdot \exp\left[-\left(\frac{\theta}{B_D}\right)^2\right] + A_{CP} \cdot \exp\left[-\left(\frac{\theta}{B_P}\right)^2\right] + A_H \cdot \exp\left[-\left(\frac{\theta}{B_H}\right)^2\right]\right\}. \quad (18.11)$$

It is now convenient to normalize the core-related quantities, $A_{CD}$ and $A_{CP}$, with respect to the halo-related quantity, $A_H$. The normalized quantities, which we denote by $A_D$ and $A_P$, are defined as follows (c.f., 13.52).

$$A_D = \frac{A_{CD}}{A_H} \tag{18.12}$$

and

$$A_P = \frac{A_{CP}}{A_H}. \tag{18.13}$$

By re-writing Eq. 18.11 in terms of $A_D$ and $A_P$, we find that the $A_H$ terms cancel from the equation, enabling $\langle I(\theta)\rangle$ to be expressed in terms of the five remaining parameters:

$$\begin{aligned}\langle I(\theta)\rangle = {} & \frac{1}{(A_D + A_P + 1)} \\ & \times \left\{ A_D \cdot \exp\left[-\left(\frac{\theta}{B_D}\right)^2\right] + A_P \cdot \exp\left[-\left(\frac{\theta}{B_P}\right)^2\right] \right. \\ & \left. + \exp\left[-\left(\frac{\theta}{B_H}\right)^2\right]\right\}. \end{aligned} \tag{18.14}$$

The intensity PSF section, $I(\theta)$, may also be written in a form (c.f., 13.55) where, upon rotation, the enclosed volume—and thus the enclosed light energy—are both unity:

$$\begin{aligned}\langle I(\theta)\rangle = {} & \frac{1}{\pi \cdot \left(A_D \cdot B_D^2 + A_P \cdot B_P^2 + B_H^2\right)} \\ & \times \left\{ A_D \cdot \exp\left[-\left(\frac{\theta}{B_D}\right)^2\right] + A_P \cdot \exp\left[-\left(\frac{\theta}{B_P}\right)^2\right] \right. \\ & \left. + \exp\left[-\left(\frac{\theta}{B_H}\right)^2\right]\right\} \end{aligned} \tag{18.15}$$

From Fig. 18.3 (Right), it is clear that the maximum possible resolution obtainable from an ELT instrument is ultimately set by the angular width of the central disc portion of the core. When this feature contains a significant fraction of the total light energy, it affords diffraction-limited resolution ($\sim 1.22 \cdot \lambda/\mathrm{D}$) which can be a factor of 50- to 100-times higher than provided by the halo.

At shorter wavelengths, where there may be negligible light energy in the central disc, resolution is then determined by the angular width of the core pedestal. At even shorter wavelengths, where there may be negligible energy in either the central disc or the pedestal, resolution falls back to the angular width of the halo, the star image having now degenerated into a seeing disc.

### 18.7.3 Fourier Transform Relationships Between the Intensity PSF and the Residual OPD Fluctuation

As discussed in Chap. 13, telescope central obstructions cause narrowing, or sharpening, of the central disc portion of the core, with the excess light off-loaded into the core pedestal (c.f., Fig. 13.3). For an ELT of diameter, D, and central obstruction diameter, d, the angular radius of the central disc shrinks in the same proportion as the central disc produced by an unobstructed instrument with a slightly larger effective diameter, $D_{Ef}$, given by

$$D_{Ef} \approx \sqrt{(D^2 + d^2)}. \tag{18.16}$$

For central obstructions in the size range, $0 \leq d \leq 0.4 \cdot D$, the above approximation for $D_{Ef}$ gives the radius of the central disc ($\sim 1.22 \cdot \lambda/D_{Ef}$) to within 0.5% accuracy. For $d = 0.5 \cdot D$, the accuracy is still within 2%. For the 3.8 m Mayall instrument (Page 371) where $d \approx 0.43 \cdot D$, 18.16 gives an effective diameter of 4.5 m. For the 39.3 m E-ELT instruments with its 4.2 m central obstruction, $D_{Ef} \approx 39.5$ m.[8]

The three-Gaussian function approximation to the intensity PSF provides sufficient flexibility to assign one of these Gaussian functions to represent the central disc, a second to represent the core pedestal, and the third to represent the halo. In this way, a more accurate and physically realistic model of the intensity PSF is obtained than was possible with the Chap. 13 two-Gaussian function approximation. Using an analysis similar to that in Sect. 13.1.5, the following relationships can be established between the parameter set describing the best-fit three-Gaussian function approximation, $A_D$, $A_P$, $B_C$, $B_P$ and $B_H$, and the physical parameter set, D, d, $\lambda$, $\sigma_F$, $\sigma_D$, $w_{oF}$ and $w_{oD}$, which describe the telescope aperture geometry, the imaging wavelength, and the fixed and dynamical OPD fluctuation residuals in the AO-corrected imaging waves (c.f., 13.78)

$$B_D = \frac{2 \cdot \lambda}{\pi \cdot \sqrt{(D^2 + d^2)}}, \tag{18.17}$$

$$B_P = \frac{2 \cdot \sigma_F}{w_{oF}}, \tag{18.18}$$

$$B_H = 2 \cdot \sqrt{\frac{\sigma_F^2}{w_{oF}^2} + \frac{\sigma_D^2}{w_{oD}^2}}, \tag{18.19}$$

[8] It might be noted that, by increasing the central obstruction diameter of the E-ELT to 7.5 m—which could readily be accomplished by placing a suitably scaled annular mask in a secondary pupil plane—the E-ELT could again reclaim its status as a '40-m' instrument, at least in terms of the angular subtense of the core's central disc. Taking such a strategy to its limit, $d \to D$, the angular diameter of the core's central disc formed by the 39.3-m E-ELT would be the same as that of an Airy pattern formed by a 62.5-m unobstructed instrument.

$$A_D = \frac{\pi^2 \cdot (D^2 + d^2)}{\lambda^2} \cdot \left( \frac{\sigma_F^2}{w_{oF}^2} + \frac{\sigma_D^2}{w_{oD}^2} \right) \times \frac{\exp\left[ -\frac{4 \cdot \pi^2 \cdot (\sigma_D^2 + \sigma_F^2)}{\lambda^2} \right]}{1 - \exp\left( -\frac{4 \cdot \pi^2 \cdot \sigma_D^2}{\lambda^2} \right)} \quad (18.20)$$

and

$$A_P = \frac{w_{oF}^2}{\sigma_F^2} \cdot \left( \frac{\sigma_F^2}{w_{oF}^2} + \frac{\sigma_D^2}{w_{oD}^2} \right) \times \frac{\exp\left( -\frac{4 \cdot \pi^2 \cdot \sigma_D^2}{\lambda^2} \right) \cdot \left[ 1 - \exp\left( -\frac{4 \cdot \pi^2 \cdot \sigma_F^2}{\lambda^2} \right) \right]}{1 - \exp\left( -\frac{4 \cdot \pi^2 \cdot \sigma_D^2}{\lambda^2} \right)}. \quad (18.21)$$

Angle $B_D$ may be calculated directly from 18.17 from the parameters, $\lambda$, $D$ *and* $d$. Since these parameters are known at the outset, there is no compelling reason to measure this angle from the actual star image. Assuming that the reference star image motion was properly corrected when the reference star's intensity PSF was obtained—that is, by continuous real-time re-centering of the core's central disc feature (while entirely ignoring the light energy centroid)—the 1/e angular half-width of the central disc in the intensity PSF should precisely match that given by 18.17. With $B_D$ effectively pre-determined, now only four parameters, $A_D$, $A_P$, $B_P$ *and* $B_H$, are needed to specify the three-Gaussian function intensity PSF approximation.

### *18.7.4 Inverse Fourier Transform Relationships*

The inverse Fourier transform relationships can readily be established from 18.17 to 18.21. These inverse relationships enable direct calculation of the parameter set, $\sigma_F$, $\sigma_D$, $w_{oF}$ and $w_{oD}$, from the parameter set, $D$, $d$,, $A_D$, $A_P$, $B_D$, $B_P$, $B_H$:

$$\sigma_F = \frac{\lambda}{2 \cdot \pi} \times \sqrt{-\ln\left( \frac{4 \cdot \lambda^2 \cdot A_D}{\pi^2 \cdot (D^2 + d^2) \cdot A_P \cdot B_P^2 + 4 \cdot \lambda^2 \cdot A_D} \right)}, \quad (18.22)$$

$$\sigma_D = \frac{\lambda}{2 \cdot \pi} \times \sqrt{-\ln\left( \frac{\pi^2 \cdot (D^2 + d^2) \cdot A_P \cdot B_P^2 + 4 \cdot \lambda^2 \cdot A_D}{4 \cdot \lambda^2 \cdot A_D + \pi^2 \cdot (D^2 + d^2) \cdot (A_P \cdot B_P^2 + B_H^2)} \right)}, \quad (18.23)$$

$$w_{oF} = \frac{\lambda}{\pi \cdot B_P} \times \sqrt{-\ln\left( \frac{4 \cdot \lambda^2 \cdot A_D}{\pi^2 \cdot (D^2 + d^2) \cdot A_P \cdot B_P^2 + 4 \cdot \lambda^2 \cdot A_D} \right)}, \quad (18.24)$$

$$w_{oD} = \frac{\lambda}{\pi \cdot \sqrt{\left(B_H^2 - B_P^2\right)}} \times \sqrt{-\ln\left(\frac{\pi^2 \cdot \left(D^2 + d^2\right) \cdot A_P \cdot B_P^2 + 4 \cdot \lambda^2 \cdot A_D}{4 \cdot \lambda^2 \cdot A_D + \pi^2 \cdot \left(D^2 + d^2\right) \cdot \left(A_P \cdot B_P^2 + B_H^2\right)}\right)}. \quad (18.25)$$

Though $B_{\mathrm{C}}$ does not appear in any of these four equations, it is implicit (c.f., 18.17) in the D, d and λ values.

**Numerical check of** 18.17–18.21 **and** 18.22–18.25

As a final check that equations 18.17–18.21 and the inverse equations 18.22–18.25 are circularly self- consistent, we input the following arbitrarily chosen parameter values into 18.17–18.21:

D = 39.3 *mm*, d= 4.2 *mm*, $\lambda = 1.65\ \mu m$, $\sigma_{\mathrm{F}} = 0.18\ \mu m$, $\sigma_{\mathrm{D}} = 0.3\ \mu m$, $w_{oF} = 0.8$ *m*, and $w_{o\mathrm{D}}= 0.3\ m$. The resulting parameter values, $A_{\mathrm{D}}$, $A_{\mathrm{P}}$, $B_{\mathrm{D}}$, $B_{\mathrm{P}}$, and $B_H$ are as follows:

$A_{\mathrm{D}} = 1383.67$, $A_{\mathrm{P}} = 2.89443$, $B_{\mathrm{D}} = 0.0054819$ arcsec, $B_{\mathrm{P}} = 0.09282$ arcsec, and $B_H = 0.42284$ arcsec.

When the latter parameter values are entered into 18.22–18.25 (while maintaining D = 39.3 *m*, d= 4.2 *m*, and $\lambda = 1.65\ \mu m$) these equations identically reproduce—as they should—the starting parameter value set, in this illustrative case, $\sigma_F = 0.18\ \mu m$, $\sigma_{\mathrm{D}} = 0.3\ \mu m$, $w_{oF} = 0.8\ m$ and $w_{o\mathrm{D}} = 0.3\ m$.

### *18.7.5 Calculation of Star Image Intensity PSFs at Other Wavelengths*

Once a reference star image has been obtained by an AO-equipped ELT instrument (at a wavelength consistent with a core and halo image) and the intensity PSF envelop has been expressed in terms of the three-Gaussian function approximation, we then possess sufficient information to calculate the intensity PSF (formed by the same AO-assisted ELT instrument observing over the same atmospheric path in the same seeing conditions) at all wavelengths in the extended optical range, 0.3–1000 $\mu m$.

Star images formed by large telescopes are strongly wavelength dependent and this is especially true of ELT instruments. To indicate that dependence as we go forward, we now re-label $A_{\mathrm{D}}$, $A_{\mathrm{P}}$, $B_{\mathrm{D}}$, $B_{\mathrm{P}}$, and $B_H$, by $A_{\mathrm{D}}(\lambda)$, $A_{\mathrm{P}}(\lambda)$, $B_{\mathrm{D}}(\lambda)$, $B_{\mathrm{P}}(\lambda)$ and $B_H(\lambda)$. If the initial reference star image was obtained at wavelength, $\lambda_M$, the parameter values that describe the best-fit intensity PSF approximation can be denoted by $A_{\mathrm{D}}(\lambda_M)$, $\mathrm{A_P}(\lambda_M)$, $B_{\mathrm{D}}(\lambda_M)$, $\mathrm{B_P}(\lambda_{\mathrm{M}})$, and $B_{\mathrm{H}}(\lambda_M)$. The corresponding parameter values describing the intensity PSF at any other arbitrarily chosen wavelength, λ, can be similarly denoted, $A_{\mathrm{D}}(\lambda)$, $A_{\mathrm{P}}(\lambda)$, $B_{\mathrm{D}}(\lambda)$, $B_{\mathrm{P}}(\lambda)$ and $B_H(\lambda)$.

By using 18.17–18.21, we can set up two equation sets, one for wavelength, $\lambda_M$, the other for some other wavelength, $\lambda$. By combining these equation sets, the following relationships can be established.[9]

$$A_D(\lambda) = A_D(\lambda_M) \cdot \frac{\lambda_M^2}{\lambda^2} \cdot \exp\left[-4 \cdot \pi^2 \cdot (\sigma_D^2 + \sigma_F^2) \cdot \left(\frac{1}{\lambda^2} - \frac{1}{\lambda_M^2}\right)\right] \times \left[\frac{1 - \exp\left(-\frac{4 \cdot \pi^2 \cdot \sigma_D^2}{\lambda_M^2}\right)}{1 - \exp\left(-\frac{4 \cdot \pi^2 \cdot \sigma_D^2}{\lambda^2}\right)}\right] \quad (18.26)$$

$$A_P(\lambda) = A_P(\lambda_M) \cdot \exp\left[-4 \cdot \pi^2 \cdot \sigma_D^2 \cdot \left(\frac{1}{\lambda^2} - \frac{1}{\lambda_M^2}\right)\right] \times \frac{\left[1 - \exp\left(-\frac{4 \cdot \pi^2 \cdot \sigma_F^2}{\lambda^2}\right)\right] \cdot \left[1 - \exp\left(-\frac{4 \cdot \pi^2 \cdot \sigma_D^2}{\lambda_M^2}\right)\right]}{\left[1 - \exp\left(-\frac{4 \cdot \pi^2 \cdot \sigma_D^2}{\lambda^2}\right)\right] \cdot \left[1 - \exp\left(-\frac{4 \cdot \pi^2 \cdot \sigma_F^2}{\lambda_M^2}\right)\right]}, \quad (18.27)$$

$$B_D(\lambda) = \frac{\lambda}{\lambda_M} \cdot B_D(\lambda_M), \quad (18.28)$$

$$B_P(\lambda) = B_P(\lambda_M), \quad (18.29)$$

$$B_H(\lambda) = B_H(\lambda_M). \quad (18.30)$$

## 18.8 Representative Intensity PSFs Delivered by AO-Assisted ELTs

Once the best-fitting parameter set, $A_D(\lambda_M)$, ..... $B_H(\lambda_M)$, has been calculated for a reference star image obtained at wavelength, $\lambda_M$, Eqs. 18.26–18.30 can also be used to calculate the corresponding values, $A_D(\lambda)$, ..... $B_H(\lambda)$, for any other arbitrarily chosen wavelength, $\lambda$. A central section through the intensity PSF at wavelength, $\lambda$, may then be calculated from either of the following two expressions (c.f., 18.14 and 18.15):

[9] In cases where $\lambda$ and $\lambda_M$ are widely separated, the refractive index dispersion of ai (Chap. 3, Sect. 3.1.) would have to be taken into consideration. This could be accomplished by recasting the rms residual dynamical OPD fluctuation, $\sigma_D$, used in Eqs. 18.28 –18.32 as a wavelength dependent parameter. At any arbitrarily chosen wavelength, $\sigma_D$ could be calculated as per Chap. 6 (Eq. 6.61). No similar modifications are needed for the quantity, $\sigma_F$. Its value is determined entirely by the residual aberrations of the telescope\AO system. Since the air path variation caused by these aberrations is of the order of microns, and the refractive index of air is approximately unity (to within 1 part in 3500) the effect of dispersion in this instance is entirely negligible.

$$\langle I(\theta,\ \lambda)\rangle = \frac{1}{(A_D(\lambda) + A_P(\lambda) + 1)} \times \left\{ A_D(\lambda) \cdot \exp\left[-\left(\frac{\theta}{B_D(\lambda)}\right)^2\right] + A_P(\lambda) \cdot \exp\left[-\left(\frac{\theta}{B_P(\lambda)}\right)^2\right] + \exp\left[-\left(\frac{\theta}{B_H(\lambda)}\right)^2\right]\right\} \tag{18.31}$$

$$\langle I(\theta, \lambda)\rangle = \frac{1}{\pi \cdot [A_D(\lambda) \cdot B_D(\lambda)^2 + A_P(\lambda) \cdot B_P(\lambda)^2 + B_H(\lambda)^2]} \left\{ A_D(\lambda) \cdot \exp\left[-\left(\frac{\theta}{B_D(\lambda)}\right)^2\right] + A_P(\lambda) \cdot \exp\left[-\left(\frac{\theta}{B_P(\lambda)}\right)^2\right] + \exp\left[-\left(\frac{\theta}{B_H(\lambda)}\right)^2\right]\right\} \tag{18.32}$$

Equation 18.31 provides the central sections in unit-normalized form (c.f., 18.14), while 18.32 provides central sections which, upon rotation, enclose unit volume and hence unit light energy (c.f., 18.15). Star image intensity PSF sections calculated from 18.32 are shown in Fig. 18.4 for the 39.3 m E-ELT instrument for several visible and IR wavelengths. They show the enormous variations in star image angular size and shape likely to be delivered by such a massive instrument.

The sections are based on the same $\sigma_D$, $\sigma_F$, $w_{oD}$, and $w_{oF}$ values (0.3 μm, 0.18 μm, 0.3 m and 0.8 m) previously used to generate the 10 m Keck II star images shown in Figs. 14.19 and 14.20. The $A_D$ (λ), $A_P(\lambda)$, $B_H$ (λ), $B_p(\lambda)$, and $B_D(\lambda)$ values inserted into 18.32 were generated from 18.17 to 18.21 with D and *d* set to 39.3 m and 4.2 m, respectively. The optimum wavelength, $\lambda_{Opt}$, for these $\sigma_D$ and $\sigma_F$ values is about 2.2 μm, with the sweet-spot wavelength range roughly lying between 1.5 μm to 3.5 μm.

To apply $\sigma_D$, $\sigma_F$, $w_{oD,}$and $w_{oF}$ values to a 39.3 m instrument that have so far only been demonstrated with a 10 m instrument may be slightly optimistic, at least in the near term. Consequently, Fig. 18.5 shows a second set of intensity PSF sections obtained by a 39.3 m instrument where the $\sigma_D$, $\sigma_F$, $w_{oD}$ and $w_{oF}$ values have all been doubled in size to 0.6μm, 0.36 μm, 0.6 m and 1.6 m, respectively.

The image sections in Fig. 18.5 are similar to those in Fig. 18.4. Again, huge variations are seen in the angular sizes and shapes of the intensity PSFs at different wavelengths. The optimum wavelength has now increased to 4.4 μm, and the sweet-spot wavelength range has moved out to 3.0–7.0 μm.

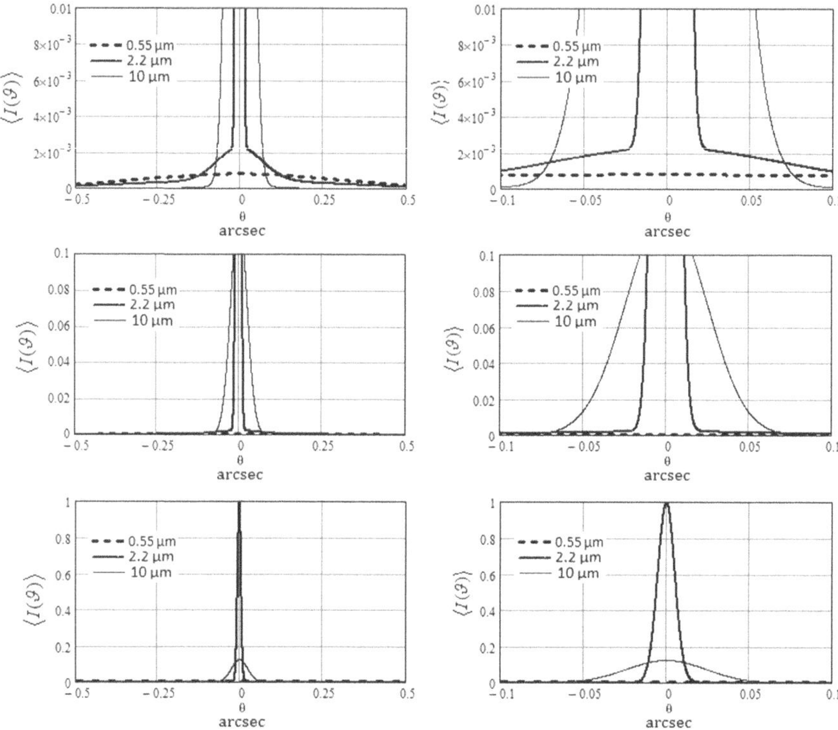


**Fig. 18.4** (Left Bottom) Star image intensity PSFs of the sort likely to be obtained by the 39.3 m E-ELT instrument at visible and IR wavelengths in good seeing conditions. (Left, Middle and Top) Same intensity PSFs but with 10 × and 100 × times vertical scale expansions. (Right Bottom) Same intensity PSFs but with a 5 × horizontal scale expansion. (Right, Middle and Top) Same as Right Bottom but with 10 × and 100 × vertical scale expansions. The intensity PSF at the visible wavelength, 0.55 μm, appears as a diffuse halo (seeing disc). The intensity PSF at 2.2 μm displays the full image feature compliment—core central disc, core pedestal, and halo. The intensity PSF at 10 μm almost entirely consists of a central core. (Parameter values used to calculate the plots: D = 39.3 m, d = 4.2 m, $\lambda_M$ = 1.65 $\mu m$, $\sigma_D$= 0.3 $\mu m$, $\sigma_F$ = 0.18 $\mu m$, $w_{oD}$ = 0.3 m, $w_{oF}$ = 0.8 m)

## Logarithmic Plots of the Intensity PSF Envelops

The extreme variations of the star image central intensities and angular widths seen in Figs. 18.4 and 18.5 may be better appreciated by plotting the intensity PSFs against a logarithmic intensity scale, $\log_{10}(I)$. Figures 18.6 and 18.7 show 2-D surface plots of AO-sharpened 39.3 m E-ELT star image intensity envelops plotted in this way. The intensity envelops are shown for four different wavelengths spanning the visible to far-IR wavelength range, 0.55–30 μm. Figure 18.6 shows images consistent with a 0.7-arcsec FWHM seeing disc image at the visible wavelength, 0.55 μm. Figure 18.7 shows images consistent with a 1.4-arcsec FWHM seeing disc image at the same visible wavelength.

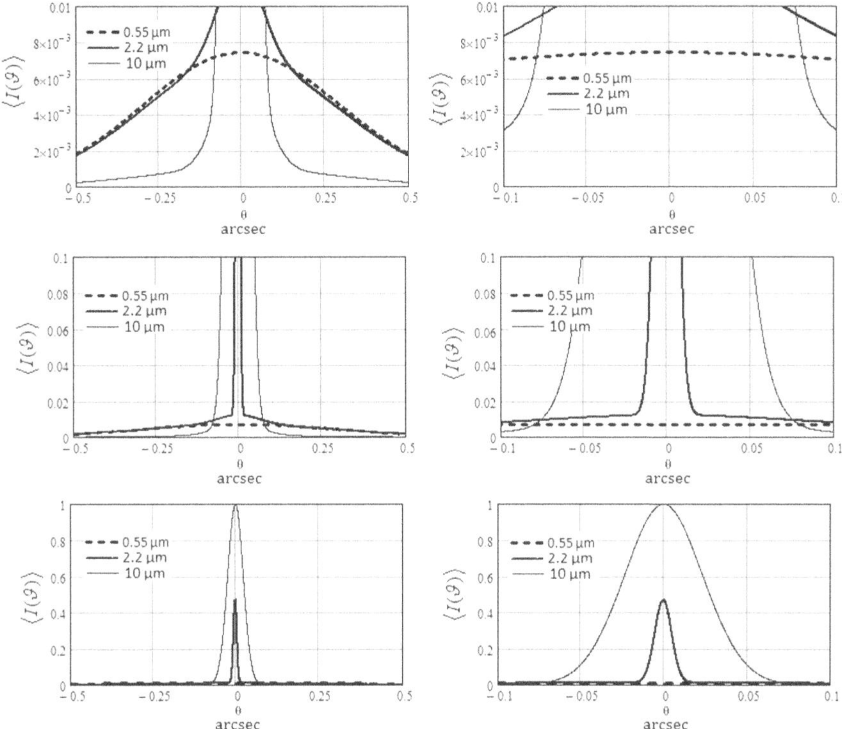


**Fig. 18.5** Intensity PSFs similar to those shown in Fig. 18.4 but in slightly less favorable seeing conditions where the residual fixed and dynamical rms variations take values twice as large as those used in Fig. 18.4. (Left Bottom) Star image intensity PSFs similar to those likely to be obtained by the 39.3 m E-ELT instrument at visible and IR wavelengths in good seeing conditions. (Left, Middle and Top) Same intensity PSFs with 10 × and 100 × vertical scale expansions. (Right Bottom) Same intensity PSFs with a 5 × horizontal scale expansion. (Right, Middle and Top) Same as Right Bottom but with 10 × and 100 × vertical scale expansions. The intensity PSF at the visible wavelength, 0.55 μm, again appears as a diffuse halo (seeing disc). The intensity PSF at 2.2 μm largely consists of a halo with only a small fraction of the light energy remaining in the core. The intensity PSF at 10 μm again largely consists of a central core. (Parameter values used to calculate the plots: D = 39.3 m, d = 4.2 m, $\lambda_M$ = 1.65 μm, $\sigma_D$ = 0.6 μm, $\sigma_F$ = 0.36 μm, $w_{oD}$ = 0.6 m, $w_{oF}$ = 1.6 m)

The parameter values used to calculate the star image intensity envelops in Fig. 18.6 are the same as those used previously to calculate the star image intensity sections in Fig. 18.4 (D = 39.3 m, d = 4.2 m, $\sigma_D = 0.3\ \mu m$, $\sigma_F = 0.18\ \mu m$, $w_{oD}$ = 0.3 m, $w_{oF}$ = 0.8 m). The values used to produce the Fig. 18.7 intensity envelops correspond to larger wave disruptions, where $\sigma_D$ and $\sigma_F$ have both been doubled in size to 0.6 $\mu m$ and 0.36 $\mu m$, the other parameter values remaining unchanged.

The rationale for the above parameter value choices was given in the text above Fig. 18.4. Since guesswork was involved—albeit somewhat informed guesswork—it would be unwise to attach too much credence to the chosen values, at least not

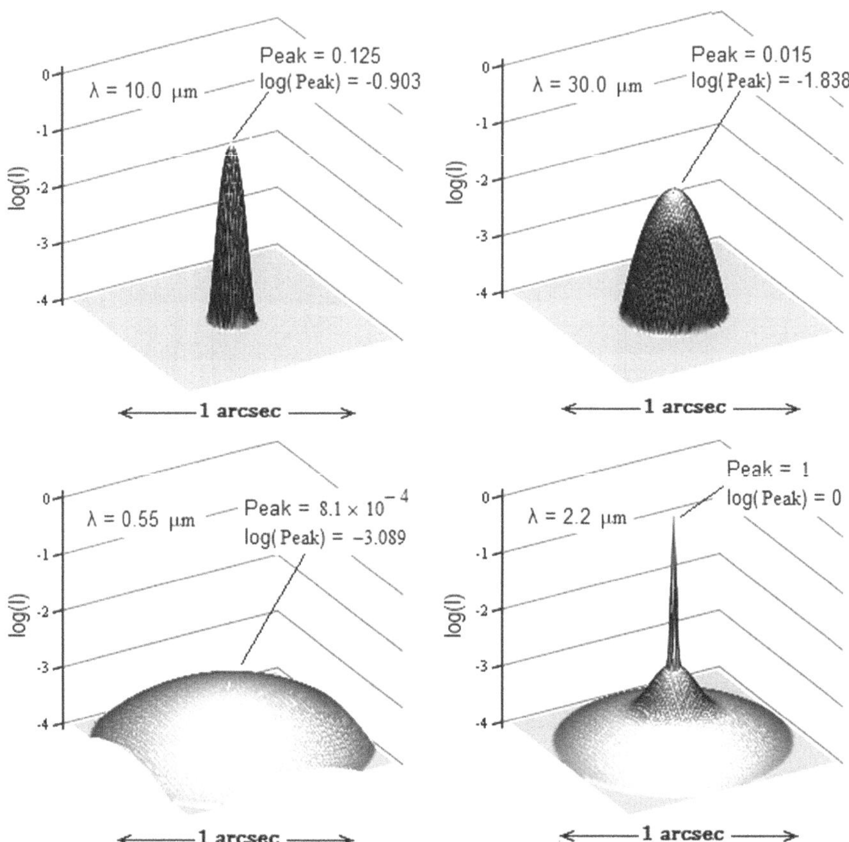


**Fig. 18.6** Star image intensity PSF envelopes plotted against a $\log_{10}(I)$ intensity scale for the 39.3 m E-ELT instrument at the four wavelengths indicated in the figure. The following parameter values were used to calculate the intensity envelops: $D = 39.3$ m, d $= 4.2$ m, $\sigma_D = 0.3\,\mu m$, $\sigma_F = 0.18\,\mu m$, $w_{oD} = 0.3\,m$, and $w_{oF} = 0.8\,m$. Since 18.32 was used to calculate the intensity envelops, all four enclose the same amount of light energy: unity. The logarithmic scale on the vertical axis captures the intensity variation over the range, $10^{-4}$–1. The central intensity attained by the 0.55-$\mu m$ visible image (lower left) is seen to be about 1250 times less than the central intensity at the optimum wavelength, 2.2 $\mu m$. Note that the latter central intensity has been (arbitrarily) normalized to unity. Also note that $\log_{10}(1) = 0$. The light energy fractions in the cores at wavelengths, 10 $\mu m$ and 30 $\mu m$ (upper left and upper right images) are 0.65 and 0.86, respectively. Cores comprising such large light fractions dominate star images

until after the various ELTs actually begin delivering star images. When that moment comes, just one single long-exposure, core-and-halo star image, obtained with image wander removed and analyzed in the manner described in this, and earlier, chapters, should instantly provide more meaningful values for the AO-improved imaging wave front parameters, $\sigma_D$, $\sigma_F$, $w_{oD}$, and $w_{oF}$.

The imaging wavelength, or wavelengths, chosen to produce a suitable core-and-halo star image for the above purpose will likely lie in the near-IR wavelength region.

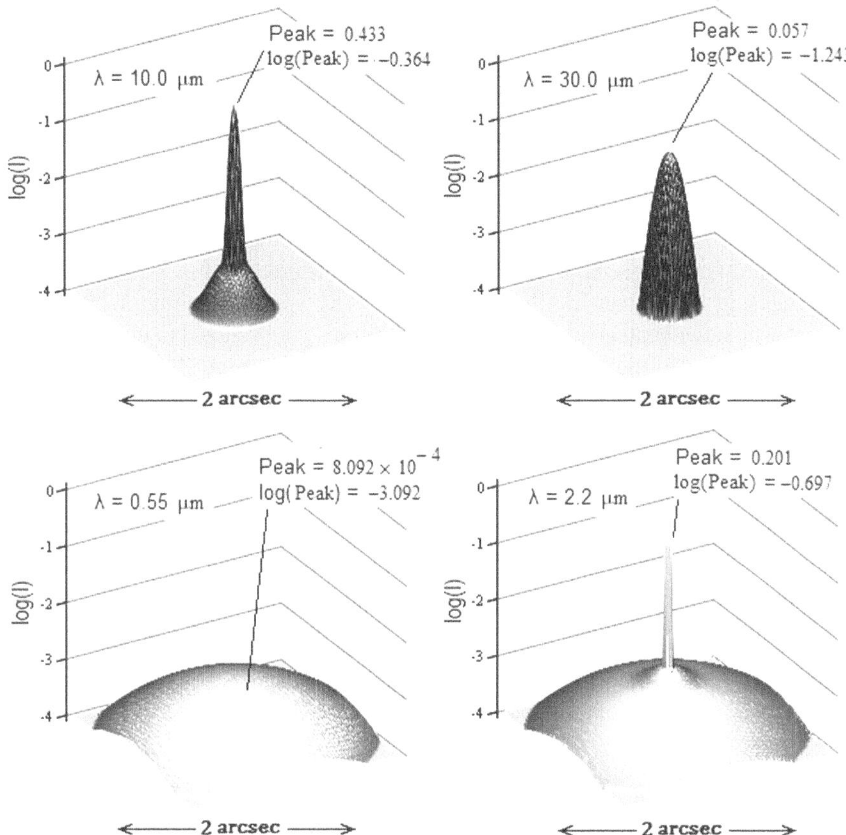


**Fig. 18.7** Star image intensity PSF envelopes plotted against a $\log_{10}(I)$ intensity scale for the 39.3 m E-ELT instrument. The envelops correspond to the same wavelength set used in Fig. 18.6, namely 0.55 μm, 2.2 μm, 10 μm, and 30 μm. With the exception of $\sigma_D$ and $\sigma_F$ whose values have now been doubled to $\sigma_D = 0.6\mu m$ and $\sigma_F = 0.36\mu m$, the other parameter values remain the same as those used in Fig. 18.6, namely $D = 39.3$ m, d = 4.2 m, $w_{oD} = 0.3\,m$, and $w_{oF} = 0.8\,m$. Since the same normalization scheme was used to produce the intensity envelop images in this figure as was used to produce the Fig. 18.6 images, all eight plots in the two figures enclose the same amount of light energy: unity. The relative central intensities and angular widths of all eight star images shown in Figs. 18.6 and 18.7 may therefore be gauged directly from the various images. Doubling the $\sigma_D$ and $\sigma_F$ values has the effect of doubling the FWHM angular width of the seeing disc image at 0.55 $\mu m$. It also doubles the ' optimum wavelength to 4.4 $\mu m$. To accommodate the increased angular widths of the images in Fig. 18.7, the field of view has been increased from 1-arcsec to 2-arcsec

As the various ELTs are brought into ever-improved states of adjustment, the $\sigma_D$ and $\sigma_F$ values should naturally shrink. The values taken by $w_{oD}$ and $w_{oF}$ are more difficult to predict. As discussed and illustrated in Chap. 13 (Sect. 13.2 and Fig. 13.12) their values sharply depend on the response characteristics of the individual AO systems.

To produce the Fig. 18.6 intensity envelops, Eqs. 18.17–18.21 were used to calculate values for $A_D$ ($\lambda$), $A_P(\lambda)$, $B_D(\lambda)$, $B_p(\lambda)$ and $B_H(\lambda)$, and 18.32 was used to calculate the final envelops. A consequence of using 18.32 is that all four images in Fig. 18.6 enclose the same amount of light energy: unity. The optimum wavelength for these particular parameter value choices is about 2.2 $\mu m$ (c.f., later Eq. 18.36). The intensity envelop at the optimum wavelength is shown at lower right in the figure. The intensity envelop at 0.55 $\mu m$ (lower left) is a halo-only seeing disc image. Equation 13.46 gives the FWHM angular width of this image as 0.7 arcsec.

The central intensity attained by the 0.55-$\mu m$ visible image in Fig. 18.6 (lower left) is a staggering 1250-times less than the central intensity at the optimum wavelength, 2.2 $\mu m$. The latter central intensity has been arbitrarily normalized to unity, with the central intensities of the other three images scaled in proportion. (Note that log(1.0) = 0.) The core light fractions at 10 $\mu m$ and 30 $\mu m$ in Fig. 18.6 (upper left and upper right) are 0.65 and 0.86, respectively. As can be seen in these two images, cores containing such large light energy fractions dominate the image centers.

To generate the intensity envelops in Fig. 18.7, the following, slightly different set of parameter values was used: $D = 39.3$ m, d = 4.2 m, $\sigma_D = 0.6\mu m$, $\sigma_F = 0.36\mu m$, $w_{oD} = 0.3$ m, and $w_{oF} = 0.8$ m. Note that $\sigma_D$ and $\sigma_F$ have now doubled in size while the other parameter values remain the same as in Fig. 18.6. Equations 18.17–18.21 were again used to calculate values for $A_D$ ($\lambda$), $A_P(\lambda)$, $B_D(\lambda)$, $B_p(\lambda)$ and $B_H(\lambda)$, and since 18.32 was again used to calculate the final intensity envelops, again the four intensity envelops in Fig. 18.7 each enclose the same amount of light energy: unity. The optimum wavelength for this new parameter set increases to about 4.4 $\mu m$ (c.f., later Eq. 18.36).

If the intensity envelop at the optimum wavelength, 4.4 $\mu m$, had been shown in Fig. 18.7, the peak intensity would have taken the value 0.25, with the logarithm of this intensity taking the value, -0.602. The 4-times intensity reduction factor reflects the $1/\lambda_{opt}^2$ dependence of star image central intensity on optimum wavelength (c.f., 18.10.17).

To obtain fully resolved star images in the field from the 39.3 m E-ELT instrument—images that accurately display the narrow spikey core central disc features seen in Fig. 18.6 (lower right) and Fig. 18.7 (lower right)—precise removal of image wander is absolutely essential. A suitable procedure for realizing the desired precision was described previously in Chap. 17, Sect. 17.4.

The FWHM angular width of the core central disc shown in Fig. 18.6 (lower right) is a mere 0.007 arcsec. The FWHM of the core central disc shown in Fig. 18.7 (lower right) is about 0.014 arcsec. To resolve these extremely narrow central disc features in actual star images, the detector pixel spacing would have to be no greater than the Nyquist sampling interval (c.f., Sect. 11.6.4, Footnote 4). In terms of the applicable angular spatial frequencies dealt with here, the Nyquist sampling interval is given by $\lambda/2D$. Inserting $\lambda = 2.2 \cdot \mu m$ and $D = 39.3$ m, gives the maximum allowed (angular) pixel spacing, 0.006 arcsec.

It is common practice to over-sample and thus obtain cleaner, more satisfying looking images (though it should be emphasized that this does not in fact provide additional image information). In this instance, star images clearly depicting the core central disc, and thus might be more pleasing to the eye, could be obtained using a 0.001 arcsec pixel spacing, spreading the central disc feature over a smooth ~ 12-pixel span. But is also be observed that such profligate over-sampling would greatly reduce the overall field of view captured by the imaging FPA. Some $10^3$ pixels arranged edge-to-edge would be needed to span a 1-arcsec field width, with $10^6$ pixels needed to cover a 1-arcsec × 1-arcse field of view.

### *18.8.1 Intensity Attained in the Center of the Average Short-Exposure PSF*

The intensity in the center of the average short-exposure intensity PSF at an arbitrarily chosen wavelength, λ, can be obtained from 18.32 by setting $\vartheta = 0$:

$$\langle I(0,\lambda)\rangle = \frac{A_D(\lambda) + A_P(\lambda) + 1}{\pi \cdot \left[A_D(\lambda)\cdot B_D(\lambda)^2 + A_P(\lambda)\cdot B_P(\lambda)^2 + B_H(\lambda)^2\right]}. \tag{18.33}$$

By using 18.17–18.21 (and considerable algebraic manipulation) equation 18.33 can be expressed in the more explicit form,

$$\langle I(0,\lambda)\rangle = \frac{w_{oF}{}^2 \cdot \left[\sigma_F{}^2 \cdot w_{oD}{}^2 + w_{oF}{}^2 \cdot \sigma_D{}^2 \cdot \exp\left(-\frac{4\cdot\pi^2\cdot\sigma_D{}^2}{\lambda^2}\right)\right]}{4\cdot\pi\cdot\sigma_F{}^2\cdot\left(w_{oD}{}^2\cdot\sigma_F{}^2 + w_{oF}{}^2\cdot\sigma_D{}^2\right)} + \frac{\left[\pi^2\cdot\left(D^2+d^2\right)\cdot\sigma_F{}^2 - \lambda^2\cdot w_{oF}{}^2\right]\cdot\exp\left[-\frac{4\cdot\pi^2\cdot\left(\sigma_D{}^2+\sigma_F{}^2\right)}{\lambda^2}\right]}{4\cdot\pi\cdot\lambda^2\cdot\sigma_F{}^2}. \tag{18.34}$$

Evaluations of 18.34 shown in Fig. 18.8 for the 39.3 m E-ELT instrument indicate enormous wavelength-dependent variations of the central intensities in star images formed by that instrument—the larger the instrument the larger the variation. The different plots correspond to the different rms residual OPD fluctuation values, $\sigma_{Tot}$, indicated in the figure. Wavelengths corresponding to the highest central intensities indicate intensely bright central cores. These wavelengths are therefore ideally-suited for distinguishing small faint objects, such as exoplanets. The platform intensity level seen at shorter wavelengths—which regrettably would likely include visible wavelengths—indicates halo-only seeing disc star images and low resolution. The ratio of the maximum central intensity to the platform level increases as the square of the telescope diameter. For the 39.3 m E-ELT, the ratio is about 16-times greater than for a 10 m instrument and about 100-times greater than for a 4 m instrument(c.f., Figs. 10.12, 10.13, and 10.14).

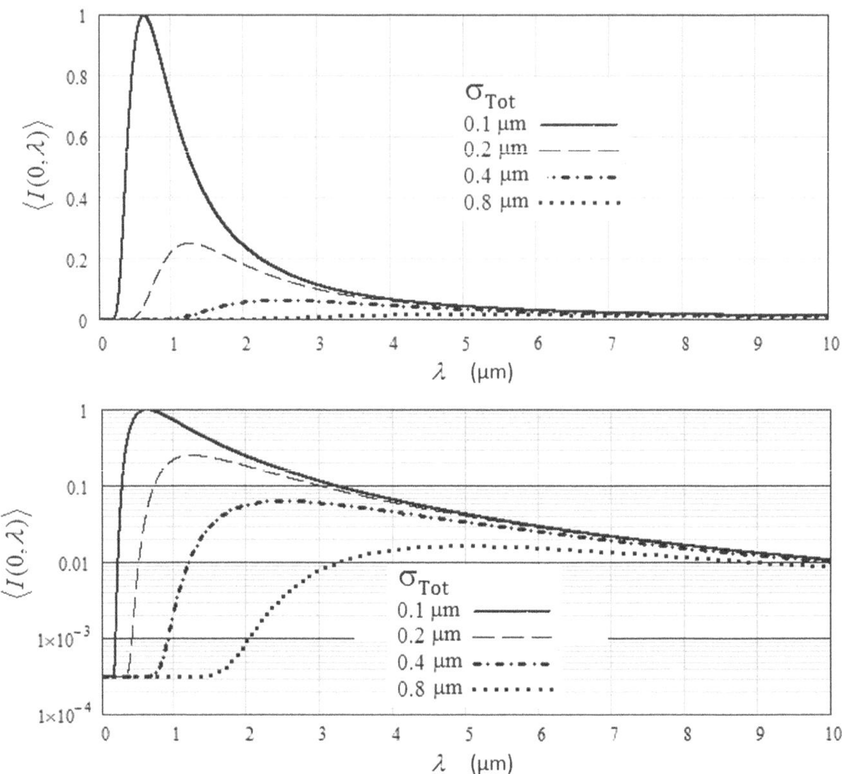


**Fig. 18.8** Intensity variation in the center of a star image, $I(0, \lambda)$, plotted against wavelength for the 39.3 m E-ELT instrument for the various $\sigma_{\mathrm{Tot}}$ (c.f., 18.35) values indicated. Top: Variation plotted against linear scale. Bottom: Variation plotted against logarithmic scale. At short wavelengths, all plots converge to the same level. This contrasts with plots shown previously in Fig. 14.14 because the $w_{oF}$ and $w_{\mathrm{oD}}$ values used here, instead of being held fixed, are scaled in proportion to $\sigma_{\mathrm{Tot}}$

## 18.8.2 Sweet-Spot Wavelength Regions for ELT Instruments

The optimum wavelengths in Fig. 18.8 correspond to those wavelengths where the central intensities take their maximum values (where it is assumed that the same amount of light energy is present at all wavelengths). Optimum wavelengths shift towards longer wavelengths in proportion to the total residual rms OPD fluctuation, $\sigma_{Tot}$, where,

$$\sigma_{Tot} = \sqrt{\sigma_D^2 + \sigma_F^2}. \tag{18.35}$$

The optimum wavelength, $\lambda_{opt}$, is given approximately by

$$\lambda_{opt} = 2 \cdot \pi \cdot \sqrt{\sigma_D^2 + \sigma_F^2}. \tag{18.36}$$

By differentiating 18.34 with respect to λ, we obtain

$$\frac{\partial}{\partial\lambda}\langle I(0,\lambda)\rangle = \frac{2\cdot\pi\cdot w_{oF}^4\cdot\sigma_D^4}{\lambda^3.\sigma_F^2\cdot\left(w_{oD}^2\cdot\sigma_F^2+w_{oF}^2\cdot\sigma_D^2\right)}\cdot\exp\left[-\frac{4\cdot\pi^2\cdot\sigma_D^2}{\lambda^2}\right]$$
$$+\left\{\frac{\pi\cdot\left[\sigma_F^2\cdot(D^2+d^2)\cdot(4\cdot\pi^2\cdot\sigma_D^2+4\cdot\pi^2\cdot\sigma_F^2-\lambda^2)-4\cdot\lambda^2\cdot w_{oF}^2\cdot(\sigma_D^2+\sigma_F^2)\right]}{2\cdot\lambda^5\cdot\sigma_F^2}\right\}$$
$$\times\exp\left[-\frac{4\cdot\pi^2\cdot(\sigma_D^2+\sigma_F^2)}{\lambda^2}\right]. \tag{18.37}$$

A more exact value of $\lambda_{opt}$ may be acquired by solving 18.37 for the case,

$$\frac{\partial}{\partial\lambda}\langle I(0,\lambda)\rangle = 0. \tag{18.38}$$

The Author was unable to find an analytic solution for 18.37 and 18.38. However, solutions can readily be obtained using numerical methods. Solutions can also be obtained by plotting 18.37 as in Fig. 18.9. The plots in the figure correspond to the 39.3 m E-ELT instrument for various $\sigma_{Tot}$ values. The optimum wavelengths, $\lambda_{opt}$, correspond to the wavelength locations where the plots cross the x-axis.

An approximate value for the intensity in the center of a star image at the optimum wavelength, $\langle I(0,\lambda)\rangle$, can be obtained by inserting the approximate expression for $\lambda_{opt}$ given by 18.36 into 18.34. The result simplifies to

$$\langle I(0,\lambda_{opt})\rangle = \frac{w_{oF}^2\cdot\left[w_{oD}^2\cdot\sigma_F^2+w_{oF}^2\cdot\sigma_D^2\cdot\exp\left(-\frac{\sigma_D^2}{\sigma_D^2+\sigma_F^2}\right)\right]}{4\cdot\pi\cdot\sigma_F^2\cdot\left(w_{oD}^2\cdot\sigma_F^2+w_{oF}^2\cdot\sigma_D^2\right)}$$
$$+\left[\frac{\left(D^2+d^2\right)\cdot\sigma_F^2-4\cdot w_{oF}^2\cdot(\sigma_D^2+\sigma_F^2)}{16\cdot\pi\cdot e\cdot\sigma_F^2\cdot\left(\sigma_D^2+\sigma_F^2\right)}\right] \tag{18.39}$$

## 18.9 Resolution Versus Wavelength for the E-ELT Instrument

The intensity PSF plots in Figs. 18.4, 18.5, 18.6 and 18.7 illustrate the huge variation in angular resolution anticipated for the 39.3 m AO-equipped E-ELT instrument. The variations are attributable to two main factors: the imaging wavelength and the level of phase correction achieved by the AO system. In this section, we examine in more detail the resolution from this, the largest of the ELTs, using the generalized form of the Rayleigh criterion (Sect. 14.1.1) as the resolution metric.

According to this criterion, a binary star consisting of equally bright components is considered just-resolved when the intensity midway between the twin image peaks is 26.5% less than in the peaks themselves. Star images obtained by the E-ELT at

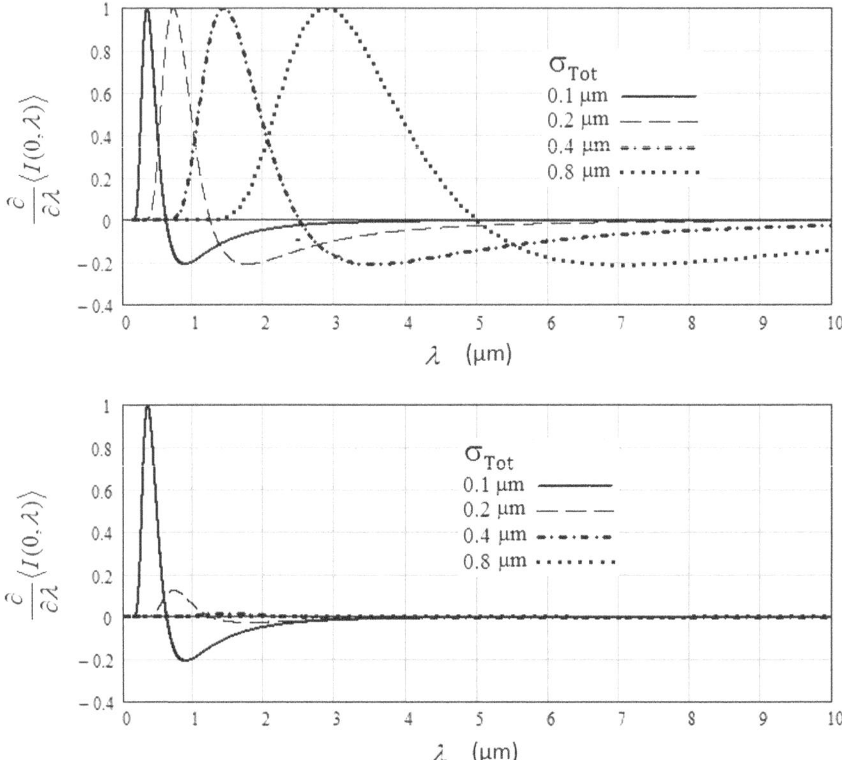


**Fig. 18.9** $\partial I(0, \lambda)/\partial\lambda$ versus wavelength for the 39.3 m E-ELT for various $\sigma_{Tot}$ values. (Bottom) Plots drawn to the same scale.) (Top) Each plot normalized to unity for easier identification of the optimum wavelengths (where the plots cross the x-axis)

long enough IR wavelengths will ultimately appear as near-perfect Airy patterns (as is true of any telescope with a circular aperture). Ignoring the small effect of the central obstruction, the required 26.5% dip occurs at angular separation $1.22 \cdot \lambda /D$.

At shorter wavelengths, core and halo star images begin to appear. When the residual dynamical rms OPD fluctuation, $\sigma_D$, and the imaging wavelength, λ, satisfy the condition, $\sigma_D \leq 0.25 \cdot \lambda$, cores can account for significant fractions of the light energy in star images, potentially offering extremely high resolution. Even if the above condition is not satisfied, resolution can still show enormous variations, depending on the residual fixed rms OPD fluctuation, $\sigma_F$.

For small $\sigma_F$ values, $\sigma_F \leq 0.13 \cdot \lambda$, the central disc contains more than 50% of the core's light energy so that this feature closely resembles (in terms of angular size and shape) the central disc of an Airy pattern formed by a diffraction-limited instrument of the same diameter. Because of the sheer size of the E-ELT instrument, the twin central discs in binary star images will tower above their respective core pedestals and halos. A 26.5% dip between the two central discs will again correspond to angular resolution $\sim 1.22 \cdot \lambda/D$. For larger $\sigma_F$ values, the central disc gradually disappears, with all of the core's light energy now transferring into the pedestal. The resolution provided by star images of this type is likely to be only marginally better than the coarse resolution provided by halo-only seeing disc star images.

To comfortably realize the required 26.5% intensity dip between the twin central discs in core and halo binary star images, the intensity level attained by these twin features must be at least five times greater than the intensity level attained by the twin overlapping halos. With reference to Fig. 18.3 and Eq. 18.12, to meet this 'five-times' requirement demands $A_D \geq 10$. Inspection of 18.20, indicates that there are several ways of adjusting $A_D$ to meet this condition:

(1) Minimize $\sigma_D$ and $\sigma_F$;
(2) Minimize $w_{oD}$ and $w_{oF}$;
(3) Use a suitably long imaging wavelength, $\lambda$;
(4) Use largest possible telescope diameter, D.

To achieve diffraction-limited resolution ( $1.22 \cdot \lambda/D$) in a binary star image at visible wavelengths ($\lambda \approx 0.55\ \mu m$) requires $\sigma_F \leq 0.07\ \mu m$ and $\sigma_D \leq 0.13\ \mu m$, which combine to give $\sigma_{Tot} \leq 0.15\ \mu m$. At the high end of this range, $0.15\ \mu m$, visible image cores contain a mere 0.05 fraction of the light energy in the star image. Yet, despite this modest allotment, the central disc features in such cores should appear as stunningly bright points of light compared to the background halos. To increase the core light energy fraction to 0.5 would require $\sigma_{Tot} \leq 0.075\ \mu m$. While such a level of AO correction is theoretically possible, it represents a level presently well beyond the reach of the 10 m Keck instruments, for which $\sigma_{Tot}$ is rarely less than $0.25\ \mu m$.

Achieving $\sigma_{Tot} \leq 0.25\ \mu m$ with any of the ELT instruments represents a formidable challenge. To go further and achieve $\sigma_{Tot} \leq 0.075\ \mu m$ would represent a staggering achievement, even given the most pristine observing conditions. Such a level of AO correction would enable the E ELT instrument to achieve diffraction- limited resolution, $1.22 \cdot \lambda/D$, throughout the entire visible to far-IR wavelength spectrum.

Achieving that objective could prove illusive for decades to come. It may ultimately require developments not yet on the horizon and the maturation of technologies, such as quantum computing, that are presently in their infancy. Figure 18.10 shows the Rayleigh resolution limits anticipated for the 39.3 m E-ELT for various $\sigma_{Tot}$ residuals, where it is assumed that the $\sigma_D$ residuals (which are more difficult to control) are about a factor of two larger than the $\sigma_F$ residuals. Thus, for the various $\sigma_{Tot}$ values indicated, $\sigma_F \approx \sigma_{Tot}/\sqrt{5}$ and $\sigma_D \approx 2 \cdot \sigma_{Tot}/\sqrt{5}$. Star images attain their maximum central intensity at the optimum wavelength, $\lambda_{opt} \approx 2 \cdot \pi \cdot \sigma_{Tot}$. For $\sigma_{Tot} = 0.1\ \mu m$, $\lambda_{opt} \approx 0.6\ \mu m$; for $\sigma_{Tot} = 3.2\ \mu m$, $\lambda_{opt} \approx 20\ \mu m$.

At optimum wavelengths, star image cores contain the fraction 1/e (~ 0.38) of the total light energy (Sect. 10.4). However, for a telescope as large as the E-ELT, even star image cores containing much smaller light energy fractions could still tower high above their respective halos. In such cases, the 26.5% central dip in a binary star image can be delivered at wavelengths significantly shorter than the optimum wavelength, as may be confirmed by examining the resolution plots in Fig. 18.10.

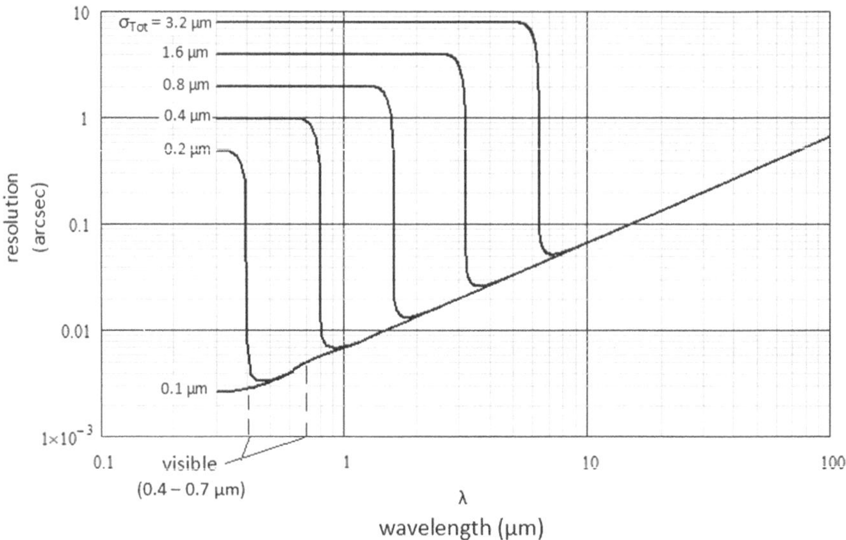


**Fig. 18.10** Resolution versus wavelength according to the Rayleigh criterion for the 39.3 m E-ELT. The level of AO-correction is indicated by the residual $\sigma_{Tot}$ values indicated. The straight line portion of the plots (upper left) corresponds to diffraction-limited resolution (1.22 $\lambda$/D) delivered by the central disc portions of the image cores. The abrupt departures from this line correspond to the absence of image cores when resolution abruptly deteriorates to the much coarser levels provided by seeing disc star images. To obtain substantially diffraction limited resolution over all visible and IR wavelengths requires $\sigma_{Tot} \leq 0.1$ µm

## 18.10 Detection of Faint Objects

The combined effect of the residual fixed and dynamical OPD fluctuations scatters light out of the core's central disc, decimating the central intensity in star images. The effect is most dramatic at (shorter) visible wavelengths. Consequently, the intensity in the center of star images formed by ELT instruments at visible wavelengths might often be less than that obtained in star images formed by smaller AO-equipped instruments capable of diffraction-limited performance. The plots in Fig. 18.11 show how the intensity level in the center of a star image formed by the 39.3 m diameter E-ELT at visible wavelengths could appear more than 10 magnitudes dimmer than a star image produced by an ideal 39.3 m telescope with diffraction-limited performance. To obtain substantially diffraction-limited imaging performance at visible wavelengths in the classical sense—Strehl intensity $\geq$0.8—requires $\sigma_{Tot} \leq 0.04$ µm.

### *18.10.1 Detection of Exoplanets*

Exoplanets are typically a billion times fainter (22.5 magnitudes) than their parent stars. To have any real chance of observing them directly, we would be obliged to

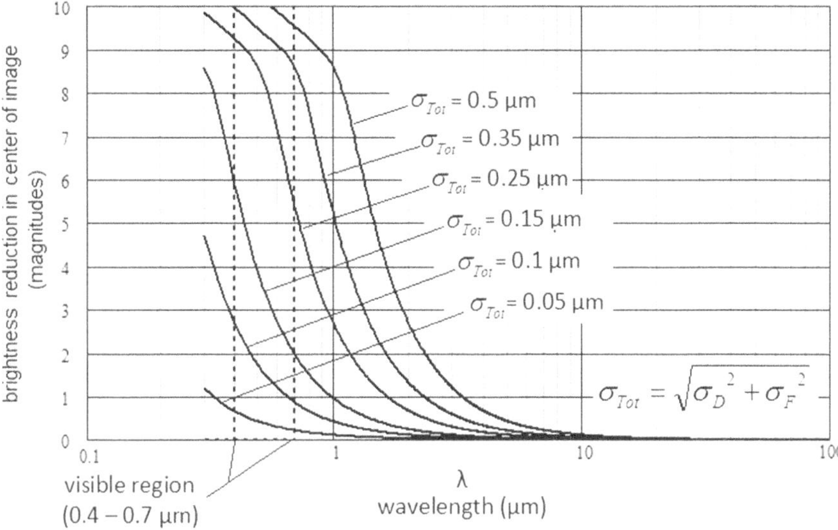


**Fig. 18.11** The intensity reduction in the center of star images formed by the 39.3 m E-ELT plotted against wavelength for various levels of rms uncorrected OPD fluctuation, $\sigma_{Tot}$. The logarithmic vertical axis scale directly expresses the reduction in magnitudes. At visible wavelengths, the intensity reduction could amount to 10 or more magnitudes

observe in the sweet-spot wavelength region. Fig. 18.12 shows computer-generated E-ELT images of a hypothetical exoplanet and parent star at three well-separated wavelengths, 0.7, 2.2, and 5 μm. Equation 18.34 was used to calculate the images.

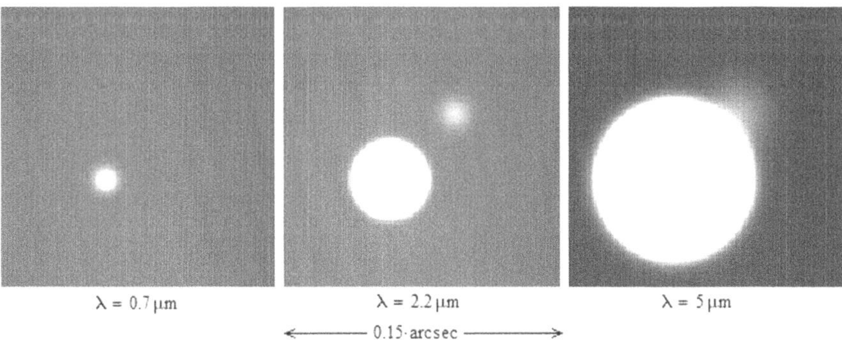


**Fig. 18.12** Images at three different wavelengths of a hypothetical parent star and exoplanet (0.05-arcsec separation) formed by the 39.3 m AO-equipped E-ELT. (Left) Image at wavelength 0.7 μm showing the over-exposed halo of the parent star, yet still no sign of the exoplanet. (Center) Image at the optimum wavelength, 2.2 μm. By over-exposing the parent star (at the cost of blooming), the core in the exoplanet image has become clearly visible. (Right) Image at wavelength 5 μm. The parent star has again been deliberately over-exposed, causing significant blooming, to reveal just a bare hint of the exoplanet at top right

With the angular separation between the parent star and the exoplanet set at 0.05-arcsec, the exoplanet lies well within the 0.4-arcsec (FWHM) halo of the parent star. While an admittedly unrealistically small, 8-magnitude brightness difference was used to generate the images, use of a more realistic (far greater) brightness difference risked exceeding the dynamic range of the book's image reproduction process, thereby undermining the entire exercise. For the level of AO- correction assumed here, $\sigma_D = 0.3$ μm, $\sigma_F = 0.15$ μm, and $\sigma_{Tot} = 0.335$ μm, star images achieve maximum central irradiance at wavelength 2.2 μm. The 2.2 μm image does indeed seem to show the exoplanet most clearly.

### 18.10.2 Star Cluster Images Formed by the 25 m GMT Instrument at Visible and IR Wavelengths

Figures 18.13 and 18.14 show computer-generated images of a star cluster formed by the 25 m Giant Magellan Telescope (GMT) for two levels of AO performance.

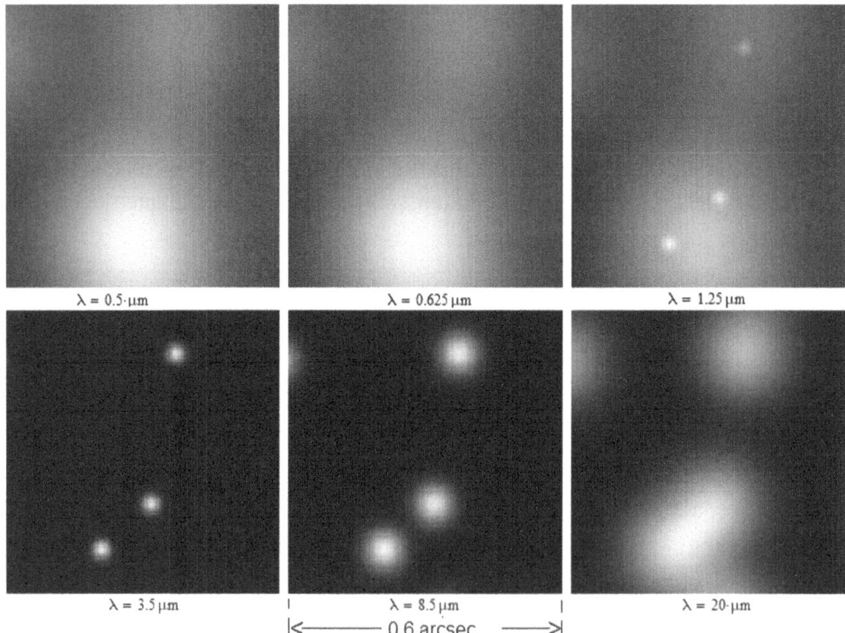


**Fig. 18.13** Computer-generated star cluster as might be observed by the 25 m GMT. The star images at wavelengths 0.5 μm and 0.625 μm are halo-only (seeing disc) images. At 1.25 μm they show typical core and halo structure. At 3.5 μm and 8.5 μm they again show core and halo structure, with substantial energy fractions invested in the halos even if the cores dominate the star image centers. At 20 μm the cores contain more than 97% of the total light energy. (OPD parameter values: $\sigma_F = 0.24$ μm, $\sigma_D = 0.48$ μm, $\sigma_{Tot} = 0.54$ μm and $w_{oD} = w_{oF} = 1.0$ m.)

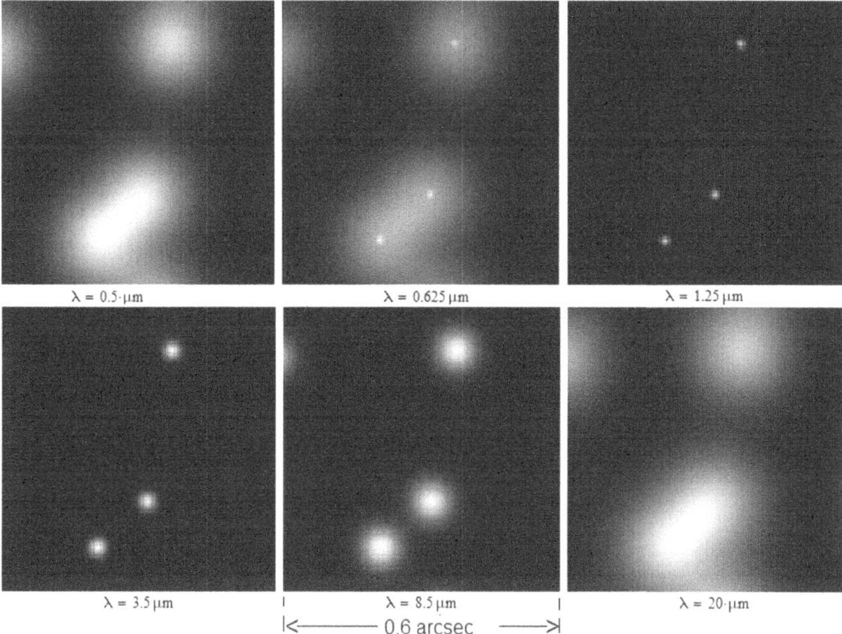


**Fig. 18.14** Computer-generated star cluster as observed by a 25 m GMT with improved AO correction. The star images at 0.5 μm are halo-only (seeing disc) images. The images at 0.625 μm, 1.25 μm, and 3.5 μm are core and halo images with substantial energy fractions invested in the halos, while cores dominate the star image centers. The images at 8.5 μm and 20 μm closely approximate core-only images. (OPD parameter values: $\sigma_F = 0.12$ μm, $\sigma_D = 0.24$ μm, $\sigma_{Tot} = 0.27$ μm and $w_{oD} = w_{oF} = 1.0$ m.)

Fig. 18.13 shows images corresponding to $\sigma_{Tot} = 0.54\ \mu m$, an AO performance level that could be realized early in the life of the instrument in quiescent atmospheric conditions. Fig. 18.14 shows the cluster imaged at the same wavelengths but where AO performance has improved by a factor of two ($\sigma_{Tot} = 0.27\ \mu m$). One could reasonably expect such imaging performance at some point in the GMT's lifetime in calm observing conditions. While the core central discs at the visible wavelength, 0.625 μm, contain no more than a 0.001 light energy fraction, they are nonetheless conspicuous enough to provide resolution of about 4 mas.

## 18.11 Mathematical Notation Used in This Chapter

The mathematical notation used in this chapter is indicated in Table 18.1.

**Table 18.1** Mathematical notation used in this chapter along with the SI dimensional units of the individual quantities

| Symbol | Quantity | Dimensions |
|---|---|---|
| $\lambda$ | Wavelength | $m$ |
| $\lambda_M$ | Star image measurement wavelength | $m$ |
| $\lambda_{opt}$ | Optimum wavelength | $m$ |
| $\lambda_{X0}$ | Cross-over wavelength | $m$ |
| $D$ | Telescope diameter | $m$ |
| $d$ | Telescope central obstruction | $m$ |
| $D_{EF}$ | Effective telescope diameter | $m$ |
| $SI(\lambda)$ | Telescope Strehl intensity at wavelength $k$ | "1" |
| $E_C$ | Light energy fraction in core | "1" |
| $E_{CD}$ | Light energy fraction in core central disc | "1" |
| $E_{CP}$ | Light energy fraction in core pedestal | "1" |
| $E_H$ | Light energy fraction in halo | "1" |
| $w_{0F}$ | 1/*e* half-width of Gaussian approximation to autocorrelation of residual dynamical OPD fluctuation | $m$ |
| $w_{0D}$ | 1/*e* half-width of Gaussian approximation to autocorrelation of residual fixed OPD fluctuation | $m$ |
| $\rho_F(\xi)$ | Gaussian approximation to autocorrelation of residual fixed OPD fluctuation | "1" |
| $\rho_D(\xi)$ | Gaussian approximation to autocorrelation of residual dynamical OPD fluctuation | "1" |
| $\sigma_F$ | rms variation of residual fixed OPD fluctuation | $m$ |
| $\sigma_D$ | rms variation of residual dynamical OPD fluctuation | $m$ |
| $\sigma_{Tot}$ | rms variation of total residual OPD fluctuation | $m$ |
| $A_{CD}$ | Central height of Gaussian core central disc | "1" |
| $A_{CP}$ | Central height of Gaussian core pedestal | "1" |
| $A_H$ | Central height of Gaussian halo | "1" |
| $A_D$ | Normalized height of core central disc | "1" |
| $A_P$ | Normalized height of core pedestal | "1" |
| $B_D$ | Angular 1/*e* half-width of Gaussian core central disc | "1" |
| $B_P$ | Angular 1/*e* half-width of Gaussian core pedestal | "1" |
| $B_H$ | Angular 1/*e* half-width of Gaussian halo | "1" |
| $B_C$ | Angular 1/*e* half-width of Gaussian core | "1" |
| $I(\vartheta)$ | Normalized star image intensity for radially symmetric images | "1" |
| $\vartheta$ | Angular coordinate in telescope image plane | "1" |

# References

Levenberg, K. (1944). A method for the solution of certain non-linear problems in least squares. *Quarterly of Applied Mathematics.*, *2*(2), 164–168.

Marquardt, D. W. (1963). An algorithm for least-squares estimation of nonlinear parameters. *J Soc for Industrial and Applied Mathematics, 11*(2), 431–441.

Wynne, C. G. (1959). Lens designing by electronic digital computer: I. *Proceedings of the Physical Society of London, 73*(5), 777–787.

McKechnie, T.S. (2016). Wavelength dependence of star images formed by large ground-based telescopes including ELTs. In Ground-based and Airborne Instrumentation for Astronomy VI, edited by Christopher J. Evans, Luc Simard, Hideki Takami, Proceedings of SPIE Vol. 9908 Article CID # 9908AC.

# Appendix A
# James Clerk Maxwell and the Electromagnetic Field Equations

**James Clerk Maxwell—His Life and Contributions to Science**

At the age of ten, Maxwell (Mahon 2004) enrolled at Edinburgh Academy[1] only a short distance from his India Street birthplace. He soon acquired the dubious nickname "Dafty," his peers unable to relate to the new boy's odd Galloway[2] accent and the curious shoes and clothes he wore. The name stuck, but Maxwell's genial manner, athleticism, and invigorating intellect soon made him one of the most popular students and no less popular in later life. Maxwell completed undergraduate studies in mathematics and physics at the University of Edinburgh from 1847 to 1850. The University was founded in 1583.

Maxwell's contributions to science are considered by many to be of the same magnitude as those of Newton and Einstein. His profound influence on the physical sciences is evident in every branch he touched (Campbell and Garnett 1882). His analysis in 1857 of the stability of Saturn's rings, a problem that had puzzled astronomers for over 200 years, showed that the rings must consist of separate bodies orbiting the primary. The stability of Saturn's rings was chosen in that year by St. John's College, Cambridge, as the topic of the prestigious Adam's prize.[3] Such was the difficulty of the task, Maxwell, with the only entry, was awarded the prize. Sir George Airy, the Astronomer Royal, declared James' submission "one of the most remarkable applications of mathematics to physics that I have ever seen." No one

---

[1] Edinburgh Academy is a prestigious private school, accepting students between the ages of 4 and 19. The author attended a rival Edinburgh school, George Watson's College. Differences between the two schools are settled in the annual cricket and rugby matches.

[2] Galloway is a rural county in the southwest of Scotland where Maxwell's father, John, had inherited a sizable estate.

[3] The Prize is named after the mathematician, John Couch Adams. It was endowed by members of St. John's College, Cambridge University, and approved by the senate of the university in 1848 to commemorate Adams' contribution to the discovery of the planet Neptune.





T. S. McKechnie, *General Theory of Light Propagation and Imaging Through the Atmosphere*, Progress in Optical Science and Photonics 20,
https://doi.org/10.1007/978-3-030-98828-9

since Maxwell has taken the understanding much further. Voyager I and II fly-bys in the 1980s showed the ring structure to be exactly as Maxwell had predicted.

His paper on gas theory in 1860 was profoundly important. At a time when most scientists favored Newton's conjecture that pressure was the result of static repulsion, Maxwell's intuition drew him to kinetic theory. Reasoning that it would be impossible to track the individual motions and collisions of every molecule, in a stroke of genius he attacked the problem statistically, developing a statistical model of the velocity distribution of gas molecules. The model said nothing about individual velocities but instead described the more meaningful probability density function of the velocities, known as Maxwell's distribution of molecular velocities.[4] This was the first statistical law of physics and precursor to many more applications of this powerful approach, including the application in this book to the problem of light scattering and imaging through the atmosphere.

Although Maxwell's kinetic theory of gases resolved most of the issues, his predicted value for the ratio of the specific heats of air at constant volume and constant pressure differed significantly from measured values. Finding himself no closer to resolving the problem after years of trying, he concluded that "something essential to the complete understanding of the physical theory of molecular encounters must have hitherto escaped us." The mystery was solved some 50 years later by Planck's quantum theory.

Maxwell took it upon himself to submit a paper (1863) to the British Association in which he recommended a complete system of units defining all physical quantities.[5] Going considerably beyond his original proposal, he developed a systematic way of defining all units in terms of mass, length, and time. Nowadays, under the name dimensional analysis, this field is now considered a natural part of the physical science and remains unaltered from the form originally presented by Maxwell.

In 1868, in a paper titled "On governors" Maxwell gave the first mathematical analysis of control systems. This paper is now considered the foundation of modern control theory. His work in this area attracted little attention until the 1940s when, with the urgency of the Second World War, its importance was finally recognized. For many years, the subject of color vision fascinated and intrigued Maxwell.

Building on an idea first suggested 50 years earlier by Thomas Young, Maxwell demonstrated the validity of the red, green, and blue color triangle and gave it proper mathematical expression. Maxwell reduced his color theory to practice in 1861 when he demonstrated the first color photograph in a lecture given to the Royal Institution.

Any one of the above contributions to science would have been enough to place Maxwell among the world's great scientists, but his electromagnetic equations placed him an altogether higher level—a level shared only by Newton and Einstein. An assessment of Maxwell's status in the grand scheme of things was given by an

[4] Maxwell's approach caught the attention of the Austrian physicist, Ludwig Boltzmann (1844–1906) who, along with Maxwell, further developed the theory of gases, their names forever linked by the so-called Maxwell–Boltzmann distribution.

[5] At that time, there was considerable confusion about units especially, but not just confined to, electrical and magnetic units, such as the units of the electric and magnetic fluxes.

acerbic individual whose criticisms could be withering but who had an entirely different regard for Maxwell:

> A part of us lives after us, diffused through all humanity - more or less - and through all nature. This is the immortality of the soul. There are large souls and small souls ... That of a Shakespeare or Newton is stupendously big. Such men live the best part of their lives after they are dead. Maxwell is one of these men. His soul will live and grow for long to come, and hundreds of years hence will shine as one of the bright stars of the past, whose light takes ages to reach us.
>
> Oliver Heaviside

Maxwell's Equations

Maxwell's equations in their original form consisted of eight equations. Maxwell was aware that their number could be reduced but that "to eliminate a quantity which expresses a useful idea would be a loss rather than a gain at this stage of our enquiry." The above-mentioned Englishman, Oliver Heaviside (1850–1925), later condensed the equations into a set with no redundant quantities, similar to those given in Sect. 3.9.1. In free space, in the absence of charges and currents where all quantities are expressed in Gaussian units [for which Maxwell rather than Gauss was primarily responsible (Mahon 2004)], we find that $D = E$ and $B = H$. Maxwell's equations then display a remarkable combination of elegance and simplicity:

$$\text{curl } E = -\frac{1}{c} \cdot \frac{\partial H}{\partial t}, \tag{A.1}$$

$$\text{curl } H = \frac{1}{c} \cdot \frac{\partial E}{\partial t}, \tag{A.2}$$

$$div \quad E - 0, \tag{A.3}$$

$$div \quad H = 0. \tag{A.4}$$

Even without mathematical training, one can sense the power and beauty of these equations. In time, the exploitation of electromagnetic waves would lead to wireless telegraphy, wireless transmission of radio and television, satellite communications, radar, X-rays, IR astronomy, radio astronomy, gamma-ray astronomy, microwave ovens, and more. All electromagnetic waves, including light waves, are bound by Maxwell's equations as they travel through space. Remarkably, quantum theory and relativity theory—two revolutionary theories that were to come some 50 years after Maxwell had developed the electromagnetic equations—did not affect the primacy of these equations.

# Appendix B
# Coherence Terminology

Throughout the book, various coherence functions are used, including several different complex coherence factors and two different two-point two-wavelength correlation functions. In this appendix, we list some of the more important coherence functions and show how they relate to one another. We also show how they are related to the complex amplitude and intensity.

In Chap. 3 (Sect. 3.8), we saw that for quasi-monochromatic light, the mutual coherence function, $\Gamma\left(u', v', u, v, \lambda, \tau\right)$,[6] is for all practical purposes identical to $\Gamma\left(u', v', u, v, \lambda, 0\right)$. We also saw that the latter function can be expressed in terms of the analytic signal (Sect. 3.7), $V(u, v, \lambda, t)$, by

$$\Gamma\left(u', v', u, v, \lambda, 0\right) = \left\langle V\left(u', v', \lambda, t\right) \cdot V^*(u, v, \lambda, t)\right\rangle. \tag{B.1}$$

For notational brevity, the function $\Gamma\left(u', v', u, v, \lambda, 0\right)$ is usually denoted by $J\left(u', v', u, v, \lambda\right)$. Thus,

$$J\left(u', v', u, v, \lambda\right) = \Gamma\left(u', v', u, v, \lambda, 0\right). \tag{B.2}$$

Function $J\left(u', v', u, v, \lambda\right)$ is referred to as the mutual intensity function at wavelength $\lambda$; it may be expressed in terms of the complex amplitude as follows:

$$J\left(u', v', u, v, \lambda\right) = \left\langle U\left(u', v', \lambda\right) \cdot U^*(u, v, \lambda)\right\rangle. \tag{B.3}$$

[6] $\Gamma\left(u', v', u, v, \lambda, \tau\right)$ Is essentially the same as the mutual coherence function which Born and Wolf denote by $\Gamma_{1,2}(\tau)$. These authors denote the mutual intensity function by $J_{1,2}$ and the complex degree of coherence by $\mu_{1,2}$



T. S. McKechnie, *General Theory of Light Propagation and Imaging Through the Atmosphere*, Progress in Optical Science and Photonics 20,
https://doi.org/10.1007/978-3-030-98828-9

In the case where the two points coalesce (i.e., $u' = u$ and $v' = v$) the mutual intensity function degenerates to simply become the average intensity,

$$J(u, v, u, v, \lambda) = \big\langle U(u, v, \lambda) \cdot U^*(u, v, \lambda)\big\rangle = \langle I(u, v, \lambda)\rangle. \tag{B.4}$$

Frequently, the unit-normalized form of the mutual intensity function arises. We denote this function by $\mu\big(u', v', u, v, \lambda\big)$; it may be defined in two equivalent ways:

$$\begin{aligned}\mu\big(u', v', u, v, \lambda\big) &= \frac{J\big(u', v', u, v, \lambda\big)}{[J(u', v', u', v', \lambda) \cdot J(u, v, u, v, \lambda)]^{\frac{1}{2}}} \\ &= \frac{\langle U\big(u', v', \lambda\big) \cdot U^*(u, v, \lambda)\rangle}{[\langle I(u', v', \lambda)\rangle \cdot \langle (u, v, \lambda)\rangle]^{\frac{1}{2}}}.\end{aligned} \tag{B.5}$$

We refer to function $\mu\big(u', v', u, v, \lambda\big)$ as the complex coherence factor; in some books, however, this function is referred to as the complex degree of coherence.

In the degenerate case, $u' = u$ and $v' = v$, the complex coherence function, $\mu(u, v, u, v, \lambda)$ takes its highest value, unity. Thus,

$$\mu(u, v, u, v, \lambda) = 1. \tag{B.6}$$

For waves at two different wavelengths, $\lambda_1$ and $\lambda_2$, we may write a more generalized form of the complex coherence factor. Denoting this function by $\mu\big(u', v', u, v, \lambda_1, \lambda_2\big)$, we refer to it in this book as the (unit-normalized) two-point two-wavelength correlation function of the complex amplitudes. This function is defined in terms of the complex amplitudes at the two wavelengths by

$$\mu\big(u', v', u, v, \lambda_1, \lambda_2\big) = \frac{\big\langle U\big(u', v', \lambda_1\big) \cdot U^*(u, v, \lambda_2)\big\rangle}{[\langle I(u', v', \lambda_1)\rangle \cdot \langle I(u, v, \lambda_2)\rangle]^{\frac{1}{2}}}. \tag{B.7}$$

A function of this type can be found in Chap. 6 for light waves arriving in the telescope pupil after having propagated over an extended atmospheric path. That particularly important function is denoted in its most general form by $S\big(x', y', x, y, \lambda_1, \lambda_2\big)$. However, because the statistical properties of atmospheric turbulence are assumed in that chapter (and elsewhere in the book) to be spatially stationary, we can simplify the spatial arguments as follows: $S(\xi, \eta, \lambda_1, \lambda_2)$, where $x' - x = \xi$ and $y' - y = \eta$.

A second function of the above type defined by (B.7) arises in Chap. 7 (Sect. 7.5.2) where it is used to describe the degree of correlation of the complex amplitudes at two different wavelengths and at two different locations in a telescope image. In the limiting case where the two points coalesce (i.e., $u' = u$ and $v' = v$), the function defined by (B.7) simplifies to the function $\mu(u, v, u, v, \lambda_1, \lambda_2)$ which we may then denote by the simpler from, $\mu(u, v, \lambda_1, \lambda_2)$. This function may be referred to as the (unit-normalized) spectral correlation function; it is defined in terms of the complex amplitudes by

$$\mu(u, v, \lambda_1, \lambda_2) = \frac{\langle U(u, v, \lambda_1) \cdot U^*(u, v, \lambda_2) \rangle}{[\langle I(u, v, \lambda_1) \rangle \cdot \langle I(u, v, \lambda_2) \rangle]^{\frac{1}{2}}}. \tag{B.8}$$

In the center of the image, i.e., where $u = 0$ and $v = 0$, the above form reduces to

$$\mu(0, 0, \lambda_1, \lambda_2) = \frac{\langle U(0, 0, \lambda_1) \cdot U^*(0, 0, \lambda_2) \rangle}{[\langle I(0, 0, \lambda_1) \rangle \cdot \langle I(0, 0, \lambda_2) \rangle]^{\frac{1}{2}}}. \tag{B.9}$$

Function $\mu(0, 0, \lambda_1, \lambda_2)$ arises in Chap. 7 (Sect. 7.5.4.1) where it describes the correlation coefficient for the complex amplitudes in the center of a star image at two arbitrary wavelengths, $\lambda_1$ and $\lambda_2$.

# Appendix C
# Turbulence Outer-Scale Limits Measured by Coulman et al.

Coulman et al. (1988) made extensive measurements of the atmospheric turbulence outer-scale limit, $L_0$, using balloon-borne temperature probes and scintillometry techniques at sites in three different countries—France, the United States, and Chile. Whereas these authors found some turbulence structure measuring 4 m across at high altitudes (8 km), at other altitudes the largest turbulence structures were generally found to be less than 1 m. When due account is taken of the high, $C_n^2$, weighting factors that apply at mid- and low-altitudes, their measurement data indicate an effective $L_0$ value (i.e., the average value over the entire atmospheric depth) of less than 1 m. Coulman et al. conclude that "a small $L_0$ value implies, in principle, a restriction of the range within which Kolmogorov theory of turbulence may be applied to estimate atmospheric seeing limitations to large-aperture telescopes."

In this appendix, we analyze the $L_0$ data gathered by Coulman et al. and deduce from it a possible form of the overall wavefront structure function averaged over the entire depth of the atmosphere. The rollover behavior of this function (i.e., when it departs from the 5/3-power law) establishes the effective outer-scale limit, $L_0$. The value calculated from the Coulman et al.'s data lies somewhere in the range, 30–40 cm; this range is consistent with the outer-scale limits deduced by the Author in 1975–1976 at the Royal Greenwich Observatory, Herstmonceux Castle (Sect. 8.3.).

Coulman et al. found that the $L_0$ variation with altitude can be modeled empirically by the relation,

$$L_0(z) = \frac{4}{1 + \left(\frac{z-8500}{2500}\right)^2}, \tag{C.1}$$

where both $L_0(z)$ and the altitude, $z$, are expressed in meters. Figure C.1 shows the plot of $L_0(z)$ against altitude, $z$, generated from this equation.



T. S. McKechnie, *General Theory of Light Propagation and Imaging Through the Atmosphere*, Progress in Optical Science and Photonics 20,
https://doi.org/10.1007/978-3-030-98828-9

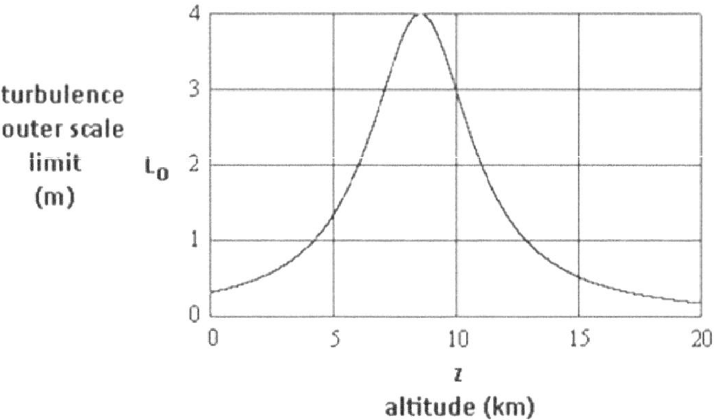


**Fig. C.1** The variation of the turbulence outer-scale limit, $L_0$, with altitude as measured by Coulman et al. Since most of the atmospheric turbulence strength lies in the first 4 or 5 km (see Fig. C.2), the average $L_0(z)$ value over the entire atmospheric path turns out to be smaller than 0.5 m

In turbulent weather conditions and/or when cloud formations are present, one might have expected to find turbulence structures significantly larger than the (largest) 4-m structures observed by Coulman et al. However, in calmer conditions that might be more suitable for astronomical viewing, the absence of large structure is not surprising; it is certainly consistent with the image cores discussed in Chap. 10. The comparative lack of large turbulence structure in the measurement data obtained by Coulman et al. and the consequences of this in regard to the atmospheric isoplanatic angle are discussed in Chap. 17, Sect. 17.2.3.3.

For simplicity, we choose a one-dimensional representation here. Otherwise, we use the same notation used in Chap. 6 (Sect. 6.2). The variance and autocorrelation of the optical path difference (OPD) fluctuation introduced by the individual random phase screens comprising a phase screen stack representing an atmospheric path, $\sigma_j^2$ and $\rho_j(\varepsilon)$, can be expressed by

$$\sigma_j^2 = \langle h_j(x)^2 \rangle, \tag{C.2}$$

and

$$\rho_j(\varepsilon) = \frac{\langle h_j(x+\varepsilon) \cdot h_j(x) \rangle}{\langle h_j(x)^2 \rangle}. \tag{C.3}$$

As indicated previously (Sect. 6.5.4), the integrated turbulence strength, $\sigma^2$, and the average autocorrelation function of the integrated OPD fluctuation over the entire atmospheric path, $\rho(\varepsilon)$, are given by

$$\sigma^2 = \sum_{j=1}^{n} \sigma_j^2, \tag{C.4}$$

$$\rho(\varepsilon) = \frac{1}{\sigma^2} \cdot \sum_{j=1}^{n} \sigma_j^2 \cdot \rho_j(\varepsilon). \tag{C.5}$$

We now assume that a Kolmogorov, 5/3-power law wavefront structure function is associated with each of the uncorrelated random phase screens in the path model.

We also assume that this functionality abruptly rolls-off for each of the phase screens at the outer-scale limits measured by Coulman et al., the limit set for each individual phase screen by the altitude of that phase screen (cf., C.1). With these assumptions, the autocorrelation function of the OPD fluctuation, $\rho_j(\varepsilon)$, introduced by the turbulence structure at the arbitrary altitude, $z_j$, may be written (cf., 12.2)

$$\begin{aligned} \rho_j(\varepsilon) &= 1 - \left(\frac{\varepsilon}{L_0(z_j)}\right)^{\frac{5}{3}} \quad \text{for} \quad \varepsilon \le L_0(z) \\ &= 0 \qquad\qquad\qquad \text{otherwise} \end{aligned} \tag{C.6}$$

where $z_j$ is the altitude of the $j$th layer.

Since the variance of the OPD fluctuation introduced by the individual random phase screens (each of which represents the turbulence contained in an associated atmospheric layer) is proportional to the $C_n{}^2$ value at the layer altitude, we may write

$$\sigma_j^2 \propto \int_{z_{j-1}}^{z_j} C_n{}^2(z) \cdot dz, \tag{C.7}$$

and

$$\sigma^2 \propto \int_{z_1}^{z_n} C_n{}^2(z) \cdot dz. \tag{C.8}$$

The average autocorrelation function of the integrated OPD fluctuation over the entire propagation path, $\rho(\varepsilon)$, which was given previously by (C.5) may then be written in the integral form,

$$\rho(\varepsilon) = \frac{\int_0^{z_0} C_n^2(z) \cdot \rho(\varepsilon, z) \cdot dz}{\int_0^{z_0} C_n^2(z) \cdot dz}, \tag{C.9}$$

where $z_0$ is the effective height of the atmosphere, which we consider here to be about 20,000 m.

To evaluate the autocorrelation function of the integrated OPD fluctuation from (C.9) we use (C.6) generate the $\rho_j(\varepsilon)$ corresponding to each individual layer and

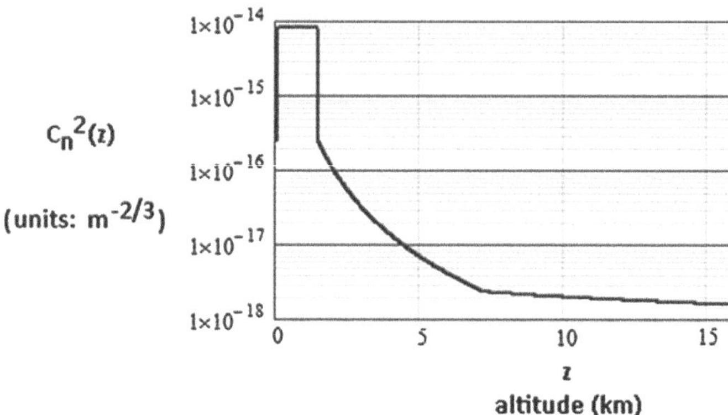


**Fig. C.2** Variation of the atmospheric refractive index structure constants, $C_n{}^2(z)$, with altitude according to the SLC night model. Evidently, most of the turbulence strength lies in the first 4 or 5 km

the submarine laser communication (SLC) night model to generate the $C_n^2(z)$ values. The SLC night model has been given by Good et al. (1981) in the form,

$$
\begin{aligned}
C_n^2(z) &= 8.4 \times 10^{-15} \cdot \mathrm{m}^{-\frac{2}{3}} \quad \text{for} \quad 0\,\mathrm{m} \le z \le 18.5\,\mathrm{m} \\
&= \left(2.87 \times 10^{-12}/z^2\right) \cdot \mathrm{m}^{-\frac{2}{3}} \quad \text{for} \quad 18.5\,\mathrm{m} \le z \le 110\,\mathrm{m}, \\
&= 8.4 \times 10^{-15}\,\mathrm{m}^{-\frac{2}{3}} \quad \text{for} \quad 110\,\mathrm{m} \le z \le 1500\,\mathrm{m} \\
&= \left(8.87 \times 10^{-7}/z^3\right) \cdot \mathrm{m}^{-\frac{2}{3}} \quad \text{for} \quad 1500\,\mathrm{m} \le z \le 7200\,\mathrm{m} \\
&= \left(2.0 \times 10^{-16}/\sqrt{z}\right) \cdot \mathrm{m}^{-\frac{2}{3}} \quad \text{for} \quad 7200\,\mathrm{m} \le z \le 20{,}000\,\mathrm{m}.
\end{aligned} \tag{C.10}
$$

A plot of the SLC night model is shown in Fig. C.2.

The autocorrelation function, $\rho(\varepsilon)$, of the integrated OPD fluctuation, calculated from the data of Coulman et al. using (C.5–C.10), is shown in Fig. C.3. For comparison purposes, the figure also shows two autocorrelation functions consistent with the 1975–1976 Royal Greenwich Observatory (RGO) measurement data described in Chap. 8. Because seeing was not specifically monitored during the RGO experiments, the two autocorrelation functions shown (which were calculated previously in Sect. 8.3) correspond to 0.75- and 1.5-arcsec full-width half-maximum (FWHM) visible seeing (cf., 13.43), these two seeing values being considered the likely best and worst seeing estimates at the time the RGO measurements were made. The two autocorrelation functions that arise from these seeing estimates can be seen as entirely consistent with the autocorrelation function that arises from Coulman et al.'s measured data. Also shown in Fig. C.3 is an autocorrelation for a much larger outer-scale limit, $L_0 = 5$ m. This function is plainly inconsistent with the autocorrelation functions arising from either Coulman et al.'s measurements or those calculated from the RGO data.

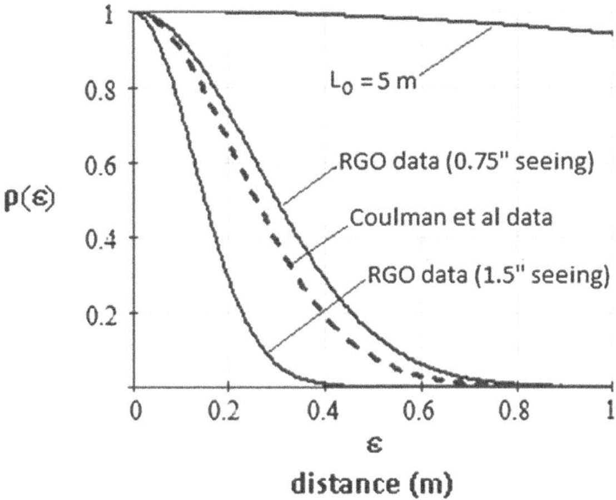


**Fig. C.3** Average turbulence structure size distribution in the atmosphere as described by the function $\rho(\varepsilon)$. The plots shown are for the data of Coulman et al. and the Author's RGO data and each of them seems to indicate relatively small outer-scale limits ($L_0 < 1$ m). All of these plots clearly contradict the Kolmogorov assumption that the outer-scale limit is larger than the largest telescope aperture

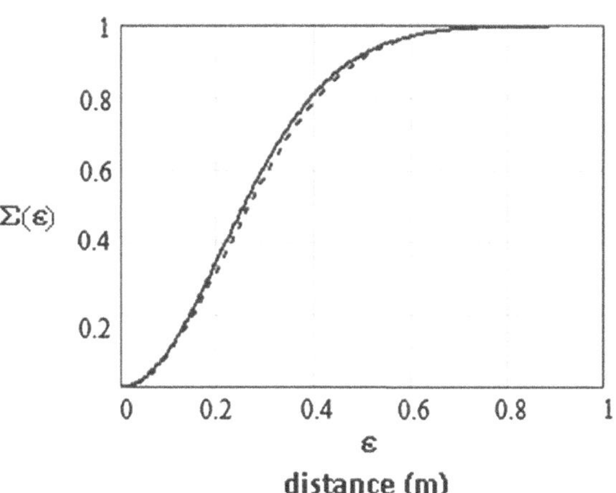


**Fig. C.4** Wavefront structure function (*solid line*) averaged over the entire atmospheric path based on the data of Coulman et al. The *dotted line* shows the best-fit Gaussian-based function given by (C.11) with $L_0 = 0.4$ m

The wavefront structure function, $\Sigma(\varepsilon)$, can be obtained directly from $\rho(\varepsilon)$ by using the relation previously given by (8.32), namely $\Sigma(\varepsilon) = 1 - \rho(\varepsilon)$. Function $\Sigma(\varepsilon)$ is plotted in Fig. C.4. For comparison, a best-fit Gaussian wavefront structure function is also shown in the figure (cf., 12.3) calculated from the equation

$$\Sigma(\varepsilon) = 1 - \exp\left[-\left(\frac{\varepsilon}{0.8 \cdot L_0}\right)^2\right], \tag{C.11}$$

where $L_0$ was set to the value 0.4 m to obtain the best-fit. This function can be seen as remarkably similar to the function derived from Coulman et al.'s measured data.

The roll-off in the structure function shown in Fig. C.4 occurs at about 0.4 m, which implies a much smaller effective outer-scale limit $L_0$ than assumed in Kolmogorov

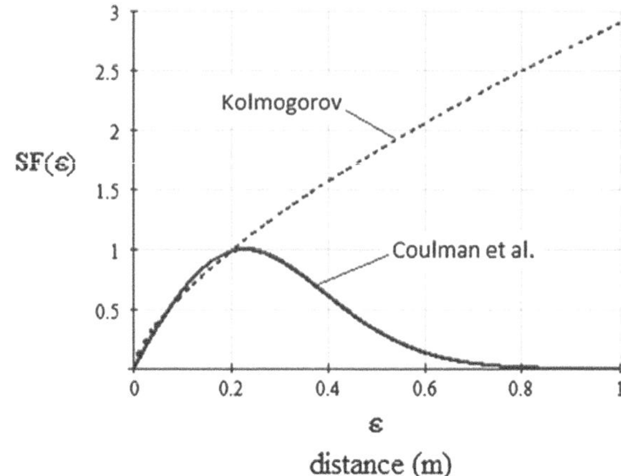


**Fig. C.5** The refractive index structure function for the turbulence measured by Coulman et al., calculated from (C.12). A best-fit 2/3-power law (as assumed in Kolmogorov formulations) is shown for comparison

theory. It is also evident that, although a 5/3-power wavefront structure function was assumed at each layer in the atmosphere (C.6), the large range of variation of the outer-scale limit with altitude causes the average wavefront structure function (averaged over the entire atmospheric depth) to depart significantly from the 5/3-power law. Unless we suppose that the outer-scale limit remains constant with altitude—a supposition that would be entirely at odds with the measurements of Coulman et al.—the average wavefront structure function is obliged to depart from the 5/3-power law. It also follows that the first derivative of the average wavefront structure function—which is simply the refractive index structure function—also departs significantly from the 2/3-power law.

By differentiating the approximate form of the wavefront structure function provided by (C.11), the refractive index structure function can be established in the form,

$$SF(\varepsilon) = \frac{2 \cdot \varepsilon}{(0.8 \cdot L_0)^2} \cdot \exp\left[ -\left( \frac{\varepsilon}{0.8 \cdot L_0} \right)^2 \right]. \tag{C.12}$$

The unit-normalized refractive index structure function calculated from the above equation with $L_0 = 0.4$ m is shown in Fig. C.5. The plot shown (solid line) roughly follows a 2/3-power law, but only for a limited distance. At about $\varepsilon = 0.2m$, the function rolls off and ultimately falls back to its initial value, zero.

# Appendix D
# Optical Path Characterization Using Scintillometry

Scintillometry is the name given to techniques for characterizing atmospheric paths by studying the intensity fluctuations in twinkling images, usually images of point objects. For astronomical paths, stars of course make near-ideal point-objects. For communication beam paths and high-energy laser (HEL) beam paths, artificial point sources can be used, located at either end of the beam path, as convenient.

Venet (1998) has given a useful and creditable account of the technique as it was used to characterize a 51.4-km atmospheric path between two neighboring mountain peaks in New Mexico, USA; a several-hundred-meter-deep valley separated the two peaks. To create a mathematical relationship between the intensity scintillation and the turbulence properties in the atmospheric path, Venet makes a number of assumptions:

(a) Kolmogorov turbulence is assumed (i.e., a 2/3-power law refractive index structure function);
(b) The outer-scale turbulence limit, $L_0$, is assumed larger than the telescope aperture;
(c) Uniform turbulence strength is assumed over the propagation path (i.e., the $C_n^2$ values are assumed constant over the path).

The fact that not all of these assumptions are entirely realistic draws attention to certain deficiencies in this technique. We have already seen evidence that assumptions (a) and (b) may not be valid. But even if they are valid in this particular instance, assumption (c) is unlikely to be valid due to the height of the propagating beam relative to the valley floor varying significantly over the path.

No less problematic are the following known shortcomings of scintillometry:

(a) For the convergent beam path considered by Venet (cf., Chap. 16), the technique has no sensitivity whatsoever to turbulence structure at either end of the path. (For astronomical imaging paths, the technique has zero sensitivity to turbulence near the ground, which paradoxically is where most of the turbulence strength actually resides);



T. S. McKechnie, *General Theory of Light Propagation and Imaging Through the Atmosphere*, Progress in Optical Science and Photonics 20,
https://doi.org/10.1007/978-3-030-98828-9

(b) Over very long paths, scintillation ultimately saturates into Gaussian speckle, at which point the technique loses any ability to discriminate turbulence structure characteristics;
(c) The formulation used by Venet makes use of a "semi-empirical result."

In regard to the last mentioned item, an empirically based result that applies to one specific atmospheric path does not necessarily reliably extend to other atmospheric paths. The three shortcomings just listed, together with the three uncertain assumptions, make it unlikely that definitive path characterizations can be obtained using scintillometry techniques. As a general rule, it is best to avoid using scintillation techniques for characterizing atmospheric paths.

# Appendix E
# Radiometry of the Sun and Stars

The spectral irradiance, $E(\lambda)$, of the surface of a blackbody emitter (i.e., the irradiance per unit area of the blackbody per unit wavelength interval) is given by (Electro-Optics Handbook 1974)

$$E(\lambda) = \frac{2 \cdot \pi \cdot h \cdot c^2}{\lambda^5 \cdot \left(\exp\left(\frac{h \cdot c}{\lambda \cdot k \cdot T}\right) - 1\right)} \quad \left(\mathrm{W} \cdot \mathrm{m}^{-2} \cdot \mathrm{m}^{-1}\right), \tag{E.1}$$

where $T$ is the black body temperature expressed in Kelvin, and $h$ is Planck's constant ($6.6260755 \times 10^{-34}$ J s) $c$ is the speed of light in vacuum ($2.99792458 \times 10^8$ m s$^{-1}$) and $k$ is Boltzmann's constant ($1.3806505 \times 10^{-23}$ J K$^{-1}$). Ramsey (1962) has given spectral irradiance data for stars and planets above the atmosphere from 0.1 to 100.0 μm.

The Sun's mean distance from Earth is about $1.496 \times 10^{11}$ m and its average total radiated emission is about $3.83 \times 10^{26}$ W (Ridpath 1989). These two numbers combine to give the solar constant (the average irradiance of solar radiation at the top of Earth's atmosphere) 1362 W/m$^2$ (Kopp and Lean 2011). To a reasonable approximation, the Sun radiates as a perfect blackbody; the effective temperature over its radiative surface, the photosphere, is about 5780 K [Handbook of Chemistry and Physics, 88th edition (Lide 2007–2008)].

The apparent visual magnitude of a celestial object, $m_v$, is a measure of the object's brightness as perceived by the human eye (i.e., the object's apparent brightness at wavelength 0.55 μm). The visual magnitude of the Sun is about −26.72 (Ridpath 1989). The visual magnitude of the brightest star, Sirius, is about −1.47. With only a few exceptions, stars have positive magnitudes, the larger the magnitude the fainter



T. S. McKechnie, *General Theory of Light Propagation and Imaging Through the Atmosphere*, Progress in Optical Science and Photonics 20,
https://doi.org/10.1007/978-3-030-98828-9

the star appears to the eye. The absolute magnitude of a celestial object, $M_v$, gives the object's brightness when observed at a standard distance of 10 parsecs (about 33 light years).[7] Absolute and apparent magnitudes are related by the formula (Ridpath 1989),

$$M_v = m_v + 5 + 5 \cdot \log(\pi), \tag{E.2}$$

where $\pi$ is used here (as is customary) to denote the object's (trigonometric) parallax expressed in arcsec. The parallax of an object is defined as the angle subtended by one astronomical unit located at the same distance as the object. In the case of the Sun, the parallax angle is naturally one radian, or 206,265 arcsec. Inserting this value for the Sun's parallax into the above formula gives the absolute magnitude of the Sun as 4.85.

The spectral irradiance of solar emissions at the top of the atmosphere, $E_{Sun}(\lambda)$, may be obtained from (E.1) by creating a version appropriate to the Sun that is consistent with both the effective blackbody surface temperature of the Sun, 5780 K, and the solar constant, 1362 W/m$^2$. Thus, we can write

$$E_{\mathrm{Sun}}(\lambda) = 2.115 \times 10^{-5} \cdot \frac{2 \cdot \pi \cdot h \cdot c^2}{\lambda^5 \cdot \left(\exp\left(\frac{h \cdot c}{\lambda \cdot k \cdot 5780}\right) - 1\right)} \quad \left(\mathrm{W} \cdot \mathrm{m}^{-3}\right). \tag{E.3}$$

It might be noted that, by obtaining the integral of $E_{Sun}(\lambda)$ over the entire wavelength range, we do indeed obtain the anticipated result, 1362 W/m$^2$.

**Solar Radiant Flux Collected By a Telescope**

The radiant flux, $P_{\mathrm{Sun}}(\lambda,\Delta\lambda)$ (expressed in Watts) in an image of the Sun formed by a telescope with a circular aperture of diameter, $D$, and central obstruction of diameter, $d$, over a small wavelength band, $\Delta\lambda$, centered at wavelength, $\lambda$, may readily be deduced from the above expression for $E_{\mathrm{Sun}}(\lambda)$ in the form,

$$\begin{aligned} P_{Sun}(\lambda, \Delta\lambda) \\ &= \frac{\left(6.3243 \times 10^{-21} \cdot \mathrm{W} \cdot \mathrm{m}^2\right)}{\lambda^5 \cdot \left(\exp\left(\frac{h \cdot c}{\lambda \cdot k \cdot 5780}\right) - 1\right)} \\ &\quad \cdot \Delta\lambda \cdot \left(D^2 - d^2\right) \cdot T_o(\lambda) \cdot T_A(\lambda)^{\left(\frac{1}{\cos(ZA)}\right)} \ (\mathrm{W}). \end{aligned} \tag{E.4}$$

where $T_o(\lambda)$ is the intensity transmission factor of the telescope ($0 \leq T_o(\lambda) \leq 1$), $T_A(\lambda)$ is the intensity transmission factor of the atmosphere for a vertical path ($0 \leq T_A(\lambda) \leq 1$), ZA is the Sun's zenith angle, and all multiplier constants have been consolidated into the single constant, $6.3243 \times 10^{-21}$ W m$^2$ (including the

[7] One parsec is the distance at which one astronomical unit subtends an angle of one arc second. One parsec is about 3.26 light years or 31 trillion ($3.1 \times 10^{13}$) km. One light year is about 9.5 trillion ($9.5 \times 10^{12}$) km. The distance to the nearest star, Proxima Centauri, is about 1.3 parsecs, or 4.2 light years.

term, $\pi/4$, that converts the quantity, $\left(D^2 - d^2\right)$, to the light collection area of the telescope).

**Stellar Radiant Flux Collected By a Telescope**

The radiant flux (in Watts) in the image of a star of visual magnitude, $m_V$, and color temperature, $T$, formed by the above telescope in the same small wavelength band, $\Delta\lambda$, centered at an arbitrary wavelength, $\lambda$, may be obtained by scaling the quantity $P_{Sun}(0.55\ \mu m, \Delta\lambda)$ (which may be obtained from (E.4) above by inserting $\lambda = 0.55\ \mu m$) in proportion to the ratio of the apparent visual brightness of the Sun and the star, $2.51189^{(-26.72-m_v)}$, and by using a correction factor to account for the different relative outputs of the Sun (color temperature, 5780 K) and the star (color temperature, $T$) at the two wavelengths, 0.55 μm and $\lambda$. In this way, we obtain

$$P_{star}(\lambda, \Delta\lambda) = P_{Sun}(0.55 \cdot \mu m, \Delta\lambda) \cdot 2.51189^{(-26.72-m_v)} \times \left(\frac{0.55 \cdot \mu m}{\lambda}\right)^5 \cdot \frac{\exp\left(\frac{h \cdot c}{0.55 \cdot \mu m \cdot k \cdot T}\right) - 1}{\exp\left(\frac{h \cdot c}{\lambda \cdot k \cdot T}\right) - 1} W. \tag{E.5}$$

Using (E.4) to substitute for $P_{Sun}(0.55 \cdot \mu m, \Delta\lambda)$ allows us to write the above expression in the form,

$$P_{Star}(\lambda, \Delta\lambda) = 2.51189^{-m_v} \cdot \left(\frac{1}{\lambda}\right)^5 \cdot \frac{\exp\left(\frac{h \cdot c}{0.55 \cdot \mu m \cdot k \cdot T}\right) - 1}{\exp\left(\frac{h \cdot c}{\lambda \cdot k \cdot T}\right) - 1} \times \Delta\lambda \cdot \left(D^2 - d^2\right) \cdot T_o(\lambda) \cdot T_A(\lambda)^{\left(\frac{1}{\cos(ZA)}\right)} \cdot \left(1.41958 \cdot 10^{-33} \cdot W \cdot m^2\right) W. \tag{E.6}$$

where the various multiplier constants have been consolidated into the single constant, $(1.41958 \times 10^{-33}\ W\ m^2)$.

**Photoelectrons Liberated By a Star Image Falling on a Detector**

In the arbitrary integration period, $\Delta t$, the number of photons corresponding to the radiant flux given above by (E.6) is given by Electro-Optics Handbook (1974)

$$photons = \frac{P_{star}(\lambda, \Delta\lambda) \cdot \Delta t \cdot \lambda}{h \cdot c}, \tag{7}$$

where the energy associated with each photon is given by $h \cdot c/\lambda$.

By using (E.6) to substitute for $P_{star}(\lambda, \Delta\lambda)$ in (E.7), and where the quantum efficiency of the detector is denoted QE($\lambda$), the number of photoelectrons liberated by the detector, PE, is given by

$$PE = 2.51189^{-m_v} \cdot \left(\frac{1}{\lambda}\right)^4 \cdot \frac{\exp\left(\frac{h \cdot c}{0.55 \cdot \mu m \cdot k \cdot T}\right) - 1}{\exp\left(\frac{h \cdot c}{\lambda \cdot k \cdot T}\right) - 1} \times QE(\lambda) \cdot \Delta t \cdot \Delta\lambda \cdot \left(D^2 - d^2\right) \cdot T_o(\lambda)$$

$$\times\ T_A(\lambda)^{\left(\frac{1}{\cos(ZA)}\right)} \cdot \left(7.4675 \times 10^{-9} \cdot \frac{m}{s}\right), \tag{E.8}$$

where the various multiplier constants have again been consolidated into the single constant, $7.4675 \times 10^{-9}$ m/s.

**Example Calculation of Photoelectrons Liberated by a Star Image Falling on a Detector**

The following parameter values are used in this example calculation:

| | |
|---|---|
| Telescope diameter | $D = 1$ m, |
| Central obstruction diameter | $d = 0$ m |
| Star visual magnitude | $m_V = 5$, |
| Star surface temperature | $T = 4700$ K, |
| Imaging wavelength | $\lambda = 2.2$ μm, |
| Bandwidth | $\Delta\lambda = 0.2$ μm |
| Detector integration time | $\Delta t = 0.01$ s, |
| Transmission through telescope | $T_o(2.2\ \mu m) = 0.5$, |
| Atmospheric transmission for vertical path | $T_A(2.2\ \mu m) = 0.78$ (see Fig. 3.9), |
| Star zenith angle | $ZA = 38.51^0$ |
| Detector quantum efficiency | $QE(2.2\ \mu m) = 0.5$, |

Inserting these values into (E.8) indicates that $10^5$ photoelectrons are liberated during the $\Delta t = 0.01$ s integration period.

# Appendix F
# Intensity Correlation Coefficient Estimates and Photon Noise Compensation

A measurement method was described in Chap. 8 (Sect. 8.2) for obtaining an estimate of the rms optical path difference (OPD) fluctuation, $\sigma$, associated with the image-forming light waves in the pupil of a telescope. The method requires making a large number of simultaneous intensity measurements in the center of a star image at two different wavelengths, $\lambda_1$ and $\lambda_2$. To obtain adequate statistical accuracy, each of the measured intensity data sequences described in Chap. 8 comprised 60,000 intensity measurements, obtained at a 1-kHz collection rate over a one-minute measurement period.

Figure F.1 shows the typical appearances of the unit-normalized temporal autocorrelation functions of the measured intensity sequences at each of the two wavelengths. These autocorrelation functions, which we denote here by $A_1$(Lag) and $A_2$(Lag), can be calculated as follows:

$$A_1(Lag) = \frac{\sum_{i+1}^{N-Lag} I_{i+Lag}(0, 0, \lambda_1) \cdot I_i(0, 0, \lambda_1)}{\sum_{i+1}^{N-Lag} [I_i(0, 0, \lambda_1)]^2} \tag{F.1}$$

and

$$A_2(Lag) = \frac{\sum_{i+1}^{N-Lag} I_{i+Lag}(0, 0, \lambda_2) \cdot I_i(0, 0, \lambda_2)}{\sum_{i+1}^{N-Lag} [I_i(0, 0, \lambda_2)]^2}, \tag{F.2}$$

where the Lag number takes the integer values 0, 1, 2, 3, .... Both autocorrelation functions are normalized to unity at zero Lag. Because the sampling interval used to obtain the intensity data was exactly 1 ms, the Lag interval indicated in Fig. F.1 is simply the Lag number expressed in milliseconds. Both autocorrelations were observed to fall to zero within about 100 Lag intervals, indicating that the measured intensities in each of the two data sequences largely decorrelate in a time interval of approximately 0.1 s.



T. S. McKechnie, *General Theory of Light Propagation and Imaging Through the Atmosphere*, Progress in Optical Science and Photonics 20,
https://doi.org/10.1007/978-3-030-98828-9

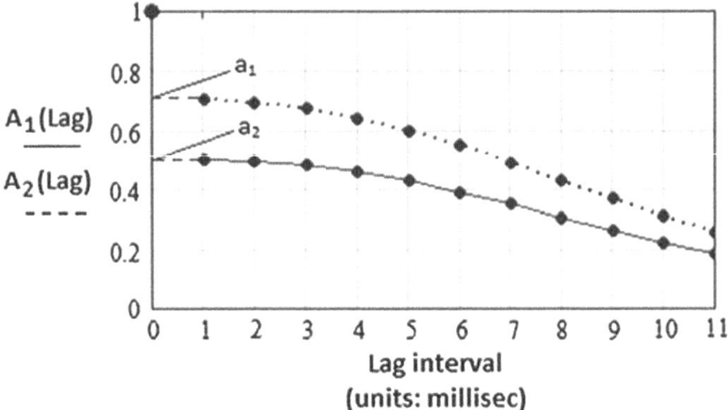


**Fig. F.1** Temporal autocorrelation functions for the intensity data obtained in the two detection channels when photon noise is present. Photon noise causes both functions to appear to intercept the vertical axis at values (indicated by $a_1$ and $a_2$) of less than unity. In the absence of photon noise, both functions would have converged to the point (0,1)

In the absence of photon noise, the two plots shown would smoothly approach the ordinate value, unity, exactly attaining that value at zero Lag interval. However, because there was in fact significant photon noise, as can be seen in the figure the two plots appear to converge to points on the vertical axis, indicated by $a_1$ and $a_2$, that are significantly less than unity. This behavior, if not compensated, would cause significant bias in the σ estimates.

Bias-corrected estimates can be obtained once accurate values have been established for both $a_1$ and $a_2$. Such values can be obtained by curve-fitting the discrete points on each plotted function for the first several Lag intervals and then using extrapolation to find the $a_1$ and $a_2$ intercepts on the vertical axis.

Once $a_1$ and $a_2$ have been established in this way, a corrected version of the intensity correlation coefficient—one free of the biasing effects caused by photon noise—may be calculated by scaling the (biased) intensity correlation coefficient arising directly from the measured data, $\left[\frac{I(0,0,\lambda_1)\cdot I(0,0,\lambda_2)}{I(0,0,\lambda_1)\cdot I(0,0,\lambda_2)} - 1\right]$, by the correction factor, $1/\sqrt{a_1 \cdot a_2}$. Thus, the photon noise bias-corrected estimate of the intensity correlation coefficient, $|\mu(0, 0, 0, 0, \lambda_1, \lambda_2)|^2$, may be obtained as follows:

$$|\mu(0, 0, 0, 0, \lambda_1, \lambda_2)|^2 = \frac{1}{\sqrt{a_1 \cdot a_2}} \cdot \left[\frac{\langle I(0, 0, \lambda_1) \cdot I(0, 0, \lambda_2)\rangle}{\langle I(0, 0, \lambda_1)\rangle \cdot \langle I(0, 0, \lambda_2)\rangle} - 1\right]. \quad \text{(F.3)}$$

Use of this unbiased form of $|\mu(0, 0, 0, 0, \lambda_1, \lambda_2)|^2$ yields the unbiased σ estimate given in Chap. 8 by (8.9).

# Appendix G
# Image Core Correspondence from Roger F. Griffin

Letter received. November, 1989:
University of Cambridge Institute of Astronomy The Observatories Madingley Road Cambridge, CB3 0HA
England November 27, 1989
Dear Dr. McKechnie
Thank you very much for your letter of October 20; I have since obtained your paper on loan from your mother, copied it and read it.

Naturally I am pleased to find that an in-depth theoretical treatment reproduces so exactly my empirical observations, which were laughed out of court when I set them out in 'Observatory' seventeen years ago. (See also Observatory 93, 138, 1973.) It was rather curious that when I received your letter I had only just revived my assertions of 1973 by writing a letter to Sky and Telescope on the subject—it has been accepted for publication, probably in February—having been prompted by my experience of the NTT which, being accurately figured, gave very good images straight away without having to wait for specially good seeing, just as I would expect.

The cores shown in your graphs are smaller than I ever saw at the 100 or 200-in., but that is obviously because they were calculated for perfect telescopes, whereas the core sizes I actually observed would have been limited by the optical aberrations. I am interested to see the wavelength dependence of the images in your models; it shows me why the most amazing images I ever saw at the 100-in. (whose optics are better than those of the 200-in.) were when I undertook with my wife a limited programme of infrared photographic spectroscopy and we were guiding through a filter that only transmitted the far-red and infrared. The image of Aldebaran (a very red first-magnitude star), in particular, seemed to be largely concentrated into a searing point of light, just as if one were looking straight at a laser.

You can see some core and halo structure in the image very frequently—of the order 25–50% of the time—on the two large telescopes with which I was familiar.


T. S. McKechnie, *General Theory of Light Propagation and Imaging Through the Atmosphere*, Progress in Optical Science and Photonics 20,
https://doi.org/10.1007/978-3-030-98828-9

(I can't use them any more: the 100-in. has been closed since 1985, and the Palomar Observatory won't give me observing time nowadays.) There are few photographic data on images, but nowadays there are groups who do 'speckle interferometry,' making runs of exposures of a few milliseconds each with CCDs, and I would think that they would be of interest for your purposes. In my experience, angular excursions of the image due to quiver were virtually nil: the better the image the more smoothly did it glide along the slit. (My purpose in looking at images, all night, night after night, was simply to trail them along the spectrograph slit, which served as a fixed reference line against which to view them.)

With good wishes,

Yours sincerely, Roger Griffin.

Email received, June 2, 2009:

Dear Stewart,

Thank you for your enquiry, and sorry not to have responded more promptly. Yes, I am still busy whenever the sky is clear. The exceptionally good images that I saw at the 100-in. were when I was taking spectra in the red part of the spectrum of certain first-magnitude stars—I particularly remember an observation of Aldebaran. To cut out the overlapping higher grating orders it was necessary to use a red filter, and it was convenient and appropriate to put the filter ahead of the guiding eyepiece too, so that one was guiding on the same part of the spectrum as one was trying to observe, otherwise there would be a chromatic offset of the wanted image relative to the visually observed one owing to atmospheric dispersion. (That was especially the case at the 100-in., where the slit was projected onto the sky as a 'horizontal' line when the object was on the meridian.) The filter used was Wratten 29, which is opaque to wavelengths shorter than 6300 Angstrom but rapidly becomes transparent above that wavelength. It looks to the eye to be a full deep red colour.

Best wishes

Roger

(The above letter and email are published by courtesy of Professor Roger F. Griffin.)

# Appendix H
# Light Scattering by Spherical Turbulence Structures

In the idealized model considered here (Fig. H.1) we consider spherical turbulence structures of fixed diameter, d, surrounded by air of constant refractive index, n. The refractive index inside the turbulence structures is also assumed to be constant, but taking the slightly different value n + Δn. We first analyze how plane waves behave as they propagate through and beyond these spherical structures. Our ultimate objective is to discover the extent to which scintillation develops in the waves, both during propagation through the structures and during subsequent propagation beyond the structures.

In effect, the structures behave as though they were spherical ball-lenses, of positive or negative power depending on whether $\Delta n$ is positive or negative. For visible light at sea level where we recall that the refractive index of air is about 1.00028. At altitude, the refractive index takes smaller values. Thus, for present purposes, we can assume that the refractive index of the air surrounding the spherical 'turbulence' structures is restricted to the range, $n \leq 1.00028$.

We also assume (without loss of generality) that $\Delta n$ is positive so that the structures behave as positive lenses, which then naturally image incident plane light waves into focused spots, albeit aberrated spots.

By applying paraxial optics principles, we obtain a first-order approximation of the lens-like behavior of these spherical lens structures. Their focal length can be expressed in terms of the lens diameter, d, as follows,

$$f = \frac{d}{4 \cdot \Delta n}. \tag{H.1}$$

Irrespective of the value of the local refractive index of air, it is always safe to assume that the refractive index difference, $\Delta n$; is small compared to the quantity ($n$ − 1). Thus, we write



T. S. McKechnie, *General Theory of Light Propagation and Imaging Through the Atmosphere*, Progress in Optical Science and Photonics 20,
https://doi.org/10.1007/978-3-030-98828-9

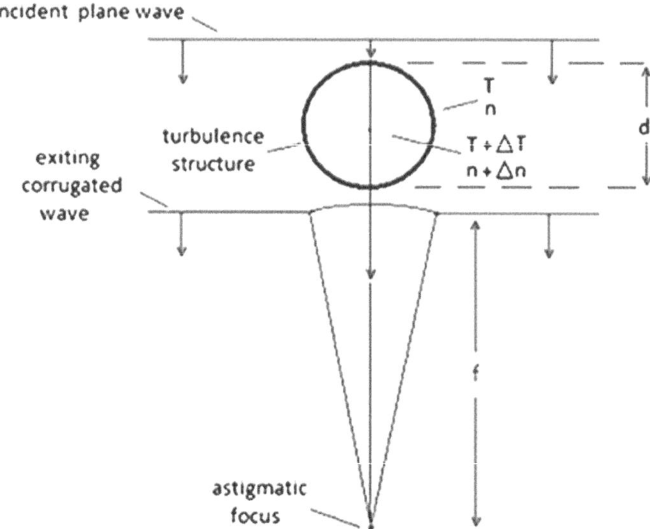


**Fig. H.1** Schematic showing the focusing effect of a spherical turbulence structure. The refractive index and temperature of the air surrounding the spherical structure are both constants, $n$ and $T$, respectively. The refractive index and temperature within the structure are also constants, $n + \Delta n$ and $T + \Delta T$, respectively (for the positive lens behavior illustrated)

$$\varepsilon = \frac{\Delta n}{(n-1)}, \tag{H.2}$$

where $0 < \varepsilon \ll 1$ and, for later use, we note here that $\varepsilon^2$ and higher powers are approximately zero.

Denoting the convergence angle, or scatter angle, of the incident plane waves as they emerge from the turbulence structure by $A_{Con}$ this angle is approximately given by

$$A_{Con} \approx \frac{d}{f}. \tag{H.3}$$

Using (H.1) and (H.2), angle $A_{Con}$ may be expressed in the alternative form,

$$A_{Con} \approx 4 \cdot \Delta n. \tag{H.4}$$

Convergence behavior begins to develop as soon as the light waves enter the structure. The ratio of the diameter of the converging wave portions as they exit the structure, $d_E$, to the diameter of the incident plane wave portions entering the structure, $d$, is given approximately by

$$\frac{d_E}{d} \approx 1 - \frac{A_{Con}}{2}. \tag{H.5}$$

From energy conservation considerations, beam brightening naturally accompanies beam convergence. Since beam convergence occurs in two dimensions, the beam brightening factor that occurs within the structure, which we denote by $B_f$, is given approximately by

$$B_f \approx \left(\frac{d}{d_E}\right)^2 \approx \left(\frac{1}{1-\frac{A_{Con}}{2}}\right)^2. \quad \text{(H.6)}$$

Using (H.4), the above relation may be written in the alternative form

$$B_f \approx \frac{1}{(1-2\cdot\Delta n)^2}, \quad \text{(H.7)}$$

which, by using (H.2) coupled with a binomial expansion, may be expressed in the form

$$B_f \approx 1 + 4\cdot(n-1)\cdot\varepsilon. \quad \text{(H.8)}$$

According to Hufnagel and Stanley (1964), temperature is the dominant mechanism that determines the refractive index within any given turbulence structure. If indeed this is so, the parameter, ε (cf., H.2), may be expressed in terms of the temperature parameters, $T$ and $\Delta T$; as follows:

$$\varepsilon \approx \frac{-\Delta T}{T}, \quad \text{(H.9)}$$

where $T$ and $\Delta T$ are both expressed in degrees Kelvin (K), and we note that, since ε is positive for a positive lens, $\Delta T$ must be negative for such a lens. Thus, for the turbulence structure to behave like a positive lens, we require the temperature inside the structure to be less than the temperature of the surrounding air.

Using the above expression for $\varepsilon$, the beam brightening factor given by (H.8) may be written in the desired form

$$B_f \approx 1 - 4\cdot(n-1)\cdot\frac{\Delta T}{T}. \quad \text{(H.10)}$$

Interestingly, this relation indicates that the amount of beam brightening that occurs within the extent of the turbulence structure is independent of the structure diameter.

The scatter angle caused by the structure (H.4) can also be expressed in terms of $T$ and $\Delta T$ by

$$A_{Con} \approx -4\cdot(n-1)\cdot\frac{\Delta T}{T}, \quad \text{(H.11)}$$

from which we see that scatter angle is also independent of the diameter of the structure.

The focal length of the structure (cf., H.1) may also be expressed in terms of the refractive index and temperature parameters, $n$, $T$, and $\Delta T$:

$$f \approx \frac{-d}{4 \cdot (n-1) \cdot \frac{\Delta T}{T}}. \tag{H.12}$$

The above expression indicates that, for any given combination of $n$, $T$, and $\Delta T$; the focal length of the structure grows in proportion to the structure diameter. When the structure diameter exceeds a certain size, the focal length of the structure must ultimately far exceed the length of even the longest physically possible atmospheric path. For such large structures, scintillation simply has no opportunity to develop over actual atmospheric paths. The effect of these large structures on propagating light waves may then be seen as limited solely to the introduction of phase alone.

**Example Evaluations for a Typical "Air Lens" Turbulence Structure**

In this illustrative example, we suppose that our spherical turbulence structure lies at an altitude of 10 km where $n \approx 1.00008$.[8] It is noted that 10 km is typical of jet stream altitudes where strong turbulence is often found. Recalling previous Fig. 3.2, we set $T$ to the average air temperature at 10-km altitude, 220 K. We also assume relatively strong turbulence by setting $\Delta T = -0.22$ K, which corresponds to $\Delta T/T = -0.001$.

With the above values of $n$, $T$, and $\Delta T$, (H.10) gives the beam brightening factor, $B_f \approx 1.0000003$. Because this value is so close to unity, the amount of scintillation that develops as the light waves propagate over the extent of the turbulence structure is, for all practical purposes, zero.

Although we only give one example evaluation here, further examples would readily demonstrate that, for all physically possible turbulence structures, the beam brightening factor, $B_f$, over the extent of these turbulence structure always takes values close to unity—to within six or more decimal places. This unique behavior is extremely convenient; it has crucial importance in the Chap. 6 wave propagation analysis.

Evaluation of (H.11) with the above parameter value choices gives a scatter angle of about 0.066 arcsec. In nighttime seeing conditions, a scatter angle of this magnitude (from a single turbulence structure) could be considered extremely large, given that the entire atmospheric path comprises many other similar turbulence structures randomly distributed over the path. The scatter angle that develops over the path increases roughly as the square root of the number of individual turbulence structures encountered in the path.

---

[8] Ignoring the effects of temperature variation with altitude, the quantity $(n-1)$ at a given altitude varies in proportion to the air pressure at that altitude (cf. Fig. 4.4).

Equation (H.12) indicates that the focal length associated with any given spherical turbulence structure of the type considered here varies linearly as the structure diameter. For $d \approx 0.05$ m, (H.12) gives the focal length as 156 km.

Although this focal length might seem quite large, at a zenith angle of 70°, the slant angle distance to the ground from an altitude of 10 km is about 30 km, which amounts to a significant fraction of 156 km. Thus, at large zenith angles, significant amounts of scintillation can be expected to develop from the relatively small (0.05 m diameter) turbulence structures. Consistent with this expectation, the lateral scale of the scintillation features shown previously in Fig. 3.11 seems to indicate that at least some of the upper atmosphere structures responsible for these patterns must indeed have measured about 0.05 m (5 cm) across.

For $d \approx 10$ m, (H.12) gives the focal length as 31,000 km—a distance far larger than the length of even the longest conceivable atmospheric path. Consequently, turbulence structures measuring more than about 10 m across produce virtually zero scintillation over any physically possible atmospheric path. This conclusion is entirely consistent with previous findings described in Chap. 5 (Sect. 5.7).

# Appendix I
# A Critique of Kolmogorov Theory as Applied to Atmospheric Turbulence Modeling

In applying Kolmogorov assumptions to atmospheric turbulence, Hufnagel and Stanley (1964) forewarned of problems: "Furthermore, there appears to be no satisfactory theory for predicting their [the atmospheric parameters[9] that determine the $C_n^2$] functional interdependence except for the lowest few hundred meters of the atmosphere. While some slight progress has been made in relating these quantities, this paper will bypass this problem …" But, as well as this particular problem, there were a number of others.

**Dimensional Inconsistencies**

The dimensional inconsistencies in Kolmogorov formulations have been discussed previously (McKechnie 2003). The choice of the Kolmogorov 2/3-power law for the refractive index structure function leads (by integration) to a 5/3-power law wavefront structure function. It may readily be demonstrated that the radii of curvature of both of these power law functions tend to zero for limitingly small separations. Since such behavior is physically unrealistic, in an attempt to restore realism, Kolmogorov formulations assume an inner-scale limit of turbulence size at which both the 2/3- and 5/3-power law functions are allowed to modify. Similarly, as discussed in Sect. 6.1.2.1, Kolmogorov formulations must also assume an outer-scale limit, usually denoted by $L_0$, where again the 2/3- and 5/3-power law functions are allowed to modify; in the case of the wavefront structure function, the 5/3- power law function rolls off and flattens out.

The interval between the inner and outer-scale limits in which the 2/3- and 5/3-power laws are assumed valid is referred to as the inertial subrange. Outside of this range, Kolmogorov theory anticipates the development of power law functionalities

[9] The altitude-dependent parameters referred to by Hufnagel and Stanley are as follows: δ, the rate of energy per unit mass dissipated by viscous friction, $\beta$, the average shear rate of the wind, and $\gamma$, the average vertical gradient of the potential temperature. Each is a function of the other two.



T. S. McKechnie, *General Theory of Light Propagation and Imaging Through the Atmosphere*, Progress in Optical Science and Photonics 20,
https://doi.org/10.1007/978-3-030-98828-9

other than 2/3 and 5/3. While all of this might seem reasonable enough on the surface, in fact it adds up to dimensional chaos and absurdity.

Hufnagel and Stanley give the following expression (7.6 in their paper) for the atmospheric modulation transfer function (MTF),

$$M(\varepsilon, \lambda) = \exp\left[-\frac{2 \cdot \pi^2}{\lambda^2} \cdot \langle S^2 \rangle\right], \tag{I.1}$$

where the quantity $\langle S^2 \rangle$ is defined (7.2 in the Hufnagel and Stanley paper) by

$$\langle S^2 \rangle = 2.91 \cdot \varepsilon^{\frac{5}{3}} \cdot \int_0^z C_n^2(z') \cdot dz'. \tag{I.2}$$

Plainly, the exponent term in square brackets on the right-hand side of (I.1) must be a dimensionless number. Noting that the parameters, λ, ε, and z, all have the dimensions of length, it follows from (I.1) and (I.2) that the dimensions of $C_n^2$ must be that of *length*$^{-2/3}$. Similarly, the dimensions of the refractive index structure constants, $C_n$, must be that of *length*$^{-1/3}$. However, suppose now that, in the vicinities of the inner- and outer-scale limits, the wavefront structure function (indicated in (I.2) by the function $\varepsilon^{-5/3}$) rolls-off as required and thus departs from the 5/3-power law functionality. To allow such altered functionality and yet still preserve a dimensionless exponent in (I.1) plainly requires corresponding modifications to the dimensions of $C_n^2$. However, because these dimensions are fixed at the outset in Kolmogorov formulations to those of *length*$^{-2/3}$, their subsequent modification is simply not allowed. Thus, outside of the inertial subrange, Kolmogorov formulations are dimensionally inconsistent and consequently mathematically untenable.

The alternative general formulation developed for the atmospheric MTF in Chap. 6 is set up in a way that safeguards against dimensional inconsistencies of the sort just described. Integrated turbulence strength is described in the formulation by the variance of the optical path difference (OPD) fluctuation of the light waves arriving in the telescope pupil, $\sigma^2$; this quantity was defined previously by (6.29) and has the dimensions of *length*$^2$. This allows the corresponding wavefront structure function to be expressed in the form, $[1 - \rho(\varepsilon)]$. Since function $\rho(\varepsilon)$ is defined in dimensionless form (c.f., 6.30) the wavefront structure function must itself be dimensionless. Thus, this function can be assigned any conceivable shape or form without causing dimensional inconsistencies.

In principle, Kolmogorov theory could be re-formulated in terms of a dimensionless function of the type $[1 - \rho(\varepsilon)]$ that conforms to the Kolmogorov 5/3-power law in the inertial subrange while taking other functionalities outside of this range (cf., Appendix C, C.6). The general formulation developed in this book readily allows the structure function characteristics postulated in Kolmogorov theory to be expressed over the entire range of turbulence structure scale sizes in a dimensionally consistent form. However, proper expression of such a formulation would oblige us to abandon the familiar Kolmogorov structure constants $C_n^2$ (with their dimensions of length$^{-2/3}$) and replace them with some other measurable quantity that has the dimensions of length$^2$.

**Troublesome consequences of the dimensional inconsistencies**
As previously mentioned in Chap. 2 an inevitable, yet absurd, practical consequence of the dimensional problems of Kolmogorov theory is that measured values of the structure constants, $C_n$, now depend on the length of the baseline over which the measurements are obtained. Smaller measured $C_n$ values arise when longer measurement baselines are used; larger values arise when shorter baselines are used. Such behavior of course results in anomalous inconsistent atmospheric path characterizations which, in turn, lead to anomalous and erroneous star image property predictions.

**Experimental Evidence Contradicting Kolmogorov Assumptions**
Our objective here is show that substantial material exists in the literature that is not consistent with Kolmogorov assumptions. In this way, we demonstrate that Kolmogorov theory has no general validity, thus identifying the need for a more soundly-based and more precise theory of light propagation and imaging through the atmosphere—one that has general applicability.

In Chap. 2, mention was made of instances where observed image properties are not consistent with Kolmogorov assumptions, the cited examples being the central cores seen in star images at visible wavelengths by Griffin (1973) and too little image motion observed by Woolf et al. (1982). Meinel (1960) has described image cores observed with the Mt. Palomar 200-in., the Mt. Wilson 100-in., and the McDonald 82-in. telescopes with properties closely similar to those seen by Griffin. Dozens more examples could have been cited.

Beckers and Williams (1982) describe star image features observed using the multi mirror telescope (MMT) that are intriguingly similar to image cores. They do not use the word "core"—a phenomenon unrecognized and eschewed by the overwhelming majority of astronomers at that time. Instead they refer to "elements" in the image "which remain stationary" in good seeing conditions. This description plainly suggests a core formed by a telescope with less than diffraction-limited performance. For severely aberrated telescopes, where the Strehl intensity at visible wavelengths is significantly less than 0.1, the telescope point-spread function (PSF), which determines core shape, usually arises as a fixed speckle pattern, spread over an angular diameter roughly given to the product of the angular size of individual speckles at the observation wavelength and the inverse square root of the Strehl intensity at that wavelength.

Griffin gave particularly detailed descriptions of image cores (Sect. 10.8), stating that "A big telescope often gives an image with a small bright core ... If the telescope were optically diffraction limited, the core would be smaller and brighter; the resolution of most existing large telescopes is set by the optics." Griffin's core observations were widely discounted for many years (Griffin 1990) for no reason other than that they were not consistent with Kolmogorov theory. According to standard scientific practice, when theoretical predictions do not agree with observations, it is concluded that the theory must be wrong and at the very least requires modification. In the case of Griffin's observations, the rule was reversed.

In their descriptions of images obtained with the MMT and 2.3-m telescopes in which much less image motion was observed than predicted by conventional theory, Woolf et al. (1982) state that "in data where the mean image size changed by a factor 2.5, no trend in image motion was discernible." They further state that "a number of astronomers have expressed doubts as to actual seeing following the predictions of seeing theory." Griffin (1973) also commented on lack of motion in observed images, finding no "bodily motions of the image ... through angular distances comparable to the angular diameter of the image core."

Analysis of star image measurements made by the Author at the Royal Greenwich Observatory using the 36-in. Yapp reflector (McKechnie 1991) indicated that the largest turbulence structures carrying any significant strength measured less than 0.5-m across. Structure size estimates by Meinel (1960) based on knife-edge measurements with the Mt. Palomar 200-in. telescope led him to observe, "the frequency of occurrence with size diminishes rapidly above "20–30 cm." For his part, Griffin (1973) speaks of 4 in. [10 cm] as being a "size corresponding to the spatial frequencies predominant in atmospheric turbulence."

More recently, Coulman et al. (1988) used balloon-borne temperature probes and scintillometry techniques to measure the outer-scale turbulence limit, $L_0$, as a function of altitude. Their extensive measurements, described in Appendix C, were made over a multi-year period at observatory sites in three different countries—France, the United States, and Chile. Their measurements consistently indicated relatively small $L_0$ values, largely confined to the range, "30–40 cm.". Although these authors measured some 4-m turbulence structure at high altitudes (8 km), the energy fraction invested at these large scales (as weighted by the tiny $C_n^2$ values at these altitudes) was simply not enough to significantly influence the $L_0$ value averaged over the entire depth of the atmosphere (Appendix C). Coulman et al. conclude that "a small $L_0$ value implies, in principle, a restriction of the range within which Kolmogorov theory of turbulence may be applied to estimate atmospheric seeing limitations to large-aperture telescopes."

Bufton (1973), using balloon-borne temperature probes, also found smaller than expected turbulence structure sizes. His measurement data led him to observe, "the limited vertical extent of some observed structure indicates a small outer-scale $L_0$ and truncation of the inertial subrange. The major effect of this error source would be to alter the quantitative prediction of optical effects."

Experiments using long-baseline interferometers have also produced evidence of smaller $L_0$ values. Many of the path length differences measurements made on Mt. Wilson using an 11-m baseline interferometer by a team led by Charles H. Townes (Bester et al. 1992) were not consistent with Kolmogorov assumptions. The flat dismissal by W.C. Danchi of Kolmogorov assumptions in typical astronomical viewing conditions (quoted previously in Chap. 2) was based on these, and other, measurements made by the Townes' measurement team.

Mariotti and Di Benedetto (1984) used an interferometer with an adjustable, 8– to 16-m baseline to measure path length differences and also found their measured data inconsistent with Kolmogorov assumptions. Phase differences measured between the two pupils were found to be five times smaller than expected. Finding it "impossible to fit them [the phase differences] with the theory," Mariotti and Di Benedetto speculate that the cause might have been "saturation of the phase" at distances smaller than the baseline.

All of this evidence—taken together with other evidence such as the near-IR cores captured by the 3.8-m Mayall telescope (Sect. 10.3.5)—contradicts the basic Kolmogorov assumption that $L_0$ is larger than the largest telescope aperture and that roll-off departures in the 2/3-power law structure function occur only beyond the extent of the telescope aperture. At the time most of the above evidence was obtained, the diameter of the world's largest telescope was only about 5 m. Since then, with more than a dozen 8- to 10-m class instruments now operational, and with several even larger, extremely large telescopes (ELTs) due to be completed within the next five years, the long-cherished assumption that Lo is larger than the largest telescope aperture should play no further part in the conversation.

**The Coherence Parameter**

The coherence parameter, $r_0$, introduced by Fried (1966) quantifies the coherence patch diameter associated with light waves that have propagated over extended atmospheric paths. The angular resolution obtained by a large telescope observing through the atmosphere can be approximately quantified in terms of $r_0$ through the angular quantity, $\lambda/r_0$. The coherence parameter, $r_0$, has the dimensions of *length* and is useful—both practically and conceptually—for quantifying the coherence patch size of light waves at visible wavelengths where geometrical optics often provides reasonably accurate image descriptions. However, at IR wavelengths, where the rms OPD fluctuation of the image forming wavefronts in the telescope pupil is often found to be smaller than the dimensions of the wavelength itself, geometrical optics loses validity.

When Kolmogorov assumptions are combined with geometrical optics, the $r_0$ parameter can be shown to vary with wavelength as $\lambda^{6/5}$, and resolution can be shown to improve in proportion to $\lambda^{-1/5}$. Thus, for example, if 1-arcsec resolution were obtained at the visible wavelength, 0.55 $\mu$m, the $\lambda^{-1/5}$ resolution dependence on wavelength indicates that at the near-IR wavelength, 2.2 $\mu$m, resolution would improve marginally to about 0.76 arcsec. If image wander were also corrected, thus obtaining the average short-exposure image, resolution would further improve by a factor no larger than 1.9 (but, depending on the size of the telescope, perhaps considerably less) to a final value not less than 0.4 arcsec.[10]

[10] According to the paper by Fried (1966), "Optical resolution through a randomly inhomogeneous medium for very long and very short exposures," depending on the telescope aperture size, resolution obtained in the average short-exposure image can be as much as 3.8 times higher (cf., Fig. 1 in Fried's paper) than resolution obtained in long-exposure images. However, resolution is defined Fried's paper in terms of "cycles squared per unit area." In terms of linear resolution—the more

However, if roll-off departures of the Kolmogorov 5/3-power law wavefront structure function occur at distances smaller than the telescope aperture size—as is the case for the type of turbulence measured by Coulman et al. (1988)—the $\lambda^{6/5}$ and $\lambda^{-1/5}$ wavelength dependencies no longer apply, especially at IR wavelengths. As described previously in Chap. 14, because of the prominent image cores delivered by the 6.5-m MMT instrument at 2.2 μm, 0.1-arcsec resolution was achieved even though visible seeing at the time the images were recorded was no better than 1 arcsec.

In addition to not taking account of image cores, the $\lambda^{-1/5}$ resolution dependence on wavelength takes no account of the fact that resolution ultimately decreases in images obtained at far-IR wavelengths. At these longer wavelengths, typical telescope wavefront aberrations of just a few waves (HeNe) generally have little effect on image quality; star images formed at these wavelengths by large telescopes with circular apertures must ultimately appear as near-perfect Airy patterns. The angular resolution obtained from these Airy patterns, instead of improving with increasing wavelength in proportion to $\lambda^{-1/5}$, now actually reduces in proportion to $\lambda$ (as per 1.22 $\lambda/D$).

The paper by Chanan et al. (1998) which deals with the phasing of the primary mirror segments in the state-of-the-art Keck 10-m telescopes may possibly have been over-reliant on the coherence parameter, $r_0$, and the associated $\lambda^{-1/5}$ resolution dependence on wavelength. When the individual 1.8-m hexagonal mirror segments are all correctly phased, Fig. 1 in that paper indicates that, as wavelength ranges from 0.5 to 10 μm, only marginal improvements in resolution are anticipated. At the same time, the star image intensity envelops show only marginal shape changes over this wavelength range.

In stark contrast, the diffraction analysis of Keck telescope images given in Chap. 14 and illustrated in Fig. 14.20 indicates entirely different behavior. As wavelength ranges from 0.5 to 10 μm, image shape gradually morphs from halo-only image characteristics, through a core and halo stage, until ultimately settling into a core-only form. The angular resolution obtained by the Keck instruments from these different types of image, instead of showing only marginal variations over the wavelength range, 0.5–10 μm, now shows variations spanning more than an order of magnitude, with highest resolution of a mere 0.03 arcsec (30 mas) anticipated in the J atmospheric window ( ~1.25 μm).

**Legacy of Kolmogorov Telescope Optical Tolerance Prescriptions**

As discussed in Chap. 2, an entire generation of telescopes constructed in the 1970s and 1980s was built to coarse, Kolmogorov, optical tolerance prescriptions. The UKIRT 3.8-m instrument (Mauna Kea, HI) is possibly the most shocking example. The optical specification of that instrument (Griffin 2011) which saw first light in 1979, required it, in the absence of atmosphere, to concentrate 80% of the light in a star image into a 1-arcsec patch diameter. (This specification translates to a telescope point spread function with the FWHM width of about 0.66 arcsec.) However, as

common way of indicating resolution used in this book and elsewhere—the maximum improvement factor that can be obtained from a short-exposure image is about $\sqrt{3.8}$, i.e., a factor < 2.

described previously (Sect. 10.9) some twenty years after its construction—but only after extensive optics upgrades had been carried out—the UKIRT instrument was finally able to deliver 0.13-arcsec images at the near-IR wavelength 2 μm (Hawarden 1998).[11] It is recalled here that Hawarden's superb images were obtained without use of adaptive optics. To obtain the images, it was only necessary to correct image motion.

The huge improvement in resolution ultimately achieved by the UKIRT instrument can now be seen as not only a refutation of Kolmogorov optical prescriptions for large ground-based telescopes but a refutation of Kolmogorov theory itself. The full resolution potential of large ground-based astronomical telescopes, first suspected by the Author in 1976 (McKechnie 1976), only became clear to astronomers around 1990 when they began equipping their instruments with (newly developed) infrared focal plane array (IR-FPA) detectors. With these new detectors, central cores were suddenly appearing routinely in star images formed by large telescopes at near-IR and longer wavelengths.

Previous detector types (such as single-point bolometer detectors where the image had to be sequentially scanned by the detector) required significantly longer image acquisition times, providing abundant opportunity for image motion[12] to smear out and obscure the crucial core features. Using the new IR-FPA detectors, short-exposure images could be recorded with image motion effectively frozen. By eliminating image smear, short-exposure images at 2.2 μm obtained by the Multi Mirror Telescope (MMT)[13] on Mt. Hopkins (AZ) and by the 3.8-m Mayall telescope on Kitt Peak (AZ) were suddenly providing angular resolution of 0.10 and 0.14 arcsec, respectively, even though visible seeing was no better than 1 arcsec.[14]

But why, around 1990, were optical tolerances for large ground-based telescopes suddenly being tightened by almost an order of magnitude when Kolmogorov theory had previously seemed to provide—at least at visible wavelengths—reasonably

[11] The Hubble Space Telescope, launched by the Space Shuttle in 1990, was initially found to have a defective primary mirror (Allen, Angel, Mangus, Rodney, Shannon, and Spoelhof, NASA-TM-103443, Nov 1990). However, following a servicing mission in 1993, it has spectacularly achieved its intended performance.

[12] Image motion is caused by the combined effects of telescope shake, tracking errors, and wavefront tilts caused by atmospheric turbulence.

[13] Circa 1990, the MMT comprised six separate 1.8-m diameter mirrors, the effective diameter of the instrument being about 5.5 m. In 2010, the six individual mirrors were replaced by a single lightweight mirror measuring 6.5 m across.

[14] Remarkably, the resolution delivered by these two instruments was significantly sharper than was being achieved by the adaptive optics telescope system under development at the US Air Force Starfire Optical Range (SOR) in New Mexico (Fugate et al. 1994). At that time, the SOR AO imaging platform was attached to a 1.5-m telescope. This system could not match the resolution that was now being obtained by larger telescopes in Arizona, even though these instruments were not equipped with AO and were known to have significant aberrations. The construction of a 3.5-m telescope at SOR (1994 first light) helped restore competitivity, although its intrinsic imaging performance specification in the absence of atmosphere merely required it to concentrate 50% of the light in a 0.33-arcsecond patch diameter. The 3-5-m SOR telescope was possibly the last large ground- based instrument to be built to coarse Kolmogorov, geometrical-based optical specifications.

adequate image descriptions? And why had Kolmogorov theory not predicted the highly resolved image cores that were now being seen routinely at near-IR wavelengths? The plain answer is that—just as observed by Danchi, Coulman et al., and McKechnie—Kolmogorov assumptions do not apply in atmospheric conditions suitable for observational astronomy. Yet, if this is so, why did this anomalous situation remain hidden for so many years?

The angular size of the "seeing-disk" images frequently delivered by large telescopes at visible wavelengths in average visible seeing conditions—say about 1 arcsec—can readily be explained by geometrical optics. To correctly predict the angular size of this type of image, it is only necessary to correctly model the average gradients of the image-forming light waves in the telescope pupil—which naturally, in 1-arcsec seeing, is also about 1 arcsec. To a first-order approximation, the magnitude of the peak-to-valley OPD fluctuation associated with the light wave portions collected by large telescopes is proportional to the product of the average wavefront gradient and the average lateral size of the wavefront structures arriving in the pupil of the telescope. Because the size of the latter naturally scales in proportion to the average turbulence structure size, for Kolmogorov turbulence the average peak-to-valley OPD fluctuation turns out significantly larger than the corresponding peak-to-valley fluctuation for the much smaller average turbulence structure sizes measured by Coulman et al.[15]

It is nontheless still possible to 'explain away' the extremely large peak-to-valley OPD fluctuations that arise from Kolmogorov turbulence in a way that allows these large fluctuations to be consistent with the correct average image size. One must simply choose, albeit retrospectively, a proportionately large value for the integrated turbulence strength over the atmospheric path![16] But sadly, by doing this we have simply fudged the correct answer; in effect, two wrongs have been combined to makes things seem right.

The combined effect of the two choices—even if they are both incorrect—results in a more or less correct value for the average wavefront gradient in the telescope pupil, which then translates to a seemingly correct angular size prediction for the star image. In this way we see that, even if the turbulence structure sizes in the atmosphere are in fact far smaller than postulated by Kolmogorov theory, as far as visible images are concerned it would still be possible for this problem to remain hidden.

However, at near-IR (and longer) wavelengths where the OPD fluctuations of the image-forming waves in the telescope pupil are frequently smaller than the wavelengths themselves, it is no longer enough to simply match the average wavefront gradients. With diffraction now being the determining mechanism (rather than geometrical optics) it is now imperative to base calculations on a correct value for

[15] For average wavefront gradients of the order 1 arcsec (~0.000005 radians) even 1-m average wavefront structure size can produce peak-to-valley OPD variations of the order 5 μm—an amount inconsistent with the appearance of image cores at wavelengths shorter than about 5 μm.

[16] The rms OPD fluctuation is roughly proportional to the square root of the integral, $\int C_n^2(z) \cdot dz$ over the atmospheric path. In practical applications, for Gaussian distributed OPD fluctuations the peak-to-valley (P—V)) OPD variation can generally be considered about 3 times larger than the rms variation.

the rms OPD fluctuation of the image-forming waves in the telescope pupil. If this value happens to be larger than the near-IR wavelengths considered—as is generally the case when Kolmogorov turbulence is assumed—it then becomes impossible to explain the presence of image cores at these wavelengths.

To simultaneously provide a correct description of star image appearance at both visible and near-IR wavelengths, there is now no room for error. To account for image cores at near-IR wavelengths, the rms OPD fluctuation of the image-forming waves must necessarily be smaller than the imaging wavelength. In this regime, the geometrical optics approach collapses. A diffraction analysis becomes obligatory and to execute it correctly obliges us to use both the correct turbulence structure function and the correct rms OPD fluctuation.

## References


Beckers, J. M., & Williams, J. T. (1982). Performance of the MMT III: Seeing experiments with the MMT. In L. D. Barr & G. Burbidge (Eds.), *Advanced technology optical telescopes, SPIE* (Vol. 332, pp. 16–23). Bellingham, WA.

Bester, M., Danchi, W. C., Degiacomi, C. G., Greenhill, L. J., & Townes, C. H. (1992). Atmospheric fluctuations: Empirical structure functions and projected performance of future instruments. *The Astrophysical Journal, 392*, 357–374.

Bufton, J. L. (1973). Comparison of vertical profile turbulence structure with stellar observations. *Applied Optics,12*, 1785–1793.

Campbell, L., & Garnett, W. (1882). *The life of James Clerk Maxwell with a selection from his correspondence and occasional writings, and a sketch of his contributions to science*. Macmillan.

Chanan, G., Troy, M., Dekens, F., Michaels, S., Nelson, J., & Mast, T. (1998). Phasing the mirror segments of the Keck telescopes: the broadband phasing algorithm. *Applied Optics, 37*(1), 140–155.

Coulman, C. E., Vernin, J., Coqueugniot, Y., & Caccia, J. L. (1988). Outer scale of turbulence appropriate to modeling refractive index structure profiles. *Applied Optics, 27*, 155–160.

Electro-Optics Handbook. (1974). *Electro-optics handbook*. Burle Industries, Inc.

Fried, D. L. (1966). Optical resolution through a randomly inhomogeneous medium for very long and very short exposures. *JOSA, 56*, 1372–1379.

Fugate, R., Ellerbroek, B., Stewart, E., Colucci, D., Ruane, R., Spinhirne, J. et al. (1994). First observations with the Starfire Optical Range 3.5-m telescope. In *Proceedings of SPIE 2199, Seeing and Telescope Improvement, Advanced Technology Optical Telescopes V* (p. 481).

Good, R. E., Beland, R. R., Murphy, E. A., Brown, J. H., & Dewan, E. M. (1981). Atmospheric models of optical turbulence. In L. S. Rothman (Ed.), *Modeling of the Atmosphere (Critical Reviews), Proceedings of SPIE* (Vol. 928, pp. 165–186). *Bellingham, WA*.

Griffin, R. F. (1973). On image structure, and the value and challenge of very large telescopes. *Observatory, 93*, 3–8.

Griffin, R. F. (1990, May). Giant telescopes, tiny images. *Sky & Telescope*, p. 469.

Griffin, R. F. (2011, January). Private email communication. In R. F. Griffin (Ed.), *Member of the 8-person UKIRT steering committee*.

Hawarden, T. (1998, 20 August). *The United Kingdom infrared telescope*. Retrieved from http://www.jach.hawaii.edu/UKIRT/public/tele-descrip.html.

Hufnagel, R. E., & Stanley, N. R. (1964). Modulation transfer function associated with image transmission through turbulent media. *JOSA, 54*, 52–61.

Kopp, G., & Lean, J. L. (2011). A new, lower value of total solar irradiance: Evidence and climate significance. *Geophysical Research Letters, 38*, L01706.
Lide, D. R. (2007–2008). *Handbook of chemistry and physics* (88th ed.). CRC Press.
Mahon, B. (2004). *The man who changed everything (The life of James Clerk Maxwell).* Wiley.
Mariotti, M., & Di Benedetto, G. P. (1984). Pathlength stability of synthetic aperture telescopes. The case of the 25 cm Cerga interferometer. In H. Ulrich & K Jaer (Eds.), *Proceedings of International Astronomical Union Colloquium* (V'ol. 79, pp. 257–265). Garching bei Miinchen, Germany: European Southern Observatory.
McKechnie, T. S. (1976). Cores in star images. *JOSA, 66*(6), 635.
McKechnie, T. S. (1991). Light propagation through the atmosphere and the properties of images formed by large ground-based telescopes. *JOSA A, 8*, 346–365.
McKechnie, T. S. (2003). General formulation for light propagation and imaging through atmospheric turbulence, and naturally derived turbulence measurement procedures. In C. Y. Young & J. S. Stryjewski (Eds.), *Atmospheric Propagation, Proceedings of SPIE* (Vol. 4976(4), pp. 34–46).
Meinel, A. B. (1960). Astronomical seeing and observatory site selection. In G. P. Kuiper & B. M. Middlehurst (Eds.), *Telescopes*. University of Chicago Press.
Ramsey, R. C. (1962). Spectral irradiance from stars and planets above the atmosphere from 0.1 to. 100.0 microns. *Applied Optics, 1*(4), 465–471.
Ridpath, I. (1989). *Norton's 2000.0, star atlas and reference handbook* (18th ed.). Longman Group UK Ltd.
Venet, B. (1998). Optical scintillometry over a long, elevated horizontal path. *SPIE, 3381*, 212–219.
Woolf, N. J., McCarthy, D. W., & Angel, J. R. (1982). Performance of the MMT VII: Image shrinkage in sub-arc second seeing at the MMT and 2.3 m telescopes. In L. D. Barr & G. Burbridge (Eds.), *Proceedings of SPIE Engineering, Advanced Technology Optical Telescopes, Bellingham, WA* (Vol. 332, pp. 50–56).

# Appendix J
# The Importance of HEL Wavelength Choice IN HEL Weapon Systems

The paper, "Focusing laser beams over atmospheric paths and identifying wavelengths that deliver maximum irradiance in the target plane" was submitted by the Author to the Optical Society of America (OSA) conference on "Propagation Through and Characterization of Atmospheric and Oceanic Phenomena," Washington, D.C., USA, June 27–29, 2016. In somewhat odd circumstances, it was not accepted for presentation. Because of the paper's direct relevance to the development of effective HEL laser weapon systems, and especially after the embarrassing failure and subsequent axing of the flagship ABL Program in 2012, one might reasonably have expected the paper to be of considerable interest—even paramount interest—when one notes the military affiliations of the majority of the Paper Selection Committee. For posterity, the three-page abstract and summary requested by, and submitted to, this committee is reproduced in full below.

**Paper Selection Committee**

**Chair**

Steven Fiorino, Air Force Institute of Technology, United States.

Denis Oesch, Guidestar Optical Systems, United States.

Stacie Williams, Air Force Office of Scientific Research, United States.

**Members**

Terry Brennan, Prime Plexus, United States.

Julian Christou, Large Binocular Telescope Observatory, United States.

Thomas Farrell, US Air Force Research Laboratory, United States.

Szymon Gladysz, Fraunhofer Institute IOSB, Germany.

Venkata Gudimetla, US Air Force, United States.

Dan Herrick, US Air Force Research Laboratory, United States.

Pat Kelly, US Air Force Research Laboratory, United States.





T. S. McKechnie, *General Theory of Light Propagation and Imaging Through the Atmosphere*, Progress in Optical Science and Photonics 20,
https://doi.org/10.1007/978-3-030-98828-9

Kent Miller, US Air Force Office of Scientific Res, United States.
Steve Miller, Colorado State University, United States.
Julie Moses, US Air Force, United States.

Andreas Muschinski, NorthWest Research Associates, United States.
Douglas Nelson, US Army, United States.
Troy Rhoadarmer, Guidestar Optical Systems, Inc., United States.
Darryl Sanchez, US Air Force Research Laboratory, United States.
Jason Schmidt, MZA Associates Corporation, United States.
Julie Smith, US Air Force Research Laboratory, United States.
Jeremy Solbrig, Colorado State University, United States.
Knut Solna, University of California Irvine, United States.
Mark Spencer, Air Force Research Laboratory, United States.
David Voelz, New Mexico State University, United States.
Mikhail Vorontsov, University of Dayton, United States.

**Title: "Focusing laser beams over atmospheric paths and identifying wavelengths that deliver maximum irradiance in the target plane"**
**Author: T. Stewart. McKechnie**

**Abstract**: The irradiance at the focus of laser beams projected through the atmosphere by large telescopes critically depends on the laser wavelength. A path characterization procedure is described that identifies wavelengths that deliver maximum irradiance.

**Summary**: Historically, when large telescopes have been used to project laser beams through the atmosphere at distant targets, the extent to which the irradiance (Watts/m$^2$) delivered at beam focus varies with the laser wavelength has not been fully appreciated. A measurement procedure is described for characterizing laser beam paths of this type and identifying the 'optimum' wavelength that delivers maximum irradiance at the beam focus for the same amount of laser beam power (Watts). The procedure is based on a rigorous light propagation analysis (Ref. 1), but makes some additional simplifying approximations whose accuracy increases with telescope diameter. While these approximations do not compromise the underlying physics, they greatly simplify the mathematics, making the procedure less complicated and more readily implementable in practice.

For any given telescope/atmospheric-path combination, the use of a laser wavelength significantly shorter than, or significantly longer than the optimum wavelength can result in a catastrophic irradiance reduction at the beam focus compared to the level attainable at the optimum wavelength.

Path characterization is carried out using the actual beam projection telescope itself. Ordinarily, the point-object would be located at the intended focus of the laser beam, with the image formed in the primary image plane of the telescope; however, the imaging could equally be accomplished in the reverse direction. Either way, to characterize the path, multiple short-exposure images of the point-object are captured (perhaps ten) at a judiciously chosen wavelength, as discussed shortly. The images are subsequently summed together (while correcting image motion) to provide a suitable estimate of the average short-exposure image intensity envelope.

To obtain fully definitive path characterization by the procedure, it is crucially important to form the images at a wavelength consistent with the images exhibiting core and halo structure. The average intensity envelope for such an image not only enables characterization of the integrated Optical Path Difference (OPD) over the imaging path, it also enables characterization of the fixed OPD contribution from the telescope optics. (Halo-only images formed at shorter wavelengths and core-only images formed at longer wavelengths represent degenerate cases of the more general core-and-halo image; neither contains the full information complement necessary for complete path characterization.)

Core-and-halo images obtained by large telescopes can be sufficiently approximated for our purposes as the sum of two Gaussian functions (Ref. 1), a narrow one representing the core portion and a wider one representing the halo portion. Such an image approximation can be specified in terms of just three parameters: the 1/e-half widths of the two Gaussian functions (Fig. 1), denoted respectively by $B_C$ ($\lambda_M$) and $B_H$ ($\lambda_M$), where $\lambda_M$ is the measurement wavelength, and the ratio of the intensities in the centers of the two Gaussian functions, denoted by $A_{CH}$ ($\lambda_M$) ( $= A_C$ ($\lambda_M$)/$A_H$ ($\lambda_M$)). The appropriate values of these parameters are those that provide a best-fit to the average short-exposure image envelope obtained from the telescope images. These values can be used directly to quantify the three fundamental parameters that characterize the end-to-end imaging path: the rms OPD fluctuation associated with the image-forming waves in the telescope pupil, $\sigma_P$, the 1/e-half width of the autocorrelation function of that fluctuation, $w_o$, and the Strehl intensity of the telescope, $S(\lambda_M)$. The required mapping equations are given by Eqs. 13.70–13.72 in Ref. 1.

To characterize an atmospheric path that has been corrected by Adaptive Optics (AO), the AO system would have to operate during image acquisition. However, in this case, rather than characterizing the OPD contributions from the atmospheric path and telescope optics, per se, more fittingly the procedure characterizes the residual OPD fluctuation of the AO-corrected image-forming waves, separately quantifying the residual dynamical OPD fluctuation and the residual fixed OPD contribution from the telescope optics.

Assuming that the OPD fluctuation associated with the image-forming waves arriving in the telescope pupil is Gaussian distributed, the light energy fraction in the core, $E_C$, is given by

$$E_C = \exp\left[-\left(\frac{2 \cdot \pi \cdot \sigma_P}{\lambda}\right)^2\right], \tag{J.1}$$

with the halo light energy fraction given by $(1 - E_C)$. To obtain core-and-halo images suitable for our purposes, the imaging wavelength, $\lambda$, is chosen so that the core energy fraction falls in the approximate range, $0.2 \leq E_C \leq 0.8$, with the halo energy fraction falling into the same range. For reflector telescopes where the rms OPD fluctuation contribution from telescope aberrations, $\sigma_T$, is independent of wavelength, and assuming that the fluctuation is Gaussian distributed, the Strehl intensities at the measurement wavelength, $\lambda_M$, and any other arbitrary wavelength, $\lambda$, are given by

$$S(\lambda_M) = \exp\left[-\left(\frac{2 \cdot \pi \cdot \sigma_T}{\lambda_M}\right)^2\right] \tag{J.2}$$

and

$$S(\lambda) = \exp\left[-\left(\frac{2 \cdot \pi \cdot \sigma_T}{\lambda}\right)^2\right]. \tag{J.3}$$

The average intensity envelope at wavelength $\lambda$ (Fig. 2) can be calculated (Ref. 1, Eq. 13.55) using

$$\langle I(\vartheta, \lambda)\rangle = \frac{1}{\pi \cdot \left(A_{CH}(\lambda) \cdot B_C(\lambda)^2 + B_H(\lambda)^2\right)} \times \left\{A_{CH}(\lambda) \cdot \exp\left[-\left(\frac{\vartheta}{B_C(\lambda)}\right)^2\right] + \exp\left[-\left(\frac{\vartheta}{B_H(\lambda)}\right)^2\right]\right\}, \tag{J.4}$$

where $B_C$ $(\lambda)$, $B_H$ $(\lambda)$ and $A_{CH}$ $(\lambda)$, can be calculated from $B_C$ $(\lambda_M)$, $B_H$ $(\lambda_M)$ and $A_{CH}$ $(\lambda_M)$ using:

$$B_C(\lambda) = B_C(\lambda_M) \cdot \frac{\lambda}{\lambda_M} \cdot \exp\left[-2 \cdot \pi^2 \cdot \sigma_T^2 \cdot \left(\frac{1}{\lambda_M^2} - \frac{1}{\lambda^2}\right)\right], \tag{J.5}$$

$$B_H(\lambda) = B_H(\lambda_M), \tag{J.6}$$

and

$$A_{CH}(\lambda) = A_{CH}(\lambda_M) \cdot \frac{\lambda_M^2}{\lambda^2} \times \exp\left[-4 \cdot \pi^2 \cdot \left(\sigma_P^2 + \sigma_T^2\right)\right] \times \left\{\frac{1 - \exp\left[-\left(\frac{2 \cdot \pi \cdot \sigma_P}{\lambda_M}\right)^2\right]}{1 - \exp\left[-\left(\frac{2 \cdot \pi \cdot \sigma_P}{\lambda}\right)^2\right]}\right\}. \tag{J.7}$$

For a telescope of diameter, $D$, Eqs. H.4–H.7 combine to show that maximum irradiance in the center of the image ($\vartheta = 0$) occurs at a wavelength, $\lambda_{optI}$, obtained by solving the transcendental equation,

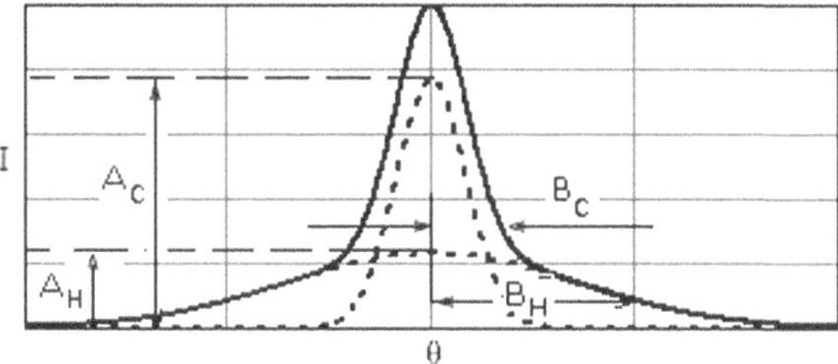


**Fig. J.1** (Solid) Point-object image separated into (Dots) best-fit Gaussian core and halo portions

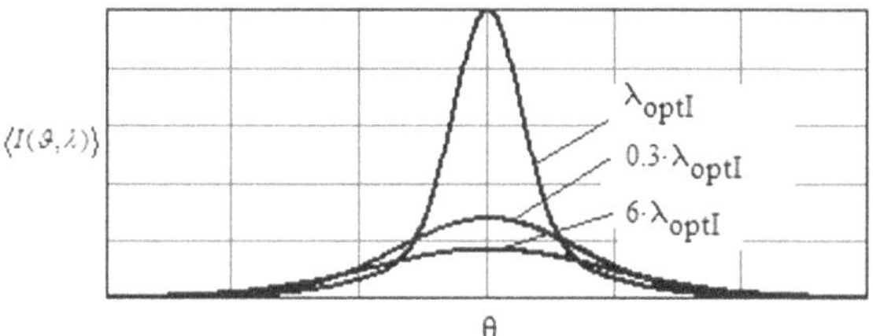


**Fig. J.2** Average intensity envelopes at the laser beam focus. Upon rotation, all enclose the same volume, i.e., the same total light energy. The large irradiance differences seen at $\vartheta = 0$ underline the importance of using optimum wavelengths. As telescope size increases, irradiance variation with wavelength is likely to be significantly larger than shown

**Fig. J.3** Typical wavelength dependence of the irradiance at the focus of a laser beam projected through the atmosphere by a large telescope

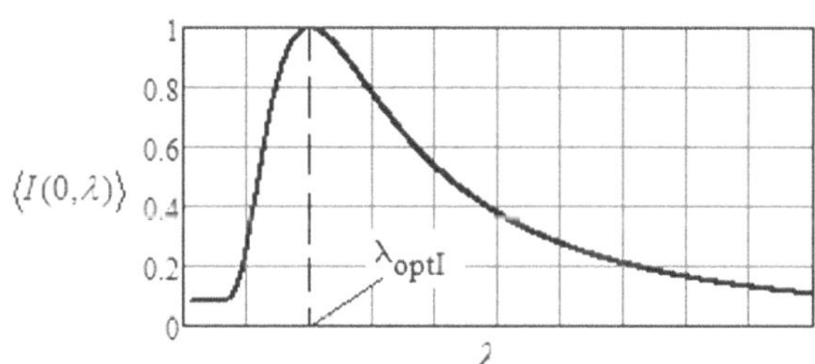


$$\lambda_{optI} = 2 \cdot \pi \cdot \sqrt{\frac{\sigma_P^2 + \sigma_T^2}{1 + 4 \cdot \frac{w_0^2}{D^2} \cdot \exp\left(\frac{4 \cdot \pi^2 \cdot \sigma_T^2}{\lambda_{optI}}\right)}} \approx 2 \cdot \pi \cdot \sqrt{\sigma_P^2 + \sigma_T^2} \quad \text{(J.8)}$$

Figure J.2 illustrates how wavelength choices either longer or shorter than the optimum wavelength can result in significantly lower irradiances at beam focus. Figure J.3 particularly underscores how wavelength choices only marginally shorter than optimum can result in catastrophic irradiance reductions

**References**: (1) McKechnie, T. S. (2015). *General theory of light propagation and imaging through the atmosphere* (Vol. 196). Springer series in optical sciences. Springer International Publishing.

# Appendix K
# Three-Gaussian Function Least Mean Squares Best-Fit Approximation of Star Images

This appendix specifically relates to Chap. 18 (Sect, 18.1) where the objective is to establish a least mean squares best fit approximation of a star image intensity PSF in terms of the sum of three Gaussian functions. Suppose the star image is obtained by an AO-equipped ELT fitted with a Focal Plane Array (FPA) detector, the latter consisting of a square $n \times n$ pixel array, where the pixel locations are denoted by ($x_i$, $y_j$) for i = 1,.... $n$ and j = 1,.... $n$.

A long exposure AO-corrected star image recorded by such an ELT/camera combination directly provides a suitable intensity PSF for our purposes, assuming that the star is unresolved and the image is obtained at a wavelength consistent with the most general type of star image, a core and halo image. It is further assumed that the entire star image falls within the pixel array, so that the intensity PSF is fully represented by the recorded $n \times n$ intensity array,$\langle I\left(x_i, y_j\right)\rangle$. It is noted that n must be large enough to sample the star image at an angular frequency not less than the Nyquist sampling frequency, $\lambda/D$, where λ is the imaging wavelength and D is the telescope diameter.

A central section through the most general three-Gaussian function approximation of a star image intensity PSF was given previously (18.11). The central section in the x-direction is given by $\langle I\left(x_i, 0\right\rangle$ which, for the one-dimensional analysis given below, we abbreviate to $\langle I\left(x_i\right\rangle$. The essential elements of 18.11 can be expressed in terms of six independent parameters, $a_1$,... $a_6$, as follows

$$\langle I\left(x_i\right)\rangle \approx \left\{a_1 \cdot \exp\left[-\left(\frac{x_i}{a_2}\right)^2\right] + a_3 \cdot \exp\left[-\left(\frac{x_i}{a_4}\right)^2\right] + a_5 \cdot \exp\left[-\left(\frac{x_i}{a_6}\right)^2\right]\right\}. \tag{K.1}$$

The task now it to find a suitable method for calculating values of the six parameters, a1, ... a6, so as to minimize the sum of the squares of the deviations between the measured intensity values , $\langle I(\mathrm{x})\rangle$, and the three-Gaussian function approxima-



T. S. McKechnie, *General Theory of Light Propagation and Imaging Through the Atmosphere*, Progress in Optical Science and Photonics 20,
https://doi.org/10.1007/978-3-030-98828-9

tion. Denoting the said sum by $S(a_1, a_2, a_3, a_4, a_5, a_6)$, we may calculate this sum as follows:

$$
\begin{aligned}
&S(a_1, a_2, a_3, a_4, a_5, a_6) = \\
&\sum_{k=1}^{n}\left\{\langle I(x_i)\rangle - a_1 \cdot \exp\left(-\left(\frac{x_i}{a_2}\right)^2\right)\right. \\
&\left. -a_3 \cdot \exp\left(-\left(\frac{x_i}{a_4}\right)^2\right) - a_5 \cdot \exp\left(-\left(\frac{x_i}{a_6}\right)^2\right)\right\}
\end{aligned}
\tag{K.2}
$$

The least mean squares best fits occurs when each of the six partial derivatives of the right hand side of the above equation takes the value zero. This requirement can be expressed by

$$
\frac{\partial}{\partial a_k} S(a_1, a_2, a_3, a_4, a_5, a_6) = 0, \quad \text{for} \quad \mathrm{k} = 1, \ldots 6
\tag{K.3}
$$

For obvious reasons, the partial derivatives arise in two distinct forms. One has the form

$$
\begin{aligned}
&\frac{\partial}{\partial a_1} S(a_1, a_2, a_3, a_4, a_5, a_6) = \\
&2 \cdot a_1 \cdot \sum_{k=1}^{n} \exp^{-2\frac{x_i^2}{a_2^2}} - 2 \cdot \sum_{k=1}^{n} \langle I(x_i)\rangle \cdot \exp^{-\frac{x_i^2}{a_2^2}} \\
&+ 2 \cdot a_3 \cdot \sum_{k=1}^{n} \exp^{-\frac{x_i^2}{a_2^2} - \frac{x_i^2}{a_4^2}} \\
&+ 2 \cdot a_5 \cdot \sum_{k=1}^{n} \langle I(x_i)\rangle \cdot \exp^{-\frac{x_i^2}{a_2^2} - \frac{x_i^2}{a_6^2}} = 0,
\end{aligned}
\tag{K.4}
$$

with closely similar expressions for $\frac{\partial}{\partial a_3} S(a_1, a_2, a_3, a_4, a_5, a_6)$ and $\frac{\partial}{\partial a_5} S(a_1, a_2, a_3, a_4, a_5, a_6)$. The other has the form,

$$
\begin{aligned}
&\frac{\partial}{\partial a_2} S(a_1, a_2, a_3, a_4, a_5, a_6) = \frac{4 \cdot a_1}{a_2^3} \\
&\times \left\{a_1 \cdot \sum_{k=1}^{n} x_i^2 \cdot \exp^{-2\frac{x_i^2}{a_2^2}} - \sum_{k=1}^{n} x_i^2 \cdot \langle I(x_i)\rangle \cdot \exp^{-\frac{x_i^2}{a_2^2}}\right\} \\
&+ \frac{4 \cdot a_1}{a_2^3} \cdot \left\{a_3 \cdot \sum_{k=1}^{n} x_i^2 \cdot \exp^{-\frac{x_i^2}{a_2^2} - \frac{x_i^2}{a_4^2}} + a_5 \cdot \sum_{k=1}^{n} \exp^{-\frac{x_i^2}{a_2^2} - \frac{x_i^2}{a_6^2}}\right\} = 0,
\end{aligned}
\tag{K.5}
$$

with closely similar expressions for $\frac{\partial}{\partial a_4} S(a_1, a_2, a_3, a_4, a_5, a_6)$ and $\frac{\partial}{\partial a_6} S(a_1, a_2, a_3, a_4, a_5, a_6)$

Solutions for parameter values, $a_1, \ldots, a_6$, simultaneously satisfying all six equations can be carried out by inputting sensible starting values and then allowing the least mean squares best fit algorithm to iteratively refine these values until the six equations are satisfied to the desired level of accuracy.

Levenberg-Marquardt Algorithms (LMAs) and derivative algorithms are well-suited for the above task. They are robust with respect to the starting value set and most likely to obtain the global minimum, as required, rather than some local minimum.

# Appendix L
# Acronyms

| | |
|---|---|
| (AO) | Adaptive optics |
| (ABL) | Air borne laser |
| (AOI) | Angle of incidence |
| (AFRL) | *United States* Air Force Research Laboratory |
| (BRDF) | Bidirectional reflectance distribution function |
| (COIL) | Chemical oxygen iodide laser |
| (CTE) | Coefficient of thermal expansion |
| (DoD) | *Unites States* Department of Defense |
| (DM) | Deformable mirror |
| (ESO) | European Southern observatory |
| (E-ELT) | European extremely large telescope |
| (EPS) | Equivalent phase screen |
| (FWHM) | Full width half maximum |
| (FPA) | Focal plane array |
| (FEL) | Free electron laser |
| (FSM) | Fast steering mirror |
| (GMT) | Giant Magellan telescope |
| (HEL) | High energy laser |
| (HeNe) | Helium neon |
| (HST) | Hubble space telescope |
| (IR) | InfraRed |
| (IR-FPA) | InfraRed focal plane array |
| (JWT) | James Webb Telescope |
| (LGS) | Laser guide star |
| (LMA) | Levenberg Marquardt algorithm |
| (MTF) | Modulation transfer function |
| (MEMS) | Micro electro mechanical systems |
| (mas) | Milli arc second |
| (MDA) | *United States* Missile Defense Agency |


T. S. McKechnie, *General Theory of Light Propagation and Imaging Through the Atmosphere*, Progress in Optical Science and Photonics 20, https://doi.org/10.1007/978-3-030-98828-9

| | |
|---|---|
| (MMT) | Multi mirror telescope |
| (NTT) | New technology telescope |
| (NOT) | Nordic optical telescope |
| (NOAO) | National optical astronomy observatory |
| (NASA) | National Aeronautics and Space Administration |
| (OPD) | Optical path difference |
| (OTF) | Optical transfer function |
| (PSF) | Point spread function |
| (PDF) | Probability density function |
| (PMT) | Photo multiplier tube |
| (P–V) | Peak to valley |
| (RMS) | Root mean square |
| (RGO) | Royal Greenwich Observatory |
| (ROE) | Royal observatory, Edinburgh |
| (SI) | International System *of units* |
| (SNL) | Sandia National Laboratories |
| (SNR) | Signal to Noise Ratio |
| (SLC) | Submarine laser communication |
| (SOR) | Starfire optical range |
| (TMT) | Thirty meter telescope |
| (TBM) | Theater ballistic missile |
| (TPTWCF) | Two-point two-wavelength correlation function |
| (UV) | Ultra violet |
| (UKIRT) | United Kingdom infra red telescope |
| (USAF) | United States air force |
| (VLT) | Very large telescope |

# Index




T. S. McKechnie, *General Theory of Light Propagation and Imaging Through the Atmosphere*, Progress in Optical Science and Photonics 20,
https://doi.org/10.1007/978-3-030-98828-9

The manufacturer's authorised representative in the EU is Springer Nature Customer Service Centre GmbH, Europaplatz 3, 69115 Heidelberg, Germany. If you have any concerns regarding our products, please contact ProductSafety@springernature.com

Printed and bound by CPI Group (UK) Ltd, Croydon, CR0 4YY
15/07/2026
02167620-0008